Introduction to Optimization with Matlab® Examples

Jose Maria Giron-Sierra

Introduction to Optimization with Matlab® Examples

Springer

Jose Maria Giron-Sierra
Systems Engineering and Automatic Contr.
Universidad Complutense de Madrid
Madrid, Spain

ISBN 978-3-032-05047-2 ISBN 978-3-032-05048-9 (eBook)
https://doi.org/10.1007/978-3-032-05048-9

This Springer imprint is published by the registered company Springer Nature Switzerland AG
The registered company address is: Gewerbestrasse 11, 6330 Cham, Switzerland

To my family

Preface

The application of Optimization in different fields has an important economic and social impact, like, for instance, in industrial processes, financial decisions, well-tuned products, commerce, transportation, and engineering solutions. Consequently, many academic studies include optimization matters, in the context of engineering, management, economy, information processing, and others.

It is normally desired to get the better of our resources. This could require a mathematical treatment, for even simple-looking problems. In response to this need, a wealth of analytic optimization methods has been developed along time. More recently, digital algorithms have entered the scene, giving impressive power for handling complex real-life optimization problems, like in the case of logistics.

Optimization is not far from our daily life. Actually, we are frequently using optimization methods and algorithms, without noticing it. For example, there are apps helping you to get to a certain place, offering on a map several optimized route alternatives (by car, train, walking, etc.). What happens here is that the machine—using its optimization algorithms—is helping you to *take good decisions*: a matter of man-machine collaboration.

Along this book, the main Optimization theoretical and practical developments are introduced, from well-established mathematical methods to contemporary algorithms.

This book connects with our teaching experience, including theory classes and laboratory exercises. MATLAB tools for Optimization were successfully used, based on a gradual progress in the concepts and methods being studied.

The author's experience on Optimization covers many years of graduate-level teaching at the University Complutense of Madrid, Spain, and the participation in several research activities. An important part of the research was devoted to the optimization of several industrial objectives, like Fuel and Gas distribution in Spain and other countries, part of the Pharma distribution in Spain, the production of beer by a well-known company, and other applications related to transportation.

The book is implicitly divided into several parts, according to the following schemes:

- Begin with fundamental concepts and basic optimization techniques.
- Introduction of important algorithms (for computers), habitually used in practical applications, like Linear Programming and Dynamic Programming.
- Classic analytic methods, due to Lagrange and Euler, which empower optimization for a large variety of problems.
- Return to algorithms, covering now different optimization scenarios: quadratic, conic, convex, integer, etc.
- Another relatively recent approach is the interior point methodology that had a profound impact on available computational tools; this is introduced in Chapter 10.
- In the last years, a new evolutionary methodology (Global Optimization) has been created that expands the capabilities of optimization for complex problems. This is considered in three chapters.
- Other important matters are then considered, in Chapters 14 and 15, like multi-objective optimization, game theory, proximal methods, etc.
- The book ends with a chapter devoted to image processing, which is a very intuitive optimization field.

The book has an appendix for long programs, in order to simplify the chapters.

Concerning the MATLAB programs, the programming style is purposely simple, illustrative. The programs work in various MATLAB versions, with perhaps some possible changes for some functions (in this case, MATLAB itself suggests appropriate changes).

There are traditional books that could be used for consultation. The chapters include references to those books, and to pertinent scientific papers. Most of the papers are available from the Internet.

The reader is invited to typeset the programs included in the book, since it would help in noticing coding details. All programs are available from the book web page:

www.dacya.ucm.es/giron/OptBook/Programs

For teaching or self-exploration of the book topics, a series of Lectures in PowerPoint is available from the web page:

www.dacya.ucm.es/giron/OptBook/Lectures

Please send feedback and suggestions for further improvements and support.

Madrid, Spain Jose Maria Giron-Sierra

Acknowledgments Thanks to my University, my colleagues, and my students. Since this book required a lot of time taken from nights, weekends, and holidays, I have to sincerely express my gratitude to my family.

Competing Interests The author has no competing interests to declare that are relevant to the content of this manuscript.

Contents

List of Figures

Chapter 1
Fundamental Aspects of Optimization

1.1 Introduction

During the Olympic Games one sees the athletes trying to run faster, to jump higher, to throw farther, etc. There are measures to be **minimized or maximized**, such as time, distance, velocity, etc. In every case, they want to reach **an optimum**. This is just one example of optimization, as something that happens in our usual circumstances. More examples—perhaps nearer to our daily life—could be added, concerning cost minimization, or profit maximization, or using resources as best as possible.

Just to illustrate how optimization is present in many common scenarios, Fig. 1.1 visualizes the case of airplanes. They should **optimize fuel consumption** for a number of important reasons.

In some cases, there are a number of alternatives, and one has to find the optimum. For example, suppose you want to find the highest tree in a certain small forest. With patience it would be possible to measure the height of each tree, 32 m, 15 m, 73 m, etc., and then put the data as the following set:

$$\{32, 15, 73, \ldots\}$$

This set of data can be seen as an array or a vector. This book pays special attention to the use of MATLAB in optimization problems. MATLAB is precisely quite appropriate for handling arrays and matrices. As it will be shown in this chapter, MATLAB provides simple solutions for finding the maximum and the minimum in a data set.

Consider now another example, in which we face a continuum of data. This is the classical ballistic trajectory of a cannon ball (Fig. 1.2). One wishes to find **h**, the maximum height of the trajectory. As it is well-known, derivatives are helpful for obtaining the solution.

J. M. Giron-Sierra, *Introduction to Optimization with Matlab® Examples*,
https://doi.org/10.1007/978-3-032-05048-9_1

Fig. 1.1 A notable optimization instance: airplanes and fuel consumption

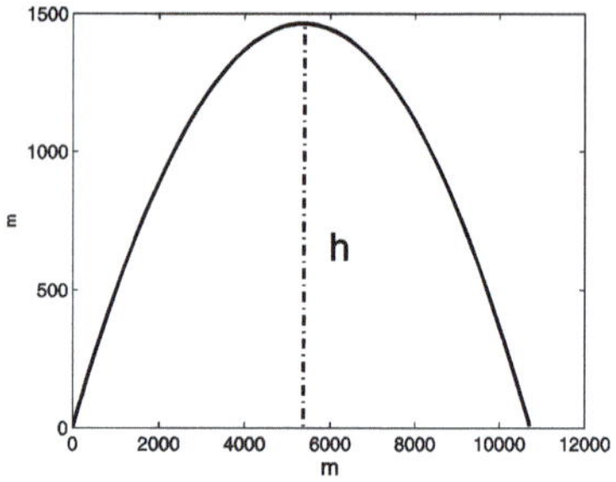

Fig. 1.2 A ballistic trajectory

Frequently optimization matters could be expressed with equations and functions. There is a rich heritage of mathematical theory that helps us to analyze and solve optimization questions. In this first chapter, some fundamental aspects of this theory will be introduced. Next chapters would give a more complete view of the optimization methodology.

The chapter starts with a concise account of basic analysis in one dimension, which is extended in the next section to multivariate functions. Then, some MATLAB functions especially suitable for optimization are introduced.

In continuation, the chapter presents some basic examples of optimization. As an important particular case, we shall focus on least squares fitting. To conclude, the chapter includes a summary of the MATLAB Symbolic Toolbox and some convenient resources.

Later on in this book, we shall consider optimization aspects of special economic importance, concerning industrial production and logistics. Figure 1.3 visualizes two examples: chemical industry and intelligent material handling (like in Amazon and other distribution services).

There are many textbooks on Optimization. In particular, for this initial chapter, the following publications would be recommended as background [1, 2, 4, 5], and [3].

Fig. 1.3 Examples of industrial optimization areas

1.2 Basic Analysis in One Dimension

Let us begin with a rapid overview of the optimization-oriented analysis of univariate functions.

1.2.1 *Points of Interest*

Supposing a continuous function $f(x)$, if there are points where the derivative is zero, these points could be local maxima, local minima, or inflection points. Notice that several local maxima may exist; it would be necessary to compare them to determine which is the global maximum. Figure 1.4 shows an example of function with several maxima and several minima.

Notice that a sinusoidal signal—as the one depicted in Fig. 1.5—has many equal local maxima, so none is a global maximum. Similarly, it would be said so about local and global minima of the sine signal.

Figure 1.6 shows an example of a piecewise function. Notice that the derivative does not exist at the vertex (cusp).

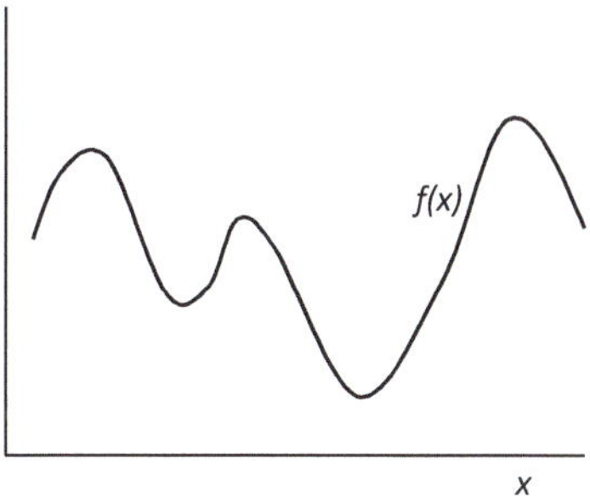

Fig. 1.4 Example of function with several maxima and several minima

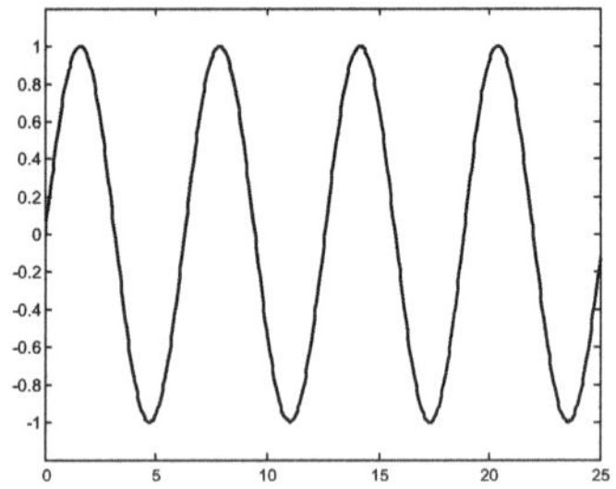

Fig. 1.5 The sinusoidal signal has many equal local maxima and minima

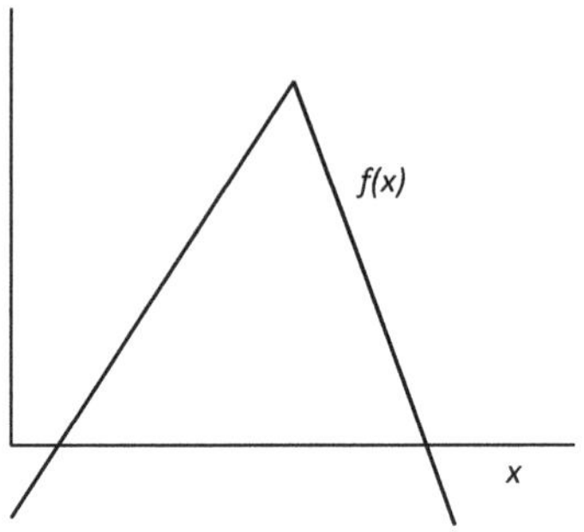

Fig. 1.6 Example of point where the derivative does not exist

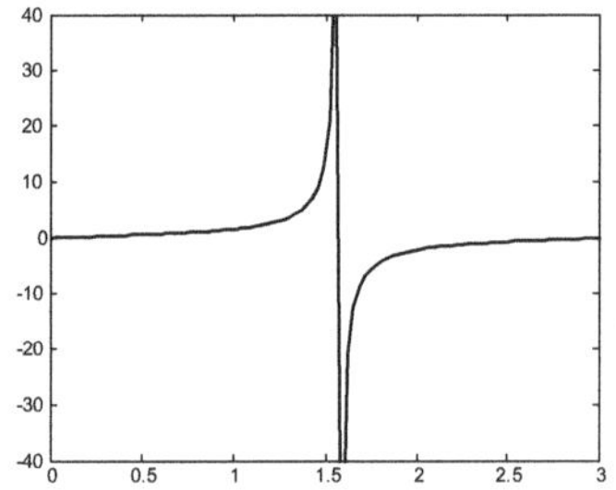

Fig. 1.7 Example of vertical asymptote

Points where the derivative is zero, or does not exist, are called critical points. Figure 1.7 shows a function with a vertical asymptote, this is another example of critical point where the derivative does not exist.

The local (or global) maximum or minimum values of a function are called extrema.

The extrema occur at critical points (Fermat's theorem). The inverse is not always true, for instance, an inflection point is a critical point but there is no extremum at this point.

Critical points where the derivative is defined and equal to zero are called stationary points.

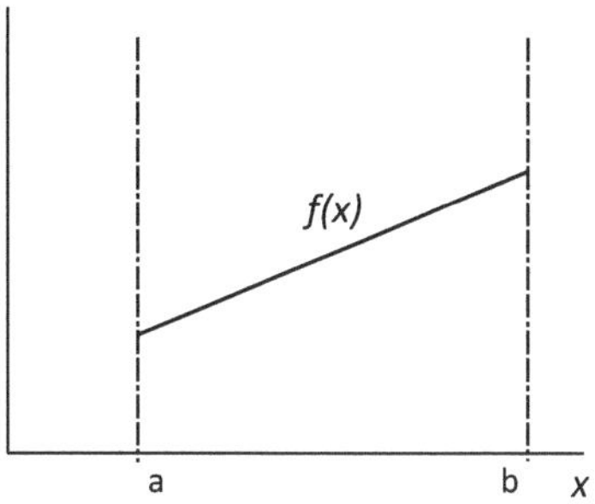

Fig. 1.8 Example of a line in an interval [a,b]

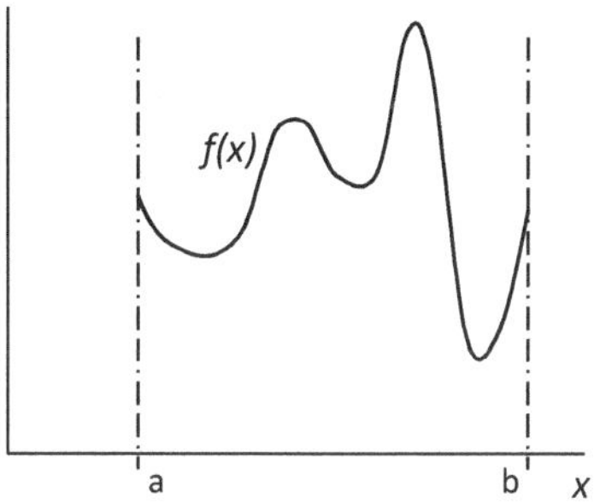

Fig. 1.9 Example of global maximum and global minimum in an interval [a,b]

1.2.2 Bounds

Roads are plenty of speed limit signals. The day has only 24 hours. Water boils at 100 °C. Many more examples of bounds can be noticed in real life. The reason of mentioning this is that optimization scenarios change a lot when one considers functions defined in a closed interval $[a, b]$. For instance, an inclined line has no maximum, but *there is one* in case of having the line in the interval $[a, b]$. Figure 1.8 depicts this case, the maximum being at b.

The extreme value theorem establishes that if $f(x)$ is continuous in $[a, b]$, then $f(x)$ has a global maximum at some point of the interval, and a global minimum at some point of the interval. Figure 1.9 shows a simple example of it (of course, Fig. 1.8 also shows an example of this).

1.2.3 Derivatives

The derivative of $f(x)$ at c (if it exists) is the slope of the line tangent to $f(x)$ at c. A curve follows the direction of its tangent line at any given point. Positive derivatives correspond to $f(x)$ increasing and negative derivatives to $f(x)$ decreasing. Derivatives equal to zero correspond to critical points.

The second derivative could be used to determine if there is an extremum at the critical point c where the first derivative is zero. If the second derivative is zero at c, the point is an inflection point. If it is negative, there is a maximum at c. If it is positive, there is a minimum at c.

If a curve in $[a, b]$ is such as the second derivative is always positive, the curve is concave up on the interval (the curve has a U shape). If the second derivative is always negative, the curve is concave down. Inflection points are points where a curve changes concavity.

1.2.4 Derivative Approximations Using Differences

Consider a set of equally spaced values of a variable x, for example, 0, 0.1, 0.2, ...100. The corresponding values of $f(x)$ would be also a set of numbers (an array, or vector). The *plot( )* function of MATLAB could be used for producing a plot of $f(x)$. Now, consider the following short program in MATLAB:

Program 1.1 Sine signal

```
x=0:0.05:12;
y=sin(x);
plot(x,y,'k');
```

The first sentence creates a vector of 241 equi-spaced values of x:

$$0, 0.05, 0.10, 0.15, \ldots, 12.$$

The second sentence obtains a vector of 241 values of the sine function at the values of x. Figure 1.10 shows the result of the third sentence, which uses the *plot( )* function.

The *plot( )* function automatically inserts lines to join successive points, in order to give an impression of continuity. This is the usual way for representing functions or curves in MATLAB. Of course, if the space between successive values of x was too large, the visual trick would become evident. The reader is invited to change 0.05 in the first sentence for a larger value. For example, Fig. 1.11 shows what happens with an interval of 1.

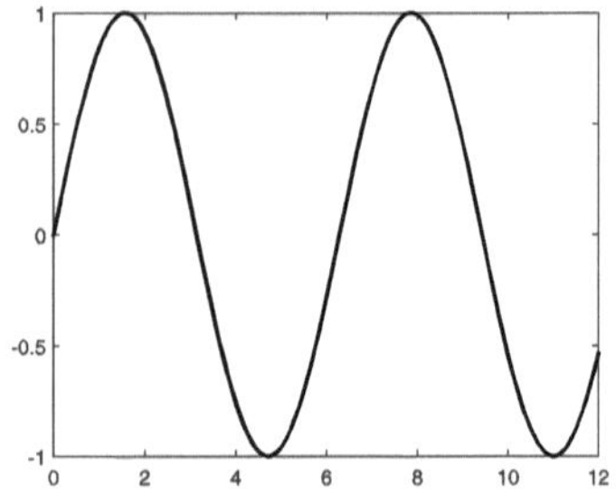

Fig. 1.10 Result of our first MATLAB program

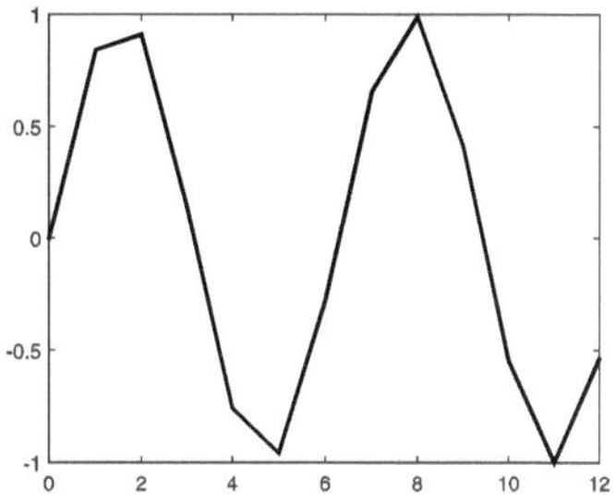

Fig. 1.11 Result of increasing the interval between values of x

Typically, digital computers use derivative approximations based on differences. There are three common approaches: the backward difference, the forward difference, and the central difference. Here are the mathematical expressions of these approaches:

- Backward difference:

$$f'(x_k) = \frac{f(x_k) - f(x_{k-1})}{x_k - x_{k-1}}$$

- Forward difference:

$$f'(x_k) = \frac{f(x_{k+1}) - f(x_k)}{x_{k+1} - x_k}$$

- Central difference:

$$f'(x_k) = \frac{f(x_{k+1}) - f(x_{k-1})}{x_{k+1} - x_{k-1}}$$

Figure 1.12 helps to situate the alternatives we have for approximating the slope of the tangent at point x_k:

The second derivative can be approximated based on differences of first derivatives.

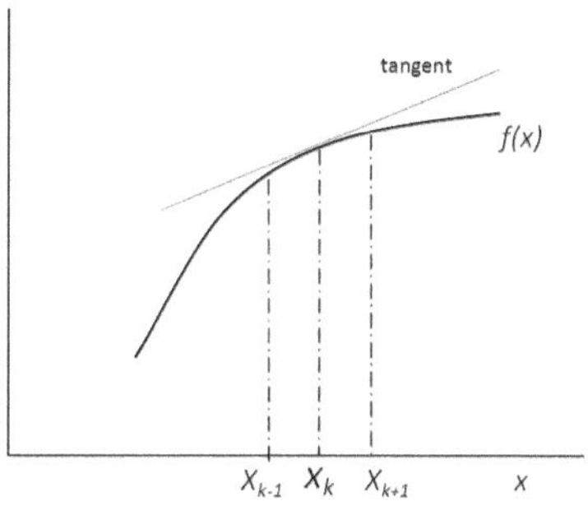

Fig. 1.12 Concept of derivative approximation based on differences

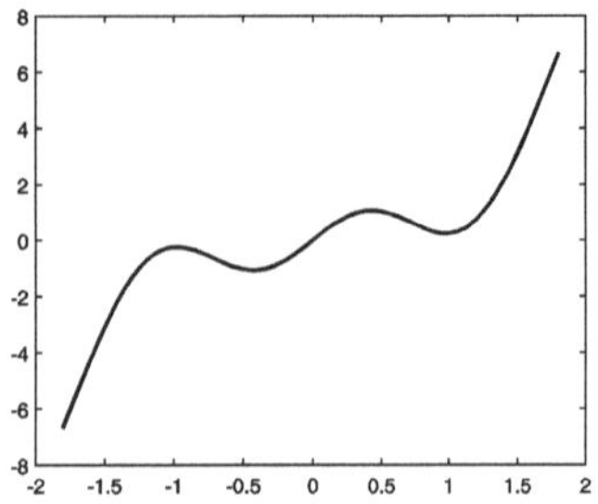

Fig. 1.13 The function considered in Program 1.2

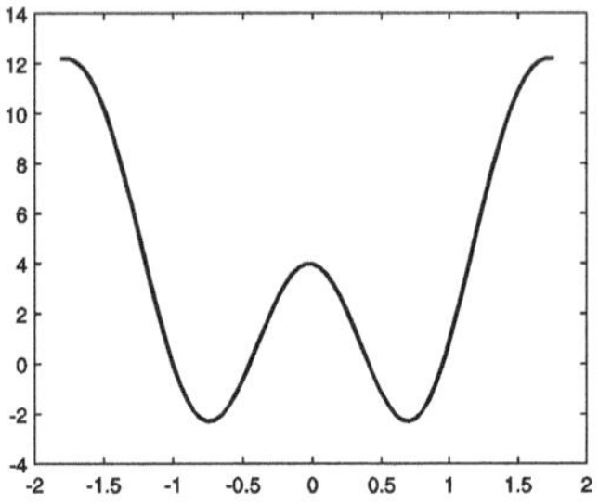

Fig. 1.14 The approx. derivative of the function

MATLAB provides the *diff( )* function which computes differences. Given a vector x of N values, *diff(x)* returns another vector with N-1 values corresponding to the differences between adjacent values of x.

It is possible to compute derivative approximations using *diff( )*. An example of this is given by Program 1.2, which plots a function and its derivative. The function chosen was:

$$y = \sin(4x) + x^3$$

Figure 1.13 presents a plot of the function.

And the approximated derivative is shown in Fig. 1.14.

Notice that the derivative crosses the zero value four times. These crossings correspond to stationary points.

Both Figs. 1.13 and 1.14 have been generated by the Program 1.2. The program applies forward differences.

Program 1.2 Function and derivative

```
x=-1.8:0.05:1.8;
y=sin(4*x)+(x.^3);
deriv=diff(y)./diff(x); % derivative approximation
figure(1)
plot(x,y,'k');
figure(2)
plot(x(1:end-1),deriv,'k');
```

1.3 Multivariate Functions

When climbing a hill it is possible to choose different directions in each step. Similarly, in the case of a function of several variables, the optimization-oriented analysis could move in several directions within the multi-dimensional domain. Typically, the mathematical study would rely on partial derivatives. Taking this into consideration, the concepts already introduced for univariate functions can be naturally extended.

Let us introduce now a series of mathematical concepts and tools. Then, we will consider again candidates for optimal solutions and some examples in MATLAB.

1.3.1 Mathematical Concepts and Tools

1.3.1.1 Gradient and Hessian

The gradient of a function of n variables is defined as follows:

$$\nabla f(x_1, x_2, \ldots, x_n) = \begin{pmatrix} \frac{\partial f}{\partial x_1} \\ \frac{\partial f}{\partial x_2} \\ \vdots \\ \frac{\partial f}{\partial x_n} \end{pmatrix}$$

The Hessian matrix of a function of n variables is defined as follows:

$$\nabla^2 f(x_1, x_2, \ldots, x_n) = \begin{pmatrix} \frac{\partial^2 f}{\partial x_1 \partial x_1} & \frac{\partial^2 f}{\partial x_1 \partial x_2} & \cdots & \frac{\partial^2 f}{\partial x_1 \partial x_n} \\ \frac{\partial^2 f}{\partial x_2 \partial x_1} & \frac{\partial^2 f}{\partial x_2 \partial x_2} & \cdots & \frac{\partial^2 f}{\partial x_2 \partial x_n} \\ \vdots & \vdots & \vdots & \vdots \\ \frac{\partial^2 f}{\partial x_n \partial x_1} & \frac{\partial^2 f}{\partial x_n \partial x_2} & \cdots & \frac{\partial^2 f}{\partial x_n \partial x_n} \end{pmatrix}$$

1.3.1.2 Directional Derivative

The directional derivative of a function $f()$ in the direction $\vec{v}$ (a vector) is the dot product:

$$\nabla f \cdot \frac{\vec{v}}{|\vec{v}|}$$

Notice that the gradient shows the direction with maximal variation of the function.

For example, let us find the directional derivative of

$$f(x, y, z) = \sqrt{x \cdot y \cdot z}$$

in the direction of $\vec{v} = (-1, -2,\ 2)$ at the point (3,2,6).
The magnitude of $\vec{v}$ is $|\vec{v}| = \sqrt{1 + 4 + 4} = 3$. The gradient components are:

$$\frac{\partial f}{\partial x} = \frac{y \cdot z}{2\sqrt{x \cdot y \cdot z}}$$

$$\frac{\partial f}{\partial y} = \frac{x \cdot z}{2\sqrt{x \cdot y \cdot z}}$$

$$\frac{\partial f}{\partial z} = \frac{x \cdot y}{2\sqrt{x \cdot y \cdot z}}$$

The gradient at (3,2,6) would be:

$$\nabla f = \left[\frac{12}{2 \times 6}, \frac{18}{2 \times 6}, \frac{6}{2 \times 6}\right] = \frac{1}{2}(2,\ 3,\ 1)$$

And therefore, the directional derivative is:

$$D_v f = \frac{\nabla f \cdot \vec{v}}{|\vec{v}|} = \frac{1}{3 \times 2}(2, 3, 1) \cdot (-1,\ -2,\ 2) = -1$$

1.3.1.3 Level Curves

Let $z = f(x_1, x_2)$ be the function under study. The curves on the plane (x_1, x_2) corresponding to different constant values of z (say for example 1, 2, 3…) are called level curves. For example, Fig. 1.15 shows level curves corresponding to $z = x_1^2 + x_2^2$.

Figure 1.15 has been generated with Program 1.3, which uses the MATLAB *contour( )* function.

Program 1.3 Level curves

```
m=-3:0.05:3;
[vx1,vx2]=meshgrid(m);
z=(vx1.^2)+(vx2.^2);
figure(1)
contour(vx1,vx2,z,'LineWidth',2);
xlabel('x1'); ylabel('x2');
```

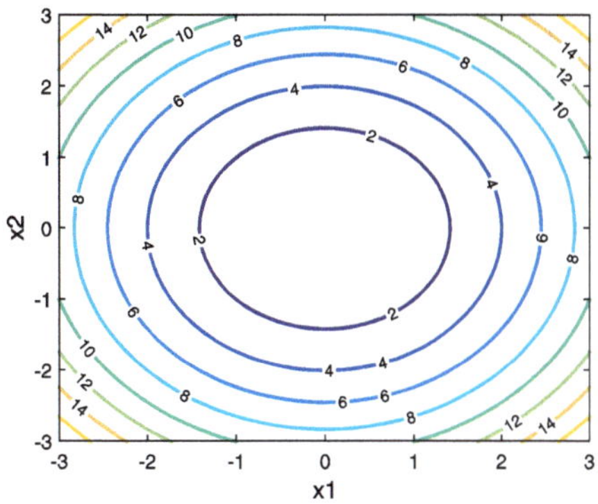

Fig. 1.15 Level curves for $z = x_1^2 + x_2^2$

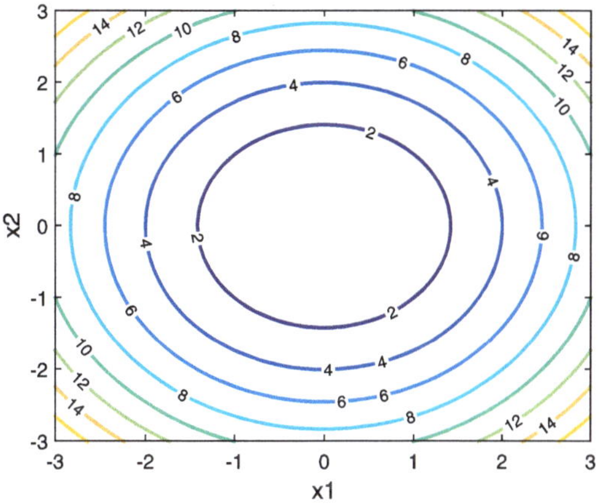

Fig. 1.16 Level curves with labels for $z = x_1^2 + x_2^2$

It is possible to add labels to the contours, using the *clabel( )* function. Program 1.4 provides an example of it, as shown in Fig. 1.16.

Program 1.4 Level curves with labels

```
m=-3:0.05:3;
[vx1,vx2]=meshgrid(m);
z=(vx1.^2)+(vx2.^2);
[C,H]=contour(vx1,vx2,z);
figure(1)
clabel(C,H,'FontSize',12);
xlabel('x1'); ylabel('x2');
```

Take any point P belonging to a level curve, the gradient of $z = f(x_1, x_2)$ is perpendicular to the level curve through that point. For example, Fig. 1.17 shows level curves and gradients corresponding to $z = x_1^2 + x_2^2$. For evident reasons, ideas concerning electric fields and potentials come to mind here.

Figure 1.17 has been obtained with the Program 1.5. Notice the use of the *quiver( )* function for adding arrows to the plot.

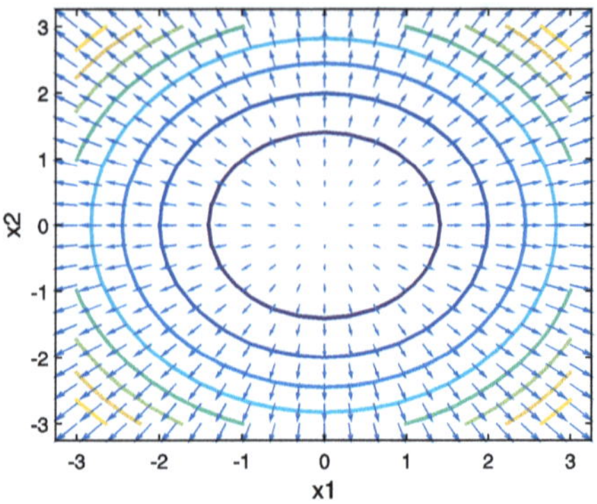

Fig. 1.17 Level curves and gradients for $z = x_1^2 + x_2^2$

Program 1.5 Level curves and gradients

```
m=-3:0.3:3;
[vx1,vx2]=meshgrid(m);
z=(vx1.^2)+(vx2.^2);
[g1,g2]=gradient(z,0.3,0.3);
figure(1)
contour(vx1,vx2,z,'LineWidth',2); hold on;
quiver(vx1,vx2,g1,g2);
xlabel('x1'); ylabel('x2');
```

In case the function domain had three dimensions, the gradient would be perpendicular to level surfaces. This fact is useful for computing tangent planes. In particular, given a point $P(x_0, y_0, z_0)$ and the gradient through this point $\nabla f(P) = (F_x, F_y, F_z)$ then the tangent plane at P would be:

$$F_x \cdot (x - x_0) + F_y \cdot (y - y_0) + F_z \cdot (z - z_0)$$

1.3.2 Points of Interest

An interior point of the domain of a multivariate function where all first partial derivatives are zero or where one or more of these derivatives does not exist is a critical point of the function.

Critical points where the partial derivatives are defined and equal to zero are called stationary points. Obviously, the gradient at stationary points is zero.

The extrema occur at critical points.

A critical point that is not a local extremum is called a *saddle point*. Figure 1.18 (generated by Program 1.6) depicts in 3D the function $z = x_1^2 - x_2^2$ which has a saddle point at (0, 0).

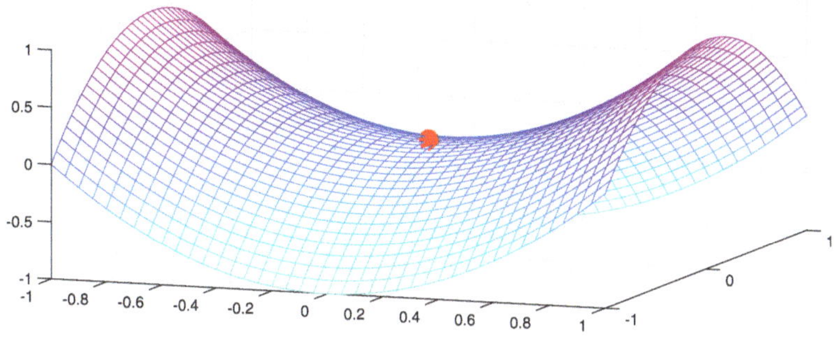

Fig. 1.18 Example of saddle point, function $z = x_1^2 - x_2^2$

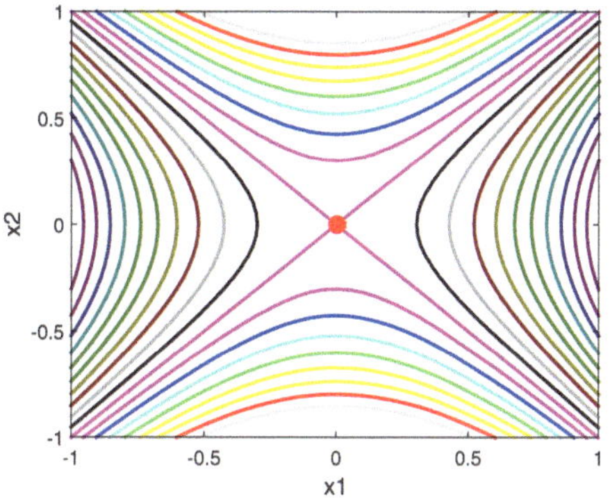

Fig. 1.19 Contour plot of $z = x_1^2 - x_2^2$; the saddle point is in the center

Program 1.6 Saddle point in 3D scenario

```
[x1,x2]=meshgrid(-1:0.05:1);
z=x1.^2-x2.^2;
figure(1)
mesh(x1,x2,z); hold on;
plot(0,0,'ro','LineWidth',4);
colormap('cool');
view(20,20);
```

The Hessian can be used to determine whether a stationary point is a local maximum or minimum. If the Hessian matrix is positive definite (the matrix has all eigenvalues positive), then it is a local minimum. If the Hessian matrix is negative definite (all eigenvalues negative), it is a local maximum. If there are both positive and negative eigenvalues, it is a saddle point.

Figure 1.19 shows a contour plot corresponding to $z = x_1^2 - x_2^2$, with the saddle point in the center. This figure has been obtained with the Program 1.7.

Program 1.7 Contours and saddle point

```
[x1,x2]=meshgrid(-1:0.05:1);
z=x1.^2-x2.^2;
```

```
figure(1)
contour(x1,x2,z,21); hold on;
plot(0,0,'ro','LineWidth',4);
colormap('vga');
xlabel('x1');ylabel('x2');
```

The Hessian matrix of $z = x_1^2 - x_2^2$ is:

$$H = \nabla^2 f = \begin{bmatrix} \frac{\partial^2 f}{\partial x_1 \partial x_1} & \frac{\partial^2 f}{\partial x_1 \partial x_2} \\ \frac{\partial^2 f}{\partial x_2 \partial x_1} & \frac{\partial^2 f}{\partial x_2 \partial x_2} \end{bmatrix} = \begin{bmatrix} 2 & 0 \\ 0 & -2 \end{bmatrix}$$

Now, by using the following MATLAB code,

Program 1.8 Square signal

```
H=[2 0; 0 -2];
eig(H)
```

it is possible to compute the eigenvalues of the Hessian, which are −2, 2. Then, there are positive and negative eigenvalues, and so the stationary point (0,0) is a saddle point.

Let us add another example of analysis. This is the following function:

$$f(x_1, x_2) = x_1^3 + x_2^3 + 2x_1^2 + 4x_2^2 + 6$$

The gradient components are:

$$\frac{\partial f}{\partial x_1} = x_1 (3x_1 + 4)$$

$$\frac{\partial f}{\partial x_2} = x_2 (3x_2 + 8)$$

There are four points where both gradient components are zero:

$$p_1 = (0, 0);\ p_2 = (0, -8/3);\ p_3 = (-4/3, 0);\ p_4 = (-4/3, -8/3).$$

The Hessian is:

$$H = \begin{bmatrix} 6x_1 + 4 & 0 \\ 0 & 6x_2 + 8 \end{bmatrix}$$

The evaluation of the Hessian at the four critical points, gives:

$$H(p_1) = \begin{bmatrix} 4 & 0 \\ 0 & 8 \end{bmatrix} = def + \qquad H(p_2) = \begin{bmatrix} 4 & 0 \\ 0 & -8 \end{bmatrix} = indef$$

$$H(p_3) = \begin{bmatrix} -4 & 0 \\ 0 & 8 \end{bmatrix} = indef \qquad H(p_4) = \begin{bmatrix} -4 & 0 \\ 0 & -8 \end{bmatrix} = def-$$

Points p_2 and p_3 are saddle points, p_1 is a local minimum, and p_4 a local maximum.

1.4 Useful MATLAB Functions

1.4.1 Maximum and Minimum Values

Given an ordered set of values (a vector), the location and value of the maximum and the minimum can be found with the functions *max( )* and *min( )*. Take for example the following set:

D = [0, 1, 2, 3, 4, 3, 2, 1];

Then, the next MATLAB program:

Program 1.9 Maximum and minimum values

```
D=[0, 1, 2, 3, 4, 3, 2, 1];
[Vmax,imax]=max(D)
[Vmin,imin]=min(D)
```

gives you the maximum value (4) and its location (5) and the minimum value (0) and its location (1).

1.4.2 Finding Values

By using *find*( ), it is possible to determine the locations of data satisfying a specified condition. For example, the locations of values higher than 2 in data set D can be found with the following short program:

Program 1.10 Locations of values higher than 2

```
D=[0, 1, 2, 3, 4, 3, 2, 1];
ix=find(D>2)
```

which gives you the locations 4, 5, and 6.

There is a specific MATLAB function for determining local peaks, called *findpeaks( )*. The next section gives an example of using it.

1.4.3 Gradients and Hessians

MATLAB provides the function *gradient( )* for numerical computation of the gradient. Program 1.5 has already shown an example of this.

It is also possible to use the MATLAB Symbolic Math Toolbox. In particular, the function *jacobian( )* can be used for gradients and Hessians and the functions *gradient( )* and *hessian( )* in recent MATLAB versions. The following code examples were taken from the MATLAB Documentation site:

```
syms x y z
f = x*y + 2*z*x;
hessian(f,[x,y,z])
```

Equivalently, the Hessian can also be computed with:

```
jacobian(gradient(f))
```

An example of symbolic gradient is:

```
syms x y z
f = 2*y*z*sin(x) + 3*x*sin(z)*cos(y);
gradient(f, [x, y, z])
```

There is a recent specific function of the MATLAB Image Processing Toolbox for computing image gradients, called *imgradient( )*.

1.5 Examples

1.5.1 A Basic Exercise of Finding Local Maxima and Minima

The following example is devoted to use *diff( )* and find for detecting the local peaks of a given function (example taken from Dr. Wickert; see Sect. 1.9). For illustration purposes, a combination of trigonometric and polynomial functions has been built, with the following expression:

$$f(x) = 3\ (\cos(11\,x)) + x^3 - x^2 - 4x + 12$$

A representation of this function is shown in Fig. 1.20, which has been generated by Program 1.11. Several local maxima and minima can be clearly observed.

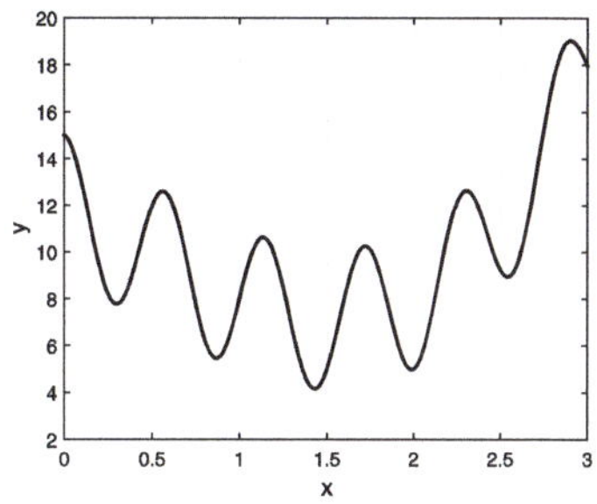

Fig. 1.20 The poly-trig function example

Program 1.11 Example of poly-trig function

```
%poly-trig function example
x=0:0.005:3;
y=3*cos(11*x)+x.^3-x.^2-4*x+12;
%display-------------------------
figure(1)
plot(x,y,'k');
axis([0 3 2 20]);
grid;
xlabel('x'); ylabel('y');
```

Figure 1.21 shows an approximation of the derivative of the function. It has been computed using *diff( )*. Obviously our interest focuses on the points where the derivative crosses the zero value. These points can be detected by taking note of changes in the sign of the derivative. Program 1.12 adopts this approach and obtains the points marked on Fig. 1.21.

Figure 1.22 shows the peaks that have been detected by determining zero-crossings of the derivative. The result was satisfactory. However, the reader is invited to change the value 0.005 in the first active sentence (establishing a range of values of x); higher values would mean lower density of function samples and less precision in the determination of the peaks.

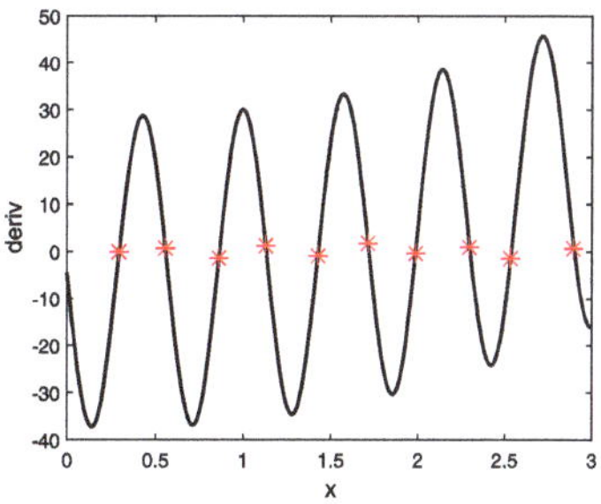

Fig. 1.21 Derivative approximation and zero-crossings

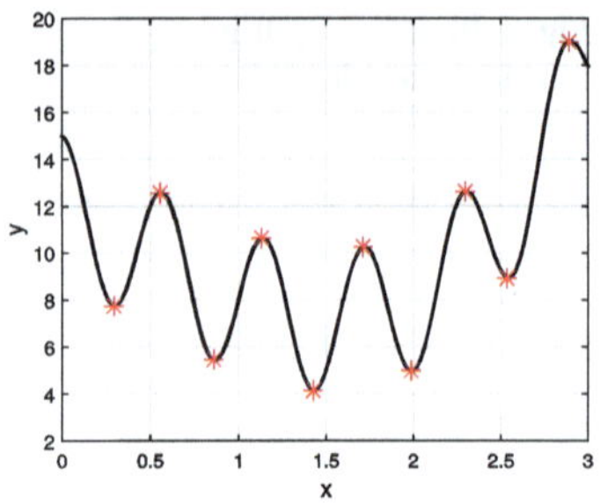

Fig. 1.22 The function under study and the peaks that have been found

Program 1.12 Finding of extrema

```
%finding of extrema
%poly-trig function example
x=0:0.005:3;
y=3*cos(11*x)+x.^3-x.^2-4*x+12;
deriv=diff(y)./diff(x); %derivative approximation
xb=x(1:end-1); %backward diff. points
%derivative zero crossings
N=length(xb);
prods=zeros(1,N-1);
for nn=1:N-1,
prods(nn)=deriv(nn)*deriv(nn+1);
end;
xc=xb(find(prods<0)); %the zero-crossing points
derivc=deriv(find(prods<0));
yc=y(find(prods<0));
%display-------------------------
figure(1)
plot(xb,deriv,'k'); hold on;
plot(xc,derivc,'r*');
grid
xlabel('x'); ylabel('deriv');
figure(2)
plot(x,y,'k'); hold on;
plot(xc,yc,'r*');
axis([0 3 2 20]);
grid
xlabel('x'); ylabel('y');
```

1.5.2 Using findpeaks( )

As said before, MATLAB provides a specific function for cases like the example above. It is called *findpeaks( )*.

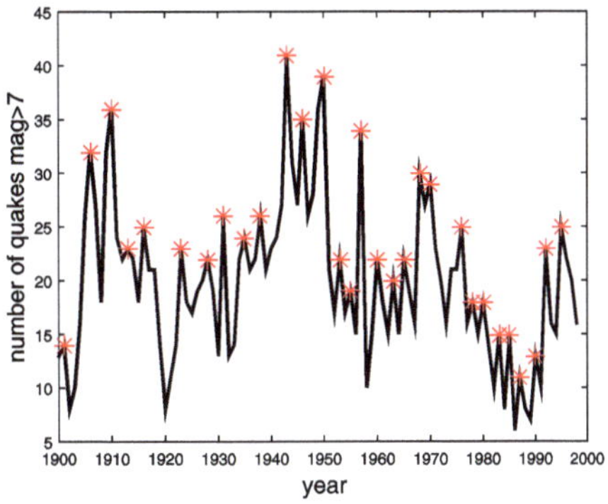

Fig. 1.23 Number of main Earthquakes per year

For the next example, we take from Internet a data set with the **number of main Earthquakes** per year, in the years 1900 to 2000. Only quakes with magnitude equal or higher than 7 were selected. Figure 1.23 depicts the contents of the data set and marks on the local peaks that have been detected with *findpeaks( )*. This figure has been obtained with the Program 1.13.

The local minima can be also determined by a simple trick: just use *findpeaks(-dat)*.

Program 1.13 Finding of peaks

```
%Example of findpeaks
%Number of quakes
fer=0;
while fer==0,
fid2=fopen('Num_of_quakes.txt','r');
if fid2==-1, disp('read error')
else
  dat=fscanf(fid2,'%f \r\n');
  fer=1;
end;
end;
fclose('all');
T=1900:1:1998; %years
[yp,ix]=findpeaks(dat);
%display---------------
figure(1)
plot(T,dat,'k'); hold on;
plot(T(ix),yp,'r*');
xlabel('year');
ylabel('number of quakes mag>7');
```

Since it is quite easy to employ the Program 1.13 for other data sets, we included two more figures. Figure 1.24 corresponds to the **annual flow of the river Nile** at Aswan (formerly Assuan), from 1871 to 1970; a changepoint can be observed near 1898.

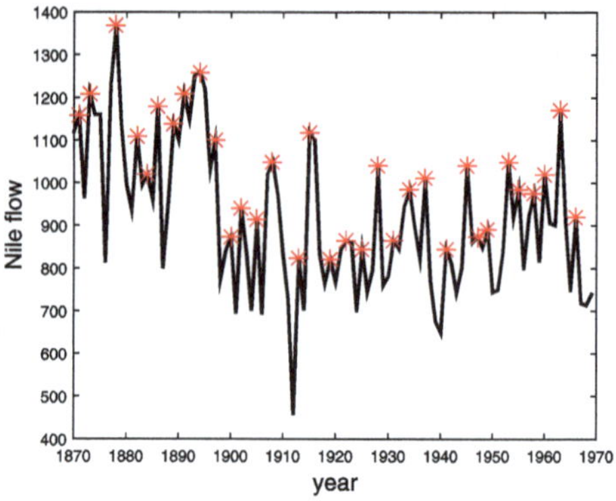

Fig. 1.24 Yearly flow of the river Nile

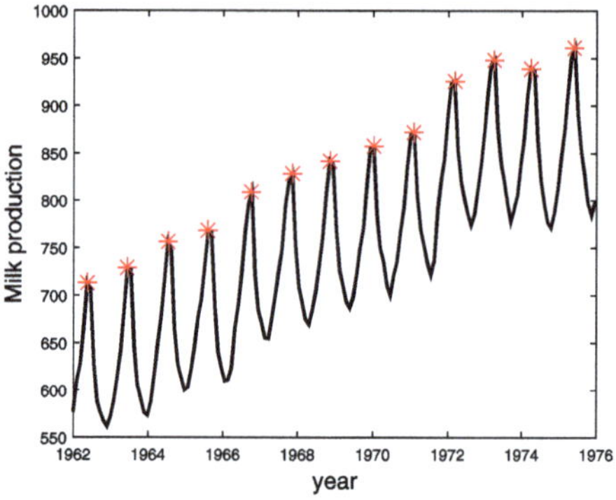

Fig. 1.25 Monthly production of milk

Figure 1.25 shows monthly data corresponding to **milk production**, pounds per cow from Jan 62 to Dec 75. There is an increasing trend.

People involved in time series data enjoy a lot of public-access data bases. The data used for the previous figures were taken from Internet sites, which are mentioned in the section on resources, at the end of the chapter.

The *findpeaks( )* function has many options, which is an indication that the subject of finding peaks could be really complicated, due to noisy data and other factors.

1.5.3 Example of Function Analysis

There are many examples of optimization-oriented analysis of functions that can be reached from Internet. Let us choose one of these examples, which refer to the following function:

$$f(x_1, x_2) = -x_1 x_2 \cdot \exp\left(\frac{-(x_1^2 + x_2^2)}{2}\right)$$

Fig. 1.26 3D representation of the function under study

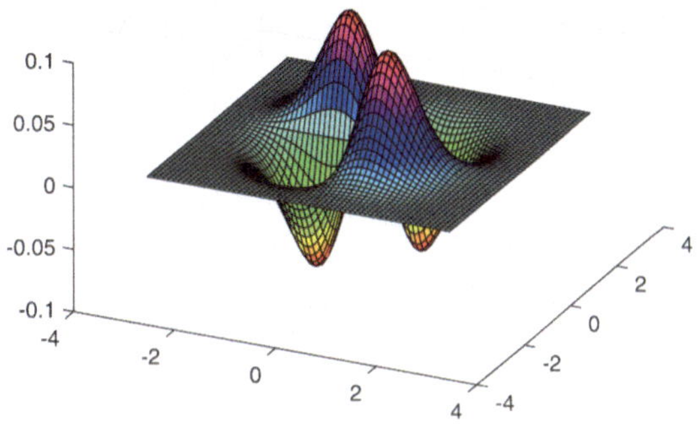

By means of Program 1.14 a 3D plot of the function was obtained, as depicted in Fig. 1.26. The function has two hills and two deep valleys. The reader is invited to move the perspective (using the sentence *view( )* or just a button provided by MATLAB plot windows for this purpose).

Program 1.14 3D plot of a function

```
[x1,x2]=meshgrid(-3:0.1:3);
z=-x1.*x2.*(exp(-(x1.^2+x2.^2))/2);
figure(1)
surf(x1,x2,z);
colormap('hsv');
view(25,35);
```

It is found that the gradient components are zero at five points: two at the top of hills, two at the bottom of valleys, and the point (0, 0). For easiest analysis of this situation is convenient to plot the set level of the function around (0, 0), as it is done in Fig. 1.27. This representation has been obtained with the Program 1.15, which is a simple variation of the previous program.

Fig. 1.27 Level set of the function under study

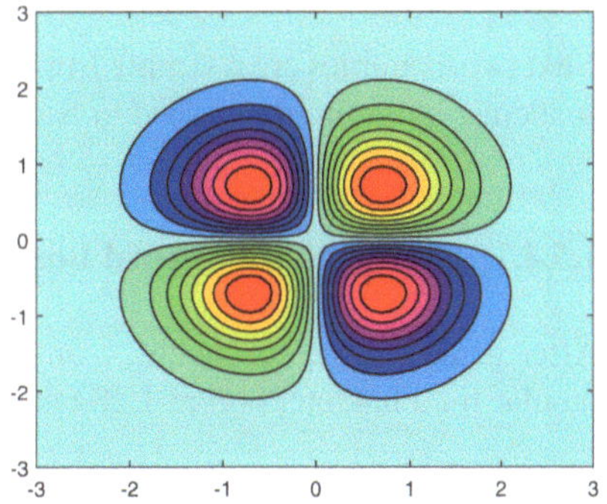

Program 1.15 Contours of the 3D function

```
[x1,x2]=meshgrid(-3:0.1:3);
z=-x1.*x2.*(exp(-(x1.^2+x2.^2))/2);
figure(1)
contourf(x1,x2,z,16);
colormap('hsv');
```

For simplification purposes, denote:

$$Q = \exp\left(\frac{-(x_1^2 + x_2^2)}{2}\right)$$

Then, the components of the Hessian would be:

$$\frac{\partial^2 f}{\partial x_1 \partial x_1} = Q\ (3\,x_1 x_2 - x_1^3\, x_2)$$

$$\frac{\partial^2 f}{\partial x_2 \partial x_2} = Q\ (3\,x_1 x_2 - x_1\, x_2^3)$$

$$\frac{\partial^2 f}{\partial x_1 \partial x_2} = \frac{\partial^2 f}{\partial x_2 \partial x_1} = Q\ (-1 + x_1^2 + x_2^2 - x_1^2 \cdot x_2^2)$$

Therefore, the Hessian at (0,0) would be:

$$H = \begin{bmatrix} 0 & -1 \\ -1 & 0 \end{bmatrix}$$

The reader can easily check that the eigenvalues of this Hessian are -1, 1, and so the point (0,0) is a saddle point.

1.5.4 Examples from Ordinary Life

Next two examples are a sample from popular optimization problems included in the academic literature mentioned in Sect. 1.9 (in particular, from Kiryl Tsishchanka).

1.5.4.1 Making a Cardboard Box

After getting some cardboard, many types of boxes can be fabricated: flat or not, tubular for a hat, etc. Figure 1.28 takes a view of this situation.

Fig. 1.28 Transforming a cardboard sheet into a box

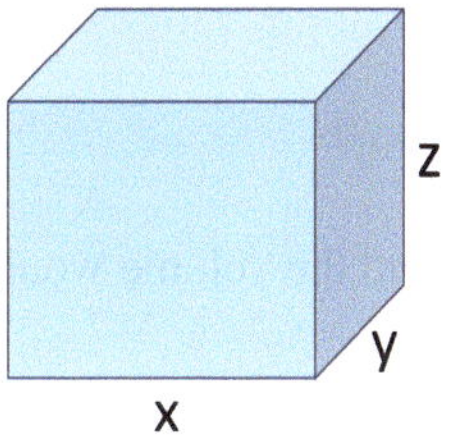

Fig. 1.29 A sketch of the box

One wants to construct a box with a square base, using at most $10\,\mathrm{m}^2$ of cardboard. The question is to determine the maximum volume that the box can have.

Figure 1.29 depicts a sketch of the box. Let us write equations. The area of the base is $A = x \cdot y$, and the volume of the box is $V = x \cdot y \cdot z$. The surface of the box is $S = 2xy + 2xz + 2yz = 10$.

Since $x = y$, then:

$$V = x^2 \cdot z, \text{ and } S = 2x^2 + 4xz = 10.$$

Hence,

$$V = x^2 \cdot z = x^2 \cdot \left(\frac{5 - x^2}{2x}\right) = \frac{1}{2} \cdot (5x - x^3)$$

Let us compute the derivative of V and equal it to zero:

$$\frac{dV}{dx} = \frac{1}{2} \cdot (5x - 3x^2) = 0$$

Therefore, the maximum is obtained with $x = \sqrt{\frac{5}{3}}$. The value of z would be,

$$z = \left(\frac{5 - x^2}{2x}\right) = \sqrt{\frac{5}{3}}$$

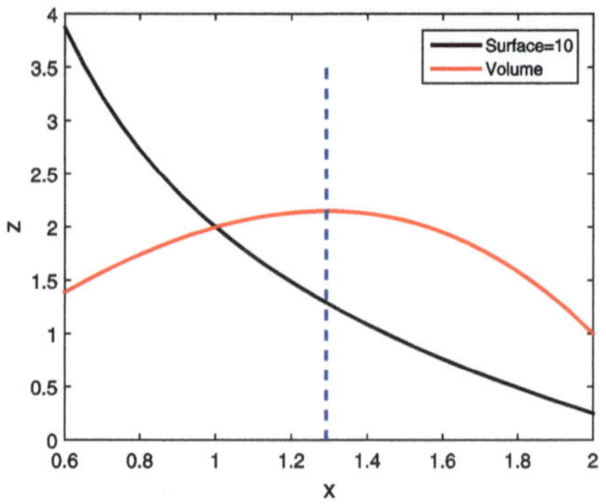

Fig. 1.30 Main aspects of the optimization problem

and the volume would be:

$$V = x^2 \cdot z = \left(\sqrt{\frac{5}{3}}\right)^3 \approx 2.1517$$

Figure 1.30 shows the main aspects of the problem: that the surface must be $10\,\text{m}^2$ and that the volume must be maximized.

The Fig. 1.30 has been generated with the Program 1.16.

Program 1.16 Box with maximum volume

```
%box with maximum volume
x=0.6:0.05:2;
z=(5-x.^2)./(2*x);
vol=z.*(x.^2);
xopt=sqrt(5/3);
figure(1)
plot(x,z,'k'); hold on;
plot(x,vol,'r');
plot([xopt xopt],[0 3.5],'b--');
xlabel('x'); ylabel('z');
legend('Surface=10','Volume');
```

1.5.4.2 Crossing a River

Figure 1.31 represents a river. The flow is marked with an arrow. The problem is a time minimization case.

A man wants to reach a point B as quickly as possible. He starts from A, launches a boat, and rows with 6 km/h speed. Having reached the point D, he runs toward point B with 8 km/h speed. Where is D?

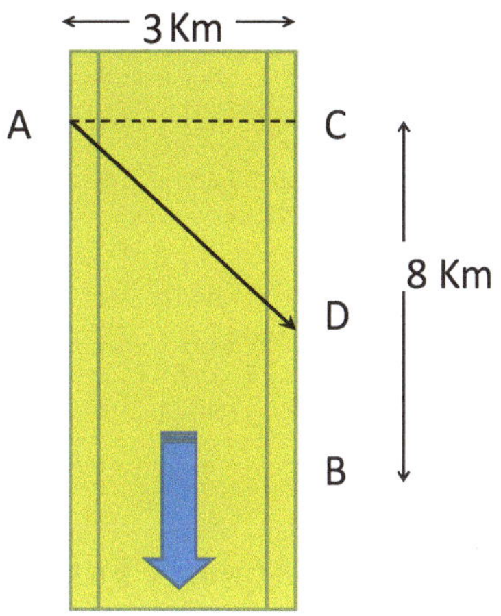

Fig. 1.31 From A to B, crossing a river

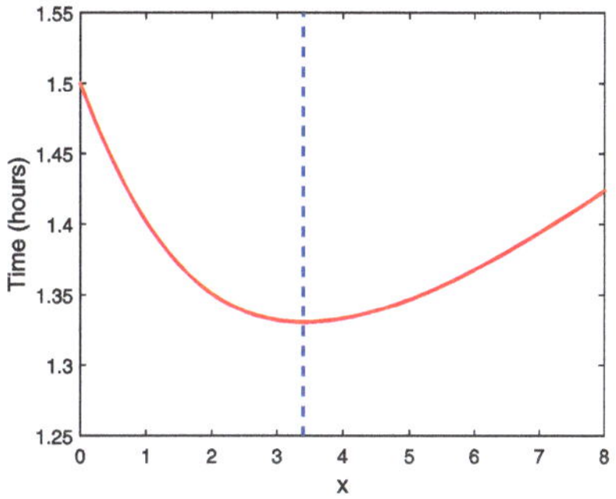

Fig. 1.32 The optimization problem

Let x be the distance between C and D. The total time from A to B would be:

$$T = \frac{\sqrt{x^2+9}}{6} + \frac{8-x}{8}$$

The derivative of T is:

$$\frac{dT}{dx} = \frac{x}{6\sqrt{x^2+9}} - \frac{1}{8}$$

Then,

$$\frac{dT}{dx} = 0 \Rightarrow \frac{x}{6\sqrt{x^2+9}} = \frac{1}{8} \Rightarrow 16x^2 = 9\,(x^2+9) \Rightarrow 7x^2 = 81$$

Therefore $x = \frac{9}{\sqrt{7}}$.

A representation of this time minimization problem is presented in Fig. 1.32, which has been obtained with the Program 1.17.

Program 1.17 Crossing a river

```
%crossing a river
x=0:0.02:8;
T=(sqrt(x.^2+9)/6)+((8-x)/8);
xopt=9/sqrt(7);
figure(1)
plot(x,T,'r'); hold on;
plot([xopt xopt],[1.25 1.55],'b--');
axis([0 8 1.25 1.55]);
xlabel('x'); ylabel('Time (hours)');
```

1.6 Least Squares Fitting

An important example of optimization is least square fitting. Figure 1.33 presents a typical case: there is a set of data points, and the problem is to fit a straight line to the data as best as possible. Given a candidate line, the errors (also called residuals) between the data and the line can be considered (the figure marks for instance one of these errors, e_3).

In least squares fitting one chooses as criterion to optimize a non-negative quantity, which is the sum of squared errors. This criterion should be minimized.

The mathematics of this approach starts by writing the sum of squared errors:

$$E = \sum e_i^2$$

If one wants to try a straight line for the fitting, then:

$$E = \sum (y_i - (a\,x_i + b))^2$$

The minimization of E implies that: $\frac{\partial E}{\partial a} = 0$ and $\frac{\partial E}{\partial b} = 0$ Therefore, for the straight line:

$$\frac{\partial E}{\partial a} = \sum 2\,(y_i - (a\,x_i + b)) \cdot (-x_i) = 0$$

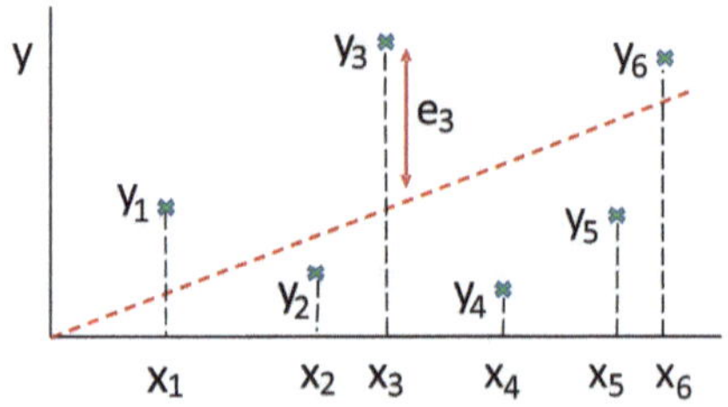

Fig. 1.33 Least squares fitting example

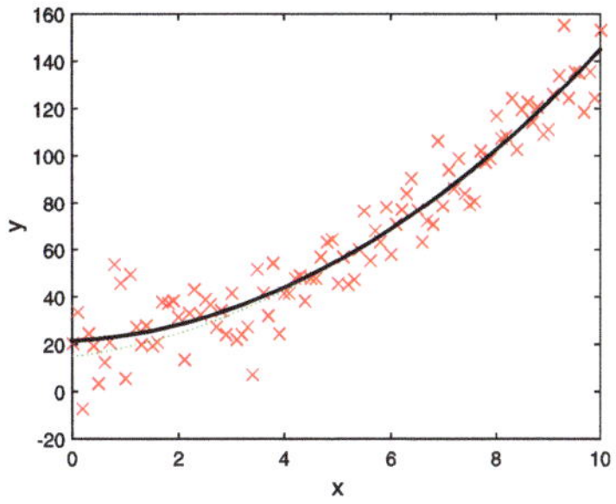

Fig. 1.34 Least squares fitting example

$$\frac{\partial E}{\partial b} = \sum 2\,(y_i - (a\,x_i + b)) \cdot 1 = 0$$

In order to simplify the work, the following expressions are introduced:

$$S_{xx} = \sum x_i^2\,; \quad S_x = \sum x_i$$

$$S_{xy} = \sum x_i\, y_i\,; \quad S_y = \sum y_i$$

Then, the equations corresponding to $\frac{\partial E}{\partial a} = 0$ and $\frac{\partial E}{\partial b} = 0$ can be written as follows:

$$\begin{bmatrix} S_{xx} & S_x \\ S_x & \sum_{i=1}^{N} 1 \end{bmatrix} \begin{pmatrix} a \\ b \end{pmatrix} = \begin{bmatrix} S_{xy} \\ S_y \end{bmatrix}$$

And the solution of the fitting can be obtained with:

$$\begin{pmatrix} a \\ b \end{pmatrix} = \begin{bmatrix} S_{xx} & S_x \\ S_x & \sum_{i=1}^{N} 1 \end{bmatrix}^{-1} \cdot \begin{bmatrix} S_{xy} \\ S_y \end{bmatrix}$$

The standard MATLAB command for least-squares fitting by a polynomial is *polyfit( )*. Figure 1.34 shows an example where *polyfit( )* has been used for a data fitting with a second order polynomial. The figure has been generated with the Program 1.18.

Program 1.18 Polynomial fitting

```
% polynomial fitting example
x=0:0.1:10; %independent variable data set
y=x.^2+3*x+15; %example of y=f(x)
N=length(x); %number of data points
di=randn(1,N); %random disturbance
```

```
ym=y+(10*di); %adding some disturbance to data
%polinomial fitting of disturbed data:
  [P,S]=polyfit(x,ym,2);
ye=polyval(P,x); %estimated y values
%display of results
figure(1)
plot(x,y,':g'); hold on; %plots original data
plot(x,ym,'xr'); %plots disturbed data
plot(x,ye,'k'); %plots the fitting
xlabel('x'); ylabel('y');
cor=corrcoef(ym,ye); %fit quality checking
```

There is a MATLAB Curve Fitting Toolbox that contains an extensive consideration of curve fitting methods.

1.7 A Summary of the MATLAB Symbolic Toolbox

Depending on the type of approach and tool that would be more convenient in each case, it could be opportune to use the MATLAB Symbolic Toolbox, or any other coding alternative. This section is devoted to provide a concise introduction to the Symbolic Toolbox.
– For declaring symbolic variables and expressions:

- For variables, use *syms*
- For expressions and equations, use *sym( )*

(it is also assumed that all strings may be manipulated as symbolic expression or equations)

Example:

```
syms x y z
r = sqrt(x^2+y^2+z^2)
```

The variable r is a symbolic variable, since it was defined based on at least one symbolic variable.
Other examples:

```
L = sym(`x*cos(y)') %L contains a symbolic expression
A= sym(`[a, b; c, d]') %symbolic matrix
```

– Some commands:
- *diff( )*, differentiation
- *int(*), integration
- *jacobian( )*, the Jacobian

- *determ( ),* or *det( )*, the determinant of a symbolic matrix
- *subs(function, old, new)*, replaces old with new in function
- *simplify( )*, collects terms
- *pretty( )*, for nice printouts
- *eval( )*, evaluate a symbolic expression
- *sum, prod, mean, std, del2, conj, etc.*

Example:

```
syms x y z
r = sqrt(x^2+y^2+z^2)
dr = diff(r,x)
```

The last sentence would give:
$dr = (1/sqrt(x^2 + y^2 + z^2)) * x$
If you add:
x=3; y=4; z=12;
Then you can obtain results with:
eval(r) (the answer is ans = 13)
eval(dr) (the answer is ans = 0.230769230769231)
Another example:

```
syms x a t
y = cos(t)
dy = diff(y)
Y = int(y)
```

Then, you will have $dy = -sin(t)$, and $Y = sin(t)$.

More information on the Symbolic Toolbox can be obtained by typing *'help sym'* or *'help syms'* on the MATLAB prompt.

1.8 Resources

1.8.1 MATLAB

1.8.1.1 Tutorials

To find extrema with symbolic MATLAB
http://www.math.umd.edu/~petersd/241/matlabextrn.html

1.8.1.2 Toolboxes

MATLAB Optimization Toolbox
https://www.mathworks.com/products/optimization.html
MATLAB Curve Fitting Toolbox
https://www.mathworks.com/products/curvefitting.html
EZYFIT Curve Fitting Toolbox for MATLAB
http://www.fast.u-psud.fr/ezyfit/
MATLAB Symbolic Toolbox
https://es.mathworks.com/products/symbolic.html

1.8.1.3 MATLAB Code

Find local min and max values (University of Cambridge)
http://www-h.eng.cam.ac.uk/help/tpl/programs/Matlab/minmax.html
Critical points of multivariable functions (KTH)
http://www.math.kth.se/math/student/courses/5B1148/ME/200607/lab/5B1148 matlab/lab3/KarlssonLehtinenLab3.pdf
Eli Billauer, Peak Detection
http://www.billauer.co.il/peakdet.html
Curve Fitting (University of Cambridge)
http://www-h.eng.cam.ac.uk/help/tpl/programs/Matlab/curve_fitting.html

1.8.2 Web Sites

Github, caesar0301, Time Series Data Sets
https://github.com/caesar0301/awesome-public-datasets
Rdatasets, Time Series Data Sets
https://vincentarelbundock.github.io/Rdatasets/datasets.html
TerpConnect, University of Maryland, Peak Finding
https://terpconnect.umd.edu/~toh/spectrum/PeakFindingandMeasurement.htm
CHEBFUN Developers, Oxford University, Least Squares Fitting:
http://www.chebfun.org/examples/stats/LeastSquares.html

1.8.3 Academic and Scientific Literature

Dr. Wickert, (2017), University of Colorado, Colorado Springs, ECE 1010, Lecture Notes
https://ece.uccs.edu/~mwickert/ece1010/lecture_notes/1010n7b.PDF

Kiryl Tsishchanka, (2010), The University of Texas at Austin, Lecture Notes https://mrsfucsko.weebly.com/uploads/9/1/6/6/91664946/4_-_optimization_notes_from_book.pdf

1.9 Exercises

1- The following function:

$$f(x_1, x_2) = x_1^3 + x_2^3 + 2x_1^2 + 4x_2^2 + 6$$

has been studied in Sect. 1.3.2. Now, visualize this function in 2D (contours) and 3D. Mark on the visualizations the critical points.

2- The Rosenbrock function is:

$$f(x_1, x_2) = 100(x_2 - x_1^2)^2 + (1 - x_1)^2$$

visualize the function in 2D (contours) and 3D. Add arrows indicating the gradient at points (−1,0), (0,0), (1,0), and (1,2). What about critical points?

3- Take the monthly production of milk represented in Sect. 1.5.2. Could you find a reasonable model of the evolution of year averages?

4- [1.1] A milk store produces x units of whole milk and $y$units of skim milk. The price for whole milk is $p(x) = 80 - x$ and the price for skim milk is $q(y) = 80 - y$. The cost of production is $C(x, y) = x^2 + x\,y + y^2$. How the store should choose x and y to maximize profits? Are there critical points? Visualize the problem.

5- [1.2] The yield of an agricultural crop could be modeled in function of nitrogen in the soil, as follows:

$$Y(N) = \frac{N}{1 + N^2}$$

where N is nitrogen concentration and $Y(N)$ is the yield. Visualize the yield in function of N. Find the maximum yield.

6- [1.3] It is the policy of an airline that baggage must be like a box, with the sum of length, width, and height not exceeding 64 in. Width and length being equal. Find the maximum volume baggage dimensions. Visualize the volume in function of width.

7- [1.4] A concert promoter knows that 4,000 people will attend with tickets at $45. For each dollar less in the ticket price, an additional 800 people will attend. Find a ticket price for maximum profit. Visualize the problem.

1.9.1 Exercise References

[1.1] https://www.eco.uc3m.es/docencia/matematicasii/english/Exercises/I- problemas5-sol.pdf
[1.2] https://www.ms.uky.edu/~ma137/Fall15/Lectures/Lecture_32.pdf
[1.3] https://www.csun.edu/~hcmth008/150a/bccalclt03_0405.pdf
[1.4] https://colleencondra.weebly.com/uploads/1/3/9/7/13973661/ab_solutions_p111-139.pdf

References

1. Bakr, M. (2013). *Nonlinear optimization in electrical engineering with applications in Matlab*. Institution of Engineering and Technology.
2. Lopez, C. (2014). *MATLAB optimization techniques*. Apress.
3. Messac, A. (2015). *Optimization in practice with MATLAB®: For engineering students and professionals*. Cambridge University Press.
4. Rao, S. S., & Rao, S. S. (2009). *Engineering optimization: theory and practice*. John Wiley & Sons.
5. Venkataraman, P. (2009). *Applied optimization with MATLAB programming*. John Wiley & Sons.

Chapter 2
Searching Methods

2.1 Introduction

Although searching is a natural human reaction when looking for optimization, it is most convenient to use a certain methodology, at least for not exploring the same point twice.

In favor of searching is the fact that some optimization cases resist to analytical treatment. However, also searching could fail if not adequate.

As a familiar example of searching, Fig. 2.1 displays a man in search of archaeologic pieces with a metal detector.

This chapter focuses on iterative searching methods. There is a good number of iterative methods that have been proposed along history (before computers existed).

First, the chapter introduces searching in one dimension. And then, several dimensions are considered, like when you climb a hill not following a trail: a direction should be now selected, and then try some steps, and then another direction should be decided, and so on. This is not the only alternative, as we shall see when introducing "*trust regions*".

In this chapter we shall remain deterministic. There are families of methods based on dice throwing (and some heuristics) that will be treated in other chapters.

Being to some extent simple, the methods to be introduced next are quite relevant in contemporary advanced algorithms, in the context of neural networks for deep learning, and other methodologies related to artificial intelligence and big data, [1, 34].

J. M. Giron-Sierra, *Introduction to Optimization with Matlab® Examples*,
https://doi.org/10.1007/978-3-032-05048-9_2

Fig. 2.1 Searching

2.2 Searching in One Dimension

This section will consider the following problem: determine the minimum (or maximum) of a single variable function over a closed interval $[a_0\ b_0]$using a searching procedure.

Searching methods assume that the function is *unimodal* in the interval considered: that is, the function has exactly one point where a maximum occurs in the interval. Denoting that point as x*, then the function is increasing in $[a_0\ x*]$ and decreasing in $[x*\ b_0]$. If the function is multimodal, then it must be subdivided into separate unimodal functions. Figure 2.2 shows an example of unimodal function. Analogously the definition of unimodal function could refer to having just one minimum, etc.

Normally the searching starts with a bounding phase, in which one finds a lower and an upper bound of the maximum (or minimum), and a second phase where a convenient algorithm is chosen to obtain a satisfactory approximation of the solution.

There are several bracketing methods that can be applied for the bounding phase [5]. The most elementary approach is to use a fixed step size L, as depicted in Fig. 2.3.

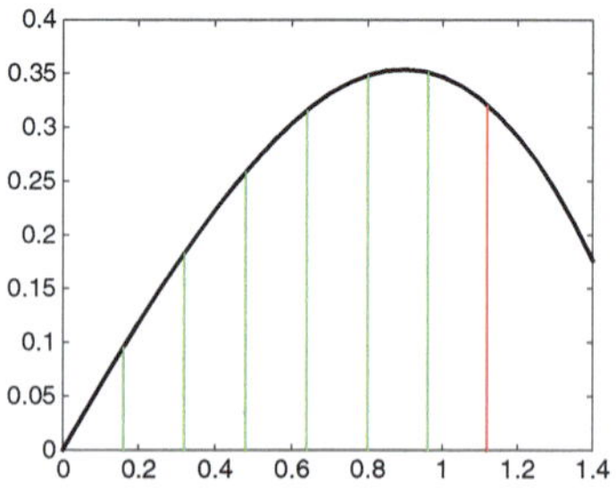

Fig. 2.2 Example of unimodal function

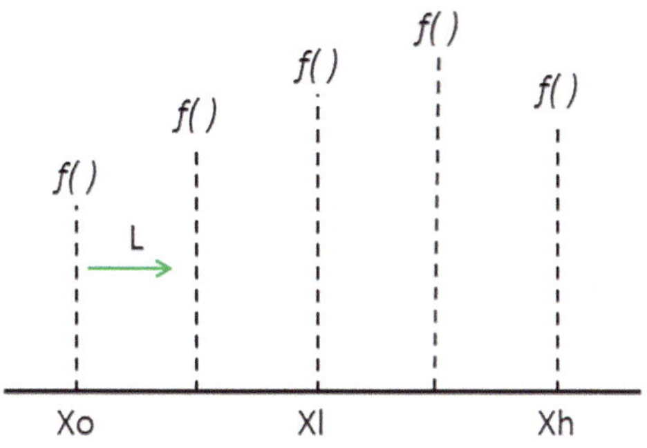

Fig. 2.3 Fixed step size bracketing

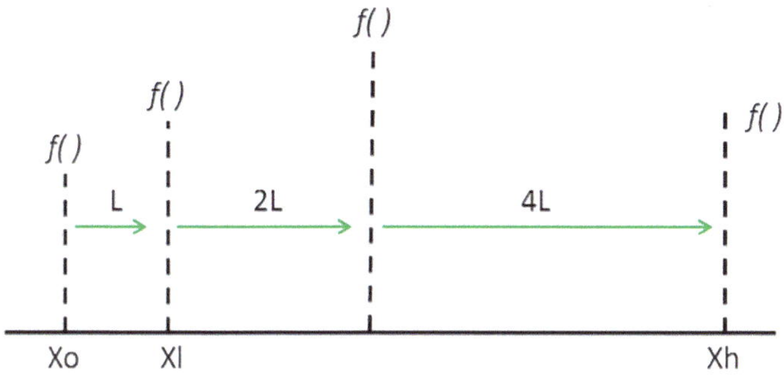

Fig. 2.4 Accelerated version

According with Fig. 2.3 one moves from x_0 to $x_0 + L$, then to $x_0 + 2L$, then to $x_0 + 3L$, and so on while the value of the function keeps increasing, until it starts to decrease. Figure 2.3 shows that the solution has been bracketed inside $[x_l \; x_h]$.

An accelerated version could be applied, in which the search step size is doubled in each iteration. Figure 2.4 illustrates the concept.

As simple example, let us apply the fixed step size approach for bracketing the following function, in order to find its maximum in the interval [0 1.4]:

$$f(x) = 1.6 \; \sin(x) \; - x$$

Figure 2.5 shows the sequence of steps for a step size of 0.16. This figure has been obtained with a direct coding in MATLAB, the Program 2.1. The last step of the search has been highlighted in darker color.

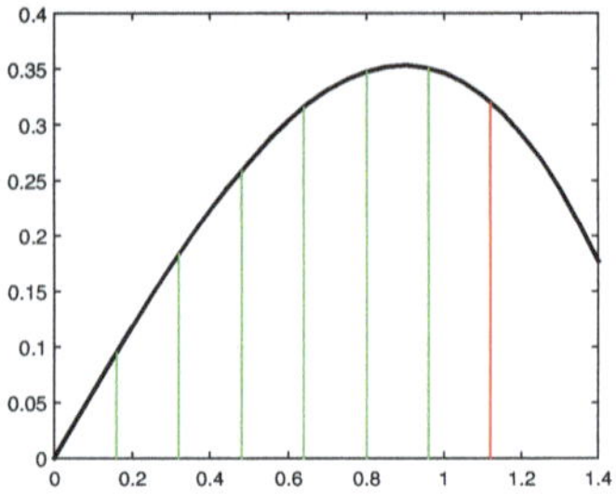

Fig. 2.5 Example of fixed step size bracketing

Program 2.1 Fixed step size bracketing

```
% Fixed step size bracketing
%
% function: y = (1.6*sin(x))-x
%
% First, we plot the function
% in the considered interval
x=0:0.05:1.4;
y=(1.6*sin(x))-x;
figure(1)
plot(x,y,'k');
hold on;
%
% Getting bounds with fixed length steps
L=0.16; %step size
x=0; %initial value of x
flag=0;
while flag==0,
  yold= (1.6*sin(x))-x;
  plot([x x],[0 yold],'g');
  x=x+L;
  ynew=(1.6*sin(x))-x;
  if ynew<yold, flag=1; end;
end;
% the solution is between x and x-2L
plot([x x],[0 ynew],'r');
```

Once the solution has been bracketed, one starts the second phase. Let us introduce three types of approaches that could be chosen for this: region elimination, interpolation, and derivative-based methods.

2.2.1 *Region Elimination*

The interval where the solution must live can be gradually narrowed, by discarding non-pertinent regions. Recall that one assumes that the function is unimodal. It would be readily noticed that the four methods to be introduced now have a classic, even historical flavor.

For comparison purposes, the same example would be used in the four methods; which is to find the maximum of:

$$f(x) = 1.6 \ \sin(x) \ - x$$

Recall that the bracketing result for this function was that the solution must be inside the interval [0.8 1.12].

2.2.1.1 Dichotomy

One departs from having bracketed the interval $[x_l \ x_h]$. This interval is divided into three equal sub-intervals by considering the points a and b. The function is evaluated at these points and the corner points. Figure 2.6 depicts this scheme. Now, one takes into account that the function is unimodal and discards one of the three sub-intervals. For example, in the case shown in Fig. 2.6, the maximum of the function cannot be in the leftmost sub-interval. Therefore, a new interval becomes the subject of study, namely, the interval $[a \ x_h]$. Again, this narrower interval would be divided into three equal sub-intervals, one of these sub-intervals will be discarded, and so on.

For example, let us apply dichotomy to the chosen example. Figure 2.7 shows the results of applying four times the dichotomy. The interval—left and right bounds painted as dashed lines—becomes more and more narrow. The point in the middle of each interval was represented with a continuous line. Eventually a satisfactory approximation of the solution can be found after not too many steps.

The reasoning associated to this method was implemented in the Program 2.2, which presents the results in Fig. 2.7, where the plot on top left corner corresponds to the first iteration, the second iteration is on top right, the third on bottom left, and the fourth on bottom right.

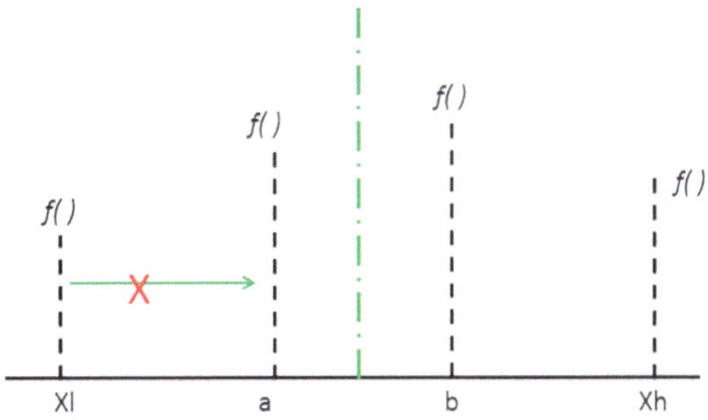

Fig. 2.6 Dichotomy method

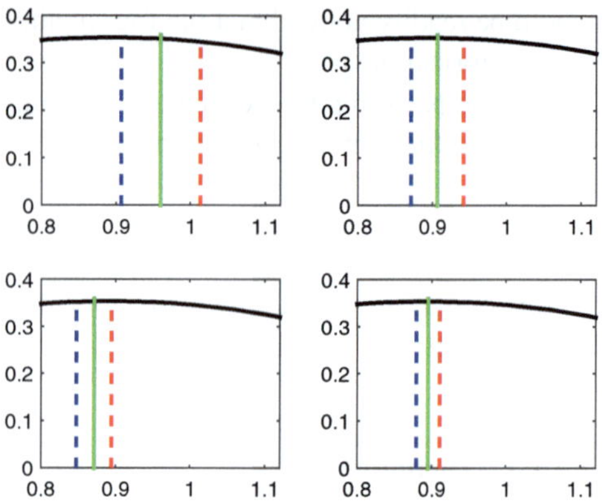

Fig. 2.7 First four steps of the dichotomy

Program 2.2 Dichotomy

```
% Now we apply dichotomy for refining
%
% function: y = (1.6*sin(x))-x
%
% interval determined by bracketing:
L=0.16;
xh=1.12; xl=xh-(2*L);
% for plotting the function in the bracketed interval
x=xl:0.01:xh;
y=(1.6*sin(x))-x;
%
for n=1:4, % we limit the searching to 4 steps
  xm=(xh-L); %middle point
  a=xm-(L/3); %left point
  ya=(1.6*sin(a))-a;
  b=xm+(L/3); %right point
  yb=(1.6*sin(b))-b;
  if yb<ya,
    xh=b; %eliminate third interval
  end;
  L=2*L/3;
  % display
  figure(1)
  subplot(2,2,n)
  plot(x,y,'k');
  axis([0.8 1.12 0 0.4]);
  hold on;
  plot([a a],[0 ya],'b--');
  plot([b b],[0 yb],'r--');
  plot([xm xm],[0 0.36],'g');
end;
```

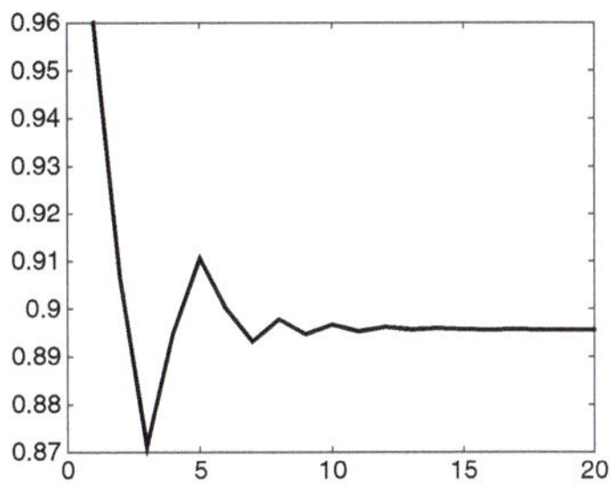

Fig. 2.8 Convergence of x_m toward the solution

One of the variables handled in the Program 2.2 is '*xm*'. This variable contains the value of the middle point of the current interval. As the intervals change (becoming narrower) during the searching process, also the value of *xm* changes (see Fig. 2.7).

With simple modifications of Program 2.2 it is possible to keep a record of the successive positions of *xm*. Figure 2.8 shows how *xm* converges toward the solution. This kind of graphic representation can help to determine when to stop the searching. Normally one stops when no significant precision gain is obtained; for example, it is clear that after the 15th iteration, the evolution becomes horizontal, so one could say that the solution was reached.

Program 2.3 Convergence of dichotomy

```
% Convergence of the dichotomy
% function: y = (1.6*sin(x))-x
% interval determined by bracketing:
L=0.16;
xh=1.12; xl=xh-(2*L);
% for plotting the function in the bracketed interval
x=xl:0.01:xh;
y=(1.6*sin(x))-x;
%
vxm=zeros(1,20); %reserve memory for a vector
%
for n=1:20, % we limit the searching to 20 steps
  xm=(xh-L); %middle point
  a=xm-(L/3); %left point
  ya=(1.6*sin(a))-a;
  b=xm+(L/3); %right point
  yb=(1.6*sin(b))-b;
  if yb<ya,
    xh=b; %eliminate third interval
  end;
  L=2*L/3;
  vxm(n)=xm; %save value in vector
end;
xm
%
```

```
%display:
figure(1)
plot(vxm,'k');
title('evolution of xm');
```

2.2.1.2 Interval Halving

The interval halving method divides the current interval into four equal sub-intervals. The method departs from the bracketed interval $[x_l\ x_h]$. Figure 2.9 illustrates the idea. The function is evaluated at the bounds *xl* and *xh* and at points a, b, and c. There are three half intervals to consider:

1. the left half $[x_l\ a]$.
2. the central half $[a\ b]$.
3. the right half $[b\ x_h]$.

Based on unimodality of the function, one discards the left half, or the right half, or keeps the central half. For example, in the case shown in Fig. 2.9, the solution cannot be in the right half. Thus, the next interval to study is $[x_l\ b]$, which then would be divided into four sub-intervals, and so on.

Let us apply the interval halving method to the chosen example. Figure 2.10 shows the results of applying four times the method. The interval—between dashed lines—becomes more and more narrow. The figure has been generated by the Program 2.4, which includes the interval halving reasoning.

Program 2.4 Interval halving

```
% Interval halving
%
% function: y = (1.6*sin(x))-x
%
% interval determined by bracketing:
L=0.16;
xh=1.12; xl=xh-(2*L);
% for plotting the function in the bracketed interval
x=xl:0.01:xh;
y=(1.6*sin(x))-x;
%
for n=1:4, % we limit the searching to 4 steps
  b=(xh-L); %middle point
  yb=(1.6*sin(b))-b;
  a=b-(L/2); %left point
  ya=(1.6*sin(a))-a;
  c=b+(L/2); %right point
  yc=(1.6*sin(c))-c;
  if ya>yb,
    xh=b; %eliminate 3 and 4 intervals
```

```
  elseif yc>yb,
    xl=b; %eliminate 1 and 2 intervals
  else
    xl=a; xh=c; %reserve 2 and 3 intervals
  end;
  L=L/2;
  yl=(1.6*sin(xl))-xl;
  yh=(1.6*sin(xh))-xh;
  % display
  figure (1)
  subplot(2,2,n)
  plot(x,y,'k');
  axis([0.79 1.12 0 0.4]);
  hold on;
  plot([xl xl],[0 yl],'b--');
  plot([xh xh],[0 yh],'r--');
end;
```

In order to visualize how the method converges toward the solution, a simple modification of Program 2.4 was done to keep a record of the successive positions of the right bound. Figure 2.11 shows the evolution of this bound, which becomes horizontal after the 8th iteration.

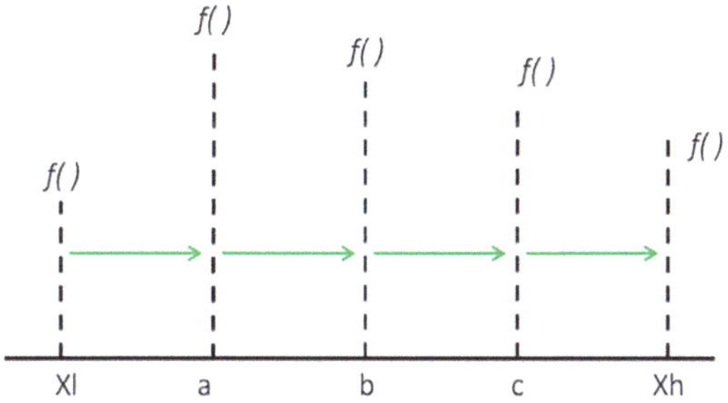

Fig. 2.9 Interval halving method

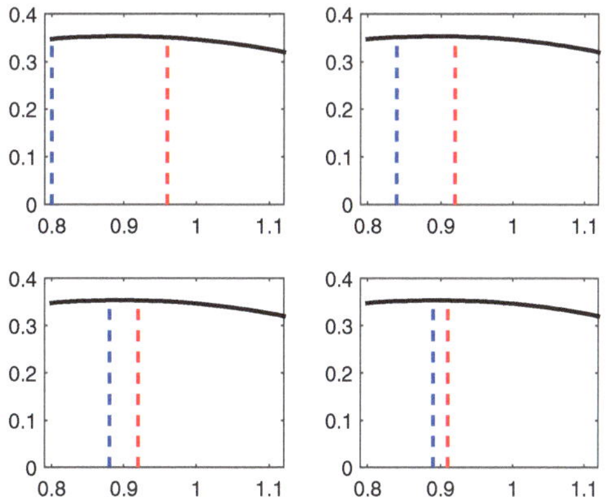

Fig. 2.10 First four steps of the interval halving

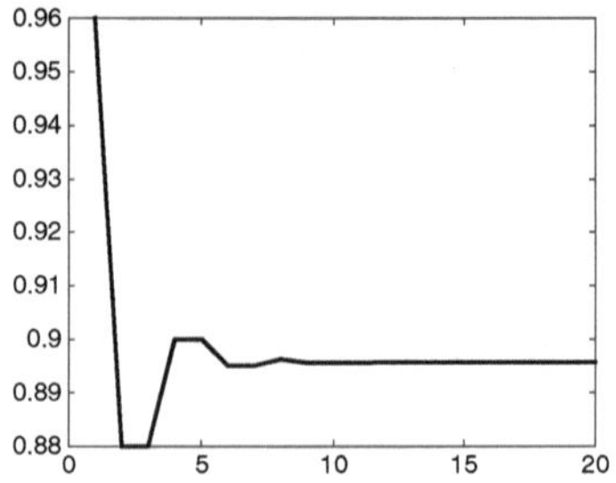

Fig. 2.11 Convergence of b toward the solution

Program 2.5 Convergence of interval halving

```
% Convergence of interval halving
% function: y = (1.6*sin(x))-x
% interval determined by bracketing:
L=0.16;
xh=1.12; xl=xh-(2*L);
% for plotting the function in the bracketed interval
x=xl:0.01:xh;
y=(1.6*sin(x))-x;
%
vb=zeros(1,20); %reserve memory for a vector
%
for n=1:20, % we limit the searching to 20 steps
  b=(xh-L); %middle point
  yb=(1.6*sin(b))-b;
  a=b-(L/2); %left point
  ya=(1.6*sin(a))-a;
  c=b+(L/2); %right point
  yc=(1.6*sin(c))-c;
  if ya>yb,
    xh=b; %eliminate 3 and 4 intervals
  elseif yc>yb,
    xl=b; %eliminate 1 and 2 intervals
  else
    xl=a; xh=c; %reserve 2 and 3 intervals
  end;
  L=L/2;
  vb(n)=b;
  % display
  figure (1)
  plot(vb,'k');
  title('evolution of b');
end;
```

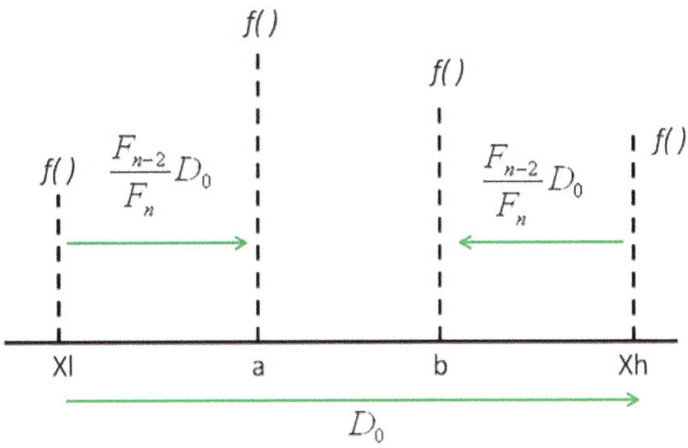

Fig. 2.12 Scheme of Fibonacci and golden section methods

2.2.1.3 Fibonacci

One of the most famous sequences of numbers is 1, 1, 2, 3, 5, 8, 13, 21, 34, 55, …It is the sequence of Fibonacci, which was published in the book *'Liber abaci'* in 1202. The author of this book was Leonardo Pisano (Leonardo from Pisa), who was the son of Bonacci (*'filius Bonacci'*). One of the mathematical problems studied in the book was about rabbit's reproduction; this study led to the following generation rule:

$$x(n+1) = x(n) + x(n-1),$$

which is the formula to generate the Fibonacci numbers.

Coming now to the problem of dividing the bracketed interval into sub-intervals, one could use the Fibonacci sequence for building the division. Figure 2.12 illustrates the idea, where F_n, F_{n-1}, F_{n-2} … are the Fibonacci numbers. In the case depicted in Fig. 2.12, the interval was divided based on F_n and F_{n-2}. The solution cannot be in $[b\ x_h]$, so this sub-interval is discarded and a new, narrower interval $[x_l\ b]$ is considered. This interval would be divided into three, based on F_{n+1} and F_{n-1}, and so on.

Figure 2.13 shows the results of using this method four successive times. This figure has been generated by the Program 2.6, which takes for the initial step the numbers 5 and 8 from the Fibonacci sequence, and then generates the next Fibonacci numbers as the iterations go on.

Program 2.6 Fibonacci

```
% Fibonacci
%
% function: y = (1.6*sin(x))-x
%
% interval determined by bracketing:
L=0.16;
xh=1.12; xl=xh-(2*L);
% for plotting the function in the bracketed interval
```

```
x=xl:0.01:xh;
y=(1.6*sin(x))-x;
%
D=2*L;
F1=8; F0=5; %initial Fibonacci numbers (edit)
%
for n=1:4, % we limit the searching to 4 steps
  F2=F1+F0;
  r=F0/F2;
  a=xl+(r*D);
  ya=(1.6*sin(a))-a;
  b=xh-(r*D);
  yb=(1.6*sin(b))-b;
  if yb<ya,
    xh=b; %eliminate third interval
  else
    xl=a; %eliminate first interval
  end;
  F0=F1;
  F1=F2;
  D=xh-xl;
  %
  yl=(1.6*sin(xl))-xl;
  yh=(1.6*sin(xh))-xh;
  %
  % display
  figure(1)
  subplot(2,2,n)
  plot(x,y,'k');
  axis([0.79 1.12 0 0.4]);
  hold on;
  plot([xl xl],[0 yl],'b--');
  plot([xh xh],[0 yh],'r--');
end;
```

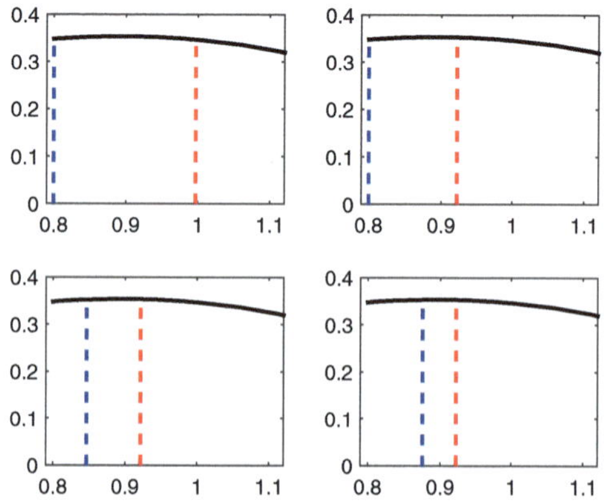

Fig. 2.13 First four steps of Fibonacci

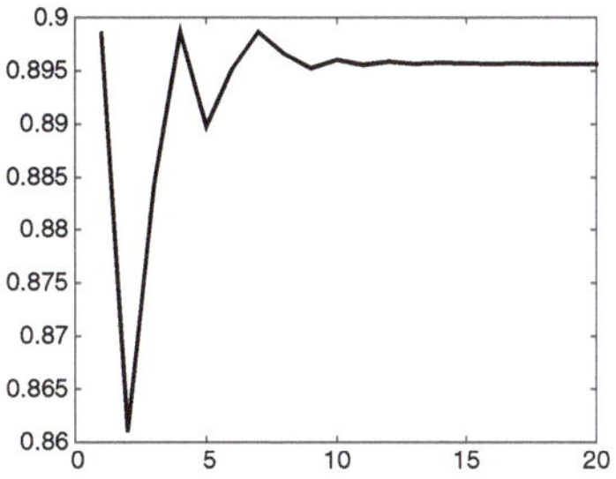

Fig. 2.14 Convergence of middle point toward the solution

Using again simple modifications, a visualization of the convergence of the middle point of each interval was obtained, Program 2.7. It becomes horizontal by the 13th iteration (Fig. 2.14).

Program 2.7 Convergence of Fibonacci

```
% Convergence of Fibonacci
% function: y = (1.6*sin(x))-x
% interval determined by bracketing:
L=0.16;
xh=1.12; xl=xh-(2*L);
% for plotting the function in the bracketed interval
x=xl:0.01:xh;
y=(1.6*sin(x))-x;
%
vm=zeros(1,20); %reserve memory for a vector
%
D=2*L;
F1=8; F0=5; %initial Fibonacci numbers (edit)
%
for n=1:20, % we limit the searching to 20 steps
  F2=F1+F0;
  r=F0/F2;
  a=xl+(r*D);
  ya=(1.6*sin(a))-a;
  b=xh-(r*D);
  yb=(1.6*sin(b))-b;
  if yb<ya,
    xh=b; %eliminate third interval
  else
    xl=a; %eliminate first interval
  end;
  F0=F1;
  F1=F2;
  D=xh-xl;
  vm(n)=(xh+xl)/2;
  %
  % display
  figure(1)
```

```
  plot(vm,'k');
  title('evolution of middle point');
end;
```

2.2.1.4 Golden Section

The golden ratio has old roots. Around 2550 years ago, the Pythagoreans discovered that the diagonal and the side of a pentagon were incommensurable (the ratio of these numbers is an irrational number, which was named '*Phi*'). This number, 1.618..., was renamed as "*golden ratio*" by Martin Ohm, brother of Georg Simon Ohm (author of the Ohm's law), not so long ago, in the 1830s. The golden ratio can be observed in nature, like in the angle of rotation of the leaves of certain plants (Fig. 2.15).

Using a scheme similar to what has been illustrated in Fig. 2.12, the golden section method takes sub-intervals of $r\, D_{n-1}$ width, using the ratio:

$$r = \frac{F_{n-1}}{F_n} = 0.618,$$

Actually, the numbers of the Fibonacci sequence tend to fulfill this ratio.

Figure 2.16, which has been generated with the Program 2.8, shows the results of using the golden section method four successive times.

Program 2.8 Golden section

```
% Golden section
%
% function: y = (1.6*sin(x))-x
%
% interval determined by bracketing:
L=0.16;
xh=1.12; xl=xh-(2*L);
% for plotting the function in the bracketed interval
x=xl:0.01:xh;
y=(1.6*sin(x))-x;
%
D=2*L;
r=(3-sqrt(5))/2; %golden rule
%
for n=1:4, % we limit the searching to 4 steps
  a=xl+(r*D);
  ya=(1.6*sin(a))-a;
  b=xh-(r*D);
  yb=(1.6*sin(b))-b;
  if yb<ya,
    xh=b; %eliminate third interval
```

```
    else
      xl=a; %eliminate first interval
    end;
    D=xh-xl;
    %
    yl=(1.6*sin(xl))-xl;
    yh=(1.6*sin(xh))-xh;
    %
    % display
    figure(1)
    subplot(2,2,n)
    plot(x,y,'k');
    axis([0.79 1.12 0 0.4]);
    hold on;
    plot([xl xl],[0 yl],'b--');
    plot([xh xh],[0 yh],'r--');
end;
```

Like before, with simple modifications a visualization of the convergence of the middle point of each interval was obtained, Program 2.9. As expected, it is similar to the convergence of Fibonacci method (Fig. 2.17).

Fig. 2.15 Examples of golden ratio spirals

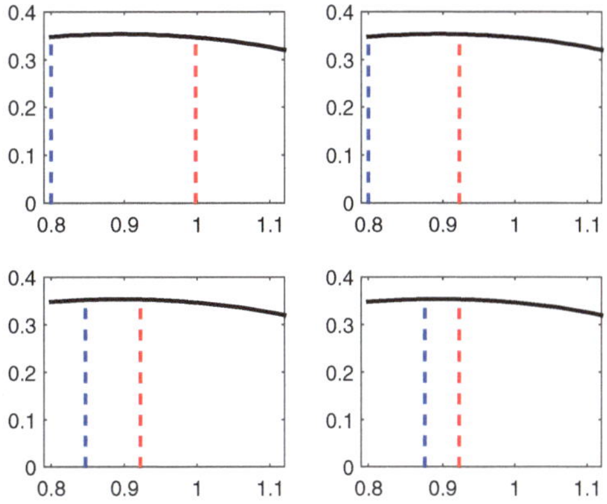

Fig. 2.16 First four steps of the golden section method

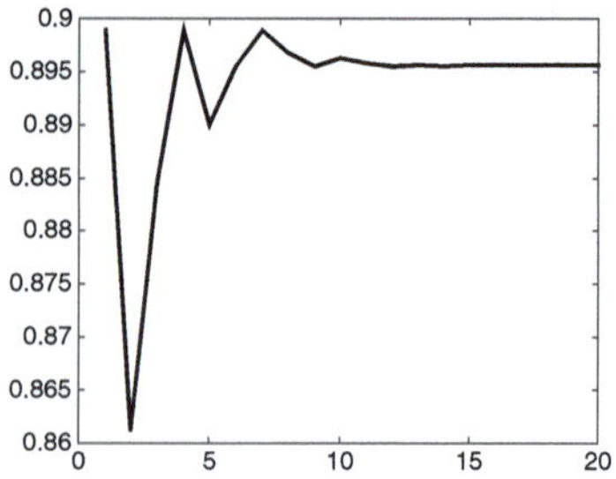

Fig. 2.17 Convergence of the middle point toward the solution

Program 2.9 Convergence of the golden section

```
% Convergence of the Golden section
% function: y = (1.6*sin(x))-x
% interval determined by bracketing:
L=0.16;
xh=1.12; xl=xh-(2*L);
% for plotting the function in the bracketed interval
x=xl:0.01:xh;
y=(1.6*sin(x))-x;
%
vm=zeros(1,20); %reserve memory for a vector
%
D=2*L;
r=(3-sqrt(5))/2; %golden rule
%
for n=1:20, % we limit the searching to 20 steps
  a=xl+(r*D);
  ya=(1.6*sin(a))-a;
  b=xh-(r*D);
  yb=(1.6*sin(b))-b;
  if yb<ya,
    xh=b; %eliminate third interval
  else
    xl=a; %eliminate first interval
  end;
  D=xh-xl;
  vm(n)=(xh+xl)/2;
  %
  % display
  figure(1)
  plot(vm,'k');
  title('evolution of middle point');
end;
```

2.2.2 *Interpolation*

In certain cases, it could be convenient to use an interpolation approach. The idea would be to devise a simple mathematical model to approximate the function $f(x)$ in the bracketed interval. Once the model was fit to $f(x)$, one computes the extremum of the model as an approximation to the solution.

Illustrative examples of this method are quadratic models, and cubic models. Let us describe in more detail the application of these models.

2.2.2.1 Quadratic Models

The quadratic model to be used has three coefficients, a, b, and c:

$$M(x) = a + bx + c\,x^2$$

Let us take three points of $f(x)$ in the interval of interest, f_1, f_2, f_3. Figure 2.18 shows an example of this.

It is not difficult to determine the coefficients of $M(x)$, such that the model crosses these three points. Just solve the following equation:

$$\begin{pmatrix} 1 & x_1 & x_1^2 \\ 1 & x_2 & x_2^2 \\ 1 & x_3 & x_3^2 \end{pmatrix} \begin{pmatrix} a \\ b \\ c \end{pmatrix} = \begin{pmatrix} f_1 \\ f_2 \\ f_3 \end{pmatrix}$$

This equation can be easily solved with MATLAB. The solution would be the values of the three coefficients of the model, a, b, and c.

Once a quadratic model was established, the position of the extremum of the model can be obtained using the derivative:

$$\frac{dM(x)}{dx} = b + 2\,c\,x = 0 \;\Rightarrow\; x = -\frac{b}{2\,c}$$

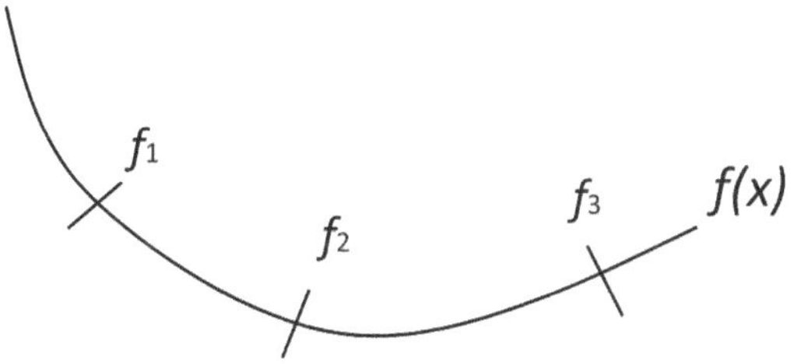

Fig. 2.18 Choosing three points of f(x)

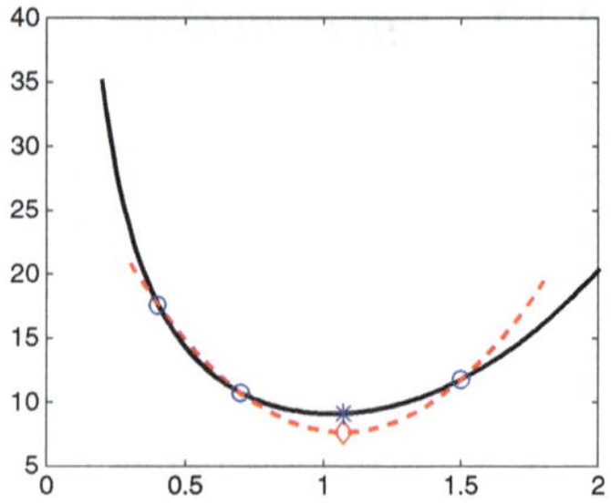

Fig. 2.19 Example of quadratic model

The second derivative is:

$$\frac{d^2M(x)}{dx^2} = 2c$$

For the extremum to be a minimum, c should be positive.

Let us put an example. Consider the following function:

$$y = 2.1\ x^3 + (7/x)$$

Figure 2.19 depicts this function in an interval of interest. Three points were chosen, and then a quadratic model was found, such that it crosses the three points. The function was drawn in black color and the model in lighter color. The minimum of the model was marked with a diamond. This value was the result of the method, which is not exactly the minimum of the function. In order to notice the difference, the minimum of the function was marked with an asterisk.

Figure 2.19 was obtained with the Program 2.10, which includes the determination of the model and finding its minimum.

The reader is invited to change the position of the three points. It would be easy to obtain a better solution, nearer to the actual minimum of the function.

Program 2.10 Using a quadratic model

```
% Using a quadratic model
%
% function: y = 2.1(x^3) + 7/x
% in the interval [0.2 2]
%
% First, we plot the function
x=0.2:0.01:2;
y=(2.1*(x.^3))+(7./x);
figure(1)
plot(x,y,'k');
hold on;
%
% Taking 3 points
```

```
x1=0.4; y1=(2.1*(x1^3))+(7/x1);
x2=0.7; y2=(2.1*(x2^3))+(7/x2);
x3=1.5; y3=(2.1*(x3^3))+(7/x3);
%
% Getting the model from Ac=b
A=[1 x1 x1^2; 1 x2 x2^2; 1 x3 x3^2];
b=[y1;y2;y3];
c=A\b;,
%
% we plot the model and the 3 points
x=0.3:0.01:1.8; % a shorter range of values
M=c(1)+(c(2).*x)+(c(3)*x.^2);
plot(x,M,'r--');
plot(x1,y1,'bo','MarkerSize',10);
plot(x2,y2,'bo','MarkerSize',10);
plot(x3,y3,'bo','MarkerSize',10);
%
% The minimum of the model is:
mx=-c(2)/(2*c(3));
my=c(1)+(c(2)*mx)+(c(3)*mx^2);
plot(mx,my,'rd','MarkerSize',10);
%
% The value of the function at mx is:
qy=(2.1*(mx^3))+(7/mx);
plot(mx,qy,'b*','MarkerSize',10);
mx
```

2.2.2.2 Cubic Models

In the case of cubic models, the first task is to fit the following cubic model,

$$M(x) = a + bx + cx^2 + dx^3$$

to the function in the interval of interest. A possible approach is to choose two points of the function and solve the following equation:

$$\begin{pmatrix} 1 & x_1 & x_1^2 & x_1^3 \\ 0 & 1 & 2x_1 & 3x_1^2 \\ 1 & x_2 & x_2^2 & x_2^3 \\ 0 & 1 & 2x_2 & 3x_2^2 \end{pmatrix} \begin{pmatrix} a \\ b \\ c \\ d \end{pmatrix} = \begin{pmatrix} f_1 \\ f_1' \\ f_2 \\ f_2' \end{pmatrix},$$

where f_1', f_2', are derivatives.

Again, this type of equations can be easily solved with MATLAB. The solution would be the values of the four coefficients of the model. The extremum of the

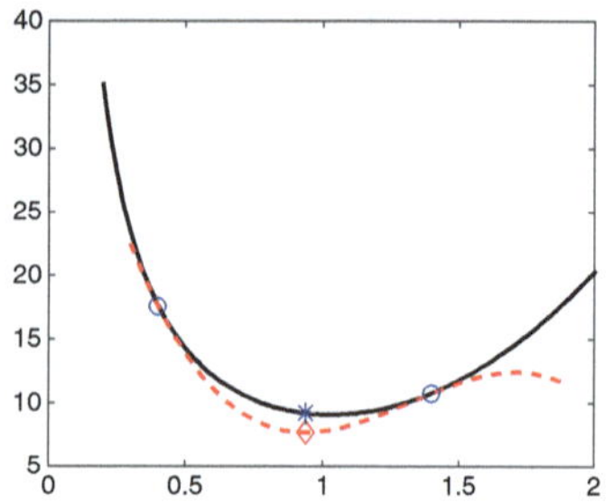

Fig. 2.20 Example of cubic model

model would take place at:

$$\frac{dM(x)}{dx} = b + 2c\,x + 3d\,x^2 = 0 \;\Rightarrow\; x = -\frac{-c \pm \sqrt{c^2 - 3b\,d}}{3d}$$

The second derivative is:

$$\frac{d^2M(x)}{dx^2} = 2c + 6d\,x$$

which should be positive for the extremum to be a minimum.

Continuing with the same example as before, let us choose two points and try the cubic model. The result is shown in Fig. 2.20, which was generated with the Program 2.11.

Again, the reader is invited to get better results by choosing different points.

Program 2.11 Using a cubic model

```
% Using a cubic model
% function: y = 2.1(x^3) + 7/x
% in the interval [0.2 2]
%
% First, we plot the function
x=0.2:0.01:2;
y=(2.1*(x.^3))+(7./x);
figure (1)
plot(x,y,'k');
hold on;
%
% Taking 2 points
x1=0.4; y1=(2.1*(x1^3))+(7/x1);
yd1=(6.3*(x1^2))-(7/(x1^2));
x2=1.4; y2=(2.1*(x2^3))+(7/x2);
yd2=(6.3*(x2^2))-(7/(x2^2));
%
% Getting the model from Ac=b
A=[1 x1 x1^2 x1^3; 0 1 2*x1 3*(x1^2);...
```

```
 1 x2 x2^2 x2^3; 0 1 2*x2 3*(x2^2)];
b=[y1;yd1;y2;yd2];
c=A\b;,
%
% we plot the model and the 3 points
x=0.3:0.01:1.9; % a shorter range of values
M=c(1)+(c(2).*x)+(c(3)*(x.^2))+(c(4)*(x.^3));
plot(x,M,'r--');
plot(x1,y1,'bo','MarkerSize',10);
plot(x2,y2,'bo','MarkerSize',10);
%
% The minimum of the model is:
mx=(-c(3)+sqrt(c(3)^2-(3*c(2)*c(4))))/(3*c(4));
my=c(1)+(c(2)*mx)+(c(3)*mx^2)+(c(4)*mx^3);
plot(mx,my,'rd','MarkerSize',10);
%
% The value of the function at mx is:
qy=(2.1*(mx^3))+(7/mx);
plot(mx,qy,'b*','MarkerSize',10);
mx
```

2.2.3 *Derivative-Based Methods*

There are some important searching methods that are based on derivatives, enjoying in general faster convergence rates. Although the historical details are intricate, one could point at Newton as a main reference for this type of methods.

2.2.3.1 The Newton-Raphson Method

Consider the following expression, obviously related to a truncated Taylor expansion:

$$f'(x) = f'(x_1) + f''(x_1)\,(x - x_1)$$

In an extremum point, it would happen that $f'(x) = 0$ and so:

$$x_{j+1} = x_j - \frac{f'(x_j)}{f''(x_j)}$$

(where a slight notation change has been used, to suggest successive iterations)

The iterations could not converge to the solution x* if the starting point was relatively far from x*.

This method is frequently named as the "*method of tangents*". Others prefer to cite it as the Newton-Raphson method.

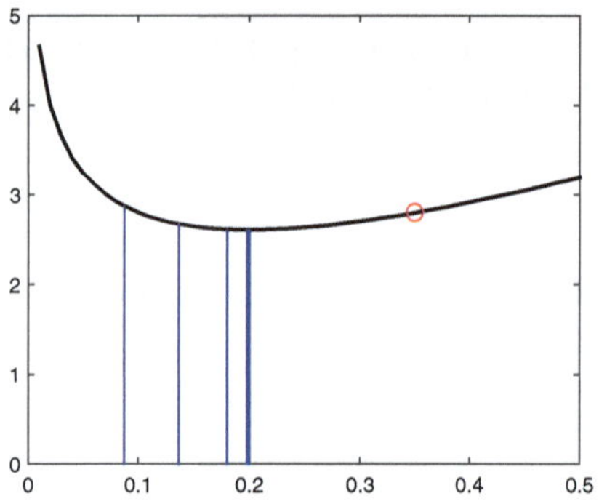

Fig. 2.21 Example of Newton-Raphson method

In order to illustrate the method, let us try to find, for example, the minimum of the following function:

$$y = 5x - \log(x)$$

Figure 2.21 shows the successive results of the method iterations, starting from $x = 0.35$ (marked with a circle). From the 5th iteration the results become quite similar (convergence). The reader could try other starting points.

Figure 2.21 has been obtained with the Program 2.12, which is not too long.

Program 2.12 Using the Newton-Raphson method

```
% Using the Newton-Raphson method
%
% function: y = 5x - log(x);
%
% First, we plot the function
x=0.01:0.01:0.5;
y=5*x-log(x);
figure (1)
plot(x,y,'k');
hold on;
%
% We start from:
x=0.35;
plot(x,(5*x-log(x)),'ro');
%
for n=1:5, %we limit the searching
  % the derivative values are:
  dy=5-(1/x);
  %(from dy=0 we know the solution is x=0.2)
  hy=1/(x^2);
  % the next value of x is
  xnew=x-(dy/hy);
  x=xnew;
```

```
    y=5*x-log(x);
    plot([x x],[0 y],'b');
end
x
```

2.2.3.2 The Secant Method

Suppose that the solution is in an interval of interest $[x_a\ x_b]$. Then $f'(x_a)$ and $f'(x_b)$ have different sign.

Let us approximate the second derivative as follows:

$$f''(x_j) = \frac{f'(x_a) - f'(x_b)}{(x_a - x_b)}$$

Now, by substitution in the Newton-Raphson iteration, one obtains:

$$x_{j+1} = x_j - \frac{f'(x_a)(x_b - x_a)}{f'(x_b) - f'(x_a)}$$

The new point x_{j+1} replaces x_a or x_b depending on the sign of $f'(x_{j+1})$.

Consider, for example, that one wants to find the minimum of the following function:

$$y = -2x\ \sin(0.8\,x) + \exp(-2\,x)$$

Figure 2.22 shows the successive results of the method iterations, starting from $x_a = 1$ (marked as a diamond on the left) and $x_b = 5.3$ (marked as another diamond, on the right). The color of the lines corresponding to iterations is lighter for positive derivative and darker for negative derivative.

Figure 2.22 has been obtained with the Program 2.13. The program shows how the sign of the derivative is considered for replacing x_a or x_b.

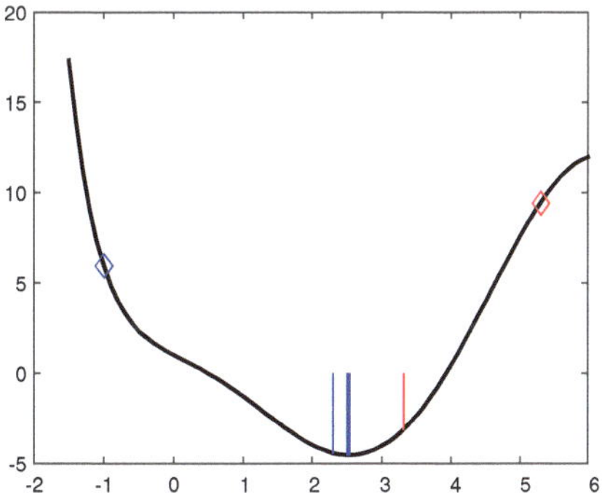

Fig. 2.22 Example of secant method

Program 2.13 Using the secant method

```
% Using the secant method
%
% function: y = -2x sin(0.8x)+exp(-2x);
%
% First, we plot the function
x=-1.5:0.1:6;
y=-2*x.*sin(0.8*x)+exp(-2*x);
figure (1)
plot(x,y,'k');
title('Secant method');
hold on;
%
% We start from:
a=-1; b=5.3;
% display of initial bounds
ya=-2*a*sin(0.8*a)+exp(-2*a);
yb=-2*b*sin(0.8*b)+exp(-2*b);
plot(a,ya,'bd');
plot(b,yb,'rd');
%
% Start of iterations
for n=1:5, %we limit the searching
  % the derivative values are:
  k=2*0.8;
  dya=-2*sin(0.8*a)-k*a*cos(0.8*a)-2*exp(-2*a);
 dyb=-2*sin(0.8*b)-k*b*cos(0.8*b)-2*exp(-2*b);
  % the quotient is:
  q=((b-a)*dya)/(dyb-dya);
  %the new point is:
  c=a-q;
  %the derivative at c is:
  dyc=-2*sin(0.8*c)-k*c*cos(0.8*c)-2*exp(-2*c);
  %replacing a or b:
  if dyc>0,
    b=c;
    yb=-2*b*sin(0.8*b)+exp(-2*b);
    plot([b b],[0 yb],'r');
  else
    a=c;
    ya=-2*a*sin(0.8*a)+exp(-2*a);
    plot([a a],[0 ya],'b');
  end;
  %
end
b
yb
```

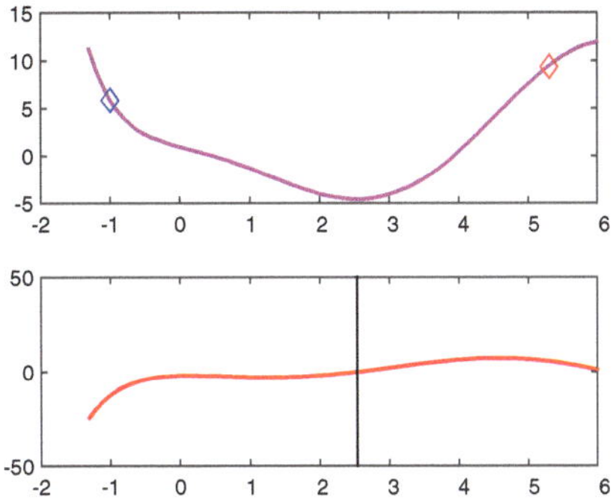

Fig. 2.23 The function under study and its derivative

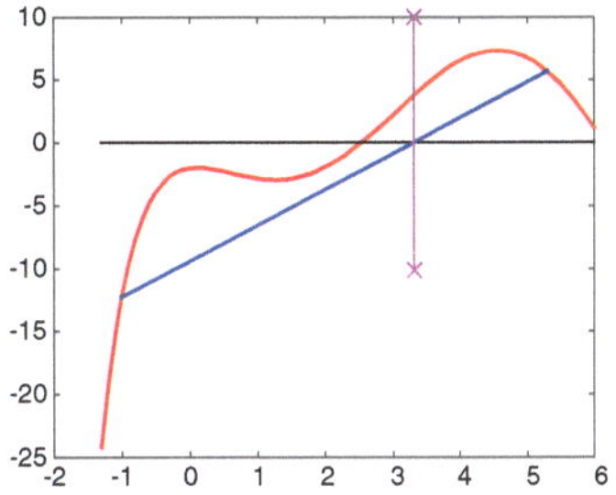

Fig. 2.24 Example of secant method, first iteration

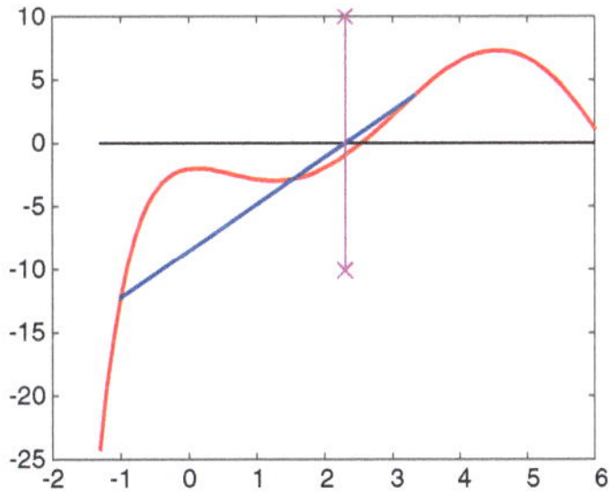

Fig. 2.25 Example of secant method, second iteration

Again, the reader could move the initial values of x_a and x_b to see what happens. It could become somewhat surprising, that's why a detailed description of what occurs with this example is now given.

Figure 2.23 includes two plots. On top, the function subject to study. Below, the derivative of the function, with a black vertical line indicating where the derivative is zero.

Figure 2.24 corresponds to the first iteration. The gray curve is the derivative. A dark line is drawn from $f'(x_a)$ to $f'(x_b)$. The intersection of this line with the horizontal line (value = 0) determines the new x_b (since the derivative is positive at this intersection).

In the second iteration, Fig. 2.25, the dark line joining $f'(x_a)$ and $f'(x_b)$ gives a new intersection. The derivative at this intersection is negative and therefore the intersection determines the new x_a.

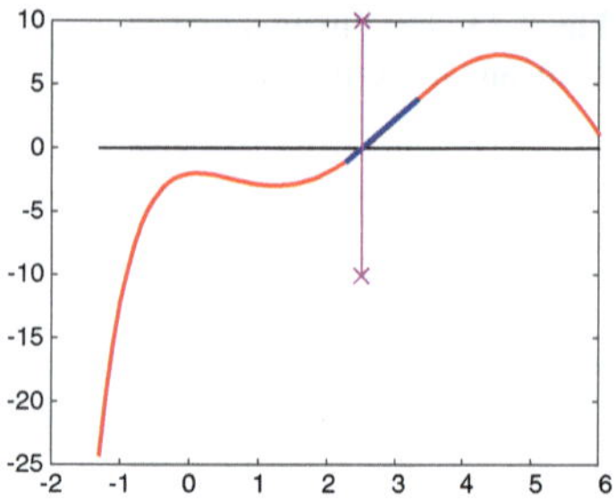

Fig. 2.26 Example of secant method, third iteration

In the third iteration, Fig. 2.26, there is a new intersection. Although it is difficult to see, the derivative at the intersection is negative, and so the intersection determines a new x_a.

Let us stop here. More figures could be obtained by simple changes in the Program 2.14.

Program 2.14 Step by step secant method

```
% Using the secant method
% step by step illustration
%
% function: y = -2x sin(0.8x)+exp(-2x);
%
% First, we plot the function
% and its derivative
x=-1.2:0.1:6;
y=-2*x.*sin(0.8*x)+exp(-2*x);
k=2*0.8;
dy=-2*sin(0.8*x)-k*x.*cos(0.8*x)-2*exp(-2*x);
%
figure (1)
subplot(2,1,1)
plot(x,y,'k');
title('the function')
hold on;
subplot(2,1,2)
plot(x,dy,'r');
title('the derivative; dy=0 black line');
hold on;
% place where dy=0
plot([2.5388 2.5388], [-50 50],'k');
%
% We start from:
a=-1; b=5.3;
% Note: the sign of the derivative must be different
% at a and at b
% display of initial bounds
ya=-2*a*sin(0.8*a)+exp(-2*a);
```

```
yb=-2*b*sin(0.8*b)+exp(-2*b);
subplot(2,1,1)
plot(a,ya,'bd');
plot(b,yb,'rd');
%
% First iteration -----------------------------
% the derivative values at a an b, are:
dya=-2*sin(0.8*a)-k*a*cos(0.8*a)-2*exp(-2*a);
dyb=-2*sin(0.8*b)-k*b*cos(0.8*b)-2*exp(-2*b);
% the quotient is:
q=((b-a)*dya)/(dyb-dya);
%the new point is:
c=a-q % for user info
% see what happens with derivatives and the new point
figure(2)
plot(x,dy,'r'); hold on;
plot([x(1) x(end)], [0 0],'k'); % horizontal axis
plot([a b],[dya dyb],'b');
plot([c c],[-10 10],'mx-'); %point c
title('derivative, and first iteration');
% Note1: the point c is the intersection of line
% between a and b (blue),
% and the horizontal axis (black line)
% Note2: the minimum is at the intersection of the
% derivative curve (red) and the horizontal axis
%
% Second iteration -----------------------------
% the new point b would be:
b=c;
% then:
yb=-2*b*sin(0.8*b)+exp(-2*b);
dyb=-2*sin(0.8*b)-k*b*cos(0.8*b)-2*exp(-2*b);
% the quotient is:
q=((b-a)*dya)/(dyb-dya);
%the new point is:
c=a-q % for user info
% see what happens with derivatives and the new point
figure(3)
plot(x,dy,'r'); hold on;
plot([x(1) x(end)], [0 0],'k'); % horizontal axis
plot([a b],[dya dyb],'b');
plot([c c],[-10 10],'mx-'); %point c
title('derivative, and second iteration');
% the derivative at c is:
dyc=-2*sin(0.8*c)-k*c*cos(0.8*c)-2*exp(-2*c)
% which is negative, and so the new point a is:
a=c;
%
% Third iteration -----------------------------
ya=-2*a*sin(0.8*a)+exp(-2*a);
dya=-2*sin(0.8*a)-k*a*cos(0.8*a)-2*exp(-2*a);
```

```
% the quotient is:
q=((b-a)*dya)/(dyb-dya);
%the new point is:
c=a-q % for user info
% see what happens with derivatives and the new point
figure(4)
plot(x,dy,'r'); hold on;
plot([x(1) x(end)], [0 0],'k'); % horizontal axis
plot([a b],[dya dyb],'b');
plot([c c],[-10 10],'mx-'); %point c
title('derivative, and third iteration');
% the derivative at c is:
dyc=-2*sin(0.8*c)-k*c*cos(0.8*c)-2*exp(-2*c)
% which is negative, and so the new point a is:
a=c;
% And so on
```

It should be said that several variants of the methods just introduced can be obtained by using different approximation of the derivatives.

2.3 Basic Multidimensional Searching

When searching in one-dimensional problems, it is only possible to move to the left or to the right. Obviously, this is not the case in multidimensional problems; one has to choose searching directions among infinite possibilities.

In this section a set of basic methods for multidimensional searching will be introduced. Most of them use gradients.

Some of the methods mentioned in this chapter would look familiar to readers with background in numerical calculus. A main reason is that if you consider the following function to be minimized:

$$f(\vec{x}) = -\frac{1}{2}\vec{x}^T A\,\vec{x} - \vec{x}^T\vec{b}$$

then (theorem), if A is a symmetric positive definite matrix, there is exactly one $\vec{x}^*$ for which:

$$f(\vec{x}^*) = \min_x f(\vec{x})$$

and that $\vec{x}^*$ is the solution of the linear system $A\,\vec{x} = \vec{b}$.

Hence, the searching methods presented in this chapter can be used as iterative methods for the solution of equations.

Notice that in general $A\,\vec{x} = \vec{b}$ can have one solution, infinite solutions, or no solution at all, depending on A. This also concerns the optimization context.

2.3.1 Steepest Ascent

An intuitive example for introducing steepest ascent is hill climbing. One tries to reach the top, step by step. Supposing there are no paths to follow, you should decide the direction of each step. It seems efficient to always choose the steepest one. Of course, this example deals with a multidimensional scenario.

It is opportune to consider contours, which have been already used in the previous chapter. One is tempted now, to excurse a little bit on the connection of contours and topographic maps. These maps are useful when you wander through difficult mountains. Typically, such maps use isolines to represent the relief. Figure 2.27 illustrates this feature.

Figure 2.28 shows an example of topographic map with isolines. It corresponds to Hawaii.

Now, let us analyze the consequences of taking steepest steps. The direction of largest variation of a function is given by its gradient. If you use minus the gradient, you are choosing the steepest descent sense: while the gradient leads to steepest ascent.

Consider the contours depicted in Fig. 2.29. They are concentric circumferences. Then, departing from any initial position, the gradient would point to the center,

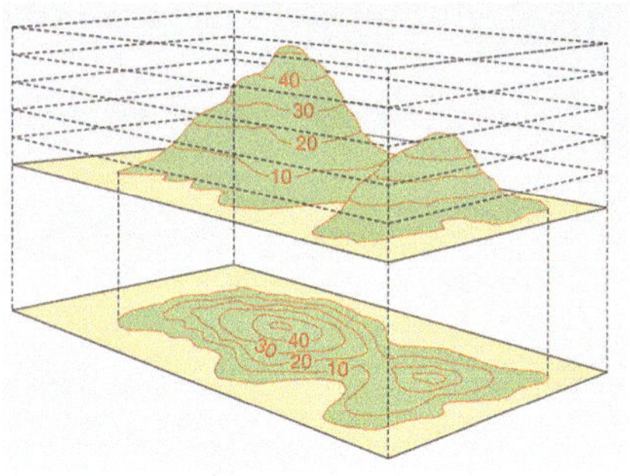

Fig. 2.27 Using isolines for relief representation

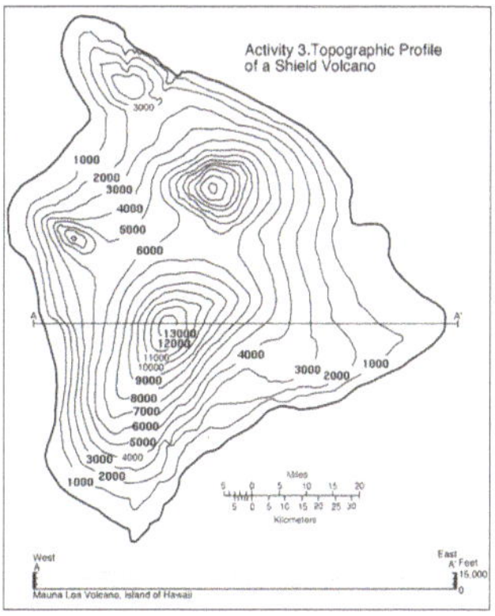

Fig. 2.28 Example of topographic map

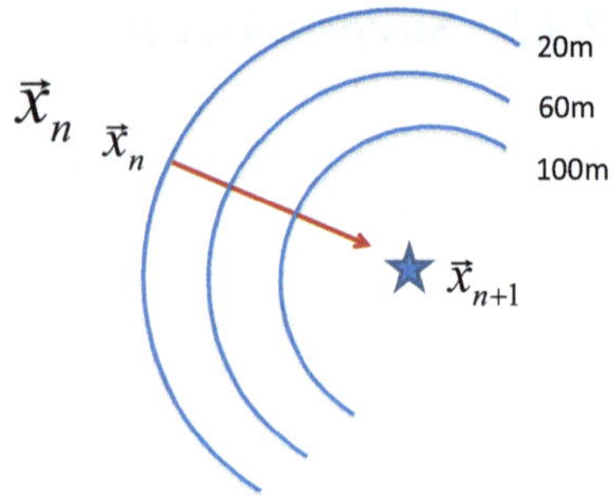

Fig. 2.29 Steepest ascent in case of circular concentric contours

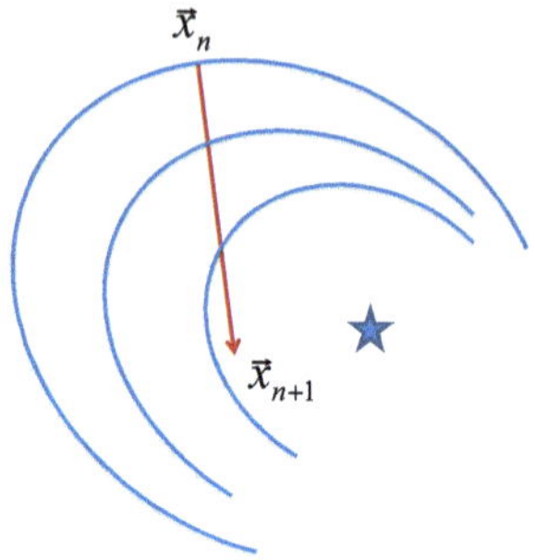

Fig. 2.30 A case of non-circular contours

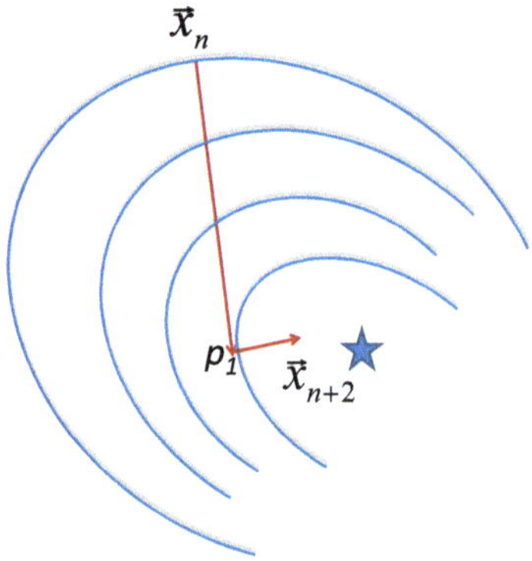

Fig. 2.31 After the first step, a second one

where the maximum is. Any unidimensional search along the line in the gradient direction will give you the optimal solution. In other words, the solution would be reached in just one step (with the proper size).

However, if it was not the case of concentric circumferences, more steps would be needed. Suppose, for example, the contours depicted in Fig. 2.30.

In most cases the gradient would not point to the maximum. A line search along the gradient direction would take you to a point $p1$, with maximal value. At $p1$, the line in the gradient direction would be tangent to a certain contour. There, you should take another step. Keeping the decision rule, one takes now the direction of the gradient at $p1$. The new line would be perpendicular to the contour passing through $p1$. See Fig. 2.31.

Usually, the steepest ascent rule consequence is a zigzag trajectory of successive steps toward the maximum.

In mathematical terms, the procedure is the following. Given a function $f(\vec{x}_n)$, the steepest step would be:

$$\vec{x}_{n+1} = \vec{x}_n + \beta \cdot \nabla f(\vec{x}_n)$$

where β would be the solution of:

$$\min_{\beta} f(\vec{x}_n + \beta \cdot \nabla f(\vec{x}_n))$$

A case of particular interest is when the function is quadratic, that is:

$$f(\vec{x}) = \frac{1}{2}\vec{x}^T Q \vec{x} + \vec{q}^T \vec{x}$$

where Q is a positive definite symmetric matrix.

The minimum of this function is located at:

$$\vec{x}* = -Q^{-1}\vec{q}$$

So, the minimum would be:

$$f(\vec{x}*) = -\frac{1}{2}\vec{q}^T Q^{-1}\vec{q}$$

The steepest descent would be based on minus the gradient:

$$\vec{d} = -Q\vec{x} - \vec{q}$$

The optimal step size from the point $\vec{x}$ would be:

$$\beta = \frac{\vec{d}^T \vec{d}}{\vec{d}^T Q \vec{d}}$$

The number of zigzags required to reach the solution depends on the ratio A/a, where A is the largest eigenvalue of Q and a the smallest eigenvalue of Q. For example, if this ratio was 3 there would be 6 steps, if the ratio was 100 then 58 steps would be needed.

Program 2.15 considers a simple quadratic function:

$$y = 2x_1^2 + 7x_2^2$$

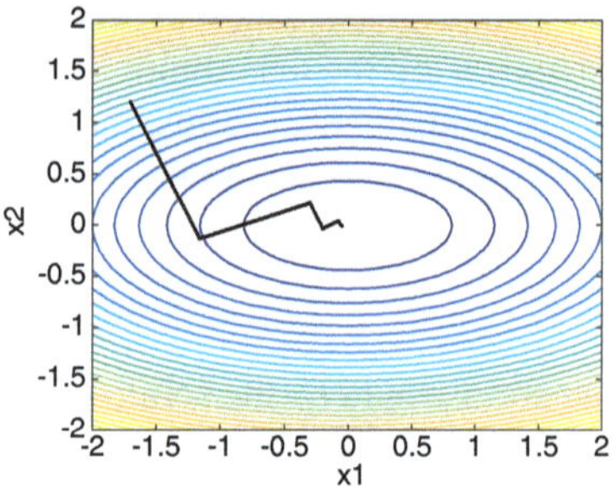

Fig. 2.32 Example of steepest descent

The minimum of this function is obviously at (0, 0). Figure 2.32 shows the result of the steepest descent procedure, which is a series of zigzags. The reader is invited to change aspects of the problem, like the initial point or the function.

Program 2.15 has been coded in a simple form for readability. It includes basic line searching to determine the points for starting new steps. The sign of the gradient component on axis $x1$ was employed for determining the sense of the searching (in this case from left to right).

Program 2.15 Steepest descent

```
% steepest descent example
x1=-2:0.05:2;
x2=-2:0.05:2;
N=length(x1);
y=zeros(N,N); %reserve memory
%
%set of function values
for i=1:N,
  for j=1:N,
    y(j,i)=2*(x1(i)^2)+7*(x2(j)^2);
  end;
end;
%function display--------
figure (1)
contour(x1,x2,y,26); hold on;
xlabel('x1'); ylabel('x2');
% searching-------------
% initial point
x1i=-1.7; x2i=1.2;
yi=2*(x1i^2)+7*(x2i^2);
g=[(4*x1i),(14*x2i)]; %gradient value
% sign of g along x1:
s=1;
if g(1)<0, s=-1; end;
%
% first search segment
m=g(2)/g(1); %slope
```

```
flag=1; yref=yi;
x1=x1i;
while flag==1,
  x2=m*(x1-x1i)+x2i;
  y=2*(x1^2)+7*(x2^2);
  if y>yref,
    flag=0;
  else
    yref=y;
    plot(x1,x2,'k.');
    x1o=x1; x2o=x2;
    x1=x1-(s*0.01);
  end;
end;
% more search segments
for M=1:4,
  x1i=x1o; x2i=x2o;
  x1=x1i;
  g=[(4*x1i),(14*x2i)]; %gradient value
  m=g(2)/g(1); %slope
  flag=1;
  while flag==1,
    x2=m*(x1-x1i)+x2i;
    y=2*(x1^2)+7*(x2^2);
    if y>yref,
      flag=0;
    else
      yref=y;
      plot(x1,x2,'k.');
      x1o=x1; x2o=x2;
      x1=x1-(s*0.01);
    end;
  end;
end;
```

For comparison purposes it is also convenient to see what happens with a non-quadratic function, like the following example:

$$y = \exp(-(x_1 - 3)/2) + \exp((x_1 + 4x_2)/10) + \exp((x_1 - 4x_2)/10)$$

Figure 2.33 depicts the behavior of the steepest descent method, which obtains the result after a series of zigzags.

Figure 2.33 has been generated with the Program 2.16, which—compared with Program 2.15—has been coded in a more compact form.

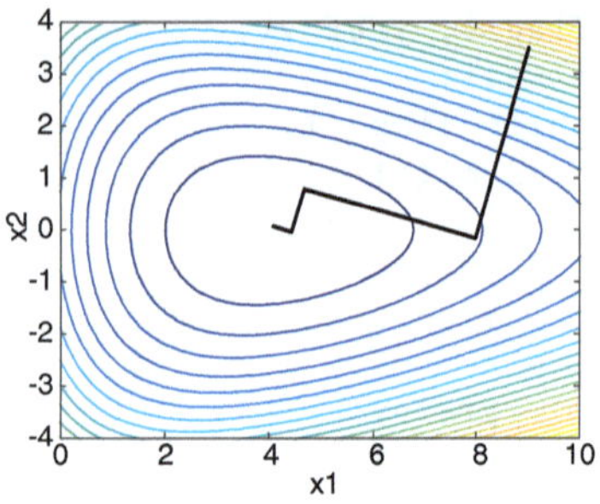

Fig. 2.33 Another example of steepest descent

Program 2.16 Steepest descent example

```
%example of steepest descent
%
[vx1,vx2]=meshgrid(0:0.1:10, -4:0.1:4);
y=(exp(-(vx1-3)/2))+(exp((vx1+4*vx2)/10))...
+(exp((vx1-4*vx2)/10));
%
% function display-------------
figure (1)
contour(vx1,vx2,y,20); hold on;
xlabel('x1'); ylabel('x2');
%
% searching ------------------
% initial point
x1i=9; x2i=3.5;
% segments
for M=1:4,
  %
  yi=(exp(-(x1i-3)/2))+(exp((x1i+4*x2i)/10))...
  +(exp((x1i-4*x2i)/10));
  % gradient
  aux1=(-0.5*exp(-(x1i-3)/2))...
  +(0.1*exp((x1i+4*x2i)/10))...
  +(0.1*exp((x1i-4*x2i)/10));
  aux2=(0.4*exp((x1i+4*x2i)/10))...
  -(0.4*exp((x1i-4*x2i)/10));
  g=[aux1,aux2];
  m=g(2)/g(1); % slope
  % sign of g along x1:
  s=1;
  if g(1)<0, s=-1; end;
  %
  flag=1; yref=yi;
  x1=x1i;
  while flag==1,
    x2=m*(x1-x1i)+x2i;
    y=(exp(-(x1-3)/2))+(exp((x1+4*x2)/10))...
    +(exp((x1-4*x2)/10));
    if y>yref,
```

```
      flag=0;
    else
      yref=y;plot(x1, x2, 'k.');
      x1o=x1; x2o=x2;
      x1=x1-(s*0.01);
    end;
  end;
  %
  x1i=x1o; x2i=x2o;
end;
```

The non-quadratic example will be used again in the next subsections.

2.3.2 Methods Based on Conjugate Vectors

A number of methods has been proposed in order to decrease the zigzags. An important part of these methods is based on conjugate vectors. The directions of these vectors can be built in several ways, like, for instance, taking into account gradients.

2.3.2.1 Conjugate Vectors

It is said that two vectors $\vec{v}_1$ and $\vec{v}_2$ are *conjugate*, or *Q-orthogonal*, with respect to a symmetric matrix Q if:

$$\vec{v}_1^T\, Q\, \vec{v}_2 = 0$$

Orthogonal vectors are a particular case, for $Q = I$.

A finite set of vectors $\vec{v}_1, \vec{v}_2, \cdots, \vec{v}_n$ is said to be a *Q-orthogonal set* if $\vec{v}_i^T\, H\, \vec{v}_j = 0$ for all $i \neq j$.

It can be shown that if Q is symmetric positive definite, the vectors belonging to the Q-orthogonal set are linearly independent [13]. Then, these vectors can be used as a basis for $\Re^n$.

Consider the minimization of a quadratic function:

$$\min\left(\frac{1}{2}\vec{x}^T\, Q\, \vec{x} + \vec{q}^T \vec{x}\right)$$

with Q symmetric positive definite.

This problem is equivalent to solving:

$$Q\, \vec{x} + \vec{q} = 0$$

Let us put the solution $\vec{x}*$ in terms of the linear basis:

$$\vec{x}* = \alpha_1 \vec{v}_1 + \alpha_2 \vec{v}_2 + \ldots + \alpha_n \vec{v}_n$$

Therefore:

$$Q\,(\alpha_1 \vec{v}_1 + \alpha_2 \vec{v}_2 + \ldots + \alpha_n \vec{v}_n) + \vec{q} = 0$$

And then:

$$\vec{v}_i^T\, Q\,(\alpha_1 \vec{v}_1 + \alpha_2 \vec{v}_2 + \ldots + \alpha_n \vec{v}_n) + \vec{v}_i^T \vec{q} = 0$$

Since the vectors are conjugate, the values of α_i can be obtained with:

$$\alpha_i = -\frac{\vec{v}_i^T\, \vec{q}_i}{\vec{v}_i^T\, Q\, \vec{v}_i}$$

In order to propose an iterative procedure, the following fact can be demonstrated. Suppose one starts from an initial $\vec{x}_0$ and then one uses the following rule for generating a series of steps:

$$\vec{x}_{k+1} = \vec{x}_k + \alpha_k \vec{v}_k \qquad (1)$$

Then, the sequence $\vec{x}_0,\ \vec{x}_1,\ \ldots,\ \vec{x}_k$ converges to $\vec{x}*$ in n steps.

Denote as A_k the subspace generated by $\{\vec{x}_0,\ \vec{x}_1,\ \ldots,\ \vec{x}_{k-1}\}$. It can be shown that the gradient at $\vec{x}_k$ is orthogonal to A_k.

2.3.2.2 Powell's Method (Conjugate Directions)

The method of Powell does not use derivatives. It was proposed in a scientific article in 1964 [50]. This article has been cited thousands of times.

Suppose a problem in three dimensions. Choose an initial point $\vec{x}_0$ and explore on the x-axis for a maximum (you can apply any line search method), which would happen at certain point $\vec{p}_x$. Then, from $\vec{p}_x$ explore on the y-axis for a maximum, getting a new point, $\vec{p}_y$. Then, from $\vec{p}_y$ explore on the z-axis for a maximum, obtaining the point $\vec{p}_z$. Now one forms the direction given by $\vec{d}_{\vec{1}} = (\vec{p}_z - \vec{x}_0)$, and explores on this direction for a maximum, obtaining the point $\vec{x}_1$. After that, one prepares a new set of directions D for the next step. The old set was $D_0 = \{\vec{d}_x, \vec{d}_y,\ \vec{d}_z\}$; the new one would be $D_1 = \{\vec{d}_y,\ \vec{d}_z,\ \vec{d}_1\}$.

In other terms: one discards the first direction that was explored (the x-axis), left-shift the other directions (y-axis, z-axis) and adds the direction $\vec{d}_1$ (which would be conjugate of the other two directions). Then, next four explorations would take place from the point $\vec{x}_1$. The procedure is repeated (including discarding the first direction, shifting and adding) until a satisfactory solution is reached.

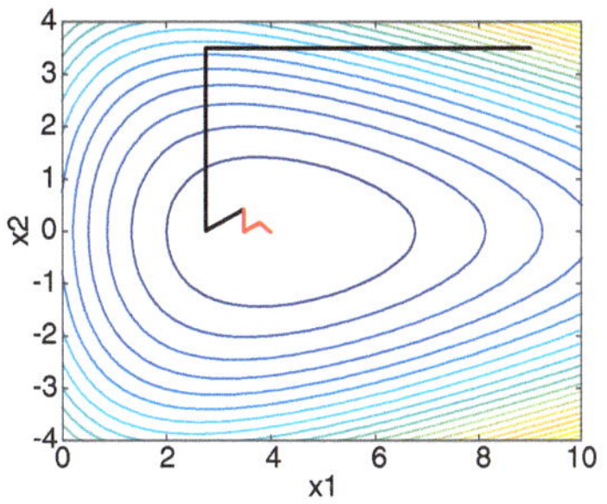

Fig. 2.34 Example of Powell's method

The generalization for problems in n-dimensions is evident.

Figure 2.34 shows the behavior of the method in the non-quadratic example employed before. The figure has been obtained with the Program 2.17. It shows only two epochs of the procedure, the first epoch in black and the second in gray. Here we mean with epoch a complete exploration of a set of directions D.

Notice in Program 2.17 that the sense of each line search was specified by hand according with observation of exploration progress (this could be done in software by testing both senses before doing a line search)

Program 2.17 Powell's method

```
% Example of Powell's method
%
% function y=(exp(-(x1-3)/2))+(exp((x1+4*x2)/10))...
% +(exp((x1-4*x2)/10));
%
[vx1,vx2]=meshgrid(0:0.1:10, -4:0.1:4);
y=(exp(-(vx1-3)/2))+(exp((vx1+4*vx2)/10))...
+(exp((vx1-4*vx2)/10));
%
% function display-------------
figure (1)
contour(vx1,vx2,y,20); hold on;
xlabel('x1'); ylabel('x2');
%
% First epoch-----------------
% initial point
x1i=9; x2i=3.5;
yi=(exp(-(x1i-3)/2))+(exp((x1i+4*x2i)/10))...
+(exp((x1i-4*x2i)/10));
%
x1=x1i; x2=x2i;
yref=yi;
for M=1:3, % segments counter
  %
  flag=1;
  if M==3,
    m1=(x2-x2i)/(x1-x1i); %slope
  end;
```

```
  %
  % three explorations
  while flag==1,
    y=(exp(-(x1-3)/2))+(exp((x1+4*x2)/10))...
    +(exp((x1-4*x2)/10));
    if y>yref,
      flag=0; x1=x1o; x2=x2o;
    else
      yref=y; plot(x1,x2,'k.');
      x1o=x1; x2o=x2;
      switch M,
        case 1,
          x1=x1-0.01; %from right to left
        case 2
          x2=x2-0.01; %downwards
        case 3,
          %from left to right:
          x1=x1+0.01; x2=m1*(x1-x1i)+x2i;
      end;
    end;
  end;
end;
%
% Second epoch------------------
x1i=x1; x2i=x2;
for M=1:3, % segments counter
  %
  flag=1;
  if M==3,
    m2=(x2-x2i)/(x1-x1i); %slope
  end;
  %
  % three explorations
  while flag==1,
    y=(exp(-(x1-3)/2))+(exp((x1+4*x2)/10))...
    +(exp((x1-4*x2)/10));
    if y>yref,
      flag=0; x1=x1o; x2=x2o;
    else
      yref=y; plot(x1,x2,'r.');
      x1o=x1; x2o=x2;
      switch M,
        case 1,
          x2=x2-0.01; %downwards
          x1p=x1; x2p=x2;
        case 2,
          %from left to right
          x1=x1+0.001; x2=m1*(x1-x1p)+x2p;
        case 3
          %from left to right
          x1=x1+0.01; x2=m2*(x1-x1i)+x2i;
      end;
    end;
  end;
end;
```

2.3.2.3 Conjugate Gradients

Consider a slight notation change in (1):

$$\vec{x}_{k+1} = \vec{x}_k + \alpha_k \vec{d}_k$$

where $\vec{d}_k$ belongs to a set of conjugate directions.

Conjugate gradient methods use the following rule for iterative generation of the directions:

$$\vec{d}_{k+1} = g_k + \beta_k \vec{d}_k$$

where $\vec{g}_k$ is the gradient at the current location (a negative sign would be used in case of descent).

The essence of this methodology is to combine the current gradient with the gradient in the previous step (keeping some memory of the searching history).

Define $\vec{y}_k = g_{k+1} - g_k$. In the original proposal on using conjugate gradients [32], the value of β_k was computed as follows:

$$\beta_k = -\frac{\vec{g}_{k+1}^T \vec{y}_k}{\vec{d}_k^T \vec{y}_k}$$

Generally, conjugate gradient methods compare favorably with respect to steepest ascent.

2.3.2.4 Fletcher-Reeves Method (Conjugate Gradients)

The scientific article in which Fletcher and Reeves introduced their method was published in 1964 [18], in the same journal as the article of Powell. It is also a heavily cited article. A main feature of the proposed method was that it contemplates its use for nonlinear optimization.

The algorithm of Fletcher-Reeves is based on the general outline of the conjugate gradient methodology, with two changes. One is that the computation of α_k is replaced by a line search. The other is that β_k is computed as follows:

$$\beta_k = -\frac{\vec{g}_{k+1}^T \vec{g}_{k+1}}{\vec{g}_k^T \vec{g}_k}$$

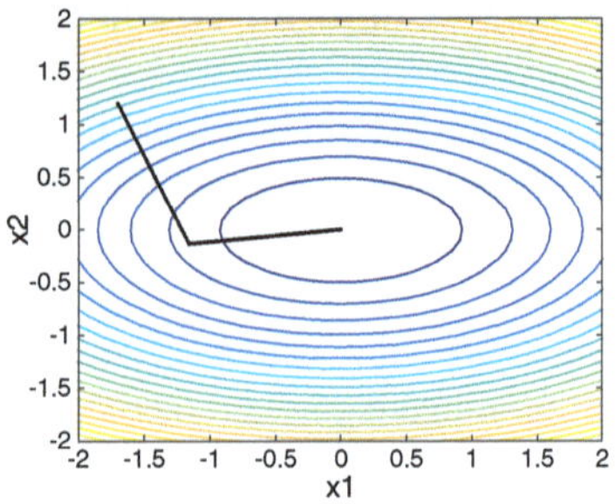

Fig. 2.35 Example of Fletcher-Reeves method

Here is the algorithm of Fletcher-Reeves for minimization:

1. Choose an initial point $\vec{x}_0$
2. Find by line search the minimum along the direction $\vec{d}_0 = -\nabla f(\vec{x}_0)$, which will be located at certain $\vec{x}_1^*$
3. From $\vec{x}_1^*$ find by line search the minimum along the direction:

$$\vec{d}_1 = -\nabla f(x_1^*) + \frac{\left\| \nabla f(x_1^*) \right\|^2}{\left\| \nabla f(x_0^*) \right\|^2} \vec{d}_0$$

so a new point $\vec{x}_2^*$ is determined

And then loop back to 3, taking $\vec{x}_2^*$ as initial point.

Figure 2.35 shows the good result of using Fletcher-Reeves for the quadratic function example. The figure has been obtained with the Program 2.18. It suffices with five steps for the algorithm to get close to the solution.

Program 2.18 Fletcher-Reeves method

```
% Example of Fletcher-Reeves method
%
v=-2:0.1:2;
[vx1,vx2]=meshgrid(v);
y=(2*vx1.^2)+(7*vx2.^2);
%
% function display-------------
figure (1)
contour(vx1,vx2,y,20); hold on;
xlabel('x1'); ylabel('x2');
% initial point
x1i=-1.7; x2i=1.2;
yi=2*(x1i^2)+7*(x2i^2);
% gradient
g=[(4*x1i),(14*x2i)];
% sign of g along x1:
s=1;
```

```
if g(1)<0, s=-1; end;
%
% first search segment
m=g(2)/g(1); %slope
flag=1; yref=yi;
x1=x1i;
while flag==1,
  x2=m*(x1-x1i)+x2i;
  y=2*(x1^2)+7*(x2^2);
  if y>yref,
    flag=0;
  else
    yref=y;
    plot(x1,x2,'k.');
    x1o=x1; x2o=x2;
    x1=x1-(s*0.01);
  end;
end;
%
% more search segments
x1i=x1o; x2i=x2o;
go=g; po=-g;
%
for M=1:4,
  yi=2*(x1i^2)+7*(x2i^2);
  g=[(4*x1i),(14*x2i)]; % gradient at current corner
  %
  q=(norm(g))^2/(norm(go)^2);
  p=-g+(q*po); %new search direction
  % sign for searching along x1:
  s=1;
  if p(1)>0, s=-1; end;
  m=p(2)/p(1); %slope
  flag=1; yref=yi;
  x1=x1i;
  while flag==1,
    x2=m*(x1-x1i)+x2i;
    y=2*(x1^2)+7*(x2^2);
    if y>yref,
      flag=0;
    else
      yref=y;
      plot(x1,x2,'k.');
      x1o=x1; x2o=x2;
      x1=x1-(s*0.01);
    end;
  end;
  x1i=x1o; x2i=x2o;
  go=g; po=p;
end;
```

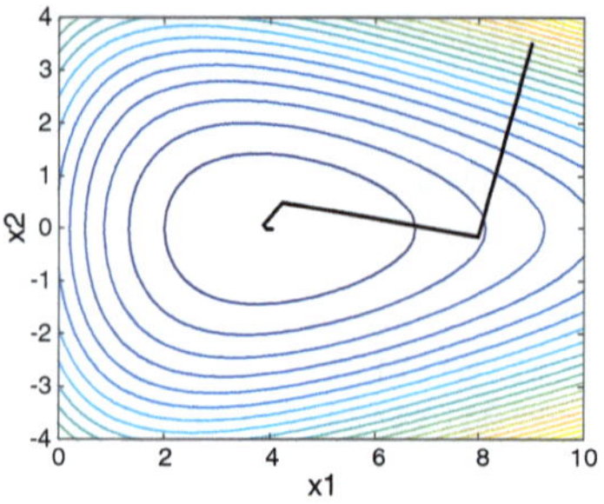

Fig. 2.36 Another example of Fletcher-Reeves method

Let us check the result of the Fletcher-Reeves method in the non-quadratic function example. As shown in Fig. 2.36, a good solution approximation was reached in five steps. Program 2.19 contains the MATLAB coding details for this case.

Program 2.19 Fletcher-Reeves method example

```
% Example of Fletcher-Reeves method
%
% function y=(exp(-(x1-3)/2))+(exp((x1+4*x2)/10))...
% +(exp((x1-4*x2)/10));
%
[vx1,vx2]=meshgrid(0:0.1:10, -4:0.1:4);
y=(exp(-(vx1-3)/2))+(exp((vx1+4*vx2)/10))...
+(exp((vx1-4*vx2)/10));
%
% function display-------------
figure (1)
contour(vx1,vx2,y,20); hold on;
xlabel('x1'); ylabel('x2');
%
% initial point
x1i=9; x2i=3.5;
yi=(exp(-(x1i-3)/2))+(exp((x1i+4*x2i)/10))...
+(exp((x1i-4*x2i)/10));
%
% gradient
aux1=(-0.5*exp(-(x1i-3)/2))...
+(0.1*exp((x1i+4*x2i)/10))+(0.1*exp((x1i-4*x2i)/10));
aux2=(0.4*exp((x1i+4*x2i)/10))...
-(0.4*exp((x1i-4*x2i)/10));
g=[aux1,aux2];
% sign of g along x1:
s=1;
if g(1)<0, s=-1; end;
%
% first search segment
m=g(2)/g(1); %slope
flag=1; yref=yi;
```

```
x1=x1i;
while flag==1,
  x2=m*(x1-x1i)+x2i;
  y=(exp(-(x1-3)/2))+(exp((x1+4*x2)/10))...
  +(exp((x1-4*x2)/10));
  if y>yref,
    flag=0;
  else
    yref=y;
    plot(x1,x2,'k.');
    x1o=x1; x2o=x2;
    x1=x1-(s*0.01);
  end;
end;
%
% more search segments
x1i=x1o; x2i=x2o;
go=g; po=-g;
%
for M=1:4,
  yi=(exp(-(x1i-3)/2))+(exp((x1i+4*x2i)/10))...
  +(exp((x1i-4*x2i)/10));
  aux1=(-0.5*exp(-(x1i-3)/2))+...
  (0.1*exp((x1i+4*x2i)/10))+...
  (0.1*exp((x1i-4*x2i)/10));
  aux2=(0.4*exp((x1i+4*x2i)/10))...
  -(0.4*exp((x1i-4*x2i)/10));
  g=[aux1,aux2]; % gradient at current corner
  %
  q=(norm(g))^2/(norm(go)^2);
  p=-g+(q*po); %new search direction
  % sign for searching along x1:
  s=1;
  if p(1)>0, s=-1; end;
  m=p(2)/p(1); %slope
  flag=1; yref=yi;
  x1=x1i;
  while flag==1,
    x2=m*(x1-x1i)+x2i;
    y=(exp(-(x1-3)/2))+(exp((x1+4*x2)/10))...
    +(exp((x1-4*x2)/10));
    if y>yref,
      flag=0;
    else
      yref=y;
      plot(x1,x2,'k.');
      x1o=x1; x2o=x2;
      x1=x1-(s*0.01);
    end;
  end;
  x1i=x1o; x2i=x2o;
  go=g; po=p;
end;
```

2.3.2.5 A More General View of Conjugate Gradients

Nowadays, there exists a family of conjugate gradient methods. Several alternatives for updating the searching direction have been proposed. The seminal paper of Hestenes and Stiefel, in 1952 [32], uses

$$\beta_k = -\frac{\vec{g}_{k+1}^T \, \vec{y}_k}{\vec{d}_k^T \, \vec{y}_k} \quad \text{(HS method)}$$

for β_k update. The algorithm focused on symmetric, positive-definite linear systems.

The algorithm introduced in 1964 by Fletcher and Reeves, [18], is considered as the first nonlinear conjugate gradient algorithm. It uses

$$\beta_k = -\frac{\vec{g}_{k+1}^T \, \vec{g}_{k+1}}{\vec{g}_k^T \, \vec{g}_k} \quad \text{(FR method)}$$

for β_k update.

In 1969, Polak and Ribiere [46], and Polyak [47], the following update was proposed:

$$\beta_k = -\frac{\vec{g}_{k+1}^T \, \vec{y}_k}{\vec{g}_k^T \, \vec{g}_k} \quad \text{(PRP method)}$$

In 1977 Powell pointed out that the FR method with exact line search could incur into jamming problems (too many short steps for a poor convergence progress) [49]. According to the interesting survey of [26], the methods with $\vec{g}_{k+1}^T \, \vec{g}_{k+1}$ in the numerator are susceptible to jamming, while having $g_{k+1}^T \, \vec{y}_k$ in the numerator prevents from this problem (short steps drive the new searching direction to the steepest descent direction, which is a kind of algorithm re-start).

Later on, in 1984, Powell showed that the PRP method for nonconvex minimization may not converge [53].

Other β_k updates have been proposed, (see [HG2]), choosing $\vec{g}_{k+1}^T \, \vec{g}_{k+1}$ or $g_{k+1}^T \, \vec{y}_k$ for the numerator, and $\vec{d}_k^T \, \vec{y}_k$, or $-\vec{d}_k^T \, \vec{g}_k$, or $\vec{g}_k^T \, \vec{g}_k$ for the denominator.

See [22] and [31] for historical details of conjugate gradient methodology.

An interesting new conjugacy condition has been proposed in the year 2000 by Dai and Liao [24]. This idea has been followed by [4] and [38].

2.3.3 *Aspects Related with Searching*

In practice, searching algorithms can be improved in several ways, like preconditioning, adequate selection of step size, and acceleration means.

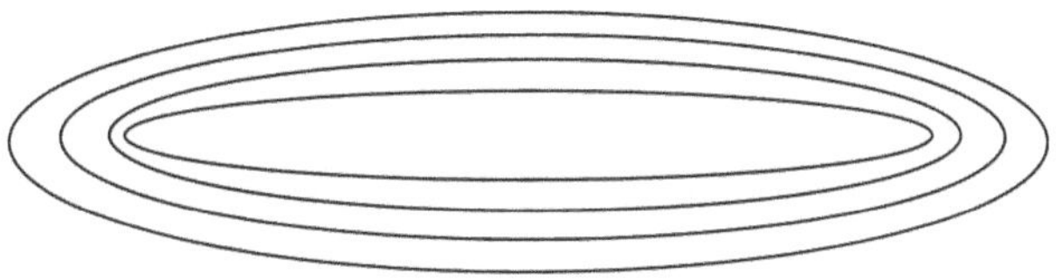

Fig. 2.37 Contours of a 2D quadratic case

2.3.3.1 Preconditioning

There are problems that require a lot of iterations for solving them. This situation could be greatly improved by means of preconditioning, which transforms the problem to an equivalent version requiring much less searching steps.

Consider for example a simple 2D quadratic case:

$$J = \vec{x}^T Q \vec{x} = \frac{x_1^2}{a^2} + \frac{x_2^2}{b^2}$$

The contours will be ellipses. If the ellipses were quite elongated, as it is represented in Fig. 2.37, most probably there would be a lot of zigzags for reaching the solution.

The preconditioning idea would be in this case to try a change of variables, so the ellipses became circles and then the solution could be obtained in just one step. This can be done with a simple scaling of the variables, like, for instance, $z_1 = x_1/a$ and $z_2 = x_2/b$.

In general, bad situations are characterized by having a matrix Q with large condition number. The condition number of a matrix Q is:

$$\kappa(Q) = \left\| Q^{-1} \right\| \cdot \| Q \|$$

Any consistent norm could be used. In case that the norm is $\| \|_2$, if Q is normal, then the condition number can be expressed in terms of the maximum and minimum eigenvalues of the matrix:

$$\kappa(Q) = \frac{|\lambda_{max}|}{|\lambda_{min}|}$$

In the simple 2D quadratic example, Fig. 2.37, the ratio of the lengths of the major and minor axes of the ellipses is given by:

$$\frac{a}{b} = \sqrt{|\lambda_2|/|\lambda_1|} = \sqrt{\kappa_2(Q)}$$

The case depicted in Fig. 2.37 corresponds to a large value of the condition number. It can be seen from the example that the rate of convergence of the searching would be slow if the condition number is high.

With the advent of large dimension problems, preconditioning is becoming an important issue. A common way for introducing the topic is by considering the classical linear system of equations:

$$A\vec{x} = \vec{b}$$

A first observation is that if you multiply both sides by A^{-1} the result is a simpler equation:

$$I\vec{x} = A^{-1} \cdot \vec{b}$$

Therefore, one could say that the best preconditioner would be A^{-1}. Note that the condition number of I is one.

Obviously if A was easily invertible, there is no need of preconditioning. What we are considering are more the situations with difficulties for matrix inversion. The desire is to improve the scenario with a matrix M easy to invert and being a good approximation of A^{-1}.

The conditioning could be applied from the left,

$$M^{-1}A\vec{x} = M^{-1}\vec{b}$$

or from the right,

$$A\,M^{-1}\vec{y} = \vec{b} \quad \text{and} \quad \vec{x} = M^{-1}\vec{y}$$

A simple approach (Jacobi) is to take a matrix $M = diag(Q)$, then apply a Cholesky decomposition $M = R^T R$, and then use the following change of variables:

$$\hat{x} = R\vec{x}$$

The transformed matrix would be:

$$\hat{Q} = R^{-T}\,Q\,R^{-1}$$

which will be a symmetric matrix, enjoying a good condition number.

The survey [9] gives a lot of information on preconditioning, including relevant references. Algorithmic implementation details for conjugate gradients were given by [6].

2.3.3.2 Step Size Conditions

One of the steps of the Fletcher-Reeves method involves line searching. Some other methods also include this step. It has been found that trying to get close to the exact minimizer in the line is not worth the computational expense. It is more efficient just to take a sufficient decrease of the objective function and continue with the algorithm.

During line search iterations, one proceeds as follows:

$$\vec{x}_{k+1} = \vec{x}_k + \alpha_k \, \vec{p}_k$$

where $\vec{p}_k$ gives the direction of the search. The problem is to choose a convenient step size α_k. This step could be too large or too short.

The so-called Armijo's rule gives a criterion to avoid too large step sizes. One takes as reference the expected decrease for a step size, which would be:

$$g(\vec{x}_{k+1}) = f(\vec{x}_k) + \alpha_k \, f'(\vec{x}_k)$$

Actually, the value of the objective function for a given step would be:

$$f(\vec{x}_{k+1}) = f(\vec{x}_k + \alpha_k \, \vec{p}_k)$$

One could compare $g(\vec{x}_{k+1})$ and $f(\vec{x}_{k+1})$ to see how well was the actual decrease. If this was small, it would be an indication of a too large step. The Armijo's rule proposes a line for deciding whether the decrease was acceptable or not. Figure 2.38 depicts $f(\vec{x}_{k+1})$ in function of the step size, and two lines: one corresponds to $g(\vec{x}_{k+1})$, and the other corresponds to the Armijo's rule.

The Armijo's line is the following:

$$L(\vec{x}_{k+1}) = f(\vec{x}_k) + c_1 \, \alpha_k \, f'(\vec{x}_k)$$

where the constant c_1 is usually a small value.

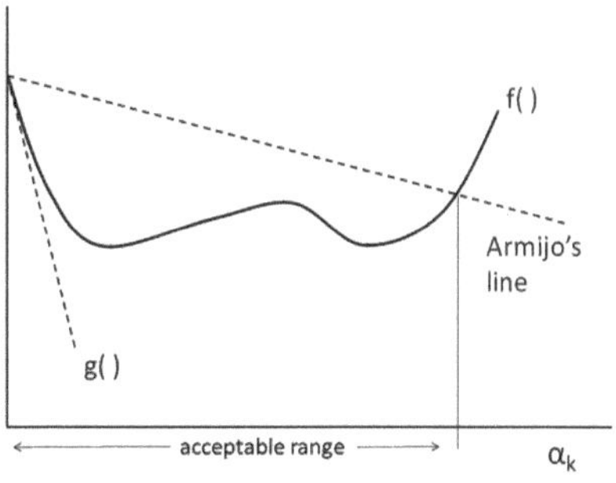

Fig. 2.38 Armijo's rule representation

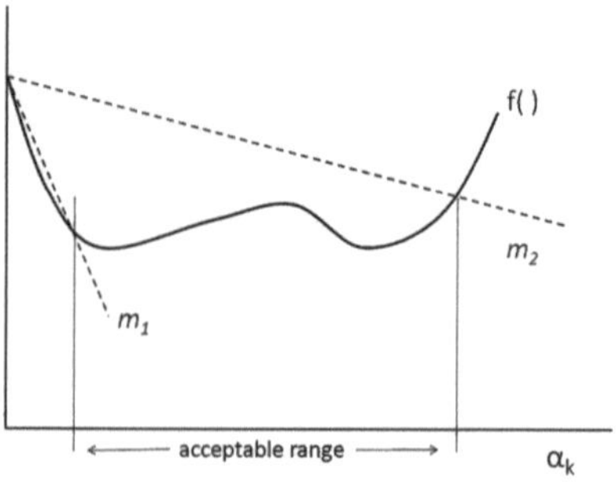

Fig. 2.39 Goldstein and Price conditions

Then, the criterion for a step size to be acceptable is:

$$f(\vec{x}_k + \alpha_k \vec{p}_k) \leq f(\vec{x}_k) + c_1 \, \alpha_k \, f'(\vec{x}_k)$$

Notice in Fig. 2.38 that very short step sizes would be also acceptable. However, cases exist where a too small step size has undesirable consequences of no convergence to the solution.

Goldstein and Price extended the conditions by considering two numbers m_1 and m_2, such that $0 < m_1 < m_2 < 1$, and defining with them two lines, one for too short step size, and the other for too large step size. Figure 2.39 visualizes the concept. The line corresponding to m_2 is the Armijo's line. The condition could be written as follows:

$$m_1 \, f'(\vec{x}_k) \leq \frac{f(\vec{x}_k + \alpha_k \vec{p}_k) - f(\vec{x}_k)}{\alpha_k} \leq m_1 \, f'(\vec{x}_k)$$

In addition to the comparison between $g(\vec{x}_{k+1})$ and $f(\vec{x}_{k+1})$, it is convenient to consider what happens with $f'(\vec{x}_{k+1})$. One could ensure that the step size is not too short, if the derivative of the objective function $f'(\vec{x}_{k+1})$ has some increase with respect to $f'(\vec{x}_k)$. Then (*'curvature condition'*):

$$f'(\vec{x}_k + \alpha_k \vec{p}_k) \geq c_2 \, f'(\vec{x}_k)$$

with $0 < c_2 < 1$ (notice that the derivatives are negative).

This curvature condition, combined with the Armijo's rule, is the Wolfe's conditions.

There is a line search algorithm, called "*backtracking algorithm*," which is a direct application of the Armijo's rule. The algorithm starts with a relatively large size step, and then:

1. Check if the Armijo's rule is satisfied
2. If not, decrease the step size, and go back to (1)
   Else, accept the step size

2.3.3.3 Accelerated Methods

In response to the growing interest on searching methods, the research has proposed a number of accelerated methods, achieving faster convergence to the solution.

One of the useful concepts for better convergence is to capture and use information on the curvature. For instance, this could be done by taking into account $\vec{x}_k$ and $\vec{x}_{k-1}$. The so-called "*heavy-ball*" method proposes the following searching iteration:

$$\vec{x}_{k+1} = \vec{x}_k - \alpha \nabla f(\vec{x}_k) + \beta (\vec{x}_k - \vec{x}_{k-1})$$

The heavy-ball metaphor considers the state $\vec{x}_k$ as a point mass with *momentum* [48]. Then, it tends to move in the direction $\vec{x}_k - \vec{x}_{k-1}$.

Suggested values for the constants were:

$$\alpha = \frac{4}{L} \frac{1}{(1 + 1/\sqrt{\kappa})^2}$$

$$\beta = \left(1 - \frac{2}{1 + \sqrt{\kappa}}\right)^2$$

where L is a constant related to the Lipschitz condition; a function is said to be Lipschitz if a constant L>0 exists such that:

$$\| f(\vec{x}) - f(\vec{y}) \|_2 \leq L \, \| \vec{x} - \vec{y} \|_2 \, , \forall \vec{x}, \, \vec{y} \in \Re$$

Nesterov has introduced several acceleration methods (see [30]). The first one, published in 1983, appears in the literature in different versions, like the following [44]:

$$\vec{y}_k = \vec{x}_k + \beta_k (\vec{x}_k - \vec{x}_{k-1})$$

$$\vec{x}_{k+1} = \vec{y}_k - \alpha \nabla f(\vec{y}_k)$$

with the following constants:

$$\beta_k = \frac{t_{k-1}}{t_{k+1}}$$

$$t_{k+1} = \frac{1 + \sqrt{1 + 4\, t_k^2}}{2}$$

$$\alpha = \frac{1}{L}$$

Another contribution was the method proposed by Barzilai and Borwein in 1988 [7]. It was expressed as follows:

$$\vec{x}_{k+1} = \vec{x}_k - \alpha_k \nabla f(\vec{x}_k)$$

$$\alpha_k = \arg\min_{\alpha} \|\vec{s}_k - \alpha\, \vec{z}_k\|$$

where:

$$\vec{s}_k = \vec{x}_k - \vec{x}_{k-1}$$

$$\vec{z}_k = \nabla f(\vec{x}_k) - \nabla f(\vec{x}_{k-1})$$

2.4 Multidimensional Second-Order Methods

The Hessian gives information on the curvature that can be used for better convergence to the optimal solution.

Since the Hessian could be hard to use for computational reasons, a series of methods have been proposed that employ Hessian approximations. These methods are now quite popular.

In honor to the classical roots, this section starts considering the Newton's method, and then it continues with the introduction of the methods based on Hessian approximations.

It is appropriate to include here a portrait of Sir Isaac Newton, University of Cambridge, taken from the site [39], which also contains an interesting reading (Fig. 2.40).

Fig. 2.40 Sir Isaac Newton

2.4.1 Second-Order Newton's Method

Consider the following quadratic Taylor expansion:

$$f(\vec{x}) \approx q(\vec{x}) = f(\vec{x}_0) + \nabla f(\vec{x}_0)^T (\vec{x} - \vec{x}_0) + \frac{1}{2} (\vec{x} - \vec{x}_0)^T H(\vec{x}_0) (\vec{x} - \vec{x}_0)$$

The minimum of the quadratic function $q(\vec{x})$ can be found with:

$$\nabla q(\vec{x}) = 0 \Rightarrow \nabla f(\vec{x}_0) + H(\vec{x}_0) (\vec{x} - \vec{x}_0)$$

Therefore:

$$\vec{x} = \vec{x}_0 - H(\vec{x}_0)^{-1} \nabla f(\vec{x}_0)$$

This expression provides the basis for step-by-step searching. In case of $f(\vec{x})$ being a quadratic function, the solution would be reached in just one step. It can be seen as a direct generalization of the Newton-Raphson method for functions of several variables.

It would be not necessary to explicitly invert the Hessian matrix. Instead, solve the linear system:

$$H(\vec{x}_0) \vec{\delta} = -\nabla f(\vec{x}_0)$$

(the unknown variable to be determined is $\vec{\delta} = \vec{x} - \vec{x}_0$)

When far from the solution, the method may not converge. The so-called *Damped Newton method* uses line search along the direction of $\vec{\delta}$ for a more robust behavior (see [27]).

Let us employ the same two examples that were used in the previous section. In the case of the first example, which is a quadratic function, it is expected to attain the solution in just one step. Somewhat dramatically, Fig. 2.41 confirms this expectation.

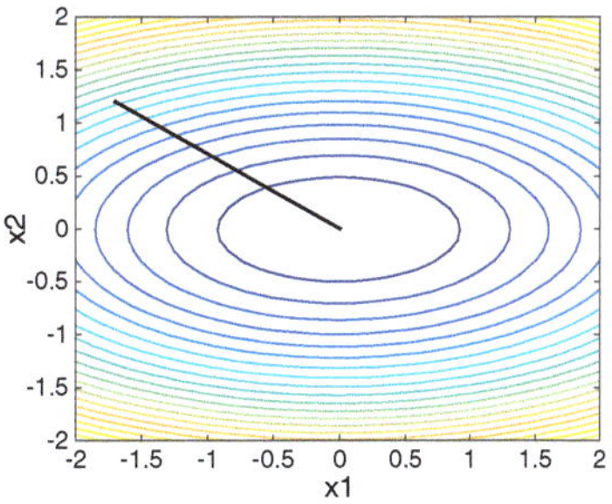

Fig. 2.41 Example of Newton's method

Program 2.20, which was used to obtain Fig. 2.41, presents a compact coding of the Newton's method. Notice the use of the MATLAB backslash operator \ for computing the single step in just one line of code.

Program 2.20 Newton's method example

```
% Example of Newton's method
%
v=-2:0.1:2;
[vx1,vx2]=meshgrid(v);
y=(2*vx1.^2)+(7*vx2.^2);
%
% function display-------------
figure (1)
contour(vx1,vx2,y,20); hold on;
xlabel('x1'); ylabel('x2');
%
% initial point
x1i=-1.7; x2i=1.2;
yi=2*(x1i^2)+7*(x2i^2);
% gradient
g=[(4*x1i),(14*x2i)];
% Hessian
H=[4 0;0 14];
%
delta=H\-g'; %step
%
x1f=x1i+delta(1);
x2f=x2i+delta(2);
plot([x1i x1f], [x2i x2f],'k');
```

Figure 2.42 depicts the behavior of the Newton's method in the non-quadratic function example. A satisfactory solution approximation was found in five steps (the last steps were quite short).

Figure 2.42 was obtained with the Program 2.21, which is similar to Program 2.20 although now it provides a loop for several steps and not only one.

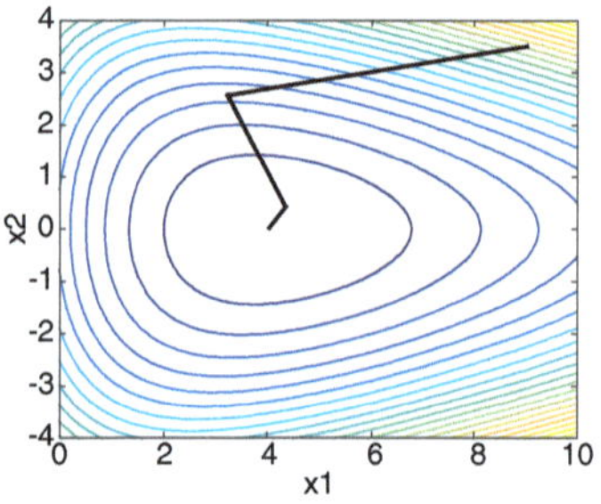

Fig. 2.42 Another example of Newton's method

Program 2.21 Newton's method example

```
% Example of Newton's method
%
% function y=(exp(-(x1-3)/2))+(exp((x1+4*x2)/10))...
% +(exp((x1-4*x2)/10));
%
[vx1,vx2]=meshgrid(0:0.1:10, -4:0.1:4);
y=(exp(-(vx1-3)/2))+(exp((vx1+4*vx2)/10))...
+(exp((vx1-4*vx2)/10));
%
% function display-------------
figure (1)
contour(vx1,vx2,y,20); hold on;
xlabel('x1'); ylabel('x2');
%
% initial point
x1i=9; x2i=3.5;
% steps
for M=1:4,
  a=(exp(-(x1i-3)/2));
  b=(exp((x1i+4*x2i)/10));
  c=(exp((x1i-4*x2i)/10));
  yi=a+b+c;
  %
  % gradient
  aux1=(-0.5*a)+(0.1*b)+(0.1*c);
  aux2=(0.4*b)-(0.4*c);
  g=[aux1,aux2];
  %
  % Hessian
  qH11=(0.25*a)+(0.01*b)+(0.01*c);
  qH12=(0.04*b)-(0.04*c);
  qH21=(0.04*b)-(0.04*c);
  qH22=(0.16*b)+(0.16*c);
  H=[qH11 qH12; qH21 qH22];
  %
  delta=H\-g'; %step
  %
  x1f=x1i+delta(1);
  x2f=x2i+delta(2);
  plot([x1i x1f], [x2i x2f],'k');
  %
  x1i=x1f; x2i=x2f;
end
```

For more mathematical details about the second-order Newton's method, the lecture notes [60] by Ryan Tibshirani would be recommended.

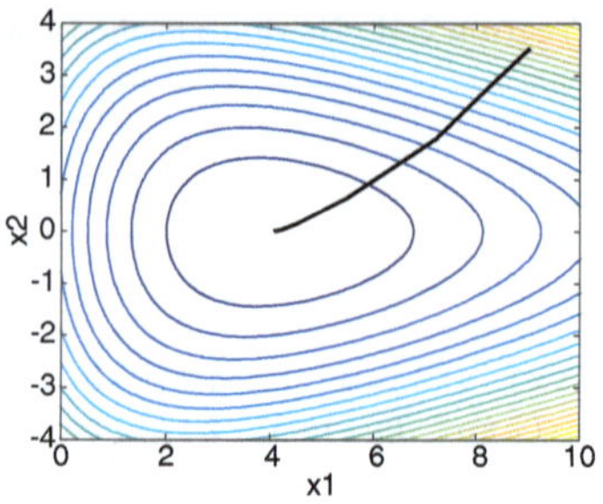

Fig. 2.43 Example of Levenberg-Marquardt method

2.4.2 Levenberg-Marquardt Method

Since the information given by the Hessian far from the solution could be misleading, Levenberg [36] proposed a blending of the Newton and the steepest descent methods, by using the following step:

$$\vec{x}_{n+1} - \vec{x}_n = -(H(\vec{x}_n) + \lambda I)^{-1} \nabla f(\vec{x}_0)$$

In this way, the value of λ is used to modify the eigenvalues of the matrix inside the parentheses, and so it would be possible to ensure that it is semi-definite positive. Besides, a larger value of λ makes the step to be smaller, more prudent.

As shown in Fig. 2.43, the Levenberg method exhibits a good, smooth behavior in the non-quadratic example (compare with Fig. 2.42).

Program 2.22, which generates Fig. 2.43, is similar to Program 2.21 with some added lines for including the gradient.

Program 2.22 Levenberg-Marquardt method

```
% Example of Levenberg-Marquardt method
%
% function y=(exp(-(x1-3)/2))+(exp((x1+4*x2)/10))...
% +(exp((x1-4*x2)/10));
%
[vx1,vx2]=meshgrid(0:0.1:10, -4:0.1:4);
y=(exp(-(vx1-3)/2))+(exp((vx1+4*vx2)/10))...
+(exp((vx1-4*vx2)/10));
%
% function display-------------
figure (1)
contour(vx1,vx2,y,20); hold on;
xlabel('x1'); ylabel('x2');
%
% initial point
x1i=9; x2i=3.5;
% steps
for M=1:5,
```

```
    a=(exp(-(x1i-3)/2));
    b=(exp((x1i+4*x2i)/10));
    c=(exp((x1i-4*x2i)/10));
    yi=a+b+c;
    %
    % gradient
    aux1=(-0.5*a)+(0.1*b)+(0.1*c);
    aux2=(0.4*b)-(0.4*c);
    g=[aux1,aux2];
    %
    % Hessian
    qH11=(0.25*a)+(0.01*b)+(0.01*c);
    qH12=(0.04*b)-(0.04*c);
    qH21=(0.04*b)-(0.04*c);
    qH22=(0.16*b)+(0.16*c);
    H=[qH11 qH12; qH21 qH22];
    %
    mhu=0.1; %parameter (edit)
    I=eye(2);
    delta=-(inv(H+(mhu*I)))*g'; %step
    %
    x1f=x1i+delta(1);
    x2f=x2i+delta(2);
    plot([x1i x1f], [x2i x2f],'k');
    %
    x1i=x1f; x2i=x2f;
end
```

Marquardt improved the method with the following modification [40]:

$$\vec{x}_{n+1} - \vec{x}_n = -(H(\vec{x}_n) + \lambda\ diag(H(\vec{x}_n))^{\ 1}\ \nabla f(\vec{x}_0)$$

The purpose of the modification is to get some benefit from the Hessian even when the dominant action was steepest descent, [57].

Generally, the Levenberg-Marquardt method works very well, so it has become popular indeed. The paper [40] has been cited more than 41374 times. A motivating historical view is offered by [45].

Many improvements and implementations have been proposed, like, for instance, [62]. See [19] for some interesting examples of coding the method in MATLAB.

2.4.3 Methods Based on Hessian Approximation

In order to avoid the numerical inversion of the Hessian, the research has proposed a number of methods that use approximations of the Hessian or the inverse of the Hessian. This kind of methods is called quasi-Newton methods. The approximations are updated in each iteration.

Denote as B_k the approximation to the Hessian at kth iteration. In the next iteration, the process comes from $\vec{x}_k$ and $\vec{g}_k$, to new values $\vec{x}_{k+1}$ and $\vec{g}_{k+1}$, and

$$B_{k+1} = F(B_k, \Delta\vec{x}_k, \Delta\vec{g}_k)$$

Given the increment $\Delta\vec{x}_k$, an increment $\Delta\vec{g}_k$ is observed, and one expects that B_{k+1} would predict this gradient change as follows:

$$B_{k+1}\,\Delta\vec{x}_k = \Delta\vec{g}_k,$$

which is called the *"secant condition"*.

In 1967, Broyden suggested that it was numerically convenient to choose matrices such that the rank of $(B_{k+1} - B_k)$ was small [56]. This could be achieved with updates having the form:

$$B_{k+1} = B_k + U \cdot V,$$

with $U \cdot V$ being a low-rank matrix.

In this context, it is opportune to consider the *"Sherman-Morrison" formula.* Suppose $B \in \Re^{n\times n}$ is an invertible matrix and u, v are column vectors with n components, then:

$$(B + \vec{u}\,\vec{v}^T)^{-1} = B^{-1} - \frac{B^{-1}\vec{u}\,\vec{v}^T B^{-1}}{1 + \vec{v}^T B^{-1}\,\vec{u}},$$

A theorem by Sherman, Morrison, and Woodbury [25] generalizes this inversion formula to the case in which $U, V \in \Re^{n\times p}$:

$$(B + UV^T)^{-1} = B^{-1} - B^{-1}U\,(I + V^T B^{-1}U)^{-1}\,V^T B^{-1}$$

2.4.3.1 Broyden's Method

Let us try a symmetric rank-one update of the form:

$$B_{k+1} = B_k + \vec{u}\,\vec{u}^T,$$

This update must satisfy the secant condition, and so:

$$(B_k + \vec{u}\,\vec{u}^T)\Delta\vec{x}_k = \Delta\vec{g}_k,$$

Which could be written as:

$$(\vec{u}^T\,\Delta\vec{x}_k)\,\vec{u} = \Delta\vec{g}_k - B_k\Delta\vec{x}_k,$$

It follows [29] that:

$$\vec{u} = \frac{\Delta\vec{g}_k - B_k \Delta\vec{x}_k}{(\vec{u}^T \Delta\vec{x}_k)},$$

Therefore, the update would be:

$$B_{k+1} = B_k + \frac{(\Delta\vec{g}_k - B_k \Delta\vec{x}_k)(\Delta\vec{g}_k - B_k \Delta\vec{x}_k)^T}{(\Delta\vec{g}_k - B_k \Delta\vec{x}_k)^T \Delta\vec{x}_k},$$

This update was suggested by a number of authors in 1967–1969, [29]. However, it is possible to use the Sheridan-Morrison formula for a more numerically convenient update, [29]. Denote as A_k the inverse of B_k , then:

$$A_{k+1} = A_k + \frac{(\Delta\vec{g}_k - A_k \Delta\vec{x}_k)(\Delta\vec{g}_k - A_k \Delta\vec{x}_k)^T}{(\Delta\vec{g}_k - A_k \Delta\vec{x}_k)^T \Delta\vec{x}_k},$$

A possible algorithm could be the following:

1. Choose an initial point $\vec{x}_0$
2. Compute the gradient $\vec{g}_0$ at $\vec{x}_0$. Initialize: $A_0 = I$
3. Obtain a direction: $\vec{s}_n = -A_n \, \vec{g}_n$
4. Find by line search the minimum along that direction, which will be located at certain $\vec{x}_{n+1} = \vec{x}_n + \beta \, \vec{s}_n$
5. Compute: $\vec{g}_{n+1} = \nabla f(\vec{x}_{n+1})$; $\Delta x_n = \vec{x}_{n+1} - \vec{x}_n$; $\Delta g_n = \vec{g}_{n+1} - \vec{g}_n$
6. Use (Broyden's update):

$$A_{n+1} = A_n + \frac{(\Delta\vec{x}_n - A_n \Delta\vec{g}_n)\,(\Delta\vec{x}_n - A_n \Delta\vec{g}_n)^T}{(\Delta\vec{x}_n - A_n \Delta\vec{g}_n)^T \cdot \Delta\vec{g}_n} = A_n + \Delta A_n$$

to compute A_{n+1}

And then loop back to 3.

Thus, one never has to invert the Hessian.

The method may fail if the denominator of ΔA_n was too small. Another drawback is that, it does not preserve positive definiteness of A_{n+1}.

Let us see the result of the Broyden's method for the non-quadratic function example. Figure 2.44 shows the behavior of the algorithm described before. It gets close to the solution in five steps (the last one is very short).

Figure 2.44 was obtained with the Program 2.23. Notice that the first step takes the direction of the gradient (compare with Fig. 2.42, in which the first step takes another direction based on the Hessian).

Note that in Program 2.23 a specific code fragment is marked, in between comment %QA and comment %QB. This fragment will be changed in the two methods to be introduced next.

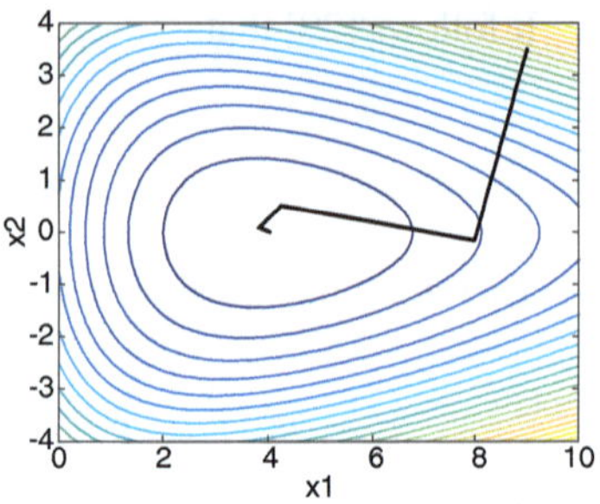

Fig. 2.44 Example of Broyden's method

Program 2.23 Broyden's method

```
% Example of Broyden's method
%
% function y=(exp(-(x1-3)/2))+(exp((x1+4*x2)/10))...
% +(exp((x1-4*x2)/10));
[vx1,vx2]=meshgrid(0:0.1:10, -4:0.1:4);
y=(exp(-(vx1-3)/2))+(exp((vx1+4*vx2)/10))...
+(exp((vx1-4*vx2)/10));
%
% function display-------------
figure (1)
contour(vx1,vx2,y,20); hold on;
xlabel('x1'); ylabel('x2');
%
% initial point
x1i=9; x2i=3.5;
%
a=(exp(-(x1i-3)/2));
b=(exp((x1i+4*x2i)/10));
c=(exp((x1i-4*x2i)/10));
yi=a+b+c;
%
% gradient
aux1=(-0.5*a)+(0.1*b)+(0.1*c);
aux2=(0.4*b)-(0.4*c);
g=[aux1;aux2];
%
% matrix A
A=eye(2); %matrix I
%
% first step -------------------------
delta=-A*g;
% sign of delta along x1:
s=1;
if delta(1)<0, s=-1; end;
%
% first search segment
m=delta(2)/delta(1); %slope
```

```
flag=1; yref=yi;
x1=x1i;
while flag==1,
  x2=m*(x1-x1i)+x2i;
  a=(exp(-(x1-3)/2));
  b=(exp((x1+4*x2)/10));
  c=(exp((x1-4*x2)/10));
  y=a+b+c;
  if y>yref,
    flag=0;
  else
    yref=y;
    plot(x1,x2,'k.');
    x1o=x1; x2o=x2;
    x1=x1+(s*0.01);
  end;
end;
%
x1=x1o; x2=x2o;
%
% next steps -------------------------
for M=1:4,
  a=(exp(-(x1-3)/2));
  b=(exp((x1+4*x2)/10));
  c=(exp((x1-4*x2)/10));
  aux1=(-0.5*a)+(0.1*b)+(0.1*c);
  aux2=(0.4*b)-(0.4*c);
  go=g; %old value of the gradient
  g=[aux1;aux2]; %new gradient
  dg=g-go;
  dx=[x1 x1i; x2 x2i];
  Ao=A; %old value of A
  %QA (CODE FRAGMENT BEG)
    %new value of A
    A=Ao+((dx-A*dg)*(dx-A*dg)')/((dx-A*dg)'*dg);
  %QB (CODE FRAGMENT END)
  x1i=x1; x2i=x2; %new start point
  %
  % line search
  delta=-A*g;
  % sign of delta along x1:
  s=1;
  if delta(1)<0, s=-1; end;
  %
  m=delta(2)/delta(1); %slope
  flag=1;
  x1=x1i;
  while flag==1,
    x2=m*(x1-x1i)+x2i;
    a=(exp(-(x1-3)/2));
    b=(exp((x1+4*x2)/10));
```

```
    c=(exp((x1-4*x2)/10));
    y=a+b+c;
    if y>yref,
      flag=0;
    else
      yref=y;
      plot(x1,x2,'k.');
      x1o=x1; x2o=x2;
      x1=x1+(s*0.01);
    end;
  end;
  x1=x1o; x2=x2o;
end;
```

2.4.3.2 Method of Davidon-Fletcher-Powell (DFP)

In order to preserve positive definiteness one could introduce more degrees of freedom, by considering a rank-two update of the form:

$$A_{k+1} = A_k + \vec{u}\,\vec{u}^T + \vec{v}\,\vec{v}^T,$$

From the secant condition, it is possible to obtain:

$$A_{n+1} = A_n + \frac{\Delta\vec{x}_n\,\Delta\vec{x}_n^T}{\Delta\vec{x}_n^T\,\Delta\vec{g}_n} - \frac{A_n\,\Delta\vec{g}_n\,\Delta\vec{g}_n^T\,A_n}{\Delta\vec{g}_n^T\,A_n\,\Delta\vec{g}_n} = A_n + \Delta A_n \quad (2)$$

This formula has a peculiar history. Davidon was a physicist at Argonne National Lab., he had not the computational power we enjoy today, and had difficulties for solving certain large optimization problems. He then devised the first rank-two quasi-Newton method. It was introduced in 1959 as internal report for the Lab. A revised version was published by Fletcher and Powell in 1963, being a revolutionary contribution [17]. In 1991, the original paper of Davidon was published in the inaugural issue of the SIAM Journal of Optimization [15].

There is an alternative way for obtaining the DFP formula. It is convenient to have B_{k+1} as close as possible to B_k. The difference between these matrices could be measured using the Frobenius norm, which for a given matrix D is defined as follows:

$$\|D\|_F = \left(\sum_{i\,j} d_{i\,j}^2\right)^{1/2}$$

The weighted Frobenius norm of D with respect to a weighting matrix W is:

$$\|D\|_W = \left\| W^{1/2} D\; W^{1/2} \right\|_F$$

Now, in order to get the B_{k+1} closest to B_k one has to solve the following minimization problem:

$$\arg\min_{B} \; \|B - B_k\|_W$$

where B has to satisfy the secant condition.

If one chooses W to be the average Hessian, which is:

$$W = \int_0^1 \nabla^2 f(x_k + \tau \, \Delta x_k) d\tau$$

Then it can be shown that the solution of the minimization problem is:

$$B_{k+1} = (I - \rho_k \Delta\vec{g}_k \Delta\vec{x}_k^T) \, B_k (I - \rho_k \Delta\vec{x}_k \Delta\vec{g}_k^T) + \rho_k \Delta\vec{g}_k \Delta\vec{g}_k^T$$

where,

$$\rho_k = \frac{1}{\Delta\vec{g}_k^T \, \Delta\vec{x}_k}$$

Then, the inverse A_{k+1} can be obtained with the Sherman-Morrison-Woodbury formula, which gives the DFP update expression (2).

An algorithm for using the DFP method can be simply be built by changing the matrix update in the algorithm described before for Broyden's method. It is just a matter of changing the step 6 by the DFP update.

If one continues using the non-quadratic function example, the result of the DFP method is quite similar to the behavior shown in Fig. 2.44. Besides, the corresponding program for generating the figure is almost the same as Program 2.23, with only a difference: substitute the code fragment between %QA and %QB, with the following fragment:

```
%QA (CODE FRAGMENT BEG)
%new value of A
A=Ao+((dx*dx')/(dx'*dg))...
   -((Ao*dg*dg'*Ao)/(dg'*Ao*dg));
%QB (CODE FRAGMENT END)
```

Note (Program 2.23-DFP)

The complete listing of the program 2.23-DFP, corresponding to the DFP method and the non-quadratic example has been included in the Appendix for Long Programs.

2.4.3.3 Method of Broyden-Fletcher-Goldfarb-Shanno (BFGS)

Another way of preserving positive definiteness is to consider a rank-two update of the form:

$$B_{k+1} = B_k + \vec{u}\,\vec{u}^T + \vec{v}\,\vec{v}^T,$$

From the secant condition, one obtains:

$$B_{n+1} = B_n + \frac{\Delta\vec{g}_n\,\Delta\vec{g}_n^T}{\Delta\vec{g}_n^T\,\Delta\vec{x}_n} - \frac{(B_n\Delta\vec{x}_n)(B_n\Delta\vec{x}_n)^T}{(B_n\Delta\vec{x}_n)\Delta\vec{x}_n} = B_n + \Delta B_n \quad (3)$$

As before, it is also possible to derive this update formula by considering the following minimization problem:

$$\arg\min_B \|A - A_k\|_W$$

The minimization solution is:

$$A_{k+1} = (I - \rho_k\Delta\vec{x}_k\Delta\vec{g}_k^T)\,A_k(I - \rho_k\Delta\vec{g}_k\Delta\vec{x}_k^T) + \rho_k\Delta\vec{x}_k\Delta\vec{x}_k^T$$

And, again, the update expression (3) can be obtained with the Sherman-Morrison-Woodbury formula.

The BFGS update was independently published in the same year, 1970, by Broyden [10], Fletcher [16], Goldfarb [21], and Shanno [58]. The acronym BFGS is then quite justified. There is a famous photograph of the four authors all together that we happily reproduce in Fig. 2.45.

Fig. 2.45 From left to right, Broyden, Fletcher, Goldfarb, and Shanno

An algorithm for using the BFGS method can be simply be built by changing the matrix update in the step 6 of the Broyden's algorithm by the following formula:

$$A_{n+1} = A_n + \frac{\Delta\vec{x}_n\,\Delta\vec{x}_n^T}{\Delta\vec{x}_n^T\,\Delta\vec{g}_n}\left(1 + \frac{\Delta\vec{g}_n^T\,A_n\,\Delta\vec{g}_n}{\Delta\vec{x}_n^T\,\Delta\vec{g}_n}\right) - \frac{A_n\,\Delta\vec{g}_n\,\Delta\vec{x}_n^T}{\Delta\vec{x}_n^T\,\Delta\vec{g}_n} - \frac{\Delta\vec{x}_n\,\Delta\vec{g}_n^T\,A_n}{\Delta\vec{x}_n^T\,\Delta\vec{g}_n} = A_n + \Delta A_n$$

For the non-quadratic function example, the result of the BFGS method is quite similar to Fig. 2.44. Likewise, the corresponding program for generating the figure is like Program 2.23, with only a difference: substitute the code fragment between %QA and %QB, with the following fragment:

```
%QA (CODE FRAGMENT BEG)
%new value of A
A=Ao-((Ao*dg*dx')/(dx'*dg))...
+(((dx*dx')/(dx'*dg))*(1+(dg'*Ao*dg)/(dx'*dg)))...
-((dx*dg'*Ao)/(dx'*dg));
%QB (CODE FRAGMENT END)
```

Note (Program 2.23-BFGS)

The complete listing of the program 2.23-BFGS, corresponding to the BFGS method and the non-quadratic example has been included in the Appendix for Long Programs.

2.4.3.4 The Broyden's Family of Methods

The most general symmetric rank-two update for the inverse Hessian can be written as:

$$A_{k+1} = A_k + [\Delta\vec{x}_k\ A_k\Delta\vec{g}_k]\begin{pmatrix} a & b \\ b & c \end{pmatrix}\begin{bmatrix} \Delta\vec{x}_k^T \\ \Delta\vec{g}_k^T A_k \end{bmatrix}$$

The update should satisfy the secant condition, which is:

$$A_{k+1}\,\Delta\vec{g}_k = \Delta\vec{x}_k$$

After some algebra, one obtains the following system:

$$a\,\Delta\vec{x}_k^T\,\Delta\vec{x}_k \;+\; b\,\Delta\vec{g}_k^T A_k \Delta\vec{g}_k = 1$$

$$b\,\Delta\vec{x}_k^T\,\Delta\vec{g}_k \;+\; c\,\Delta\vec{g}_k^T A_k \Delta\vec{g}_k = -1$$

Therefore, there are two equations and three unknowns. Then, one has one degree of freedom, and this would mean a complete family of methods. After more algebra, again, one reaches the following update expression (Broyden's family):

$$A_{k+1} \;=\; A_k + \frac{\Delta\vec{x}_n\,\Delta\vec{x}_n^T}{\Delta\vec{x}_n^T\,\Delta\vec{g}_n} - \frac{A_n\,\Delta\vec{g}_n\,\Delta\vec{g}_n^T A_n}{\Delta\vec{g}_n^T A_n\,\Delta\vec{g}_n} + \phi_k\,\vec{v}_k\,\vec{v}_k^T$$

where:

$$\phi_k \;=\; b\,\Delta\vec{x}_k^T\,\Delta\vec{g}_k$$

and,

$$\vec{v}_k \;=\; (\Delta\vec{g}_k^T\,A_k\,\Delta\vec{g}_k)^{1/2}\left(\frac{\Delta\vec{x}_k}{\Delta\vec{x}_k^T\,\Delta\vec{g}_k} - \frac{A_k\Delta\vec{g}_k}{\Delta\vec{g}_k^T\,A_k\,\Delta\vec{g}_k}\right)$$

Special cases of the Broyden's family are DFP for $\phi_k = 0$, and BFGS for $\phi_k = 1$

References [2, 3] offer an extensive overview of quasi-Newton methods, including the Broyden's family.

2.4.3.5 Comments on Methods Based on the Hessian

Quasi-Newton methods have fast convergence compared to gradient-based methods. In general, BFGS is the preferred option compared to DFP, since DFP is less effective than BFGS at self-correcting of the Hessian, and DFP could fail for general nonlinear problems, could stop at a saddle point, and could have numerical problems. The study of Powell [54] with a quadratic objective function concludes in favor of BFGS.

Concerning the history of nonlinear optimization methods, it would be recommended to read [55] and [11].

2.5 Trust-Region Methods

The methods already described first decide the direction and then the step size. Another alternative is the opposite, select a step size and then decide a direction. Trust-region methods use an approximate model to predict $f(\vec{x}_k)$ in a region near $\vec{x}_k$; if the model fits well the region can be enlarged; otherwise, it can be reduced. Once the model can be trusted, a convenient direction inside the region would be selected, [14, 64], and [42].

Commonly, a quadratic model is used:

$$m(\vec{x}_k + \vec{p}) = f(\vec{x}_k) + \nabla f(\vec{x}_k)^T \vec{p} + \frac{1}{2} \vec{p}^T B_k \vec{p}$$

which obviously resembles a Taylor expansion.

The question now is to solve the following minimization problem:

$$\min_{\vec{p} \in T} m(\vec{x}_k + \vec{p}) = \min_{\vec{p} \in T} \left\{ f(\vec{x}_k) + \nabla f(\vec{x}_k)^T \vec{p} + \frac{1}{2} \vec{p}^T B_k \vec{p} \right\}$$

where T is the trust region. Usually, T is a ball with radius r:

$$T(r) = \{ \vec{x} \mid \|\vec{x} - \vec{x}_k\| \leq r \}$$

This aspect of the trust-region method is usually called the minimization sub-problem.

Denote as $\vec{x}_+$ the solution of the minimization sub-problem. The trust region method will accept or not this solution for a new step. Define the actual reduction as:

$$R_a = f(\vec{x}_k) - f(\vec{x}_+)$$

and the reduction predicted by the model:

$$R_p = m(\vec{x}_k) - m(\vec{x}_+)$$

Then, in order to judge the profit of the proposed step, one uses the ratio:

$$\rho = R_a / R_p$$

and two parameters, $0 \leq \lambda_L < \lambda_H$.

If $\rho \geq \lambda_L$ the step is accepted, otherwise $\vec{x}_{k+1} = \vec{x}_k$ [23].

And the size of the trust region is updated as follows:

- If $\rho < \lambda_L$ the radius of T is decreased
- If $\lambda_L \leq \rho < \lambda_H$ the size remains equal
- If $\rho > \lambda_H$ the radius of T is increased according with $\max(\alpha \cdot step,\ r)$

In most cases, the radius is divided by 2 when decreasing, or using $\alpha = 2$ when increasing.

Clearly, it is convenient to use a simple and fast method for solving the sub-problem, or at least obtain an approximation. One of the alternatives is to minimize $m(\vec{x}_k + \vec{p})$ in the steepest descent direction; the result is the so-called "*Cauchy point*". In case of using the quadratic model, the Cauchy point is reached with:

$$\vec{p}^C = -\tau \frac{r}{\|\nabla f(\vec{x}_k)\|} \nabla f(\vec{x}_k)$$

where:

$$\tau = \begin{cases} 1, & if \quad \nabla f(\vec{x}_k)^T B_k \nabla f(\vec{x}_k) \le 0 \\ \min(1, \rho)), & \text{otherwise} \end{cases}$$

and,

$$\rho = \frac{\|\nabla f(\vec{x}_k)\|^3}{r \, \nabla f(\vec{x}_k)^T B_k \nabla f(\vec{x}_k)}$$

A better than steepest descent step would in general be possible. There is a lemma which is most opportune here. The lemma says that, being B a symmetric matrix, a vector $\vec{d}$ is a solution of the problem:

$$\min_{\vec{d}} \left\{ \vec{g}^T \vec{d} + \frac{1}{2} \vec{d}^T B \vec{d} \right\}$$

subject to $\left\| \vec{d} \right\|_2 \le r$

if and only if there exists $\lambda^* \ge 0$ such that:

$$(B + \lambda^* I) \vec{d} = -g$$

and that $(B + \lambda^* I)$ is positive definite, $\left\| \vec{d} \right\|_2 \le r$, and $\lambda^* (r - \left\| \vec{d} \right\|_2) = 0$.

Except for a specific case called the "*hard case*", λ^*can be found by solving:

$$\left\| (B + \lambda I)^{-1} g \right\|_2^{-1} - (1/r) = 0$$

In the hard case one has that $\lambda^* = -\sigma_n(B)$, where $\sigma_n(B)$ is the least eigenvalue of B. If this eigenvalue was zero, then the solution would be $\vec{d}^* = -B^{-1} g$ (the Newton step). In case that $\lambda^* = -\sigma_n(B) > 0$, then there are difficulties which require specific treatment: see [20] for a detailed study of this aspect.

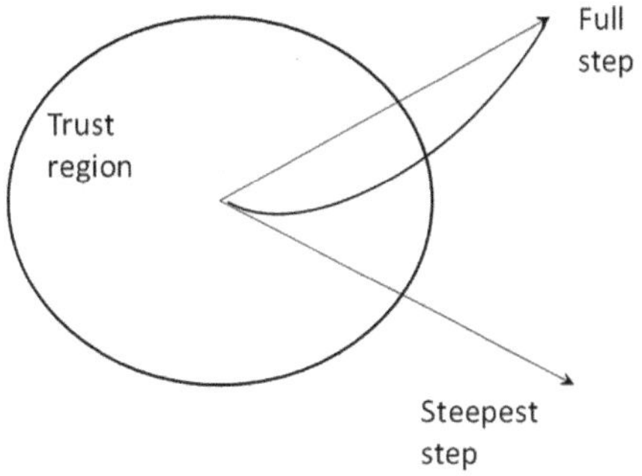

Fig. 2.46 The path of the minimum as trust region increases

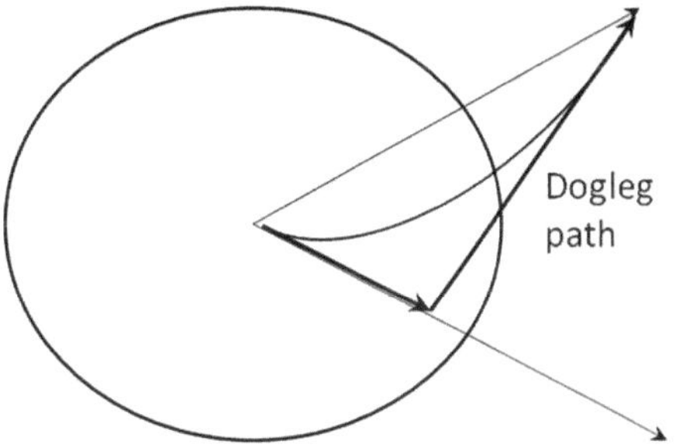

Fig. 2.47 The dog-leg path

In addition to alternatives like taking the steepest descent direction, or using the Newton step (also called the *"full step"*), some other possibilities could be considered.

An interesting approach for dealing with the minimization sub-problem, when B_k is positive definite, is the so-called *"dog-leg method"*, which was proposed by Powell in 1970 [51, 52]. Notice that when the trust region is small, the quadratic term in:

$$\min_{\vec{p}\in T}\left\{f(\vec{x}_k) + \nabla f(\vec{x}_k)^T\,\vec{p} + \frac{1}{2}\,\vec{p}^T\,B_k\,\vec{p}\right\}$$

is small, so the minimum would be close to the steepest descent direction. On the other hand, if the trust region grows, the minimum will move to the full step. Figure 2.46 shows the curve that the minimum would follow as the trust region increases.

The dog-leg method uses the full step if it is in the trust region. Otherwise, it obtains the Cauchy point; if this point is outside the trust region, the method comes to the intersection of steepest descent direction and the border of the trust region. If the Cauchy point is inside the trust region, then the dog-leg path is used.

The dog-leg path starts with a line segment reaching the Cauchy point. The second part of the path is a line segment reaching the full step tip. Figure 2.47 shows the dog-leg path.

In mathematical terms, the dog-leg path can be expressed as follows:

$$\vec{p}(\tau) = \begin{cases} \tau\, \vec{p}_k^{\,C}, & 0 \leq \tau \leq 1 \\ \vec{p}_k^{\,C} + (\tau - 1)(\vec{p}_k^{\,FS} - \vec{p}_k^{\,C}), & 1 \leq \tau \leq 2 \end{cases}$$

where $\vec{p}_k^{\,FS}$ is the full step.

The minimum along the dog-leg path is found at the intersection of this path with the trust region border. This intersection takes place at a value of τ such as:

$$\left\| \vec{p}_k^{\,C} + (\tau - 1)(\vec{p}_k^{\,FS} - \vec{p}_k^{\,C}) \right\|^2 = r^2$$

Another method for the minimization sub-problem was introduced by [61] and [59] and is now quite popular. It is a truncated conjugate gradient method, which uses conjugate gradient steps as long as they inside the trust region. In case these steps lead to a termination point inside the trust region, this would be the solution. Otherwise, if the new step comes out of the trust region, or a negative curvature direction is computed, a truncated step is taken to the intersection with the trust region border.

Some other methods have been proposed, considering, for instance, a two-dimensional search in the zone inside the trust region and between the steepest descent and the Newton step directions. See [64] for pertinent references.

A number of searching methods, like the Levenberg-Mardquardt, have trust-region versions [41].

In order to provide an example, the trust-region method using Cauchy points has been used to find the minimum in the non-quadratic example. Program 2.24 gives an example of simple coding of this method in MATLAB. Figure 2.48 shows the result obtained by the program.

Program 2.25 gives an example of dog-leg method coding in MATLAB. The case considered is again the non-quadratic problem. Figure 2.49 shows the result of this algorithm. Notice that the very end of the steps trajectory has a pronounced (and small) zigzag.

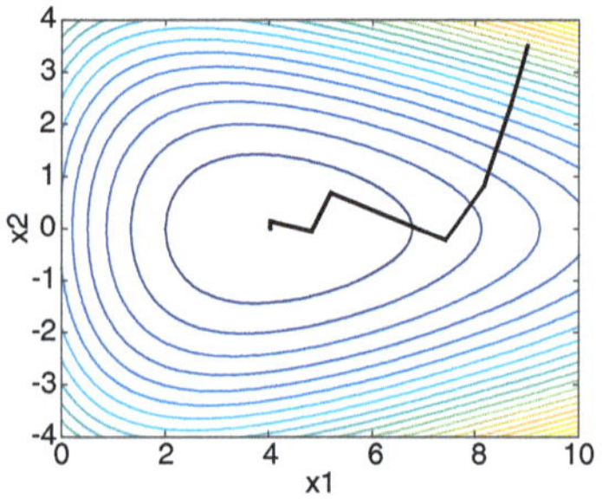

Fig. 2.48 Example of trust-region method

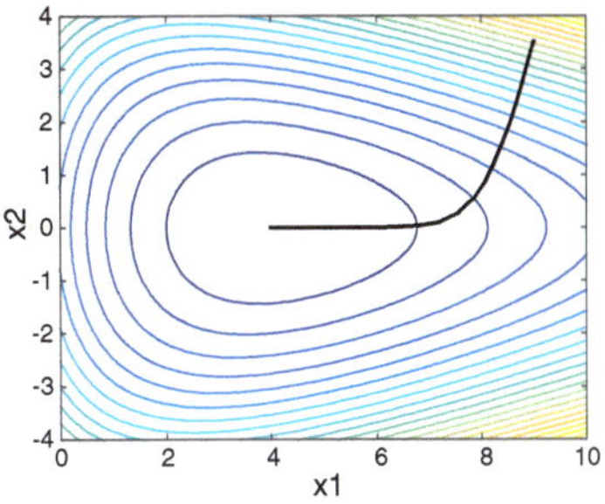

Fig. 2.49 Example of trust-region dog-leg method

Note (Programs 2.24 and 2.25)

The complete listings of the Programs 2.24 and 2.25 have been included in the Appendix for Long Programs.

2.6 The Case of Vector Functions: Gauss-Newton

It is not uncommon to have optimization problems involving vector functions. Actually, an important application is the fitting of models to data. This section will consider a simple example of such application.

The line of reasoning for the Gauss-Newton method is again based on a Taylor expansion. In particular, for a vector function, one considers the following first-order approximation:

$$\vec{f}(\vec{x}) = \vec{f}(\vec{x}_n) + J_n(\vec{x})(\vec{x} - \vec{x}_n)$$

where $J_n(\vec{x})$ is a Jacobian.

In the case of model fitting, the vector function is a set of errors (residuals) between data points and the model. One wants to have $f(\vec{x}) = 0$ for a certain value of $\vec{x}$, being $\vec{x}$ the parameters of the model.

Setting $f(\vec{x}) = 0$ in the Taylor expansion, one can write:

$$J_n(\vec{x})(\vec{x} - \vec{x}_n) = -\vec{f}(\vec{x}_n)$$

The Jacobian is a rectangular matrix that cannot be inverted. Looking for a more convenient expression, one multiplies both sides of the expression by J_n^t. And then:

$$J_n^T J_n \cdot (\vec{x} - \vec{x}_n) = -J_n^T \cdot \vec{f}(\vec{x}_n) \Rightarrow \Delta\vec{x}_n = -(J_n^T J_n)^{-1} \cdot J_n^T \vec{f}(\vec{x}_n)$$

In practice no matrix inversion is computed. Instead, one considers that:

$$\vec{x}_{n+1} = \vec{x}_n + \vec{\delta}_n$$

And the value of $\vec{\delta}_n$ is obtained by solving the following linear equation:

$$(J_n^T \cdot J_n)\, \vec{\delta}_n = -\, J_n^T \cdot \vec{f}(\vec{x}_n)$$

Let us give a simple example. The following data corresponds to the population evolution of bears in a certain region

$$t_i = 1,\ 2,\ 4,\ 5,\ 8$$
$$y_i = 3,\ 4,\ 6,\ 11,\ 20$$

The following model was suggested in order to fit the data:

$$y = x_1 \exp(x_2\, t)$$

There are 5 functions (residuals) to be minimized:

$$r_i = y_i - x_1 \exp(x_2\, t_i), \quad i = 1, \ldots, 5$$

The Jacobian would be:

$$J = \begin{pmatrix} \frac{\partial r_1}{\partial x_1} & \frac{\partial r_1}{\partial x_2} \\ \frac{\partial r_2}{\partial x_1} & \frac{\partial r_2}{\partial x_2} \\ \ldots & \ldots \\ \ldots & \ldots \\ \frac{\partial r_5}{\partial x_1} & \frac{\partial r_5}{\partial x_2} \end{pmatrix}$$

where:

$$\frac{\partial r_i}{\partial x_1} = -\exp(x_2\, t_i)$$

$$\frac{\partial r_i}{\partial x_2} = -\, x_1\, t_i \exp(x_2\, t_i)$$

Now, one applies the Gauss-Newton method. Figure 2.50 shows the data (cross marks) and the model (continuous curve) that has been fitted with the method

Figure 2.50 has been created with the Program 2.26, which includes a compact MATLAB coding of the Gauss-Newton step. The algorithm converged to satisfactory model parameter approximations in six iterations.

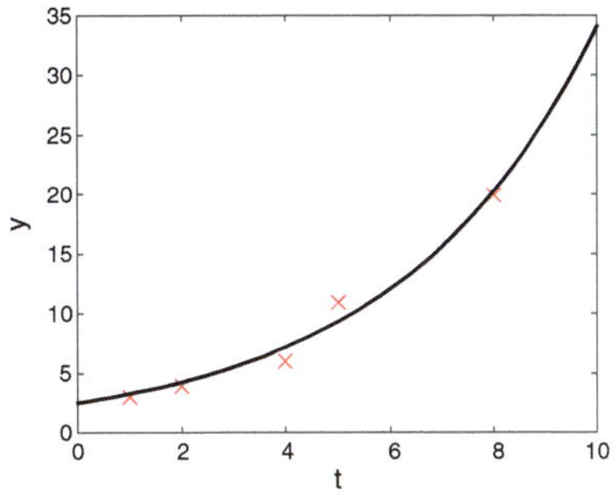

Fig. 2.50 Example of Gauss-Newton method

Program 2.26 Gauss-Newton method

```
% Example of Gauss-Newton method
% model fitting
%
% data to be fitted (column vectors)
t=[1 2 4 5 8]';
y=[3 4 6 11 20]';
%
% reserve memory
N=length(t);
f=zeros(N,1);
J=zeros(N,2);
% model is y = x1 * exp(x2*t)
% initial values of model parameters
x1=6; x2=0.1;
%
% steps -----------------------
for M=1:6,
  % value of function
  for n=1:N,
    f(n)= y(n)-(x1*exp(x2*t(n))); % residuals
  end;
  % value of the Jacobian
  for n=1:N,
    J(n,1)= -exp(x2*t(n));
    J(n,2)= -(x1*t(n))*exp(x2*t(n));
  end;
% Gauss-Newton step:
  delta=(J'*J)$-J'*f);
  x1=x1+delta(1);
  x2=x2+delta(2);
end;
% model fitting display -------------------
figure (1)
%the data:
plot(t,y,'rx'); hold on;
% the model:
```

```
tm=0:0.2:10;
ym=x1*exp(x2*tm);
plot(tm,ym,'k');
xlabel('t'); ylabel('y');
```

2.7 Direct, Derivative-Free Methods

In some problems, derivative-based approaches have difficulties in certain points or entire regions. Still, some methodologies exist for direct searching, [35, 37].

Basic derivative-free methods can be random jumping, or choosing directions at random, or following a systematic search on a grid, etc. In the case of random directions, you depart from a point, choose at random a direction and then run a line-search: if it determines a better point, the point is accepted, and a random direction from this new point is followed and a line search is run, etc. (if a direction was not fruitful, another direction is tried at random).

There are some other methods that should be mentioned.

2.7.1 *The Method of Hooke and Jeeves*

Also called *Pattern Search method*, [28]. It was proposed in 1961 by R. Hooke and T.A. Jeeves. From an initial point x_0 of the search in n-dimensions, exploratory moves (with a fixed step) on each dimension are taken. For instance, in 2D four points are checked, as shown in Fig. 2.51:

If the exploration determines a best move, with a certain direction, then the method continues to move farther in this direction until no improvement is obtained: this is called a *pattern move*.

If the exploration does not find a better point, then a smaller step size is tried.

The iterative process terminates when reaching a too small step size.

Some MATLAB implementations of the method can be obtained from the MathWorks File Exchange site and from GitHub.

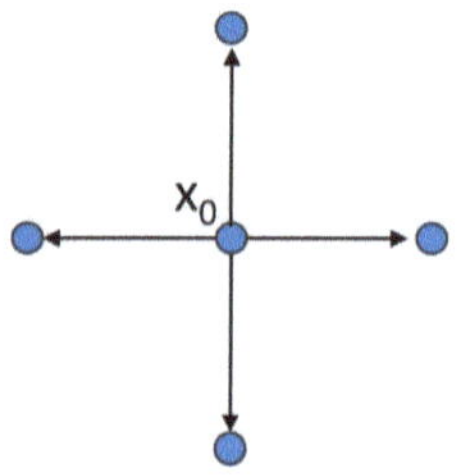

Fig. 2.51 Exploratory moves in the Hooke and Jeeves method

Several variations of the algorithm have been proposed, as well as hybridization with heuristic approaches.

2.7.2 The Method of Nelder and Mead

In 1965, J.A. Nelder and R. Mead introduced what they called a *Simplex Method* [43]. The method constructs a special polyhedron and then moves and modifies it to perform the search.

A simplex in n dimensions is a polyhedron with n+1 vertices. For instance, in two dimensions is a triangle; and in three dimensions is a tetrahedron, etc.

The method of Nelder and Mead is **not** related to the simplex method employed in Linear Programming (next chapter), which dates back to 1947.

Once a first simplex polyhedron is formed, with n+1 points, the first task of the Nelder and Mead method is ordering these points from worst to best. Then, excluding the worst point, a centroid (mean) of the rest of points is computed. Denote this centroid as c.

Then, the method considers a series of transformations of the simplex, [33]:

- Reflection:

A reflection of the worst point is tried, forming a new simplex. For instance, in 2D we have (Fig. 2.52):

If the new point is better than the second-worst point, but no better than the current best, the worst point is replaced by the new point.

- Expansion:

If in the reflection above, the new point was better than the current best, an expanded point is tried, as shown in the next figure (Fig. 2.53):

The worst point is replaced by the better of the reflection point or the expanded point (variants of the method use other alternatives).

- Contraction:

In case the reflection point was worst than the second-worst point, the method tries a contraction, which can be outward or inward. For example, in 2D here is an inward contraction (Fig. 2.54):

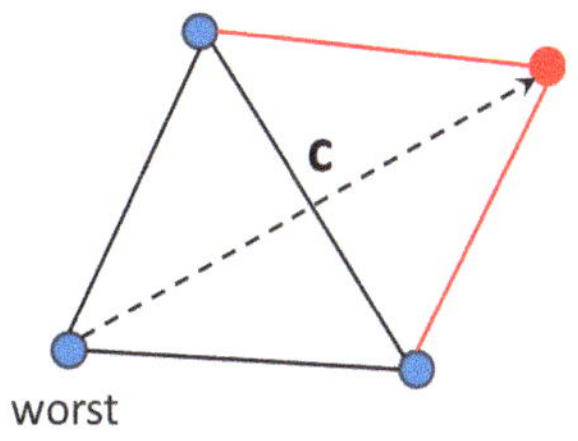

Fig. 2.52 A reflection move of the simplex

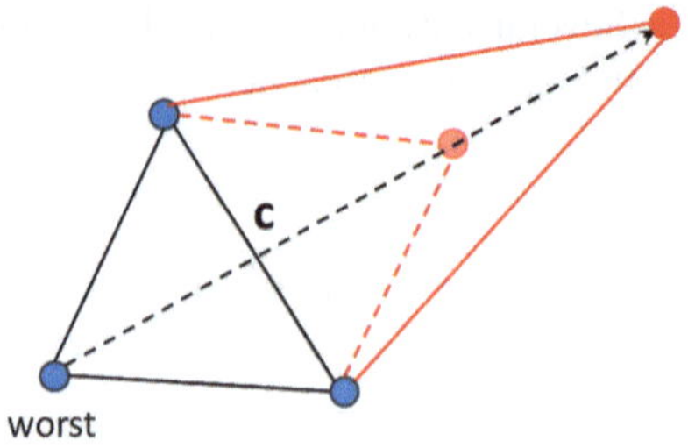

Fig. 2.53 An expansion move of the simplex

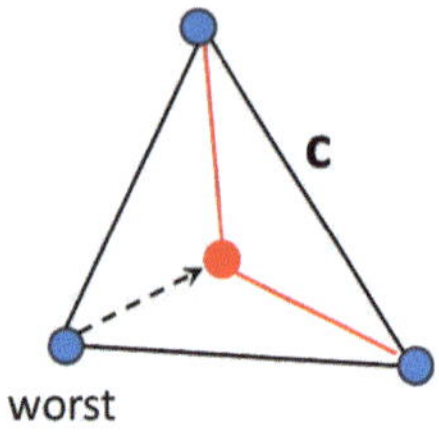

Fig. 2.54 An inward contraction of the simplex

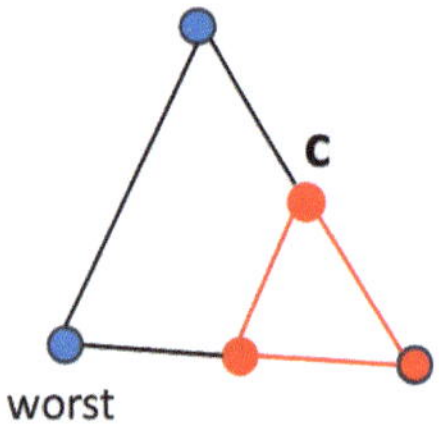

Fig. 2.55 A shrink of the simplex

If the contracted point is better than the worst point, the worst point is replaced by the contracted point. If not, then we go to a shrink of the simplex.

- Shrink:

The simplex is redefined, moving the points toward the current best point. For example, in 2D (Fig. 2.55):

In practice, shrinking is not usually required.

The idea of the simplex method is to adapt the search to the data in n-dimensions. The simplex motions are iterated until a satisfactory convergence is reached. Figure 2.56 shows an example of simplex searching, where you can see how the simplex moves and adapts toward the optimum.

The Nelder and Mead article [43] is a "Science Citation Classic," with more than 40,000 cites. An interesting publication on the method is [63] , revealing historical details and the scientific discussion on convergence (in some scenarios, the original method may fail, and some variations have been proposed). An extensive document on the Nelder and Mead method, including some improvements, is [8].

Some MATLAB implementations of the Nelder and Mead Simplex method can be obtained from the MathWorks File Exchange site and from GitHub (and from Academia, like [12]).

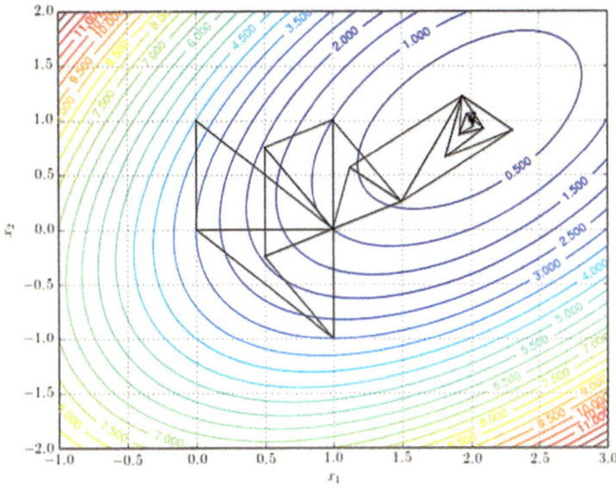

Fig. 2.56 Evolution of a simplex search

2.8 Exercises

1- Use the unidimensional searching methods explained in Sect. 2.2.1, to approximate the minimum of:

$$f(x) = x^5 - 5x^3 - 20x + 5$$

Visualize the iterations. Compare convergence behaviors of each method.

2- Use the unidimensional searching methods explained in Sect. 2.2.2, to approximate the minimum of the same function above. Visualize iterations; compare convergences (also with respect to problem 1).

3- Use the methods in Sect. 2.2.1 to approximate the minimum of:

$$f(x) = 0.65 - (0.75/(1 + x^2)) - 0.65 \operatorname{arctg}(1/x)$$

inside interval (0,3)

Visualize the iterations. Compare convergence behaviors of each method.

4- Use the methods in Sect. 2.2.2, to approximate the minimum of the same function above. Visualize iterations; compare convergences (also with respect to problem 3).

5- Use the methods in Sect. 2.2.1, to approximate the minimum of:

$$f(x) = -3\alpha \sin(0.75\alpha) + \exp(-2\alpha)$$

inside interval $(0, 2\pi)$

Visualize the iterations. Compare convergence behaviors of each method.

6- Use the methods in Sect. 2.2.2, to approximate the minimum of the same function above. Visualize iterations; compare convergences (also with respect to problem 5).

7- Given the function:

$$f(\vec{x}) = (x_1 - 1)^2 + (x_2 - 5)^2 + (x_3 - 4)^2$$

Starting from (3, -3, 5), use steepest descent to approximate the minimum. Visualize how the distance to the solution decreases along iterations.

8- Given the function:

$$f(\vec{x}) = 4(x_1 - 3)^2 + 3(x_2 - 5)^2 + (x_3 - 4)^2$$

Starting from (0, 0, 0), use Powell's method to approximate the minimum. Visualize how the distance to the solution decreases along iterations.

9- Use Fletcher-Reeves for the same function above. Compare convergence behavior.

10- Given the function:

$$f(\vec{x}) = x_1^2 \exp\{1 - x_2^2 - 10(x_1 - x_2)^2\}$$

Starting from (0.1, 0.1), use Powell's method to approximate the minimum. Visualize in 2D the iterations. What could happen if another initial point was chosen?

11- Given the function:

$$f(\vec{x}) = x_1^2 + 4x_2^2 + x_1 x_2 + 1$$

Starting from (0, 2), use the second-order Newton's method to find the minimum. Visualize in 2D the behavior of the method.

12- Given the function:

$$f(\vec{x}) = 1.5x_1^2 + 2x_2^2 + 1.5x_3^2 + x_1 x_3 + 2x_2x_3 - 3x_1 - x_3$$

Starting from (3, -7, 0), use the Levenberg-Marquardt method to approximate the minimum. Visualize the evolution of distance to the solution along iterations.

13- Employ the Broyden's method for problem 12. Use also the DFP and the BFGS methods, and compare.

14- Given the function:

$$f(\vec{x}) = (x_1 + 5)^2 + (x_2 + 8)^2 + (x_3 + 7)^2 + 2x_1^2x_2^2 + 4x_1^2x_3^2$$

Starting from (1, 1, 1), use the dog-leg method to approximate the minimum. Visualize the evolution of distance to the solution along iterations.

15- [2.1] [2.2] [2.3] Consider the following objective function:

$$J = x^2 - 4x + y^2 - y - xy$$

Apply the Nelder-Mead simplex method to find the minimum. Visualize the behavior of the simplex during the method iterations. Try several initializations; add comments on convergence.

15- [2.4] The following objective function has two minima:

$$J = x_1^4 + x_2^4 - \frac{5}{4}x_1^2 + \frac{1}{4}.$$

Apply Nelder-Mead departing from different initializations, in order to find the two minima. Visualize in 2D the behavior of the algorithm.

2.8.1 Exercise References

[2.1] https://jasoncantarella.com/downloads/NelderMeadProof.pdf
[2.2] https://people.sc.fsu.edu/~jburkardt/m_src/nelder_mead/nelder_mead.html
[2.3] https://es.mathworks.com/matlabcentral/fileexchange/69636-nelder-and-mead-simplex-algorithm
[2.4] https://www.jakubkonka.com/2013/10/22/nelder-mead-simplex.html

References

1. Abdulkadirov, R., Lyakhov, P., & Nagornov, N. (2023). Survey of optimization algorithms in modern neural networks. *Mathematics, 11*(11), 2466. Available from: https://www.mdpi.com/2227-7390/11/11/2466
2. Al-Baali, M., & Khalfan, H. (2007). An overview of some practical quasi-Newton methods for unconstrained optimization. *Sultan Qaboos University Journal for Science [SQUJS], 12*(2), 199–209.
3. Al-Baali, M., Spedicato, E., & Maggioni, F. (2014). Broyden's quasi-Newton methods for a nonlinear system of equations and unconstrained optimization: A review and open problems. *Optimization Methods and Software, 29*(5), 937–954.
4. Babaie-Kafaki, S., Ghanbari, R., & Mahdavi-Amiri, N. (2010). Two new conjugate gradient methods based on modified secant equations. *Journal of Computational and Applied Mathematics, 234*(5), 1374–1386.
5. Bakr, M. (2013). *Nonlinear optimization in electrical engineering with applications in Matlab*. Institution of Engineering and Technology.
6. Barlow, J. L. (2006). The preconditioned conjugate gradient method. Available from: https://www.cse.psu.edu/~b58/cse456/lecture20.pdf
7. Barzilai, J., & Borwein, J. M. (1988). Two-point step size gradient methods. *IMA Journal of Numerical Analysis, 8*(1), 141–148.
8. Baudin, M. (2010). Nelder-mead user's manual. *Consortium Scilab-Digiteo*. Available from: https://www.scilab.org/sites/default/files/neldermead.pdf
9. Benzi, M. (2002). Preconditioning techniques for large linear systems: A survey. *Journal of Computational Physics, 182*(2), 418–477.

10. Broyden, C. G. (1970). The convergence of a class of double-rank minimization algorithms 1. General considerations. *IMA Journal of Applied Mathematics, 6*(1), 76–90.
11. Broyden, C. G. (2000). On the discovery of the "good Broyden" method. *Mathematical Programming, 87*(2), 209–213.
12. Burkardt, J. (2019). Nelder-mead. Available from: https://people.sc.fsu.edu/~jburkardt/m_src/nelder_mead/nelder_mead.html
13. Burke, J. (2020). The conjugate gradient algorithm. Available from: https://sites.math.washington.edu/~burke/crs/408f/notes/nlp/cg.pdf
14. Conn, A. R., Gould, N. I. M., & Toint, P. L. (2000). *Trust region methods*. SIAM.
15. Davidon, W. C. (1991). Variable metric method for minimization. *SIAM Journal on Optimization, 1*(1), 1–17.
16. Fletcher, R. (1970). A new approach to variable metric algorithms. *The Computer Journal, 13*(3), 317–322.
17. Fletcher, R., & Powell, M. J. D. (1963). A rapidly convergent descent method for minimization. *The Computer Journal, 6*(2), 163–168.
18. Fletcher, R., & Reeves, C. M. (1964). Function minimization by conjugate gradients. *The Computer Journal, 7*(2), 149–154.
19. Gavin, H. (2011). The Levenberg-Marquardt method for nonlinear least squares curve-fitting problems. Available from: https://people.duke.edu/~hpgavin/ExperimentalSystems/lm.pdf
20. Geyer, C. J. (2020). Trust regions. Available from: https://mirrors.nics.utk.edu/cran/web/packages/trust/vignettes/trust.pdf
21. Goldfarb, D. (1970). A family of variable-metric methods derived by variational means. *Mathematics of Computation, 24*(109), 23–26.
22. Golub, G. H., & O'Leary, D. P. (1989). Some history of the conjugate gradient and lanczos algorithms: 1948–1976. *SIAM Review, 31*(1), 50–102.
23. Gould, N. I. M., Orban, D., Sartenaer, A., & Toint, P. L. (2005). Sensitivity of trust-region algorithms to their parameters. *4OR: A Quarterly Journal of Operations Research, 3*(3), 227–241.
24. Dai, Y. H., & Liao, L. Z. (2001). New conjugacy conditions and related nonlinear conjugate gradient methods. *Applied Mathematics & Optimization, 43*(1), 87–101.
25. Hager, W. W. (1989). Updating the inverse of a matrix. *SIAM Review, 31*(2), 221–239.
26. Hager, W. W., & Zhang, H. (2006). A survey of nonlinear conjugate gradient methods. *Pacific Journal of Optimization, 2*(1), 35–58.
27. Hanzely, S., Kamzolov, D., Pasechnyuk, D., Gasnikov, A., Richtárik, P., & Takác, M. (2022). A damped Newton method achieves global $\mathcal{O}\left(\frac{1}{k^2}\right)$ and local quadratic convergence rate. *Advances in Neural Information Processing Systems, 35*, 25320–25334. Available from: https://labmmo.ru/upload/000/u8/3/f/2211-00140.pdf
28. Harder, D. W. (2021). Hooke-jeeves method for finding extrema in n dimensions. Available from: https://ece.uwaterloo.ca/~dwharder/nm/Lecture_materials/pdfs/8.2.1%20Hooke-Jeeves%20method.pdf
29. Hauser, R. (2005). Quasi-Newton methods. Available from: https://www.numerical.rl.ac.uk/people/nimg/oupartc/lectures/raphael/lectures/lecture4.pdf
30. He, N. (2016). Gradient descent and acceleration. Available from: https://github.com/niaohe/Big-Data-Optimization-Course/blob/main/lecture_scribe/IE598-lecture9-gradient-descent-and-acceleration.pdf
31. Hestenes, M. R. (2012). *Conjugate direction methods in optimization* (Vol. 12). Springer Science & Business Media.
32. Hestenes, M. R., & Stiefel, E. (1952). Methods of conjugate gradients for solving linear systems. *Journal of Research of the National Bureau of Standards, 49*(1), 409–436.
33. Joglekar, S. (2016). Nelder-mead optimization. Available from: https://codesachin.wordpress.com/2016/01/16/nelder-mead-optimization/
34. Kashyap, R. (2022). A survey of deep learning optimizers–first and second order methods. arXiv preprint arXiv:2211.15596. Available from: https://arxiv.org/pdf/2211.15596

35. Kramer, O., Ciaurri, D. E., & Koziel, S. (2011). Derivative-free optimization. In *Computational optimization, methods and algorithms* (pp. 61–83). Springer.
36. Levenberg, K. (1944). A method for the solution of certain non-linear problems in least squares. *Quarterly of Applied Mathematics, 2*(2), 164–168.
37. Lewis, R. M., Torczon, V., & Trosset, M. W. (2000). Direct search methods: Then and now. *Journal of computational and Applied Mathematics, 124*(1–2), 191–207.
38. Li, G., Tang, C., & Wei, Z. (2007). New conjugacy condition and related new conjugate gradient methods for unconstrained optimization. *Journal of Computational and Applied Mathematics, 202*(2), 523–539.
39. Macardel, J. (2010). 18th century mezzotint portrait of sir Isaac Newton (1643–1727). Available from: https://www.tregeaglefineart.com/en-GB/pictures/18th-century-mezzotint-portrait-of-sir-isaac-newton-1643-1727-/prod_10182
40. Marquardt, D. W. (1963). An algorithm for least-squares estimation of nonlinear parameters. *Journal of the Society for Industrial and Applied Mathematics, 11*(2), 431–441.
41. Moré, J. J. (1978). The Levenberg-Marquardt algorithm: Implementation and theory. In *Numerical analysis* (pp. 105–116). Springer.
42. More, J. J. (1983). Recent developments in algorithms and software for trust region methods. In *Mathematical programming The state of the art* (pp. 258–287). Springer.
43. Nelder, J. A., & Mead, R. (1965). A simplex method for function minimization. *The Computer Journal, 7*(4), 308–313.
44. Nocedal, J., & Wright, S. J. (2006). *Numerical optimization* (2nd ed.). Springer.
45. Paul, D. (1993). Levenberg–Marquardt methods and nonlinear estimation. *SIAM News, 8*(1).
46. Polak, E., & Ribiere, G. (1969). Note sur la convergence de méthodes de directions conjuguées. *Revue Française d'Informatique et de Recherche Opérationnelle. Série rouge, 3*(16), 35–43.
47. Polyak, B. T. (1969). The conjugate gradient method in extremal problems. *USSR Computational Mathematics and Mathematical Physics, 9*(4), 94–112.
48. Polyak, B. T. (1987). *Introduction to optimization.* Optimization Software Inc.
49. Powell, M. J. D. (1977). Restart procedures for the conjugate gradient method. *Mathematical Programming, 12*(1), 241–254.
50. Powell, M. J. D. (1964). An efficient method for finding the minimum of a function of several variables without calculating derivatives. *The Computer Journal, 7*(2), 155–162.
51. Powell, M. J. D. (1970). A hybrid method for nonlinear equations. In *Numerical methods for nonlinear algebraic equations* (Vol. 7, pp. 87–114). Gordon and Breach.
52. Powell, M. J. D. (1970). A new algorithm for unconstrained optimization. In *Nonlinear programming* (pp. 31–65). Academic Press.
53. Powell, M. J. D. (1984). Nonconvex minimization calculations and the conjugate gradient method. In *Numerical analysis* (pp. 122–141). Springer.
54. Powell, M. J. D. (1986). How bad are the BFGS and DFP methods when the objective function is quadratic? *Mathematical Programming, 34*(1), 34–47.
55. Powell, M. J. D. (2010). *On nonlinear optimization since 1959.* World Scientific.
56. Rheinboldt, W. C. (1998). *Methods for solving systems of nonlinear equations*. SIAM.
57. Roweis, S. (1996). Levenberg-Marquardt optimization. *Notes, University Of Toronto.*
58. Shanno, D. F. (1970). Conditioning of quasi-Newton methods for function minimization. *Mathematics of Computation, 24*(111), 647–656.
59. Steihaug, T. (1983). The conjugate gradient method and trust regions in large scale optimization. *SIAM Journal on Numerical Analysis, 20*(3), 626–637.
60. Tibshirani, R. (2019). Newton's method. *Notes for convex optimization: Machine learning* (pp. 10–725). Available from: https://www.stat.cmu.edu/~ryantibs/convexopt-S15/scribes/14-newton-scribed.pdf
61. Toint, Ph. L. (1981). Towards an efficient sparsity exploiting Newton method for minimization. In *Sparse matrices and their uses* (pp. 57–88). Academic Press.
62. Transtrum, M. K., & Sethna, J. P. (2012). Improvements to the Levenberg-Marquardt algorithm for nonlinear least-squares minimization. arXiv preprint arXiv:1201.5885.

63. Wright, M. H. et al. (2010). Nelder, mead, and the other simplex method. *Documenta Mathematica, 7*, 271–276. Available from: https://www.emis.de/journals/DMJDMV/vol-ismp/42_wright-margaret.pdf
64. Yuan, Ya-X. (2000). A review of trust region algorithms for optimization. In *ICIAM* (Vol. 99, pp. 271–282).

Chapter 3
Linear Programming

3.1 Introduction

Many practical optimization applications, in industry, finance, commerce, etc. can be expressed as linear programming problems. To get an impression of the importance of this topic, you could just search for linear programming on the Internet and find an avalanche of millions of results. One of the reasons for being so relevant is that in many daily life cases the objective to be optimized, like cost, time, energy, etc., is directly proportional to the values of several factors; in other words, a linear relation exists between objective and variables. Although academic examples use not many variables, to keep things simple and clear, real linear programming problems may involve thousands of variables.

A typical way of solving linear programming problems is by using an algorithm called ***"simplex"***. This algorithm is considered as one of the ten most important algorithms of the twentieth century.

The first contributions to linear programming trace back to the work of L. Kantorovich in the 1930s, for optimization problems in economics. Like other important theory developments written in Russian, it remained internationally unknown for some years. It was applied to shipping problems by T. Koopmans, at the beginning of the 1940s. In 1947, G.B. Dantzig published the simplex algorithm for solving linear programming problems. The name "linear programming" was proposed by Koopmans in a discussion with Dantzig in 1948.

Figure 3.1 presents a photograph of George B. Dantzig, who worked at RAND Corporation and several research centers and universities, like Berkeley and Stanford.

Curiously, the term "programming" did not mean at early 1940s anything related to computer programs. Instead, it refers to "planning." Digital computers were not present until 1946, when the ENIAC computer was introduced on the market. This computer was programmed using around 6000 switches. Actually, an important

J. M. Giron-Sierra, *Introduction to Optimization with Matlab® Examples*,
https://doi.org/10.1007/978-3-032-05048-9_3

Fig. 3.1 George B. Dantzig

activity of Dantzig, after 1947, was to implement the simplex algorithm on a digital computer, obtaining successful results in 1953.

More historical details can be found in papers by Dantzig himself [7], and articles like [10] or the more recent [4]. A new edition of a most cited book [8] on linear programming is [6].

Linear programming is also a motivating avenue for introducing the following important topic: optimization in the presence of constraints (*constrained optimization*).

Consider the following optimization problem:

Maximize:

$$z = c_1 x_1 + c_2 x_2 + \ldots + c_n\, x_n$$

Subject to (constraints):

$$a_{11} x_1 + a_{12} x_2 + \ldots + a_{1n}\, x_n \le b_1$$

$$a_{21} x_1 + a_{22} x_2 + \ldots + a_{2n}\, x_n \le b_2$$

$$a_{m1} x_1 + a_{m2} x_2 + \ldots + a_{m\,n}\, x_n \le b_m$$

(the variables take continuous real values; inequalities can also be $\ge$)

This is an example of *'linear programming' problems*. This type of problems includes an objective function to optimize and a set of constraints to be obeyed. What is characteristic of linear programming is that both the objective function and the constraints are linear.

Internally, the chapter is divided into two parts. The first is devoted to the exposition of the method, and the second presents some illustrative applications, like diets, scheduling, logistics, etc.

It is not difficult to get from the Internet good tutorials and documentation. Likewise, there are many books, more or less oriented to the theory, like [27, 32], or to specific application fields [21, 34].

Most books on operations research devote ample space to linear programming and related problems.

3.2 Polyhedral Analysis

An intuitive way for analyzing linear programming problems is to adopt the point of view of geometry. In this context, the following expression:

$$a_1x_1 + a_2x_2 + \ldots + a_n x_n = b$$

corresponds to a hyperplane in n-dimensions. In the case of $n = 2$, this is a line. An equivalent expression of the hyperplane is the following:

$$\vec{a}^T \vec{x} = b$$

Two hyperplanes with the following equations:

$$\vec{a}^T \vec{x} = b_1$$

$$\vec{a}^T \vec{x} = b_2$$

are parallel.

A hyperplane separates the n-dimensional space into two half spaces.

In the case of an inequality:

$$\vec{a}^T \vec{x} \leq b$$

this also separates the n-dimensional space into two half spaces, one of which satisfies the inequality.

The constraints of the linear programming problem define a set of points $\vec{x}$ inside an n-dimensional region formed by the intersection of hyperplanes and half spaces.

For example, suppose the following constraints:

$$8x_1 + 10x_2 \geq 80$$

$$18\,x_1 + 14\,x_2 \leq 252$$

$$x_1\,,\ x_2 \geq 0$$

Let us plot in Fig. 3.2 the region defined by these constraints. This region is called the feasible solution set.

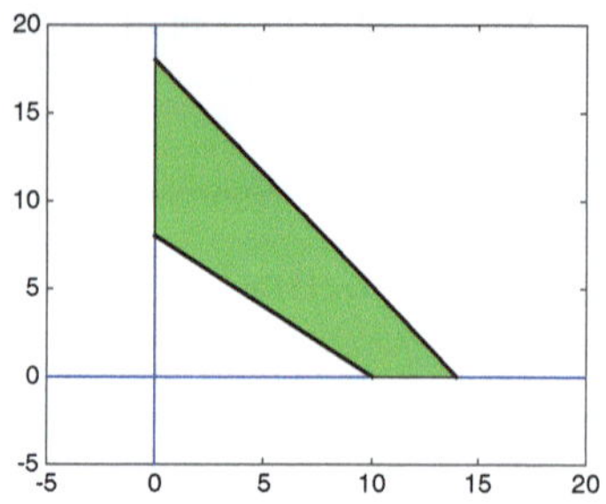

Fig. 3.2 A feasible solution set

Usually, the feasible solution region is bounded, in which case it is a polygon (2D), or a polyhedron (3D), or a polytope (nD).

Figure 3.2 has been obtained with the Program 3.1. The feasible region has been painted using the *patch( )* function.

Program 3.1 Displaying a 2D feasible set

```
%display of 2D feasible set
% example 1
% ax + by =c
a1=8; b1=10; c1=80;
a2=18; b2=14; c2=252;
%
% constraint corner points:
X1=c1/a1; Y1=c1/b1;
X2=c2/a2; Y2=c2/b2;
figure(1)
%axes:
plot([-5 20],[0 0],'b'); hold on;
plot([0 0],[-5 20],'b');
%feasible set:
patch([0 0 X2 X1],[Y1 Y2 0 0],'g');
%constraints
plot([0 X1],[Y1 0],'k');
plot([0 X2],[Y2 0],'k');
```

Another example of constraints is the following:

$$10x_1 + 18x_2 \leq 180$$

$$14x_1 + 7x_2 \leq 98$$

$$x_1, \ x_2 \geq 0$$

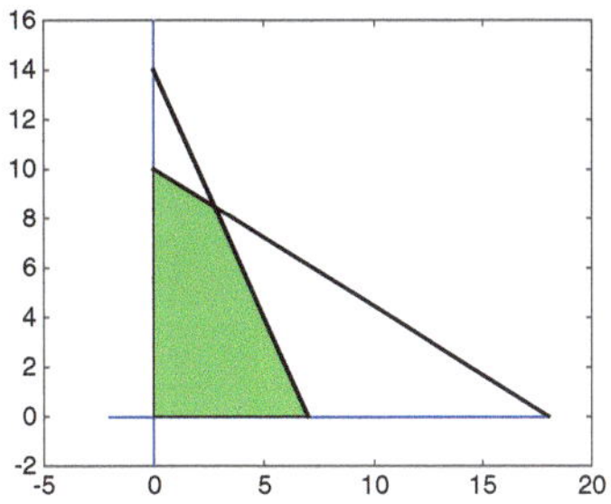

Fig. 3.3 A feasible solution set

Figure 3.3 visualizes the region defined by these constraints. This figure has been obtained with Program 3.2, which includes the calculation of the crossing point of constraint border lines.

Program 3.2 Another example of feasible set

```
%display of 2D feasible set
% example 2
% ax + by =c
a1=10; b1=18; c1=180;
a2=14; b2=7; c2=98;
% constraint corner points:
X1=c1/a1; Y1=c1/b1;
X2=c2/a2; Y2=c2/b2;
% intersection:
aux1=(Y2-Y1); aux2=(Y2*X1)-(Y1*X2);
xi=(X1*X2*aux1)/aux2;
yi=Y1-((Y1*xi)/X1);
figure(1)
%axes:
plot([-2 18],[0 0],'b'); hold on;
plot([0 0],[-2 16],'b');
%feasible set:
patch([0 0 xi X2],[0 Y1 yi 0],'g');
%constraints
plot([0 X1],[Y1 0],'k');
plot([0 X2],[Y2 0],'k');
```

Now, suppose that the function to maximize is:

$$z = 3x_1 + 2x_2$$

Figure 3.4 shows some parallel lines obtained for different values of z. This figure has been generated by the Program 3.3. The function *compass( )* has been used for drawing the arrow.

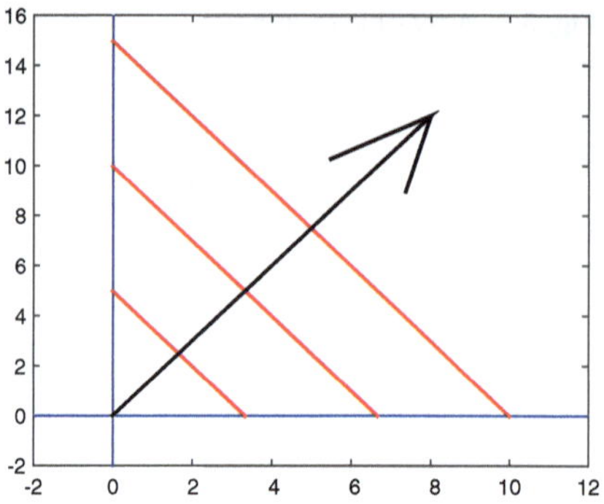

Fig. 3.4 Representation of the objective function

Program 3.3 Displaying the objective function

```
%display of objective function
% (3 values)
%ax+by=z;
a=3; b=2;
% axis intersection points
z=10; X1=z/a; Y1=z/b;
z=20; X2=z/a; Y2=z/b;
z=30; X3=z/a; Y3=z/b;
%
figure(1)
%axes:
plot([-2 12],[0 0],'b'); hold on;
plot([0 0],[-2 16],'b');
%objective function
plot([0 X1],[Y1 0],'r');
plot([0 X2],[Y2 0],'r');
plot([0 X3],[Y3 0],'r');
compass(b*4,a*4,'k');
```

3.2.1 *Key Observation*

The important observation is that the maximum of z is attained at a vertex of the feasible solution region, as it is shown in Fig. 3.5. The figure represents a set of lines, corresponding to some values of the objective function, upon the feasible set. One of these lines (in different color), intersecting with a vertex, gives the solution of the optimization problem: it is the maximum value of z obeying to the constraints,

Figure 3.5 has been obtained with Program 3.4. The data for the feasible zone gray (green) patch were inherited from Program 3.2.

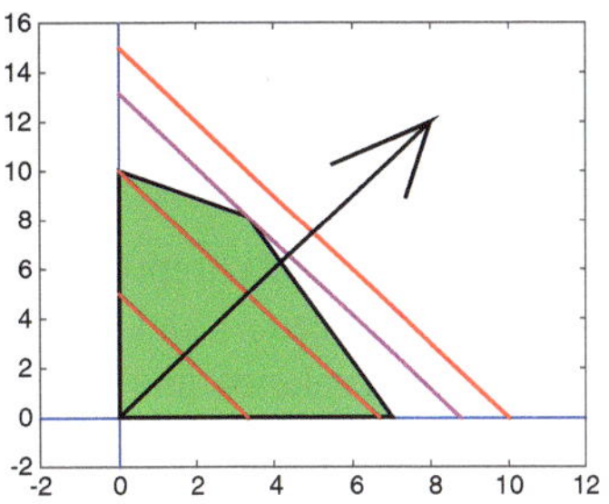

Fig. 3.5 The feasible set and some values of the objective function

Program 3.4 Displaying both the feasible set and the objective function

```
%display of objective function
% (3 values)
%ax+by=z;
a=3; b=2;
% axis intersection points
z=10; X1=z/a; Y1=z/b;
z=20; X2=z/a; Y2=z/b;
z=30; X3=z/a; Y3=z/b;
z=26.323; Xo=z/a; Yo=z/b;
%
figure(1)
%axes:
plot([-2 12],[0 0],'b'); hold on;
plot([0 0],[-2 16],'b');
%feasible set:
xi=576/172; xf=7; yf=10; yi=yf-((yf*xi)/18);
patch([0 0 xi xf],[0 yf yi 0],'g');
%objective function
plot([0 X1],[Y1 0],'r');
plot([0 X2],[Y2 0],'r');
plot([0 X3],[Y3 0],'r');
plot([0 Xo],[Yo 0],'m','LineWidth',2);
compass(b*4,a*4,'k');
```

3.2.2 *Vertices and Segments*

Let us say in advance that the simplex algorithm tests adjacent vertices in sequence, so that in maximization problems at each new vertex the objective function increases, or at least remains unchanged (in minimization problems, it decreases or do not change). Therefore, it is convenient to study in more detail vertices and segments.

A segment between two points $\vec{x}_1$ and $\vec{x}_2$ is the set of points $\vec{x}$ given by:

$$\vec{x} = \lambda x_2 + (1 - \lambda) x_1 \ , \quad 0 \leq \lambda \leq 1$$

And now a very important concept. A set C is ***convex*** if for all $\vec{x}_1, \vec{x}_2 \in C$ the segment between $\vec{x}_1$ and $\vec{x}_2$ is composed entirely of points of C.

The hyperplanes and the half-spaces are convex. The intersection of a finite number of convex sets is convex.

A point $\vec{x} \in C$ is an *extreme point* of C if there is no segment between two points of C containing $\vec{x}$.

A point $\vec{x} \in C$ is an *interior point* of C if around it there is a ball made entirely of points of C.

As it will be detailed soon, by adding artificial variables, all inequalities can be converted to equalities. This way, the linear programming problem can be compactly expressed as follows:

Maximize:

$$z = \vec{c}^T \cdot \vec{x}$$

Subject to:

$$A\vec{x} = \vec{b} \ , \ \vec{x} \geq \vec{0}$$

Already one has the following facts. The constraints define a set F of feasible solutions. It is a convex set. The maximum of the objective function takes place on an extreme point of F.

Now, it can be shown that if there is a set of linearly independent columns of A, $\vec{a}_1 , \vec{a}_2, \ , \vec{a}_k , \ (k \leq m)$, such that:

$$x_1 \vec{a}_1 + x_2 \vec{a}_2 + \ldots + x_k \vec{a}_k = \vec{b}, \ \ \forall x_j \geq 0$$

then the point $\vec{x} = (x_1, x_2, \ldots, x_k, 0, \ldots, 0)$ is an extreme point of F.

There are at most $\binom{n}{m}$ sets of m linearly independent columns of A.

One could check all sets, to determine the optimum value of the objective function, but this is ordinarily an excessive effort, especially in case of hundreds or even thousands of variables. Fortunately, the simplex algorithm provides a computational cheaper way.

3.2.3 Conversion to Equality Form

The original work of Dantzig considered the following linear programming problem:

Minimize $z = \vec{c}^T \cdot \vec{x}$, subject to $A\vec{x} = \vec{b}$, $\vec{x} \geq \vec{0}$, $\vec{b} \geq \vec{0}$

Dantzig called this *"the standard form"*. However, for clarity reasons we would prefer to call it the equality minimum form. It was the case that some classical books (mostly related to economic fields) denote as standard, a form with $A\vec{x} \leq \vec{b}$. For such expression, we would like to call it instead an inequality form (that could be further specified as $\leq$ or $\geq$).

If you browse a number of papers on linear programming, you could find still more mixing of terms, since some authors denote as canonical the form with $A\vec{x} \leq \vec{b}$. And this is not all, but let us not get distracted.

The point we wish to highlight is that inequalities can be transformed to equalities by adding artificial variables. For instance, suppose one has:

$$3x_1 + 5x_2 \geq 15$$

$$4x_1 + 9x_2 \leq 36$$

$$x_1, x_2 \geq 0$$

This can be converted to equalities as follows:

$$3x_1 + 5x_2 - x_3 = 15$$

$$4x_1 + 9x_2 + x_4 = 36$$

$$x_1, x_2, x_3, x_4 \geq 0$$

It was a matter of adding the artificial variables x_3, x_4. Variable x_3 is called a "*surplus*" variable and x_4 a "*slack*" variable.

In case of having a variable that can be as well positive or negative, this variable could be expressed as the difference of two nonnegative variables.

3.3 The Simplex Algorithm

After introducing the mathematics underlying the simplex algorithm, this section addresses the popular way in which the algorithm is usually applied, which is by using a specific table format called "*tableau*" (a French word). An easy example will be used to describe step-by-step the application of the algorithm.

A last part of the section expands the view to other types of examples, which motivate a general method for the automatic initialization of the algorithm.

3.3.1 Mathematical Description

Suppose that the problem comes in equality maximum form

Maximize $z = \vec{c}^T \cdot \vec{x}$, subject to $A\vec{x} = \vec{b}$, $\vec{x} \geq \vec{0}$, $\vec{b} \geq \vec{0}$

The idea is to move from one vertex to another, using columns of A.

The basis of a matrix is a maximal linearly independent set of columns of that matrix. Suppose that a basis of the matrix A has been found and form the matrix B with the columns $\vec{a}_1$, $\vec{a}_2$, , $\vec{a}_m$ of this basis. Any other column $\vec{a}_j$ of A can be written in function of the basis:

$$\vec{a}_j = y_{1j}\,\vec{a}_1 + y_{2j}\,\vec{a}_2 + \ldots + y_{m\,j}\,\vec{a}_m$$

Or, equivalently:

$$\vec{a}_j = B\,\vec{y}_j$$

As seen before, there is an extreme point associated to B, and in this point the objective function has the value z_B. Now, we want to substitute one of the columns of B by another column of A, so a new basis B' was obtained and such that $z'_B \geq z_B$ (z'_B is the value of the objective function at the extreme point associated to B'). There are two questions, what column to extract from B, and what column to take from A.

Consider a partition of the constraint equations:

$$A\vec{x} = [B, N]\vec{x} = [B, N] \cdot \begin{bmatrix} \vec{x}_B \\ \vec{x}_N \end{bmatrix} = \vec{b}$$

Basic feasible solutions of the constraint equations are feasible solutions with no more than m positive entries. One of these solutions will be such that:

$$\vec{x}_N = \vec{0}$$

$$B\,\vec{x}_B = \vec{b}$$

This second equation can be also written as follows:

$$x_{B1}\,\vec{a}_1 + x_{B2}\,\vec{a}_2 + \ldots + x_{Bm}\,\vec{a}_m = \vec{b}$$

After an algebraic study, the simplex algorithm concludes that the column $\vec{a}_r$ to extract from B should be such that:

$$\frac{x_{Br}}{y_{rj}} = \min_i \left\{ \frac{x_{Bi}}{y_{i\,j}},\ y_{i\,j} > 0 \right\}$$

This was the answer to the first question. For the second question, denote:

$$z_j = \sum_{i=1}^{m} c_i\, y_{i\,j}$$

The column $\vec{a}_j$ to take from A should maximize the following expression:

$$\frac{x_{Br}}{y_{r\,j}}\,(c_j - z_j)$$

The term y_{rj} is usually named as *"pivot"*.

In order to establish for the algorithm a suitable tabular form, it is convenient to derive a series of expressions. If one focuses on the exchange of a column $\vec{a}_r$ by the column $\vec{a}_j$, the mutual relationship is:

$$\vec{a}_j = y_{rj}\,\vec{a}_r + \sum_{i \neq r} y_{ij}\vec{a}_i$$

From which,

$$\vec{a}_r = \frac{1}{y_{rj}}\,\vec{a}_j - \sum_{i \neq r} \frac{y_{ij}}{y_{rj}}\,\vec{a}_i$$

Any column $\vec{a}_k$ can be expressed in the old basis as follows:

$$\vec{a}_k = y_{rk}\,\vec{a}_r + \sum_{i \neq r} y_{ik}\,\vec{a}_i = \frac{y_{rk}}{y_{rj}}\,\vec{a}_j + \sum_{i \neq r} (y_{ik} - \frac{y_{ij}\,y_{rk}}{y_{rj}})\,\vec{a}_i$$

This same column can be expressed in the new basis as:

$$\vec{a}_k = \sum_i y^{;}_{ik}\,\vec{a}^{;}_i$$

with:

$$\vec{a}^{;}_i = \vec{a}_i, \quad \forall i \neq r$$

$$\vec{a}^{;}_r = \vec{a}_j$$

$$y'_{ik} = y_{ik} - \frac{y_{ij}y_{rk}}{y_{rj}} \vec{a}_i, \ \forall i \neq r$$

$$y'_{rk} = \frac{y_{rk}}{y_{rj}}$$

These last expressions are quite useful for computing the y'_k terms in function of the y_k terms.

Another aspect of interest is what happens with $z_k - c_k$ in the new basis. From definition, one knows that:

$$z'_k - c_k = \sum c'_{Bi} y'_{ik} - c_k$$

with:

$$c'_{Bi} = c_{Bi}, \ \forall i \neq r$$

$$c'_{Br} = c_j$$

Using the y'_k terms, it could be written that:

$$z'_k - c_k = \left[\sum_{i \neq r} c_{Bi} (y_{ik} - \frac{y_{ij}y_{rk}}{y_{rj}}) + \frac{y_{rk}}{y_{rj}} c_j \right] - c_k$$

Include now a dummy term:

$$c_{Br} (y_{rk} - \frac{y_{rj}y_{rk}}{y_{rj}}) = 0$$

Then,

$$z'_k - c_k = \left[\sum_{i} c_{Bi} (y_{ik} - \frac{y_{ij}y_{rk}}{y_{rj}}) + \frac{y_{rk}}{y_{rj}} c_j \right] - c_k =$$

$$= \sum c_{Bi} y_{ik} - \frac{y_{rk}}{y_{rj}} \sum c_{Bi} \, y_{ij} + \frac{y_{rk}}{y_{rj}} c_j - c_k$$

Therefore:

$$z'_k - c_k = z_k - c_k - \frac{y_{rk}}{y_{rj}} (z_j - c_j)$$

It can also be established that:

$$z' = z - \frac{x_{Br}}{y_{rj}}(z_j - c_j)$$

3.3.2 *The Tableau Format*

In practice, linear programming problems are handled with a specific tableau format, so the algorithm can be applied in a clear systematic way.

Let us present an example of using the tableau format. The problem to solve is:

Maximize:

$$z = 8x_1 + 6x_2$$

Subject to:

$$4x_1 + x_2 \leq 160$$

$$2x_1 + 3x_2 \leq 240$$

$$x_1, x_2 \geq 0$$

Notice that the problem is given in inequality maximum form. Slack variables are added for obtaining the equality form:

$$4x_1 + x_2 + s_1 = 160$$

$$2x_1 + 3x_2 + s_2 = 240$$

Figure 3.6, which was generated with the Program 3.5, shows the feasible region corresponding to this problem.

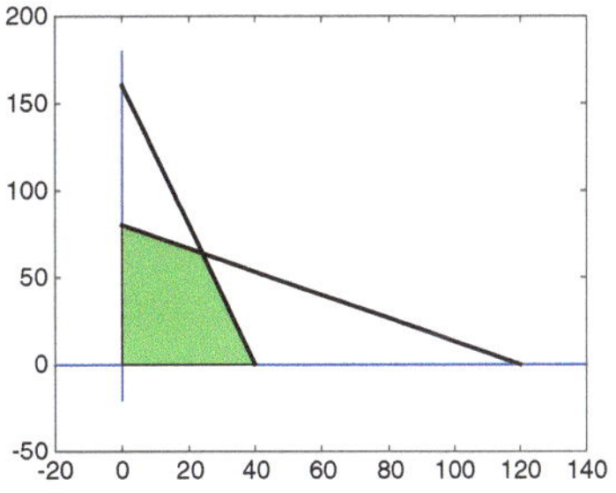

Fig. 3.6 The feasible set for the problem considered

Program 3.5 Displaying the feasible set of a case to be studied

```
%display of 2D feasible set
% example for tableaux
% ax + by =c
a1=4; b1=1; c1=160;
a2=2; b2=3; c2=240;
%
% constraint corner points:
X1=c1/a1; Y1=c1/b1;
X2=c2/a2; Y2=c2/b2;
% intersection:
aux1=(Y2-Y1); aux2=(Y2*X1)-(Y1*X2);
xi=(X1*X2*aux1)/aux2;
yi=Y1-((Y1*xi)/X1);
figure(1)
%axes:
plot([-20 140],[0 0],'b'); hold on;
plot([0 0],[-20 180],'b');
%feasible set:
patch([0 0 xi X1],[0 Y2 yi 0],'g');
%constraints
plot([0 X1],[Y1 0],'k');
plot([0 X2],[Y2 0],'k');
```

The first tableau is easily built as follows:

$c_j \rightarrow$		8	6	0	0	
$\downarrow$	Basis	x_1	x_2	s_1	s_2	b
0	s_1	4	1	1	0	160
0	s_2	2	3	0	1	240
	z_j	0	0	0	0	0
	$c_j - z_j$	8	6	0	0	

Notice that the initial tableau is filled with just the problem statement: the objective function and the constraints. If all numbers in the last row were zero or negative, the optimum was reached; but this is not the case, the algorithm should continue. The second column indicates, below the tittle "Basis," the variables chosen for the current basis, which are s_1 and s_2. The values of these variables (the values of x_{Br}) are shown in the last column: 160 and 240.

Now, the effort concentrates in selecting *a pivot*. The pivot column is where $c_j - z_j$ is maximum, so it is the x_1 column. Now, compute 160/4 and 240/2 (the numbers 160 and 240 belong to the b column; the numbers 4 and 2 belong to the pivot column). The smallest non-negative result tells us the pivot row, which in this case is the s_1 row. Therefore, s_1 abandons the basis and is substituted by x_1. The value of the pivot was 4. The next tableau is the following:

$c_j \rightarrow$		8	6	0	0	
$\downarrow$	Basis	x_1	x_2	s_1	s_2	b
8	x_1	1	0.25	0.25	0	40
0	s_2	0	2.50	−0.50	1	160
	z_j	8	2	2	0	320
	$c_j - z_j$	0	4	−2	0	

The x_1 row has been obtained from the s_1 row in the previous tableau, dividing it by the pivot.

Take the kth entry of the new s_2 row. It has been obtained from the kth entry of the old s_2 row minus a quantity u_k, where u_k is the product of the number of the old s_2 row which is below the pivot (2), and the kth entry of the new x_1 row. That is:

$$s_2(1) = 2 - 2 \cdot 1;\; s_2(2) = 3 - 2 \cdot 0.25;\; s_2(3) = 0 - 2 \cdot 0.25;$$
$$s_2(4) = 1 - 2 \cdot 0;\; s_2(5) = 240 - 2 \cdot 40$$

The values of the z_j row are obtained from the column of c_j and the x_1, x_2,s_1,s_2, columns as follows:

$$z_j(1) = 8 \cdot 1 + 0 \cdot 0;\; z_j(2) = 8 \cdot 0.25 + 0 \cdot 2.50;$$
$$z_j(3) = 8 \cdot 0.25 + 0 \cdot -0.50;\; z_j(4) = 8 \cdot 0 + 0 \cdot 1;\; z_j(5) = 8 \cdot 40 + 0 \cdot 160$$

The values of x_1 and s_2 in the new vertex are 40 and 160.

The next pivot would be the 2.50 at the intersection of the x_2 column and the s_2 row. A third tableau should be computed, which results to be the final one with the optimum.

$c_j \rightarrow$		8	6	0	0	
$\downarrow$	Basis	x_1	x_2	s_1	s_2	b
8	x_1	1	0	0.30	−0.10	24
6	x_2	0	1	−0.20	0.40	64
	z_j	8	6	1.20	1.60	576
	$c_j - z_j$	0	0	−1.20	−1.60	

The last column gives the solution, $x_1 = 24$, $x_2 = 64$. The maximum thus attained is 576.

A word of caution: the term "simplex" is also used to designate a set of $n + 1$ points in a n-dimensional space. There are iterative optimization methods based on the use of simplexes. These methods should not be confused with the simplex algorithm just described.

Program 3.6 provides an example of MATLAB coding of the simplex method. It could be used for studying problems like the previous case, with two variables. The program generates the series of tableaux, which imply coding efforts for output formatting. Instead of $z(5)$, the program uses q for the last entry of the z_j row. The program is not protected against all possible difficulties (some of these will be considered later on).
Note (Program 3.6)

The complete listing of the Program 3.6, which presents a coding of the simplex method, has been included in the Appendix for Long Programs.

3.3.3 The Initial Solution

Consider again the example just used for introducing the tableau format. In abstract terms, it could be written as follows:

Maximize $z = \vec{c}^T \cdot \vec{x}$,
Subject to $D\,\vec{x}_D \leq \vec{b}\,,\ \vec{x}_D \geq \vec{0},\ \vec{b} \geq \vec{0}$

Adding now slack variables, one has:

$$D\,\vec{x}_D + I\,\vec{x}_S = \vec{b}\,,\ \vec{x}_D \geq \vec{0}, \vec{x}_S \geq \vec{0}$$

Clearly, if one takes $\vec{x}_D \equiv \vec{0}$, then

$$I\,\vec{x}_S = \vec{b}$$

gives us a basic feasible solution.

Actually, this was the initial basis chosen for the first tableau in the previous example.

Suppose now a change in the constraint $4\,x_1 + x_2 \leq 160$, so it would be $4\,x_1 + x_2 \geq 160$ instead. A surplus variable would be added, and if one takes $\vec{x}_D \equiv \vec{0}$, then

$$\begin{bmatrix} -1 & 0 \\ 0 & 1 \end{bmatrix} \vec{x}_S = \vec{b}$$

which means that the corresponding solution has $s_1 < 0$, and so infeasible.

In order to have an initial basic feasible solution for any case (with slack and surplus variables), one could add to the problem with $D\,\vec{x}_D + I\,\vec{x}_S = A\,\vec{x} = \vec{b}$ a set of artificial variables $\vec{x}_a$, so that:

$$A\,\vec{x} + I\,\vec{x}_a = \vec{b}$$

and then, one uses $\vec{x} \equiv \vec{0}$, $\vec{x}_a = \vec{b}$ as a basic feasible solution, which would not be so for $A\vec{x} = \vec{b}$. Now, the trick is to assign large negative values (say a $-M$) to the contribution of artificial variables to the objective function,

$$z = c_1 x_1 + c_2 x_2 + \ldots + 0 s_1 + 0 s_2 + \ldots - M x_{a1} - M x_{a2} - \ldots$$

Therefore, if one starts the simplex algorithm, the process will arrive to $\vec{x}_a \equiv \vec{0}$, because maximization of z would imply the expulsion of the $-M$ terms. From this stage on, the algorithm will obtain solutions for $A\vec{x} = \vec{b}$.

3.3.4 Dealing with Minimization

In case of a minimization problem, like the following:

Minimize $z = \vec{c}^T \cdot \vec{x}$,
Subject to $D\vec{x}_D \leq \vec{b}$, $\vec{x}_D \geq \vec{0}$, $\vec{b} \geq \vec{0}$

there are a few changes to be applied concerning the procedure just described. They can be summarized with three actions:

Use $+M$ terms instead of $-M$ terms in the extended version of z.
Choose the column where $c_j - z_j$is minimum to select the pivot.
The solution is reached when all $c_j - z_j$ terms are nonnegative.

Of course, there are other ways for adapting the general algorithm for different cases, like, for instance, using sign changes in the expression of the problem, etc.

In a next section on illustrative applications, there would be a diet example, which refers to cost minimization while keeping adequate nutrition. The example has a program that includes the three changes just mentioned.

3.3.5 Degeneracy

Recall that the basic feasible solutions have no more than m positive entries. If any of these solutions has less than m positive entries, then it is a degenerate basic feasible solution. In general, degeneracy is caused by redundant constraints.

It happens that a change of basis modifies the value of z as follows:

$$z' = z + \frac{x_{Br}}{y_{rj}}(c_j - z_j)$$

In case of degeneracy, it would be possible that a change of basis makes $x_{Br} = 0$, and then $z' = z$. Therefore, the algorithm iterations may re-visit again and again a basis already explored. This would be a bad cyclic behavior. Fortunately, it is very infrequent in real applications, at least because of numerical errors.

The problem of cyclic behavior has been tackled by the research in several ways. See, for instance, [33] on the Bland's rule for pivoting or the perturbation method to avoid zeros [31]. An interesting survey of the research on pivoting rules, is [29].

3.4 Special Situations

In practical optimization application, some special circumstances may appear. Frequently, they are related to the feasible solution region F.

Of course, using the simplex coding given in this chapter would be not recommended for the bad situations to be described next.

For instance, a bad specification of constraints may lead to an empty feasible solution set; usually this would be due to inconsistent (contradictory) constraints. In this case, there is no solution (it is said that the problem is *infeasible*). Next specifications give an example of this:

Subject to:

$$- x_1 + 2x_2 \leq 2$$

$$- x_1 + x_2 \geq 2$$

$$x_1 , x_2 \geq 0$$

Figure 3.7 depicts the scenario corresponding to the given constraints: two independent zones without overlapping. There is no feasible region.

Figure 3.7 has been obtained with Program 3.7. The MATLAB function *quiver()* has been used for drawing the arrows.

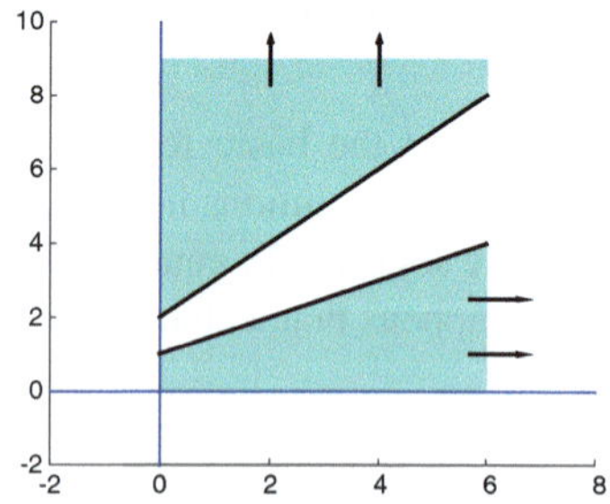

Fig. 3.7 A case with inconsistent constraints

Program 3.7 Displaying of a case with inconsistent constraints

```
%display of 2D feasible set
% special case 2
% ax + by =c
a1=-1; b1=2; c1=2;
a2=-1; b2=1; c2=2;
%
% constraint corner points:
X1=c1/a1; Y1=c1/b1;
X2=c2/a2; Y2=c2/b2;
% rightmost point:
xr=6;
yr1=(c1-(a1*xr))/b1;
yr2=(c2-(a2*xr))/b2;
figure(1)
%feasible set:
patch([0 0 xr xr],[0 Y1 yr1 0],'c',...
'EdgeColor','c'); hold on;
patch([0 0 xr xr],[Y2 yr2+1 yr2+1 yr2],'c',...
'EdgeColor','c');
%axes:
plot([-2 8],[0 0],'b');
plot([0 0],[-2 10],'b');
%constraints
plot([0 xr],[Y1 yr1],'k');
plot([0 xr],[Y2 yr2],'k');
% arrows:
quiver([5.7,5.7],[1,2.5],[1,1],[0,0],'k');
quiver([2,4],[8.3,8.3],[0,0],[1,1],'k');
```

It is also possible that, according with the given constraints, the feasible region was unbounded. An example of this situation is the following:

Subject to:

$$-3x_1 + 2x_2 \leq 3$$

$$-x_1 + 2x_2 \leq 8$$

$$x_1,\ x_2 \geq 0$$

Figure 3.8 shows the corresponding feasible region. The arrow indicates that the region is open on the right side. Figure 3.8 has been obtained with Program 3.8.

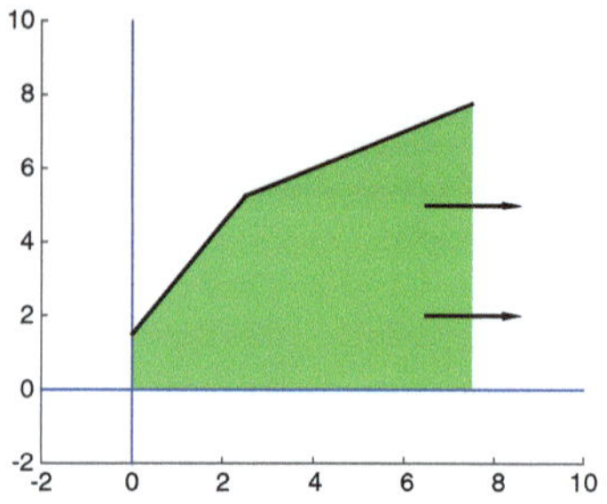

Fig. 3.8 A case with unbounded feasible set

Program 3.8 A case with unbounded constraints

```
%display of 2D feasible set
% special case 1
% ax + by =c
a1=-3; b1=2; c1=3;
a2=-1; b2=2; c2=8;
%
% constraint corner points:
X1=c1/a1; Y1=c1/b1;
X2=c2/a2; Y2=c2/b2;
% intersection:
aux1=(Y2-Y1); aux2=(Y2*X1)-(Y1*X2);
xi=(X1*X2*aux1)/aux2;
yi=Y1-((Y1*xi)/X1);
% rightmost point (on second constraint):
xr=3*xi; yr=(c2-(a2*xr))/b2;
figure(1)
%feasible set:
patch([0 0 xi xr xr],[0 Y1 yi yr 0],'g',...
'EdgeColor','g'); hold on;
%axes:
plot([-2 10],[0 0],'b');
plot([0 0],[-2 10],'b');
%constraints
plot([0 xi],[Y1 yi],'k');
plot([xi xr],[yi yr],'k');
% arrows
quiver([6.5,6.5],[2,5],[1,1],[0,0],'k');
```

An unbounded feasible region may represent, or not, a difficulty. Suppose, for example, that the objective function is:

Maximize:

$$z = 2x_1 + x_2$$

Figure 3.9 depicts several line segments corresponding to increasing values of the objective function. Clearly these values can increase with no limit, yet obeying the given constraints. The solution is thus unbounded.

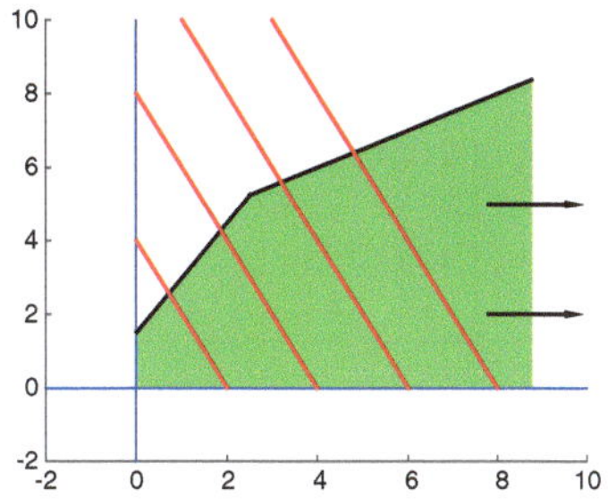

Fig. 3.9 A case with unbounded solution

Figure 3.9 has been generated by the Program 3.9, which, obviously, inherits many lines from Program 3.8.

Program 3.9 Case with unbounded constraints and unbounded solution

```
%display of 2D feasible set and objective function
% special case 1, unbounded
% ax + by =c
a1=-3; b1=2; c1=3;
a2=-1; b2=2; c2=8;
%
% constraint corner points:
X1=c1/a1; Y1=c1/b1;
X2=c2/a2; Y2=c2/b2;
% intersection:
aux1=(Y2-Y1); aux2=(Y2*X1)-(Y1*X2);
xi=(X1*X2*aux1)/aux2;
yi=Y1-((Y1*xi)/X1);
% rightmost point (on second constraint):
xr=3.5*xi; yr=(c2-(a2*xr))/b2;
figure(1)
%feasible set:
patch([0 0 xi xr xr],[0 Y1 yi yr 0],'g',...
'EdgeColor','g'); hold on;
%axes:
plot([-2 10],[0 0],'b');
plot([0 0],[-2 10],'b');
%constraints
plot([0 xi],[Y1 yi],'k');
plot([xi xr],[yi yr],'k');
% arrows
quiver([7.8,7.8],[2,5],[1,1],[0,0],'k');
% objective function values
%2x+y=z;
a=2; b=1;
% segment corners
z=4; X1=z/a; Y1=z/b;
z=8; X2=z/a; Y2=z/b;
z=12; X3=z/a; xl3=X3-5; yl3=z-(a*xl3);
```

```
z=16; X4=z/a; xl4=X4-5; yl4=z-(a*xl4);
%
plot([0 X1],[Y1 0],'r');
plot([0 X2],[Y2 0],'r');
plot([xl3 X3],[yl3 0],'r');
plot([xl4 X4],[yl4 0],'r');
```

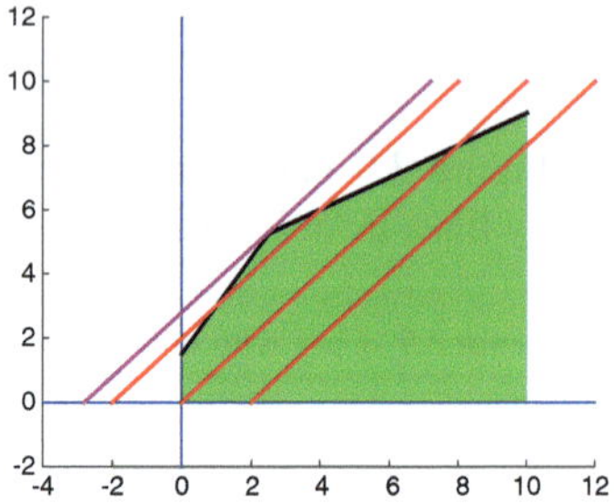

Fig. 3.10 A case with unbounded feasible set, however with unique solution

Figure 3.10 shows another case with the same unbounded feasible region, but with a unique solution at a vertex.

Maximize:

$$z = -x_1 + x_2$$

Figure 3.10 has been obtained with Program 3.10, quite similar to Program 3.9

Program 3.10 Case with unbounded constraints but having a solution

```
%display of 2D feasible set and objective function
% special case 1, single vertex solution
% ax + by =c
a1=-3; b1=2; c1=3;
a2=-1; b2=2; c2=8;
%
% constraint corner points:
X1=c1/a1; Y1=c1/b1;
X2=c2/a2; Y2=c2/b2;
% intersection:
aux1=(Y2-Y1); aux2=(Y2*X1)-(Y1*X2);
xi=(X1*X2*aux1)/aux2;
yi=Y1-((Y1*xi)/X1);
% rightmost point (on second constraint):
xr=4*xi; yr=(c2-(a2*xr))/b2;
figure(1)
%feasible set:
patch([0 0 xi xr xr],[0 Y1 yi yr 0],'g',...
'EdgeColor','g'); hold on;
```

```
%axes:
plot([-4 12],[0 0],'b');
plot([0 0],[-2 12],'b');
%constraints
plot([0 xi],[Y1 yi],'k');
plot([xi xr],[yi yr],'k');
% objective function values
%-x+y=z;
a=-1; b=1;
% segment corners
z=-2; X1=z/a; xr1=X1+10; yr1=z-(a*xr1);
z=0; X2=z/a; xr2=X2+10; yr2=z-(a*xr2);
z=2; X3=z/a; xr3=X3+10; yr3=z-(a*xr3);
z=2.8; X4=z/a; xr4=X4+10; yr4=z-(a*xr4);
%
plot([X1 xr1],[0 yr1],'r');
plot([X2 xr2],[0 yr2],'r');
plot([X3 xr3],[0 yr3],'r');
plot([X4 xr4],[0 yr4],'m');
```

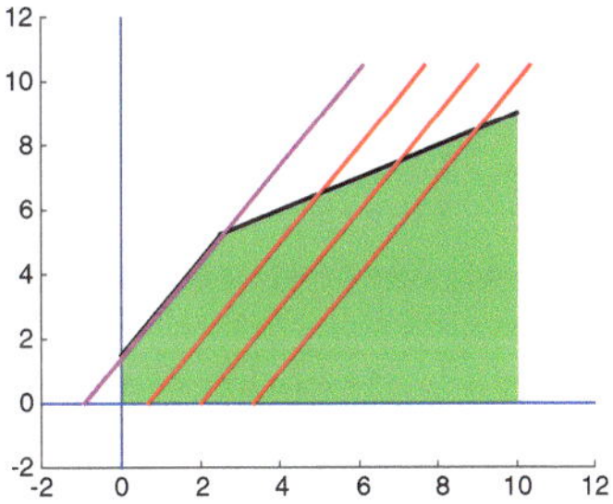

Fig. 3.11 A case with infinite solutions

Finally, Fig. 3.11 shows a case with infinite (good) solutions, since the optimal objective function is coincident with a line segment joining two vertices. The figure has been generated with Program 3.11, again similar to Program 3.9.

Maximize:

$$z = -3x_1 + 2x_2$$

Program 3.11 Case with infinite (good) solutions

```
%display of 2D feasible set and objective function
% special case 1, infinite number of solutions
% ax + by =c
a1=-3; b1=2; c1=3;
a2=-1; b2=2; c2=8;
%
```

```
% constraint corner points:
X1=c1/a1; Y1=c1/b1;
X2=c2/a2; Y2=c2/b2;
% intersection:
aux1=(Y2-Y1); aux2=(Y2*X1)-(Y1*X2);
xi=(X1*X2*aux1)/aux2;
yi=Y1-((Y1*xi)/X1);
% rightmost point (on second constraint):
xr=4*xi; yr=(c2-(a2*xr))/b2;
figure(1)
%feasible set:
patch([0 0 xi xr xr],[0 Y1 yi yr 0],'g',...
'EdgeColor','g'); hold on;
%axes:
plot([-2 12],[0 0],'b');
plot([0 0],[-2 12],'b');
%constraints
plot([0 xi],[Y1 yi],'k');
plot([xi xr],[yi yr],'k');
% objective function values
%-3x+2y=z;
a=-3; b=2;
% segment corners
z=-10; X1=z/a; xr1=X1+7; yr1=(z-(a*xr1))/b;
z=-6; X2=z/a; xr2=X2+7; yr2=(z-(a*xr2))/b;
z=-2; X3=z/a; xr3=X3+7; yr3=(z-(a*xr3))/b;
z=2.8; X4=z/a; xr4=X4+7; yr4=(z-(a*xr4))/b;
%
plot([X1 xr1],[0 yr1],'r');
plot([X2 xr2],[0 yr2],'r');
plot([X3 xr3],[0 yr3],'r');
plot([X4 xr4],[0 yr4],'m');
```

When using tableaux for the simplex algorithm, it may turn out in a certain stage that every entry in the pivot column is either zero or negative. In this case, it is not possible to determine what variable leaves the basis. The algorithm cannot continue. This corresponds to unbounded solution.

If the optimal solution reached by the simplex algorithm contains one or more artificial variables, then the problem was infeasible.

In order to confirm that a unique optimum has been obtained, and this was the last stage, one could try to continue with the method and see if the next value of z remains the same. Two equal optima mean that there are infinite good solutions. An indication of multiple optimal solutions is to have a zero in a $c_j - z_j$ corresponding to a non-basic variable.

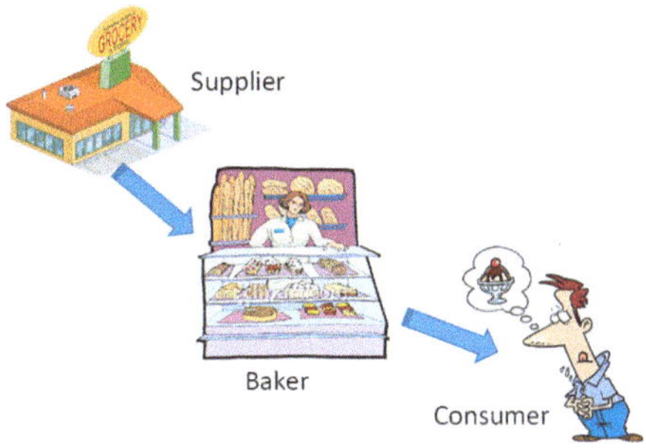

Fig. 3.12 Example of duality in normal life

3.5 Duality

There is a classic example that is quite useful for introducing duality. A bakery offers brownies and cheesecake. The baker buys from a supplier the ingredients for cooking the products. Clients want to purchase these products in order to fit certain nutritional requirements. Figure 3.12 offers a sketch of the situation.

There are several facets of this situation, what the supplier wants—a good profit—and what the consumer wishes, a good price. And, indeed, the baker also wants some profit.

Let us specify a scenario. The bakery products contain chocolate, sugar, and cream cheese. Each baker's product has a price. The next table summarizes the relevant data.

	Brownie	Cheesecake	Requirements
Chocolate	3 oz	0 oz	6 oz
Sugar	2 oz	4 oz	10 oz
Cream cheese	2 oz	5 oz	8 oz
Price (cents)	50	80	

The consumer wants to satisfy a set of nutritional requirements, as written on the last column of the table. A minimum cost is desired. Therefore, the consumer has the following minimization problem:

Minimize:

$$z = 50x_1 + 80x_2$$

Subject to:

$$\begin{aligned} &3x_1 \geq 6 \quad (chocolate) \\ &2x_1 + 4x_2 \geq 10 \quad (sugar) \\ &2x_1 + 5x_2 \geq 8 \quad (cream\, cheese) \end{aligned}$$

$$x_1,\ x_2 \geq 0$$

where x_1 , x_2 are quantity of brownies and cheesecake. The solution is $x_1 = 2$; $x_2 = 3/2$, and the optimal cost is 2.20.

This is a simple example of the well-known diet problem, which will be considered in more detail in the next section.

Now, let us adopt the point of view of the supplier. Denote as y_1, y_2, y_3 the cost of each ounce of chocolate, sugar, and cream cheese. The supplier needs that the baker had some profit by selling the brownies and cheesecakes at the prices of 50 and 80. Therefore:

$3\,y_1 + 2\,y_2 + 2\,y_3 \leq 50 \quad (Brownie)$
$4\,y_2 + 5\,y_3 \leq 80 \quad (Cheesecake)$

The supplier wants to maximize the revenue of selling the ingredients, which are in turn the nutritional components the client gets. Then,

Maximize:

$$p = 6\,y_1 + 10\,y_2 + 8\,y_3$$

Subject to:

$$3\,y_1 + 2\,y_2 + 2\,y_3 \leq 50$$

$$4\,y_2 + 5\,y_3 \leq 80$$

$$y_1,\; y_2 \geq 0$$

The solution of this problem is $y_1 = 10/3$; $y_2 = 20$; $y_3 = 0$, and the optimal revenue is 2.20. Both problems get the same optimum value.

More formally, this example showed that there was a "primal" problem with the following formulation:

Minimize:

$$z = \vec{c}^T \cdot \vec{x}$$

Subject to:

$$A\vec{x} \geq \vec{b}\,,\; \vec{x} \geq \vec{0}$$

And there was another related "dual" problem, with the following expression:

Maximize:

$$p = \vec{b}^T \cdot \vec{y}$$

Subject to:

$$A^T \vec{y} \leq \vec{c}\ ,\ \vec{y} \geq \vec{0}$$

Commonly in economics literature the y_i variables are called "*shadow prices*" or "*marginal values*".

Returning to the example, notice that the solution of the dual implies no profit for the baker. This would be not fair. That aspect should be considered in the problem, with the consequence that a gap would appear between the minimum cost for the consumer and the maximum revenue for the supplier. Another interesting point is that it turns out that a zero price for the cream cheese was convenient for the supplier.

Supposing that the primal problem was in equality minimum form (the Dantzig standard form):

Minimize $z = \vec{c}^T \cdot \vec{x}$, subject to $A\vec{x} = \vec{b}\ ,\ \vec{x} \geq \vec{0},\ \vec{b} \geq \vec{0}$

The dual would be:

Maximize $p = \vec{b}^T \cdot \vec{y}$, subject to: $A^T \vec{y} \leq \vec{c}\ ,\ \vec{y}\ free$

In cases with mixed constraints, the rules for duality are summarized in the next table.

Min problem	Max problem
Type of constraint	Type of variable
$\vec{a}_j^T \vec{x} \geq b_j$	$y_j \geq 0$
$\vec{a}_j^T \vec{x} \leq b_j$	$y_j \leq 0$
$\vec{a}_j^T \vec{x} = b_j$	y_j unrestricted
Type of variable	Type of constraint
$x_j \geq 0$	$\vec{y}^T A_j \leq c_j$
$x_j \leq 0$	$\vec{y}^T A_j \geq c_j$
x_j unrestricted	$\vec{y}^T A_j = c_j$

where A_j denotes the *j-th* column of the matrix A.

Duality is an involution: the dual of the dual is the primal.

Suppose that the problem was given in equality minimum form. The "*weak duality*" theorem establishes that if $\vec{x}$ and $\vec{y}$ are feasible for the primal and the dual problems, respectively, then:

$$\vec{c}^T \vec{x} \geq \vec{b}^T \vec{y}$$

Corollaries:

- If $\vec{x}$ and $\vec{y}$ are feasible and $\vec{c}^T \vec{x} = \vec{b}^T \vec{y}$ then $\vec{x}$ and $\vec{y}$ are optimal for their respective problems.

- If $\vec{x}$ and $\vec{y}$ are feasible and have the same objective function values, then they are optimal for their respective problems.
- If either problem has unbounded solution, the other problem is infeasible. The converse is not true, since both problems could be infeasible.

The *"strong duality"* theorem establishes that if either the primal or the dual has a finite optimal solution, then the other has a finite optimal solution and the corresponding optimal objective function values are equal.

According with these theorems, the dual problem could be useful for several reasons. For instance, in case of having successfully obtained a solution for the primal, the result can be confirmed by obtaining the same solution with the dual. Also, sometimes finding an initial feasible solution to the dual is easier that finding one for the primal. Moreover, it may happen that the dual is easier to solve, in particular when having many constraints and few variables.

More details on duality can be found in books like [3] and [30]. There are also many tutorials on the Internet, like for instance the lecture [23].

An interesting historical note is that in 1947 Dantzig visited J. von Neumann, who rapidly noticed during their conversation that there were analogies with game theory, pertaining in particular to duality. This aspect will appear in a next chapter.

3.6 Illustrative Applications

Linear programming is a quite successful and popular method, and so many types of applications could have a place in this moment of the chapter. Being not possible to introduce them all, we just did a small selection. In any case, the applications considered have special aspects of interest.

3.6.1 Diets

Diets are important for hospitals, fast food chains, merchandise offered by stores, etc., and our quotidian life.

Indeed, what is represented in Fig. 3.13 seem exaggerated, but it serves to introduce the atmosphere of this optimization scenario.

The diet problem has a history. It was one of the first optimization problems studied in the 1930s and 1940s, [2]. A famous version of the problem was formulated in 1982 by George Stigler (Nobel Laureate in Economics), considering 77 foods ([13] offers a review of this problem). The article [14] (year 2018) extensively considers the optimization diet question, with an eye on sustainability.

Fig. 3.13 Diet optimization

3.6.1.1 A Simple Diet

You have a repertory of foods. Each one has certain nutritional contents, and a cost. Take, for example, the simple case represented in the following table:

	Cheese	Potato
Protein (gr)	7.0	1.3
Calories	106	97
Cost	8.0	1.5

The question is to find a minimum-cost diet that offers at least 300 calories and not more than 10 gr of protein.

The problem can be mathematically expressed as follows:

Minimize:

$$z = 8x_1 + 1.5x_2$$

Subject to:

$$7x_1 + 1.3x_2 \leq 10$$

$$106x_1 + 97x_2 \geq 300$$

$$x_1, \; x_2 \geq 0$$

Figure 3.14, which was generated with the Program 3.12, shows the feasible region corresponding to this simple diet.

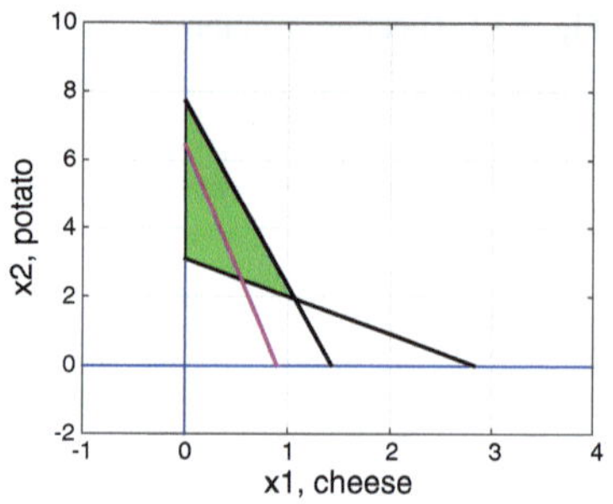

Fig. 3.14 The feasible set for the problem considered

Program 3.12 Feasible set of the simple diet

```
%display of 2D feasible set
% example 1
% ax + by =c
a1=7; b1=1.3; c1=10;
a2=106; b2=97; c2=300;
% z=ax+by
a=10; b=1.4;
%
% constraint corner points:
X1=c1/a1; Y1=c1/b1;
X2=c2/a2; Y2=c2/b2;
% objective function corner points:
z=9; Zx=z/a; Zy=z/b;
% intersection:
aux1=(Y2-Y1); aux2=(Y2*X1)-(Y1*X2);
xi=(X1*X2*aux1)/aux2;
yi=Y1-((Y1*xi)/X1);
figure(1)
%axes:
plot([-1 4],[0 0],'b'); hold on;
plot([0 0],[-2 10],'b');
xlabel('x1, cheese'); ylabel('x2, potato');
%feasible set:
patch([0 0 xi],[Y2 Y1 yi],'g');
%constraints
plot([0 X1],[Y1 0],'k');
plot([0 X2],[Y2 0],'k');
% objective function:
plot([0 Zx],[Zy 0],'m');
grid;
```

Using slack and surplus variables, the problem can be easily transformed to equality minimum form:

Minimize:

$$z = 8x_1 + 1.5x_2$$

Subject to:

$$7x_1 + 1.3x_2 + s_1 = 10$$

$$106x_1 + 97x_2 - s_2 = 300$$

$$x_1,\ x_2,\ s_1,\ s_2 \geq 0$$

Two artificial variables x_{a1} and x_{a2} are added in order to have an initial solution. Then, it is possible to start the simplex algorithm with the following tableau:

$c_j \rightarrow$		8	1.5	0	0	1000	1000	
$\downarrow$	Basis	x_1	x_2	s_1	s_2	x_{a1}	x_{a2}	b
1000	x_{a1}	7	1.3	1	0	1	0	10
1000	x_{a2}	106	97	0	−1	0	1	300
	z_j	113000	98300	1000	−1000	1000	1000	310000
	$c_j - z_j$	−112992	−98298	−1000	1000	0	0	

Notice that values of 1000 were used in place of the two Ms. The initial values of the z_j row are obtained from the column of c_j and the x_1, x_2, s_1, s_2, x_{a1}, x_{a2} columns.

The first pivot chosen corresponds to leaving of x_{a1} to be replaced by x_1 for the new basis.

$c_j \rightarrow$		8	1.5	0	0	1000	1000	
$\downarrow$	Basis	x_1	x_2	s_1	s_2	x_{a1}	x_{a2}	b
8	x_1	1	0.19	0.14	0	1	0	1.4
1000	x_{a2}	0	77.31	−15.14	−1	0	1	148.6
	z_j	8	77315.77	−15141.71	−1000	1000	1000	148582.86
	$c_j - z_j$	0	−77314.27	15141.71	1000	0	0	

Next pivot is 77.31, in the second row and second column. The second artificial variable x_{a2} abandons the basis, letting the place for x_2.

$c_j \rightarrow$		8	1.5	0	0	1000	1000	
$\downarrow$	Basis	x_1	x_2	s_1	s_2	x_{a1}	x_{a2}	b
8	x_1	1	0	0.18	0	1	0	1.1
1.5	x_2	0	1	−0.20	−0.01	0	1	1.9
	z_j	8	1.50	1.14	−0	1000	1000	11.46
	$c_j - z_j$	0	0	−1.14	0	0	0	

With a little surprise, the solution was not reached. There is a negative value in the last row. Next pivot is 0.18, in the first row. Then, s_1 enters the basis, and x_1 leaves.

$c_j \rightarrow$		8	1.5	0	0	1000	1000	
$\downarrow$	Basis	x_1	x_2	s_1	s_2	x_{a1}	x_{a2}	b
0	s_1	5.58	0	1	0.01	1	0	6
1.5	x_2	1.09	1	0	−0.01	0	1	3.1
	z_j	1.64	1.50	0	−0.02	1000	1000	4.64
	$c_j - z_j$	6.36	0	0	0.02	0	0	

This was the last tableau. The solution is $x_2 = 3.1$ and $z = 4.64$. The optimal diet is only potato.

The series of tableaux has been obtained with Program 3.13, which includes the opportune changes for a minimization problem.

Note (Program 3.13)

The complete listing of the Program 3.13, corresponding to a simple diet, has been included in the Appendix for Long Programs.

3.6.1.2 A More Complete Diet

A more extended diet problem could be the following:

	Bread	Cheese	Potato	Fish
Protein (gr)	4.0	7.0	1.3	8.0
Fat (gr)	1.0	9.0	0.1	7.0
Carbohydrates (gr)	15.0	0.4	22.6	0.0
Calories	90	106	97	130
Cost	2.0	8.0	1.5	11.0

Now, the question is to find a minimum-cost diet that offers at least 300 calories, not more than 10 gr of protein, not less than 10 gr of carbohydrates, and not less than 8 gr of fat.

Based on Program 3.13, a new program—Program 3.14—has been made with extended capabilities. For reasons of screen space, no tableaux are shown (resulting in a shorter program). The program includes more measures to prevent difficulties related to degeneration and pivot selection or pivot being zero. These difficulties were not found in the diet example but will arise in other examples in this chapter.

Note (Program 3.14)

The complete listing of the Program 3.14, corresponding to a more complete diet, has been included in the Appendix for Long Programs.

The solution for the extended diet is $x_1 = 0.5183$ (bread), $x_2 = 0.8121$ (cheese), $x_3 = 1.7244$ (potato), with an optimal cost $z = 10.1203$

It is convenient to have a confirmation of the optimal result. This can be done by solving the dual problem. Program 3.15 offers an example of this approach, playing again with the extended diet.

The result of Program 3.15 is $z = 10.1415$, which confirms (with some numerical error) the result of the primal approach.

Note (Program 3.15)

The complete listing of the Program 3.15, corresponding to dual method for obtaining optimal extended diet, has been included in the Appendix for Long Programs.

3.6.2 Blending

Blending or mixing is a frequent process in several industries (oil&gas, paint, foods, etc.). See the book [26], which can freely be accessed on the Internet, for an extensive treatment of this type of optimization application.

Let us introduce the topic with the following example.

A company makes two blends of fuel by mixing three oils. Figure 3.15 gives a sketch of the example.

The next table shows the cost and daily availability of each oil.

Oil	Cost ($/liter)	Amount available (liters)
A	0.30	6,000
B	0.40	10,000
C	0.48	12,000

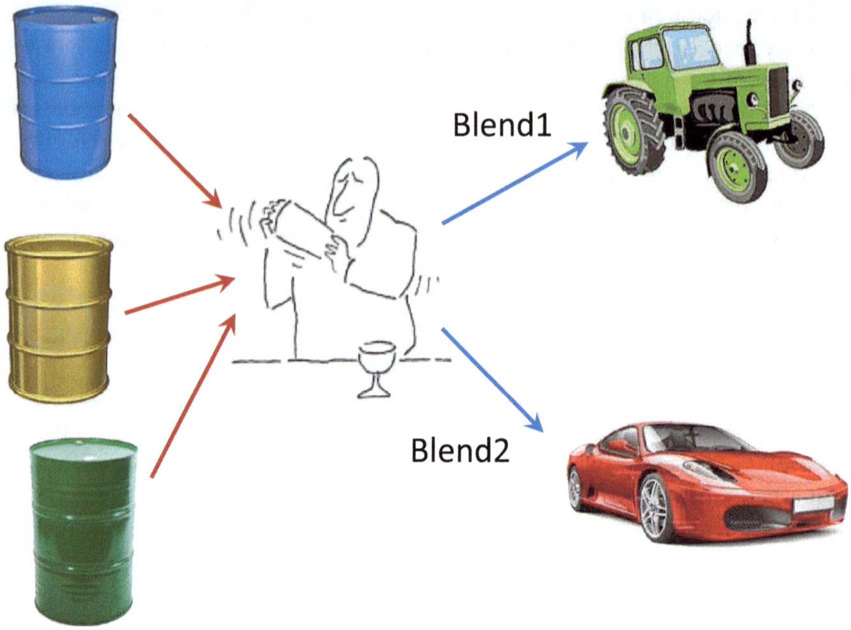

Fig. 3.15 A sketch of the blending example

The blends are specified according with the next table:

Blend 1	At least 30% of A At most 50% of B At least 30% of C
Blend 2	At most 40% of A At least 35% of B At most 40% of C

The price of blend 1 is 1.10 $ per litter, and for blend 2 is 1.20 $ per litter.

At least 10,000 litters of each blend should be produced.

Let us denote as $x_1,\ x_2,\ x_3$ the amounts of A, B, and C oils in blend 1. Likewise, $x_4,\ x_5, x_6$ the amounts of A, B, and C oils in blend 2.

According with the problem statement, the aspects concerning the objective function are:

Revenue = $1.10\,(x_1\ +\ x_2 +\ x_3)\ +\ 1.20(x_4\ +\ x_5 +\ x_6)$
Cost = $0.3\ (x_1\ +\ x_4)\ +\ 0.4\ (x_2\ +\ x_5)\ +\ 0.48\ (x_3\ +\ x_6)$
And then: *Profit* = *Revenue* – *Cost*.

Regarding the constraints, there will be six about the mixing, one of them for example with the form:

$$x_1 / (x_1 + x_2 + x_3) \geq 0.3$$

This can be transformed to:

$$x_1 \geq 0.3\ (x_1 + x_2 + x_3)$$

And then to:

$$-0.7x_1 + 0.3x_2 + 0.3x_3 \leq 0$$

So, these six constraints are:

$$-0.7x_1 + 0.3x_2 + 0.3x_3 \leq 0$$

$$-0.5x_1 + 0.5x_2 - 0.5x_3 \leq 0$$

$$0.3x_1 + 0.3x_2 - 0.7x_3 \leq 0$$

$$0.6x_4 - 0.4x_5 - 0.4x_6 \leq 0$$

$$0.35x_4 - 0.65x_5 + 0.35x_6 \leq 0$$

$$-0.4x_4 - 0.4x_5 + 0.6x_6 \leq 0$$

There are also constraints due to amount of oil available:

$$x_1 + x_4 \leq 6{,}000$$

$$x_2 + x_5 \leq 10{,}000$$

$$x_3 + x_6 \leq 12{,}000$$

Two more constraints express that at least 10,000 litters of each blend should be produced.

$$x_1 + x_2 + x_3 \geq 10{,}000$$

$$x_4 + x_5 + x_6 \geq 10{,}000$$

In total, there are 11 constraints. Some of the constraints have a zero in the right-hand side. This fact implies cycling problems when applying the simplex algorithm. That is why some measures were taken to avoid pivoting difficulties. In particular, according with one of the recommended techniques for escaping cycling situations, a perturbation was added to the zeros.

By modifying the initial part of Program 3.15 (corresponding to the dual of the extended diet), another program (Program 3.16) was built to solve the blending problem.

The code fragment that has been deleted from Program 3.15 was:

Fragment 3.1 Fragment FrA

```
% simplex with invisible tableaux
% diet 1:
% dual problem
%
% Problem specification:
N=4; %number of variables
L=4; %number of constraints
%total number of variables+added variables
W=N+(2*L);
Oc=[2; 8; 1.5; 11];
OA=[4 7 1.3 8;
1 9 0.1 7;
15 0.4 22.6 0;
90 106 97 130];
Ob=[-10; 8; 10; 300]; %notice the minus sign
c=zeros(W,1); A=zeros(L,W); p=zeros(L,1);
%
% (L slack and L surplus variables are added)
M=-1000; %large negative value (edit)
c(1:N)=Ob; c(N+L+1:end)=M*ones(L,1); %dual
A(:,1:N)=OA'; %dual
A(:,N+1:N+L)=[-1 0 0 0;
0 1 0 0;
0 0 1 0;
0 0 0 1]; %(edit according with constraints)
A(:,N+L+1:end)=eye(L);
b=Oc; %dual
```

The new code fragment that has been used for Program 3.16, in substitution of the deleted fragment, is:

Fragment 3.2 Fragment FrB

```
% simplex with invisible tableaux
% blend0
%
% Problem specification:
N=6; %number of variables
L=11; %number of constraints
%total number of variables+added variables
W=N+(2*L);
```

```
Oc=[0.8; 0.7; 0.62; 0.9; 0.8; 0.72];
OA=[-0.7 0.3 0.3 0 0 0;
-0.5 0.5 -0.5 0 0 0;
0.3 0.3 -0.7 0 0 0;
0 0 0 0.6 -0.4 -0.4;
0 0 0 0.35 -0.65 0.35;
0 0 0 -0.4 -0.4 0.6;
1 0 0 1 0 0;
0 1 0 0 1 0;
0 0 1 0 0 1;
1 1 1 0 0 0;
0 0 0 1 1 1];
Ob=[0; 0; 0; 0; 0; 0; 6000; 10000; 12000;...
 10000; 10000];
c=zeros(W,1); A=zeros(L,W); p=zeros(L,1);
%
% (L slack and L surplus variables are added)
M=-1000; %large negative value (edit)
c(1:N)=Oc; c(N+L+1:end)=M*ones(L,1);
A(:,1:N)=OA;
%(edit according with constraints):
A(:,N+1:N+L)=eye(L);
A(L-1,N+1:N+L)=[0 0 0 0 0 0 0 0 0 -1 0];
A(L,N+1:N+L)=[0 0 0 0 0 0 0 0 0 -1];
A(:,N+L+1:end)=eye(L);
b=Ob;
```

The solution of the blending problem is *profit= 21040*, with:

$x_1 = 3000,\ x_2 = 2200,\ x_3 = 4800,\ x_4 = 3000,\ x_5 = 7800,\ x_6 = 7200$

Note (Program 3.16)

The complete listing of the Program 3.16, which deals with a blending problem, has been included in the Appendix for Long Programs.

3.6.3 Scheduling

In research or development projects, a planning should be done and be respected along time. Part of the planning is to decompose the work into tasks with a certain duration. There are tasks that should wait for others to be finished. To build a bicycle, you need wheels, pedals, chain, etc. Scheduling is the art of ordering tasks along time. A typical tool for elaborating a schedule is the Gantt chart (see [16] for more details).

From our optimization point of view, the peculiar aspect of scheduling is *time*.

Figure 3.16 intends to illustrate our subject. Project managers know that scheduling is a difficult exercise, which usually requires several trials and the council of participants and experts.

Fig. 3.16 Activity scheduling scene

There are a lot of books, articles, and videos about project management that include scheduling. What we do next is to consider a very simple example, just to give the flavor of the question.

Here is the example. This is a small residence for students. A fast cleaning up in the morning was desired. There are enough people for the work, only that tasks should be well organized. It happens, like in many other scheduling scenarios, that some tasks cannot start until other precedent tasks were finished. The next table gives an idea of the problem.

Preceding	Name	Task	Duration (min)
n/a	A	Clean fridges	10
n/a	B	Wash dishes	25
n/a	C	Make up beds	15
A,B	D	Clean sinks	7
C,D	E	Take dust	18
E	F	Vacuum carpets	12
D	G	Take out garbage	3
F,G	H	Tidy up the area	14

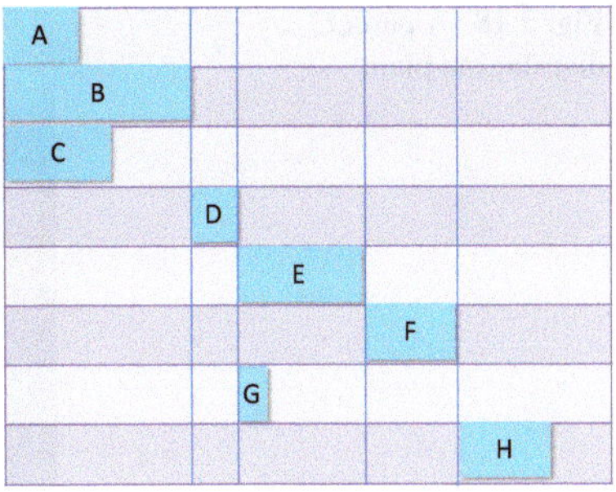

Fig. 3.17 Timing chart corresponding to the example

Denote as t_j the start time of task j. The problem can be specified as to finally have task H finished soon:

Minimize:

$$J = t_H + 14$$

The constraints would correspond to precedence relations. For instance, sink cleaning must wait for dishes to be washed, and so:

$$t_D \geq t_B + 25$$

Also, fridges should be cleaned before, and then the sink:

$$t_D \geq t_A + 10$$

Based on the rest of precedence relations, the series of constraints is completed as follows:

$$t_E \geq t_D + 7$$

$$t_F \geq t_E + 18$$

$$t_G \geq t_D + 7$$

$$t_H \geq t_F + 12$$

$$t_H \geq t_G + 3$$

Figure 3.17 shows schematically the solution.

The example has been taken from [5]. This reference would be recommended for it shows through many examples, how to formulate them as linear programming problems.

This type of scheduling problems is called "*activity scheduling*." There are methods, apart from the simplex algorithm, that can be used specifically for scheduling, like the "*critical path method*" (CPM) (see [1]).

Fig. 3.18 A parcel distribution plant

3.7 Special Methods for Logistics

This section will consider some important applications, related to logistics, for which specific methods are relevant. Indeed, these applications can be treated with the simplex algorithm, but the procedure would require large tableaux. Instead, the specific methods need much less mathematical manipulation.

The world of logistics has a decisive impact in modern economies and social habits, including our use of Amazon and other services. Figure 3.18 presents a view of a typical parcel medium-size distribution plant

A main reference for this section is [28] (our reference includes a link to the part of interest in that big book).

3.7.1 *Transportation*

Suppose there are m factories manufacturing a certain product and n warehouses that demand these products. There are $m \times n$ routes connecting factories and warehouses. Denote as c_{ij} the cost of sending one unit of the product from factory i to warehouse j. Also, denote as x_{ij} the size of a shipment (how many units of the product) from factory i to warehouse j.

Assuming linear relationships, the total cost of transportation from factories to warehouses would be:

$$J = \sum_i \sum_j c_{ij} x_{ij}$$

There are constraints. Each factory can manufacture a certain maximum number a_i of units; then,

$$\sum_j x_{ij} \leq a_i$$

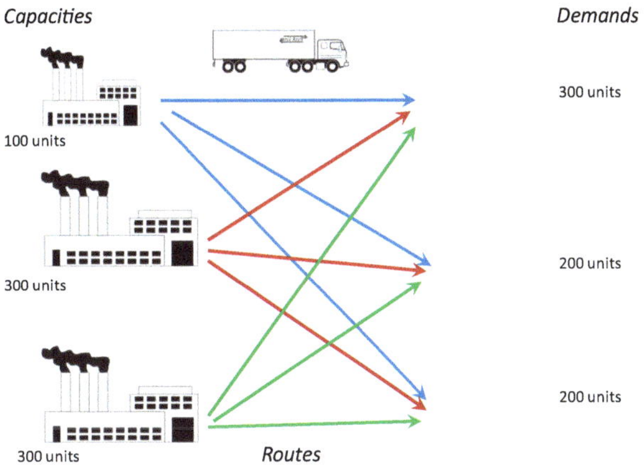

Fig. 3.19 A sketch of the transportation example

In addition, each warehouse wants no less than a certain minimum number b_j of units. Then:

$$\sum_i x_{ij} \geq b_j$$

The problem is to minimize J.

In many cases it is assumed that the total supply is equal to the total demand:

$$\sum_i a_i = \sum_j b_j$$

A simple example would be now considered. A company that manufactures desks has three factories, from which the products should be transported to three warehouses. Figure 3.19 presents a sketch of the problem, which is to minimize the transportation costs.

The next table gives information of the cost ($) of each route. It also contains how many desks are produced by each factory and how many desks are demanded by each warehouse. It is a "*balanced problem*": the demand equals the capacity.

From \ To	A	B	C	Capacity
D	5	4	3	100
E	8	4	3	300
F	9	7	5	300
Demand	300	200	200	700

In order to solve the problem, one follows two steps: first an initial solution and then improving that solution.

3.7.1.1 Initial Solution

There is a number of methods for obtaining an initial solution. For its simplicity, let us choose the *"North West corner rule"*. One starts in the upper left-hand entry. The procedure exhausts the supply of each row before moving to the row below. Also, regarding the columns, the procedure moves to the right only when a demand was completed.

In this example, the initial entry takes all production from factory A, 100 units. One continues with the second row, with 200 units the demand of first column was satisfied, and moving to the right, 100 units were put in the second column. This exhausts the production of factory E, 300 units. The third row is filled by completing with 100 units the demand in the second column, and then taking 200 units for the third column, exhausting the offer from factory F, 300 units.

There is a specific tableau for handling this type of problems. The next table shows an example of it, filled with the initial solution that has been found, together with the pertinent problem information: items and costs.

From \ To	A	B	C	Capacity
D	[5] 100	[4]	[3]	100
E	[8] 200	[4] 100	[3]	300
F	[9]	[7] 100	[5] 200	300
Demand	300	200	200	700

By using the tableau contents, it is easy to compute the costs of any solution. In this case—the initial solution—the total transportation cost would be 4200 $. The next table summarizes the cost calculation.

Route	Units shipped	Unit cost	Total cost
DA	100	5	500
EA	200	8	1600
EB	100	4	400
FB	100	7	700
FC	200	5	1000
			Total: 4200

3.7.1.2 Finding the Optimal Solution

There are also several methods for taking you from the initial solution to the optimal solution. A popular method is the so-called *"stepping stone method"*. Perhaps you may recall a time when you crossed a river or a pond using stepping stones, that's the name. One starts on any unused entry and then builds from here a closed path with only vertical or horizontal moves. You can change directions only on occupied entries.

For example, start from first row-second column. Observe the next table. A plus (+) was placed in the start entry. Move left (moving right would take you nowhere). A minus (−) was placed on this entry. As a general rule, one marks the path with alternating + or − signs. Then one moves down, and place a (+). Then to the right and a (−). And finally, back to the start.

From \ To	A	B	C	Capacity
D	[5] 100 − ←	START [4] +	[3]	100
E	↓ [8] 200 + →	↑ [4] 100 −	[3]	300
F	[9]	[7] 100	[5] 200	300
Demand	300	200	200	700

Now, an improvement index I_{ij} is computed as follows:

$$I_{DB} = +4 - 5 + 8 - 4 = +3$$

Another path can be built from the entry in the crossing of first row and third column. The next table shows this path, which is a bit more complicated.

From \ To	A	B	C	Capacity
D	[5] 100 − ←	[4] ← ← ←	START [3] +	100
E	↓ [8] 200 + →	[4] 100 −	↑ [3] ↑	300
F	[9]	↓ [7] 100 + →	↑ [5] 200 −	300
Demand	300	200	200	700

The improvement index of this path is:

$$I_{DC} = +3 - 5 + 8 - 4 + 7 - 5 = +4$$

There are two more paths, which have the following improvement indexes:

$$I_{EC} = +3 - 4 + 7 - 5 = +1$$

$$I_{FA} = +9 - 7 + 4 - 8 = -2$$

For our purposes the good news is that a negative index exists (if several appear, one chooses the more negative). The next table shows the FA path:

From \ To	A	B	C	Capacity
D	[5] 100	[4]	[3]	100
E	[8] 200 ↓ − ←	[4] 100 +	[3]	300
F	START [9] + →	[7] ↑ 100 −	[5] 200	300
Demand	300	200	200	700

The idea would be to use the route F-A at maximum, since it saves money. For this aim, one looks at F-A path entries with negative signs and selects the smallest number in there (in this case, 100). Following the path, that number is added to the entries with + sign and subtracted from entries with − sign. The result is shown in the next table.

From \ To	A	B	C	Capacity
D	[5] 100	[4]	[3]	100
E	[8] 100	[4] 200	[3]	300
F	[9] 100	[7]	[5] 200	300
Demand	300	200	200	700

The shipping cost of this new solution is 4000 \$. Now there are four empty entries, and the improvement indexes of the corresponding closed paths are:

$$I_{DB} = +4 - 5 + 8 - 4 = +3$$

$$I_{DC} = +3 - 5 + 9 - 5 = +2$$

$$I_{EC} = +3 - 8 - 9 - 5 = -1$$

$$I_{FB} = +7 - 4 + 8 - 9 = +2$$

Therefore, a better solution can be found by taking advantage of route E-C. The next table shows the closed path from entry EC.

From \ To	A	B	C	Capacity
D	[5] 100	[4]	[3]	100
E	[8] 100 − ↓	← ← [4] 200	START [3] ← +	300
F	↓ [9] 100 + →	→ → → [7]	↑ [5] 200 −	300
Demand	300	200	200	700

Using the rules for updating the table with the smallest number of negative entries along the path (which is 100), one gets the solution shown in the next table.

From \ To	A	B	C	Capacity
D	[5] 100	[4]	[3]	100
E	[8]	[4] 200	[3] 100	300
F	[9] 200	[7]	[5] 100	300
Demand	300	200	200	700

The shipping cost of this new solution is 3900 \$. No negative improvement indexes can be found among all possible closed paths; therefore, this is the optimal solution.

For a detailed survey of transportation optimization, land, sea, and air, it would be opportune the article [9]. A most cited contribution is [24] (1326 cites) on transportation between customers and depot. In fact, the field of transportation optimization is quite large and varied; the literature review [19] and references therein help to explore the topic.

3.7.2 Assignment

A basic assignment scenario is having several tasks, to be distributed among several people, one task per person. In order to describe the assignment, one could use the following variables:

x_{ij} = 1 if person i is assigned to task j
x_{ij} = 0 if person i is not assigned to task j
c_{ij} = 0 the cost of assigning a person to a task

Since the cost should be minimized, the problem is:

Minimize:

$$\sum_i \sum_j c_{ij} x_{ij}$$

Subject to:

$$\sum_j x_{ij} = 1, \quad \forall i$$

$$\sum_i x_{ij} = 1, \quad \forall j$$

Figure 3.20 presents a simple task assignment scenario, in which three persons should be assigned to three tasks. It may be that this figure could suggest preferences due to different aptitudes, which is not unusual. This example will be treated in economic terms, but it could be also treated in other more suitable terms.

The information considered pertinent for the assignment problem is contained in the following table, which gives the costs ($) of each person assignment to a task.

TASK	1	2	3
PERSON			
Fabius	11	14	6
Frank	8	10	11
Humbert	9	12	7

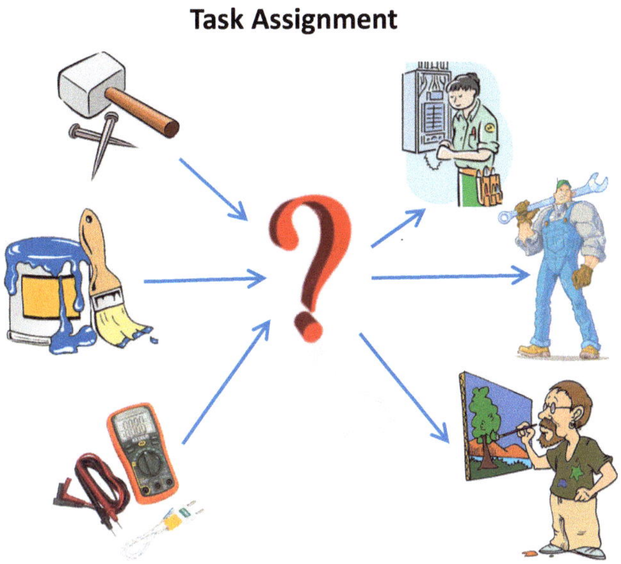

Fig. 3.20 A sketch of the assignment example

Since this problem is quite simple, it is possible to directly compute all alternatives, as shown in the next table. The best solution corresponds to the last row.

TASK			COST TERMS	TOTAL COST
1	2	3		
Fabius	Frank	Humbert	11+10+7	28
Fabius	Humbert	Frank	11+12+11	34
Frank	Fabius	Humbert	8+14+7	29
Frank	Humbert	Fabius	8+12+7	26
Humbert	Fabius	Frank	9+14+11	34
Humbert	Frank	Fabius	9+10+6	25

In general, assignment problems are really more complicated, and so the *brute force* method was not recommended. There are a number of other methods that have been proposed. One of the most popular methods is the so-called *"Hungarian method"*, which will be introduced next.

Consider the table give before, with the essential information. Row reduction is obtained by subtracting the smallest number in each row. Next table shows the result.

TASK	1	2	3
PERSON			
Fabius	5	8	0
Frank	0	2	3
Humbert	2	5	0

Then, column reduction is computed by subtracting the smallest number in each column. The next table shows the result. It is called an "*opportunity cost table*."

TASK	1	2	3
PERSON			
Fabius	5	6	0
Frank	0	0	3
Humbert	2	3	0

Now, the procedure continues by drawing lines for zero elimination. If the minimum number of lines is equal either to the number of row or the number of columns, an optimal solution was reached. If it is less, the procedure continues.

Then, let us draw vertical and horizontal lines to cross all zeros in the table. The result is shown in the next table. There were two lines, so we must go on.

TASK	1	2	3
PERSON			
Fabius	5	6	0
Frank	0	0	3
Humbert	2	3	0

Look at not crossed numbers in the table. Select the smallest number (2). Subtract this number from each not crossed number. Also, add this number to the intersection of the lines. The result is shown in the next table.

TASK	1	2	3
PERSON			
Fabius	3	4	0
Frank	0	0	5
Humbert	0	1	0

Draw vertical and horizontal lines to cross all zeros in the table. The result is shown in the next table. There were three lines, so a solution was reached.

TASK	1	2	3
PERSON			
Fabius	3	4	0
Frank	0	0	5
Humbert	0	1	0

Now, the assignment is done as follows. Fist, find a row or column with only one zero. This is indicated in the next table. Fabius is assigned to task 3.

TASK	1	2	3
PERSON			
Fabius	3	4	0
Frank	0	0	5
Humbert	0	1	0

Eliminate the row and column of the first selection. Again, find a row or column with only one zero. One of the alternatives would be task 1 for Humbert.

TASK	1	2	3
PERSON			
Fabius	3	4	0
Frank	0	0	5
Humbert	0	1	0

Eliminate the row and column of the second selection. Clearly, task 2 is for Frank.

TASK	1	2	3
PERSON			
Fabius	3	4	0
Frank	0	0	5
Humbert	0	1	0

The seminal paper on the Hungarian method was due to Harold Kuhn [20]. A short history of the topic was included in the introduction of the paper, making special mention of the work of two Hungarian scientists, and so the algorithm's name was coined. Recent articles make memory of this important contribution, like

[12]. With occasion of the 50th anniversary of the topic, [25] offered an interesting survey on the assignment problem.

3.8 Network Flows

Many linear programming problems, as well as other problems, can be modeled as network flows. Indeed, there are problems for which network flow models are quite suitable, as, in particular, real flows of water, oil, electricity, etc. Figure 3.21 visualizes part of a gas processing & distribution plant, plenty of pipes and valves.

An advantage of networks is that they, usually, can be represented graphically. This contributes to an intuitive treatment of problems. Actually, there are optimization algorithms easy to apply in a graphical context.

The most important reference for this section is [17], which joins several lecture notes on graphs and networks.

3.8.1 *Graphs*

Figure 3.22 introduces a first example of graph. A graph is made up of vertices, or nodes, which are connected by edges, arcs, or lines. The nodes in the figure are represented with small circles. Edges may be directed (represented with an arrow), or not. When all edges are directed, the graph is said to be a *directed graph*. The graph in the figure is an *undirected* graph.

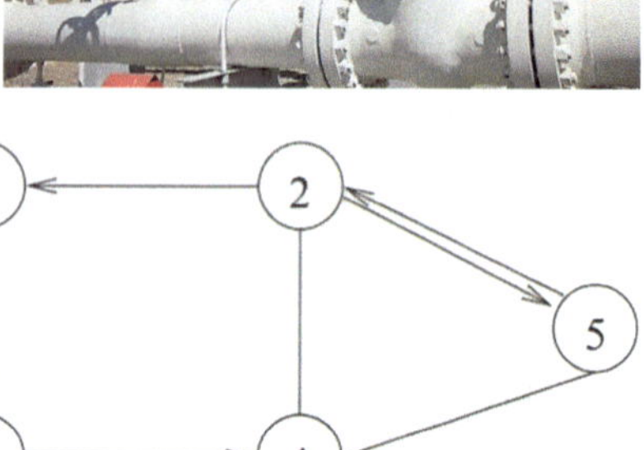

Fig. 3.21 Example of network flow control: gas system

Fig. 3.22 Example of graph

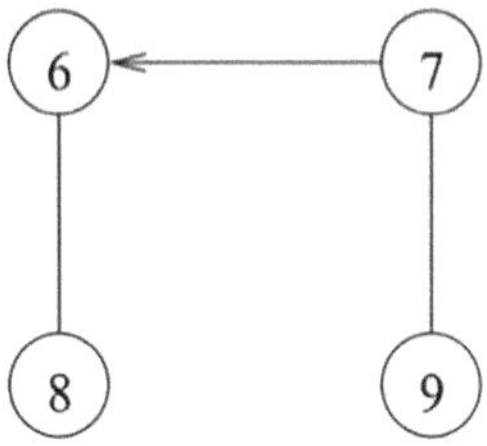

Fig. 3.23 Example of undirected, connected graph

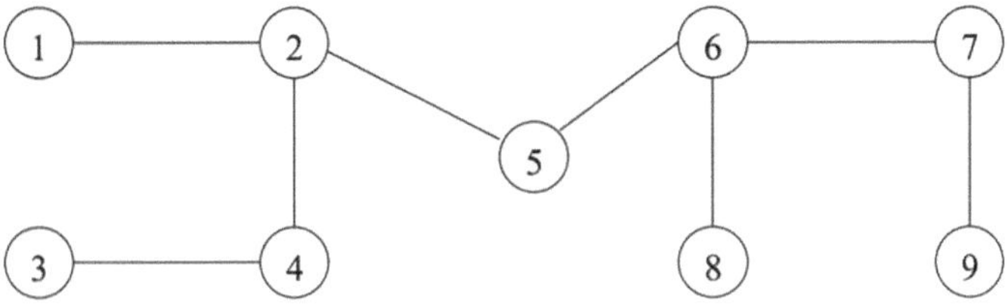

Fig. 3.24 Example of spanning tree

In an undirected graph, any two vertices u and v are connected if there is a path from u to v, no matter of the direction of directed edges. If there is no disconnected vertex, the graph is a *connected graph.*

Figure 3.23 shows, for instance, an undirected connected graph.

A directed graph is called *weakly connected* if ignoring the direction of edges, there is no disconnected vertex. It is *strongly connected* if any two vertices u and v are connected with a directed path from u to v, or from v to u.

Closed paths, cycles, may exist in a graph. Like, for instance, the path 2-4-5-2 in Fig. 3.22.

A *tree* is a collection of arcs with no cycles in it. A *spanning tree* in a graph is a collection of arcs that form a tree and that connect to every node. Figure 3.24 shows an example of spanning tree. The number of arcs in the spanning tree is equal to the number of nodes minus one.

An *end* is a node with only one arc incident to it. Spanning trees have at least two ends.

3.8.1.1 Graph Labeling

Graph labeling is the assignment of labels to the edges or vertices, or both, of a graph. One of the possible labeling ways is coloring (using several colors).

Suppose, for instance, that the graph represents flights between several airports. It would be opportune, for more complete information, to label arcs with flight costs and to label vertices with airport charges.

3.8.2 *Flows in Networks*

An important application of directed graphs is for studying flows in a network. The topology of the network is represented with a graph. There would be a flow x_{ij} associated with each arc of the graph. In many cases the flow obeys to a conservation law, so the flow that enters a node is equal to the flow leaving that node:

$$\sum_k x_{jk} - \sum_i x_{ij} = o, \quad \forall j$$

Often there are upper and lower flow constraints for each arc:

$$l_{ij} \leq x_{ij} \leq u_{ij}, \quad \forall i, \forall j$$

3.8.2.1 Maximum Flows

Sometimes the traffic of cars does not take full advantage of the capacities of available roads. In more abstract mood, consider, for example, what happens with the network represented in Fig. 3.25, where labels mean (flow, capacity):

The flow starts from the source and arrives to the sink. According with the configuration depicted in Fig. 3.25, the total flow from source to sink is 3.

However, with simple changes, the flow can increase up to 4, as depicted in Fig. 3.26.

As its name says, the maximum flow problem asks for obtaining as much as possible flow from source to sink. As a first step to tackle this problem, it is opportune to prepare some helping tools, which will be now introduced. They keep the same graphical representation context.

Notice that the differences between Figs. 3.21 and 3.22 concern to the labels of three edges. Figure 3.27 shows a path formed with these edges in particular:

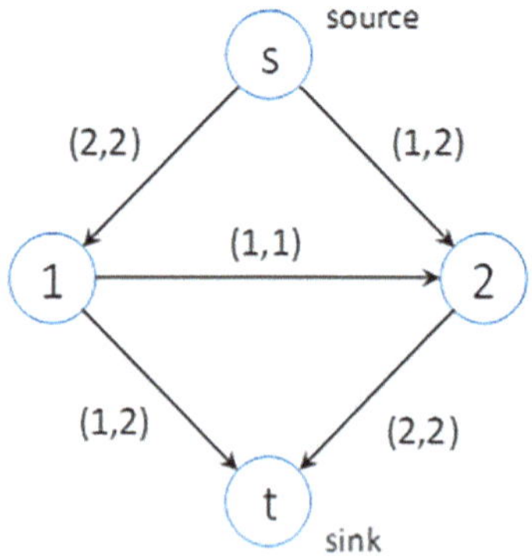

Fig. 3.25 Example of flow in a network

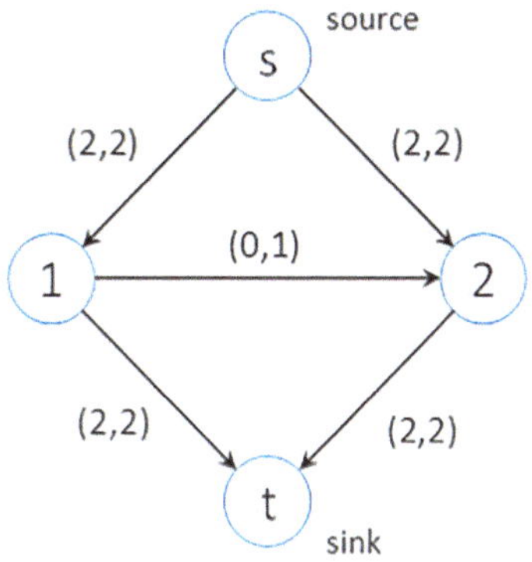

Fig. 3.26 Increasing the flow

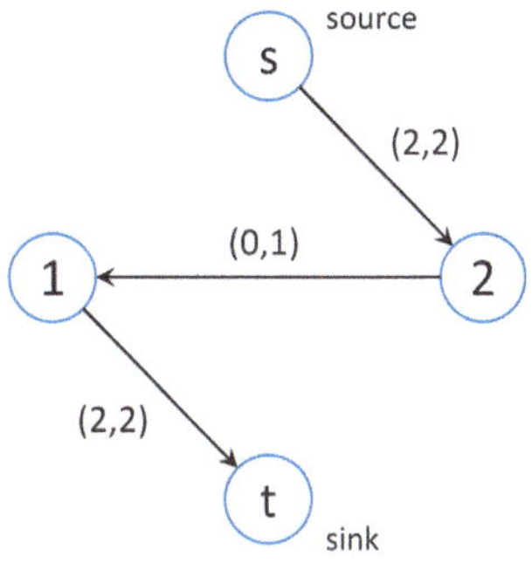

Fig. 3.27 Augmenting path

Paths like the one represented in Fig. 3.27 are called *"augmenting paths"*. Notice the change in the direction of one of the edges; this serves to indicate a decrease of flow in that edge.

In order to explore for adequate flow changes, it would be convenient to have a graphical representation of the surplus capacity of each edge (how much flow can be added before exceeding limits). Take, for example, the network in Fig. 3.21. Studying each edge, it is possible to obtain the network depicted in Fig. 3.28, which is called "*residual network*".

Notice that in Fig. 3.28, there is one directed path from source to sink and that this path is precisely the augmenting path used to obtain Fig. 3.26. This is a useful general fact: it is possible to use the residual network to find augmenting paths, as they are directed paths from source to sink.

A first method for solving the problem was proposed in 1956 by Ford and Fulkerson [11]. The method starts with zero flow in every edge. Then, the method iterates as follows:

1. A direct path is found in the residual network
2. Since it is an augmenting path, it is used in the original network to increase the flow as much as possible
3. The residual network is updated
4. back to step 1

Fig. 3.28 Residual network

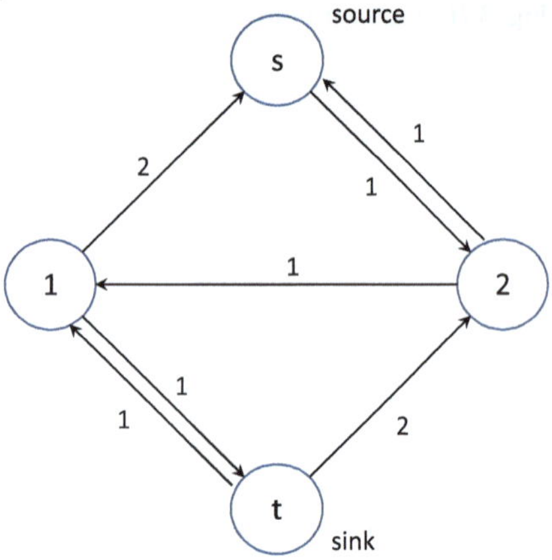

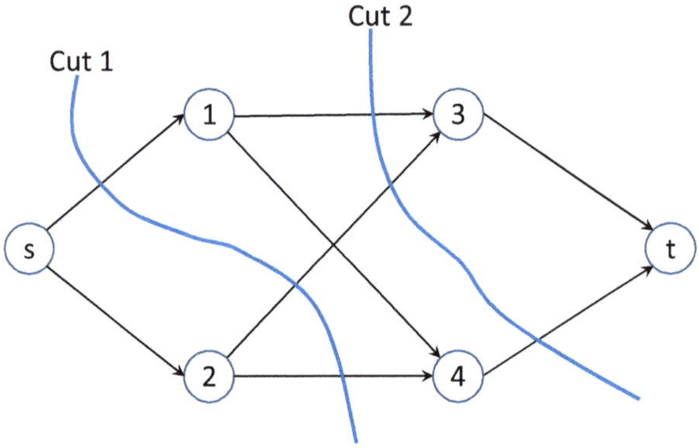

Fig. 3.29 Network with two cuts

The process is repeated until no new augmenting path can be found in step 1.

A number of improvements have been proposed, like searching in step 1 for the shortest augmenting path, or the path with higher flow. The article [15] offers a good review of classic and new algorithms for the maximum flow problem.

There is an important theorem that can be used to ascertain that an optimal flow has been achieved. It is the *max-flow min-cut* theorem. A "*cut*" is a break of a number of arcs such that all paths from source to sink are interrupted. Figure 3.29 shows an example with two cuts.

The capacity of a cut is the sum of the capacities of the arcs that have been broken by the cut. The cut (or cuts) with the smallest capacity is called the "*minimal cut*".

The theorem establishes that the capacity of the minimal cut is equal to the maximal flow between source and sink.

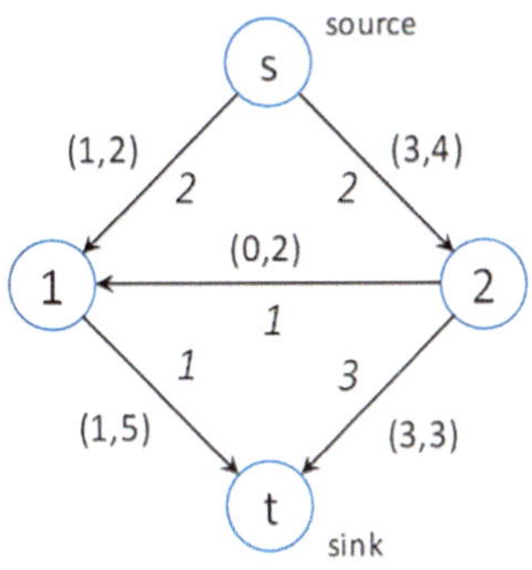

Fig. 3.30 Network with cost labels

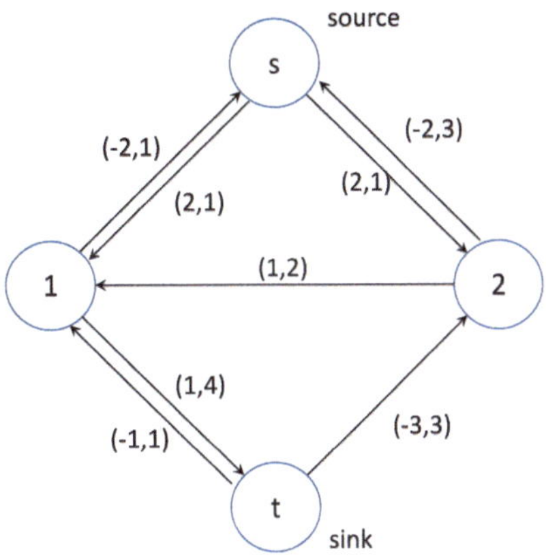

Fig. 3.31 Residual network with (cost, flow) labels

3.8.2.2 Minimum Cost Flows

There is a network connecting a source and a sink. The sink demands a certain flow. It is desired to find the minimum cost flow that satisfies that demand.

In addition to the labels (flow, capacity), another label, cost, will be used. It is a cost per unit of flow. Denote as x_{ij} the flow across the arc connecting node i and node j. Let c_{ij} be the cost label of that arc. Then, $C = x_{ij} \cdot c_{ij}$ would be the cost of the flow across the arc. Figure 3.30 depicts an example of cost labeled network. The total flow from source to sink is 4.

A basic method for solving the optimization problem is the so-called "*cycle-cancelling algorithm*". Like the Ford-Fulkerson method, it uses a residual network; only that positive and negative costs are included. Figure 3.31 shows the residual network corresponding to the network in Fig. 3.30.

Now, let us see if there are cycles. For instance, there is a cycle t-2-1-t. Adding costs one gets: $-3 + 1 + 1 = -1$. Therefore, it is a *negative cycle*. The smallest flow (among 3, 2, 4) is 2 in this cycle. Then, in order to cancel this negative cycle, the flow is augmented 2 units along the cycle. The residual network is updated, and the result is as shown in Fig. 3.32.

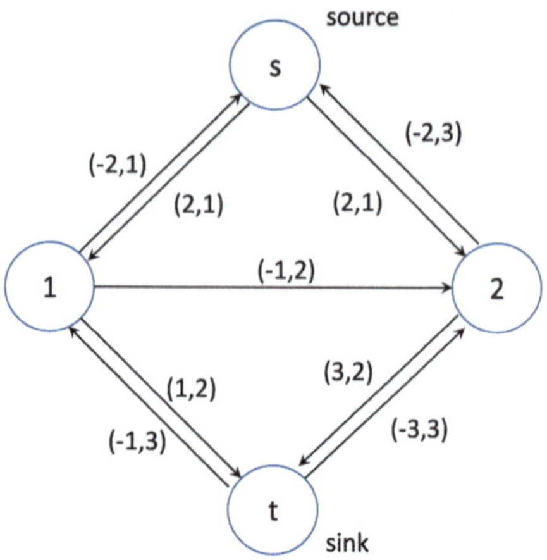

Fig. 3.32 Updated residual network

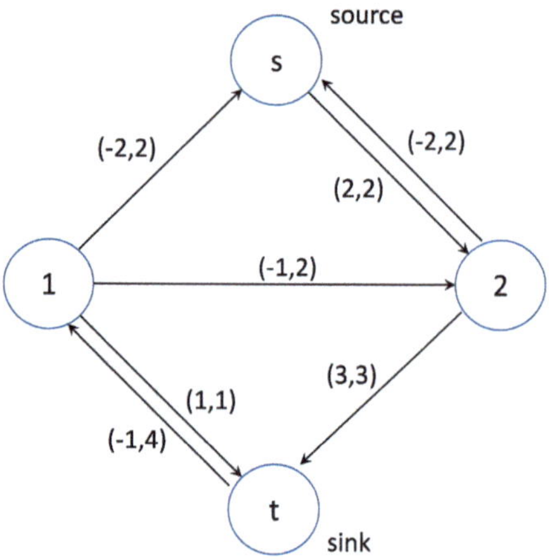

Fig. 3.33 Next updated residual network

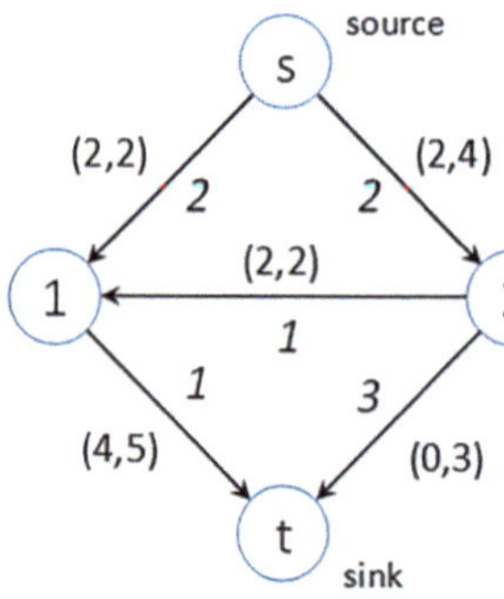

Fig. 3.34 Minimum cost flow

There is a negative cycle, t-2-s-1-t. The smallest flow in the cycle is 1. To cancel this cycle, the flow is augmented 1 unit along the cycle. Figure 3.33 shows the result.

There are no more negative cycles. The algorithm terminates. Figure 3.34 shows the optimal solution just obtained.

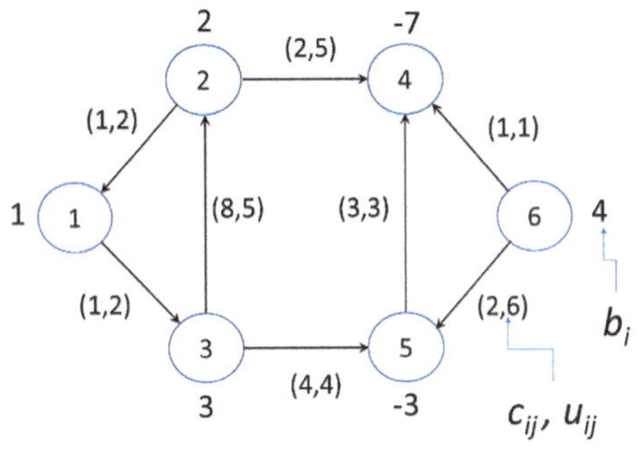

Fig. 3.35 Minimum cost flow must be determined

3.8.3 Network Simplex

The network simplex method translates the simplex algorithm to the context of graphs.

Let us introduce the network simplex by way of an example discussed in [22]. We shall start by considering a minimum cost flow problem in the following terms.

Minimize:

$$z = c_1 x_1 + c_2 x_2 + \ldots + c_n x_n$$

Subject to:

$$\sum_j x_{ij} - \sum_j x_{ji} = b_i\,,\ \forall i \in N$$

$$0 \leq x_{ij} \leq u_{ij},\ (i, j) \in A$$

The particular problem to be solved concerns the graph represented in Fig. 3.35, where (cost, flow) labels have been included.

The minimum cost flow problem for this case can be formulated as follows:

Minimize:

$$z = x_{13} + x_{21} + 2x_{24} + 8x_{32} + 4x_{35} + 3x_{54} + x_{64} + 2x_{65}$$

Subject to:

$$x_{13} - x_{21} = 1$$

$$x_{21} + x_{24} - x_{32} = 2$$

$$-x_{13} + x_{32} + x_{35} = 3$$

Node	x13	x21	x24	x32	x35	x54	x64	x65
1	1	-1						
2		1	1	-1				
3	-1			1	1			
4			-1			-1	-1	
5					-1	1		-1
6							1	1

Fig. 3.36 Constraints correspond to arcs

$$-x_{24} - x_{54} - x_{64} = -7$$

$$-x_{35} + x_{54} - x_{65} = -3$$

$$x_{64} + x_{65} = 4$$

And,

$$x_{13} \leq 2, \quad x_{21} \leq 2, \quad x_{24} \leq 5, \quad x_{32} \leq 5, \quad x_{35} \leq 4,$$

$$x_{54} \leq 3, \quad x_{64} \leq 1, \quad x_{65} \leq 6$$

It is assumed that:

$$\sum_i b_i = 0, \; i \in N$$

where N is the set of nodes.

The tabular figure above visualizes how the constraints correspond to arcs in the graph (Fig. 3.36):

Note that our example problem can be solved with standard simplex, using the numerical data. In turn, what the network simplex wants is to be directly applied on the graphs.

In matrix-vector format, the constraints can be expressed with:

$$Q\vec{x} = \vec{b}$$

where the columns of Q (columns in the table above) correspond to arcs in the graph. It turns out that at least one row is redundant and can be removed. The choice of the row to remove is irrelevant; then, one can select the row corresponding to an arbitrary root node. After removing one row, one obtains matrix P.

As seen in Sect. 3.3, the simplex algorithm starts from a basis and search for the optimum by exchanging columns of the basis. A similar procedure is used by the network simplex. In our example, a basis B would be a square matrix formed with $n - 1$ independent columns of P, such as:

$$B\,\vec{x}_B = \vec{b}_B$$

where $\vec{x}_B$ (called the *basic variables*) and $\vec{b}_B$ have $n - 1$ components.

Now, some properties of spanning trees become relevant for our algorithm:

- Every spanning tree has $n - 1$ arcs
- Every spanning tree has at least 2 end nodes
- Every pair of nodes in a spanning tree is connected by a unique path
- Addition of an arc to a spanning tree, results in a graph with exactly one cycle

Basis Lemma—Let Z be a set of arcs of our graph G, and let Bz be the corresponding set of columns of P. Then Bz is a basis for P if and only if Z **is a spanning tree** for G.

There is an algorithm, called Compute-Flows, that obtains the unique basic solution corresponding to a given spanning tree. The algorithm considers the imbalance of nodes, which is:

$$e_i = b_i - \sum_j x_{ij} + \sum_j x_{ji}$$

The algorithm starts by setting $e_i = b_i$ and $S = Z$ and then enters into the following loop:

While $S \neq 0$ do,

- Let $i \neq s$ be an end node of S, and $a = (i, j)\, or\, (j, i)$ the associated unique arc of S adjacent to i.
- If $a = (i, j)$ assign $x_{ij} = e_i$, or if $a = (j, i)$ assign $x_{ji} = -e_i$
- Remove a and i from S, and add e_i to e_j

end while

For example, taking from the graph G the following spanning tree:

$$Z = \{(2, 1),\ (3, 2),\ (2, 4),\ (3, 5),\ (6, 5)\}$$

the algorithm computes the corresponding basic solution (flows) as represented in Fig. 3.37:

Once a basic solution was obtained, its corresponding residual network can be constructed (Fig. 3.38):

With an extra work, one completes the labeling of the residual networks, with node potentials and with (cost, flow) labels on non-basic arcs. The non-basic arc with label $(-7, 1)$ has the same direction as in G, non-basic arcs with labels $(3, 3)$ and $(-10, 2)$ have the opposite direction (Fig. 3.39).

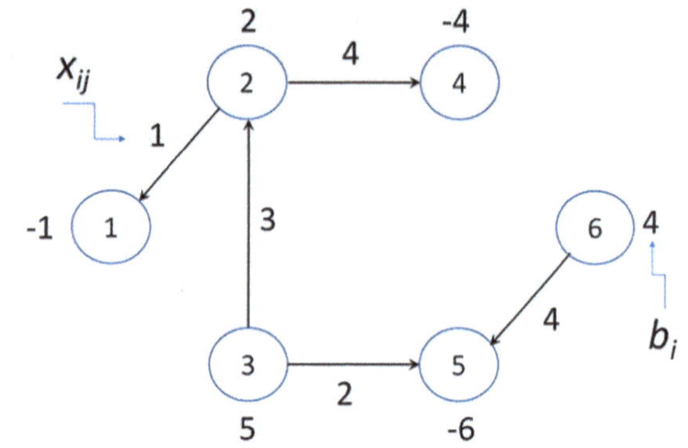

Fig. 3.37 Basic solution corresponding to spanning tree Z

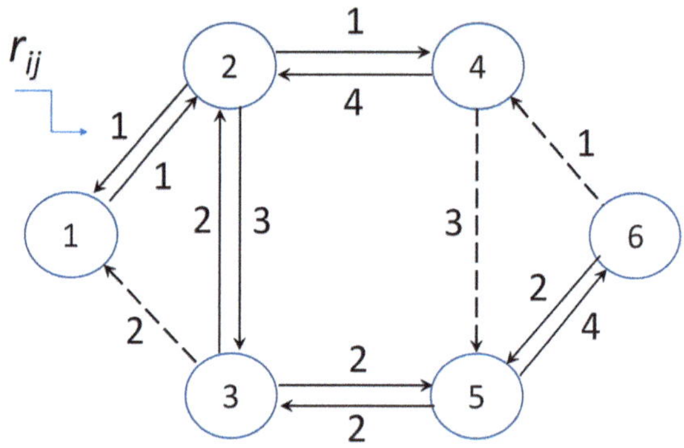

Fig. 3.38 The corresponding residual network

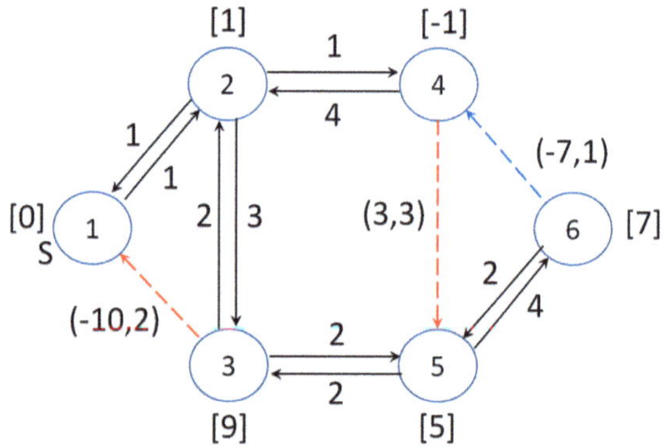

Fig. 3.39 Adding information to residual network

Now, we are in conditions to perform a pivoting step.

1. Choice of entering variable: find non-basic arc (i, j) with most negative reduced cost (if all reduced costs were non-negative, STOP, optimal solution found)
2. Insert arc (i, j) into Z. A unique directed cycle W will appear. Let (k, l) be the arc of W having minimum residual capacity, this arc will move to non-basic arcs, being coincident or opposite respect to arc directions in G.

Figure 3.40 depicts the residual network after first pivot:

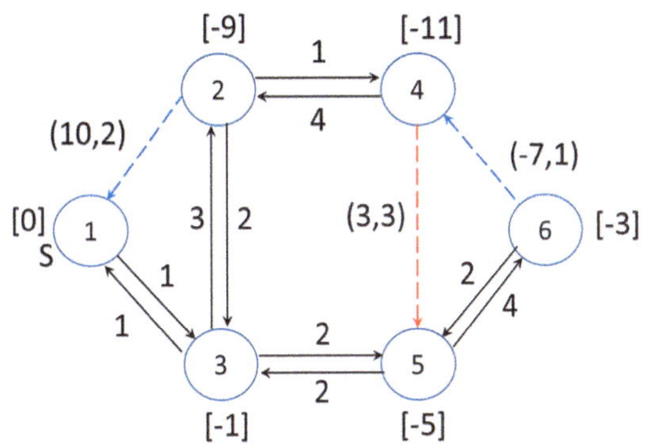

Fig. 3.40 Residual network after first pivot

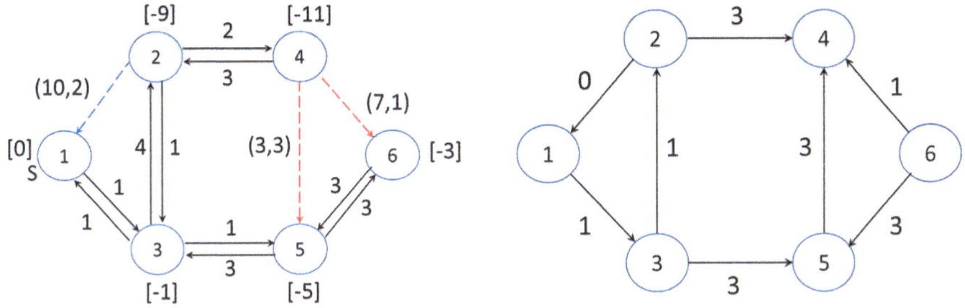

Fig. 3.41 Residual network after second pivot and final result

And, finally, Fig. 3.41 shows on the left the residual network after second pivot, which is the final, and on the right the result: the optimal flow.

An extensive treatment of the network simplex method can be found in Chapter 11 of book [18] (1357 cites). The MathWorks File Exchange site contains a MATLAB implementation of the algorithm, called "*the Network Simplex Algorithm*."

3.8.4 Networks and Graphs in MATLAB

In version 2015b, MathWorks introduced a set of functions for graph study and applications. The set involves around 60 functions. The visualization of graphs is made easy.

Function *maxflow( )* corresponds to the maximum flow problem; and function *shortestpath( )* detects shortest paths. The function *allcycles( )* detects all the cycles in a graph.

Other function are oriented to trees, like *shortestpathtree( )*, *dfsearch( )* for depth-first search, and *bfsearch( )* for breadth-first search.

Fig. 3.42 Visualization of a graph in MATLAB

Next MATLAB demo code:

Program 3.17 Graph example

```
[B,V] = bucky;
G = graph(B);
p = plot(G);
axis equal
```

produces the graph visualization shown in Fig. 3.42, which gives an idea of MATLAB visualization capabilities concerning graphs.

The MathWorks File Exchange site contains several interesting contributions, like the following:

- Network visualizer, for creating visually appealing graph plots
- grTheory-Graph Theory Toolbox, with many functions
- Network flow planning algorithm
- The Network Simplex Algorithm

Some repositories in GitHub contain contributions related to graphs and network flows. For example:

- Epanet, for hydraulic studies
- Matlab-bgl, many functions for graph theory

The Truong's weblog page at:
https://truongnghiem.wordpress.com/2008/12/10/graph_theory_in_matlab/
contains a succinct list of MATLAB packages for graphs.

One interesting example is the Bioinformatics Toolbox; there is a lot of research on biological networks, including the brain (there is a MATLAB Brain Connectivity Toolbox). We suggest a visit to the MATLAB page on Compute Functional Connectivity from Brain fMRI. Also, the package *netGraph* (MathWorks File Exchange) focuses on retinal networks.

3.9 Using MATLAB for Linear Programming

There is a MATLAB Optimization Toolbox that offers a variety of functions for several types of optimization problems. One of these functions—very popular—is *linprog( )*, which targets at linear programming problems.

The algorithms employed by *linprog( )* belongs to another category of methods, not studied in this chapter. They will be introduced later in the book. However, *linprog( )* gives us an opportunity now for checking, with alternative methods, the results obtained with simplex for the examples already considered.

This section includes a series of short programs that solve the previous examples. The programs show how to use *linprog( )* in each case.

3.9.1 The Diet

3.9.1.1 The Basic Diet

Program 3.18 The basic diet

```
% using linprog( )
% diet 0:
% minimize cost: z=8x1+1.5x2
% x1, cheese; x2, potato
% subject to:
% 7x1+1.3x2<=10
% 106x1+97x2>=300
%
% Problem specification:
f=[8; 1.5];
A=[7 1.3;
-106 -97];
b=[10; -300];
%
[X,Fval]=linprog(f,A,b,[],[],zeros(1,2));
X
Fval
```

3.9.1.2 The Extended Diet

Program 3.19 The extended diet

```
% using linprog( )
% diet 1:
```

```
% Problem specification:
f=[2; 8; 1.5; 11];
A=[4 7 1.3 8;
-1 -9 -0.1 -7;
-15 -0.4 -22.6 -0;
-90 -106 -97 -130];
b=[10; -8; -10; -300];
%
[X,Fval]=linprog(f,A,b,[],[],zeros(1,4));
X
Fval
```

3.9.1.3 The Extended Diet Dual

Program 3.20 The extended diet dual

```
% using linprog( )
% diet 1:
%
% Problem specification:
of=[2; 8; 1.5; 11];
oA=[-4 7 1.3 8;
-1 9 0.1 7;
-15 0.4 22.6 0;
-90 106 97 130];
ob=[-10; 8; 10; 300];
%
A=oA'; f=-ob; b=of;
[X,Fval]=linprog(f,A,b,[],[],zeros(1,4));
X
Fval
```

3.9.2 The Blending Problem

Program 3.21 The blending problem

```
% using linprog( )
% blending 0:
%
% Problem specification:
f=[0.8; 0.7; 0.62; 0.9; 0.8; 0.72];
A=[-0.7 0.3 0.3 0 0 0;
-0.5 0.5 -0.5 0 0 0;
```

```
0.3 0.3 -0.7 0 0 0;
0 0 0 0.6 -0.4 -0.4;
0 0 0 0.35 -0.65 0.35;
0 0 0 -0.4 -0.4 0.6;
1 0 0 1 0 0;
0 1 0 0 1 0;
0 0 1 0 0 1;
-1 -1 -1 0 0 0;
0 0 0 -1 -1 -1];
b=[0; 0; 0; 0; 0; 0; 6000; 10000; 12000;...
  -10000; -10000];
%
[X,Fval]=linprog(-f,A,b,[],[],zeros(1,6));
X
Fval
```

The solution is not completely the same as the solution found with Program 3.16.

3.10 Exercises

1- A manufacturer produces two products, A and B. The profits are $5 per unit of A, and $60 per unit of B. The following resources are required:

	Number of units	Machine hours	Labour hours	Material (kg)
Product A	x_1	4	3	1
Product B	x_2	2	5	1

Resources available per week: 100 machine hours, 150 labor hours, 50 kg of material.

Determine with Linear Programming a weekly production plan for maximum profit. Visualize in 2D the problem and its solution.

2- [3.1] An economist is considering investing in two securities P and Q. The next table summarizes risks and returns:

	Return	Risk (0 to 10)
P	9%	5
Q	15%	8

The total amount to be invested is $20,000. Total minimum return should be 12%. Maximum average risk, no more than 6.

Determine with Linear Programming how to invest. Visualize in 2D the problem and its solution.

3- [3.1] This is a case of animal breeding. There are two products, L and M, which contain the three required nutrients. The next table gives the details:

Nutrients	L, quantity/unit	M, quantity/unit	Minimum requirement
1	73	11	215
2	7	23	74
3	38	21	190

The cost of the products is $42 per unit of L, and $78 per unit of M.

Determine with Linear Programming the quantities of L and M to be given to the animals, for minimal cost. Visualize in 2D the problem and its solution.

4- [3.2] A small factory produces a high-tech gadget. The factory employs both skilled workers and apprentices. The workforce should be no more than 26 people, and it should make at least 420 gadgets a week. On average a skilled worker can make 28 gadgets a week, while an apprentice can make 13 a week.

Skilled workers are paid $300 a week, and apprentices $100 a week.

Determine with Linear Programming the number of skilled workers and apprentices to be hired, for cost minimization. Visualize in 2D the problem and its solution.

5- [3.2] It has been decided for a 700 m^2 car park to have some spaces for small cars, and the rest for large cars. For small cars 8.5 m^2 is allowed, and 12 m^2 for large cars. It was estimated that the ratio of small to large cars to park will be between 2:1 and 2:3.

Determine with Linear Programming the number of spaces in order to maximize the number of parked cars. Visualize in 2D the problem and its solution.

6- [3.1] An advertising agency is going to launch an ad campaign, on the Internet, TV, and newspapers. Next table presents data on campaign costs and reach:

Medium	Cost per unit	Reach (people) per unit
Internet	20,000	2,500,000
TV	8,000	1,000,000
Newspapers	4,000	100,000

Minimum 1,000,000 people are to be reached by the Internet. No more than 20,000 should be invested on newspaper ads. Total advertising budget is 2,000,000

Determine with Linear Programming the distribution of budget among media. Visualize in 2D the problem and its solution.

7- [3.3] The main product of ACME company is canned tomatoes. The company has three canneries in Oregon, California, and Texas. The products are shipped by truck to four distributing warehouses in Utah, Minnesota, Washington, and

Florida. The next table presents information of the shipping costs and other details:

	Warehouses				
Shipping cost/truck	1	2	3	4	Output
1 Cannery 2 3	460	511	660	875	75 125 100
	351	412	688	792	
	997	683	391	686	
Requirement	81	66	70	82	

Determine the optimal transportation (cost minimization).

8- [3.4] A manufacturer has four machines, 1, 2, 3, and 4. There are four jobs to be done, A, B, C, and D. The jobs can be performed on all machines, but with different costs. The next table presents those costs:

Job/Machine	1	2	3	4
A	19	36	29	18
B	25	31	42	11
C	22	40	37	19
D	37	40	35	16

(the above table indicates that if job A is done on machine 1, it would cost 19; etc.)

Assign jobs to machines.

3.10.1 Exercise References

[3.1] https://old.mu.ac.in/wp-content/uploads/2017/10/dormsem1linearprogramming.pdf
[3.2] https://www.cimt.org.uk/projects/mepres/alevel/discrete_ch5.pdf
[3.3] https://mat.tepper.cmu.edu/classes/QUANT/NOTES/chap11.pdf
[3.4] https://www.bbau.ac.in/dept/RM/TM/AssignmentProblem.pdf

References

1. Anonymous. (2015). The critical path: What it is and how to utilize it. Available from: https://www.imse.iastate.edu/files/2015/08/Critical-Path.pdf

2. Anonymous. (2022). The diet problem. Available from: https://neos-guide.org/case-studies/om/the-diet-problem/#:~:text=History,stillprovidingahealthydiet
3. Bachem, A., Kern, W., Bachem, A., & Kern, W. (1992). *Linear programming duality*. Springer.
4. Bixby, R. E. (2012). A brief history of linear and mixed-integer programming computation. *Documenta Mathematica*, 107–121.
5. Cakanyildirim, M. (2015). Linear programming formulation. Available from: https://personal.utdallas.edu/~metin/Or6302/Notes/formulation.pdf
6. Dantzig, G. (2016). *Linear programming and extensions*. Princeton University Press.
7. Dantzig, G. B. (1982). Reminiscences about the origins of linear programming. *Operations Research Letters, 1*(2), 43–48.
8. Dantzig, G. B. (2002). Linear programming. *Operations Research, 50*(1), 42–47.
9. Díaz-Parra, O., Ruiz-Vanoye, J. A., Bernábe Loranca, B., Fuentes-Penna, A., Barrera-Cámara, R. A., et al. (2014). A survey of transportation problems. *Journal of Applied Mathematics, 2014*. Available from: https://www.hindawi.com/journals/jam/2014/848129/
10. Dorfman, R. (1984). The discovery of linear programming. *Annals of the History of Computing, 6*(3), 283–295.
11. Ford, L. R., & Fulkerson, D. R. (1956). Maximal flow through a network. *Canadian Journal of Mathematics, 8*, 399–404. Available from: https://www.cs.yale.edu/homes/lans/readings/routing/ford-max_flow-1956.pdf
12. Frank, A. (2004). On Kuhn's hungarian method – a tribute from Hungary.
13. Garille, S. G., & Gass, S. I. (2001). Stigler's diet problem revisited. *Operations Research, 49*(1), 1–13. Available from: https://pubsonline.informs.org/doi/pdf/10.1287/opre.49.1.1.11187
14. Gazan, R., Brouzes, C. M. C., Vieux, F., Maillot, M., Lluch, A., & Darmon, N. (2018). Mathematical optimization to explore tomorrow's sustainable diets: A narrative review. *Advances in Nutrition, 9*(5), 602–616. Available from: https://www.ncbi.nlm.nih.gov/pmc/articles/PMC6140431/
15. Goldberg, A. V., & Tarjan, R. E. (2014). Efficient maximum flow algorithms. *Communications of the ACM, 57*(8), 82–89. Available from: http://doi.acm.org/10.1145/2628036.
16. Gover, D. (2002). An overview of Gantt, PERT and CPM charts. Available from: https://www.egr.msu.edu/classes/ece480/capstone/ganttv1.pdf
17. Hochbaum, D. S. (2020). Graph algorithms and network flows. Available from: https://hochbaum.ieor.berkeley.edu/files/266Notes-F2020.pdf
18. Jungnickel, D., & Jungnickel, D. (2013). *Graphs, networks and algorithms* (Vol. 3). Springer. Available from: https://link.springer.com/chapter/10.1007/978-3-642-32278-5_11
19. Kacher, Y., & Singh, P. (2021). A comprehensive literature review on transportation problems. *International Journal of Applied and Computational Mathematics, 7*, 1–49.
20. Kuhn, H. W. (1955). The hungarian method for the assignment problem. *Naval Research Logistics Quarterly, 2*, 83–97.
21. Luenberger, D. G., & Ye, Y. (2016). *Linear and nonlinear programming*.
22. Martin, J. (2021). The network simplex method. Available from: https://people.computing.clemson.edu/~jmarty/courses/commonCourseContent/common/OR-Intro/TextsAndMaterial/networkSimplexProblem.pdf
23. Newman, A. (2016). Linear programming duality. Available from: https://pagesperso.g-scop.grenoble-inp.fr/~newmana/OptApproxFall2016/OptApprox-lecture5.pdf
24. Parragh, S. N., Doerner, K. F., & Hartl, R. F. (2008). A survey on pickup and delivery problems: Part I: Transportation between customers and depot. *Journal für Betriebswirtschaft, 58*, 21–51.
25. Pentico, D. W. (2007). Assignment problems: A golden anniversary survey. *European Journal of Operational Research, 176*(2), 774–793. Available from: https://ideas.repec.org/a/eee/ejores/v176y2007i2p774-793.html
26. Schrage, L. (2002). *Optimization modelling with LINGO*. LINDO Systems Inc. Available from: https://www.lindo.com/downloads/Lingo_Textbook_5thEdition.pdf
27. Schrijver, A. (1998). *Theory of linear and integer programming*. John Wiley & Sons.

28. Taylor III, B. (2009). *Introduction to management science*. Prentice Hall. Available from: https://www.mbbcollege.in/db/notes/256.pdf
29. Terlaky, T., & Zhang, S. (1993). Pivot rules for linear programming: A survey on recent theoretical developments. *Annals of Operations Research, 46*(1), 203–233.
30. Tovey, C. A. (2020). *Linear optimization and duality: A modern exposition*. Chapman and Hall/CRC.
31. Vanderbei, R. J. (2007). Linear programming: Chapter 3, degeneracy. Available from: https://vanderbei.princeton.edu/542/lectures/lec3.pdf
32. Vanderbei, R. J., et al. (2015). *Linear programming*. Springer.
33. Williamson, D. P. (2014). ORIE6300, mathematical programming i. Available from: https://people.orie.cornell.edu/dpw/orie6300/Lectures/lec13.pdf
34. Winston, W. L., & Goldberg, J. B. (2004). *Operations research: Applications and algorithms* (Vol. 3). Thomson Brooks/Cole Belmont.

[illegible]

Chapter 4
Classic Analytical Methods

4.1 Introduction

This chapter is devoted to important classical methods that had been the cornerstone of optimization theory and practice for decades. Despite the roots being old, they are still quite relevant for many application fields, helping to gain insight into the nature of optimization problems.

While linear programming is restricted to linear problems in general, the methods to be introduced now are able to tackle **nonlinear problems**, with the nonlinearity due to the objective function, the constraints, or caused by both.

Indeed, the methods introduced in this chapter can also be used for linear problems.

The chapter is divided into three main sections: Lagrange multipliers, variational calculus, and numerical methods. The numerical section introduces some practical approaches that are important in physics and engineering.

Many books on optimization or operations research contain chapters related to the matters to be covered next. In particular, the following books can be recommended: [4] (6593 cites), which focuses on the Lagrange multiplier method, being complemented by [18], and [15] (5135 cites), more oriented to variations. A beautiful article on the history of the topic is [34].

4.2 Lagrange Multipliers

The method to be introduced in this section was devised by J.L. Lagrange in the late 1700s. A characteristic of the method is the use of certain constants as multipliers. Frequently, the method is called "*the method of multipliers*."

Figure 4.1 shows a portrait of Joseph-Louis de Lagrange (1736–1813). He made important contributions in mathematics, astronomy, and mechanics. A major

J. M. Giron-Sierra, *Introduction to Optimization with Matlab® Examples*,
https://doi.org/10.1007/978-3-032-05048-9_4

Fig. 4.1 Joseph-Louis de Lagrange

contribution was his book *Mécanique Analytique* (1788), much admired by W.R. Hamilton. The dynamics of a rigid system were represented with an elegant differential equation involving kinetic and potential energy. In order to solve this equation, the method of Lagrange multipliers was introduced. Later on, in 1804, in his treatise "*Leçons Sur le Calcul de Fonctions*," he expanded his methodology to variational calculus.

The first part of this section introduces the new concepts and describes the application of the Lagrange method for equality constraints; this is followed by a second part devoted to inequality constraints. Then, another part is devoted to duality, and a final subsection considers the use of symbolic MATLAB in this context.

4.2.1 Equality Constraints

A direct way for introducing the Lagrange method is to consider a very simple example. One wants to solve the following problem:

Maximize:

$$z = f(x, y) = y + x^2$$

Subject to (constraint):

$$g(x, y) = y - (60.63 + 40\cos(x/3)) = 0$$

Note that both the objective function and the constraint are *nonlinear functions*.

The problem can be graphically solved, as represented in Fig. 4.2. The curve on top is the constraint, and the other curves below correspond to increasing values of z. The solution is reached at the intersection of the constraint and the curve $z = 80$ (point 5.94, 44.71).

Figure 4.2 has been obtained with Program 4.1.

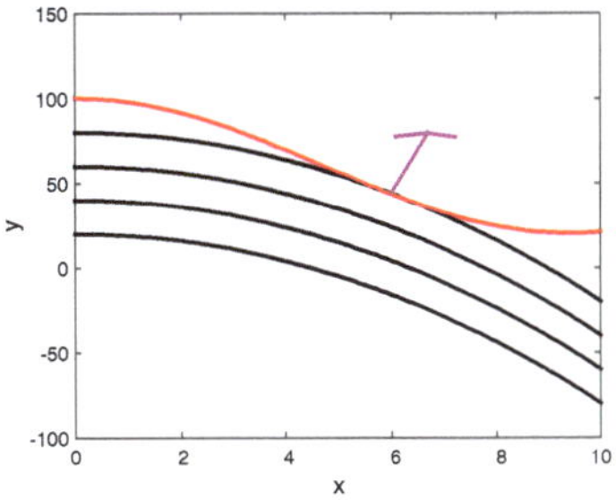

Fig. 4.2 Level curves of objective function (in black) and one constraint (dark curve on top)

Program 4.1 Level curves and a constraint

```
% level curves and one constraint
x=0:0.1:10;
y1=80-(x.^2);
y2=60-(x.^2);
y3=40-(x.^2);
y4=20-(x.^2);
yc=60.63+(40*(cos(x/3))); %constraint
figure(1) %----------------------------
plot(x,yc,'r'); hold on; %constraint
plot(x,y1,'k');
plot(x,y2,'k');
plot(x,y3,'k');
plot(x,y4,'k');
text(5,8,'\rightarrow','FontSize',50,'Rotation',...
65,'color','g');
xlabel('x'); ylabel('y');
```

What is interesting for our purpose is to realize that the gradients of g and z at the intersection point are parallel. Therefore

$$\nabla f = \lambda \, \nabla g$$

or

$$\nabla f - \lambda \, \nabla g = 0$$

where λ is a constant.

In case of several constraints, there would be intersections at which the gradient of the objective function can be expressed as a linear combination of gradients of the constraints, and so

$$\nabla f - \sum_j \lambda_j \, \nabla g_j = 0$$

Then, the intersection points can be determined with the following system of equations:

$$\nabla f(\vec{x}^*) - \sum_j \lambda_j \nabla g_j(\vec{x}^*) = 0$$

$$g_j(\vec{x}^*) = 0$$

with $j = 1, 2, \ldots, m$.

Now, consider the following function:

$$L(\vec{x}, \vec{\lambda}) = f(\vec{x}) - \sum_i \lambda_i g_i(\vec{x})$$

The derivatives of $L(\vec{x}, \vec{\lambda})$ with respect to $\vec{x}$ and λ_j will give the same system of equations for determining $\vec{x}^*$.

Based on this analysis, it is possible to formulate the following approach. Suppose a maximization problem with equality constraints. The objective function is a function of n variables. There are m constraints. The problem is

Maximize:

$$z = f(\vec{x}), \quad \vec{x} \in \Re^n$$

Subject to (constraints):

$$g_i(\vec{x}) = 0, \; i = 1, 2, \ldots, m$$

4.2.1.1 The Lagrange Function

The clever idea of Lagrange is to form an augmented function by adding the constraints to the objective function through the use of a set of multipliers. This augmented function becomes the target to be optimized. The new function is

$$L(\vec{x}, \vec{\lambda}) = f(\vec{x}) - \sum_i \lambda_i g_i(\vec{x})$$

Notice that the original constrained scenario has been transformed to an unconstrained problem, in which the use of derivatives makes sense. Also, note that for any candidate solution $\vec{x}^*$ satisfying the constraints, the summation term would be zero and so $L(\vec{x}^*, \vec{\lambda})$ and $f(\vec{x}^*)$ would become the same.

The question now is to optimize the augmented function. This can be done by solving the following system of equations:

$$\frac{\partial L}{\partial x_j} = \frac{\partial f}{\partial x_j} - \sum_i \lambda_i \frac{\partial g_i}{\partial x_j} = 0, \ j = 1, 2, \ldots, n$$

$$\frac{\partial L}{\partial \lambda_i} = -g_i(\vec{x}) = 0, \ i = 1, 2, \ldots, m$$

This system of equations provides a **necessary condition**.

The method can be applied for maximization as well as for minimization.

4.2.1.2 Example of Application (Distance)

The Lagrange method of multipliers is well suited for problems related to distances, areas, and other geometrical questions.

From time to time, news comes about a comet or another celestial object approaching our planet. We all know there are potential risks, and so it is important to estimate how far the predicted path of that object is from Earth's path in our Solar System. This is illustrated by Fig. 4.3.

Our example will be to determine the minimum distance from the following two curves on the plane x-y.

A parabola:

$$y = 1 + x^2$$

A circle:

$$(x-2)^2 + y^2 = 1$$

Fig. 4.3 Celestial object approaching us

Let us apply the Lagrange method. An augmented function is formed, and its minimum can be determined by equaling derivatives to zero:

$$\nabla L = \nabla\{(x_1 - x_2)^2 + (y_1 - y_2)^2 + \lambda_1(1 + x_1^2 - y_1) +$$

$$+ \lambda_2((x_2 - 2)^2 + y_2^2 - 1)\} = 0$$

Now, we explicit the parts embodied in the above equation:

$$\frac{\partial L}{\partial x_1} = 2(x_1 - x_2) + 2x_1\lambda_1 = 0$$

$$\frac{\partial L}{\partial x_2} = -2(x_1 - x_2) + 2\lambda_2(x_2 - 2) = 0$$

$$\frac{\partial L}{\partial y_1} = 2(y_1 - y_2) - \lambda_1 = 0$$

$$\frac{\partial L}{\partial y_2} = -2(y_1 - y_2) + 2\lambda_2 y_2 = 0$$

Combining these four equations with the equations of the parabola and the circle, it is possible to eliminate λ_1 , λ_2 , y_1 , y_2, obtaining finally

$$2x_1\left\{(1 + x_1^2) - \sqrt{1 - (x_2 - 2)^2}\right\} + (x_1 - x_2) = 0$$

$$(x_1 - x_2)\sqrt{1 - (x_2 - 2)^2} - (x_1 - x_2)\left\{(1 + x_1^2) - \sqrt{1 - (x_2 - 2)^2}\right\} = 0$$

Fortunately, computers exist, and we can devise programs to effortlessly handle such an equation system.

In our case, a MATLAB program (Program 4.2) has been developed, which systematically searches for a satisfactory approximation of the solution (in fact, the error committed is quite negligible).

Figure 4.4 shows the result, plotting in another color the segment that marks the minimum distance between the parabola and the circle.

Here is the listing of the MATLAB program.

Program 4.2 Distance between two curves

```
% distance between two 2D curves
% display of the curves
x=0:0.05:1.5;
y=1+x.^2; %parabola
plot(x,y,'r'); hold on;
x=1:0.05:3;
```

```
y=sqrt(1-(x-2).^2); %circle
plot(x,y,'m');
% search for approximate solution
sqer=10000; %initialize
for i=1:100,
  for j=1:100,
    x1=0.5+(i*0.01);
    x2=1+(j*0.01);
    aux1=1+(x1^2); aux2=x1-x2;
    auxq=sqrt(1-(x2-2)^2);
    e1=(2*x1*(aux1-auxq))+aux2;
    e2=(aux2*auxq)-(aux2*(aux1-auxq));
    M = (e1^2)+(e2^2);
    if (M < sqer)
      sqer=M; sx1=x1; sx2=x2;
    end
  end
end
sy1=1+(sx1^2);
sy2=sqrt(1-(sx2-2).^2);
plot([sx1 sx2],[sy1,sy2],'b'); %solution segment
xlabel('x'); ylabel('y');
distance = sqrt((sx1-sx2)^2 + (sy1-sy2)^2);
%print out
distance
disp('parabola point')
sx1
sy1
disp('circle point')
sx2
sy2
```

The numerical results obtained by the program were the following:

Distance = 0.9496

Point on parabola = (0.54, 1.2916)

Point on circle = (1.24, 0.6499)

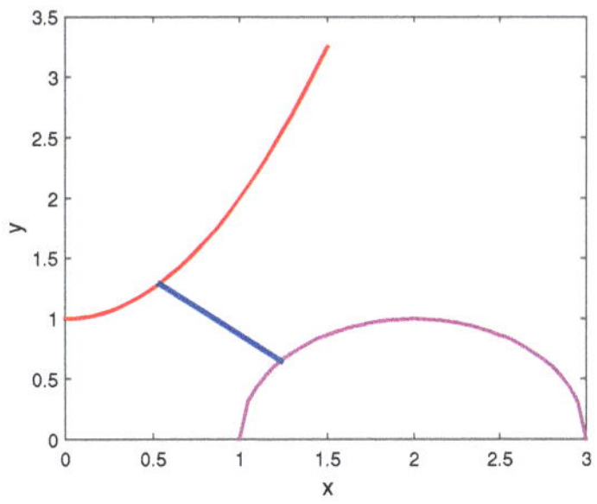

Fig. 4.4 Solution of the minimum distance problem

Fig. 4.5 Typical machinery for electricity generation

4.2.1.3 Another Example of Application (Electricity Generation)

This example confirms that the Lagrange method is of industrial interest and gives also a glimpse of practical mathematical treatment.

It is a matter of economic dispatch of energy. Three electricity generators serve a load of 952 Mw. The cost of generation (in $/hour) of the generators is

$$f_1 = x_1 + 0.0625\, x_1^2$$

$$f_2 = x_2 + 0.0125\, x_2^2$$

$$f_3 = x_3 + 0.0250\, x_3^2$$

where x_i is the output power of the *i-th* generator.

Figure 4.5 illustrates the nature of our present optimization problem.

It is desired to minimize the cost. Therefore, the problem is

Minimize:

$$J = f_1 + f_2 + f_3$$

Subject to:

$$G = x_1 + x_2 + x_3 - 952 = 0$$

Following the Lagrange method, one constructs the function $L = J - \lambda\, G$ and sets $\nabla L = 0$, which means

$$\begin{pmatrix} 0.125 & 0 & 0 & -1 \\ 0 & 0.025 & 0 & -1 \\ 0 & 0 & 0.05 & -1 \\ 1 & 1 & 1 & 0 \end{pmatrix} \begin{pmatrix} x_1 \\ x_2 \\ x_3 \\ \lambda \end{pmatrix} = \begin{pmatrix} -1 \\ -1 \\ -1 \\ 952 \end{pmatrix}$$

Solving this equation, one obtains

$$x_1 = 112\,Mw,\; x_2 = 500\,Mw,\; x_3 = 280\,Mw,\; \lambda = 15;\; J = 7616$$

4.2.2 *Inequality Constraints*

Consider the following problem, with inequality constraints:
Maximize:

$$z = f(\vec{x}) \,,\; \vec{x} \in \Re^n$$

Subject to:

$$g_i(\vec{x}) \le 0,\; i = 1, 2, \ldots, m$$

Like in the simplex algorithm, inequalities can be converted to equalities by using m positive slack variables as follows:

$$g_i(\vec{x}) + s_i^2 = 0,\; i = 1, 2, \ldots, m$$

Then, the augmented objective function would be

$$L\,(\vec{x}, \vec{\lambda}) = f(\vec{x}) - \sum_i \lambda_i\, g_i(\vec{x}) - \sum_i \lambda_i\, s_i^2$$

The constants λ_i must be nonnegative; otherwise, the term $\sum_i \lambda_i\, s_i^2$ could arbitrarily increase the value of $L()$, turning the maximization problem absurd.

Continuing with the method, one takes derivatives

$$\frac{\partial L}{\partial x_j} = \frac{\partial f}{\partial x_j} - \sum_i \lambda_i \frac{\partial g_i}{\partial x_j} = 0\,,\; j = 1, 2, \ldots, n$$

$$\frac{\partial L}{\partial y_i} = -2\,\lambda_i s_i = 0\,,\; i = 1, 2, \ldots, m$$

$$\frac{\partial L}{\partial \lambda_i} = -(g_i\,(\vec{x}) + s_i^2) = 0\,,\; i = 1, 2, \ldots, m$$

The KKT Optimality Conditions

With some analysis of these equations and considering the constraints, it is possible to write the following conditions:

$$\frac{\partial f}{\partial x_j} - \sum_i \lambda_i \frac{\partial g_i}{\partial x_j} = 0, \quad j = 1, 2, \ldots, n$$

$$\lambda_i g_i = 0, \quad i = 1, 2, \ldots, m$$

$$g_i(\vec{x}) \leq 0, \quad i = 1, 2, \ldots, m$$

$$\lambda_i \geq 0, \quad i = 1, 2, \ldots, m$$

This set of **necessary conditions** is called the "*KKT optimality conditions*."

This mysterious name corresponds to the following history: The conditions were published by Kuhn and Tucker in 1951, and later, it was found that Karush presented the conditions in his unpublished master's thesis in 1939.

Consider now the case of having a mix of equality and inequality constraints, as in the following problem:

Maximize:

$$z = f(\vec{x}), \quad \vec{x} \in \Re^n$$

Subject to:

$$h_i(\vec{x}) \leq 0, \; i = 1, 2, \ldots, m$$

$$l_j(\vec{x}) = 0, \; j = 1, 2, \ldots, r$$

The first step is to write the augmented objective function:

$$L(\vec{x}, \vec{u}, \vec{v}) = f(\vec{x}) - \sum_i u_i h_i(\vec{x}) - \sum_j v_j l_j(\vec{x})$$

Then, solve the problem (if possible) by applying the following KKT conditions:

$$\frac{\partial f}{\partial x_j} - \sum_i^m u_i \frac{\partial h_i}{\partial x_j} - \sum_i^r v_i \frac{\partial l_i}{\partial x_j} = 0, \quad j = 1, 2, \ldots, n$$

$$l_i(\vec{x}) = 0, \quad i = 1, 2, \ldots, r$$

$$h_i(\vec{x}) \leq 0, \quad i = 1, 2, \ldots, m$$

$$u_i h_i = 0, \quad i = 1, 2, \ldots, m$$

$$u_i \geq 0, \quad i = 1, 2, \ldots, m$$

Example with Inequality Constraints (Maximum Profit)

Here is an example with inequality constraints. One can buy as much as 26 oz of a certain chemical at a price of \$10 per ounce. It is possible to obtain two products A and B from that chemical. The next table shows the transforming costs and the intended prices for A and B.

Transformation cost/oz	Quantity (oz)	Intended price
3	x_1	$30 - x_1$
5	x_2	$50 - x_2$

Denote as x_3 the quantity of chemical to buy. The problem is to determine x_1, x_2, and x_3.

Let us translate the problem into equations. It would be desired to get maximum profit, therefore

Maximize:

$$z = x_1(30 - x_1) + x_2(50 - x_2) - 3x_1 - 5x_2 - 10x_3$$

Subject to:

$$x_1 + x_2 - x_3 \leq 0$$

$$x_3 \leq 26$$

$$x_1, x_2, x_3 \geq 0$$

The augmented function would be

$$L = x_1(30 - x_1) + x_2(50 - x_2) - 3x_1 - 5x_2 - 10x_3 - u_1(x_1 + x_2 - x_3) - u_2 x_3$$

Taking derivatives:

$$\frac{\partial L}{\partial x_1} = 30 - 2x_1 - 3 - u_1 = 0$$

$$\frac{\partial L}{\partial x_2} = 50 - 2x_2 - 5 - u_1 = 0$$

$$\frac{\partial L}{\partial x_3} = -10 + u_1 - u_2 = 0$$

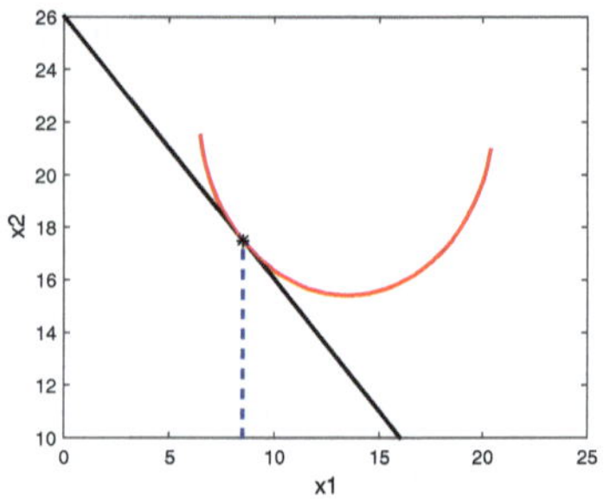

Fig. 4.6 Constraint (a line) and objective function (a curve)

The rest of the conditions is

$$u_i(x_1 + x_2 - x_3) = 0$$

$$u_2(x_3 - 26) = 0$$

$$u_1, u_2 \geq 0$$

Usually, after writing the conditions, it follows a discussion of alternatives. In our example, based on the fourth and fifth conditions, there are four cases:

1. $u_1 = 0, u_2 = 0$. This would give $-10 = 0$ in the third condition. Not a solution.
2. $u_1 = 0, x_3 = 26$. From the third condition, it would imply that $u_2 = -10$, against the last condition.
3. $x_1 + x_2 - x_3 = 0,\ u_2 = 0$. This would be a feasible solution.
4. $x_1 + x_2 - x_3 = 0,\ x_3 = 26$. This would be also a feasible solution.

If one chooses alternative 3, it leads to $u_1 = 10$ (third condition), $x_1 = 8.5$ (first condition), $x_2 = 17.5$ (second condition), and $x_3 = 26$. The corresponding value of z is 378.5.

It is easy to see that alternative 4 takes you to the same solution as alternative 3.

Figure 4.6 shows a representation of the constraint (black line) and the objective function (a dark curve), for a fixed value of $x_3 = 26$.

Figure 4.6 has been obtained with Program 4.3.

Program 4.3 Objective function and one constraint

```
% objective function and constraint
x3=26;
z=378.5;
N=140;
vx1=zeros(1,N); vx2=zeros(1,N);
for nn=1:N,
  x1=6.4+(0.1*nn);
  vx1(nn)=x1;
```

```
    aux=(27*x1)-(x1^2)-(z+260);
    x2=(45-sqrt(45^2+(4*aux)))/2;
    vx2(nn)=x2;
end;
figure(1)
%the constraint
plot([0 16],[26 10],'k'); hold on;
%the objective function for z=378.5
plot(vx1,vx2,'r');
%marking the solution
plot([8.5 8.5],[10 17.5],'b--');
plot(8.5, 17.5,'k*');
xlabel('x1'); ylabel('x2');
```

For more information on the multiplier method, the survey of Dimitri P. Bertsekas [3] would be recommended. In addition, the article [1] contains a deeper description of various aspects of the associated theory.

4.2.3 Duality

Duality has already appeared during the introduction of linear programming. It is an aspect of optimization that will also be mentioned in other moments of the book. In general, it can be basically described by two functions, A and B, which could be depicted as in Fig. 4.7.

If the problem was to minimize A, then the maximum of the dual B will act as a lower bound, so we have $\min(A) \geq \max(B)$.

If the problem was to maximize B, then the minimum of the dual A will act as an upper bound, so $\max(B) \leq \min(A)$.

In the particular context of Lagrange multipliers, suppose that the primal problem is

Minimize: $z = f(\vec{x})$
Subject to:

$$h_i(\vec{x}) \leq 0, \; i = 1, 2, \ldots, m$$

Fig. 4.7 Basic duality scenario

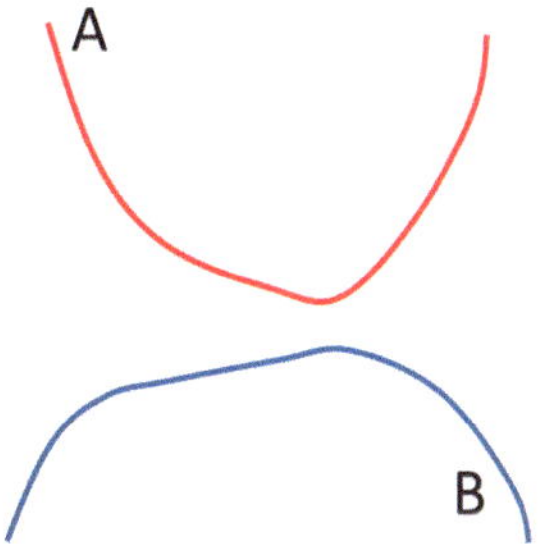

$$l_j(\vec{x}) = 0, \; j = 1, 2, \ldots, r$$

Once the Lagrange function $L(\vec{x}, \vec{u}, \vec{v})$ was built, it is possible to consider the following dual function:

$$q(\vec{u}, \vec{v}) = \min_{\vec{x}} L(\vec{x}, \vec{u}, \vec{v})$$

The dual problem will be
Maximize: $q(\vec{u}, \vec{v})$
Subject to: $\vec{u} \geq 0$
Denote f^* as the solution of the primal problem and q^* as the solution of the dual problem. Then, $q^* \leq f^*$. The difference between both solutions is called *duality gap*. If this gap is zero, one has *strong duality*. Otherwise, it is *weak duality*.

Similar considerations, *mutatis mutandis*, can be done in case the primal problem was a maximization.

The dual problem could be valuable for several purposes: to confirm a primal problem solution, for providing an easier problem to solve, etc.

Example of Dual Problem
Suppose the primal problem is
Minimize: $J = x_1^2 + x_2^2$
Subject to: $x_1 + x_2 \geq 4$
with nonnegative x_i
The inequality constraint can also be expressed as

$$g(\vec{x}) = -x_1 - x_2 + 4 \geq 0$$

Now, the dual function would be

$$q(\mu) = \min_x (J - \mu\, g(\vec{x}) = 4\,\mu + \min_x (x_1^2 + x_2^2 - \mu\, x_1 - \mu\, x_2)$$

That is

$$q(\mu) = 4\,\mu + \min_x (x_1^2 - \mu\, x_1) + \min_x (x_2^2 - \mu\, x_2)$$

For a given nonnegative value ofμ, the minimum is attained at

$$x_1 = x_2 = \frac{\mu}{2}$$

Hence

$$q(\mu) = 4\mu - \frac{\mu^2}{2}$$

(Note that $q(\mu)$ is strictly concave.)

The dual problem is

Maximize: $q(\vec{u}, \vec{v})$

Subject to: $\mu \geq 0$

From $\frac{d\,q(\mu)}{d\mu} = 0$, it is seen that the maximum is obtained at $\mu = 4$, which gives $x_1 = x_2 = 2$. Also, $J^* = q^* = 8$ (duality gap $= 0$).

Therefore, the problem has been solved in a simple way.

4.2.3.1 Using Symbolic MATLAB

Symbolic MATLAB is a good alternative for solving optimization problems with the Lagrange multiplier method.

Here is an example, based on the following link:

https://es.mathworks.com/matlabcentral/answers/1687174-finding-the-maxima-minima-using-lagrange-mutlipliers.

The problem to be solved is

Minimize: $J = \exp(x) - \sin(y)$

Subject to: $x^2 + 3\,y^2 = 1$ (an ellipse)

In this case, a most useful symbolic MATLAB function is *diff( )*. The first part of Program 4.4 is devoted to stating and solving the optimization problem via Lagrange multipliers. The second part generates the graphical representation presented in Fig. 4.8, where the curved surface corresponds to the objective function, and the constraint is drawn as a curve at the floor (the projection on the surface is painted in gray).

The result obtained by the program is that the minimization point is x = 0.689..., and y = 0.418.... This point is marked on the figure. The minimum value of J is 0.096.

Here is the listing of the program.

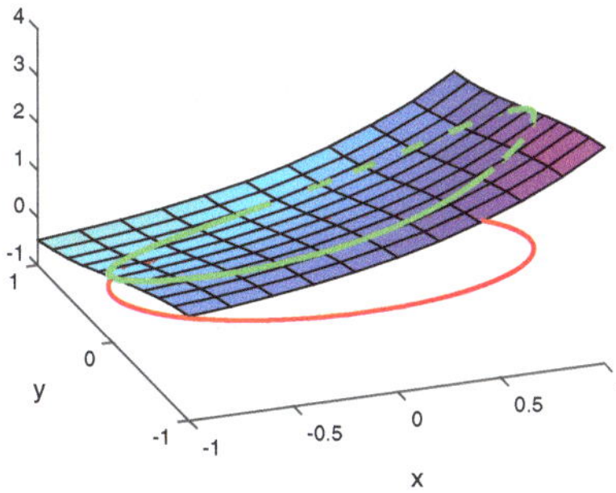

Fig. 4.8 The example being optimized

Program 4.4 Using symbolic MATLAB for Lagrange optimization

```
%example of Lagrange method in symbolic MATLAB
syms x y lambda
J = exp(x) - sin (y); %objective function
g = x^2 + 3*y^2 - 1; %equality constraint
%
L = J - lambda * g;
dL_dx = (diff(L,x) == 0);
dL_dy = (diff(L,y) == 0);
dL_dlambda = (diff(L,lambda) == 0);
system = [dL_dx; dL_dy; dL_dlambda];
[x_opt, y_opt, lambda_opt] = solve(system, ...
 [x y lambda], 'Real', true);
results_numeric = double([x_opt, y_opt,...
 lambda_opt]);
Jmin = (exp(x_opt) - sin(y_opt));
%
%display ---------------
figure(1)
m=-1:0.2:1;
[vx,vy]=meshgrid(m);
zJ=exp(vx) - sin (vy);
surf(vx,vy,zJ); hold on; % the surface
colormap('cool');
xlabel('x'); ylabel('y');
% the ellipse
q=sqrt(1/3);
y=-q:0.002:q;
x=sqrt(1-(3*y.^2));
z=0*y;
sz=exp(x)-sin(y);
plot3(x,y,z,'r'); plot3(x,y,sz,'g','LineWidth',3);
sz=exp(-x)-sin(y);
plot3(-x,y,z,'r'); plot3(-x,y,sz,'g','LineWidth',3);
% the minimum point
plot3(x_opt,y_opt,Jmin,'r*','LineWidth',4);
view(-20,35);
%numerical optimal result
Jmin
x_opt
y_opt
```

4.3 Calculus of Variations: Fundamentals

There are some optimization problems that, instead of looking for an optimal point, want to determine **an optimal function**. For instance, this function could be a suitable route for a car, in order to achieve less fuel consumption or for arriving

Fig. 4.9 Leonhard Euler

at the destination in the shortest time. This branch of study is frequently called "*dynamic optimization*".

Important fields of dynamic optimization application are orbital mechanics and space missions, civil and mechanical engineering, automatic control and robotics, management and planning, etc.

The calculus of variations provides important fundamentals for dynamic optimization. It has been studied by many famous scientists from long ago. In particular, a representative problem that will be included in this section is to determine the"*brachystochrone*" curve (more details later on). The problem was posed by Johann Bernoulli in 1696 as a challenge for intelligent people. Newton, Leibniz, L'Hôpital, and others, including Jakob Bernoulli (brother of Johann), responded with interesting contributions. Euler refined the approach of Jakob Bernoulli. In 1755, Lagrange sent to Euler a letter with an appendix introducing "*variations*." At once, Euler adhered to this approach, and renamed the subject as "*calculus of variations*" [16] and introduced the "*calculus of variations*." Lagrange continued with related work.

Figure 4.9 shows a portrait of Leonhard Euler (1707–1783). Born in Basel, Switzerland, he made decisive contributions in mathematics, physics, and observational astronomy. Although he was blind from 1766, this had almost no effect on his impressive scientific productivity.

One of the peculiar aspects of the theory to be introduced next is that the criterion to be optimized is usually an integral. Likewise, constraints can also come in the form of integrals.

This section starts with some preliminary concepts and then introduces the main result: the Euler-Lagrange equation. Then, the application of this equation to several types of problems is introduced, with examples. This is continued with more conditions (derived from second-order variations), and finally, cases with constraints will be considered.

A good, extensive reference, useful for this section, is [25].

4.3.1 Preliminaries

The calculus of variations works with "*functionals*." A functional is a mapping of functions to scalar numbers. Here is a basic example of functional:

$$J = J(y(x))$$

where the input to the functional $J()$ is the function $y(x)$ and the output J is a number.

A functional may also depend on x, or on function derivatives, for instance

$$J = J(x, y(x), y'(x))$$

where $y'(x)$ represents $dy(x)/dx$.

More generally, functionals may depend on higher derivatives and/or several independent variables $x_1, x_2, \ldots, x_n$.

A specific type of functional, called the basic one-dimensional functional, is the following:

$$J = \int_a^b F(x, y(x), y'(x))\, dx$$

with $y(a) = R$ and $y(b) = S$, which are boundary conditions.

Functionals are not strange creatures. On the contrary, there are examples of functionals that are quite familiar. One of these examples is the area under a curve:

$$A = \int_a^b y(x)\, dx$$

Another example is the length of a curve:

$$L = \int_a^b \sqrt{1 + (y'(x))^2}\, dx$$

The optimization problem is to find a function $y(x)$ so that J attains an optimal value (a maximum or minimum, depending on the particular problem).

Candidate solutions $y(x)$ should be single valued and have appropriate characteristics for the integral to exist. Such candidates are called admissible functions.

Figure 4.10 shows two examples of admissible functions.

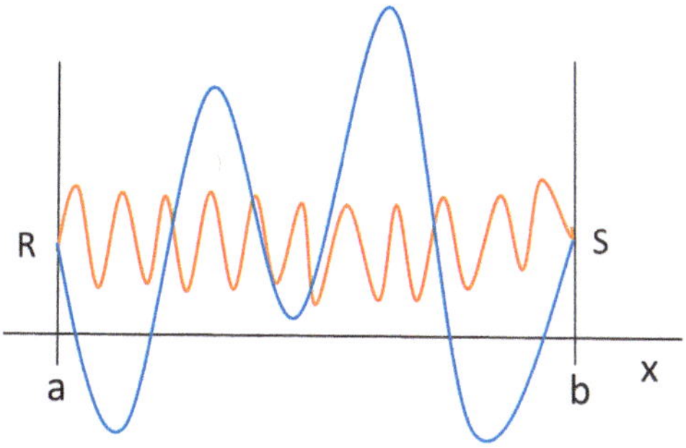

Fig. 4.10 Two examples of admissible functions

In order to check whether a given $y(x)$ could be a solution, the basic idea would be to add a perturbation $\delta\, y(x)$(a "*variation*)":

$$y(x) \; + \delta y(x)$$

and then to analyze the effect of the perturbation on the value of J.

Usually, the variation of $y(x)$ is taken to be

$$\delta\, y(x) \; = \varepsilon\, \eta(x)$$

where ε is an arbitrary parameter and $\eta(x)$ is an arbitrary continuously differentiable function.

The change of J due to the perturbation can be approximated by a Taylor expansion, as follows:

$$J[y + \delta y] \; \approx \; J[y] \, + \, \varepsilon\, \delta J[y] \, + \, \frac{1}{2}\, \varepsilon^2 \delta^2 J \, + \, \ldots$$

In which, the "*first variation*" of J is

$$\delta\, J[y] \; = \; \lim_{\varepsilon \to 0} \, \frac{1}{\varepsilon} (J[y + \delta y] \; - \; J[y])$$

The term $\delta^2\, J$ is the second variation of J and so on in the Taylor expansion. Notice the analogy between derivatives and variations.

Optimal values of J are called "*extrema*".

4.3.2 The Euler-Lagrange Equation

Let us consider the following problem: Find a function $x(t)$ such that the following criterion is minimized ($\dot{x}(t)$ represents $dx(t)/dt$):

$$J \; = \; \int_a^b \, F(x(t), \; \dot{x}(t), t)\, dt$$

where the function $x(t)$ must be such that $x(a) \, = \, x_1$, and $x(b) \, = \, x_2$.

Notice the slight change of notation, from $y(x)$ to $x(t)$, which does not modify the nature of the problem.

Suppose that, for some reason, we suspect that a certain function $x^*(t)$ is a solution of the optimization problem. Then, let us build the following function:

$$x^*(t) \;+\; \varepsilon\,\eta(t)$$

where $\eta(t)$ is an arbitrary continuously differentiable function with $\eta(a) = \eta(b) = 0$ and ε is an arbitrary parameter.

Then

$$J(\varepsilon) = \int_a^b F(x^* + \varepsilon\eta,\ \dot{x}^* + \varepsilon\dot{\eta}, t)\, dt$$

If $x^*(t)$ is really a solution, the derivative of $J(\varepsilon)$ with respect to ε would be zero for $\varepsilon = 0$:

$$\left.\frac{d\,J(\varepsilon)}{d\varepsilon}\right|_{\varepsilon=0} = 0$$

Now, let us calculate this derivative. To make things easier, denote

$$X(t) = x^*(t) \;+\; \varepsilon\,\eta(t)$$

$$\dot{X}(t) = \dot{x}^*(t) \;+\; \varepsilon\,\dot{\eta}(t)$$

Hence

$$\left.\frac{d\,J(\varepsilon)}{d\varepsilon}\right|_{\varepsilon=0} = \left[\int_a^b \left(\frac{\partial F}{\partial X}\frac{\partial X}{\partial \varepsilon} + \frac{\partial F}{\partial \dot{X}}\frac{\partial \dot{X}}{\partial \varepsilon}\right) dt\right]_{\varepsilon=0} =$$

$$= \int_a^b \left(\frac{\partial F}{\partial x^*}\eta + \frac{\partial F}{\partial \dot{x}^*}\dot{\eta}\right) dt = 0$$

Taking integration by parts

$$u = \frac{\partial F}{\partial \dot{x}^*}\;;\; v = \eta$$

$$\left.\frac{d\,J(\varepsilon)}{d\varepsilon}\right|_{\varepsilon=0} = \left.\frac{\partial F}{\partial \dot{x}^*}\eta\right|_a^b + \int_a^b \left(\frac{\partial F}{\partial x^*} - \frac{d}{dt}\left(\frac{\partial F}{\partial \dot{x}^*}\right)\right)\cdot \eta\, dt = 0$$

Since $\eta(a) = \eta(b) = 0$ and $\eta(t)$is arbitrary, then

$$\frac{\partial F}{\partial x*} - \frac{d}{dt}\left(\frac{\partial F}{\partial \dot{x}^*}\right) = 0$$

This is called **the Euler-Lagrange equation**. It is a **necessary condition** for $x*(t)$to be a solution of the variational problem.

An alternative, equivalent form of the equation is the following:

$$\frac{\partial F}{\partial t} - \frac{d}{dt}\left(F - \dot{x}^* \frac{\partial F}{\partial \dot{x}^*}\right) = 0$$

4.3.2.1 Calculation Examples

Two simple examples will be worked on next. The first one asks for the minimization of

$$J(x) = \int_0^1 \left[x^2(t) + \dot{x}^2(t)\right] dt$$

with $x(0) = 0$ and $x(1) = 1$.
Then, $F(x, \dot{x}, t) = x^2 + \dot{x}^2(t)$.

Now, proceed with the derivatives:

$$\frac{\partial F}{\partial x} = 2x(t)$$

$$\frac{\partial F}{\partial \dot{x}} = 2\dot{x}(t) \rightarrow \frac{d}{dt}\left(\frac{\partial F}{\partial \dot{x}}\right) = 2\ddot{x}(t)$$

Apply the Euler-Lagrange equation:

$$\frac{\partial F}{\partial x} - \frac{d}{dt}\left(\frac{\partial F}{\partial \dot{x}}\right) = 0 = 2x(t) - 2\ddot{x}(t)$$

The above differential equation has the solution:

$$x(t) = c_1 \exp(t) + c_2 \exp(-t)$$

The condition $x(0) = 0$ implies that $c_2 = -c_1$.
And the condition $x(1) = 1$ implies that

$$c_1 \exp(1) + c_2 \exp(-1) = 1 \rightarrow c_1 = \frac{e}{e^2 - 1}.$$

Finally, the solution is

$$x(t) = \frac{e}{e^2 - 1}(2 \sinh(t))$$

The second example considers the minimization of

$$J(x) = \int_a^b \frac{y'^2}{x^3} dx$$

(notation: $y'(x)$ means $\frac{dy(x)}{dx}$)
Then, $F(y, y', x) = \frac{y'^2}{x^3}$.
And the derivatives are

$$\frac{\partial F}{\partial y} = 0, \quad \frac{\partial F}{\partial y'} = \frac{2}{x^3} y', \quad \frac{d}{dx}\left(\frac{\partial F}{\partial y'}\right) = \frac{d}{dx}\left(\frac{2}{x^3} y'\right)$$

Using the Euler-Laplace equation, one obtains

$$\frac{d^2 y(x)}{dx^2} = \frac{3}{x} y'$$

Therefore, the solution is $y(x) = c_1 x^4 + c_2$

4.3.2.2 Using Symbolic MATLAB

Recent versions of MATLAB are offering more powerful symbolic support, well suited for the Euler-Lagrange approach.

For instance, Program 4.5 applies symbolic MATLAB to solve the first of the calculation examples just given.

Program 4.5 Using symbolic MATLAB for Euler-Lagrange variational calculus

```
%example of Euler-Lagrange equation
%in symbolic MATLAB
syms x(t)
F = x(t)^2 + (diff(x(t),t)^2); % F inside integral
%
D1 = diff(F,x(t));
Dp = diff(F,diff(x(t),t));
D2 = diff(Dp,t);
%
ode = D1-D2 == 0;
cond1 = x(0) == 0;
cond2 = x(1) == 1;
conds = [cond1 cond2];
xsol = dsolve(ode, conds);
xsol
```

It is left to the reader to apply the program, with pertinent changes, to the other calculation example.

The MathWorks File Exchange site includes some MATLAB packages related to Euler-Lagrange applications. Like the "*Euler-Lagrange tool package*."

4.3.2.3 First Integrals, Three Special Cases

There are three cases in which the functional does not depend on one of the variables. Then, at most, one integration is required. Let us briefly describe those cases.

(a) The functional does not depend on t.
A first integral of the alternative expression of the Euler-Lagrange equation can be easily obtained, and then

$$\left(F - \frac{\partial F}{\partial \dot{x}^*} \dot{x}^* \right) = \text{constant}$$

Later on, we will see two important examples, which belong to this particular case (called the "*Bertrami identity*").

(b) The functional does not depend on x.
The Euler-Lagrange equation is in this case

$$\frac{d}{dt} \left(\frac{\partial F}{\partial \dot{x}^*} \right) = 0$$

The first integral is

$$\frac{\partial F}{\partial \dot{x}^*} = \text{constant}$$

(c) The functional does not depend on $\dot{x}$.
Obviously, in this case

$$\frac{\partial F}{\partial x^*} = 0$$

4.3.2.4 More General Scenarios

There are problems with several variables, like the following:

$$J = \int_a^b F(x_1, x_2, \ldots, x_n; \dot{x}_1, \dot{x}_2, \ldots, \dot{x}_n, t)\, dt$$

By using several $\eta_i(t)$ and a parameter ε and based on reasoning similar to the basic scenario, the following necessary condition is found:

$$\frac{\partial F}{\partial x_i^*} - \frac{d}{dt}\left(\frac{\partial F}{\partial \dot{x}_i^*}\right) = 0\,, \quad i = 1,\, 2,\, \ldots, n$$

which is the Euler-Lagrange equation for this type of problem.

Another type of problem corresponds to combinations of independent and dependent variables. For example

$$F(x, y, z(x, y), \frac{\partial z}{\partial x}, \frac{\partial z}{\partial y})$$

In this case, one considers

$$Z(x, y) = z^*(x, y) \; + \; \varepsilon\, \eta(x, y)$$

and integrating by parts for computing the derivative of J with respect to ε, one obtains the condition:

$$\frac{\partial F}{\partial z} - \frac{\partial}{\partial x}\frac{\partial F}{\partial\left(\frac{\partial z}{\partial x}\right)} - \frac{\partial}{\partial y}\frac{\partial F}{\partial\left(\frac{\partial z}{\partial y}\right)} = 0$$

which is the equation of Ostrogradsky.

When there are n independent variables, the equation above is easily extended:

$$\frac{\partial F}{\partial z} - \frac{\partial}{\partial x_1}\frac{\partial F}{\partial\left(\frac{\partial z}{\partial x_1}\right)} - \frac{\partial}{\partial x_2}\frac{\partial F}{\partial\left(\frac{\partial z}{\partial x_2}\right)} - \ldots - \frac{\partial}{\partial x_n}\frac{\partial F}{\partial\left(\frac{\partial z}{\partial x_n}\right)} = 0$$

Likewise, in cases like the following

$$F\left(x, y, z(x, y), \frac{\partial z}{\partial x}, \frac{\partial z}{\partial y}, \frac{\partial^2 z}{\partial x^2}, , \frac{\partial z^2}{\partial x \partial y}, \frac{\partial z^2}{\partial y^2}\right)$$

The equation would be

$$\frac{\partial F}{\partial z} - \frac{\partial}{\partial x}\frac{\partial F}{\partial\left(\frac{\partial z}{\partial x}\right)} - \frac{\partial}{\partial y}\frac{\partial F}{\partial\left(\frac{\partial z}{\partial y}\right)} +$$

$$+ \frac{\partial^2}{\partial x^2}\frac{\partial F}{\partial\left(\frac{\partial^2 z}{\partial x^2}\right)} + \frac{\partial^2}{\partial x \partial y}\frac{\partial F}{\partial\left(\frac{\partial^2 z}{\partial x \partial y}\right)} + \frac{\partial^2}{\partial y^2}\frac{\partial F}{\partial\left(\frac{\partial^2 z}{\partial y^2}\right)} = 0$$

4.3.3 Some Geometrical Examples

The calculus of variations has been applied for a good number of optimization problems with a geometrical nature. Some of the answers obtained are now the basis of structural and constructive solutions in engineering.

The article [10] offers a historical glimpse into the topics covered by variational investigators.

4.3.3.1 The Brachistochrone Curve

This is a famous problem introduced by J. Bernoulli in 1696 and stated in the following terms: "*given two points A and B in a vertical plane, find a curve AMB such that a body gliding on it under gravity, starting from A, arrives after the shortest time to B.*"

Figure 4.11 depicts the geometry of this problem.

The motion velocity would be

$$\frac{ds}{dt} = \sqrt{2g \cdot y}$$

where ds denotes a small arc length.

With such velocity, the time for moving from A to B along a curve $y(x)$ can be calculated as follows:

$$t(y(x)) = \frac{1}{\sqrt{2g}} \int_0^{x_B} \frac{\sqrt{1 + y'^2}}{\sqrt{y}} \, dx$$

Then, the time minimization problem corresponds to a functional that does not depend on x, and so

$$F(y, y') - \frac{\partial F}{\partial y'} y' = c$$

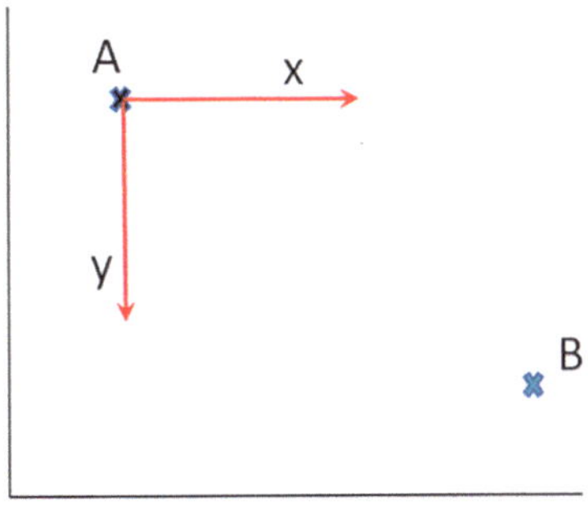

Fig. 4.11 Scenario for the brachystochrone problem

Applying this in our problem

$$\frac{\sqrt{1+y'^2}}{\sqrt{y}} - \frac{y'^2}{\sqrt{y(1+y'^2)}} = c$$

Simplifying

$$\frac{1}{\sqrt{y(1+y'^2)}} = c$$

which can be written as

$$y\,(1+y'^2) = K_1$$

Consider the following change of variable:

$$y' = ctg(t)$$

That leads to

$$y = \frac{K_1}{1+ctg^2(t)} = K\,\sin^2(t) = \frac{K_1}{2}(1 - \cos(2t))$$

and

$$dx = \frac{dy}{y'} = \frac{2\,K_1\,\sin(t)\,\cos(t)\,dt}{ctg(t)} = 2K_1\,\sin^2(t)\,dt = K_1(1 - \cos(2t))dt$$

Then

$$x = K_1\left(t - \frac{\sin(2t)}{2}\right) + K_2 = \frac{K_1}{2}(2t - \sin(2t)) + K_2$$

Therefore, the solution is the following curve:

$$\begin{cases} x = \frac{K_1}{2}(2t - \sin(2t)) + K_2 \\ y = \frac{K_1}{2}(1 - \cos(2t)) \end{cases}$$

This is a cycloid curve: the curve traced by a point on the rim of a wheel as it rolls. Figure 4.12 shows an example of a brachystochrone curve from a higher point to a lower one.

Figure 4.12 has been obtained with Program 4.6.

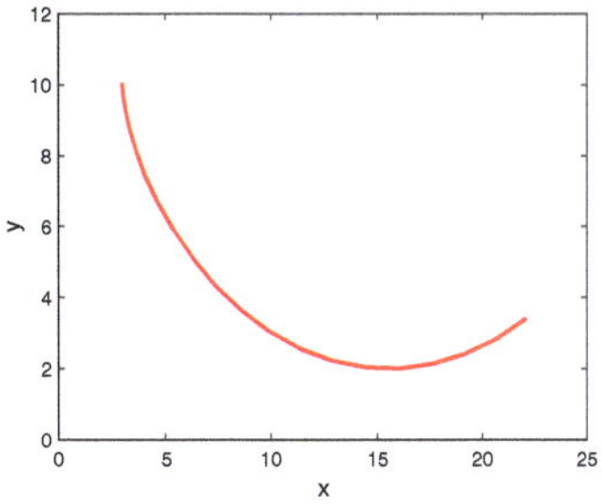

Fig. 4.12 Example of brachystochrone curve

Fig. 4.13 Example of cycloid

Program 4.6 Brachistochrone curve

```
% Brachistochrone curve
X0=1; Y0=10; %origin
t=0:0.1:2;
K1=8; K2=2;
y=(K1/2)*(1-cos(2*t));
x=(K1/2)*((2*t)-sin(2*t))+K2;
% display
plot([0 25],[0 0],'k'); hold on;
plot([0 0],[0 12],'k');
plot(X0+x,Y0-y,'r');
xlabel('x'); ylabel('y');
```

Cycloid curves are used in the design of roller coasters, ice toboggans, etc. Figure 4.13 shows a well-known example.

Using Symbolic MATLAB

In the type of problems and approaches we are treating now, it is usually convenient to employ symbolic MATLAB. Just for introducing basic aspects of how to deal with such problems, Program 4.7 obtains the differential equation corresponding to the Brachystochrone problem.

Program 4.7 Using the functionalDerivative symbolic function

```
% Finding the brachystochrone differential equation
% via symbolic functionalDerivative function
syms g y(x)
assume(g,'positive')
f = sqrt((1 + diff(y)^2)/(2*g*y));
eqn = functionalDerivative(f,y) == 0;
eqn = simplify(eqn)
```

From 2015, symbolic MATLAB offers the *functionalDerivative( )* function, which simplifies the analysis of variational problems. Program 4.7 is directly based on the documentation on this function, which includes a short program that finds rapidly the Brachistochrone differential equation.

4.3.3.2 Minimum Surface Area of Revolution

Given two separated points A and B, the problem is to find the curve $y(x)$ from A to B that generates the surface of minimum area when rotated about the x-axis. Let us say in advance that the solution is a catenary curve.

A typical application of the minimum surface area of revolution is the design of water condensation towers, like the pair shown in Fig. 4.14. Notice that in this case, the catenary rotates about the y-axis.

The area of revolution to be minimized is given by

$$J = 2\pi \int_a^b y\sqrt{1 + y'^2}dx$$

Considering again the necessary condition when the functional does not depend on x

$$F(y, y') - \frac{\partial F}{\partial y'} y' = c$$

Fig. 4.14 Example of engineering application

The following expression can be written:

$$y\sqrt{1+y'^2} - \frac{y \cdot y'^2}{\sqrt{1+y'^2}} = K_1$$

Simplifying

$$\frac{y}{\sqrt{1+y'^2}} = K_1$$

Now, with the following change of variables

$$y' = K_1 \sinh(t)$$

One obtains

$$y = K_1 \cosh(t)$$

and

$$y' = \frac{dy}{dx} \rightarrow dx = \frac{dy}{y'} = \frac{K_1 \sinh(t) \cdot dt}{\sinh(t)} = K_1 dt$$

In consequence

$$x = K_1 t + K_2$$

Therefore, the solution is

$$\begin{cases} x = K_1 t + K_2 \\ y = K_1 \cosh(t) \end{cases}$$

or by elimination of the parameter t

$$y = K_1 \cosh\left(\frac{x - K_2}{K_1}\right)$$

This is a catenary curve, having the shape of a chain or a rope hanging from two points. A scenic example is suspended bridges, like those for pedestrians in the jungle, or the more sophisticated versions for traffic, like the one shown in Fig. 4.15.

Figure 4.16 shows a catenary curve, generated according to the equations just obtained. The figure has been obtained with Program 4.8.

Fig. 4.15 A suspended bridge exhibiting catenary shapes

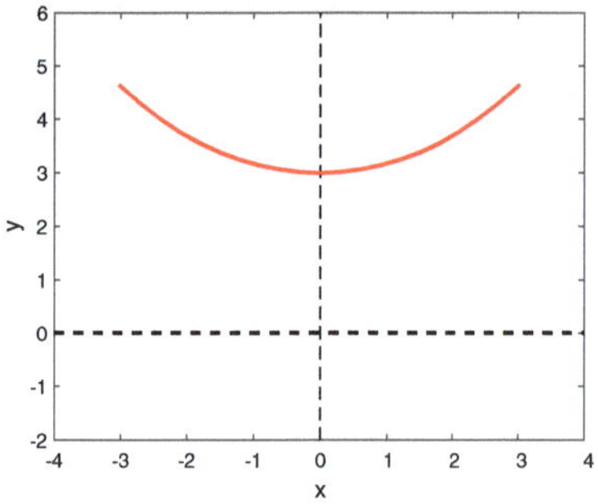

Fig. 4.16 Example of catenary curve

Program 4.8 Catenary curve

```
% Catenary curve
t=-1:0.1:1;
K1=3; K2=0;
x=(K1*t)+K2;
y=K1*cosh(t);
% display
plot([-4 4],[0 0],'k--'); hold on;
plot([0 0],[-2 6],'k--');
plot(x,y,'k');
xlabel('x'); ylabel('y');
```

Program 4.9 gives an implementation example in symbolic MATLAB, using the *functionalDerivative( )* function. The product of the program is the differential equation giving the solution.

Program 4.9 Minimum surface area of revolution

```
% Finding the minimum surface area of revolution
% via symbolic functionalDerivative function
syms y(x)
f=(2*pi*y*sqrt(1+diff(y)^2));
eqn = functionalDerivative(f,y) == 0;
eqn = simplify(eqn)
```

There is a MATLAB program called *plotCatenary.m* available from the Internet with several valuable options.

4.3.4 Applications in Physics

The Euler-Lagrange approach, together with the contributions of Sir W.R. Hamilton, is of fundamental importance in physics, especially in mechanics.

Consider the motion of a particle. Denote T as kinetic energy and U as potential energy. A function L, called *Lagrangian* by mechanical scientists, is formed in the following terms:

$$L = T - U$$

Now, **action** is defined as

$$S = \int_{t_0}^{t_1} L\,dt$$

Hamilton's principle states that the motion of the particle is such that the action is **stationary** with respect to variations in the path. That is

$$\frac{\delta S}{\delta \vec{x}(t)} = 0$$

This requirement can be satisfied if and only if

$$\frac{\partial L}{\partial \vec{x}} - \frac{d}{dt}\left(\frac{\partial L}{\partial \dot{\vec{x}}}\right) = 0$$

It can be shown that Newton's motion laws can be derived from this equation (which is the Euler-Lagrange equation).

4.3.4.1 Example of Mass-Spring System

A simple example for illustrating the use of the Euler-Lagrange equation in mechanics is the motion of the mass-spring system.

Figure 4.17 depicts, in a scheme, a mass-spring system.

The kinetic and potential energies are the following:

$$T = \frac{1}{2} m \left(\frac{dx}{dt}\right)^2$$

Fig. 4.17 Schematics of a mass-spring system

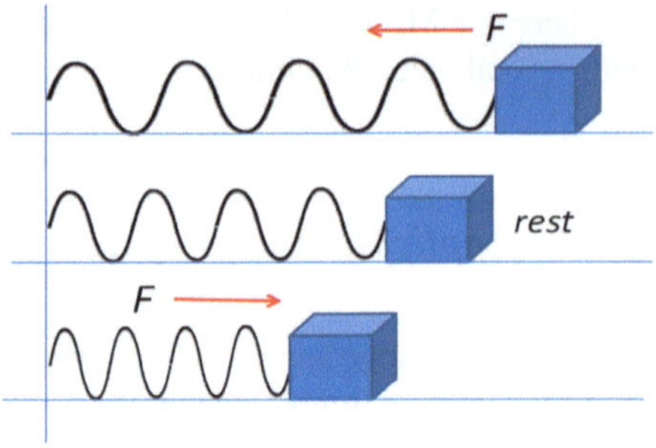

$$U = \frac{1}{2} k x$$

Program 4.10 has been mounted from the symbolic MATLAB Documentation page concerning *functionalDerivative( )*. The program obtains the oscillatory solution of the system.

Program 4.10 Mass-spring system variational study

```
% Euler-Lagrange equation example
% simple mass-spring system
syms m k x(t)
T = 1/2*m*diff(x,t)^2;
V = 1/2*k*x^2;
L = T - V
eqn = functionalDerivative(L,x) == 0
%
assume(m,'positive')
assume(k,'positive')
Dx(t) = diff(x(t),t);
xSol = dsolve(eqn,[x(0) == 10, Dx(0) == 0])
% solution print out
xSol
```

4.3.4.2 Example of a Pendulum

The pendulum oscillation is another important problem in mechanics, which can be treated with the Euler-Lagrange approach.

Figure 4.18 depicts a schematic view of a pendulum.

The kinetic and potential energies are the following:

$$T = \frac{1}{2} m l^2 \left(\frac{d\varphi}{dt} \right)^2$$

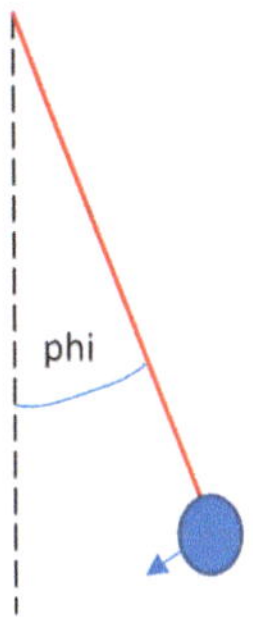

Fig. 4.18 Schematics of a pendulum

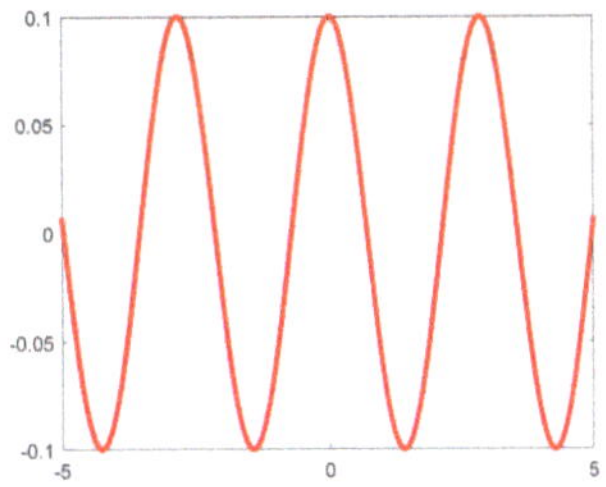

Fig. 4.19 Oscillations of the pendulum

$$U = mgl \sin(\varphi)$$

From the Euler-Lagrange equation, one obtains the equation of motion:

$$\frac{d^2\varphi}{dt^2} + \frac{g}{l} \sin(\varphi) = 0$$

For small oscillations, the approximation $\sin(\varphi) \approx \varphi$ can be used:

$$\frac{d^2\varphi}{dt^2} + \frac{g}{l}\varphi = 0$$

A program in symbolic MATLAB has been devised for the study of small oscillations of the pendulum. Figure 4.19 visualizes the result.

The first part of Program 4.11 is quite similar to the previous program. A few lines have been added at the end for plotting the result.

Program 4.11 Pendulum small osicllation variational study

```
% Euler-Lagrange equation example
% pendulum
syms m g l phi(t)
T = 1/2*m*(l^2)*diff(phi,t)^2;
V = m*g*l*(1-cos(phi));
L = T - V
eqn = functionalDerivative(L,phi) == 0
```

```
eqlin = subs(eqn,sin(phi(t)),phi(t))
assume(g,'positive')
assume(l,'positive')
Dphi(t) = diff(phi(t),t);
xSol = dsolve(eqlin,[phi(0) == 0.1, Dphi(0) == 0])
%plot the solution
g=9.8; l=2;
y=subs(xSol)
fplot(y);
```

4.3.4.3 Hamilton's Formulation

The canonical format used in mechanics is based on Hamilton's formulation, which translates the Euler-Lagrange equation to $2n$ first-order equations. This is done by the following procedure:

Depart from

$$\frac{\partial L}{\partial x_k} - \frac{d}{dt}\frac{\partial L}{\partial \dot{x}_k} = 0\ ,\quad k = 1,\ 2, \ldots n$$

Take the following change of variables

$$q_k = \frac{\partial L}{\partial \dot{x}_k},\quad k = 1,\ 2, \ldots n$$

As a result of this change, the system of equations can be written as

$$\frac{dq_k}{dt} = \frac{\partial L}{\partial x_k},\quad k = 1,\ 2, \ldots n$$

If we solve the following equations

$$\frac{\partial L}{\partial \dot{x}_k} = q_k,\quad k = 1,\ 2, \ldots n$$

with respect to $\dot{x}_k$, we obtain

$$\dot{x}_k = \omega_k\left(t, \vec{x}, \vec{q}\right)$$

Therefore, we have the following equivalent system of equations:

$$\begin{cases} \dot{x}_k = \omega_k\left(t, \vec{x}, \vec{q}\right) \\ \frac{dq_k}{dt} = \left\{\frac{\partial L}{\partial x_k}\right\} \end{cases}$$

(The braces indicate that $\dot{x}_k$ has been substituted with ω_k.)

Now, the **Hamiltonian *H*** is defined as

$$H(t, \vec{x}, \vec{q}) = \sum_k \omega_k q_k - \{L\}$$

(The total energy is $H = T + U$.)

Then, the system can be expressed as

$$\begin{cases} \dfrac{\partial x_k}{\partial t} = \dfrac{\partial H}{\partial q_k} \\ \dfrac{dq_k}{dt} = -\dfrac{\partial H}{\partial x_k} \end{cases}$$

which is the canonical system.

Some words about Sir William R. Hamilton (1805–1865), professor of astronomy at Trinity College, Dublin. The canonical formulation just described above was fundamental for further advances in physics, including quantum mechanics or the Hamiltonian formulation of general relativity. He also invented *quaternions*, so important for rotating systems (like satellites, for example).

4.3.4.4 Keplerian Orbits

This is, for example, the case of our planet orbiting around the Sun. Suppose two masses, M and m, with $M >> m$. Let us put the origin of coordinates in M. There is a gravitational attraction between both masses.

The kinetic and potential energies are the following:

$$T = \frac{1}{2} m \left(\dot{r}^2 + r^2 \dot{\theta}^2 \right)$$

$$U = \frac{-G\,M\,m}{r}$$

The Hamiltonian is

$$H = T + U = \frac{1}{2} m \left(\dot{r}^2 + r^2 \dot{\theta}^2 \right) \frac{-G\,M\,m}{r} = C$$

with C a constant.

Equivalently

$$H = \frac{(m\,\dot{r})^2}{2m} + \frac{(mr^2\dot{\theta})^2}{2mr^2} - \frac{G\,M\,m}{r} = C$$

which can be rewritten as

$$\frac{p^2}{2m} + \frac{q^2}{2mr^2} - \frac{\gamma}{r} = E$$

where p is the linear momentum and q is the angular momentum.

Let us focus on the dynamics of r. From the equations above, one has

$$\dot{r}^2 = \frac{2\,E}{m} + \frac{2\gamma}{mr} - \frac{q^2}{(mr)^2}$$

A convenient expression of $\dot{r}$ is found as follows:

$$\dot{r} = \frac{dr}{dt} = \frac{dr}{d\theta}\frac{d\theta}{dt} = \frac{dr}{d\theta}\left(\frac{q}{m\,r^2}\right) = \frac{q}{m}\frac{d(1/r)}{d\theta}$$

Then, after some algebra (see [25])

$$r = \frac{\alpha}{1 + \varepsilon\cos\theta}$$

with

$$\alpha = \frac{q^2}{\gamma\,m} \quad , \quad \varepsilon = \sqrt{\frac{2\,E}{\gamma^2 m} + \left(\frac{1}{q}\right)^2}$$

Figure 4.20 depicts a typical solution of the problem for the Sun-Earth system, which is an elliptical orbit.

There is a MATLAB package, called *Keplerian Orbit*, available from the MathWorks File Exchange site. Likewise, there are MATLAB functions for handling

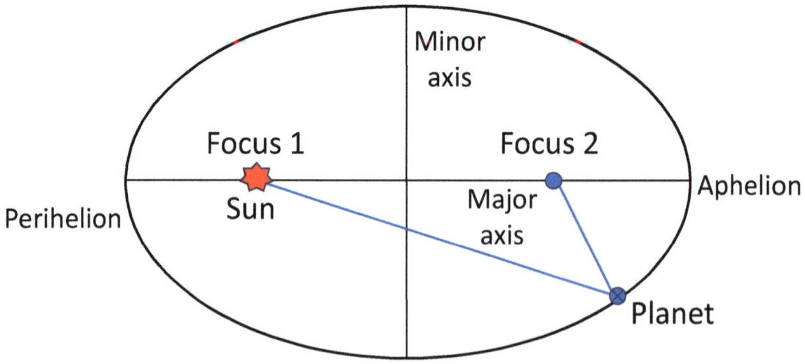

Fig. 4.20 Keplerian orbit

aspects of Kepler orbits, for instance, *keplerian2ijk( )*, *ijk2keplerian( )*, and *polarMotion( )*.

In addition, MATLAB has a GUI, called *SAT-LAB*, for simulating Keplerian orbits.

For those more oriented to professional study of orbits, File Exchange has the package *Two Body Orbit Propagation*.

GitHub offers 27 repositories of MATLAB packages related to orbital mechanics.

4.3.4.5 Other Applications in Physics and Engineering

There is an interesting set of programs for the study of diverse topics of physics, like Keplerian orbits, magnetism, waves, and quantum mechanics. These programs, produced at the Tuebingen University, are written in symbolic MATLAB and are available from the MathWorks File Exchange site under the title "*Physics Course with Symbolic Math Toolbox and Live Editor*."

Examples and introductory comments related to the application of the calculus of variations in different fields of physics and engineering can be found in

- Fermat's principle on light propagations [7, 36]
- Heat equation [24]
- Maxwell's equations [36]
- Elasticity [27]
- Quantum mechanics Schrödinger [20]
- Einstein's relativity [19]

The MathWorks site for File Exchange (MATLAB programs) has the following examples of interest:

- Simple heat equation solver
- Simulation, wave-propagation-Maxwell's differential equation
- Beam bending and deflection

(Other examples of interest are given in the *Partial Differential Equation MATLAB Toolbox*.).

In particular, the variational approach has found a prolific field of application in robotic arms (kinematic chains). A nice presentation of fundamentals of this topic is given in https://youngmok.com/lagrange-equation-by-matlab-with-examples/.

GitHub has four repositories on "*lagrange-dynamics*." Figure 4.21 reproduces one of the examples considered in the GitHub entry "*robot-manipulator-dynamics*" by G. Sensolini.

The MathWorks File Exchange site includes several packages related with robotic arms and pendula, like the "*Matlab Euler-Lagrange Library*," the "*Euler-Lagrange Solver*" (symbolic MATLAB), or the "*Dynamics Simulator for Kinematic Chains*".

The page https://pab47.github.io/robotics.html gives a table of links to robotics in MATLAB.

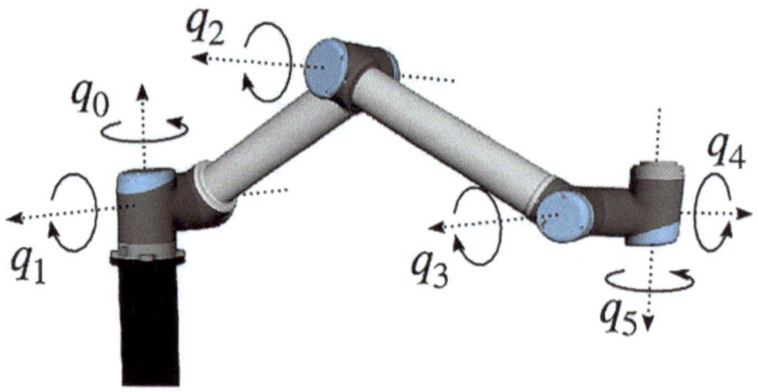

Fig. 4.21 A kinematic chain (robot arm) studied with Euler-Lagrange

4.3.5 More Conditions (Second Variation)

As it is well known, the standard way for an unconstrained single variable function optimization is to obtain candidate solutions from equaling the first derivative to zero and then to analyze the sign of the second derivative. Something similar occurs with variations, the Euler-Lagrange equation comes from equaling the first variation to zero, and then, further analysis should proceed with the second (perhaps third, etc.) variation.

To be more specific, consider the second variation:

$$\delta^2 J = \frac{d^2}{d\varepsilon^2}\,(J(x+\varepsilon\,\eta)|_{\varepsilon=0} = \int_a^b \frac{\partial^2}{\partial\varepsilon^2} F(x+\varepsilon\eta,\ \dot{x}+\varepsilon\dot{\eta},t)\,dt$$

The integrand can be expressed as follows:

$$\frac{\partial^2}{\partial\varepsilon^2} F(x+\varepsilon\eta,\ \dot{x}+\varepsilon\dot{\eta},t) = \frac{\partial}{\partial\varepsilon}(F_x\eta + F_{\dot{x}}\dot{\eta}) = F_{xx}\eta^2 + 2\,F_{x\dot{x}}\eta\dot{\eta} + F_{\dot{x}\dot{x}}\dot{\eta}^2$$

where F_x denotes $\partial F/\partial x$ and so on. All derivatives are evaluated at

$$(x+\varepsilon\eta,\ \dot{x}+\varepsilon\dot{\eta})$$

Then, for $\varepsilon = 0$

$$\delta^2 J = \int_a^b (F_{xx}\eta^2 + 2\,F_{x\dot{x}}\eta\dot{\eta} + F_{\dot{x}\dot{x}}\dot{\eta}^2)dt$$

More details about the second variation can be obtained from the book [22], chapter 6, and the lecture notes [8].

4.3.5.1 Legendre Condition

Notice that

$$2\, F_{x\dot{x}} \eta \dot{\eta} \;=\; F_{x\dot{x}} \frac{d}{dt} \eta^2$$

Also, $\eta(a) = 0$ and $\eta(b) = 0$
Therefore

$$\delta^2 J = \int_a^b \left((F_{xx} - \frac{d}{dt} F_{x\dot{x}}) \eta^2 \;+\; F_{\dot{x}\dot{x}} \dot{\eta}^2 \right) dt$$

In a more compact form

$$\delta^2 J = \int_a^b \left(Q\eta^2 \;+\; P\dot{\eta}^2 \right) \, dt$$

It can be shown that $P\,\dot{\eta}^2$ is the dominant term, and so for $\delta^2 J \geq 0$, it is necessary that $P \geq 0$ on $[a, b]$. This is the **necessary** "*Legendre condition*" for a minimum:

$$\frac{\partial^2 F}{\partial \dot{x}^2} \geq 0, \quad for \;\; x(a) \leq x \leq x(b)$$

4.3.5.2 An Example

In the Brachistochrone problem, we have

$$F(y, y') \;=\; \frac{1}{\sqrt{2g}} \int_0^{x_B} \frac{\sqrt{1 + y'^2}}{\sqrt{y}} \, dx$$

Then, the Legendre test would see whether the following expression is ≥ 0:

$$\frac{\partial^2 F}{\partial y'^2} \;=\; \frac{\partial}{\partial y'} \left(\frac{y'}{\sqrt{2\, g\, y\, ((1 \;+\; y'^2)}} \right) \;=\; \frac{1}{\sqrt{2g\, y}\;(1 \;+\; y'^2)^{3/2}}$$

Recall from the study of the Brachistochrone problem (in 4.3.3) that

$$y\, (1 + y'^2) \;=\; K_1$$

Using this, it is found that

$$\frac{\partial^2 F}{\partial y'^2} \;=\; \frac{1}{\sqrt{2gK_1}} \cdot \frac{y}{K_1}$$

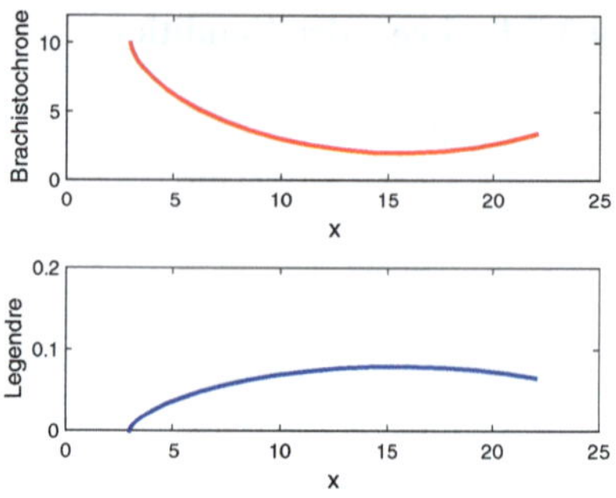

Fig. 4.22 Brachistochrone and the Legendre condition

Therefore, the Legendre test is satisfied in all points of the brachistochrone $y(x)$ between boundaries.

Figure 4.22 visualizes the evolution of the Legendre test along the brachistochrone trajectory. It confirms that the test is always nonnegative in the defined interval.

Of course, given that the condition is satisfied, it does not guarantee that the solution is a minimum (it is *only* a necessary condition). A clear utility of necessary conditions is to rapidly discard unfitting candidates.

Figure 4.22 has been generated with Program 4.12, which just adds some lines to Program 4.7.

Program 4.12 Legendre condition along the Brachistochrone

```
% Brachistochrone curve
% and Legendre condition
X0=1; Y0=10; %origin
t=0:0.1:2;
K1=8; K2=2;
g=9.8;
y=(K1/2)*(1-cos(2*t));
x=(K1/2)*((2*t)-sin(2*t))+K2;
aux=sqrt(2*g*K1)*K1;
lg=(1/aux)*y;
% display
subplot(2,1,1)
plot([0 25],[0 0],'k'); hold on;
plot([0 0],[0 12],'k');
axis([0 25 0 12]);
plot(X0+x,Y0-y,'r');
xlabel('x'); ylabel('Brachistochrone');
subplot(2,1,2)
plot([0 25],[0 0],'k'); hold on;
plot([0 0],[0 0.2],'k');
plot(X0+x,lg,'b');
xlabel('x'); ylabel('Legendre');
```

Let us continue with Legendre. He also suggested adding to the integrand the following term:

$$\frac{d}{dt}(\omega\,\eta^2)$$

It happens that

$$\int_a^b \frac{d}{dt}(\omega\,\eta^2)\,dt \;=\; \omega\,\eta^2\Big|_a^b \;=\; 0$$

Then

$$\begin{aligned}\delta^2 J &= \delta^2 J + \textstyle\int_a^b \frac{d}{dt}(\omega\,\eta^2) = \int_a^b \left(Q\eta^2 \;+\; P\dot{\eta}^2 + \frac{d}{dt}(\omega\,\eta^2)\right) dt \;= \\ &= \textstyle\int_a^b \left(P\dot{\eta}^2 + 2\omega\eta\dot{\eta} + (\dot{\omega} + Q)\,\eta^2\right) dt \;= \\ &= \textstyle\int_a^b P(\dot{\eta} + \frac{\omega}{P}\eta)^2 dt \;+\; \int_a^b (\dot{\omega} + Q - \frac{\omega^2}{P})\eta^2 dt\end{aligned}$$

Consider now the following equation:

$$\dot{\omega} + Q - \frac{\omega^2}{P} \;=\; 0$$

This expression is an instance of "*Riccati equation*," which is frequently met in optimization problems. If one finds a solution $\omega(t)$ for this equation, the second variation would simplify as follows:

$$\delta^2 J = \int_a^b P\,(\dot{\eta} \;+\; \frac{\omega}{P}\eta)^2\,dt \qquad (4.1)$$

4.3.5.3 Jacobi Condition

In order to solve the above Riccati equation, it would be convenient to use the substitution:

$$\omega \;=\; -\,\left(\frac{\dot{u}}{u}\right)\,P$$

Then

$$-\,\frac{d}{dt}\left(P\frac{du}{dt}\right) \;+\; Q\,u = 0$$

which is a second-order linear ordinary differential equation, which can be solved with a standard method. This last expression is called the "*Jacobi equation*"."

Now, let us take into consideration "*conjugate points*." Two points, $t = \alpha$ and $t = \beta$, are conjugate points of the Jacobi equation if there is a solution $u(t)$ that is nonzero between α and β and $u(\alpha) = u(\beta) = 0$.

Suppose that $P > 0$ on $[a, b]$. Then, Jacobi shows that if there are no points in $[a, b]$ conjugate to $t = a$, then the second variation has the form (1) and $\delta^2 J > 0$. The nonexistence of conjugate points is the **necessary** "*Jacobi condition*."

4.3.5.4 Weierstrass Condition

A first-order Taylor approximation of $F(t, x, \dot{x})$ can be written as follows:

$$F(t, x, \dot{x}) \approx F(t, x, p) + (\dot{x} - p)\frac{\partial F(t, x, p)}{\partial p}$$

The Weierstrass excess function is defined as

$$E(t, x, \dot{x}, p) = F(t, x, \dot{x}) - \left(F(t, x, p) + (\dot{x} - p)\frac{\partial F(t, x, p)}{\partial p}\right)$$

Supposing that $P > 0$ on $[a, b]$, the **necessary** "*Weierstrass condition*" is that $E \geq 0$ for all real values of p.

4.3.6 Boundary Freedom

Recall from the subsection where the Euler-Lagrange equation was obtained that boundaries were considered as fixed so that $x(a) = x_1$ and $x(b) = x_2$. These boundary conditions are called "*essential boundary conditions*."

However, there are problems where the boundaries are not fixed. This could concern the values of $x(a)$, $x(b)$, and even of a, b.

Figure 4.23 depicts a scenario where the same optimization problem with different boundary conditions $x(a)$, $x(b)$ originates different solution paths.

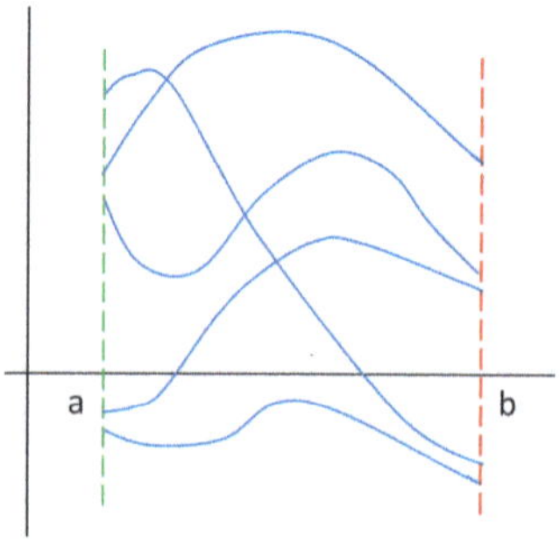

Fig. 4.23 Scenario with free boundary conditions

In case that a and b were kept, but $\eta(a)$ and/or $\eta(b)$ were nonzero, it is necessary to reconsider the expression:

$$\left.\frac{d\,J(\varepsilon)}{d\varepsilon}\right|_{\varepsilon=0} = \left.\frac{\partial F}{\partial \dot{x}^*}\eta\right|_a^b + \int_a^b \left(\frac{\partial F}{\partial x^*} - \frac{d}{dt}\left(\frac{\partial F}{\partial \dot{x}^*}\right)\right) \cdot \eta\, dt = 0$$

Then, the following boundary conditions must be added:

- If $\eta(a) \neq 0$, then

$$\left.\frac{\partial F}{\partial \dot{x}^*}\right|_{t=a} = 0$$

- If $\eta(b) \neq 0$, then

$$\left.\frac{\partial F}{\partial \dot{x}^*}\right|_{t=b} = 0$$

These conditions are called "*natural boundary conditions*."

4.3.6.1 Example of Natural Boundary Condition

Consider the following criterion to be optimized by a function [29]:

$$J = 2\pi \int_0^1 x^2\,(y')^2 - x^3\,y^2 dx$$

where $y(0) = 0$ and $y(1)$ not specified.

In order to write the Euler-Lagrange equation, obtain derivatives:

$$\frac{\partial F}{\partial y} = -2\,x^3 y$$

$$\frac{\partial F}{\partial y'} = -2\,x^2 y'$$

The Euler-Lagrange equation is

$$-\,2x^3 y - \frac{d}{dx}(2\,x^2 y') = 0$$

that is

$$x^2y'' + 2xy' + x^3y = 0$$

The natural boundary condition at $x = 1$ would be

$$\left.\frac{\partial F}{\partial y'}\right|_{x=1} = -2\,(1)^2 y'(1) = 0$$

Therefore

$$y(0) = 0, \; y'(1) = 0$$

Another possibility is that a and/or b were not fixed. Suppose for instance that b is free, then, in addition to the corresponding natural condition, the following "*transversality condition*" must be satisfied:

$$\left.\left(F - \dot{x}\frac{\partial F}{\partial \dot{x}^*}\right)\right|_{t=b} = 0$$

(similarly with a when it is free)

The case of one bound fixed and the other free is illustrated in Fig. 4.24 with a long stretch flag, although probably the length of the flag should be considered as a constraint.

Sometimes, it is required that the final point lies on a given curve $\eta(t)$. In such case, the transversality condition reads

$$\left.\left(F + (\dot{\eta} - \dot{x})\frac{\partial F}{\partial \dot{x}^*}\right)\right|_{t=b} = 0$$

Fig. 4.24 Left bound fixed, right bound free

4.3.6.2 Example of Transversality Condition

Consider the following optimization problem [v-4]:

Minimize:

$$J = \int_0^1 \sqrt{1 + \dot{x}}\, dt$$

with $x(0) = 0$ and (x_f, t_f) lying on $y(t) = -5t + 15$.

With some algebra, it is found that the Euler-Lagrange equation in this case is simply $\ddot{x} = 0$, and therefore, its solution is $x(t) = c_1 t + c_2$.

Since $x(0) = 0$ then $c_2 = 0$, and $x(t) = c_1 t$.

The transversality condition would be

$$\left(F + (\dot{\eta} - \dot{x}) \frac{\partial F}{\partial \dot{x}^*} \right) \Bigg|_{t_f} = 0$$

That is

$$-5\,\dot{x}_f + 1 = 0 \quad \rightarrow \quad -5c_1 + 10 = 0 \rightarrow \; c_1 = 1/5$$

Finally

$$x(t) = t/5; \;\; \text{and}\; t_f/5 = -5\,t_f + 15 \rightarrow \; t_f = 75/26$$

The chapter [13] describes several free boundary problems arising in industry (this chapter belongs to a complete book on the free boundary problems). The document [27] studies boundary conditions in the context of elasticity problems (bars, beams, plates, etc.). An original, surprising article is [33], which deals with natural boundary conditions in image processing; Fig. 4.25, taken from this article, gives an idea of the type of geometrical contents being considered.

Fig. 4.25 Considering boundary conditions in image processing

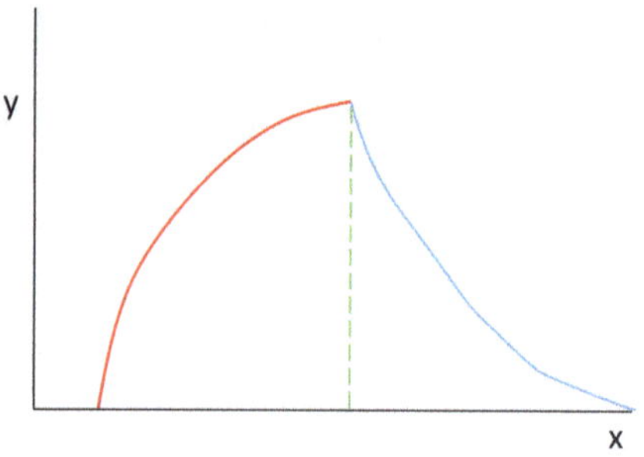

Fig. 4.26 A broken solution

4.3.7 Broken Solutions

Discontinuous solutions for the Euler-Lagrange equation are possible [26]. In such cases, the solution would be a piecewise solution, with a series of smooth arcs joining at a set of corners.

Figure 4.26 depicts a case of broken extremal.

Broken solutions must satisfy the "*Weierstrass-Erdmann conditions*" at all the corners. These conditions can be expressed in terms of limits from the left and right as follows:

$$(F - \dot{x}\, F_{\dot{x}})|_{t=t_{c-}} = (F - \dot{x}\, F_{\dot{x}})|_{t=t_{c+}}$$

and

$$(F_{\dot{x}})|_{t=t_{c-}} = (F_{\dot{x}})|_{t=t_{c+}}$$

The Weierstrass necessary condition (just introduced before this subsection) must be satisfied at every non-corner point.

4.3.7.1 Example of Extremal Broken at Unknown Point

Consider the following problem [7]:

$$F(x, y, y') = \begin{cases} a_+ y'^2 + b_+ y^2, \ if \ x \in (a, \xi) \\ a_- y'^2, \ if \ x \in (\xi, b) \end{cases}$$

with $y(a) = 0$ and $y(b) = 1$.

The Euler-Lagrange equation is

$$eqn = \begin{cases} a_+ y'' - b_+ y = 0, \ if \ x \in (a, \xi) \\ a_- y'' = 0, \ if \ x \in (\xi, b) \end{cases}$$

The solution (given the boundaries) would be

$$y_+(x) = C_1 \sinh\left(\sqrt{\frac{b_+}{a_+}}(x-a)\right), \quad if \ x \in (a, \xi)$$

$$y_-(x) = C_2\,(x-b) + 1, \quad if \ x \in (\xi, b)$$

Now, the constants ξ, C_1, C_2 should be determined from the following three conditions:

- Continuity of the extremal

$$y_+(\xi) = y_-(\xi)$$

- Weierstrass-Erdman condition

$$a_+ y_+'(\xi) = a_- y_-'(\xi)$$

- Transversality condition

$$-\,a_+(y_+'(\xi))^2 - b_+\,y(\xi)^2 = -a_-(y_-'(\xi))^2$$

This transversality condition is simplified to give

$$C_1^2 b_+ = C_2^2 a_-$$

From the second condition

$$C_1\sqrt{\frac{a_+}{b_+}}\cosh q = C_2, \text{ with } q = \sqrt{\frac{b_+}{a_+}}(\xi - a)$$

From the first condition

$$\cosh q = \sqrt{a_+ a_-} \rightarrow \xi = a + \frac{a_+}{b_+}\cosh^{-1}\sqrt{a_+ a_-}$$

also

$$C_1 \sinh q = 1 + C_2(\xi - b)$$

And transversality conditions

$$C_1 \frac{\sqrt{a_-}}{\sqrt{a_-}\ \sinh q\ -\sqrt{b_+}(\xi - b)}\quad ;\quad C_2 \frac{\sqrt{b_+}}{\sqrt{a_-}\ \sinh q\ -\sqrt{b_+}(\xi - b)}$$

4.4 Calculus of Variations: Constrained Problems

Variational problems can also have constraints to obey. This is typical, for example, of control of systems: You can determine an optimal trajectory for a car, but you have to consider the limitations of the car.

After a basic description of a general approach, this section introduces the problems of Mayer and Bolza. A large research activity has been devoted to these problems, resulting in many publications. Asking Google for the Mayer problem gets 74,700,000 answers, much less in the case of Bolza: 81,500 answers.

4.4.1 Constrained Optimization Problems

In case of variational optimization problems that include constraints, the approach of Lagrange multipliers can be used. In mathematical terms, if the problem was

Optimize:

$$J = \int_a^b F(x,\ y,\ y')\, dx$$

Subject to:

$$\Phi(x, y, y') = 0$$

Then, a Lagrangian function can be built

$$L(x,\ y,\ y',\lambda) = F(x,\ y,\ y') - \lambda\ \Phi(x, y, y')$$

and the problem would be to optimize

$$\widehat{J} = \int_a^b L(x,\ y,\ y',\lambda) dx$$

The Euler-Lagrange equation can be established and, combined with the constraint, would lead to a solution for $y(x)$ and λ.

In the case of the so-called "*isoperimetric problems*," the constraints are integrals. For instance

Optimize:

$$J = \int_a^b F(x, y, y')\, dx$$

Subject to:

$$\int_a^b \Phi(x, y, y')dx = K$$

If one takes a function $\rho(x)$ such that

$$\int_a^x \Phi(x, y, y')dx = \rho(x)$$

Then, the constraint can be expressed as

$$\Phi() - \rho'() = 0$$

And so, it is possible to write the following Lagrangian function:

$$L() = F() - \lambda\,(\Phi() - \rho'())$$

Then, proceed as in the other cases.

4.4.1.1 The Problem of Dido

When Dido, the legendary founder of Carthage, arrived in 814 B.C. on the coast of Tunisia, she asked for a piece of land. This desire was satisfied in terms of all the land that could be encompassed by an ox-hide. She then cut the ox-hide into a long, thin strip, using it to encircle an area (Fig. 4.27).

The corresponding, really old, mathematical problem is to determine the closed curve that has the maximum area for a given perimeter.

The area to be maximized is

$$J = \int_a^b y(x)\, dx$$

Subject to the length of the curve

$$\int_a^b \sqrt{1 + (y'(x))^2}dx = K$$

Fig. 4.27 Queen Dido (Carthage)

Since there is no explicit dependence on x, the Lagrange function would be

$$L(y, y') = y(x) + \lambda\sqrt{1 + (y'(x))^2}$$

The Euler-Lagrange equation is

$$\frac{\partial L}{\partial y} - \frac{d}{dx}\left(\frac{\partial L}{\partial y'}\right) = 0$$

Obviously, for our case

$$\frac{\partial L}{\partial y} = 1$$

If one takes the semicircle

$$y = \sqrt{a^2 - x^2}$$

Then, for this curve

$$\frac{\partial L}{\partial y'} = \frac{y'}{\sqrt{1 + y'^2}} = -\frac{x}{a}$$

And so, the semicircle is a solution of the Euler-Lagrange equation, taking $\lambda = a$.

4.4.2 The Problem of Mayer

The scientific activity of Christian Gustav Adolph Mayer (1839–1908) took place mainly at the University of Heidelberg. The problem, now carrying his name, was published in 1878, in the *Leipziger Berichte*. For the reader interested on historical details, the article [12] (freely available from the Internet) would be recommended.

Mayer considered a type of constrained situation in which the optimum is determined by the final extreme, for instance, a projectile trying to get close to a target.

The constraint would be given by

$$\Phi(t, x, \dot{x}) = 0$$

And the functional to optimize

$$J = P(x(b))$$

We want to start from $x(a)$ and go to a certain $x(b)$, with a fixed b, in such a way that J is optimized. The problem is to find $x(b)$.

Using the Lagrange multiplier, the problem can be stated as follows:

Optimize:

$$\widehat{J} = P(x(b)) + \int_a^b \lambda\, \Phi(t, x, \dot{x})\, dt$$

The variation of $\widehat{J}$ would be

$$\delta\, \widehat{J} = P(x(b) + \varepsilon\eta(b)) + \int_a^b \lambda\, \Phi(t, x + \varepsilon\eta, \dot{x} + \varepsilon\dot{\eta})\, dt$$

Taking

$$\left.\frac{d\widehat{J}}{d\varepsilon}\right|_{\varepsilon=0} = 0$$

And integrating by parts, the Euler-Lagrange equation is obtained again

$$\frac{\partial L}{\partial x} - \frac{d}{dt}\left(\frac{\partial L}{\partial \dot{x}}\right) = 0$$

where

$$L = \lambda\, \Phi(t, x, \dot{x})$$

In addition, the following boundary condition must be satisfied:

$$\left.\frac{\partial P}{\partial x}\eta(t)\right|_{t=b} + \left.\frac{\partial L}{\partial \dot{x}}\eta(t)\right|_{t=b} = 0$$

and also, the following transversality condition

$$\left.\frac{\partial P}{\partial x}\right|_{t=b} = -\left.\frac{\partial L}{\partial \dot{x}}\right|_{t=b} = 0$$

There are also problems for which both b and $x(b)$ are to be determined. That is

$$\widehat{J} = P(b, x(b)) + \int_a^b \lambda\, \Phi(t, x, \dot{x})\, dt$$

The Euler-Lagrange equation will be as before. The pertinent changes would be related to the transversality condition, which becomes

$$\left.\left(\frac{\partial P}{\partial x} + L - \dot{x}\,\frac{\partial L}{\partial \dot{x}}\right)\right|_{t=b} \delta\, b + \left.\left(\frac{\partial P}{\partial x} + \frac{\partial L}{\partial \dot{x}}\right)\frac{\partial L}{\partial \dot{x}}\right|_{t=b} \delta\, x(b) = 0$$

In the particular case of minimum time, the criterion to minimize would be $J = b$. Then

$$\widehat{J} = b + \int_a^b \lambda\, \Phi(t, x, \dot{x})\, dt$$

If the end value of $x(t)$ was known, so the problem is to arrive at it in minimum time, the transversality condition would be

$$\left.\left(\dot{x}\,\frac{\partial L}{\partial \dot{x}} - L\right)\right|_{t=b} = 1$$

4.4.3 The Problem of Bolza

Oskar Bolza (1857–1942), back to Freiburg (after teaching at several universities in the United States), published in 1913 his famous problem. He was influenced by the 1879 lectures of Karl Weierstrass on the calculus of variations. The link https://mathshistory.st-andrews.ac.uk/Biographies/Bolza/ offers an interesting biography of O. Bolza. See the publication of M.R. Hestenes [17], for a rigorous description of the problem.

The problem of Bolza could be seen as a mix of the problems of Lagrange and Mayer.

Optimize:

$$J = P(b, x(b)) + \int_a^b F(t, x, \dot{x})\, dt$$

Subject to:

$$\Phi(t, x, \dot{x}) = 0$$

This problem can be reduced to a Mayer problem by adding a variable z such that

$$\dot{z} = F(t, x, \dot{x})$$

Then, the functional to minimize would be

$$J = z + P(b, xb)$$

4.5 Numerical Computation

Based on variational optimization concepts and methods, several tools for computer-based analysis and simulation have been developed, with decisive impact in science, engineering, medicine, etc. In general, these tools use geometrical discretization, frequently in the form of meshes.

The computational atmosphere of this section is illustrated by Fig. 4.28. The discretized optimization, adapted to computational environments, gives interactive platforms for engineering and scientific work, facilitating sound and creative advances.

This section shall introduce representative methods, in which many numerical approaches and tools are based: the Euler method of finite differences and the methods of Ritz and Galerkin that use approximation by series of functions. A final part is devoted to the finite element methodology.

Fig. 4.28 Computational use of the optimization concepts

4.5.1 Euler Method: Finite Differences

Before computers existed, the intuition of Euler considered how to discretize variational problems. As is the case, scientific curiosity about aerodynamics and hydrodynamics (where the form of surfaces is decisive), as well as other problems, invites discretization-which, nowadays, is a well-established approach. The *finite difference method* proposed by Euler for variational problems is a convenient method *per se* and, at the same time, is an important precedent of modern finite element methodologies.

In order to introduce Euler's method, consider again the following problem:

Optimize:

$$J = \int_a^b F(x, y, y')\,dx$$

Let us evaluate J along a piecewise trajectory, like the one represented in Fig. 4.29.

In this way, the value of J becomes a function of the ordinates of the trajectory corner points:

$$J = J(y_1, y_2, \ldots, y_n)$$

Our work would be to choose values of $y_1, y_2, \ldots, y_n$ such as $J(y_1, y_2, \ldots, y_n)$ was optimized, and so

$$\frac{\partial J}{\partial y_1} = 0; \; \frac{\partial J}{\partial y_2} = 0; \; \ldots \frac{\partial J}{\partial y_n} = 0; \qquad (2)$$

This would become a system of equations providing the solution. Accepting successive increasings of n, a series of optimal values of J would be obtained, which may converge to a certain limit. Let us study this aspect in more detail. Suppose we

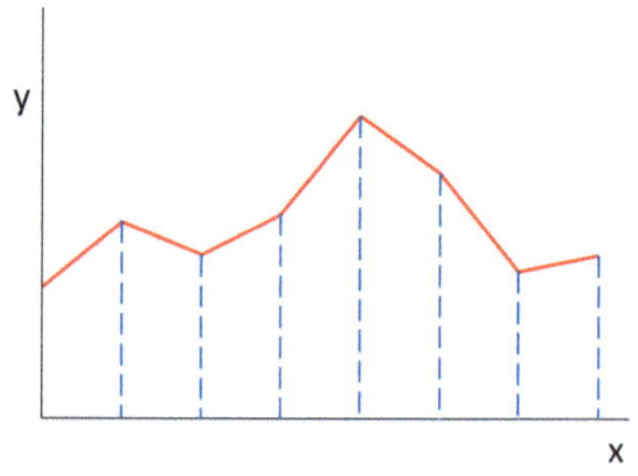

Fig. 4.29 Piecewise solution

try an approximation:

$$J = \int_a^b F(x, y, y')\, dx \approx \sum_k^{n-1} \int_{a+k\Delta x}^{a+(k+1)\Delta x} F\left(x, y, \frac{y_{k+1} - y_k}{\Delta x}\right) dx$$

With some elaboration, one gets a sum of many small areas:

$$\sum_i^{n-1} F\left(x_i, y_i, \frac{\Delta y_i}{\Delta x}\right) \Delta x$$

Notice that for any y_i, only two terms of the sum depend on its value:

$$F\left(x_i, y_i, \frac{y_{i+1} - y_i}{\Delta x}\right) \Delta x \text{ , and } F\left(x_{i-1}, y_{i-1}, \frac{y_i - y_{i-1}}{\Delta x}\right) \Delta x$$

Coming now to the equations

$$\frac{\partial J}{\partial y_i} = 0, \quad i = 1, 2, \ldots, n-1$$

Substitute

$$\frac{\partial F}{\partial y}\left(x_i, y_i, \frac{y_{i+1} - y_i}{\Delta x}\right) \Delta x + \frac{\partial F}{\partial y'}\left(x_i, y_i, \frac{y_{i+1} - y_i}{\Delta x}\right)\left(\frac{-1}{\Delta x}\right) \Delta x +$$

$$+ \frac{\partial F}{\partial y'}\left(x_{i-1}, y_{i-1}, \frac{y_i - y_{i-1}}{\Delta x}\right)\left(\frac{1}{\Delta x}\right) \Delta x = 0 , \quad i = 1, 2, \ldots, n-1$$

which can be expressed as follows

$$\frac{\partial F}{\partial y}\left(x_i, y_i, \frac{\Delta y_i}{\Delta x}\right) - \left(\frac{\frac{\partial F}{\partial y'}(x_i, y_i, \frac{\Delta y_i}{\Delta x}) - \frac{\partial F}{\partial y'}(x_{i-1}, y_{i-1}, \frac{\Delta y_{i-1}}{\Delta x})}{\Delta x}\right) = 0$$

or equivalently

$$\frac{\partial F}{\partial y}\left(x_i, y_i, \frac{\Delta y_i}{\Delta x}\right) - \frac{\left(\Delta \frac{\partial F}{\partial y'}\right)}{\Delta x} = 0$$

Going to the limit for $n \to \infty$, the equations become

$$\frac{\partial F}{\partial y} - \frac{d}{dx}\frac{\partial F}{\partial y'} = 0$$

And so, one arrives at the Euler-Lagrange equation.

In general, one would accept a sufficient approximation by choosing a suitable value of n, solving the system of equations (2), and finding the ordinates y_i.

4.5.2 Ritz Method

In 1907, the Academy of Science in Paris published a challenge (with a prize of 4000 fr), about the deformation of elastic plates under external forces. In French, it was the *Prix Vaillant.*

Being informed of the prize by Paul Ehrenfest, W. Ritz worked on the problem and published his famous approximation method in 1909.

Walther H.W. Ritz (1878–1909) published his method during his last academic activity at University of Göttingen (previously, he had periods in Zürich, Göttingen, Leiden, Bonn, Paris, and Tübingen). In 1900, he contracted tuberculosis, which caused him to move from place to place. He was a fellow student of Einstein in Zürich. His doctoral thesis (Göttingen) dealt with spectral lines of atoms and was influential on the atomic model of Niels Bohr and Ernest Rutherford.

Given a differential equation to be solved, the main idea of Ritz is to translate that equation to a variational expression and then to approximate the solution based on a series of functions.

Consider a simple example, with the following equation to be solved:

$$y'' = f(x)$$

with $y(0) = \sigma, \; y(1) = \varphi$

Now, it can be shown that the Euler-Lagrange equation corresponding to

$$J = \int_0^1 \left[\frac{1}{2}(y')^2 + f(x)\,y\right] dx$$

is $y'' - f(x) = 0$.

Hence, both the original differential equation and the variational study of J have the same solution.

Let us build the following solution candidate:

$$Y(x) = \sum_{k=1}^{N-1} \alpha_k W_k$$

where the functions $W_k(x)$ act as a basis.

Then, the task is to optimize J in function of the parameters α_i. For this purpose, a system of equations is established:

$$\frac{\partial J(\alpha_1, \alpha_2, \ldots, \alpha_n)}{\partial \alpha_i} = 0, \quad i = 1, 2, \ldots, n$$

One could take this to the limit for $n \to \infty$, trying to get an exact solution. But in most cases, it would be sufficient to use a moderate value of n.

More in detail, the candidate solution is introduced as follows:

$$J(\vec{\alpha}) = \int_0^1 \left[\frac{1}{2} \left(\sum_{k=1}^{N-1} \alpha_k (W_k)' \right)^2 + f(x) \sum_{k=1}^{N-1} \alpha_k W_k \right] dx$$

Derivatives would be taken for optimization, giving a system of equations:

$$0 = \frac{\partial J(\alpha_j)}{\partial \alpha_j} = \int_0^1 \left[\left(\sum_{k=1}^{N-1} \alpha_k (W_k)' \right) \cdot (W_j)' + f(x)\, W_j \right] dx,$$

$$j = 1, 2, \ldots, N-1$$

equations that become

$$0 = \sum_{k=1}^{N-1} \alpha_k \int_0^1 (W_k)' \cdot (W_j)' dx + \int_0^1 f(x)\, W_j dx,$$

$$j = 1, 2, \ldots, N-1$$

And then, they can be expressed in matrix-vector format, as follows:

$$A\vec{x} = \vec{b}$$

with

$$A_{kj} = \int_0^1 (W_K)' \cdot (W_j)'\, dx$$

$$b_j = -\int_0^1 f(x)\, W_j dx$$

Of course, it will be advantageous to select appropriate functions, for instance, the trigonometric functions used in Fourier transforms. This would depend on the type of problem at hand.

4.5.2.1 An Example

Let the problem be

$$y'' = -x^2$$

with $y(0) = 0,\ y(1) = 0.$

Choose, for example, $\{W_j\} = \{1,\ x,\ x^2,\ x^3\}.$

And mount $Y(x) = c_0 + c_1 x + c_2x^2 + c_3x^3.$

As a first step, look at boundary conditions:

$$Y(0) = c_0 = 0,\ \text{and}\ Y(1) = c_1 + c_2 + c_3 = 0$$

Then, the candidate solution can be expressed as

$$Y(x) = a\,x(1-x) + b\,x^2(1-x)$$

To simplify, it is possible to take $\Psi_1 = x(1-x)$ and $\Psi_2 = x^2(1-x)$ as the functional basis.

We can now begin to develop the matrix-vector format:

$$A_{11} = \int_0^1 (1-2x)^2 dx = \int_0^1 (1 - 4x + 4x^2)\,dx = \left(x - 2x^2 + \frac{4}{3}x^3\right)\Big|_0^1 = \frac{1}{3}$$

$$A_{22} = \int_0^1 (2x - 3x^2)^2 dx = \int_0^1 (4x^2 - 12x^3 + 9x^4)\,dx = \frac{2}{15}$$

$$A_{12} = A_{21} = \int_0^1 (1-2x)(2x-3x^2)dx = \frac{1}{6}$$

and

$$b_1 = \int_0^1 x^2(x - x^2)dx = \frac{1}{20}$$

$$b_2 = \int_0^1 x^2(x^2 - x^3)dx = \frac{1}{30}$$

Therefore, the problem is

$$\begin{pmatrix} \frac{1}{3} & \frac{1}{6} \\ \frac{1}{6} & \frac{2}{15} \end{pmatrix} \begin{pmatrix} a \\ b \end{pmatrix} = \begin{pmatrix} \frac{1}{20} \\ \frac{1}{30} \end{pmatrix}$$

with solutions $a = 1/15$ and $b = 1/6$. So, the approximated solution for the original differential equation would be

$$Y(x) = \frac{x}{30}(1 - x)(2 + 5x)$$

that can be compared with the analytical solution, which is

$$y(x) = \frac{x}{12}(1 - x^3)$$

Figure 4.30 compares the analytical solution and its Ritz approximation. Note that boundary conditions are satisfied.

Figure 4.30 has been generated with Program 4.13.

Program 4.13 Comparison of Ritz approximation and theoretical solution

```
% Ritz approximation compared
% to theory solution
x=0:0.1:2;
yT = (x.*(1-x.^3))/12;
yA = (x.*(1-x).*(2+5*x))/30;
figure(1) %-----------------------------
plot(x,yT,'r'); hold on; %analytic solution
plot(x,yA,'b');
xlabel('x'); ylabel('y');
```

In general, the translation of the original differential equation to a variational expression is simplified with integration by parts. For example, suppose that the original problem is

$$-\nabla^2 y = f\,, \quad y \in \Omega$$

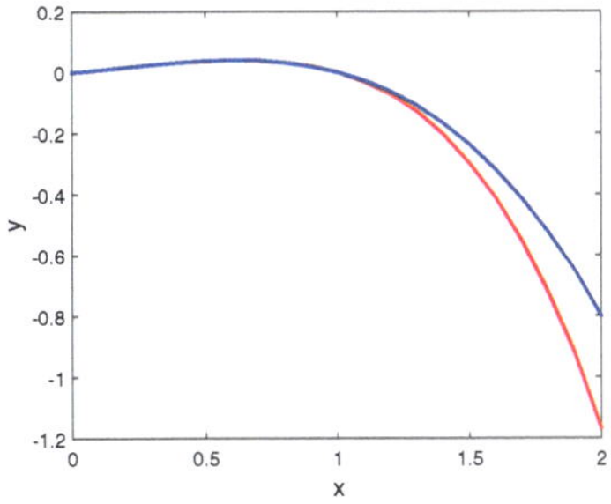

Fig. 4.30 Ritz approximation (curve on the right) to the analytic solution (curve on the left)

Then, one multiplies both sides by a test function and applies integrals:

$$\int_{\Omega} f\,\phi\,d\vec{x} = -\int_{\Omega} \nabla^2 y\,\phi\,d\vec{x} = \int_{\Omega} \nabla y \cdot \nabla\phi\,d\vec{x} - \int_{\partial\Omega} \vec{n}\,\nabla(y\phi)\,d\vec{s} =$$

$$= \int_{\Omega} \nabla y \cdot \nabla\phi\,d\vec{x}$$

Usual boundary conditions make the integral on $\partial\Omega$ to vanish.

4.5.3 Galerkin Method

Boris G. Galerkin (1871–1945), born in a poor family, started to work as calligrapher at the age of 12, and he managed to study at the St. Petersburg Technological Institute and get a job in railway companies. Due to his political activities, he was put into prison. There, he decided to switch to science. Galerkin published his approximation method in 1915. In this publication, Galerkin cited Ritz's contributions and focused on elastic equilibrium of rods and plates.

An important methodology for solving differential equations-and, in particular, boundary value problems (BVP)-is based on *weighted residuals*.

Most BVP problems can be expressed in a compact format as follows:

$$L\,y = f, \text{ with } y(a) = \alpha, \; y(b) = \beta$$

where L is a differential operator.

An approximate solution $Y(x)$ would not satisfy, in general, the equation. So, there would be a residual:

$$r(Y) = L\,Y - f, \text{ with } r(0) = 0$$

A wide class of methods intends to approximately solve the original equation by minimizing the (weighted) residual. Frequently, the target would be to obtain

$$\int_a^b r(Y)\,\phi\,dx = 0$$

where

$$Y(x) = \sum_{k=1}^{N-1} \alpha_k W_k$$

and ϕ being a test or weight function.

Popular methods in this line are collocation, the method of moments, and Galerkin.

The idea of Galerkin is to expand the candidate solution and the weight function in terms of the same basis functions:

$$Y(x) = \sum_{k=1}^{N} a_k W_k \text{ and } \phi(x) = \sum_{k=1}^{N} b_k W_k$$

Since the b_k are arbitrary

$$\int_{\Omega} r\left(\sum_{k=1}^{N} a_k(W_k)\right)\left(\sum_{k=1}^{N} b_k(W_k)\right) dx = 0 \rightarrow$$

$$\rightarrow \int_{\Omega} r\left(\sum_{k=1}^{N} a_k(W_k)\right) W_j\, dx = 0$$

4.5.3.1 An Example

Let the problem be

$$y'' + y + x = 0$$

with $y(0) = 0,\ y(1) = 0$

Choose, for example, $Y(x) = p\,x(1-x) + q\,x^2(1-x)$.

Then, $r(Y) = x + p\,(-2 + x - x^2) + q\,(2 - 6x + x^2 - x^3)$.

Build the equations

$$\int_0^1 r(Y) W_k(x)\, dx = 0 \rightarrow$$

$$\rightarrow \int_0^1 [x + p\,(-2 + x - x^2) + q\,(2 - 6x + x^2 - x^3)] \begin{Bmatrix} (x(1-x) \\ x^2(1-x) \end{Bmatrix} dx = 0$$

In matrix-vector format, we have

$$\begin{pmatrix} \frac{3}{10} & \frac{3}{20} \\ \frac{3}{20} & \frac{13}{105} \end{pmatrix} \begin{pmatrix} p \\ q \end{pmatrix} = \begin{pmatrix} \frac{1}{12} \\ \frac{1}{20} \end{pmatrix}$$

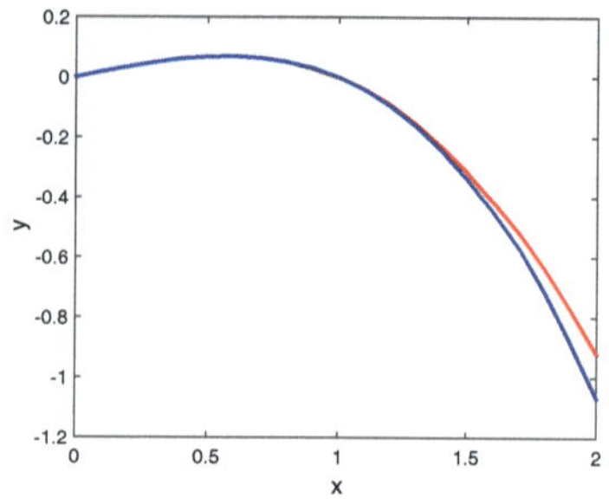

Fig. 4.31 Galerkin approximation (curve on the left) to the analytical solution (curve on the right)

with solutions $p = 7/369$ and $b = 7/41$. The approximated solution for the original differential equation would be

$$Y(x) = x(1-x)\left(\frac{71}{369} + \frac{7}{41}x\right)$$

compare with the analytical solution

$$y(x) = \frac{\sin x}{\sin 1} - x$$

Figure 4.31 depicts the analytical solution and its Galerkin approximation. Boundary conditions are satisfied.

Figure 4.31 has been generated with Program 4.14.

Program 4.14 Comparison of Galerkin approximation and theoretical solution

```
% Galerkin approximation compared
% to theory solution
x=0:0.1:2;
yT = (sin(x)/sin(1))-x;
yA = x.*(1-x).*((71/369) + (7/41)*x);
figure(1) %----------------------------
plot(x,yT,'r'); hold on; %analytic solution
plot(x,yA,'b');
xlabel('x'); ylabel('y');
```

The methods of Ritz and Galerkin paved the way for modern finite element computing. For an interesting account of how it happened, see the article [14], which includes historical and mathematical details.

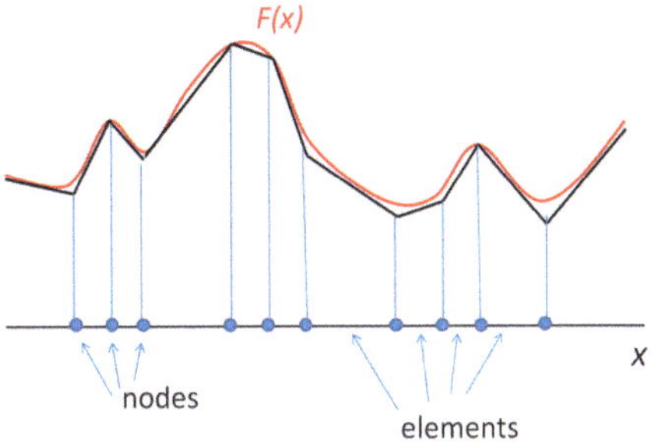

Fig. 4.32 Trying to approximate a single-variable function with finite elements

4.5.4 Finite Elements: Meshing

Finite element is a key methodology for many engineering and scientific activities. The subject is being covered by introductory books, like [6] (508 pages) or [11] (1113 cites), more theoretical books, like [35] (3096 cites), or specialized books for structural analysis, electromagnetics, heat transfer analysis, electrical engineering (covering machinery), fluids, etc. Some books include MATLAB codes.

The idea of this subsection is to briefly introduce the essentials of finite elements, adding references for personal exploration.

In general, the question is to extend a discretization approach for 2D and 3D studies. This leap commonly implies the use of partial derivatives. Here, the *MATLAB Partial Differential Equation Toolbox* is of evident interest (see [32]).

Let us enter into the topic considering a one-dimensional situation, as depicted in Fig. 4.32.

The finite element method (FEM) use a piecewise approximation based on trial functions. In Fig. 4.32, linear trial functions have been employed. The problem is to determine where to place **the nodes** for best approximation. In between the nodes are the finite elements (nodes are the boundaries of the elements).

Notice that the piecewise approximation is not a continuous curve (this is reflected in its first derivative).

There are trial functions, like quadratic curves or Hermite 3rd order polynomials, which would produce continuous approximations. However, in most real applications, the linear trial function is chosen (perhaps interspersed with curves).

The Galerkin method can be adapted to find the best node positions, as extensively detailed in [21].

4.5.4.1 Flat Geometries: FEM in 2D

The approximation of two-variable functions on a 2D space domain is usually made with trial two-variable functions and triangular finite elements. Quadrangular elements are also employed.

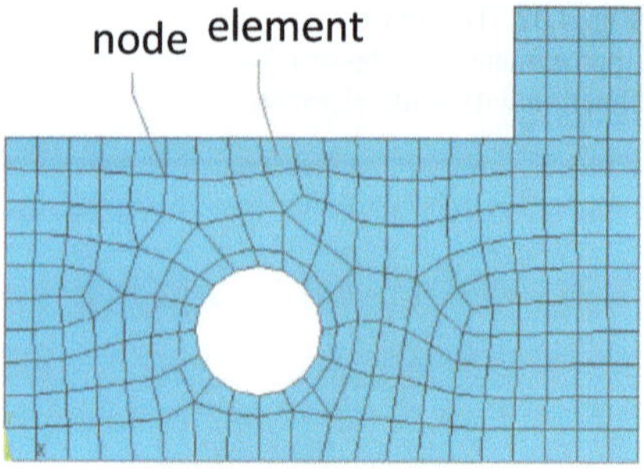

Fig. 4.33 Example of 2D mesh

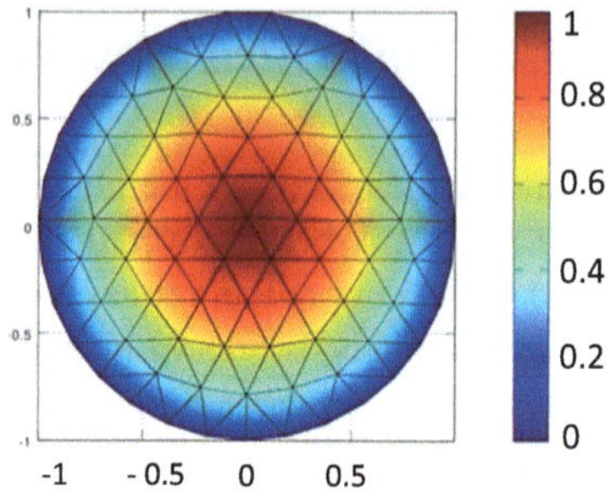

Fig. 4.34 Example of 2D approximated solution of Poisson equation

Figure 4.33 shows an example of the decomposition of a study domain in 2D. Most finite elements in this case are quadrangular (one is a triangle) [31].

The Delaunay algorithm can be used for automatic triangulation of a domain (see [30] for more details). MATLAB has a web page devoted to "*Working with Delaunay Triangulations*" and another page to document the MATLAB function *delaunayTriangulation( )* for 2D and 3D triangulations.

On top of each element, a two-variable trial function is used to approximate the solution of the 2D problem under study. The result can be visualized with pseudo-colors, like in Fig. 4.34 which shows the solution of the Poisson equation (this could correspond, for example, to the temperature of a microprocessor, which we should keep under a certain limit).

Figure 4.34 has been taken from the web page [9], which is very useful about 2D FEM in MATLAB. Notice how triangulation is used.

Figure 4.34 presents another view of the FEM approach, in the case of employing piecewise linear 2D functions defined on triangles.

Figure 4.35 belongs to a GitHub page at https://hplgit.github.io/INF5620/doc/pub/sphinx-fem/._main_fem009.html. It includes interesting illustrations related to 2D and 3D FEM.

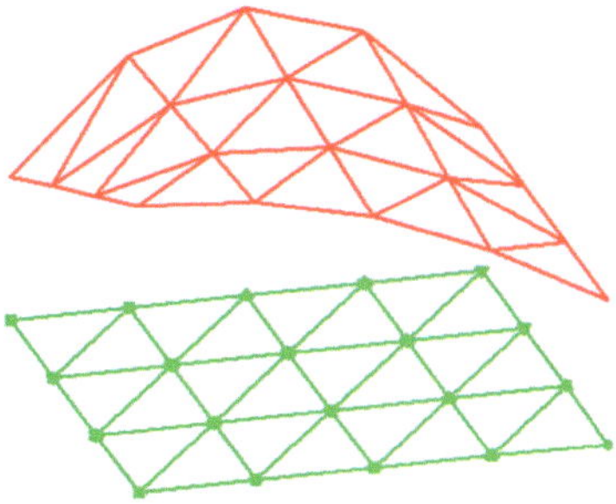

Fig. 4.35 Function approximation via linear 2D functions defined on triangles

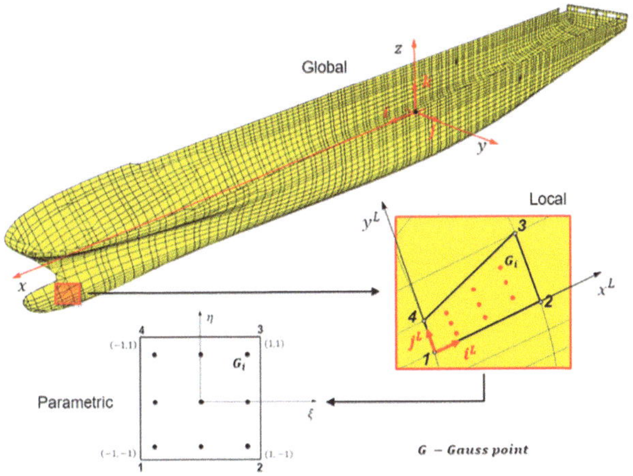

Fig. 4.36 A 3D mesh for the design and analysis of a ship

4.5.4.2 FEM in 3D

Typical finite elements for 3D applications are the tetrahedron and the octahedron, with the addition of triangles and other polygons for the surfaces (which can be curved).

Figure 4.36 shows an industrial example for the design and analysis of a ship [2]. The figure indicates two levels of work: the local study (perhaps focusing on critical regions of interest, like those related to turbulences) and the global analysis devoted to general performances.

Triangulation is most useful for rounded geometry cases, like the one depicted in Fig. 4.37.

Figure 4.37 has been taken from a very interesting presentation [5] about meshing (algorithms, tools, and problems to solve). It also mentions the proximity of the topic to the use of 3D printers (we will add some more details in a moment).

Experts consider meshing as **an art**. Given an aircraft, a ship, a turbine, an aerogenerator, a delicate piece for a machine, etc., the mesh should be designed

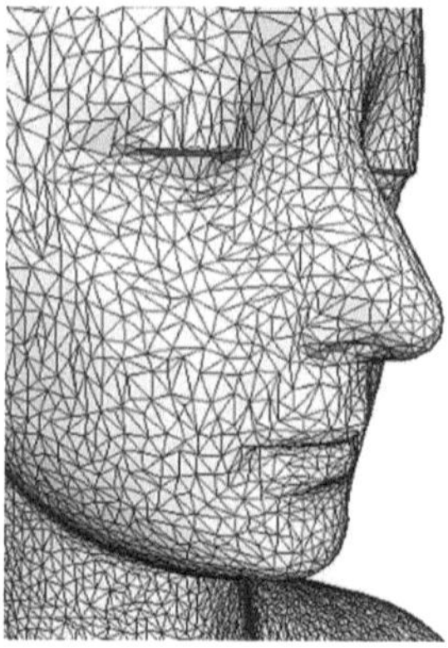

Fig. 4.37 A 3D mesh for artistic work

Fig. 4.38 Elephant in a mesh

for the best emphasis of critical phenomena (that could take place in specific small regions of the domain).

The mathematical machinery behind available FEM tools requires interpolations, optimization of the approximation, and assembly of a lot of equations derived from the mesh. All this is driven to matrix-vector format. The extensive lecture notes [28], complemented with [23], gives abundant information on this background.

Let us now include Fig. 4.38 to suggest something more. It is, clearly, an elephant. It is wrapped in a mesh.

How about animation? Let us move the elephant (some physics would be behind it). This would add another variable to the FEM, **time**. And boundaries of the body (being subject to dynamics) should be respected somehow (not to loosen the elephant).

Blogs and other means for public discussion comment, sometimes, that triangles are inherently rigid, and so it would be better to use other finite elements. Anyway, this opens to a tensorial approach. Computational efforts could become enormous and exigent.

One more figure, Fig. 4.39, shows an aircraft while flying.

The surface of the aircraft is divided into patches, according to a certain mesh. This mesh could include thousands of patches, not visible in the figure. Calculations are made for each of the patches. What you see in the figure is a pseudo-color averaged representation of the results, showing where in the aircraft important phenomena (temperature, pressure, fatigue, etc.) happen **during flight**.

Fig. 4.39 Computer-numerical study of flying aircraft

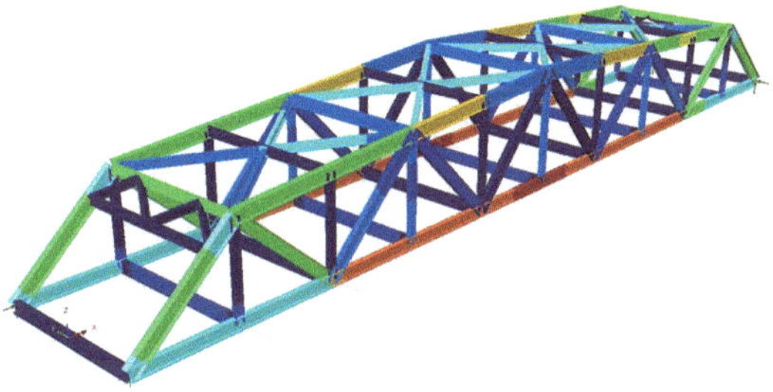

Fig. 4.40 Structural finite elements for the study of a bridge

Our treatment of the topic is being generalist up to now. Some specializations of FEM exist. For instance, there are structures were forces and other variables of interest live in material pieces (bars, springs, tiles, etc.), and so finite elements are identified with those pieces (see, for instance, https://onscale.com/meshing-in-fea-introduction-to-meshing/). This aspect concerns the field of "*Finite Element Analysis (FEA).*"

Figure 4.40 helps to recognize the peculiarities of FEM for structural analysis. It is the case of a bridge, highlighting mechanical stresses.

Modern versions of symbolic MATLAB make easy (to a certain extent) the development of FEM applications. Thanks to the guide provided by the package "*3D Finite Element Analysis with MATLAB*," available from the MathWorks File Exchange site, a simple program (Program 4.15) has been devised, which generates Figs. 4.41 and 4.42.

The first lines of Program 4.15 import an STL file, called *Block.STL*, which contains the geometry description of a block (the shape of a brick).

The page "*STL File Import*" of the MATLAB Documentation on the web shows some other geometries you can import from the MATLAB environment, like a torus, a tetrahedron, and a plate with a hole.

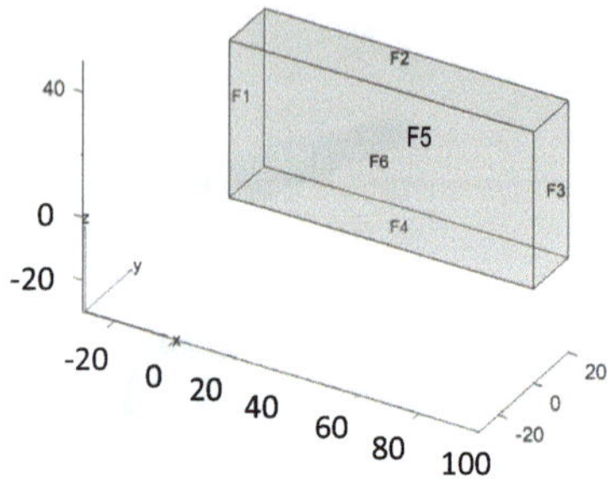

Fig. 4.41 Thermal example geometry

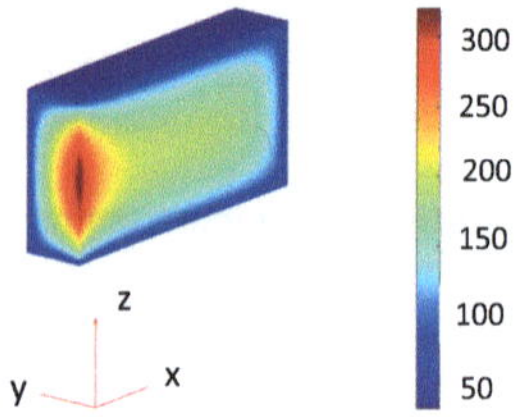

Fig. 4.42 FEM result with temperature profiles

STL files are now quite popular, since this type of file is used by 3D printers (for the description of many kinds of objects). There are several utilities you can use for reading STL files, like the packages "*STLReader*" or "*STL File Reader*" from MathWorks File Exchange or the MATLAB function "*stlread( )*. Likewise, the writing of STL files is facilitated by the package "*stlwrite*" (MathWorks File Exchange) and others. The open source environment **MeshLab** is a good example of an interactive graphical tool for editing and handling STL files (there are other commercial tools in this line). GitHub also offers several repositories related to the use of STL files.

In addition, there are programs that help you automatically generate meshes, given a certain geometry description.

The case considered in program 4.15 is the transmission of heat in a block, under the effect of heating on faces F1 and F6 (Fig. 4.41 shows the face labeling).

Figure 4.42 depicts in pseudo-colors the result of the FEM calculation, showing temperatures along the block.

Notice in Program 4.15 the sequence of steps carried on for the FEM study: geometry, boundary conditions, meshing, assembly of equations and solving, and 3D plotting.

Program 4.15 Example of FEM with symbolic MATLAB

```
% example of FEM with symbolic MATLAB
% get STL file (geometry)
oPDE = createpde(1);
importGeometry(oPDE,'Block.STL');
figure(1)
pdegplot(oPDE,'FaceLabels','on',FaceAlpha=0.3);
```

```
% PDE coeffs, Laplace eq for heat conduction
c = 1e-1;
a = 0;
f = 0;
% boundary conditions ---------------
  % (external temperature)
  extTemp = 10;
  % Convective heat transfer coefficient
  hc = 0.3;
  % Faces with convective boundary conditions
  sideFace = 2:5 ;
  protrusionFace = oPDE.Geometry.NumFaces;
% Apply convective boundary condition
oPDE.applyBoundaryCondition('Face',...
[sideFace,protrusionFace],'q',hc,'g',extTemp);
% Faces with heat source
inpFaces = [1 6];
% Enter unit flux on input heat faces
oPDE.applyBoundaryCondition('Face',inpFaces,'g',1);
% Mesh --------------------
oPDE.generateMesh('hmax',5);
% Solve ------------------
result = assempde(oPDE,c,a,f);
% Display -----------------
figure(2)
pdeplot3D(oPDE,'colormapdata',result);
```

4.6 Exercises

1- Use Lagrange multipliers for the minimization of the next function:

$$f(\vec{x}) = x_1^2 + x_2^2 - 14x_1 - 6x_2$$

Subject to:

$$x_1 + x_2 - 2 \leq 0$$

$$x_1 + 2x_2 - 3 \leq 0$$

Visualize in 2D the feasible region and the objective function (contours).

2- There is a power plant with three electricity generators. The cost of electricity production ($/hour) is the following:

$$f_1 = x_1 + 0.0625\, x_1^2$$

$$f_2 = x_2 + 0.0125\, x_2^2$$

$$f_3 = x_3 + 0.0250\, x_3^2$$

(Each x_i is generated power.)

The sum of generated power should be 952 MW.

Use Lagrange multipliers to determine the optimal values of the x_i for cost minimization.

3- Based on the Cobb-Douglas production model, given by

$$f(x, y) = 100\, x^{3/4}\, y^{1/4}$$

in which x are work units and y are capital units.

The cost of a work unit is \$200, and the cost of a capital unit is \$250.

Find how to maximize production, given that the total budget for work and capital should not exceed \$50,000.

4- Use Lagrange multipliers for the minimization of the next function:

$$f(x, y) = -2\sqrt{x} - \ln y$$

Subject to:

$$x + y \leq 16$$

$$x \geq 0,\; y \geq 0$$

Visualize in 2D the feasible region and the objective function (contours).

5- By using Lagrange multipliers, find the minimum and maximum of $f(x, y, z) = x\, y\, z$ over the sphere $x^2 + y^2 + z^2 = 12$.

Visualize the problem and the results.

6- [4.1] The director of a health clinic has an annual budget of \$750,000. He wants to maximize the number of patient visits, V, which is given as the following function:

$$V = 800\, D^{0.6}\, N^{0.3}$$

where D is the number of doctors and N is the number of nurses. A doctor's annual salary is \$55,000, and a nurses's is \$16,000.

Solve the problem with the Lagrange multiplier. Visualize the problem and the results.

7- [4.2] A manufacturer of paddle balls estimated that the profit (in thousands of dollars) is given by

$$z = 47\, x + 93\, y - x^2 - 2\, x\, y - 9\, y^2$$

where x is the number of balls (in thousands) and y is the number of hours per month of advertising.

The budget to invest is limited to

$$18\,x + 5\,y = 216$$

Maximize the profit via the Lagrange multiplier. Visualize the problem and the result.

8- [4.3] Consider a sphere $x^2+y^2+z^2 = 1$. Assume that the temperature at a point of the sphere is given by

$$T = 55\,x\,y^2\,z$$

Find the hottest and coldest points. Visualize the problem.

9- Show with Euler-Lagrange that the sphere is the surface of revolution that maximizes the contained volume. Try symbolic MATLAB.

4.6.1 Exercise References

[4.1] https://mast.queensu.ca/~math121/Assignments/Unit23_solutions.pdf
[4.2] https://math.libretexts.org/Bookshelves/Calculus/Calculus_(OpenStax)/14:_Differentiation_of_Functions_of_Several_Variables/14.08:_Lagrange_Multipliers
[4.3] https://personal.math.ubc.ca/~CLP/CLP3/clp_3_mc/sec_Lagrange.html

References

1. Arora, J. S., Chahande, A. I., & Paeng, J. K. (1991). Multiplier methods for engineering optimization. *International Journal for Numerical Methods in Engineering, 32*(7), 1485–1525. Available from: https://hal.science/hal-01352827/file/ACP.pdf
2. Bakica, A., Malenica, Š, & Vladimir, N. (2020). Hydro-structure coupling of CFD and FEM-quasi-static approach. *Ocean Engineering, 217*, 108118.
3. Bertsekas, D. P. (1976). Multiplier methods: A survey. *Automatica, 12*(2), 133–145. Available from: https://people.tamu.edu/~j-zhou/Survey_Multiplier_Methods.pdf
4. Bertsekas, D. P. (2014). *Constrained optimization and Lagrange multiplier methods*. Academic Press.
5. Burkardt, J. (2011). *Meshing for the finite element method*. Available from: https://people.sc.fsu.edu/~jburkardt/classes/fem_2011/fem_meshing.pdf
6. Chandrupatla, T., & Belegundu, A. (2021). *Introduction to finite elements in engineering*. Cambridge University Press. Available from: https://istasazeh-co.com/wp-content/uploads/2023/03/Introduction-to-Finite-Elements-in-Engineering.pdf
7. Cherkaev, A. (2019). *Basics of the theory*. Available from: https://www.math.utah.edu/%7Echerk/teach/5500-19/1-19euler.pdf

8. Cherkaev, A. (2019). *Second variation.* Available from: https://www.math.utah.edu/~cherk/teach/5500-19/sec-var19.pdf
9. Coady, J. (2024). *2d finite element method in MATLAB.* Available from: https://www.particleincell.com/2012/matlab-fem/
10. Ferguson, J. (2004). A brief survey of the history of the calculus of variations and its applications. *arXiv preprint math/0402357.* Available from: https://arxiv.org/pdf/math/0402357
11. Fish, J., & Belytschko, T. (2007). *A first course in finite elements.* Wiley. Available from: https://kgut.ac.ir/useruploads/1507881159904ygc.pdf
12. Fraser, C. (2018). The culture of research mathematics in 1860s Prussia: Adolph Mayer and the theory of the second variation in the calculus of variations. In *Research in history and philosophy of mathematics: The CSHPM 2017 annual meeting in Toronto, Ontario* (pp. 121–140). Springer.
13. Friedman, A. (1993). Free boundary problems arising in industry. In *Variational and free boundary problems* (pp. 1–10). Springer. Available from: https://link.springer.com/chapter/10.1007/978-1-4613-8357-4_1
14. Gander, M. J., & Wanner, G. (2012). From Euler, Ritz, and Galerkin to modern computing. *Siam Review, 54*(4), 627–666. Available from: https://www.vsmp.ch/cmsi/doc/attivita2011b/wanner_article.pdf
15. Gelfand, I. M., Silverman, R. A., et al. (2000). *Calculus of variations.* Courier Corporation.
16. Hanc, J. (2004). The original Euler's calculus-of-variations method: Key to Lagrangian mechanics for beginners. Submitted to *American Journal of Physics.* Available from: https://www.eftaylor.com/pub/HancEulerMethod.pdf
17. Hestenes, M. R. (1942). The problem of Bolza in the calculus of variations. *Bulletin of the American Mathematical Society, 48*(12). Available from: https://community.ams.org/journals/bull/1942-48-02/S0002-9904-1942-07600-2/S0002-9904-1942-07600-2.pdf
18. Ito, K., & Kunisch, K. (2008). *Lagrange multiplier approach to variational problems and applications.* SIAM.
19. Kaeonikhom, C. (2006). Variational principle approach to general relativity. *Naresuan University, 55*, 24–28. Available from: https://www.if.nu.ac.th/wp-content/uploads/2022/02/BS_chakkrit.pdf
20. Kang, K., Weinberger, C., & Cai, W. (2006). *A short essay on variational calculus.* Department of Mechanical Engineering, Stanford University (pp. 1–7). Available from: http://micro.stanford.edu/~caiwei/Forum/2006-05-03-VarCalc/vari_calculus_v04.pdf
21. Kelly, P. A. (2020). The (Galerkin) finite element method. Available from: https://pkel015.connect.amazon.auckland.ac.nz/SolidMechanicsBooks/FEM/index.html
22. Kot, M. (2014). *A first course in the calculus of variations* (Vol. 72). American Mathematical Society.
23. Le Dret, H., & Lucquin, B. (2016). The finite element method in dimension two. Available from: https://www.ljll.fr/~ledret/M1English/M1ApproxPDE_Chapter5-1.pdf
24. Logg, A., Langtangen, H. P., & Dokker, J. S. (2024). The heat equation. Available from: https://jsdokken.com/dolfinx-tutorial/chapter2/heat_equation.html
25. Nelson, A. (2012). *Lagrangian mechanics and optimal control.* Available from: https://www2.math.uconn.edu/~gordina/NelsonAaronHonorsThesis2012.pdf
26. Olver, P. J. (2012). The calculus of variations. *Applied Mathematics Lecture Notes. Section, 21*(4). Available from: https://www-users.cse.umn.edu/~olver/ln_/cvc.pdf
27. Olver, P. J. (2022). Boundary conditions and null Lagrangians in the calculus of variations and elasticity. *Journal of Elasticity*, 1–34. Available from: https://www-users.cse.umn.edu/~olver/a_/bcvp.pdf
28. Papadopoulos, P. (2010). Introduction to the finite element method. Available from: https://csml.berkeley.edu/Notes/ME280A.pdf
29. Persson, L.-E. (2019). Natural boundary conditions. Available from: https://www.larserikpersson.se/webcourse/iii-introduction-to-the-calculus-of-variations/9-natural-boundary-conditions/.

30. Shewchuk, J. R. (1999). Lecture notes on Delaunay mesh generation. Available from: https://people.eecs.berkeley.edu/~jrs/meshpapers/delnotes.pdf
31. Staff (2020). Introduction to finite element analysis. Available from: https://www.open.edu/openlearn/science-maths-technology/introduction-finite-element-analysis/content-section-1.5
32. Stahel, A. (2000). Calculus of variations and finite elements. *Lecture notes used at HTA Biel* (pp. 19–21). Available from: https://web.sha1.bfh.science/fem/VarFEM/VarFEM.pdf
33. Stein, O., Grinspun, E., Wardetzky, M., Jacobson, A. (2018). Natural boundary conditions for smoothing in geometry processing. *ACM Transactions on Graphics (TOG), 37*(2), 1–13. Available from: https://dl.acm.org/doi/pdf/10.1145/3186564
34. Sussmann, H. J., & Willems, J. C. (1997). 300 years of optimal control: From the brachystochrone to the maximum principle. *IEEE Control Systems Magazine, 17*(3), 32–44. Available from: https://www.uni-ulm.de/fileadmin/website_uni_ulm/mawi.inst.070/ss15/NumODE/sussmann-willems.pdf
35. Szabó, B., & Babuška, I. (2021). *Finite element analysis: Method, verification and validation*. John Wiley & Sons.
36. Townsend, P. K. (2018). Variational principles Part 1A, mathematics tripos. Available from: https://www.damtp.cam.ac.uk/user/examples/B6La.pdf

Chapter 5
Methods for Dynamic Optimization: Dynamic Programming and Pontryagin Method

5.1 Introduction

During the twentieth century, a number of new methods were proposed for optimization purposes. Some of them are related to Hamiltonian functions, as is the case with the Pontryagin principle or the dynamic programming approach. These two methods had a considerable impact on the optimization field and motivate a specific chapter for introducing them.

It could be said that the Pontryagin method belongs to the analytic atmosphere of the classic methods (Euler, Lagrange, etc.). On the other hand, dynamic programming is well suited for the use of computers, through algorithmic formulations.

The idea of this chapter is to start with the theory and then present a number of application examples that would be quite useful for grasping the practical importance of the topics considered.

Most cited books on dynamic programming are [5] (5357 cites) by R. E. Bellman and S.E. Dreyfus and [6] (16,015 cites) by D. Bertsekas (a book oriented to automatic control). A good description of the topic is the article [58].

The book [53] (16,620 cites) by L. S. Pontryagin is a main reference for his method. See the document [35] for a modern view oriented to optimal control. An opportune reading on Pontryagin's principle is the article [11], which also mentions the Mayer and Bolza problems.

5.2 Discrete Dynamic Programming

Consider a discrete system with the following dynamics:

$$x(t_{i+1}) = f(x(t_i), u(t_i), t_i)$$

J. M. Giron-Sierra, *Introduction to Optimization with Matlab® Examples*,
https://doi.org/10.1007/978-3-032-05048-9_5

It is desired to optimize an objective function, given by

$$J = \sum_{t_1}^{t_{n+1}} L(x(t_i),\ u(t_i),\ t_i) + P(x(t_n),\ t_n)$$

The problem is to find a sequence of values $u_1,\ u_2,\ \ldots,u_n$ of $u(t_i)$, such that along times $t_1,\ t_2,\ \ldots,t_n$, an optimal value of J is finally obtained. Hence, this is related to the dynamic behavior of the system.

Let us suppose that the optimal desired value is a minimum of J. In order to solve the minimization problem, it would be useful to consider the following *value function*:

$$V(x(t_i),\ t_i) = \min\ J$$

which is the minimum J obtained from a given initial condition $(x(t_i),\ t_i)$ to the final state $(x(t_n),\ t_n)$. Different initial conditions could lead to different values of V.

It is not even guaranteed that V exists for all initial conditions.

A simple example that can be helpful is represented in Fig. 5.1. It represents a series of destination places where one could go to obtain a certain reward (represented with a minus sign). During the travel, one passes through several places, incurring transportation costs. These costs are annotated with positive signs.

One desires to choose a destination and a path to get the maximum profit.

Figure 5.2 shows values of V for initial conditions only one step from final destinations:

We marked with backward arrows the optimal decisions for these initial conditions close to destinations.

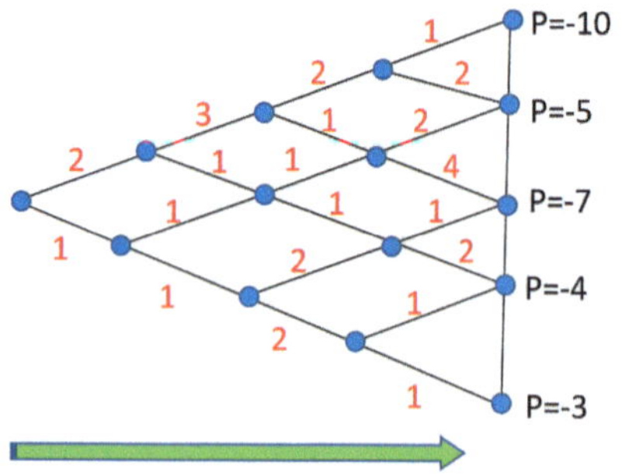

Fig. 5.1 An example for dynamic optimization

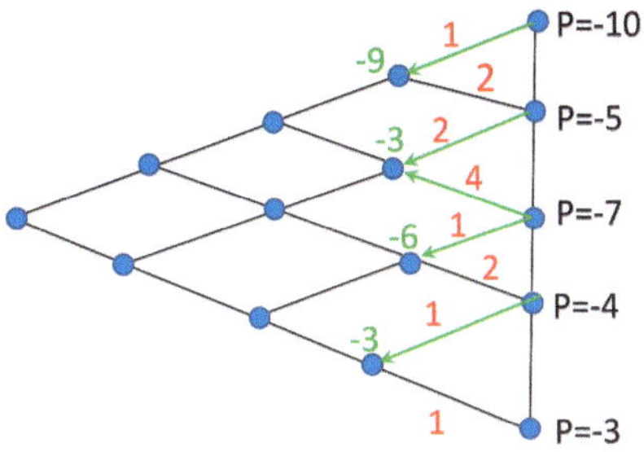

Fig. 5.2 Values of V for initial conditions close to destinations

Now, let us come right to the point. Bellman's principle of optimality says the following:

$$V(x(t_i),\ t_i) = \min_{x} \min_{u(t_i),\, u(t_{i+1}),\ldots} \left\{ \sum_{t_i}^{t_{n-1}} L + P \right\} =$$

$$= \min_{u(t_i)} \left\{ L(x(t_i),\ u(t_i),\ t_i) + \min_{u(t_{i+1}),\ldots} \left\{ \sum_{t_{i+1}}^{t_{n-1}} L + P \right\} \right\}$$

which can also be written as

$$V(x(t_i),\ t_i) = \min_{u(t_i),\, u(t_{i+1}),\ldots} \left\{ \sum_{t_i}^{t_{n-1}} L + P \right\} =$$

$$= \min_{u(t_i)} \{ L(x(t_i),\ u(t_i),\ t_i) + V(x(t_{i+1}),\, t_{i+1}) \}$$

or recalling the system dynamics

$$V(x(t_i),\ t_i) = \min_{u(t_i)} \{ L(x(t_i),\ u(t_i),\ t_i) + V(f(x(t_i),\, u(t_i),\, t_i),\ t_{i+1}) \}$$

In Bellman's words (1952), the principle of optimality is that *given an optimal sequence of decisions or choices, each subsequence must also be optimal.*

Coming back to the example, the values of V for initial conditions two steps away from the destinations are given in Fig. 5.3. Notice that these values are obtained from the values of V at one step from destinations.

Obtained with a similar procedure, the values of V for initial conditions three steps away from destinations are given in Fig. 5.4.

Finally, the optimal value of V, marked on the left vertex, is -3 (Fig. 5.5).

The optimal path is obtained by arrow reversing (-3. -4, -5, -6, -7).

Note two important points:

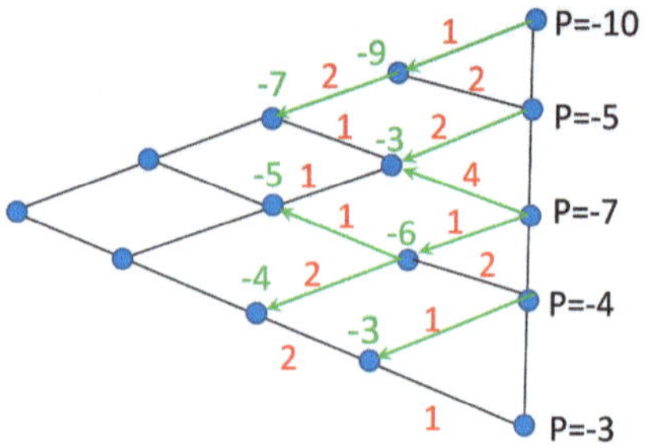

Fig. 5.3 Values of V for conditions two steps away from destinations

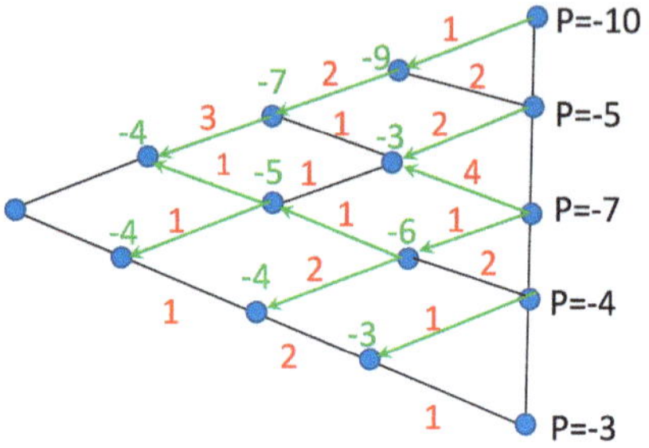

Fig. 5.4 Values of V for conditions three steps away from destinations

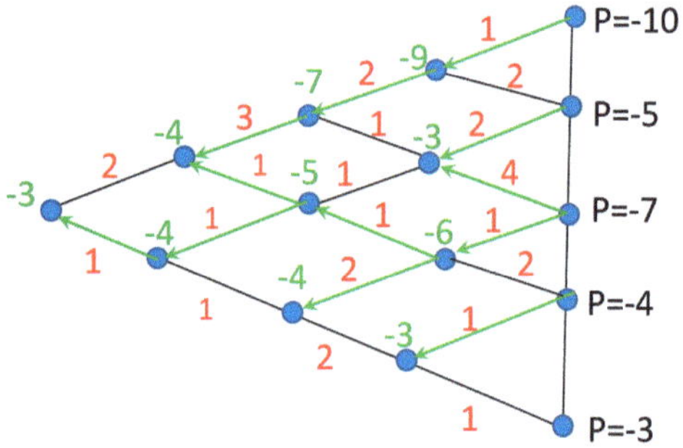

Fig. 5.5 Values of V for conditions two steps away from destinations

- This is an iterative procedure.
- It goes backwards, from destinations to origin.

Notice also that the solution is a chain of decisions (called *a policy*). In general, dynamic programming would decompose the optimization problem into subproblems.

As you can imagine from the example, there are problems in which the number of alternatives considerably increases as the sequence of decisions becomes longer. In some cases, it is a matter of combinatorial explosion. Specialists say there is a *curse of dimensionality*.

A More Involved Example

Let us now consider an example with a bit more mathematics.

It is the case of a company that tries to get the best profits from an oil well, which will be depleted after four years [80]. Figure 5.6 illustrates the example.

Fig. 5.6 Example of an oil well

The revenue each year would be

$$R_t = p_t u_t$$

where t is the time measured in years $(t = 0, 1,..,3)$, p is the price, and u is the amount of oil. The oil extraction cost would be

$$C_t = 0.05 u_t^2$$

After each year, the remaining amount of oil is

$$x_{t+1} = x_t - u_t$$

Therefore, the optimization objective would be to maximize the following quantity:

$$J = \sum_t \beta^t (p_t u_t - 0.05 u_t^2)$$

where a discount factor $\beta = 0.9$ is applied.

A boundary condition is specified as follows:

$$x_4 = 0$$

Let us solve the problem with Bellman's equation:

$$V_t = \max_{u_t} \left\{ (p_t u_t - 0.05 u_t^2) + 0.9 V_{t+1} \right\} \quad t = 0, 1, 2$$

Coming to $t = 3$, it would happen that

$$x_4 = 0 = x_3 - u_3$$

So, obviously, the last year should be devoted to selling the remaining oil:

$$u_3 = x_3$$

Bellman's equation for this stage is

$$V_3 = \max_{u_3} \left\{ (p_3 u_3 - 0.05 u_3^2) + 0.9 V_4 \right\}$$

$V_4 = 0$, and so

$$V_3 = p_3 x_3 - 0.05 x_3^2$$

Now, consider $t = 2$. Bellman's equation becomes

$$V_2 = \max_{u_2} \left\{ (p_2 u_2 - 0.05 u_2^2) + 0.9 V_3 \right\} =$$

$$= \max_{u_2} \left\{ (p_2 u_2 - 0.05 u_2^2) + 0.9 (p_3 x_3 - 0.05 x_3^2) \right\}$$

Since

$$x_3 = x_2 - u_2$$

Then

$$V_2 = \max_{u_2} \left\{ (p_2 u_2 - 0.05 u_2^2) + 0.9 \left[p_3(x_2 - u_2) - 0.05(x_2 - u_2)^2 \right] \right\},$$

In order to get the maximum

$$\frac{d \left\{ (p_2 u_2 - 0.05 u_2^2) + 0.9 \left[p_3(x_2 - u_2) - 0.05(x_2 - u_2)^2 \right] \right\}}{d u_2} = 0$$

which gives

$$p_2 - 0.19 u_2 - 0.9 p_3 + 0.09 x_2 = 0$$

Then, the amount of oil to be extracted would be

$$u_2 = 0.4737 x_2 + 5.263 p_2 - 4.737 p_3$$

With some work, one obtains

$$V_2 = 0.474 (p_2 + p_3) x_2 + 2.636 p_2^2 - 4.734 p_2 p_3 - 0.02368 x_2^2 + 2.132 p_3^2$$

For $t = 1$, Bellman's equation is

$$V_1 = \max_{u_1} \left\{ (p_1 u_1 - 0.05 u_1^2) + 0.9 V_2 \right\}$$

The mechanism would repeat the same steps as before: Insert the expression of V_2, then apply $x_2 = x_1 - u_1$, then take the derivative with respect to u_1, equal it to zero, and obtain

$$u_1 = 0.2989 x_1 + 7.013 p_1 - 2.989 p_2 - 2.989 p_3$$

Hence

$$V_1 = -0.01494 x_1^2 + 3.006 p_2^2 - 2.989 p_2 p_3 + 0.2989(p_1 + p_2 + p_3) x_1 +$$
$$+ 2.556 p_3^2 + 3.506 p_1^2 - 2.989 (p_2 + p_3) p_1$$

For $t = 0$, this value has to be now inserted into

$$V_0 = \max_{u_0} \left\{ (p_0 u_0 - 0.05 u_0^2) + 0.9 V_1 \right\}$$

And finally, repeating again the procedure steps, one obtains

$$u_0 = 0.2120 x_0 + 7.880 p_0 - 2.119 p_1 - 2.120 p_2 - 2.120 p_3$$

The symbolic MATLAB functions were really helpful for developing Program 5.1, which closely follows the procedure just described. This program displays the results of the optimization by means of two figures. Figure 5.7 shows the optimal values of u_0, u_1, u_2, u_3, and the corresponding values of remaining oil.

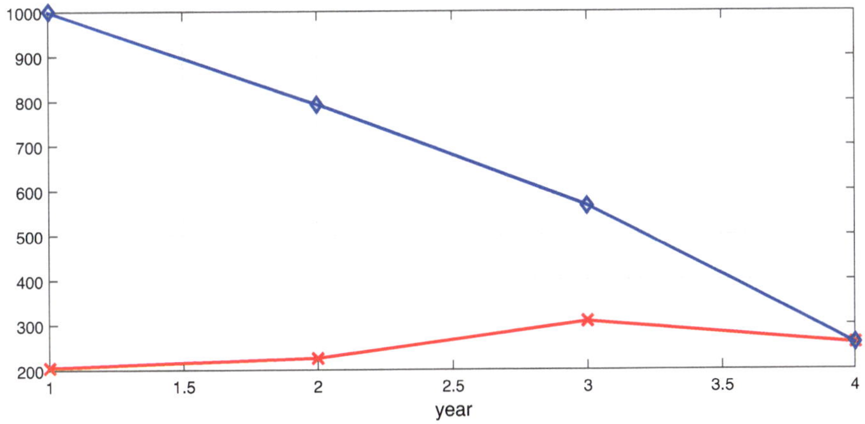

Fig. 5.7 Optimal values of u and oil quantity evolution

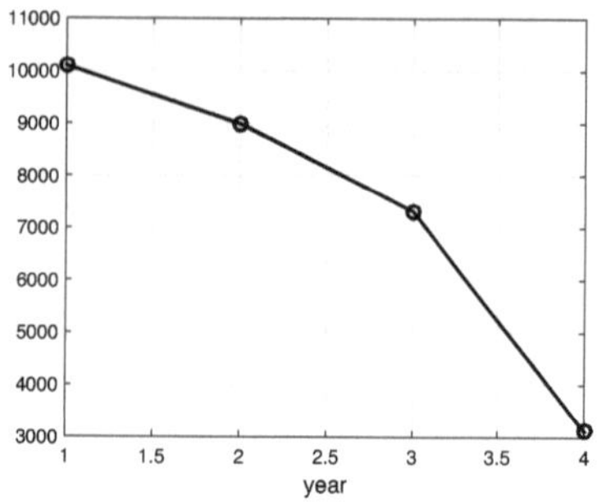

Fig. 5.8 Evolution of the value function

Figure 5.8 shows the evolution of the value function V_t. Notice that V_0 gives the total revenue after the four years.

The most interesting part of Program 5.1 is the loop in which the expressions of V_3, V_2, V_1 were found. In order to comply with MATLAB conventions, the indices in the program run from 1 to 4, corresponding to indices 0 to 3 in the mathematical equations.

Program 5.1 Oil well example

```
%oil_well example
%Note: indexes run from 1 to 4, instead of 0 to 3
%variables
u=zeros(4,1);
x=zeros(4,1);
v=zeros(4,1);
x(1)=1000; %initial oil quantity
%prices
p=[20; 22; 30; 25];
%symbolic variables and functions
syms su sx vv mxu;
sV=zeros(4,1); sV=sym(sV);
sq=zeros(3,1); sq=sym(sq);
%working with functions ------------------------
sV(4)=(p(4)*sx)-(0.05*(sx^2));
%finding sV(3)..sV(1)
for nn=3:-1:1,
  vv=subs(sV(nn+1),sx,sx-su);
  sq(nn)=(p(nn)*su)-(0.05*(su^2))+ (0.9*vv);
  dq=diff(sq(nn),su);
  mxu(nn)=solve(dq,su); %optimal su
  sV(nn)=subs(sq(nn),su,mxu(nn));
end;
%getting numerical results
for m=1:3,
  u(m)=subs(mxu(m),sx,x(m));
  v(m)=subs(sV(m),sx,x(m));
```

```
  x(m+1)=x(m)-u(m);
end;
v(4)=subs(sV(4),sx,x(4));
u(4)=x(4);
figure(1)
plot(u,'rx-'); hold on;
plot(x,'bd-');
xlabel('year');
grid;
figure(2)
plot(v,'ko-');
xlabel('year');
grid;
```

Some interesting academic examples of dynamic programming can be found in [37]. The MathWorks File Exchange site contains several packages in MATLAB related to dynamic programming, like *YADPF* (with graphical animations), *DynaProg* (there is also a book on it), and *Dynamic Programming for MATLAB*.

This last entry is also published in GitHub under the name:

Matlab_Dynamic_Programming.

Other packages in GitHub are *CompEcon* and *macrotools* (both oriented to economics). Many other contributions are also available from GitHub, which contains 146 repositories on **dynamic programming**.

5.3 Continuous Dynamic Programming

Consider a continuous system with the following dynamics:

$$\dot{x}(t) = f(x(t), u(t), t)$$

The system should arrive to a final state: $x(T), T$

It is desired to optimize an objective function, given by

$$J = \int_{t_0}^{T} L(x(t), u(t), t)\, dt + P(x(T), T)$$

As before, a value function $V(x(t), t)$ will be considered.

The principle of optimality in this case can be expressed as follows:

$$V(x(t), t) = \min_{u(t)} \{L(x(t), u(t), t)\, dt\} +$$

$$+ \min_{u(t+dt, T)} \left\{ \int_{t+dt}^{T} L(x(t), u(t), t)\, dt + P(x(T), T) \right\}$$

Likewise

$$V(x(t), t) = \min_{u(t)} \{L(x(t), u(t), t)\, dt + V(x + f(x(t), u(t), t)dt, t + dt)\}$$

Working on this last expression

$$V(x(t), t) = \min_{u(t)} \left\{ L(x(t), u(t), t)\, dt + \frac{\partial V(x, t)}{\partial x} \frac{dx}{dt} dt + \frac{\partial V(x, t)}{\partial t} dt \right\} =$$

$$= \min_{u(t)} \left\{ L(x(t), u(t), t)\, dt + \frac{\partial V(x, t)}{\partial x} f(x(t), u(t), t) dt \right\} + \frac{\partial V(x, t)}{\partial t} dt$$

(Notice that the last term does not depend on $u(t)$.)

Since dt is an arbitrary value, the minimum would be ensured with

$$\min_{u(t)} \left\{ L(x(t), u(t), t) + \frac{\partial V(x, t)}{\partial x} f(x(t), u(t), t) \right\} + \frac{\partial V(x, t)}{\partial t} = 0$$

This is the dynamic programming equation, with the condition

$$V(x(T), T) = P(x(T), T)$$

Continuing with the equations, an interesting, more synthetic expression can be obtained if one considers the following Hamiltonian:

$$H(\vec{x}, \vec{u}, \vec{y}, t) = L(\vec{x}, \vec{u}, t) + \vec{y}^T \vec{f}(\vec{x}, \vec{u}, t)$$

where $\vec{y}$ is the co-state or adjoint vector (n dimensions).

Now, denote $H^o(\vec{x}, \vec{y}, t) = \min_{\vec{u}} H$

According to this notation, the dynamic programming equation would be

$$\frac{\partial V}{\partial t} + H^o \left(\vec{x}, \frac{\partial V}{\partial \vec{x}}, t \right) = 0$$

which can also be written as

$$\frac{\partial V}{\partial t} = - H^o \left(\vec{x}, \frac{\partial V}{\partial \vec{x}}, t \right)$$

This equation is called the Hamilton-Jacobi-Bellman (HJB) equation.

Richard E. Bellman (1920–1984) received a Ph.D. from Princeton University in 1946. He developed the dynamic programming methodology and equations during his work at RAND Company, from 1949 on. Figure 5.9 shows a picture of him. He

Fig. 5.9 Richard E. Bellman

became a professor at the University of Southern California in 1965. More details of his life and scientific contributions can be found in [34].

The HJB equation constitutes the basis of many practical implementations and applications of dynamic programming. The available literature on this aspect is quite large, but we could indicate some useful readings, from relatively succinct to more extensive. The academic presentations [41] and [40] introduce important concepts (for instance, viscosity solutions) and some ways of numerical treatment and include examples. Academic documents, with more mathematics, are [47] and [63], each including two examples. The document [33] contains wise insights on HJB equation solving and comments on the Pontryagin principle. The chapter [10] is longer, being oriented to economy, and develops in detail an example. Extensive materials are covered by [64] and the thesis [20], both including finite differences and finite element numerical approaches (notice that the HJB equation uses partial derivatives).

The book [1] (year 2018) is specifically devoted to numerical methods and applications of the HJB equations in optimal control. It is a book built with contributions from 18 authors.

The MathWorks File Exchange site includes a MATLAB package named *Hamilton-Jacobi solver on unstructured triangular grids*.

GitHub includes a MATLAB package with the name *HJB-solver*, by G. DeCillia.

Quite a lot of applications with a HJB equation core have been published-in economic/financial contexts, due to the need for decision-making; in energy systems, for the same reasons; and also, in the many facets of path planning for mobile artifacts. Here are some references that could be opportune for personal exploration:

- The academic documents on Merton's portfolio problem: [27] and [57]. Both carry the topic to a stochastic context, as well as the article [28] (it includes Merton's problem and other examples).
- The article [26] in a journal of physics focuses on economic dynamics (for instance, macroeconomic growth models).
- Green energy questions are the subject of [78], coupling photovoltaic power generation and hydrogen system. The optimal management of multi-battery systems is tackled by [12].

- A search on Google about HJB equation application *for aircraft* gives 114,000 results. Concerning the flight path planning, we have, for example, [30] and [49]. The space control, with so many airplanes, needs coordination strategies, as treated in [50]. It is most convenient to devise a faster and shorter airplane descent for landing [31]. Indeed, increasing flight safety (avoiding, for example, loss of control) is a must [19].
- Nowadays, one of the key trends is to create autonomous artifacts. Advancing toward autonomous aircraft [17] proposes the use of physics-informed neural networks (PINNs) for the solution of HJB equations in closed-loop control.
- What about drones? The article [76] combines HJB and Pontryagin's principle for path planning of aerial and submarine autonomous vehicles. In [46], the reinforcement learning of deep neural networks is applied to mimic the HJB policy, being applied in autonomous drone racing. Neural networks are also utilized by [61] for approximating HJB solutions, with the following target: the path control of a massive number of drones.
- Robotics. In [69], a method based on HJB is proposed for obstacle avoidance (mobile robots). The path planning for a robotic manipulator is considered by [55], for warehousing applications. The thesis [39] considers the path planning for road vehicles. And something very peculiar: The work of [22] focuses on guided path planning of a snake robot.

Of course, this is not an exhaustive list. Also, in addition to the mentioned fields, there are others in which a series of decisions have to be taken, and the HJB approach is opportune.

5.4 Pontryagin's Minimum Principle

The minimum principle of Pontryagin is a key achievement of mathematical analysis, with applications in a wide set of optimization problems. In particular, it can be employed not only in linear problems, but also in nonlinear cases. This is important because in many scenarios, the optimal solution contains nonlinearities, like switching and saturations.

Like in the previous section, consider a continuous system with the following dynamics:

$$\dot{\vec{x}}(t) = \vec{f}(\vec{x}(t), \vec{u}(t), t)$$

The system departs from an initial state $\vec{x}(t_0)$, t_0 and should arrive to a final state: $\vec{x}(T)$, T. Denote $\vec{x}_0$ and $\vec{x}_F$ as the initial and final state, respectively.

It is desired to optimize an objective function, given by

$$J = \int_{t_0}^{T} L(\vec{x}(t), \vec{u}(t), t)\, dt + P(\vec{x}(T), T)$$

The Hamiltonian of the problem would be

$$H(\vec{x}, \vec{u}, \vec{y}, t) = L(\vec{x}, \vec{u}, t) + \vec{y}^T \vec{f}(\vec{x}, \vec{u}, t)$$

Denote $\delta \vec{u}$ as the variations of $\vec{u}(t)$. It is assumed that these variations are inside admissible bounds, such that they cause small changes in the final state. Also, it is assumed that $\delta \vec{u}$ implicitly depends on certain ε, such as $|\delta \vec{u}| \rightarrow 0$ when $\varepsilon \rightarrow 0$.

In order to introduce the minimum principle, let us suppose the problem is linear. Also, for simpler equations, let us drop the explicit time dependency. We want to analyze the impact of $\delta \vec{u}$ on the objective function.

According to Lagrange's approach, a modified objective function can be built by adding the constraint as follows:

$$\hat{J} = J + \int_{t_0}^{T} \vec{y}^T (\vec{f}(\vec{x}, \vec{u}) - \dot{\vec{x}}(t))\, dt$$

(the co-state $\vec{y}$ is used as the Lagrange multiplier)

Then

$$\hat{J} = P(\vec{x}(T)) + \int_{t_0}^{T} (H(\vec{x}, \vec{u}, \vec{y}) - \vec{y}^T \dot{\vec{x}}(t))\, dt$$

Denote $\vec{v}(t) = \vec{u}(t) + \delta\vec{u}(t)$. The state trajectory corresponding to $\vec{v}(t)$ would be $\vec{x}(t) + \delta\vec{x}(t)$. The perturbation $\delta \vec{u}$ would produce the following change:

$$\delta \hat{J} = P(\vec{x}(T) + \delta\vec{x}(T)) - P(\vec{x}(T)) +$$

$$+ \int_{t_0}^{T} (H(\vec{x}, \vec{v}, \vec{y}) - H(\vec{x}, \vec{u}, \vec{y}) - \vec{y}^T \delta\dot{\vec{x}}(t))\, dt$$

Integrating by parts, the last term becomes

$$\int_{t_0}^{T} \vec{y}^T \delta\dot{\vec{x}}(t)\, dt = \vec{y}(T)^T \delta\vec{x}(T) - \vec{y}(t_0)^T \delta\vec{x}(t_0) - \int_{t_0}^{T} \dot{\vec{y}}^T \delta\vec{x}(t)\, dt$$

Since $\delta\vec{x}(t_0) = 0$, then

$$\delta \hat{J} = P(\vec{x}(T + \delta\vec{x}(T))) - P(\vec{x}(T)) - \vec{y}(T)^T \delta\vec{x}(T) +$$
$$+ \int_{t_0}^{T} (H(\vec{x} + \delta\vec{x}, \vec{v}, \vec{y}) - H(\vec{x}, \vec{u}, \vec{y}) + \dot{\vec{y}}^T \delta\,\vec{x}(t))\, dt$$

Using a first-order approximation for the change of $P()$ and adding and subtracting the term $H(\vec{x}, \vec{v}, \vec{y})$

$$\delta \hat{J} = (\vec{\nabla}_x P|_{t=T} - \vec{y}(T))^T \delta \vec{x}(T) + \int_{t_0}^{T} (H(\vec{x} + \delta \vec{x},\, \vec{v},\, \vec{y}) -$$

$$- H(\vec{x},\, \vec{v},\, \vec{y}) + H(\vec{x},\, \vec{v},\, \vec{y}) - H(\vec{x},\, \vec{u},\, \vec{y}) + \dot{\vec{y}}^T \delta \vec{x}(t))\, dt$$

Another first-order approximation is

$$H(\vec{x} + \delta \vec{x},\, \vec{v},\, \vec{y}) - H(\vec{x},\, \vec{v},\, \vec{y}) = \frac{\partial H}{\partial \vec{x}} \delta \vec{x}$$

Then

$$\delta \hat{J} = (\vec{\nabla}_x P|_{t=T} - \vec{y}(T))^T \delta \vec{x}(T) +$$

$$+ \int_{t_0}^{T} \left(\left(\frac{\partial H}{\partial \vec{x}} + \dot{\vec{y}}^T \right) \delta \vec{x}(t) + H(\vec{x},\, \vec{v},\, \vec{y}) - H(\vec{x},\, \vec{u},\, \vec{y}) \right) dt$$

Now, if one takes the $\vec{y}$ that solves the so-called co-state (or adjoint) equation

$$\frac{\delta H}{\delta \vec{x}} + \dot{\vec{y}}^T = 0^T$$

with final condition

$$\vec{\nabla}_x P|_{t=T} = \vec{y}(T)$$

It happens that the change of the objective function is simply

$$\delta \hat{J} = \int_{t_0}^{T} (H(\vec{x},\, \vec{v},\, \vec{y}) - H(\vec{x},\, \vec{u},\, \vec{y}))\, dt$$

In case $\vec{u}(t)$ was optimal, then

$$\delta \hat{J} \geq 0, \quad \forall \delta \vec{u}$$

Consider now the following variation of $\vec{u}(t)$:

$$\delta \vec{u}(t) \begin{cases} \delta \vec{u}\,, & for\ t_i - \varepsilon < t < t_i + \varepsilon \\ 0\,, & \text{elsewhere} \end{cases}$$

(just a variation at instant t_i; this kind of variation is called spike, or needle, variation)

So, it can be seen that for $\vec{u}(t)$ being optimal

$$H(\vec{x},\, \vec{v},\, \vec{y}) \geq H(\vec{x},\, \vec{u},\, \vec{y}), \ \forall t \in [t_0, T]$$

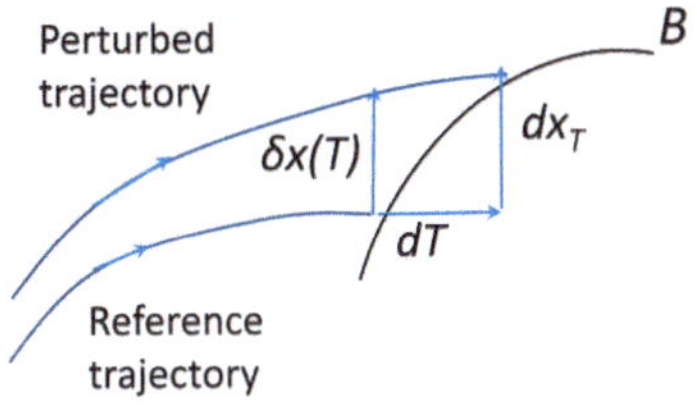

Fig. 5.10 Boundary condition vectors

In summary, Pontryagin's minimum principle says that necessary conditions for $\vec{u}(t)$ to solve the optimization problem are

$$\dot{\vec{y}} = -\left(\frac{\delta H}{\delta \vec{x}}\right)^T$$

with the final condition

$$\vec{\nabla}_x P|_{t=T} = \vec{y}(T)$$

and

$$\vec{u}(t) = \arg\min_{\vec{u}} H(\vec{x}, \vec{u}, \vec{y}), \ \forall t \in [t_0, T]$$

In case of explicit time dependency, the co-state vector is given by

- A linear differential equation:

$$\dot{\vec{y}} = -\frac{\partial \vec{f}^T}{\partial \vec{x}} \vec{y} - \frac{\partial L^T}{\partial \vec{x}}$$

- The boundary conditions:

$$\vec{y}(T)\, d\vec{x}_T - H(T)\, dT = \frac{\partial P}{\partial \vec{x}} d\vec{x}_T + \frac{\partial P}{\partial t} dT$$

The boundary conditions refer to a domain B of possible final states. The vectors $(d\vec{x}_T, dT)$ are tangents to B on the specified final state. See Fig. 5.10.

By using the spike variation, Pontryagin showed that also for nonlinear control

$$\delta \hat{J} = \int_{t_i-\varepsilon}^{t_i+\varepsilon} (H(\vec{x}, \vec{v}, \vec{y}) - H(\vec{x}, \vec{u}, \vec{y}))\, dt + o(\varepsilon) \geq 0$$

and so

$$\vec{u}(t) = \arg\min_{\vec{u}} H(\vec{x}, \vec{u}, \vec{y}), \ \forall t \in [t_0, T]$$

Fig. 5.11 Lev Semyonovich Pontryagin

Fig. 5.12 Worker and queen ants

Figure 5.11 shows a picture of L.S. Pontryagin (1908–1988). He became blind when he was 14. His first mathematical research concerned topological aspects. Then, he began to work on optimal control. His maximum principle was proposed in 1956. In 1962, he published a book on ordinary differential equations and another book on the mathematical theory of optimal processes. A history of the maximum principle can be found in [51] and [18].

Not very complicated introductions to the maximum principle can be found in [48, 75] and [16].

Example (Insects)
This is a colony of insects, with a number $w(t)$ of workers and a number $q(t)$ of queens. Figure 5.12 intends to illustrate this scenario.

A proportion $u(t)$ of workers produce more workers. The insect population evolves as follows:

$$\dot{w}(t) = A\,u(t)\,w(t) - B\,w(t)$$

$$\dot{q}(t) = (1 - u(t))\,w(t)$$

where A >B and both are positive.

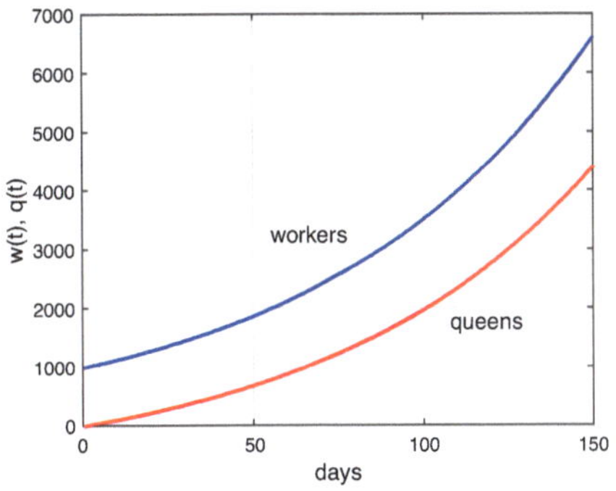

Fig. 5.13 Evolution of the number of workers and queens, with A = 0.0131, B = 0.000228, and u = 0.99

A possible choice of parameters could be $A = 0.0131$ and $B = 0.000228$. With these values and taking $u = 0.99$, the evolution of the number of workers and queens is as shown in Fig. 5.13. The time is measured in days.

Figure 5.13 has been obtained with MATLAB Program 5.2. The reader may play with this program to explore other parameter values.

Program 5.2 Worker and queen ant population. Evolution depends on model parameters

```
% model parameters and variables
A=0.0131; B=0.000228;
dw=0; dq=0;
w=zeros(150,1);
q=zeros(150,1);
u=zeros(150,1);
y=zeros(150,1);
w(1)=1000;
for nn=1:149,
  u(nn)=0.99;
  dw=(A*u(nn)*w(nn))-(B*w(nn));
  dq=(1-u(nn))*w(nn);
  w(nn+1)=w(nn)+dw;
  q(nn+1)=q(nn)+dq;
end;
aux=(1/B)*log(A/(A-B))
figure(1)
plot(w,'b'); hold on;
plot(q,'r');
xlabel('days'); ylabel('w(t), q(t)');
grid
```

Now, the question is to determine how $u(t)$ should be, in order to maximize the number of queens at time $T = 150$.

Let us proceed as proposed by Pontryagin.

The Hamiltonian is

$$H = y\,(A\,u(t) - B)\,w(t) + (1 - u(t))\,w(t)$$

and the co-state equation is

$$\dot{y} = -\frac{\partial H}{\partial w} = -[y\,(Au(t) - B) + (1 - u(t))]$$

The Hamiltonian should take its maximum value along $0 \leq t \leq T$. Let us analyze the situation. First, rewrite the Hamiltonian as follows:

$$H = ((yA - 1)\,u(t) - yB + 1)\,w(t)$$

If $(yA - 1) > 0$, the maximum value of H is reached with $u(t) = 1$. On the other hand, if $(yA - 1) < 0$, the maximum value of H is reached with $u(t) = 0$. The extreme values are

$$H = \begin{cases} (1 - yB)\,w(t), & if\ u(t) = 0 \\ (yA - yB)\,w(t), & if\ u(t) = 1 \end{cases}$$

And regarding the co-state equation

$$\dot{y} = \begin{cases} (yB - 1), & if\ u(t) = 0 \\ -y\,(A - B), & if\ u(t) = 1 \end{cases}$$

The solution of $\dot{y} = y\,B - 1$ is

$$y = (1 - e^{-B(T-t)})/B$$

This would be the solution to use for $(yA - 1) < 0$. At a certain time t_s, it will happen that $(yA - 1) = 0$, which means

$$A\,(1 - e^{-B(T-t_s)})/B = 1$$

Thus

$$t_s = T - \frac{1}{B}\ln\left(\frac{A}{(A - B)}\right)$$

For $t < t_s$, $(yA - 1) > 0$, and so $u(t) = 1$ should be used. The solution during this initial part of the insects' behavior would be

$$y = e^{-(A-B)\,t} + k$$

where

$$k = (1 - A\,\gamma)/A$$

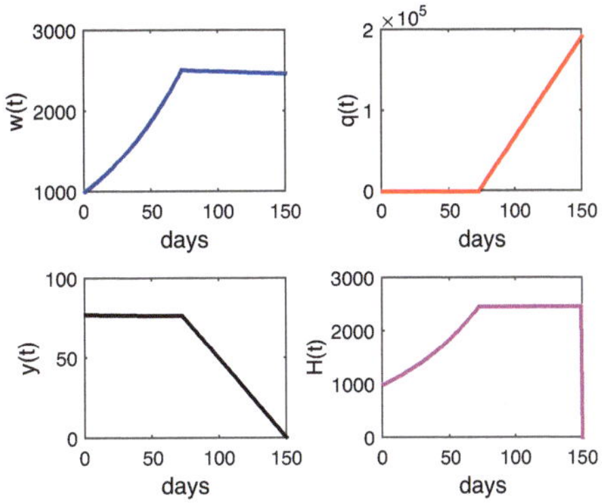

Fig. 5.14 Evolution of workers, queens, co-state, and Hamiltonian during the optimal policy

with

$$\gamma = \exp(-(A - B)\, t_s)$$

Therefore, for the optimal policy, a switching must be applied, from $u(t) = 1$ at $t < t_s$, to $u(t) = 0$ for $t > t_s$ (produce only queens).

Figure 5.14 shows the evolution of the number of workers (upper left plot), the number of queens (upper right plot), the adjoint vector, and the Hamiltonian.

By the way, the values of A and B have been chosen for the switching time to be approximately an integer value.

Figure 5.14 has been obtained with MATLAB Program 5.3.

Program 5.3 Details of ant population

```
% model parameters and variables
A=0.0131; B=0.000228;
dw=0; dq=0;
w=zeros(150,1);
q=zeros(150,1);
u=zeros(150,1);
y=zeros(150,1);
H=zeros(150,1);
w(1)=1000;
aux=(1/B)*log(A/(A-B));
ts=150-round(aux); %integer approx
kappa=exp(-(A-B)*ts);
k=(1-(A*kappa))/A;
for nn=1:ts,
  u(nn)=1;
  y(nn)=k+(exp(-(A-B)*nn));
  H(nn)=(A-B)*y(nn)*w(nn);
  dw=(A*u(nn)*w(nn))-(B*w(nn));
  dq=(1-u(nn))*w(nn);
  w(nn+1)=w(nn)+dw;
```

```
  q(nn+1)=q(nn)+dq;
end;
for nn=ts:149,
  u(nn)=0;
  y(nn)=(1-exp(-(B*(150-nn))))/B;
  H(nn)=(1-(B*y(nn)))*w(nn);
  dw=(A*u(nn)*w(nn))-(B*w(nn));
  dq=(1-u(nn))*w(nn);
  w(nn+1)=w(nn)+dw;
  q(nn+1)=q(nn)+dq;
end;
figure(1)
subplot(2,2,1)
plot(w,'b'); xlabel('days'); ylabel('w(t)');
subplot(2,2,2)
plot(q,'r'); xlabel('days'); ylabel('q(t)');
subplot(2,2,3)
plot(y,'k'); xlabel('days'); ylabel('y(t)');
subplot(2,2,4)
plot(H,'m'); xlabel('days'); ylabel('H(t)');
```

For the interested reader, it would be recommended to visit the website of J. Frédéric Bonnans at https://pages.saclay.inria.fr/frederic.bonnans/; also, see the presentation of M.J. Gander at https://www.unige.ch/gander/Preprints/LagrangeTalk.pdf and the article [65] (Lie groups and automatic control).

5.5 Some Computational Aspects

It was said, with respect to dynamic programming, that it may or may not use tables to save intermediate results, but the key is to design a suitable problem-solving procedure. This section will address this question, using some archetypical problems.

5.5.1 Fibonacci Numbers

The generation of Fibonacci numbers is a good example for the discussion of important algorithmic and computer implementation aspects. We follow here the exposition of [15].

As it is well known, the recursive definition of Fibonacci numbers is as follows:

F0 = 0, F1 = 1

Fn = Fn-1 + Fn-2 for all n >1

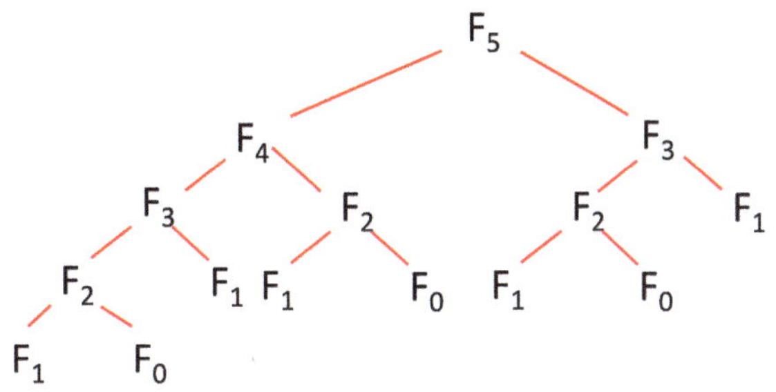

Fig. 5.15 Computing Fibonacci numbers with the naïve algorithm

This definition can be immediately translated to the following (naïve) recursive algorithm:

Naïve Algorithm

```
Fibo(n)
If (n <2)
   return n
else
   return Fibo(n-1) + Fibo(n-2)
```

Figure 5.15 visualizes the tree of computations that is behind this algorithm.

As you can see, it would be required to recompute F_2 again and again, F_3 again and again, etc. The number of mathematical operations would increase exponentially as n increases.

This type of problem, which involves overlapping subproblems, is typically suitable for a dynamic programming application. Let us now advance in this direction.

Clearly, it would be better to compute F_2 once, save it in memory, and reuse this value when required. All the same with F_3, F_4, etc. This process is called *memoization* (keeping a memo). The related algorithm would be

Algorithm with **memoization**

```
MFibo(n)
If (n <2)
   return n
else
   if F[n] is undefined
      F[n]  ← MFibo(n-1) + MFibo(n-2)
return F[n]
```

If one follows the recursive calls to MFibo( ), it is noticed that F[2] is filled first, then F[3], then F[4], and so on. The computations to obtain F[2] take place only once, the same with F[3], etc. Therefore, the number of mathematical operations increases linearly as n increases. This is much better than the naïve algorithm.

And now, once we observe how F[2], F[3], F[4], etc. are filled, a simpler procedure can be devised as follows:

Iterative Algorithm

```
IFibo(n)
F[0] ← 0
F[1] ← 1
For j ← 2 to n
   F[j] ← F[j-1] + F[j-2]
return F[n]
```

This is an iterative (nonrecursive) algorithm, which uses a table (for F[0], F[1], F[2], etc.).

It is still possible to further simplify the memory use. The idea is to use just three registers, by keeping current and previous values as follows:

Memory-Saving Algorithm

```
SFibo(n)
prev ← 0
curr ← 1
For j ← 2 to n
      next ← curr + prev
      prev ← curr
      curr ← next
return curr
```

Another computation alternative is given by the fact that

$$\begin{pmatrix} 0 & 1 \\ 1 & 1 \end{pmatrix} \begin{pmatrix} x \\ y \end{pmatrix} = \begin{pmatrix} y \\ x + y \end{pmatrix}$$

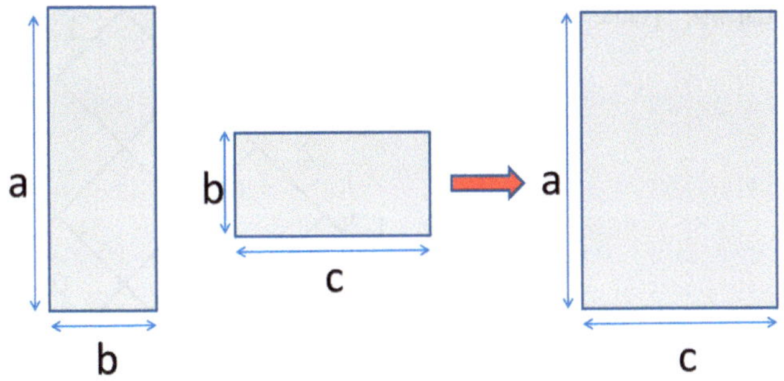

Fig. 5.16 Schematic product of two matrices: Dimensions

Then, one could write that

$$\begin{pmatrix} 0 & 1 \\ 1 & 1 \end{pmatrix}^n \begin{pmatrix} 1 \\ 0 \end{pmatrix} = \begin{pmatrix} F_{n-1} \\ F_n \end{pmatrix}$$

This expression provides a faster way to compute Fibonacci numbers.

5.5.2 *Chain Matrix Multiplication*

As it is illustrated in Fig. 5.16, the multiplication of an $a \times b$ matrix, times a $b \times c$ matrix, gives an $a \times c$ matrix. The number of multiplications involved in this operation is $a \times b \times c$.

Consider now the multiplication of three matrices, A (7 × 5), B (5 × 9), and C (9 × 3). Depending on the order of matrix multiplications, the computational cost would be

$$\cos t\ \{(A\,B)\,C\} = (7 \cdot 5 \cdot 9) + (7 \cdot 9 \cdot 3) = 504$$

$$\cos t\ \{A\,(B\,C)\} = (5 \cdot 9 \cdot 3) + (7 \cdot 5 \cdot 3) = 240$$

In this example, two different calculation order alternatives were indicated with parentheses. Some scientific papers would say there are two possible *parenthesizations*. One or another alternative leads to different computational costs.

Given a set of matrices to be multiplied, our desire is to obtain the result with a minimal computational cost. So, the problem is to determine an **optimal parenthesization**.

Let us write the problem in formal terms. There will be a set of dimensions, $p_0, p_1, \ldots, p_n$, corresponding to a matrix sequence, $A_0, A_1, \ldots, A_n$, where A_i has dimensions $p_{i-1} \times p_i$ [43].

Fig. 5.17 Initial $m[i, j]$ table

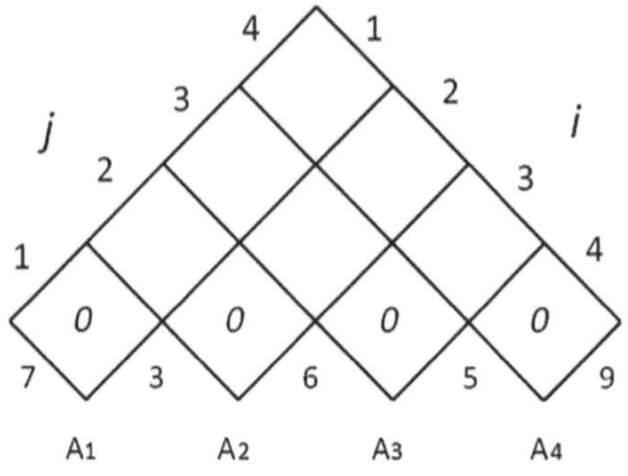

Subsequences of the matrix sequence $A_0, A_1, \ldots, A_n$ will be considered, for instance, $A_i, A_{i+1}, A_{i+2}, \ldots, A_j$. The product of these matrices would be denoted as $A_{i..j}$. The minimum number of multiplications required to compute $A_{i..j}$ would be denoted as $m[i, j]$.

Notice that, given an optimal multiplication order, the last step would be to multiply two matrices, $A_{i..k}$ and $A_{k..j}$ for some k. Then, part of the optimization problem is to determine k. In addition, the multiplication order for $A_{i..k}$ and for $A_{k..j}$ should be optimal.

Therefore, a recursive rule for obtaining $m[i, j]$would be

$$m[i, j] = \begin{cases} 0 for\ i = j \min_{i \le k < j} (m[i, k] + m[k+1, j] + p_{i-1} p_k p_j) & for\ i < j \end{cases}$$

According to this rule, in order to compute $m[i, j]$, we need to have the values of $m[i, k]$ and $m[k + 1, j]$.

A practical way to use the recursive rule would be to fill a table $m[i, j]$ in a special order, so each item would be computed using other already calculated items. Let us take, for example, four matrices, with $p_0 = 7$, $p_1 = 3$, $p_2 = 6$, $p_3 = 5$, and $p_4 = 9$. The table for $m[i, j]$ would be initialized as depicted in Fig. 5.17.

The entry $m[1, 2]$ can be obtained as follows:

$$m[1, 2] = \min_{1 \le k < 2} (m[1, k] + m[k+1, 2] + p_0 p_k p_2) = m[1, 1] + m[2, 2] + p_0 p_1 p_2$$

The table is updated accordingly, as shown in Fig. 5.18.

Likewise, $m[2, 3]$ would be

$$m[2, 3] = \min_{2 \le k < 3} (m[2, k] + m[k+1, 3] + p_1 p_k p_3) = m[2, 2] + m[3, 3] + p_1 p_2 p_3$$

The table is updated as in Fig. 5.19.

And the last entry at this level, $m[3, 4]$, would be

$$m[3, 4] = \min_{3 \le k < 4} (m[3, k] + m[k+1, 4] + p_2 p_k p_4) = m[3, 3] + m[4, 4] + p_2 p_3 p_4$$

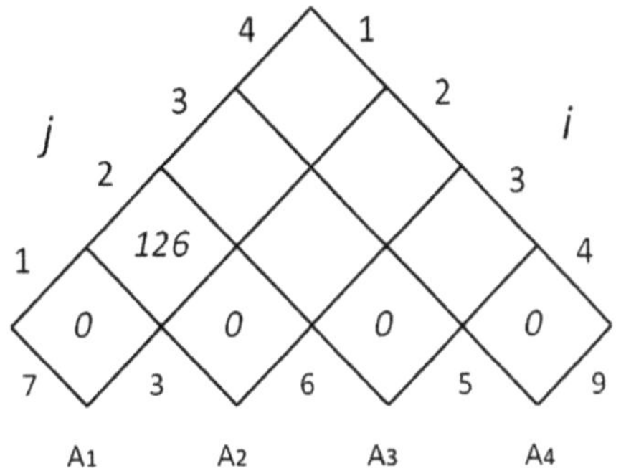

Fig. 5.18 Filling the $m[1, 2]$ entry

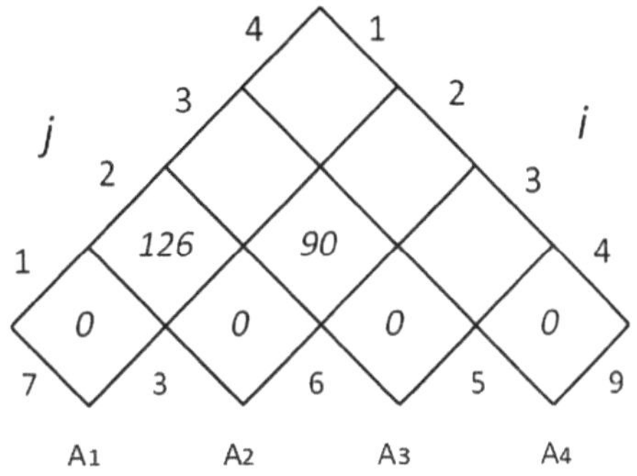

Fig. 5.19 Filling the $m[2, 3]$ entry

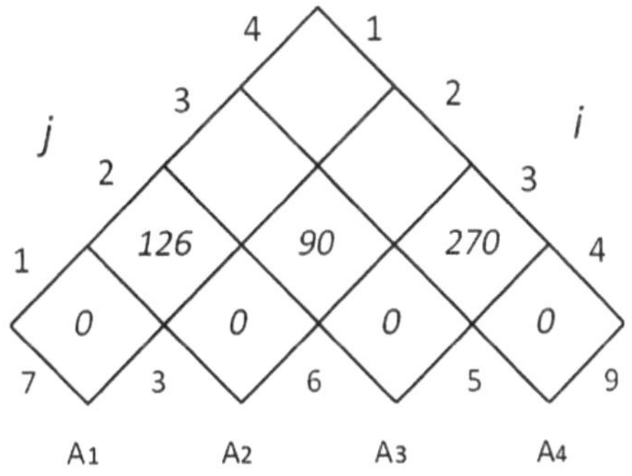

Fig. 5.20 Filling the $m[3, 4]$ entry

The table is updated as in Fig. 5.20.

Up to now, everything has been quite simple. Let us start with the next level (we are proceeding bottom-up, level by level). As you can see, the calculation of $m[1, 3]$ implies a decision:

$$m[1,\ 3] = \min_{1 \le k < 3} (m[1, k] + m[k+1, 3] + p_0 p_k p_3) =$$

$$= \min \begin{Bmatrix} (m[1,\ 1] + m[2, 3] + p_0 p_1 p_3) \\ (m[1, 2] + m[3, 3] + p_0 p_2 p_3) \end{Bmatrix}$$

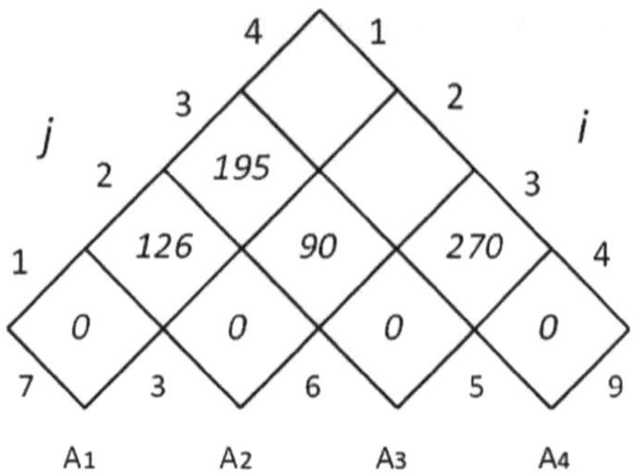

Fig. 5.21 Filling the $m[1, 3]$ entry

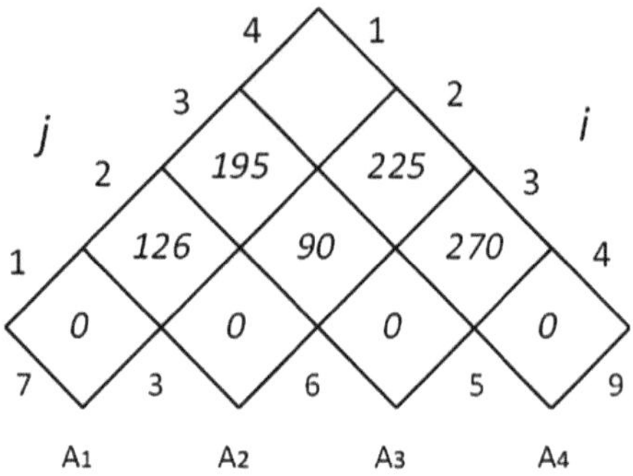

Fig. 5.22 Filling the $m[2, 4]$ entry

The first option, $k = 1$, is chosen, and so the table is updated as shown in Fig. 5.21.

The next entry, $m[2, 4]$, requires also a decision:

$$m[2,\ 4] = \min_{2 \le k < 4} (m[2, k] + m[k + 1, 4] + p_1 p_k p_4) =$$

$$= \min \begin{Bmatrix} (m[2, 2] + m[3, 4] + p_1 p_2 p_4) \\ (m[2, 3] + m[4, 4] + p_1 p_3 p_4) \end{Bmatrix}$$

The second option, $k = 3$, is chosen. The table is updated as depicted in Fig. 5.22.

It only remains to compute $m[1, 4]$. A bit more complicated decision has to be taken:

$$m[1,\ 4] = \min_{1 \le k < 4} (m[1, k] + m[k + 1, 4] + p_0 p_k p_4) =$$

$$= \min \begin{Bmatrix} (m[1, 1] + m[2, 4] + p_0 p_1 p_4) \\ (m[1, 2] + m[3, 4] + p_0 p_2 p_4) \\ (m[1, 3] + m[4, 4] + p_0 p_3 p_4) \end{Bmatrix}$$

The first option, $k = 1$, is chosen, and the result is shown in Fig. 5.23.

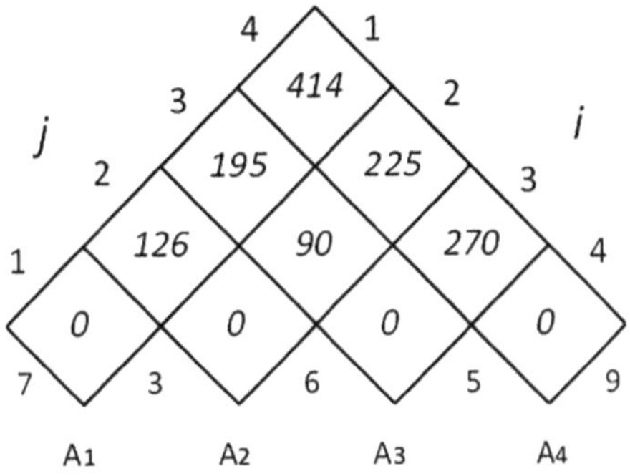

Fig. 5.23 Filling the $m[1, 4]$ entry

In order to extract from the procedure the actual solution, it is convenient to use an auxiliary array, $s[i, j]$, to keep a record of split markers. In the example just calculated, the term $s[1, 4]$ would be 1, corresponding to having chosen $k = 1$ for the $m[1, 4]$ case. Likewise, $s[2, 4]$ would be 3, and $s[1, 3]$ would be 1.

Therefore, the solution for the example is

$$A_1\,((A_2\,A_3)\,A_4)$$

In accordance with the procedure shown in the example, the bottom-up algorithm for solving the matrix-chain problem would be

Iterative Algorithm

```
Matrix-Chain (p,n)
For i ← 1 to n
      m[i,i] ← 0
   For v ←2 to n
      For i ← 1 to (n-v+1)
         j ← i+v-1
         m[i,j] ←  ∞
         For k ← i to j-1
            q ← m[i,k] + m[k+1,j] + p[i-1]*p[k]*p[j]
            If(q <m[i,j])
               m[i,j] ← q
               s[i,j] ← k
return m[1,n] and s
```

An example of MATLAB coding for the above algorithm is provided by Program 5.4.

Program 5.4 Matrix-chain iterative method

```
%Matrix-chain iterative method
n=4;
p=[7 3 6 5 9];
m=zeros(n,n);
s=zeros(n,n);
for i=1:n,
  m(i,i)=0;
end;
for v=2:n,
  for i=1:(n-v+1),
    j=i+v-1;
    m(i,j)=100000;
    for k=i:(j-1),
      q=m(i,k)+m(k+1,j)+(p(i)*p(k+1)*p(j+1));
      if (q < m(i,j)),
        m(i,j)=q;
        s(i,j)=k;
      end;
    end;
  end;
end;
m
s
```

A recursive procedure for actually doing the matrix multiplication would be the following:

Recursive Algorithm

```
Multip (i,j)
If (i == j)
        return A[i]
else
        k ← s[i,j]
        X ← Multip(i,k)
        Y ← Multip(k+1,j)
        return X * Y
```

This algorithm can be coded as the following MATLAB function:

Program 5.5 Function for multiplication of matrices in the matrix-chain problem

```
%function used for matrix_chain
function M=Multip(F,s,i,j)
if(i==j),
```

```
  M=F{i};
else
  X=Multip(F,s,i,s(i,j));
  Y=Multip(F,s,s(i,j)+1,j);
  M=X*Y;
end
```

Combining the two code pieces just suggested, it is possible to devise Program 5.6 for solving the matrix-chain problem:

Program 5.6 Matrix-chain computation example

```
%Matrix-chain computation example
A=[2 5 8; 3 4 4; 2 3 1; 15 7 1; 5 5 5; 0 3 2;...
 8 9 10];
B=[11 12 13 14 15 16; 0 1 2 3 4 5;...
 -1 -2 -3 -4 -5 -6];
C=[1 1 1 1 1; 2 2 2 2 2; 3 3 3 3 3; 4 4 4 4 4;...
 5 5 5 5 5; 6 6 6 6 6];
D=[1 2 3 4 5 6 7 8 9; 19 18 17 16 15 14 13 12 11;
-1 -2 -3 -4 -5 -6 -7 -8 -9; 1 2 1 2 1 2 1 2 1;...
 2 3 2 3 2 3 2 3 2];
%cells
F={A B C D};
%brute force test
TT=F{1}*F{2}*F{3}*F{4};
%finding optimal parenthesization -----------------
n=4; p=[7 3 6 5 9]; m=zeros(n,n); s=zeros(n,n);
for i=1:n,
  m(i,i)=0;
end;
for v=2:n,
  for i=1:(n-v+1),
    j=i+v-1;
    m(i,j)=100000;
    for k=i:(j-1),
      q=m(i,k)+m(k+1,j)+(p(i)*p(k+1)*p(j+1));
      if (q < m(i,j)),
        m(i,j)=q; s(i,j)=k;
      end;
    end;
  end;
end;
%applyng parenthesization
MM=Multip(F,s,1,n);
TT
MM
```

By using memoization, the parenthesization problem can be solved in another way, using the following recursive algorithm:

Recursive Algorithm with Memoization

```
Initialize all m[i,j] to infinity
MChain (i,j)
If (m[i,j] <  ∞)
      return m[i,j]
If (i == j)
      m[i,j] ← 0
      return 0
For k ← i to j-1
      q =MChain(i,k) + MChain(k+1,j) + p[i-1] * p[k]*p[j]
      If (q <m[i,j])
         m[i,j] ← q
      return m[i,j]
```

The first *If ()* of the algorithm avoids the recalculation of an already computed entry.

A MATLAB function can be proposed for the recursive algorithm, as follows:

Program 5.7 Function for memoized parenthesization

```
%Function for memoized parenthesization
function q=Mchain(p,i,j)
global m;
if(m(i,j)<100000),
  q=m(i,j);
  return;
end
if(i==j),
  m(i,j)=0;
  q=0;
  return;
end
q=m(i,j);
for k=i:j-1,
  u=Mchain(p,i,k)+Mchain(p,k+1,j)+...
   (p(i)*p(k+1)*p(j+1));
  if(u<q),
    q=u;
  end;
end;
m(i,j)=q;
end
```

Fig. 5.24 Example of warehouse sizing

Then, the following simple Program 5.8 can be written:

Program 5.8 Parentesization by memoization

```
%Parentesization by memoization
n=4;
p=[7 3 6 5 9];
h=100000*ones(n,n);
global m;
m=triu(h);
Mchain(p,1,n);
m
```

5.5.3 *Warehouse Expansion*

Think of a warehouse that over the last years has kept a shipment log. Part of this log could be as follows:

12	3	15	−6	14	30	−25	−3	37	4	−8	−14

Positive numbers are the number of incoming shipments, while negative numbers are the number of shipped out items [3]. Taking, for instance, the sequence from the third entry to the fifth entry, 15, −6, and 14, it requires a storage capacity of 23 items, which will be denoted as $C(3, 5)$.

The problem is to find the maximum capacity, *Cw*, which has been needed during the shipment log. This information would be useful for recommending perhaps a better warehouse sizing (Fig. 5.24).

An obvious brute-force strategy for determining the value of *Cw* is to explore all $C(i, j)$ for *i=1 to n* and $j>i$. This is what Program 5.9 does.

Program 5.9 Warehouse example, brute force

```
%warehouse example
%brute force method
```

```
Cw=0; %optimal capacity
%shipment log
L=[12 -3 15 -6 14 30 -25 -3 37 4 -8 -14];
n=length(L);
for i=1:n,
  for j=i+1:n,
    su=max([0,sum(L(i:j))]);
    if (su>Cw),
      Cw=su;
    end;
  end;
end;
%result
Cw
```

There is a lot of redundant computation in the brute-force program. For instance, $C(4,7)$ would be computed when finding $C(3,8)$, $C(2,7)$, etc. This can be avoided by considering the following recursion rule:

$$C[i,j] = \max(0,\ L(j) + C(i,j-1))$$

where $L()$ is the shipment log.

In this way, a dynamic programming procedure in MATLAB can be devised, as for example Program 5.10.

Program 5.10 Warehouse example, basic dynamic programming

```
%warehouse example
%basic dynamic prog
Cw=0; %optimal capacity
%shipment log
L=[12 -3 15 -6 14 30 -25 -3 37 4 -8 -14];
n=length(L);
C=zeros(n,n); %capacities
for i=1:n,
  C(i,i)=0;
  for j=i+1:n,
    C(i,j)=max([0, L(j-1)+C(i,j-1)]);
    if (C(i,j)>Cw),
      Cw=C(i,j);
    end;
  end;
end;
%result
Cw
```

If one defines $R(j)$ as the maximum of all capacity values ending in index j

$$R(j) = \max_{i<j}(Ci,j)$$

Then, a faster dynamic programming procedure can be written, as in Program 5.11.

Program 5.11 Warehouse example continuation, a faster dynamic programming

```
%warehouse example
%faster dynamic prog
Cw=0; %optimal capacity
R=[];
%shipment log
L=[12 -3 15 -6 14 30 -25 -3 37 4 -8 -14];
n=length(L);
R(1)=0;
for j=1:n,
  R(j+1)=max([0,L(j)+R(j)]);
  if(R(j)>Cw),
    Cw=R(j);
  end;
end;
%result
Cw
```

To give an idea of what is at stake, given a computer that could process 10^9 instructions per second, and supposing a shipment log of 10^6 transactions, it is said in [3] that the brute-force program would require 31 years of work, for giving a result. The second program (with dynamic programming) would need 17 minutes, and the third program (the faster dynamic programming) would respond in 0.001 seconds.

5.6 Representative Application Examples

In this section, several applications of great interest have been selected. They offer a glimpse into the large dimension and variety of problems in connection with this chapter.

A general, concise reference for this section is [60], which is a tutorial presentation covering several aspects of dynamic programming and related problems.

5.6.1 The Knapsack Problem

It is excursion day! You have a knapsack for carrying several items, provided the total weight does not surpass a certain limit W. You have to choose a subset of

Fig. 5.25 Example of (vintage) knapsack

items, $I_0, I_1, \ldots I_k$, from a total of n items, $I_0, I_1, \ldots I_n$. In order to establish your preferences, you could assign values to items. For example

Item i	Weight	Value
0	4	6
1	2	4
2	3	5
3	1	3
4	6	9
5	4	7

The maximum weight is $W = 10$. The problem is to maximize the total value in the knapsack, by a proper selection of items.

Figure 5.25 illustrates the case to hopefully be solved.

Many types of resource allocation problems can be studied as instances of the knapsack problem, like in the case of investments of budget, time, etc.

Of course, the problem can be solved by brute force, considering all possible subsets S_k that do not surpass W. However, a number of algorithms have been proposed that are usually better.

Now, let us introduce a procedure for solving the example [32, 74]. A table will be built, row by row. In order to fill the table, a decision will be made for each entry. Denote $f(k, g)$ as the maximum total value of a subset S_k with weight g. The decision rule will be the following:

$$f(k, g) = \begin{cases} f(k-1, g) & if\ w_k > g \\ \max\{v_k + f(k-1, g-w_k),\ f(k-1, g)\} & if\ w_k \leq g \end{cases}$$

For the first row, only the first item is considered. Here is a detailed view of how to fill this row:

Items	g = 0	g = 1	g = 2	g = 3	g = 4	g = 5	g = 6	g = 7	g = 8	g = 9	g = 10
Select					0	0	0	0	0	0	0
{0}	0	0	0	0	6	6	6	6	6	6	6

The heading of the table shows possible values of the weight (the maximum is 10). Item 0 can only be selected if the knapsack can hold a weight ≥ 4. The row $\{0\}$ is filled with the total value of the selected items.

Let us keep the first row, and fill the second one:

Items	g = 0	g = 1	g = 2	g = 3	g = 4	g = 5	g = 6	g = 7	g = 8	g = 9	g = 10
{0}	0	0	0	0	6	6	6	6	6	6	6
Select			1	1	0	0	0,1	0,1	0,1	0,1	0,1
{0,1}	0	0	4	4	6	6	10	10	10	10	10

Continue with the third row:

Items	g = 0	g = 1	g = 2	g = 3	g = 4	g = 5	g = 6	g = 7	g = 8	g = 9	g = 10
{0}	0	0	0	0	6	6	6	6	6	6	6
{0,1}	0	0	4	4	6	6	10	10	10	10	10
Select			1	2	0	1,2	0,1	0,2	0,2	0,1,2	0,1,2
{0,1,2}	0	0	4	5	6	9	10	11	11	15	15

and then the fourth row

Items	g = 0	g = 1	g = 2	g = 3	g = 4	g = 5	g = 6	g = 7	g = 8	g = 9	g = 10
{0}	0	0	0	0	6	6	6	6	6	6	6
{0,1}	0	0	4	4	6	6	10	10	10	10	10
{0,1,2}	0	0	4	5	6	9	10	11	11	15	15
Select		3	1	1,3	2,3	1,2	1,2,3	0,1,3	0,2,3	0,1,2	0,1,2,3
{0,1,2,3}	0	3	4	7	8	9	12	13	14	15	18

and so on. The final table can be expressed as follows:

Items	g = 0	g = 1	g = 2	g = 3	g = 4	g = 5	g = 6	g = 7	g = 8	g = 9	g = 10
0	0	0	0	0	6	6	6	6	6	6	6
1	0	0	4	4	6	6	10	10	10	10	10
2	0	0	4	5	6	9	10	11	11	15	15
3	0	3	4	7	8	9	12	13	14	15	18
4	0	3	4	7	8	9	12	13	14	16	18
5	0	3	4	7	8	10	12	14	15	16	19

This expression of the table does not tell what items were chosen for the best solution (value $= 19$). But this important information can be obtained with the following procedure:

Repeat for $k = n, n-1, n-2, \ldots$
If $f(k, g) \neq f(k-1, g)$, then item k was selected, $x_k = 1$; otherwise, $x_k = 0$.
The capacity for previous items would be $g \leftarrow g - w_k x_k$.

In our example

$$g = 10$$

k=5, f(5, 10) $\neq$ *f(4, 10)* then $x_5 = 1$, and $g \leftarrow 10 - 4 = 6$
k=4, f(4, 6) = f(3, 6) then $x_4 = 0$, and $g \leftarrow 6 - 0 = 6$
k=3, f(3, 6) $\neq$ *f(2, 6)* then $x_3 = 1$, and $g \leftarrow 6 - 1 = 5$
k=2, f(2, 5) $\neq$ *f(1, 5)* then $x_2 = 1$, and $g \leftarrow 5 - 3 = 2$
k=1, f(1, 2) $\neq$ *f(0, 2)* then $x_1 = 1$, and $g \leftarrow 2 - 2 = 0$
Therefore, the selection was $\{1, 2, 3, 5\}$.

One of the possible implementations of the algorithms in MATLAB is provided by Program 5.12. There are some other approaches for coding available from the Internet, for instance, [66].

Program 5.12 Knapsack example

```
%knap-sack example
%variables
N=6; %number of items
W=10; %maximum weight
w=[4 2 3 1 6 4]; %weight
v=[6 4 5 3 9 7]; %value
```

```
x=zeros(6,1); %selection
f=zeros(6,11); %table
%table filling------------
%first row
for nc=1:W+1,
  g=nc-1;
  if(w(1)>g),
    f(1,nc)=0;
  else
    f(1,nc)=v(1);
  end;
end;
%the other rows
for nr=2:N, %rows
  for nc=1:W+1, %columns
    g=nc-1;
    if(w(nr)>g),
      f(nr,nc)=f(nr-1,nc);
    else
 f(nr,nc)=max((v(nr)+f(nr-1,nc-w(nr))),f(nr-1,nc));
    end;
  end;
end;
%get selected items
g=W;
for nr=N:-1:2,
  if (f(nr,g+1) ~= f(nr-1,g+1)),
    x(nr)=1;
    g=g-(w(nr)*x(nr));
  end;
end;
f
x
```

The knapsack problem is one of those problems that initially look innocent and simple, until you dig in and start discovering things. The seminal book that pushed knapsacks into scene was [38] (year 1990, 6366 cites). Years later, another seminal book, [29] (year 2004, 4443 cites), extended the problem, adding more constraints (not only total weight).

Today, if you ask Google for **knapsack problem**, you get more than 6,230,000 results.

An extensive survey of recent research on knapsack problems is given by a series of two articles, [8] and [9]. The presentation [42] introduces some variants of the problem.

5.6.2 Sequence Alignment

Word processors usually have automatic spelling correction. In part, this feature is based on finding words similar to the misspelled word in question. Somehow, in

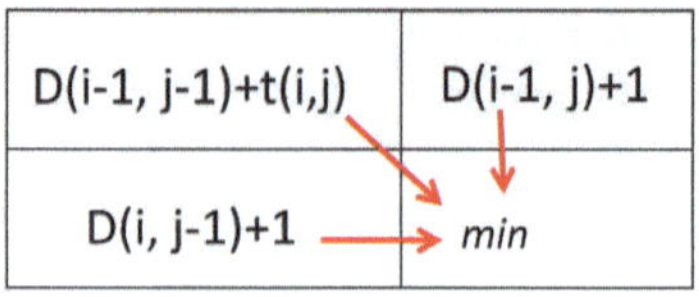

Fig. 5.26 Scheme for entry filling

order to implement this function, a measure of word similarity is needed. Notice that words are sequences (strings) of characters.

Several similarity measuring methods have been proposed. For example, a well-known method in computer science is based on the Levenshtein distance, also called *edit distance*. Suppose two strings, for instance, *kitchen* and *chicken*. The idea is to count the minimum number of edit operations, character by character, which transform one string into another. The allowed edit operations are insertion, deletion, or substitution. Then, one could try the following:

kitchen - > kitcken
kitcken - > kicken
kicken - > hicken
hicken - > chicken

This way of transforming uses four operations. Other transformation alternatives exist. Supposing that 4 is effectively the minimum number of operations, we can say the edit distance is 4.

The edit distance can be found with a dynamic programming algorithm. Denote $A[1] \ldots A[i] \ldots A[n]$ and $B[1] \ldots B[j] \ldots B[m]$ as the two strings and $D(n, m)$ as the edit distance. A recurrence relation can now be used [24]:

$$D(i,\ j) = \min \begin{cases} D(i-1,\ j)+1 \\ D(i,\ j-1)+1 \\ D(i-1,\ j-1)+\ t(i,\ j) \end{cases}$$

where

$$t(i,\ j) = \begin{cases} 1,\ if\ A[i] \neq\ B[j] \\ 0,\ if\ A[i] = B[j] \end{cases}$$

and $D(0, 0) = 0;\ \ D(i, 0)\ = i;\ D(0,\ j)\ =\ j.$

A table can be used for applying the recurrent relation. Figure 5.26 shows the scheme for filling each entry of the table.

The table is filled from left to right, top to bottom. For example

		C	H	I	C	K	E	N
	0	1	2	3	4	5	6	7
K	1	1	2	3	4	4	5	6
I	2	2	2	2	3	4	5	6
T	3	3	3	3	3	4	5	6
C	4	3	4	4	3	4	5	6
H	5	4	3	4	4	4	5	6
E	6	5	4	4	5	5	4	5
N	7	6	5	5	5	6	5	4

The bottom-most, right-most value gives the edit distance (which is 4).

Program 5.13 provides a fairly direct MATLAB implementation of the algorithm for obtaining the table.

Program 5.13 Edit distance for sequence alignment

```
%edit distance example
A='KITCHEN';
B='CHICKEN';
n=length(A);
m=length(B);
D=zeros(n+1,m+1);
%table filling---------
%first column
for i=1:n,
  D(i+1,1)=i;
end;
%first row
for j=1:m,
  D(1,j+1)=j;
end;
%the other entries
for i=1:n,
  for j=1:m;
    if (A(i)==B(j)),
      t=0;
    else
      t=1;
    end;
    D(i+1,j+1)=min([(1+D(i+1,j)),(1+D(i,j+1)),...
    (t+D(i,j))]);
  end;
end;
D
```

As you already suspect, the type of problems we are introducing now has a lot to do with genomics. It is most interesting to detect similarities in DNA sequences.

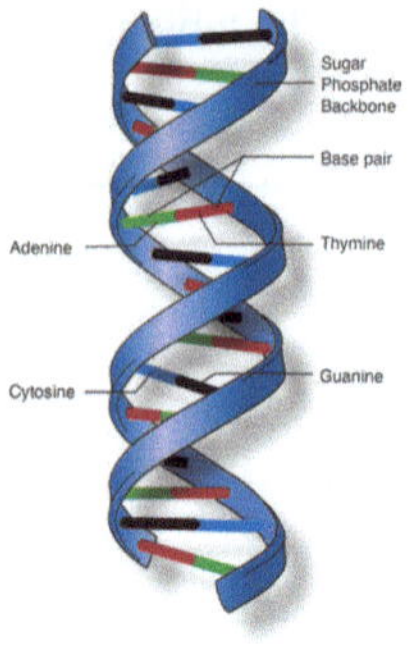

Fig. 5.27 DNA double helix (see https://www.genome.gov/genetics-glossary/Double-Helix)

Pair-wise sequence alignment is a form of representation that helps visualize similarity. For instance

work-ers
| ||
whisper

The pipe character represents a match between characters. Notice that a gap has been inserted in the first sequence, in favor of matching. It turns out that gaps are important in DNA sequences. The next example shows the alignment of two DNA sequences [45]:

CAT - TCA -C
| | | |
C- TCGCAGC

Notice that the example above only uses A, C, G, and T letters. Let us explain why. As shown in Fig. 5.27, the DNA is a double-stranded, helical molecule. Each strand is composed of a long sequence of nucleotides. There are not many nucleotides to be combined for the sequence; instead, there are only four: adenine, guanine, cytosine, and thymine. Hence, the use of only four letters for DNA sequences.

Looking at alignments, it seems opportune to introduce a certain scoring, for instance, a reward of 2 for each match, a penalty of -3 for each mismatch, and a penalty of -2 for each gap.

A scoring function could be defined, using as example "A" and "G," in the following way [70]:

$\sigma(A, A) = +2$, this is a match
$\sigma(G, A) = -3$, this is a mismatch
$\sigma(G, -) = -2$, deletion
$\sigma(-, G) = -2$, insertion

(Other numbers are possible.)

As before, a recurrence relation can be used [70]. The strings to be aligned are $A[1] \ldots A[i] \ldots A[n]$ and $B[1] \ldots B[j] \ldots B[m]$; and $S(n, m)$ is the maximum score:

$$S(i,\ j) = \max \begin{cases} S(i-1,j) + \sigma(A[i], -) \\ S(i, j-1) + \sigma(-, B[j]) \\ S(i-1, j-1) + \sigma(A[i], B[j]) \end{cases}$$

where

$$S(0,0) = 0;\ \ S(i,0) = S(i-1,0) + \sigma(A[i], -);$$

$$S(0,\ j) = S(0,\ j-1) + \sigma(-, B[j])$$

For example, let us deal with the alignment of the following sequences [45]:

ACGCATCA

ACTGATTCA

Based on the recurrence relation, one builds the following table:

		A	C	T	G	A	T	T	C	A
	0	-2	-4	-6	-8	-10	-12	-14	-16	-18
A	-2	2	0	-2	-4	-6	-8	-10	-12	-14
C	-4	0	4	2	0	-2	-4	-6	-8	-10
G	-6	-2	2	1	4	2	0	-2	-4	-6
C	-8	-4	0	-1	2	1	-1	-3	0	-2
A	-10	-6	-2	-3	0	4	2	0	-2	2
T	-12	-8	-4	0	-2	2	6	4	2	0
C	-14	-10	-6	-2	-3	0	4	3	6	4
A	-16	-12	-8	-4	-5	-1	2	1	4	8

An optimal path can be traced, beginning from the lower right corner (see next table).

		A	C	T	G	A	T	T	C	A
	0	-2	-4	-6	-8	-10	-12	-14	-16	-18
A	-2	2	0	-2	-4	-6	-8	-10	-12	-14
C	-4	0	4	2	0	-2	-4	-6	-8	-10
G	-6	-2	2	1	4	2	0	-2	-4	-6
C	-8	-4	0	-1	2	1	-1	-3	0	-2
A	-10	-6	-2	-3	0	4	2	0	-2	2
T	-12	-8	-4	0	-2	2	6	4	2	0
C	-14	-10	-6	-2	-3	0	4	3	6	4
A	-16	-12	-8	-4	-5	-1	2	1	4	8

In conformity with the optimal path, the alignment would be

AC–GCAT–CA

| | | | | | | |

ACTG–ATTCA

Looking at the last table, note that horizontal arrows correspond to gaps in the second sequence, while vertical arrows correspond to gaps in the first sequence.

The procedure for computing the score matrix S can be described as follows:

Iterative Algorithm

```
Global alignment (A,B)
S[0,0] ← 0
For i ← 1 to m
      S[i,0] ← S[i-1,0]+ σ(A[i],-)
For j ← 1 to n
      S[0,j] ← S[0,j-1]+ σ(-,B[j])
For i ← 1 to m
S[i,j]= max{ S[i-1,j-1]+ σ(A[i],B[j]) | S[i-1,j]+ σ(A[i],-) | S[i,j-1]+ σ(-,B[j])
```

Program 5.14 offers a direct MATLAB coding of the algorithm for filling the S table, including also some lines for recovering the aligned strings. The program is based on [14].

Program 5.14 DNA alignment example

```
%DNA alignment example
A='ACGCATCA';
B='ACTGATTCA';
n=length(A);
m=length(B);
sA=[]; sB=[];
sg1=2; sg2=-3; sg3=-2; sg4=-2; %sigma values
S=zeros(n+1,m+1);
%table filling---------
%first column
for i=1:n,
  S(i+1,1)=S(i,1)+sg3;
end;
%first row
for j=1:m,
  S(1,j+1)=S(1,j)+sg4;
end;
%the other entries
for i=1:n,
  for j=1:m;
    if (A(i)==B(j)),
      t=sg1;
    else
      t=sg2;
    end;
```

```
    S(i+1,j+1)=max([(sg4+S(i+1,j)),(sg3+S(i,j+1)),...
     (t+S(i,j))]);
  end;
end;
%getting the aligned strings
i=n; j=m;
%gaps will be represented with zeroes
while (i>0 | j>0),
  if (i>0 & S(i+1,j+1) == S(i,j+1)+sg4),
    sA=[A(i) sA];
    sB=[0 sB];
    i=i-1;
  elseif (j>0 & S(i+1,j+1) == S(i+1,j)+sg3),
    sA=[0 sA];
    sB=[B(j) sB];
    j=j-1;
  else
    sA=[A(i) sA];
    sB=[B(j) sB];
    i=i-1;
    j=j-1;
  end;
end;
S
sA
sB
```

The use of dynamic programming for optimal pairwise global DNA alignment was introduced in 1970 by S.B. Needleman and C.D. Wunsch. A modification of the method for local alignment of subsequences was proposed by T.F. Smith and M.S. Waterman in 1981. Since then, several improvements have been recommended [4], [67].

In addition to algorithms based on dynamic programming, other heuristic-based methods have been proposed [7].

More details of alignment can be found in extensive tutorials, like [59] on sequence alignment and [68] on sequence analysis. An interesting survey of alignment algorithms (up to the year 2010) is [36].

MathWorks offers a **MATLAB Bioinformatics Toolbox**, which includes a sequence analysis section, with a Sequence Alignment app. See the chapter [62], on how to use the toolbox for processing biological sequences.

The research and applications related to DNA sequences have grown immensely. For instance, a popular methodology is *multiple sequence alignment* (MAS) (see [13]). In 1988, a program called **Clustal** was created for MAS, and it is based on deriving phylogenetic trees and applying clustering (hence its name). Next generations of this program were ClustalV, ClustalW (very popular), ClustalX, Clustal2, and ClustalΩ (released in 2011). Today, more software tools are available for MAS, as surveyed in [79] (year 2022). A list of links for MAS resources can be found at https://tcoffee.org/Resources/aligner_list.html.

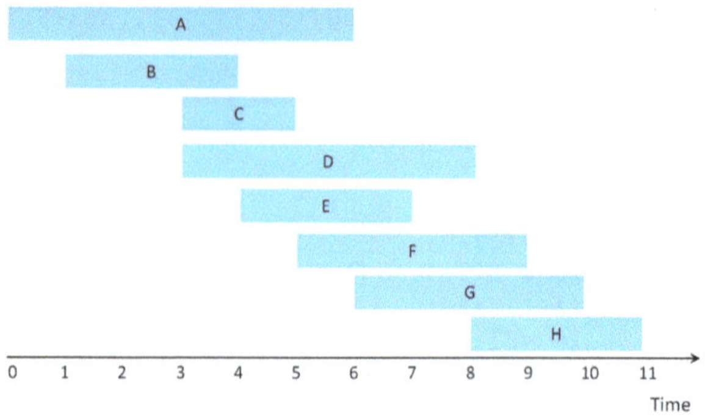

Fig. 5.28 Activities to be considered

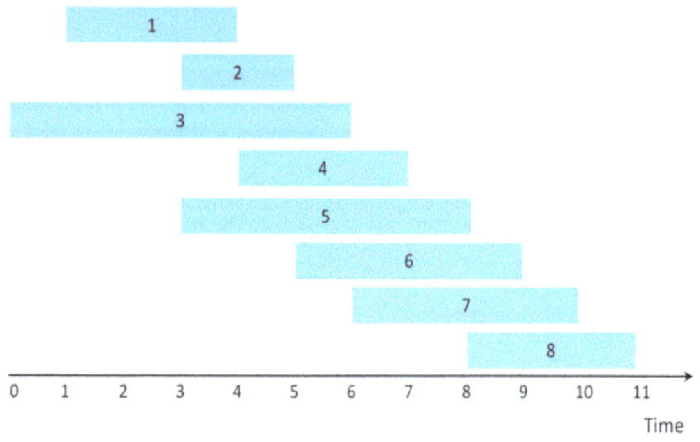

Fig. 5.29 Activities ordered by finishing time

5.6.3 Weighted Interval Scheduling

Consider the following application problem. There are several activities; each one has a start and finish time and a certain weight (profit). One wants to find a subset of nonoverlapping activities, which maximizes the sum of their profits. This optimal schedule would be denoted as S^*.

Figure 5.28 depicts a possible scenario, with a set {A, B, C…H} of activities.

The problem can be reformulated by assigning labels, 1, 2, .3,.., to the activities, corresponding to finishing times: $f_1 \leq f_2 \leq f_3, \ldots$. Figure 5.29 shows how activities were reordered for our purposes.

Denote $p(j)$ as the largest index $i < j$ such that activity i is compatible (does not overlap) with activity j. For example, in the case pictured by previous figures, $p(7) = 3$, $p(8) = 5$, etc. By convention, $p(j) = 0$ means conflict with all precedent activities.

Consider the last activity (let us call it $A(8)$). This activity may belong to S^*, or not. If not, $A(8)$ can be ignored, and proceed recursively toward the optimal solution looking now at the first $n - 1$ (seven) activities. Otherwise, $A(8)$ belongs to S^*, all other activities overlapping $A(8)$ should be ignored, and the value of $p(8)$ must be used for selecting previous compatible activities, in order to complete the optimal schedule.

Then, a recursive procedure would be suggested. Denote $opt(j)$ as the sum of profits given by the optimal solution for $1, 2, \ldots j$ activities. The recursive rule would be

$$opt(j) = \max \begin{cases} opt(j-1), & A(j)\, is\, not\ in\ S^* \\ w_j + opt(p(j)), & A(j)\, is\, in\ S^* \end{cases}$$

where w_j are the profits.

This expression can be translated to the following pseudo-code:

Recursive Algorithm

```
opt (j)
If (j == 0)
   return 0
else
   return max (opt(j-1), w[j] opt(p[j] )
```

There is a problem with this recursion. Like in the case of Fibonacci numbers, it leads to an enormous number of operations-something like 32 billion recursive calls for a schedule with 50 activities [44].

A memoized algorithm would be better. For instance, using an array $M(j)$ for storing computed values (only once for each entry), the algorithm would be

Memoized Algorithm

```
All M[j] initialized to empty
M-opt (j)
If M[j] is empty
   M[j] ←max (M-opt(j-1), w[j] M-opt(p[j] )
return M[j]
```

There is also another way, bottom-up, for computing the optimal profit. It is a non-recursive algorithm:

Bottom-Up Algorithm

```
M[0] ← 0
for j=1 to n
   M[j] ←max (M[j-1], w[j] M[p[j]] )
```

To recover the optimal schedule itself, one could apply the following procedure [73]:

```
Solution(j)
if (j==0) then return 0
else if (w[j] M[p[j] > M[j-1])
   return j ∪Solution(p[j])
else
   return Solution(j-1)
```

Program 5.15 provides a MATLAB code for the memoized algorithm, focusing on the scenario described in Fig. 5.28.

Program 5.15 Weighted scheduling example

```
%weighted scheduling example
sch=[];
global M pred;
M=10000*ones(8,1); %to indicate emptiness
pred=zeros(8,1);
%activities (already sorted)
%[init, finish, profit]
A(1,:)=[1,4,3];
A(2,:)=[3,5,2];
A(3,:)=[0,6,1];
A(4,:)=[4,7,9];
A(5,:)=[3,8,3];
A(6,:)=[5,9,5];
A(7,:)=[6,10,6];
A(8,:)=[8,11,4];
%obtaining p(j)
p=zeros(8,1);
for j=8:-1:2,
  i=j;
  nn=0;
```

```
  while (nn==0),
    if (A(i,2) > A(j,1)),
      i=i-1;
      if (i==0),
        nn=1;
      end;
    else
      p(j)=i;
      nn=1;
    end;
  end;
end;
%apply memoized algorithm
M_opt(8,A,p);
profit=max(M);
%recover schedule
j=8;
while (j>0),
  if (pred(j) == p(j)),
    sch=[j,sch]; %append to schedule
  end;
  j=pred(j);
end;
%results
profit
sch
```

The previous program recursively uses Function 5.16, which is listed below. Notice that the vector *pred[]* has been included, for easy recovery of the optimal schedule (the recovery part is based on [44]).

Program 5.16 Weighted scheduling: function for memoized parenthesization

```
%Function for memoized parenthesization
function q=M_opt(j,A,p)
global M pred;
if (j==0),
  q=0;
else
  if (M(j)==10000),
      [q,ix]=max([M_opt(j-1,A,p)...
       ,(A(j,3)+M_opt(p(j),A,p))]);
    if (ix==1),
      pred(j)=j-1; %predecessor
    else
      pred(j)=p(j); %predecessor
    end;
    M(j)=q;
  else
    q=M(j);
  end;
end;
end;
```

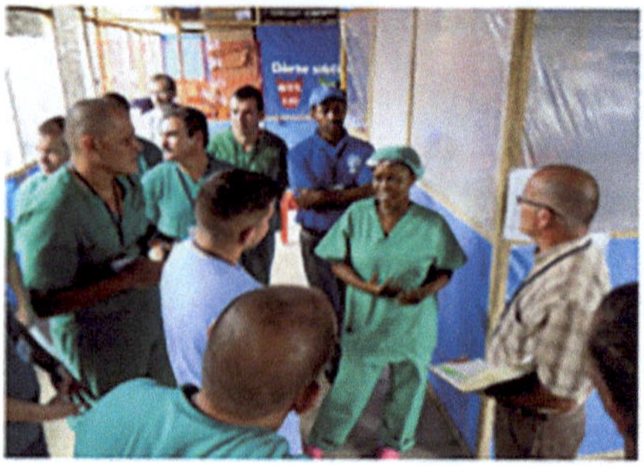

Fig. 5.30 Part of a medical team preparing for action [77]

The reader is invited to change the parameters of the scheduling problem and see what may happen.

There are several types of scheduling problems (including the weighted interval scheduling). A most cited book on scheduling is [52] (year 2012, 12,612 cites), which contemplates many aspects of the question. The main milestones of scheduling research, over the last fifty years, are described in [54].

An interesting academic work is the thesis [71] (year 2016), with contributions on dynamic programming for vehicle routing and for job shop scheduling. In particular, job shop scheduling is an important type of scheduling problems; if you ask Google about it, you get 54,800,000 results. The article [21] contains a valuable treatment of job shop scheduling using dynamic programming.

Lately, grid computing systems and cloud systems have motivated a new type of scheduling problems. There is a recent survey, [23] (year 2023), on cloud computing scheduling. The article [2] offers a survey of applying operations research methods to grid computing scheduling.

A MATLAB package, called **TORSCHE Scheduling Toolbox**, is available from GitHub, at https://github.com/CTU-IIG/TORSCHE2017.

5.6.4 Distribution of Medical Teams to Countries

A usual application of dynamic programming is the optimal distribution of resources to activities. The case to be considered now belongs to this area.

The question is to allocate five medical teams among three countries, [25] or [72], in order to contribute for humanitarian aspects, like patient care, health education, and personnel training. Figure 5.30 tries to picture the special atmosphere of this type of problem.

A measure of the effectiveness of this initiative is the increase of life expectancy, in years times the country's population (in multiples of 1000). Here is a table with data.

Medical teams	Country 1	Country 2	Country 3
0	0	0	0
1	45	20	50
2	70	45	70
3	90	75	80
4	105	110	100
5	120	150	130

The evident objective is to maximize efficiency.

The number of teams assigned to country i would be denoted as x_i. Hence, the problem is to determine x_i, x_2, x_3. Another variable that will be useful is s_i, the number of remaining teams not yet assigned. Suppose, for example, that one starts with country 1, then $s_1 = 5$; now, if it is decided to assign two teams to country 1, then $x_1 = 2$ and $s_2 = s_1 - x_1 = 3$. Focusing now on country 2, if it is decided to assign here one team, so $x_2 = 1$, then $s_3 = s_2 - x_2 = 2$ (so $x_3 = 2$).

Notice that the problem can be treated as a sequence of stages, much like the example with which this chapter opened.

In order to formalize the analysis, denote $p_i(x_i)$ as the profit corresponding to assigning x_i teams to country i. The values of this function were given by the previous table of data. Thus, the optimization problem is

Maximize:

$$J = \sum_{i=1}^{3} p_i(x_i)$$

Subject to:

$$\sum_{i=1}^{3} x_i = 5$$

The value function at stage n would be

$$V_n(s_n, x_n) = p_n(x_n) + \max \sum_{i=n+1}^{3} p_i(x_i)$$

Hence

$$V_n(s_n, x_n) = p_n(x_n) + V_{n+1}^*(s_{n+1}, x_{n+1})$$

where V^* denote maximum value.

Let us go backwards. At stage 3, one simply has $V_3^*(s_3, x_3) = p_3(x_3)$.

At stage 2, s_2 can have a value between 0 and 5. For instance, suppose that $s_2 = 3$, then there are four cases to compare:

$$x_2 = 0, \quad V_2(3,\ 0) = p_2(0) + V_3^*(3) = 0 + 80 = 80$$

$$x_2 = 1, \quad V_2(3,\ 1) = p_2(1) + V_3^*(2) = 20 + 70 = 90$$

$$x_2 = 2, \quad V_2(3,\ 2) = p_2(2) + V_3^*(1) = 45 + 50 = 95$$

$$x_2 = 3, \quad V_2(3,\ 3) = p_2(3) + V_3^*(0) = 75 + 0 = 75$$

Note that $p_2(x_i)$ and $V_3^*(s_3, x_3) = p_3(s_2 - x_2)$ are given by columns of the table of data. As a result of the study at stage 2, one obtains the following table:

$s_2\backslash x_2$	0	1	2	3	4	5	V_2*	x_2*
0	0						0	0
1	50	20					50	0
2	70	70	45				70	0 or 1
3	80	90	95	75			95	2
4	100	100	115	125	110		125	3
5	130	120	125	145	160	150	160	4

At stage 1, calculations are much simpler, since $s_1 = 5$. There are six cases to compare:

$$x_1 = 0, \quad V_1(5,\ 0) = p_1(0) + V_2^*(5) = 0 + 160 = 160$$

$$x_1 = 1, \quad V_1(5,\ 1) = p_1(1) + V_2^*(4) = 45 + 125 = 170$$

- - -

$$x_1 = 5, \quad V_1(5,\ 5) = p_1(5) + V_2^*(0) = 120 + 0 = 120$$

The results can be presented in a simple table:

$s_1\backslash x_1$	0	1	2	3	4	5	V_1*	x_1*
5	160	170	165	160	155	120	170	1

Therefore, the solution to the problem is to send one team to country 1, so $s_2 = 4$ and $x_2^* = 3$ (then send three teams to country 2), and finally one team to country 3. This would mean 170,000 additional person-years of life.

Program 5.17 just does the calculations described above. It could be useful for playing with modifications of the problem.

Program 5.17 Distribution of medical teams

```
%medical teams distribution
P=[0 0 0;
45 20 50;
70 45 70;
90 75 80;
105 110 100;
120 150 130];
V3m=P(:,3);
V2=zeros(6,6);
for s2=1:6,
  x2=1:s2;
  V2(s2,x2)=P(x2,2)+V3m(s2+1-x2);
end;
[V2m, ix2]=max(V2');
for x1=1:6,
  V1(x1)=P(x1,1)+V2m(7-x1);
end;
%result
[V,ix]=max(V1);
V
x1=ix-1
s2=5-x1;
x2=ix2(s2)
x3=5-x1-x2
```

The application of dynamic programming to the allocation of medical teams attracts a lot of interest. Asking Google about it, one obtains 7,800,000 results.

More generally, the application of dynamic programming in health care attains an impressive level of attention (Google measure), more than 60,000,000 results.

The survey [56] on the applications of operations research in health care helps grasp the variety and complexity of this field.

5.7 Exercises

1- [5.1] A salesman decided to travel from city A to city B. There are several roads that could be taken according to the following map:

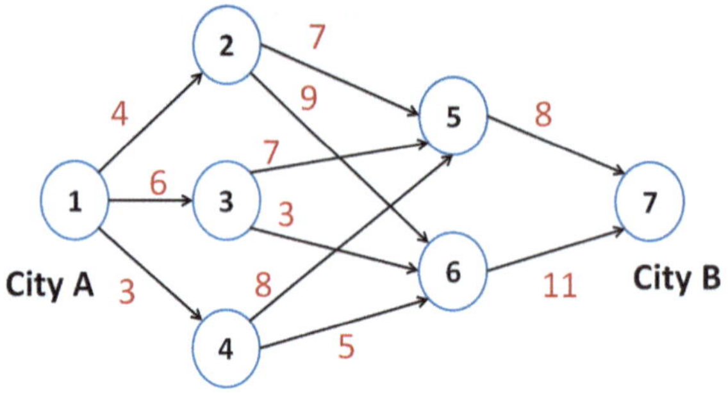

The map includes information on distances. Use dynamic programming to determine the shortest path. Explain the optimal policy.

2- [5.2] The following grid includes obstacles and rewards. Find the optimal path for a robot to get from ENTR to EXIT, collecting as many rewards as possible. The robot has to move down or left.

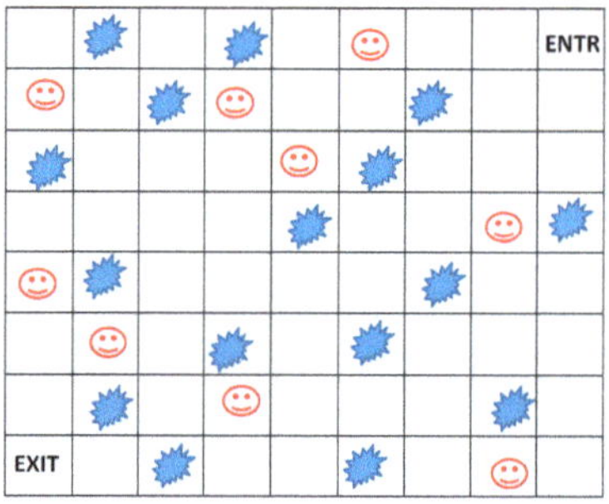

3- [5.3] Given five matrices: A1 (2 × 4), A2 (4 × 2), A3 (2 × 1), A4 (1 × 5), A5 (5 × 2), find an optimal parenthesization. Explain the result, and compare it with sub-optimal alternatives.

4- [5.4] Find the longest common subsequence among the following sequences:

TGACTAC

ACTGATGC

5- There is a row of eight coins of values **{3, 1, 5, 2, 5, 4, 2, 3}.** Pick up the maximum amount of money, subject to the following condition: No two coins adjacent in the list can be picked up. Write the history of your procedure (e.g., $C2 = 1, C5 = 5$).

6- [5.5] Space missions require careful mathematical analysis, including optimization-related considerations. In this exercise, we shall focus on just an aspect of a journey to the Moon. It is the final phase, when you try to land (not to crash) on the surface: It is technically called soft landing. The next figure illustrates the situation, supposing a vertical final trajectory (as represented on the right-hand side of the figure):

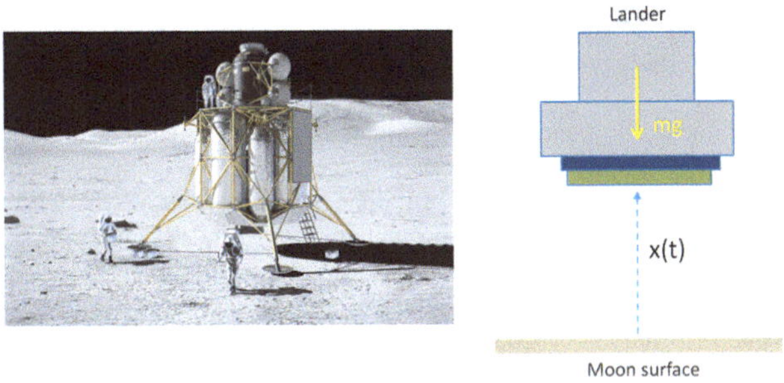

The peculiar facet of the problem is that the mass, m, decreases when you fire the rocket for braking. Therefore, one has to consider a mass flow, $\dot{m}$. The motion of the vehicle is governed by

$$\ddot{x}(t) = -\frac{K\,\dot{m}(t)}{m(t)} + g$$

where K >0 is a constant. The onboard system (or the pilot) should control the use of the rocket, so the input to consider is $u(t) = \dot{m}(t)$. We want to minimize fuel consumption, which is

$$J = \int_0^T u(t)\,dt$$

Please, apply the Pontryagin principle for determining the optimal command $u(t)$ for soft landing with minimum fuel consumption. Based on the simulation, visualize the optimal maneuver behavior along time (velocity, height, $u(t)$, fuel decreasing).

5.7.1 Exercise References

[5.1] https://universe.bits-pilani.ac.in/uploads/DP.pdf
[5.2] https://courses.cs.washington.edu/courses/cse417/18wi/lectures/lec10-dynamic-programming-1.pdf
[5.3] https://home.csulb.edu/~tebert/teaching/lectures/528/dp/dp.pdf
[5.4] https://www.jsums.edu/nmeghanathan/files/2016/01/CSC323-Sp2016-QB-Module-4-Dynamic-Programming.pdf
[5.5] https://mavmatrix.uta.edu/math_theses/13/

References

1. Akian, M., Blechschmidt, J., Botkin, N. D., Jensen, M., Kröner, A., Picarelli, A., Smears, I., Urban, K., Chekroun, M. D., Herzog, R., et al. (2018). *Hamilton-Jacobi-Bellman equations: Numerical methods and applications in optimal control* (Vol. 21). Walter de Gruyter GmbH & Co KG.
2. Attanasio, A., Ghiani, G., Grandinetti, L., Guerriero, E., & Guerriero, F. (2005). Operations research methods for resource management and scheduling in a computational grid: A survey. In *Advances in Parallel Computing* (Vol. 14, pp. 53–81). Elsevier.
3. Bafna, V. (2009). *Dynamic programming*. Available from: http://cseweb.ucsd.edu/classes/fa09/cse182/slides/dp.pdf
4. Batzoglou, S. (2016). *CS 262- lecture 2 scribe notes, sequence alignment*. Available from: https://web.stanford.edu/class/cs262/notes/lecture2.pdf
5. Bellman, R. E., & Dreyfus, S. E. (2015). *Applied dynamic programming* (Vol. 2050). Princeton University Press.
6. Bertsekas, D. (2012). *Dynamic programming and optimal control: Volume I* (Vol. 4). Athena Scientific.
7. Bucak, I., & Uslan, V. (2010). An analysis of sequence alignment: Heuristic algorithms. In *2010 Annual International Conference of the IEEE Engineering in Medicine and Biology* (pp. 1824–1827). IEEE.
8. Cacchiani, V., Iori, M., Locatelli, A., & Martello, S. (2022). Knapsack problems-An overview of recent advances. Part I: Single knapsack problems. *Computers & Operations Research, 143*, 105692. Available from: https://www.sciencedirect.com/science/article/pii/S0305054821003877
9. Cacchiani, V., Iori, M., Locatelli, A., & Martello, S. (2022). Knapsack problems-An overview of recent advances. Part II: Multiple, multidimensional, and quadratic knapsack problems. *Computers & Operations Research, 143*, 105693. Available from: https://www.sciencedirect.com/science/article/pii/S0305054821003889
10. Caputo, M. R. (2012). Foundations of dynamic economic analysis. (Dynamic programming and the Hamilton-Jacobi-Bellman equation). In *Foundations of dynamic economic analysis*, Chap. 19 (pp. 511–536). Cambridge University Press.
11. Clarke, F. (2010). The Pontryagin maximum principle and a unified theory of dynamic optimization. *Proceedings of the Steklov Institute of Mathematics, 268*(1), 58–69. Available from: https://link.springer.com/content/pdf/10.1134/S0081543810010062.pdf
12. Dokuchaev, N., & Zolotarevich, V. (2020). Optimal energy storing and selling in continuous time stochastic multi-battery setting. *Sustainable Energy, Grids and Networks, 22*, 100340. Available from: https://arxiv.org/pdf/1511.01365
13. Edgar, R. C., & Batzoglou, S. (2006). Multiple sequence alignment. *Current Opinion in Structural Biology, 16*(3), 368–373. Available from: https://www.bx.psu.edu/old/courses/bx-fall10/edgar.pdf
14. Elber, R. (2018). *CS 321- numerical methods in computational molecular biology*. Available from: http://www.cs.cornell.edu/courses/cs321/2001fa/examples/align/alignSequences.m.txt
15. Erickson, J. (2019, June). *Algorithms*. Available from: http://jeffe.cs.illinois.edu/teaching/algorithms/
16. Evans, L. C. (1983). *An introduction to mathematical optimal control theory version 0.2*. Available from: http://math.berkeley.edu/~evans/control.course.pdf
17. Furfaro, R., & D'Ambrosio, A. (2024). Increasing autonomy of aerospace systems via PINN-based solutions of HJB equation. In *AIAA SCITECH 2024 Forum* (p. 1786). American Institute of Aeronautics and Astronautics.
18. Gamkrelidze, R. V. (2019). History of the discovery of the Pontryagin maximum principle. *Proceedings of the Steklov Institute of Mathematics, 304*(1), 1–7.
19. Govindarajan, N. (2012). *An optimal control approach for estimating aircraft command margins*. Available from: https://repository.tudelft.nl/islandora/object/uuid:48dbad61-728a-4c7e-ba3e-cf8382ef1cef

20. Greif, C. (2017). *Numerical methods for Hamilton-Jacobi-Bellman equations*, Master's thesis, University of Wisconsin-Milwaukee. Available from: https://dc.uwm.edu/cgi/viewcontent.cgi?article=2485&context=etd
21. Gromicho, J. A., Van Hoorn, J. J., Saldanha-da Gama, F., & Timmer, G. T. (2012). Solving the job-shop scheduling problem optimally by dynamic programming. *Computers & Operations Research*, *39*(12), 2968–2977.
22. Guo, X., Zhu, W., & Fang, Y. (2018). Guided motion planning for snake-like robots based on geometry mechanics and HJB equation. *IEEE Transactions on Industrial Electronics*, *66*(9), 7120–7130.
23. Gupta, S., & Tripathi, S. (2023). A comprehensive survey on cloud computing scheduling techniques. *Multimedia Tools and Applications*, 1–54. Available from: https://link.springer.com/article/10.1007/s11042-023-17216-6
24. Gusfield, D. (1997). *Algorithms on strings, trees, and sequences: Computer science and computational biology*. Cambridge University Press.
25. Hillier, F. S., & Lieberman, G. J. (2016). *Introduction to operations research*. McGraw-Hill. Available from: https://www.ime.unicamp.br/~andreani/MS515/capitulo7.pdf
26. Hosoya, Y. (2023). On the basis of the Hamilton-Jacobi-Bellman equation in economic dynamics. *Physica D: Nonlinear Phenomena*, *446*, 133684. Available from: https://www.sciencedirect.com/science/article/pii/S0167278923000386
27. Jacobsson, H. (2022). *On Merton's portfolio problem: A stochastic optimal control problem*. Available from: https://www.diva-portal.org/smash/get/diva2:1667781/FULLTEXT01.pdf
28. Kafash, B., & Nadizadeh, A. (2017). Solution of stochastic optimal control problems and financial applications. *Journal of Mathematical Extension*, *11*, 27–44. Available from: https://ijmex.com/index.php/ijmex/article/viewFile/490/333
29. Kellerer, H., Pferschy, U., & Pisinger, D. (2004). *Multidimensional knapsack problems*. Springer.
30. Khardi, S. (2012). Aircraft flight path optimization. the Hamilton-Jacobi-Bellman considerations. *Applied Mathematical Sciences*, *6*(25), 1221. Available from: https://hal.science/hal-00987163/document
31. Khardi, S. (2012). Aircraft shortest and fastest continuous descent approach development. *Journal of Aircraft*, *49*(6), 1931–1939. Available from: https://arc.aiaa.org/doi/abs/10.2514/1.C031775?journalCode=ja
32. Kleinberg, J., & Tardos, E. (2006). *Algorithm design*. Pearson Education India.
33. Kohn, R. V. (2009). *A brief introduction to optimal control*. Available from: https://math.nyu.edu/~kohn/calculus-of-variations/2009/hamilton-jacobi.pdf
34. Leondes, C. T. (1980). An appreciation of Professor Richard Bellman. *Journal of Optimization Theory and Applications*, *32*(4), 399–406.
35. Lewis, A. D. (2006). *The maximum principle of Pontryagin in control and in optimal control*. Available from: https://mast.queensu.ca/~andrew/teaching/pdf/maximum-principle.pdf
36. Li, H., & Homer, N. (2010). A survey of sequence alignment algorithms for next-generation sequencing. *Briefings in Bioinformatics*, *11*(5), 473–483. Available from: https://cphg.github.io/compgen/resources/ShortReadAlignment.pdf
37. Ma, S. (2014). *Dynamic programming examples*. Available from: https://www1.se.cuhk.edu.hk/~seem3470/Handouts/3-more-examples-DP-alg
38. Martello, S., & Toth, P. (1990). *Knapsack problems: Algorithms and computer implementations*. John Wiley & Sons, Inc.
39. Micelli, P. (2018). *Path planning for road vehicles by dynamic programming*, Ph.D. thesis, Università di Parma, Dipartimento di Ingegneria ed Architettura]. Available from: https://www.repository.unipr.it/bitstream/1889/3594/4/TesiDottoratoMicelli.pdf
40. MIT OpenCourseWare. (2008). *HJB equation*. Available from: https://ocw.mit.edu/courses/16-323-principles-of-optimal-control-spring-2008/resources/lec4/
41. Moll, B. (2018). *Hamilton-Jacobi-Bellman equations*. Available from: https://benjaminmoll.com/wp-content/uploads/2019/07/Lecture3_2149.pdf

42. Monaci, M. (2019). *Knapsack problem and variants* [Conference presentation]. Available from: https://www.euro-online.org/websites/esicup/wp-content/uploads/sites/12/2019/05/esicup2019-michelemonaci.pdf
43. Mount, D. M. (2003). *CMSC 451 design and analysis of computer algorithms*. University of Maryland, Department of Computer Science.
44. Mount, D. (2017). *Cmsc 451: Lecture 10 dynamic programming: Weighted interval scheduling*. Available from: http://www.cs.umd.edu/class/fall2017/cmsc451-0101/Lects/lect10-dp-intv-sched.pdf
45. Musso, G. (2006). *CSC2427–course notes, alignment*. Available from: http://www.cs.toronto.edu/~brudno/csc2427/Lec7Notes.pdf
46. Nagami, K., & Schwager, M. (2021). HJB-RL: Initializing reinforcement learning with optimal control policies applied to autonomous drone racing. In *Robotics: Science and Systems* (pp. 1–9). Available from: https://roboticsproceedings.org/rss17/p062.pdf
47. Norris, J. (2007). *The Hamilton-Jacobi-Bellman equation*. Available from: https://www.statslab.cam.ac.uk/~james/Lectures/oc15.pdf
48. Norris, J. (2016). *Pontryagin's maximum principle*. Available from: http://www.statslab.cam.ac.uk/~james/Lectures/oc16.pdf
49. Parkinson, C., & Boyle, I. (2024). Efficient and scalable path-planning algorithms for curvature constrained motion in the Hamilton-Jacobi formulation. *Journal of Computational Physics*, *509*, 113050. Available from: https://arxiv.org/pdf/2304.12377
50. Parzani, C., & Puechmorel, S. (2018). On a Hamilton-Jacobi-Bellman approach for coordinated optimal aircraft trajectories planning. *Optimal Control Applications and Methods*, *39*(2), 933–948. Available from: https://enac.hal.science/hal-01340565/document
51. Pesch, H. J., Plail, M., & Munich, D. (2009). The maximum principle of optimal control: A history of ingenious ideas and missed opportunities. *Control and Cybernetics*, *38*(4A), 973–995.
52. Pinedo, M. L. (2012). *Scheduling* (Vol. 29). Springer. Available from: https://link.springer.com/content/pdf/10.1007/978-3-319-26580-3.pdf
53. Pontryagin, L. S. (2018). *Mathematical theory of optimal processes*. Routledge.
54. Potts, C. N., & Strusevich, V. A. (2009). Fifty years of scheduling: A survey of milestones. *Journal of the Operational Research Society*, *60*, S41–S68.
55. Prakash, R., Mohanta, J. K., & Behera, L. (2022). Closed form HJB solution for path planning of a robot manipulator with warehousing applications. In *2022 IEEE 18th international conference on automation science and engineering (CASE)* (pp. 2049–2055). IEEE. Available from: https://ieeexplore.ieee.org/abstract/document/9926505
56. Rais, A., & Viana, A. (2011). Operations research in health care: A survey. *International Transactions in Operational Research*, *18*(1), 1–31. Available from: https://onlinelibrary.wiley.com/doi/full/10.1111/j.1475-3995.2010.00767.x
57. Romsäter, T. (2023). *Merton's portfolio problem under Grezelak-Oosterlee-Van Veeren model*. Available from: https://www.diva-portal.org/smash/get/diva2:1770603/FULLTEXT01.pdf
58. Rust, J. (2008). Dynamic programming. In *The new Palgrave dictionary of economics* (Vol. 1, p. 8). Available from: https://editorialexpress.com/jrust/research/papers/dp.pdf
59. Schatz, M. (2011). *Sequence alignment*. Cold Spring Harbor Lab., NY. Bioinformatics lecture 2. Available from: http://schatzlab.cshl.edu/teaching/2011/2011.Lecture3.Sequence%20Alignment.pdf
60. Serna, M. J. (2017). *Dynamic programming* (Lecture notes T17). Universitat Politecnica de Catalunya, Barcelona Tech. Available from: https://www.cs.upc.edu/~mjserna/docencia/grauA/T17/ProgramacioDinamica.pdf
61. Shiri, H., Park, J., & Bennis, M. (2020). Communication-efficient massive UAV online path control: Federated learning meets mean-field game theory. *IEEE Transactions on Communications*, *68*(11), 6840–6857. Available from: https://mm.aueb.gr/leadingedge/pubs/2020_SPB_ToC.pdf
62. Singh, G. B. (2015). Processing biological sequences with MATLAB. In *Fundamentals of bioinformatics and computational biology* (pp. 77–93). Springer.

63. Siska, D. (2017). *Dynamic programming and Hamilton-Jacobi-Bellman equations*. Available from: https://www.maths.ed.ac.uk/~dsiska/scdaa_2017-18/03-LecNotesSCDAA-Bellman.pdf
64. Smears, I. (2019). *Hamilton-Jacobi-Bellman equations. Analysis and numerical analysis*. Available from: https://maths.dur.ac.uk/Ug/projects/highlights/PR4/Smears_HJB_report.pdf
65. Spindler, K. (2013). Optimal control on Lie groups: Theory and applications. *WSEAS Transactions on Mathematics*, *12*(5), 531–542.
66. Stevens, A. (2016). *Knapsack problem using dynamic programming*. Available from: https://es.mathworks.com/matlabcentral/answers/266697-knapsack-problem-using-dynamic-programming
67. Stojanov, D., et al. (2017). Comparing DNA applying dynamic programming: A historical review. *Tem Journal*, *6*(4), 888–893.
68. Stoye, J. (2015). *Sequence analysis* (Lecture notes). Bielefeld University. Available from: https://profs.scienze.univr.it/~liptak/FundBA/files/sequence_analysis_ws1516.pdf
69. Sundar, S., & Shiller, Z. (1997). Optimal obstacle avoidance based on the Hamilton-Jacobi-Bellman equation. *IEEE Transactions on Robotics and Automation*, *13*(2), 305–310. Available from: https://ieeexplore.ieee.org/abstract/document/351148
70. Tompa, M. (2000). *Lecture notes on biological sequence analysis*, Technical report. University of Washington, Seattle.
71. van Hoorn, J. J. (2016). *Dynamic programming for routing and scheduling: Optimizing sequences of decisions*, Ph.D. thesis, Vrije Universiteit Amsterdam. Available from: https://research.vu.nl/ws/portalfiles/portal/42163578/complete+dissertation.pdf
72. Wan, Y.-W. (2014). *Dynamic programming* (Lecture notes GSLM 52800). National Dong Hwa University, Operations Research II. Available from: http://faculty.ndhu.edu.tw/~ywan/courses/ORII/notes/dynamic_programming.pdf
73. Wayne, K. (2006). *Dynamic programming I*. Available from: https://www.cs.princeton.edu/~wayne/kleinberg-tardos/pdf/06DynamicProgrammingI.pdf
74. Wayne, K. (2006). *Lecture slides for algorithm design by Jon Kleinberg and Eva Tardos*. Available from: https://www.cs.princeton.edu/~wayne/kleinberg-tardos/
75. Weber, R. (2016). *Optimization and control*. Available from: http://www.statslab.cam.ac.uk/~rrw1/oc/oc2016.pdf
76. Weston, J., Tolić, D., & Palunko, I. (2024). Application of Hamilton-Jacobi-Bellman equation/Pontryagin's principle for constrained optimal control. *Journal of Optimization Theory and Applications*, 1–26. Available from: https://link.springer.com/article/10.1007/s10957-023-02364-4
77. World Health Organization. (2019). *Emergency medical teams*. Available from: https://www.who.int/hac/emergency-medical-teams/en/
78. Yoshioka, H., & Tsujimura, M. (2023). *The robust Orlicz risk with an application to the green photovoltaic power generation*. arXiv preprint arXiv:2307.13879. Available from: https://arxiv.org/pdf/2307.13879
79. Zhang, Y., Zhang, Q., Zhou, J., & Zou, Q. (2022). A survey on the algorithm and development of multiple sequence alignment. *Briefings in Bioinformatics*, *23*(3), bbac069. Available from: https://academic.oup.com/bib/article/23/3/bbac069/6546258
80. Zietz, J. (2007). Dynamic programming: An introduction by example. *The Journal of Economic Education*, *38*(2), 165–186.

Chapter 6
Optimal Control

6.1 Introduction

One of the challenges being confronted during these days is to substitute human drivers, by using computers, instead, on board of cars, ships, etc.

This is a clear example of automatic control. With automatic control, we could reach complete autonomy (no human pilot) in vehicles and in many other types of systems, such as power plants, smart buildings, and factories.

But let us put our feet on the ground, initiating this chapter with basic concepts, which would lead us to key developments.

Let us start with Fig. 6.1, which shows a ship and a classic helm (modern ship wheels look more functional).

The pilot of the ship should keep control of speed and heading angle. Nowadays, there are autopilots that can take charge of that.

Of course, during the voyage, obstacles—like reefs, islands, and other ships—should be avoided. And there are other questions that fall neatly in the interest of this chapter: There are *criteria to be optimized*. Time is important, and so are fuel costs.

As it will be shown in the sequel, the research has considered, from a mathematical point of view, control strategies in order to optimize energy and/or time.

A first step for this research is to devise a mathematical representation of the systems to be controlled. The next section will introduce a popular way for this purpose, based on "*state variables*."

Once we have a system representation, we can work on controlling the system in optimal ways, as it will be explained as we delve deep into the chapter. A main body of theory—as we shall see—is devoted to quadratic criteria (for example, energy).

The case of a ship, as the one depicted in Fig. 6.1, is a practical, real instance of a controlled system. Control theory considers things in abstract terms, so the ship or other controlled systems could be depicted—in general—as the block diagram shown in Fig. 6.2.

J. M. Giron-Sierra, *Introduction to Optimization with Matlab® Examples*,
https://doi.org/10.1007/978-3-032-05048-9_6

Fig. 6.1 The control of a ship

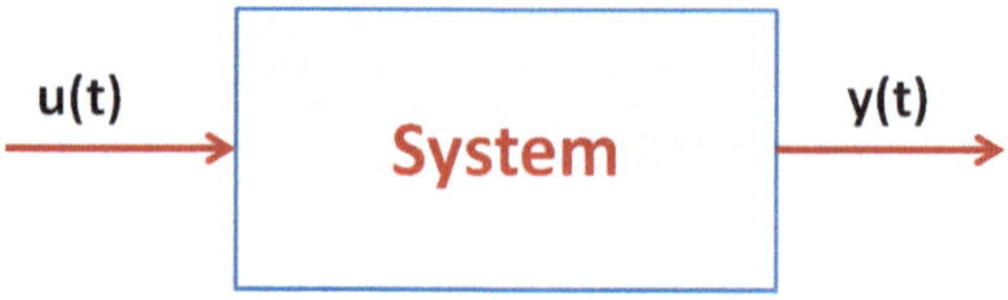

Fig. 6.2 Block diagram of a controlled system

In the diagram, u(t) represents the input control variable, and y(t) represents the output of the system. For example, in the case of a car, the input u(t) for speed control is the throttle pedal, and the output y(t) is the actual car speed.

When you push the throttle pedal, some time must be allowed for the car to reach the desired speed. It is a matter of inertia, which imposes a resistance to speed variations. In more abstract terms, one can say that the system owns certain dynamic characteristics. This fact could suppose difficulties for the control; for example, no ship, once in motion, can stop instantly (too much inertia).

In addition to ships, many other controlled systems form part of our world, including airplanes, energy generation plants, chemical factories, domestic devices, and robots.

Among the many textbooks that have been published on automatic control, we could recommend [35] and [16], which have been reedited several times. On optimal control theory, the book [30] would be useful.

For a wise perspective on the spirit and chronology of automatic control research, the article [3] would be clearly recommended.

6.2 State Variables

Before going to mathematics, it would be opportune to embrace some historic narrative about things that happened in the 1960s. The famous Sputnik satellite was put in orbit in October 1957. The space race just started, and the USA was not there. During those times, an important scientific entity, called "*International Federation of Automatic Control* (IFAC)," was born. It is a federation of national scientific societies, from all continents, involved in automatic control research. A World Congress of the IFAC takes place every three years in different places.

What is relevant for this chapter is that the first IFAC World Congress happened in Moscow, 1960. Inside the many contributions, it is easy to discover some that belong to what could be considered the *Russian school* and others that belong to, say, the *American school*. Both had responsibilities in the space race. Russians have a gift for analytic approaches (Pontryagin, etc.). The Americans had the first digital computers and preferred algorithmic approaches (Bellman, etc.).

State variables provide a simple-to-apply approach that can be implemented with matrices and vectors and is suitable for the use of digital computers.

From an aseptic perspective, state variables can be seen as a useful change of variables.

Take for example, the following differential equation:

$$\frac{d^n y}{dt^n} = \sum_{0}^{n-1} \alpha_i \frac{d^i y}{dt^i}$$

This equation could correspond, for instance, to the description of the dynamics of a system (an example will be given below).

Now, here is the change of variables:

$$\begin{aligned} y(t) &= x_0 \\ \dot{x}_0 &= x_1 \\ \dot{x}_1 &= x_2 \\ &\ldots \\ \dot{x}_{n-2} &= x_{n-1} \end{aligned}$$

Using this change, the differential equation can be expressed as follows:

$$\dot{x}_{n-1} = \sum_{0}^{n-1} \alpha_i x_i$$

In matrix-vector form

$$\dot{\mathbf{x}} = \mathbf{A}\,\mathbf{x}$$

This format is more convenient for computational purposes.

The way we made the change of variables is called the "*normal form*," which gives a series of state variables: $x_0, x_1, \ldots, x_{n-1}$.

A good introduction to state variables in the automatic control context is the book [18]. The brief tutorial [28] includes details on MATLAB implementation.

Control theory specialists use two main alternatives for the description of system dynamics. One is to consider a system as a "*black box*," so you can only apply inputs and observe the corresponding outputs. The other is to "*open the box*," by assigning to the internal pieces of the system a series of state variables.

6.3 Phase Plane

Humans can easily manage graphical representations in 2D. Some perspective tricks could also be applied for 3D.

In the case of systems represented by two state variables, we could depict in 2D the evolution of these variables over time. This is called the "*phase plane*."

Let us go directly to a classical example. This is a system whose dynamics are described with the following equation:

$$\frac{d^2y}{dt^2} + 2\,\delta\,\omega_o\frac{dy}{dt} + \omega_o^2 y \;=\; F(t)$$

where $y(t)$ is the system output.

This equation could correspond to a mass hanging from a spring. We apply the force F(t) and see what happens with the length $y(t)$ of the spring. Looking at the left-hand side of the equation, the second term (with the derivative of $y(t)$) corresponds to friction, and the third term corresponds to elastic recovery (Hooke's law).

An important aspect is that this equation is *linear*. The superposition property applies.

Let us analyze a little bit, in order to explore the system behavior. Our first target would be the case of having no forcing; that is, $F(t) = 0$.

First, let us normalize as follows:

$$\tau \;=\; \omega_o\,t \quad\rightarrow\quad d\tau \;=\; \omega_o \cdot dt$$

Hence

$$\frac{d^2y}{d\tau^2} + 2\,\delta\,\frac{dy}{d\tau} + y \;=\; 0$$

Now, choose normal state variables:

$$x_1 \;=\; y\,; \quad x_2 = \frac{dy}{d\tau}$$

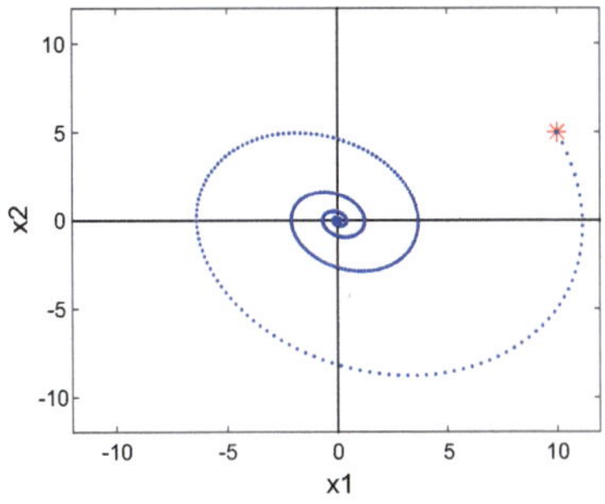

Fig. 6.3 Spring relaxing from initial forced extension (phase plane)

Then, the system dynamics equation would be

$$\frac{dx_1}{d\tau} = x_2$$

$$\frac{dx_2}{d\tau} = -2\delta \cdot x_2 \; - \; x_1$$

In matrix-vector format, we have

$$\dot{\mathbf{x}} = \mathbf{A}\,\mathbf{x}$$

with

$$\mathbf{A} = \begin{pmatrix} 0 & 1 \\ -1 & -2\delta \end{pmatrix}$$

In this way, a convenient mathematical representation (a model) of the system dynamics has been obtained.

Now, let us play with the model. For instance, let us suppose there was a weight hanging from the spring, and we take away this weight, so the spring would recover its normal length (described by $y(t) = 0$). Figure 6.3 shows on the phase plane the system behavior, from the initial condition (an asterisk) toward the origin. The axes of the phase plane are $x_1,\ x_2$.

One could interpret that the origin is attracting trajectories from distance. A point that attracts trajectories is said to be a *focus*.

Figure 6.3 has been obtained with program 6.1, which uses a simple Euler integration to compute the trajectory of the system, as described with state variables.

Program 6.1 Displaying on phase plane a second-order linear system

```
%second order linear system
%normal state variables
%the origin is a focus---------------
%damping rate
```

```
delta=0.2;
%initial point
x10=10;
x20=5;
figure(1)
%axes
plot([-12 12],[0 0],'k'); hold on;
plot([0 0],[-12 12],'k');
%initial point
plot(x10,x20,'r*','MarkerSize',10);
t=0;
dt=0.05;
x1=x10; x2=x20;
%Euler integration
while (t<30),
  plot(x1,x2,'b.');
  dx1=x2*dt;
  dx2=(-x1-(2*delta*x2))*dt;
  x1=x1+dx1;
  x2=x2+dx2;
  t=t+dt; %increment of time
end;
axis([-12 12 -12 12]);
xlabel('x1'); ylabel('x2');
```

With a simple modification of program 6.1, it is possible to depict $y(t)$ (which is equal to the first state variable) versus time, as displayed in Fig. 6.4. It is the typical behavior of a damped oscillator. More or less, damping can be studied by changing the value of delta.

If no damping exists (that is, $\delta = 0$), oscillations run forever. Figure 6.5 shows this behavior on the phase plane: a circular closed path. Closed paths are also called *cycles*.

Figure 6.5 has been obtained with program 6.2. Do not forget that Euler's approximation is by no means perfect; in particular, in the case of closed paths, you would get non-closed paths. In this program, a small integrated step has been chosen, for more precision (implying more computation effort and time). The reader is invited to use Runge-Kutta for computational improvement.

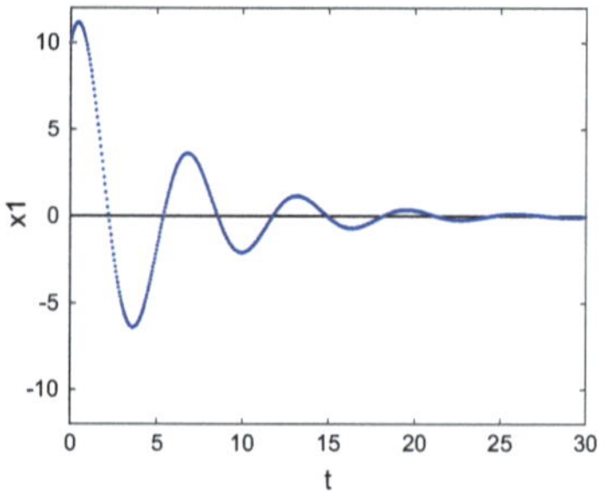

Fig. 6.4 Spring relaxing from initial forced extension $x_1(t)$ vs. t

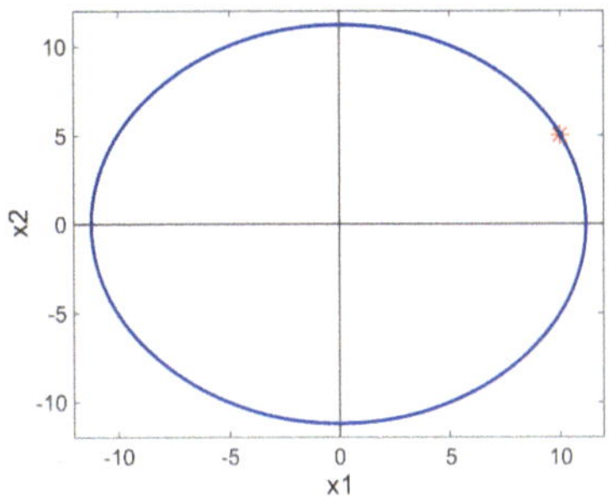

Fig. 6.5 Oscillator with no damping (phase plane)

Program 6.2 Phase plane trajectory when origin is a vortex

```
%second order linear system
%normal state variables
%the origin is a vortex---------------
%damping rate
delta=0;
%initial point
x10=10;
x20=5;
figure(1)
%axes
plot([-12 12],[0 0],'k'); hold on;
plot([0 0],[-12 12],'k');
%initial point
plot(x10,x20,'r*','MarkerSize',10);
t=0;
dt=0.0015;
x1=x10; x2=x20;
%Euler integration
while (t<6.3),
  plot(x1,x2,'b.');
  dx1=x2*dt;
  dx2=(-x1-(2*delta*x2))*dt;
  x1=x1+dx1;
  x2=x2+dx2;
  t=t+dt; %increment of time
end;
axis([-12 12 -12 12]);
xlabel('x1'); ylabel('x2');
```

Brief tutorials of the phase plane can be found in [14, 52], and [17].

Fig. 6.6 A classical pendulum clock

6.3.1 The van der Pol Oscillator

When making a mechanical clock, the key target is to obtain a sustained periodic oscillation. Energy losses due to friction (damping) can be compensated with some mechanisms involving springs. Figure 6.6 illustrates this classical scenario, with a pendulum clock.

In 1926, Balthazar van der Pol published in the Philosophical Magazine a paper on relaxation oscillators, including a mathematical model that turned out to be a motivating archetype. Since then, a lot of research has been devoted to this model, suggesting also interesting variants.

The model equations can be expressed in state variables, as follows:

$$\frac{dx_1}{dt} = x_2$$

$$\frac{dx_2}{dt} = -\mu\,(x_1^2 - 1) \cdot x_2 - x_1$$

This system **is nonlinear**, for two reasons: One of the variables is squared, and there is a product of variables.

Figure 6.7 presents a portrait of van der Pol, a Dutch scientist interested on oscillation phenomena, like in the case of heartbeat and action potentials.

Let us use the phase plane for the study of the van der Pol oscillator. Figure 6.8 displays on the phase plane two trajectories departing from two different initial conditions (marked with asterisks). What we see is that both trajectories converge to a noncircular cycle. Once the system enters the cycle, it keeps oscillating forever.

Specialists on the phase plane interpret the case of Fig. 6.8 as the cycle being an *attractor*, separating two regions with different transient behaviors. A curve separating different regions is denoted as separatrix; if it is a closed curve, it is also a limit cycle. This is the interesting scenario offered by the van der Pol model, with one limit cycle. This cycle is notably stable: All forces that appear in the phase plane drive the system toward the limit cycle (once in the cycle, you cannot escape). Indeed, this is the most suitable property for establishing a clock.

Fig. 6.7 Portrait of B. van der Pol

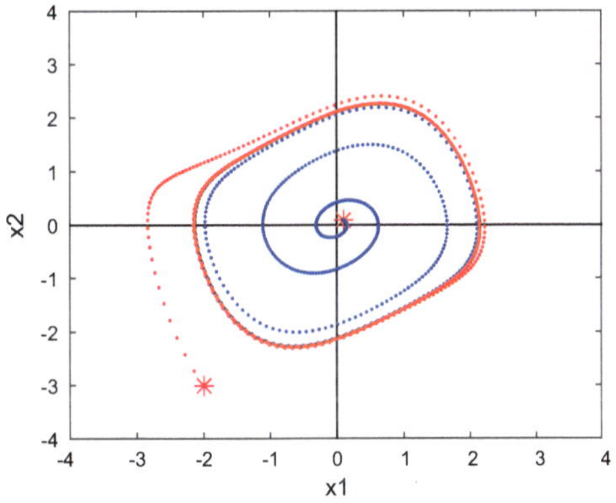

Fig. 6.8 Two trajectories of the van der Pol oscillator (phase plane)

The phase plane on Fig. 6.8 has been depicted with program 6.3, which is similar to most programs in this section. By the way, an exercise for the reader could be to display the behavior of state variables versus time.

Program 6.3 Phase plane trajectory of a van der Pol oscillator

```
%Van der Pol oscillator
%normal state variables
%there is a limit cycle---------------
%parameter
mhu=0.4;
%initial points
x10=0.1; x20=0.1;
x11=-2; x21=-3;
figure(1)
%axes
plot([-12 12],[0 0],'k'); hold on;
```

```
plot([0 0],[-12 12],'k');
%initial points
plot(x10,x20,'r*','MarkerSize',10);
plot(x11,x21,'r*','MarkerSize',10);
dt=0.05;
t=0;
x1=x10; x2=x20;
%Euler integration
while (t<30),
  plot(x1,x2,'m.');
  dx1=x2*dt;
  dx2=(-x1-(mhu*(x1^2 -1)*x2))*dt;
  x1=x1+dx1;
  x2=x2+dx2;
  t=t+dt; %increment of time
end;
t=0;
x1=x11; x2=x21;
%Euler integration
while (t<30),
  plot(x1,x2,'b.');
  dx1=x2*dt;
  dx2=(-x1-(mhu*(x1^2 -1)*x2))*dt;
  x1=x1+dx1;
  x2=x2+dx2;
  t=t+dt; %increment of time
end;
axis([-4 4 -4 4]);
xlabel('x1'); ylabel('x2');
```

One of the oldest publications of van der Pol is [50] (year 1927, 961 cites), available from the web. The article [19] considers the emergence of a concept (relaxation oscillators) and the role of van der Pol in this context.

The research around van der Pol has expanded to many unexpected aspects, as partly reviewed in [13]. For example, forced van der Pol can exhibit chaotic behavior, and this can be used for encryption [23]; coupled van der Pol oscillators show several types of behavior (depending on coupling, oscillators could converge to self-synchronization, become chaotic, etc.) (see [2]); some recent publications deal with *quantum limit cycle*s and the van der Pol oscillator [7].

Part of the research is considering the extent to what a chaotic system can be controlled [11].

The web page at http://var.scholarpedia.org/article/Van_der_Pol_model offers a motivating introduction to the van der Pol oscillator.

6.3.2 Other Limit Cycles

In order to complete the panoply of limit cycle types we can find, it would be useful to "*fabricate*" a set of illustrative models, which do not have to correspond to real systems or processes.

One of these models is the following:

$$\frac{dx_1}{dt} = x_2 - x_1(x_1^2 + x_2^2 - 1)$$

$$\frac{dx_2}{dt} = -x_1 - x_2(x_1^2 + x_2^2 - 1)$$

In favor of a direct interpretation, an equivalent model can be expressed in polar state variables, as follows:

$$\frac{d\rho}{dt} = -\rho(\rho^2 - 1)$$

$$\frac{d\theta}{dt} = -1$$

Let us choose an initial condition. If it is such $\rho < 1$, then $\left(\frac{d\rho}{dt} > 0\right)$, so the distance to the origin would tend to increase. On the contrary, if $\rho > 1$, then $\left(\frac{d\rho}{dt} < 0\right)$, so the distance to the origin would tend to decrease.

If $\rho = 1$, then $\left(\frac{d\rho}{dt} = 0\right)$, and the further trajectory would remain in a circle.

Figure 6.9 summarizes the situation. There is a stable limit cycle (all possible trajectories).

Figure 6.9 has been obtained with program 6.4. Note that a circular cycle corresponds to a sinusoidal oscillation.

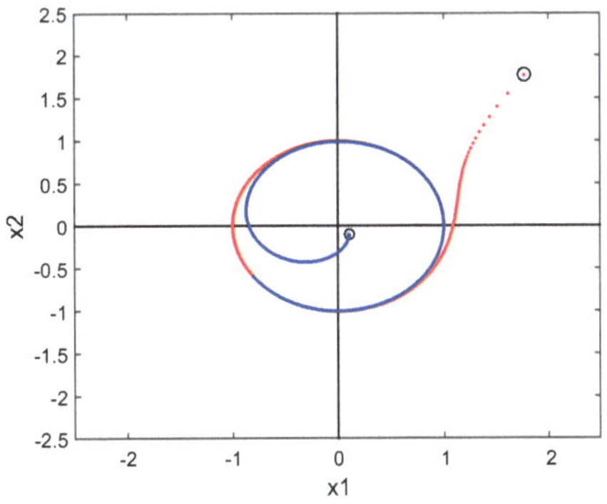

Fig. 6.9 An example of a stable limit cycle (phase plane)

Program 6.4 Stable limit cycle on phase plane

```
%Limit cycle: stable case
%in polar coordinates
%initial states---------------
rho_0 = 2.5; %radius
theta_0 = pi/4; %radians
rho_1 = 0.15; %radius
theta_1 = -pi/4; %radians
figure(1)
%axes
plot([-2.5 2.5],[0 0],'k'); hold on;
plot([0 0],[-2.5 2.5],'k');
%initial points
polar(theta_0,rho_0,'ko');
polar(theta_1,rho_1,'ko');
dt=0.02;
t=0;
rho=rho_0; theta=theta_0;
%Euler integration
while (t<8),
  polar(theta,rho,'r.');
  dtheta=-dt;
  drho=(-rho*((rho^2)-1))*dt;
  rho=rho+drho;
  theta=theta+dtheta;
  t=t+dt; %increment of time
end;
t=0;
rho=rho_1; theta=theta_1;
%Euler integration
while (t<8),
  polar(theta,rho,'b.');
  dtheta=-dt;
  drho=(-rho*((rho^2)-1))*dt;
  rho=rho+drho;
  theta=theta+dtheta;
  t=t+dt; %increment of time
end;
axis([-2.5 2.5 -2.5 2.5]);
xlabel('x1'); ylabel('x2');
```

With slight changes to the model, one obtains a radical change in the situation. Here is the new model:

$$\frac{dx_1}{dt} = x_2 + x_1(x_1^2 + x_2^2 - 1)$$

$$\frac{dx_2}{dt} = -x_1 + x_2(x_1^2 + x_2^2 - 1)$$

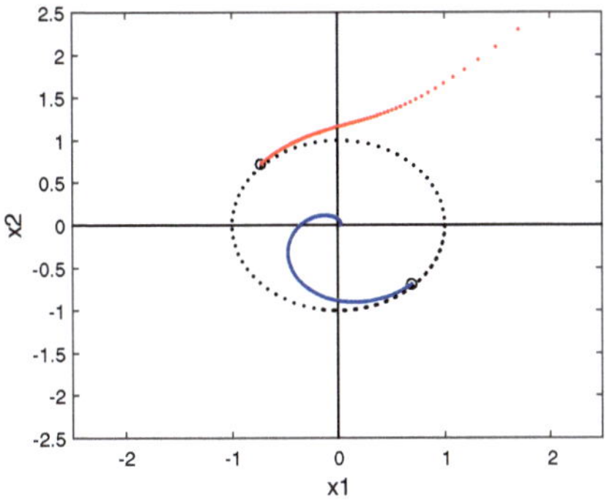

Fig. 6.10 An unstable limit cycle (phase plane)

And its version in polar state variables

$$\frac{d\rho}{dt} = \rho(\rho^2 - 1)$$

$$\frac{d\theta}{dt} = -1$$

Now, if $\rho < 1$, then $\left(\frac{d\rho}{dt} < 0\right)$, so trajectories go to the origin. If $\rho > 1$, then $\left(\frac{d\rho}{dt} > 0\right)$, so trajectories go to infinity.

Again, if $\rho = 1$, then $\left(\frac{d\rho}{dt} = 0\right)$. There is a possible cycle.

Figure 6.10 displays the phase plane of the situation. There is an unstable limit cycle.

Figure 6.10 has been generated with program 6.5, which, of course, is like the previous program.

Program 6.5 Unstable limit cycle

```
%Limit cycle: unstable case
%in polar coordinates
%initial states---------------
rho_0 = 1.03; %radius
theta_0 =3* pi/4; %radians
rho_1 = 0.97; %radius
theta_1 = -pi/4; %radians
figure(1)
%axes
plot([-2.5 2.5],[0 0],'k'); hold on;
plot([0 0],[-2.5 2.5],'k');
%initial points
polar(theta_0,rho_0,'ko');
polar(theta_1,rho_1,'ko');
```

```
%limit cycle------
dt=0.1;
t=0;
rho=1; theta=0;
%Euler integration
while (t<8),
  polar(theta,rho,'k.');
  dtheta=-dt;
  drho=(rho*((rho^2)-1))*dt;
  rho=rho+drho;
  theta=theta+dtheta;
  t=t+dt; %increment of time
end;
dt=0.02; %------------
t=0;
rho=rho_0; theta=theta_0;
%Euler integration
while (t<3),
  polar(theta,rho,'r.');
  dtheta=-dt;
  drho=(rho*((rho^2)-1))*dt;
  rho=rho+drho;
  theta=theta+dtheta;
  t=t+dt; %increment of time
end;
t=0;
rho=rho_1; theta=theta_1;
%Euler integration
while (t<5),
  polar(theta,rho,'b.');
  dtheta=-dt;
  drho=(rho*((rho^2)-1))*dt;
  rho=rho+drho;
  theta=theta+dtheta;
  t=t+dt; %increment of time
end;
axis([-2.5 2.5 -2.5 2.5]);
xlabel('x1'); ylabel('x2');
```

It is possible to devise models for one region being stable and the other unstable. For example

$$\frac{dx_1}{dt} = x_2 - x_1(x_1^2 + x_2^2 - 1)^2$$

$$\frac{dx_2}{dt} = -x_1 - x_2(x_1^2 + x_2^2 - 1)^2$$

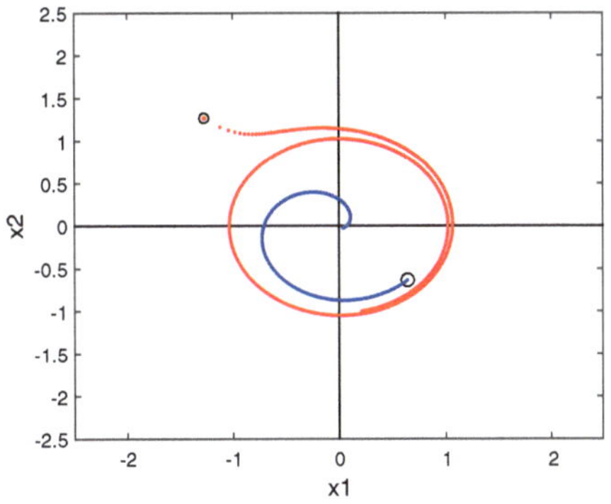

Fig. 6.11 A semistable limit cycle (phase plane)

In polar variables

$$\frac{d\rho}{dt} = -\rho(\rho^2 - 1)^2$$

$$\frac{d\theta}{dt} = -1$$

Figure 6.11 depicts what happens. There is a limit semistable cycle. Trajectories departing from outside converge to the cycle. Trajectories departing from the inside go to the origin.

Figure 6.11 has been generated by program 6.6.

Program 6.6 Semistable limit cycle in polar coordinates

```
%Limit cycle: semi-stable case
%in polar coordinates
%initial states---------------
rho_0 = 1.8; %radius
theta_0 =3* pi/4; %radians
rho_1 = 0.9; %radius
theta_1 = -pi/4; %radians
figure(1)
%axes
plot([-2.5 2.5],[0 0],'k'); hold on;
plot([0 0],[-2.5 2.5],'k');
%initial points
polar(theta_0,rho_0,'ko');
polar(theta_1,rho_1,'ko');
dt=0.02;
t=0;
rho=rho_0; theta=theta_0;
%Euler integration
while (t<10),
  polar(theta,rho,'r.');
```

```
  dtheta=-dt;
  drho=(-rho*((rho^2)-1)^2)*dt;
  rho=rho+drho;
  theta=theta+dtheta;
  t=t+dt; %increment of time
end;
t=0;
rho=rho_1; theta=theta_1;
%Euler integration
while (t<6),
  polar(theta,rho,'b.');
  dtheta=-dt;
  drho=(-rho*((rho^2)-1)^2)*dt;
  rho=rho+drho;
  theta=theta+dtheta;
  t=t+dt; %increment of time
end;
axis([-2.5 2.5 -2.5 2.5]);
xlabel('x1'); ylabel('x2');
```

A more realistic example would be a kind of *soft spring*—a type of spring that would oscillate with a small weight, but that would extend until losing any hope of recovery when the weight was too much. Figure 6.12 tries to represent a spring like that.

For this case, one could propose the following model:

$$\frac{dx_1}{dt} = x_2$$

$$\frac{dx_2}{dt} + \omega_0^2 x_1(1 - a^2 x_1^2) = 0$$

Of course, this is a nonlinear model.

Fig. 6.12 Perhaps a soft spring

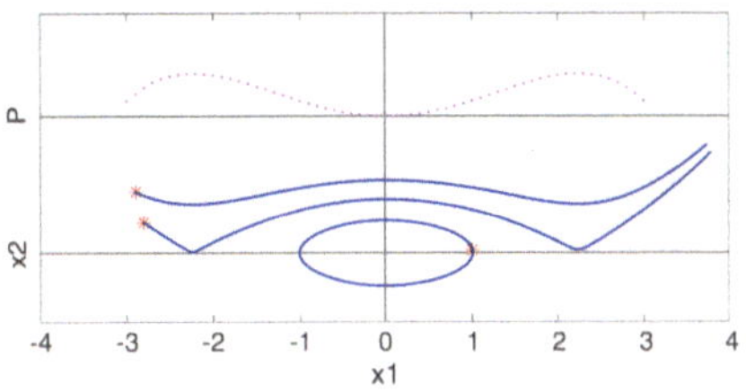

Fig. 6.13 The soft spring case

One could associate a potential energy to the spring, as follows:

$$P = \int \omega_0^2 x_1 (1 - a^2 x_1^2) dx_1 = \omega_0^2 \left(\frac{x_1^2}{2} - \frac{a^2 x_1^4}{4} \right)$$

Two plots are combined in Fig. 6.13. On top is the potential in function of spring extensión. At the bottom is the phase plane. When the weight is light, the spring would offer a sustained oscillation (a cycle). When the weight is relatively large, the spring would not recover. There is a curve separating the two distinct behaviors: It is a separatrix.

Figure 6.13 has been generated with program 6.7.

Program 6.7 Phase plane trajectories of a soft spring

```
%Soft spring
%normal state variables
%parameter
w0_2=1;
a_2=0.2;
%initial points
x10=1; x20=0.1;
x11=-2.9; x21=1.8;
x12=-2.8; x22=0.8987; %critical
figure(1)
%axes
plot([-4 4],[0 0],'k'); hold on;
plot([-4 4],[4 4],'k');
plot([0 0],[-2 7],'k');
%initial points
plot(x10,x20,'r*','MarkerSize',10);
plot(x11,x21,'r*','MarkerSize',10);
plot(x12,x22,'r*','MarkerSize',10);
set(gca,'ytick',[]); %to avoid ticks on y axis
%Potential energy--------
for x1=-3:0.1:3,
  y=w0_2*((x1^2/2)-(a_2*(x1^4)/4));
  plot(x1,y+4,'m.');
end;
dt=0.002;
```

```
%inner cycle--------
t=0;
x1=x10; x2=x20;
%Euler integration
while (t<7),
  plot(x1,x2,'b.');
  dx1=x2*dt;
  dx2=(-w0_2*x1*(1-a_2*(x1^2)))*dt;
  x1=x1+dx1;
  x2=x2+dx2;
  t=t+dt; %increment of time
end;
%outer oscillation cycle--------
t=0;
x1=x11; x2=x21;
%Euler integration
while (t<3.7),
  plot(x1,x2,'b.');
  dx1=x2*dt;
  dx2=(-w0_2*x1*(1-a_2*(x1^2)))*dt;
  x1=x1+dx1;
  x2=x2+dx2;
  t=t+dt; %increment of time
end;
%separatrix--------
t=0;
x1=x12; x2=x22;
%Euler integration
while (t<13),
  plot(x1,x2,'b.');
  dx1=x2*dt;
  dx2=(-w0_2*x1*(1-a_2*(x1^2)))*dt;
  x1=x1+dx1;
  x2=x2+dx2;
  t=t+dt; %increment of time
end;
axis([-4 4 -2 7]);
xlabel('x1');
str=['x2',' ','P'];
ylabel(str);
```

Several scientific branches are highly interested in limit cycles, since they are related to oscillations and cyclical behaviors (for instance, in economic systems [6]). In fact, we have here a kind of record: If we follow again the habit of asking Google, now about limit cycles, we obtain 326,000,000 results.

Books of interest (among others) are [10] on limit cycles of differential equations and [32] on bifurcation theory of limit cycles (this is related to chaos). Limit cycles arise in predator-prey models or population dynamics [33], power systems [41], quantum theories [20], etc. They can be used for humanoid robot walking control [25]. Also, they have been detected in oversteer vehicles [47] and are considered for the study of men's circadian rhythms [48].

The article [42] gives insight into limit cycles, following a Hamiltonian point of view.

Nobody would like a shaking car. It should be said that usually limit cycles are not welcome in controlled systems. Actually, a main drawback that may happen in control design is to have instability or unwanted oscillations, which should be avoided by all means.

6.3.3 Chaos (Duffing)

In 1963, E.N.Lorenz started to announce the existence of chaotic solutions of certain mathematical deterministic equations, which were related to weather prediction and atmospheric flows (see [46] for details). This was one of the most surprising discoveries of the twentieth century.

Extensive research on chaotic systems happened, with many publications. Among other factors that might cause chaotic behaviors in systems, one of them could be the presence of cubic nonlinearities.

Some structures, like metallic bridges or even violin strings, respond with cubic reaction forces to excitations. This is why we included Fig. 6.14 as an illustrative motivation.

A famous example of a chaotic system is the Duffing equation. This is an oscillator with a cubic term and that is subject to a sinusoidal forcing. Here is this model in state variables:

$$\frac{dx_1}{dt} = x_2$$
$$\frac{dx_2}{dt} + a\,x_2 + b\,x_1^3 = c\,(\cos\,\omega\,t)$$

Fig. 6.14 A metallic bridge

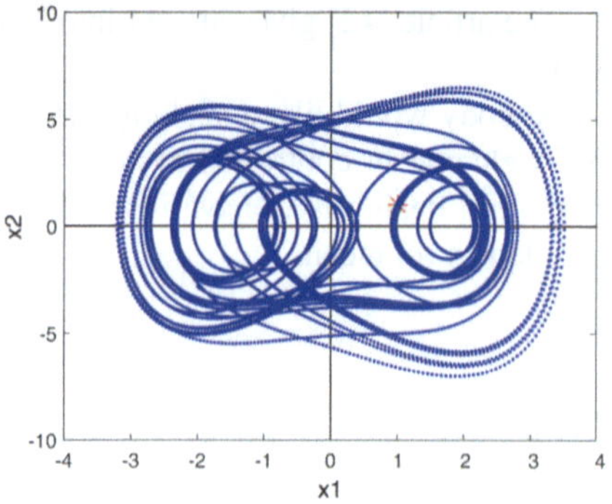

Fig. 6.15 A chaotic solution of the Duffing equation (phase plane)

The solutions of the equations are chaotic for certain combinations of parameter values. For instance, it is chaotic for the following values:

$$\omega = 1\,;\; b = 1;\; a = 0.05;\; c = 7.5$$

Figure 6.15 displays, in the phase plane, the behavior of the Duffing system with those parameter values. It may suggest a butterfly. No periodic cycling exists.

Figure 6.15 has been generated with program 6.8.

Program 6.8 Duffing chaotic oscillator on the phase plane

```
%Duffing
%initial point
x10=1; x20=1;
figure(1)
%axes
plot([-4 4],[0 0],'k'); hold on;
plot([0 0],[-10 10],'k');
%initial points
plot(x10,x20,'r*','MarkerSize',10);
dt=0.01;
t=0;
x1=x10; x2=x20;
w=1; a=0.05; b=1; c=7.5;
%Euler integration
while (t<80),
  plot(x1,x2,'b.');
  dx1=x2*dt;
  dx2=((c*cos(w*t))-((a*x2)+(b*(x1^3))))*dt;
  x1=x1+dx1;
  x2=x2+dx2;
  t=t+dt; %increment of time
end;
axis([-4 4 -10 10]);
xlabel('x1'); ylabel('x2');
```

Chaos has awakened intense attention from research. A huge quantity of publications have been devoted to this phenomenon. A partial review of the topic, focusing on nonlinear systems, is [1].

The article [12] would be specially recommended for a discussion on order and chaos (nature seems unstructured, but there are patterns).

6.4 Bang-Bang Control

After the initial review of system dynamics and the phase plane, it is time to focus on control optimization problems.

Let us start with *bang-bang control*.

Take, for example, a simple satellite maneuvering scenario. Usually, satellites should point toward certain targets, employing control systems that govern its rotational motions. The coined term for this function is *attitude control*.

In order to illustrate the example, Fig. 6.16 shows a satellite in orbit.

Consider now a simplified task. Figure 6.17 visualizes a satellite with a telescope. The satellite should point to a specific star. By means of two on/off jets, it is possible to turn the satellite (keeping in a plane).

The problem is to adequately use the jets.

First, establish a mathematical model of the system. Simple physics tells that

$$I\ddot{\theta} = \tau$$

Fig. 6.16 A satellite in orbit

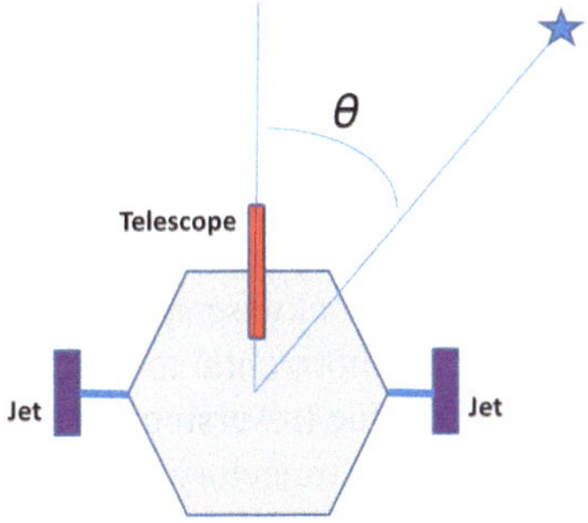

Fig. 6.17 Basic example of attitude control

where we have a torque τ that causes an angular acceleration $\ddot{\theta}$, with I being the inertia of the satellite. The torque can have three possible values:

$$\tau = \begin{cases} +I\,U \\ 0 \\ -I\,U \end{cases}$$

(Both jets should be fired, either for turning right or for turning left.)
State variables can be chosen as follows:

$$x_1 = \theta \; ; \; x_2 = \frac{d\,x_1}{dt}$$

Then, the spinning dynamics of the satellite would be described with the following equations:

$$\frac{dx_1}{dt} = x_2$$

$$\frac{dx_2}{dt} = \frac{\tau}{I} = u$$

where u could be $+\,U$, 0, or $-\,U$.
It is not difficult to determine the solutions (trajectories) of the differential equations. In case of $u = +U$, the solution is

$$\frac{x_2^2}{2} = U\,x_1 + Kte$$

And for $u = -U$, the solution is

$$\frac{x_2^2}{2} = -\,U\,x_1 + Kte$$

Now, once we have a mathematical representation, it is time to check basic behaviors of the satellite. For instance, we could start from whatever initial condition and fire the jets. Figure 6.18 presents two trajectories, departing from two different initial conditions. The vertical axis corresponds to angular velocity. For the initial condition in the right-top quadrant, we choose the negative torque ($u = -U$) to counteract the positive angular velocity of the satellite (this velocity is positive for counterclockwise spinning). Because of the velocity decreasing, the trajectory crosses the horizontal axis (zero velocity), and immediately after, velocity takes a negative value (reversing the spinning of the satellite).

The other trajectory in Fig. 6.18 corresponds to a symmetrical situation, where we choose to employ the positive torque.

Figure 6.18 has been generated with program 6.9.

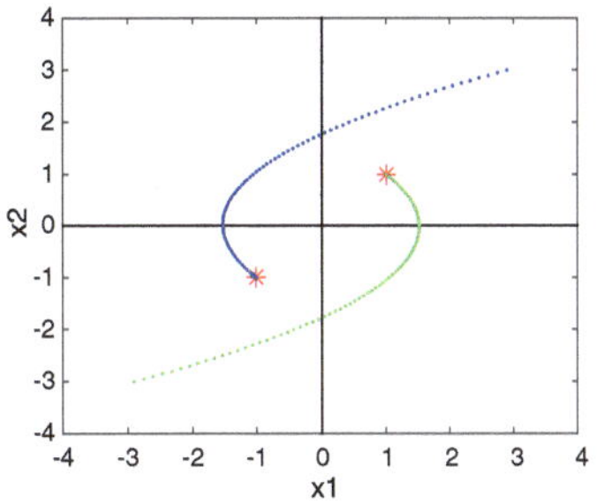

Fig. 6.18 Consequences of firing the jets, to obtain positive or negative torque (phase plane)

Program 6.9 Satellite propulsion effects on phase plane

```
%Satellite
%propulsion effects
%initial points
x10=-1; x20=-1;
x11=1; x21=1;
figure(1)
%axes
plot([-4 4],[0 0],'k'); hold on;
plot([0 0],[-4 4],'k');
%initial points
plot(x10,x20,'r*','MarkerSize',10);
plot(x11,x21,'r*','MarkerSize',10);
dt=0.05;
t=0;
x1=x10; x2=x20; U=1;
%Euler integration
while (t<4),
  plot(x1,x2,'b.');
  dx1=x2*dt;
  dx2=U*dt;
  x1=x1+dx1;
  x2=x2+dx2;
  t=t+dt; %increment of time
end;
t=0;
x1=x11; x2=x21; U=-1;
%Euler integration
while (t<4),
  plot(x1,x2,'g.');
  dx1=x2*dt;
  dx2=U*dt;
  x1=x1+dx1;
 x2=x2+dx2;
 t=t+dt; %increment of time
end;
axis([-4 4 -4 4]);
xlabel('x1'); ylabel('x2');
```

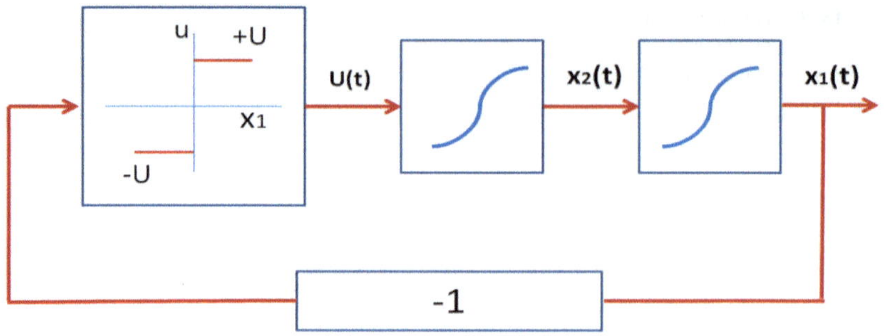

Fig. 6.19 Block diagram corresponding to a first strategy

Fig. 6.20 Results of the first strategy (phase plane)

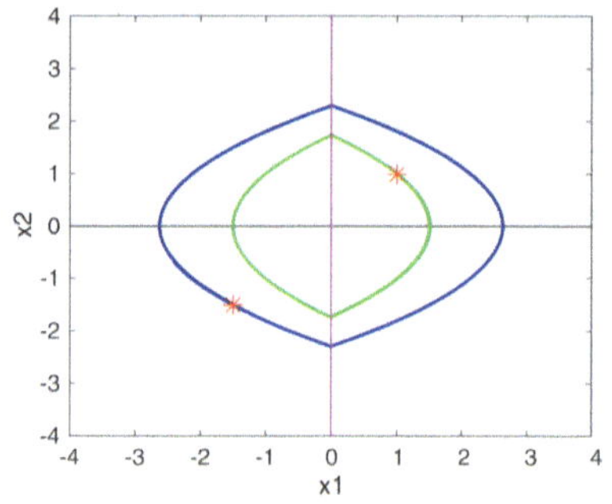

Our desire is to point the telescope at the star. That means to drive the system toward the origin of the phase plane: zero angle, zero velocity.

Well, one could devise a simple strategy: If the angle of the telescope is positive, fire the negative torque in order to make the angle decrease. If the angle is negative, do the opposite: Fire the positive torque.

It is possible to graphically express this strategy, as represented in Fig. 6.19. Two of the blocks correspond to mathematical integration. In fact, based on this diagram, a control system could be developed to put this strategy into practice. However, it would be better to try before a simulation.

According to the diagram, the strategy we are applying can be summarized as follows:

$$u(t) = -U \; sgn\,(x_1(t))$$

where *sgn(x)* is +1 or -1, as the sign of x.

Notice two main aspects of the diagram in Fig. 6.19. One is that feedback is used. The other is that a nonlinear block was included—a block that corresponds to *sgn(x)* and that, actually, functions as a commutation device.

Let us simulate the strategy. Figure 6.20 shows the results, departing from two different initial conditions. The vertical axis was depicted in color to indicate a switching line.

What happens is that the satellite will spin back and forth, forever. This is the interpretation of the cycles that result from the jet switching strategy. Let us name this strategy as naïve. Believe it or not, some of the first interplanetary missions (twentieth century) incurred in this problematic approach.

Figure 6.20 has been obtained with program 6.10.

Program 6.10 Naive conmutation strategy for satellite attitude control

```
%Satellite
%naive conmutation strategy
%initial points
x10=-1.5; x20=-1.5;
x11=1; x21=1;
figure(1)
%axes
plot([-4 4],[0 0],'k'); hold on;
plot([0 0],[-4 4],'k');
dt=0.002; %small, for more precision
U=1;
t=0;
x1=x10; x2=x20;
%Euler integration
while (t<10),
  plot(x1,x2,'b.');
  if (x1>0)
    p=-U;
  else
    p=U;
  end;
  dx1=x2*dt;
  dx2=p*dt;
  x1=x1+dx1;
  x2=x2+dx2;
  t=t+dt; %increment of time
end;
t=0;
x1=x11; x2=x21;
%Euler integration
while (t<10),
  plot(x1,x2,'g.');
  if (x1>0)
    p=-U;
  else
    p=U;
  end;
  dx1=x2*dt;
  dx2=p*dt;
  x1=x1+dx1;
  x2=x2+dx2;
  t=t+dt; %increment of time
end;
```

```
%switching line
plot([0 0],[-4 4],'m');
%initial points
plot(x10,x20,'r*','MarkerSize',10);
plot(x11,x21,'r*','MarkerSize',10);
axis([-4 4 -4 4]);
xlabel('x1'); ylabel('x2');
```

A key observation: The satellite has inertia. If you want it to stop at a certain angle, you need to act in advance with a contrary force. It is like stopping a car by braking: Some space and time are needed.

It is possible to consider some predictive action, as follows:

$$u(t) = -U\ sgn\,(x_2(t) + Kpr \cdot x_1(t))$$

Let us simulate this predictive strategy. Figure 6.21 presents the result for a certain initial condition. The switching line is depicted in color. What the system does is switch from negative torque to positive, then to negative, etc. The trajectory approaches rapidly to the origin. Similar results can be obtained from any other initial condition.

Therefore, it would be a successful strategy, with a minor problem: There could be some "*chattering*" near the origin (rapid and frequent switching for tiny, final approaching motions).

Figure 6.21 has been generated with program 6.11.

Program 6.11 Predictive conmutation strategy for satellite attitude control

```
%Satellite
%predictive conmutation strategy
%initial points
x10=1; x20=-3;
figure(1)
%axes
plot([-4 4],[0 0],'k'); hold on;
plot([0 0],[-4 4],'k');
dt=0.01;
U=1; KPR=2; aux=0;
t=0;
x1=x10; x2=x20;
%Euler integration
while (t<15),
  plot(x1,x2,'b.');
  aux=x2+(KPR*x1);
  if (aux>0)
    p=-U;
  else
    p=U;
  end;
```

```
    dx1=x2*dt;
    dx2=p*dt;
    x1=x1+dx1;
    x2=x2+dx2;
    t=t+dt; %increment of time
end;
%switching line
plot([-2 2],[2*KPR -2*KPR],'m');
%initial points
plot(x10,x20,'r*','MarkerSize',10);
axis([-4 4 -4 4]);
xlabel('x1'); ylabel('x2');
```

Our investigation has not finished yet. If one looks with more detail at the curves in Fig. 6.18, we discover that there would exist curves reaching the origin (from suitable initial conditions).

As depicted in color in Fig. 6.22, we can join two arcs driving to the origin. One of the arcs corresponds to a positive torque; the other to a negative torque. We get a curve with the following equation:

$$U\, x_1 + \frac{1}{2}\, |x_2| \cdot x_2 = 0$$

Now, the idea is to use this curve as the switching line. So the feedback control strategy would be

$$u(t) = -U\ sgn\,(U\, x_1(t) + \frac{1}{2} \cdot |x_2(t)| \cdot x_2(t))$$

which is nonlinear and predictive.

Figure 6.22 presents the result for a certain initial condition. The positive torque was applied until reaching the switching curve. Then, the torque is reversed, and we get directly to the origin—only two firings. This is why it is named *bang-bang control*.

The same behavior happens for any initial condition, except for those placed on the switching curve (only one bang is, in this case, needed to drive to the origin).

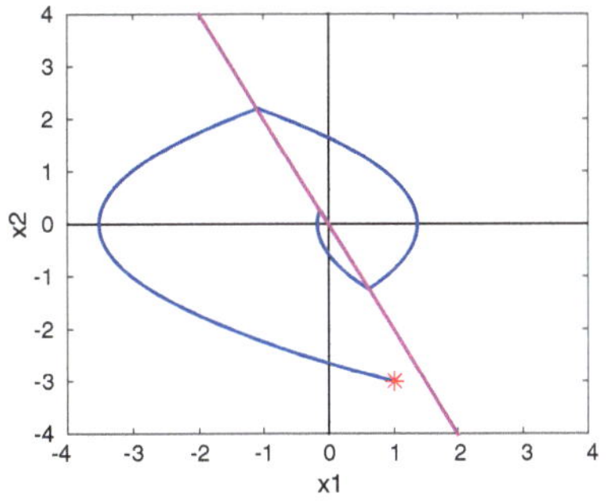

Fig. 6.21 Result of the predictive strategy (phase plane)

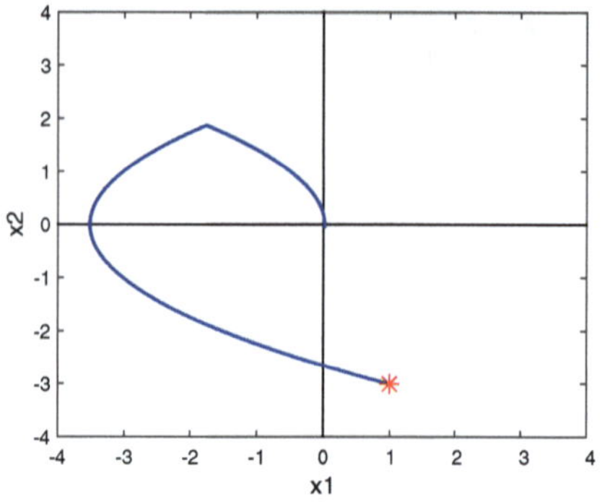

Fig. 6.22 Bang-bang example (phase plane)

This is a minimum-time strategy that can be inferred from Pontryagin's principle. Figure 6.22 has been obtained with program 6.12.

Program 6.12 Bang-bang strategy

```
%Satellite
%Bang-bang conmutation strategy
%initial points
x10=1; x20=-3;
figure(1)
%axes
plot([-4 4],[0 0],'k'); hold on;
plot([0 0],[-4 4],'k');
dt=0.01;
U=1; aux=0;
t=0;
x1=x10; x2=x20;
%Euler integration
while (t<6.8),
  plot(x1,x2,'b.');
  aux=(U*x1)+(0.5*x2*abs(x2));
  if (aux>0)
    p=-U;
  else
    p=U;
  end;
  dx1=x2*dt;
  dx2=p*dt;
  x1=x1+dx1;
  x2=x2+dx2;
  t=t+dt; %increment of time
end;
%switching line
%upper half
t=0;
x1=-3.5; x2=sqrt(abs(2*U*x1));
```

```
while (x2>0),
  plot(x1,x2,'m');
  dx1=x2*dt;
  dx2=-U*dt;
  x1=x1+dx1;
  x2=x2+dx2;
  t=t+dt; %increment of time
end;
%bottom half
t=0;
x1=3.5; x2=-sqrt(abs(2*U*x1));
while (x2<0),
  plot(x1,x2,'m');
  dx1=x2*dt;
  dx2=U*dt;
  x1=x1+dx1;
  x2=x2+dx2;
  t=t+dt; %increment of time
end;
%initial points
plot(x10,x20,'r*','MarkerSize',10);
axis([-4 4 -4 4]);
xlabel('x1'); ylabel('x2');
```

The bang-bang strategy is a consequence of the Pontryagin principle for time optimal target [15].

The application of bang-bang is very common in aerospace (see [44] related to satellite attitude control). In this context, bang-o-bang is also used; imagine a long travel: First, you fire propellers for a certain time, then you shut off for a peaceful interplanetary navigation, and finally you fire in reverse to reach a safe stop.

6.5 Linear Quadratic Regulator (LQR)

An important body of optimization theory is devoted to problems with quadratic criteria. This section starts with more details on this type of criteria, and then, it will proceed with the methods for dealing with the optimization.

6.5.1 Quadratic Criteria

Classical electric heaters are based on resistors connected to an electric current. The heat dissipated by a resistor is proportional to the square of this current. This is just an example of a typical fact: Energy is frequently related to the square of certain variables (for instance, velocity in the case of kinetic energy). In general, the use of energy has a cost, which would be also linked to squared variables.

Another different field where quadratic aspects appear is the treatment of errors. In particular, the minimization of squared errors is a popular method for polynomial fitting of data.

Supposing one is considering a discrete system, with control inputs $\boldsymbol{u}(k)$ and state variables $\boldsymbol{x}(t)$, the optimization theory usually considers the following quadratic criterion to be minimized:

$$J = \frac{1}{2}\mathbf{x}(N)^T H\,\mathbf{x}(N) + \frac{1}{2}\left(\sum_{K}^{N-1}\mathbf{x}(K)^T Q\,\mathbf{x}(K) + \mathbf{u}(K)^T V\,\mathbf{u}(K)\right)$$

Each of the three terms you can see in this expression is a quadratic term. In general, H, Q, and V are matrices. A sequence of states, for K = 0, 1, 2…N, is considered:

- The first term, with matrix H, measures final state deviation.
- The term, with matrix Q, measures state deviation.
- The last term, with matrix V, measures control input size.

An illustrative example is a missile for defence against incoming aircraft or missiles. It should reach as precisely as possible the target; this corresponds to the final state deviation, which should be minimum. During its flight, a smooth trajectory would be better (matrix Q) and not much energy consumption (matrix V), so as not to exhaust its fuel reserve.

It is the responsibility of the control designer to assign suitable values to the three matrices in this game: H, Q, and V.

6.5.2 *State Feedback*

To prepare the terrain for the next matters, it is convenient to introduce now "*state feedback*."

One could devise a feedback loop, connecting to the input of a system some information of the system output. This is a common way of building control systems.

However, if one has access to the state variables of a system, it would give you the opportunity—in general—of a more flexible feedback control.

Consider a discrete system with the following dynamics:

$$\mathbf{x}(n+1) = A\,\mathbf{x}(n) + B\,\mathbf{u}(n)$$

(A and B matrices; **x**(n) and **u**(n) vectors)

The idea of state feedback is quite simple; use as input to the system the following:

$$\mathbf{u}(n) = -R\,\mathbf{x}(n)$$

where R is a matrix to be specified.

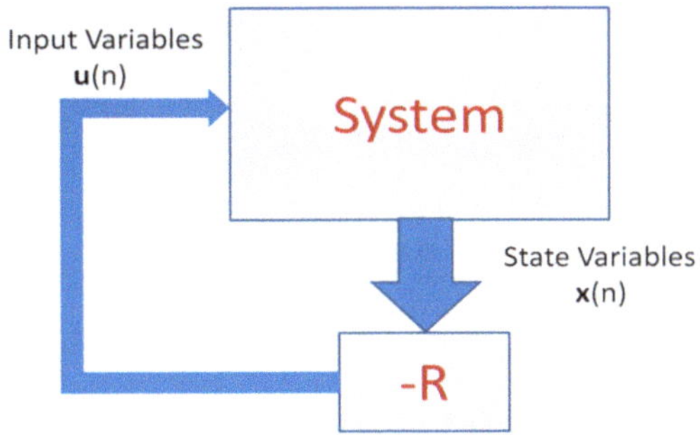

Fig. 6.23 State feedback block diagram

Figure 6.23 presents a block diagram of the state feedback.

Once the feedback loop has been established, the behavior of the complete system would be governed by the following equation:

$$\mathbf{x}(n+1) = [A - B\,R] \cdot \mathbf{x}(n)$$

Denote

$$A_R = A - B\,R$$

Then, it can be shown that the transient dynamics of the system are determined by the eigenvalues of the equation:

$$\det(zI - A_R) = 0$$

In general, the control designer wishes certain values of those eigenvalues, for instance, to guarantee convergence of the system output in a specified time range (say, for example, for a room to reach 22°C in less than five minutes, using air conditioning). Departing from the desired eigenvalues, the previous equation provides the way to compute the matrix R.

For example, suppose that the discrete system is

$$\begin{pmatrix} x_1(n+1) \\ x_2(n+1) \end{pmatrix} = \begin{pmatrix} 1 & 0.1 \\ 0 & 1 \end{pmatrix} \begin{pmatrix} x_1(n) \\ x_2(n) \end{pmatrix} + \begin{pmatrix} 0.005 \\ 0.1 \end{pmatrix} u(n)$$

And the state feedback is

$$u(n) = -R \cdot x(n) = -(r_1 \quad r_2) \cdot \begin{pmatrix} x_1(n) \\ x_2(n) \end{pmatrix}$$

Then

$$A_R = (A - B\,R) = \begin{pmatrix} 1 - 0.005\,r_1 & 0.1 - 0.005\,r_2 \\ -0.1\,r_1 & 1 - 0.1\,r_2 \end{pmatrix}$$

Therefore

$$\det(zI \; - \; A_R) \; = \; \det \begin{pmatrix} z + 0.005\, r_1 - 1 & 0.005\, r_2 - 0.1 \\ 0.1\, r_1 & z \, + \, 0.1\, r_2 \; - 1 \end{pmatrix} \; =$$

$$= z^2 \, + \, ((0.005\, r_1 \; - \; 1) \, + \, 0.1\, r_2) \cdot z \, + \, (0.005\, r_1 - 1)\,(0.1\, r_2 \; - \; 1) -$$

$$- \;\; (0.005\, r_2 \; - 0.1)\,(0.1\, r_1) \; = \; 0$$

Denote λ_1 and λ_2 as the desired eigenvalues, so

$$\det(zI \; - \; A_R) \; = \; (z - \lambda_1)\,(z - \lambda_2) \; = \; z^2 \; - (\lambda_1 \, + \, \lambda_2)\, z \, + \, \lambda_1 \cdot \lambda_2 \; = \; 0$$

Comparing term by term the two equations above, it is clear that

$$- \; (\lambda_1 \; + \; \lambda_2) \; = (0.005\, r_1 \; - \; 1) \, + \, 0.1\, r_2$$

$$\lambda_1 \cdot \lambda_2 \; = \; ((0.005\, r_1 - 1)\,(0.1\, r_2 \; - \; 1) \; - \; (0.005\, r_2 \; - 0.1)\,(0.1\, r_1))$$

For example, if one wishes $\lambda_1 = 0.6$ and $\lambda_2 = 0.8$, then it results—by solving the equations with the eigenvalues—that a state feedback with $r_1 = 8$ and $r_2 = 5.6$ will achieve the desired behavior.

In the dialect employed by control engineers, the term "*pole*" corresponds to eigenvalue. By using state feedback, it is possible *to place poles* where desired.

More information on state feedback can be obtained from [34]. The survey [45] (year 2019) reflects most of the topic extensions being considered by the research.

There is a MATLAB Documentation page devoted to pole placement (by means of state feedback). This pole placement is made using the MATLAB function called *place( )*.

6.5.3 Discrete Linear Quadratic Regulator

Let us continue considering a linear discrete system with the following dynamics:

$$\mathbf{x}(n+1) \; = \; A\,\mathbf{x}(n) \; + \; B\,\mathbf{u}(n)$$

We want to optimize the following quadratic criterion:

$$J_N \; = \; \frac{1}{2}\,\mathbf{x}(N)^T \; H\,\mathbf{x}(N) \, + \, \frac{1}{2}\left(\sum_{K}^{N-1} \mathbf{x}(K)^T \; Q\,\mathbf{x}(K) \, + \, \mathbf{u}(K)^T\, V\,\mathbf{u}(K)\right)$$

The idea will be to follow a dynamic programming strategy, working backwards. So our first step would be to look at

$$J_{N-1,\,N} = \frac{1}{2}\Big(\mathbf{x}(N)^T\,H\,\mathbf{x}(N) + \mathbf{x}(N-1)^T\,Q\,\mathbf{x}(N-1) + \\ + \mathbf{u}(N-1)^T\,V\,\mathbf{u}(N-1)\Big)$$

Let us count on the system dynamics, as follows:

$$J_{N-1,\,N} = (A\,\mathbf{x}(N-1) + B\,\mathbf{u}(N-1))^T\,H\,(A\,\mathbf{x}(N-1) + B\,\mathbf{u}(N-1)) + \\ + \mathbf{x}(N-1)^T\,Q\,\mathbf{x}(N-1) + \mathbf{u}(N-1)^T\,V\,\mathbf{u}(N-1)$$

(We got rid of the 1/2 common factor.)
We now minimize by differentiation (after some algebra):

$$\frac{\partial J_{N-1,\,N}}{\partial\,\mathbf{u}(N-1)} = 2\,B^T H\,B\,\mathbf{u}(N-1) + 2\,B^T H\,A\,\mathbf{x}(N-1) + 2\,V\,\mathbf{u}(N-1) = 0$$

That is

$$0 = (V + B^T\,H\,B)\,\mathbf{u}(N-1) + B^T H\,A\,\mathbf{x}(N-1)$$

In consequence

$$\mathbf{u}(N-1) = -\Big((V + B^T\,H\,B)^{-1}\cdot\,B^T H\,A\Big)\cdot\mathbf{x}(N-1)$$

Then, we see that the optimal **u**(N-1) is a linear state feedback. This is why we have included some previous paragraphs devoted to this kind of feedback.
For simpler expressions, let us denote

$$R(N-1) = \Big((V + B^T\,H\,B)^{-1}\cdot\,B^T H\,A\Big)$$

and so, the optimal state feedback can be written as

$$\mathbf{u}(N-1) = -\,R(N-1)\cdot\mathbf{x}(N-1)$$

Corresponding to this feedback, one has

$$J_{N-1,\,N} = (A\,\mathbf{x}(N-1) - B\,R(N-1)\,\mathbf{x}(N-1))^T\,H\,(A\,\mathbf{x}(N-1) - \\ - BR(N-1)\,\mathbf{x}(N-1)) + \mathbf{x}(N-1)^T\,Q\,\mathbf{x}(N-1) + \\ + \mathbf{x}(N-1)^T\,R(N-1)^T\,V\,R(N-1)\,\mathbf{x}(N-1)$$

which can be reorganized as follows

$$J_{N-1,\,N} = \mathbf{x}(N-1)^T\, P(N-1)\, \mathbf{x}(N-1)$$

with

$$P(N-1) = \Big((A - B\,R(N-1))^T\, H\,(A - B R(N-1) + Q +$$
$$+R(N-1)^T\, V\, R(N-1)\Big)$$

To further simplify the notation, we define *P(N)* = *H.*

After some more algebra and in consonance with the dynamic programming procedure, it is found that the solution of the quadratic optimization problem can be computed with the following algorithm:

P(N) = H
for K = N,...,1

$$R(K-1) = \Big((V + B^T\, P(K)\, B)^{-1} \cdot B^T P(K)\, A\Big)$$

$$P(K-1) = \Big((A - B\,R(K-1))^T\, P(K)\,(A - B R(K-1) + Q +$$
$$+R(K-1)^T\, V\, R(K-1)\Big)$$

This last equation, for P(K-1), can be reorganized as

$$P(K-1) = A^T\, P(K)\, A + Q - R(K-1)^T\, (V + B^T\, P(K)\, B)\, R(K-1)$$

The total cost of the optimal trajectory would be

$$J_{N-1,\,N} = \frac{1}{2}(\mathbf{x}(0)^T\, P(0)\, \mathbf{x}(0))$$

Notice that the state feedback is governed by a matrix $R(K)$ that varies along the successive stages (*K= 0, 1, 2,..*) of the system trajectory.

As you can see, linear quadratic optimization problems can be solved with a computational algorithm. This is an important motivation for the research and for practical applications.

Let us demonstrate the theory just presented, taking the same discrete system considered before (in the state feedback part). The system is

$$\begin{pmatrix} x_1(n+1) \\ x_2(n+1) \end{pmatrix} = \begin{pmatrix} 1 & 0.1 \\ 0 & 1 \end{pmatrix} \begin{pmatrix} x_1(n) \\ x_2(n) \end{pmatrix} + \begin{pmatrix} 0.005 \\ 0.1 \end{pmatrix} u(n)$$

For the quadratic optimization, we choose

$$H = \begin{pmatrix} 10 & 0 \\ 0 & 6 \end{pmatrix}; \; Q = \begin{pmatrix} 1 & 0 \\ 0 & 1 \end{pmatrix}; \; V = 2;$$

The time horizon would be N=70.

Program 6.13, in MATLAB, is devoted to implementing the linear quadratic optimization algorithm. Two figures were obtained. Figure 6.24 displays in the phase plane the evolution of the optimized from an arbitrary initial condition to the origin.

The system evolution in Fig. 6.24 corresponds to the application of a state feedback matrix R (with only two entries, in this case). Figure 6.25 depicts how

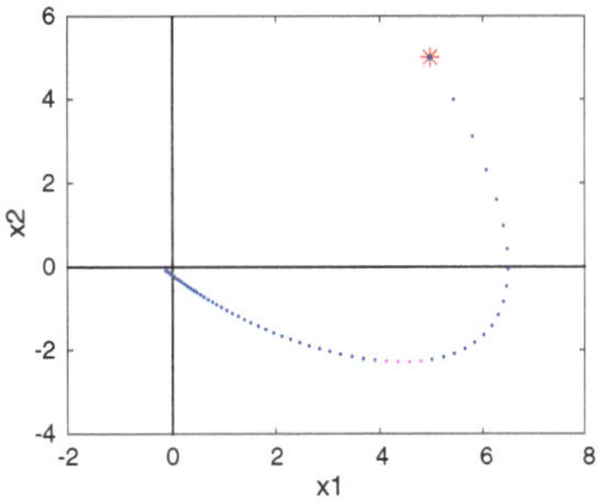

Fig. 6.24 LQR optimal trajectory (phase plane)

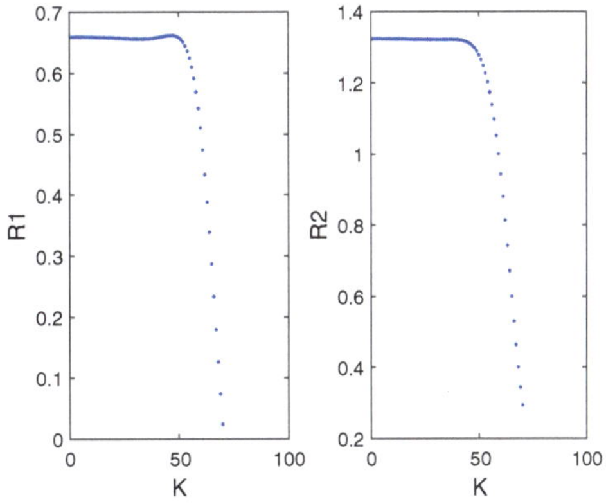

Fig. 6.25 Evolution of the state feedback matrix

the two entries of R (which we denoted as R1 and R2 in the figure) evolve along the successive iterations of the control.

The reader is invited to change the values of H, Q, V, N, and the initial condition to experiment with program 6.13 and explore different optimized results.

Program 6.13 LQR example

```
%LQR example
%simple system
%The control should drive the sysm from
%any initial condition to the origin
%while minimizing optimization criteria
%System matrices---------------
A=[1 0.1; 0 1];
B=[0.005; 0.1];
%Time horizon (in seconds)
N= 70;
%Optimization critera
H=[10 0; 0 6];
Q=[1 0; 0 1];
V=2;
%Reserve memory
X=zeros(2,1,N+1);
U=zeros(1,N+1);
P=zeros(2,2,N+1);
R=zeros(1,2,N+1);
%initial points
x10=5; x20=5; px1=0; px2=0;
%Computation of matrices
%k=1..N+1
P(:,:,N+1)=H; k=N+1;
while (k>1),
  AUX=V+(B'*P(:,:,k)*B);
  R(1,:,k-1)= (inv(AUX))*(B'*P(:,:,k)*A);
  P(:,:,k-1)= (A'*P(:,:,k)*A)+Q-(R(1,:,k-1)'*...
  AUX*R(1,:,k-1));
  k=k-1;
end;
figure(1)
%axes
plot([-2 8],[0 0],'k'); hold on;
plot([0 0],[-4 6],'k');
%initial points
plot(x10,x20,'r*','MarkerSize',10);
xlabel('x1'); ylabel('x2');
%optimal state trajectory
k=1; X(1,1,1)=x10; X(2,1,1)=x20;
while (k<N+1)
  px1=X(1,1,k); px2=X(2,1,k);
  plot(px1,px2,'b.');
  U(1,k)=-R(1,:,k)*X(:,1,k);
  X(:,1,k+1)=(A*X(:,1,k))+(B*U(1,k));
```

```
  k=k+1;
end;
figure(2)
%evolution of matrix R
k=1;
while (k<N+1)
  subplot(1,2,1)
  plot(k, R(1,1,k),'b.'); hold on;
  subplot(1,2,2)
  plot(k, R(1,2,k),'b.'); hold on;
  k=k+1;
end;
subplot(1,2,1)
xlabel('K'); ylabel('R1');
subplot(1,2,2)
xlabel('K'); ylabel('R2');
```

6.5.4 Continuous Linear Quadratic Regulator

Another important scientist to be mentioned is Rudolph E. Kalman. Now, in this subsection, we are interested in his work in the 1960s on the continuous-time quadratic optimization problem. That said, it should be added that Kalman is most famous for his developments in filtering and state estimation under stochastic conditions.

Let us summarize Kalman's approach for the quadratic optimization. A continuous-time linear system is considered:

$$\dot{\mathbf{x}}(t) = A\,\mathbf{x}(t) + B\,\mathbf{u}(t)$$

We want to optimize the following quadratic criterion:

$$J_N = \frac{1}{2}\,\mathbf{x}(t)^T\,H\,\mathbf{x}(t) + \frac{1}{2}\left(\int_{t0}^{T} (\mathbf{x}(t)^T\,Q\,\mathbf{x}(t) + \mathbf{u}(t)^T\,V\,\mathbf{u}(t))\cdot dt\right)$$

Start by considering the following Hamiltonian:

$$H_m(\mathbf{x},\,\mathbf{u},\mathbf{y}) = \frac{1}{2}\mathbf{x}^T\,Q\,\mathbf{x} + \frac{1}{2}\mathbf{u}^T\,V\,\mathbf{u} + \mathbf{y}^T\,(A\mathbf{x} + B\mathbf{u})$$

Then, by calculus of variations, the conditions for optimality say that

$$\mathbf{x}(0) = \mathbf{x}_0$$

$$\dot{\mathbf{y}} = -A^T\mathbf{y} - Q\,\mathbf{x}$$

$$0 = V\,\mathbf{u} + B^T\mathbf{y}, \text{ then: } \mathbf{u}(t) = -V^{-1}\,B^T\,\mathbf{y}(t)$$

$$\mathbf{y}(T) = H\,\mathbf{x}(T)$$

Now, one uses a mathematical result that establishes the existence of a positive semi-definite matrix P, given by the following equation (the famous Riccati equation):

$$-\dot{P} = A^T P + P\,A - P\,B\,V^{-1}B^T P + Q, \quad \text{with} \quad P(T) = H.$$

and such that

$$\mathbf{y}(t) = P(t)\,\mathbf{x}(t)$$

In order to use these expressions, one departs from the final condition $P(T) = H$ and integrates the Riccati equation backwards in time to obtain $P(t)$ from $t = 0$ to $t = T$. Most probably, it would be done numerically. Then, by using

$$\mathbf{u}(t) = -V^{-1}\,B^T\,P(t)\mathbf{x}(t)$$

in

$$\dot{\mathbf{x}}(t) = A\,\mathbf{x}(t) + B\,\mathbf{u}(t)$$

the evolution of x(t) from $t = 0$ to $t = T$ under optimum control, which is imposed by **u(t)**, can be found.

See [38, 43], and [24] for more mathematical details of the linear quadratic regulator topic.

Pertaining to the Riccati equation, it refers to one of the contributions of Jacopo Francesco Riccati (1676–1754). He was an Italian mathematician who wrote on philosophy, differential equations, and physics (he was influenced by the book of Newton *Philosophiae Naturalis Principia Mathematica* (1687)).

Being so popular and important, the Riccati equation is briefly called ARE (algebraic Riccati equation). The version we introduced for continuous time is called CARE. The version for discrete time is called DARE (it was included before, in the Discrete Linear Quadratic Regulator part).

The *Control System Toolbox* of MATLAB includes CARE and DARE functions.

Some of the published applications of the linear quadratic regulator (LQR) are the following:

- [37] (year 2019) proposes the power control of a wind turbine by using LQR.
- [36] (year 2019) designs a LQR control for a quadcopter.
- [9] (year 2022) uses LQR during the production of pharmaceutical tablets.
- [39] (year 2013) implements a LQR regulator for a continuous stirred-tank reactor.

One of the functions of the Control System Toolbox of MATLAB is *lqr( )*, specially devoted to the design of LQR control.

The MathWorks File Exchange site includes 77 entries corresponding to LQR. For example, there are interesting packages about aircraft control, submersible vehicles, robots, an inverted pendulum, a boiler-turbine system, etc. Notice that a humanoid robot *is* an inverted pendulum (to be standing up).

GitHub has 92 repositories about "*lqr-controller*" (38 repositories in MATLAB).

6.6 Observers: Kalman Filter

Kalman introduced his famous filter in a scientific article published in 1960 [27] (44,411 cites). This article has been cited by more than 39,000 publications from investigators and engineers. He doctorated in Columbia University in 1957. His work was applied to the Apollo program, NASA. Apollo 11 was the first to land on the Moon with a crew on board.

Figure 6.26 presents a photograph of Rudolph E. Kalman receiving, from President Barack Obama, the National Medal of Science in the year 2009. Kalman was born in Budapest in 1930.

6.6.1 Observers

Once again, consider the following linear discrete system:

$$\mathbf{x}(n+1) = A\,\mathbf{x}(n) + B\,\mathbf{u}(n)$$

$$\mathbf{y}(n) = C\,\mathbf{x}(n)$$

Notice that we included the output of the system $\mathbf{y}$(n).

Fig. 6.26 Rudolf E. Kalman and B. Obama

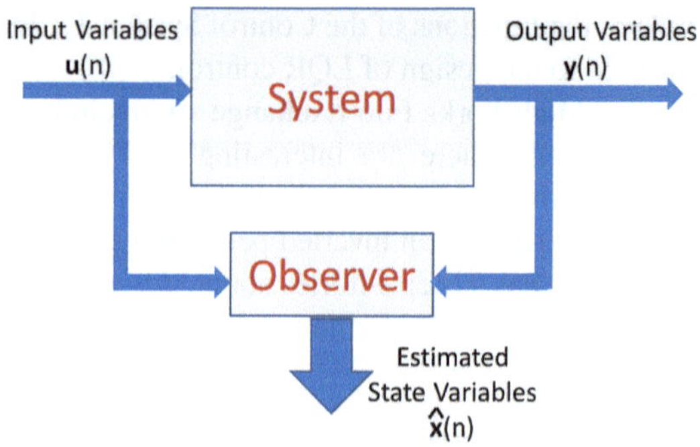

Fig. 6.27 Basic idea of an observer

Let us build an observer, according to the diagram in Fig. 6.27.

The target of the observer is to estimate the values of the system state variables, based on the inputs and outputs of the system.

In order to devise an observer, we could write the following equation:

$$\hat{\mathbf{x}}(n+1) = A\,\hat{\mathbf{x}}(n) + B\,\mathbf{u}(n) + K\,(\mathbf{y}(n) - C\,\hat{\mathbf{x}}(n))$$

where variables with a hat are estimated variables.

The structure of the observer equation combines a model of the system and a measure of estimation errors based on the system output:

$$\boldsymbol{\varepsilon}_{\mathbf{y}}(n) = \mathbf{y}(n) - C\,\hat{\mathbf{x}}(n)$$

(In the absence of estimation errors, this term would become zero.)

An important aspect of the observation is the state estimation error:

$$\boldsymbol{\varepsilon}_{\mathbf{x}}(n) = \mathbf{x}(n) - \hat{\mathbf{x}}(n)$$

This error has a dynamic behavior, which can be described with

$$\boldsymbol{\varepsilon}_{\mathbf{x}}(n+1) = [A - K\,C] \cdot \hat{\mathbf{x}}(n)$$

As in the case of state feedback, the convergence rate of this error depends on the eigenvalues of the matrix [A-KC]. Eigenvalues that can be computed from

$$\det\,(z\,I - A + K\,C) = 0$$

The designer of the observer must give suitable values to K, for a rapid but prudent convergence to zero.

The key reference for observer theory is [31] (year 1971, 3851 cites). A concise lecture is [40]. The article [8] extends the theory to continuous-time observers.

A typical application of observers is for "*software sensors*." We have some experience about it, due to our research on beer fermentation (a long time ago); it is important to get data on sugars and yeast state along the fermentation, but this requires an external laboratory with suitable analytical instruments, and it takes some hours or days for the results. However, with some sensors (pH, temperature, turbidity, etc.) and a mathematical model of the fermentative process, you can infer sugars and yeast state in real time. With observers (based on models), you can monitor a process from easy-to-obtain information.

There is a review of soft sensors [26] (year 2020), in the context of industrial processes. The article [53] (year 2020) focuses on soft sensors in fermentation processes.

6.6.2 Kalman State Estimator (Kalman Filter)

Usually, in real applications, there are perturbations and noises. In order to represent it in the context of linear discrete systems, we could add one term (with matrix J) corresponding to state perturbations and another term (normally with no matrix) for perturbations at the output:

$$\mathbf{x}(n+1) = A\,\mathbf{x}(n) + B\,\mathbf{u}(n) + J\,\omega(n)$$

$$\mathbf{y}(n) = C\,\mathbf{x}(n) + v(n)$$

Our target would be to (statistically) estimate the state variables from inputs, outputs, and the model. It is like the observer problem, but due to perturbations, we can only get expected values.

The procedure suggested by Kalman iterates two steps:

- Prediction between measurements

$$\mathbf{x}*(n) = A\,\hat{\mathbf{x}}(n-1) + B\,\mathbf{u}(n-1)$$

- Improving based on measurement of $\mathbf{y}$(n)

$$\hat{\mathbf{x}}(n) = \mathbf{x}*(n) + K(n)\,(\mathbf{y}(n) - C\,\mathbf{x}*(n))$$

This scheme leads to the following algorithm:

Start with,

$$\mathbf{x}*(0) = E(\mathbf{x}(0))$$

$$M(0) = E((\mathbf{x}(0) - \mathbf{x}*(0))(\mathbf{x}(0) - \mathbf{x}*(0))^T)$$

Then, run the following iteration loop, for n=1, 2...N,

$$K(n) = M(n)\,C^T\left[C\,M(n)\,C^T + \Phi_v\right]^{-1}$$

$$P(n) = E((\mathbf{x}(0) - \hat{\mathbf{x}}(0))(\mathbf{x}(0) - \hat{\mathbf{x}}(0))^T) = M(n) - K(n)\,C\,M(n)$$

$$\hat{\mathbf{x}}(n) = \mathbf{x}*(n) + K(n)\,(\mathbf{y}(n) - C\,\mathbf{x}*(n))$$

Predict:

$$\mathbf{x}*(n+1) = A\,\hat{\mathbf{x}}(n) + B\,\mathbf{u}(n)$$

and,

$$M(n+1) = E((\mathbf{x}(n+1) - \mathbf{x}*(n+1))(\mathbf{x}(n+1) - \mathbf{x}*(n+1))^T) =$$

$$= A\,P(n)\,A^T + J\,\Phi_w\,J^T$$

End of the loop

(Notice the presence of autocovariance matrices of the perturbations, Φ_ω, Φ_v.)

With this algorithm, one obtains the state estimation that minimizes the following quadratic criterion. In other words, an optimal state estimation is achieved:

$$J = \frac{1}{2}E((\mathbf{x}(n) - \hat{\mathbf{x}}(n))^T(\mathbf{x}(n) - \hat{\mathbf{x}}(n)))$$

A most cited introduction to the Kalman filter is [51] (year 1995, 11,251 cites). More recently, the book [22] (year 2014, 6954 cites) is devoted to the theory and practice of the Kalman filter in MATLAB.

Several publications describe the history of the Kalman filter in relation to certain fields. For instance, [49] focuses on its use in robotics, while [21] pays attention to Kalman in aerospace.

Fig. 6.28 Combination of GPS and Kalman filter for driving aid

A number of navigation applications are described in [29], which is complemented by [4] with respect to industrial applications.

Today, we use devices (the TomTom) or just a smartphone app for navigation aid while driving. This aid employs a combination of GPS and a Kalman filter, as described in [5] and references therein. Figure 6.28 illustrates this convenient helping service.

The MATLAB Control System Toolbox has the function *kalman( )*. Besides, the MathWorks File Exchange site has 283 packages related to Kalman, including "*kalman filter example*" or the "*Kalman filter virtual lab*." GitHub has 103 repositories on *kalman-filtering* (8 in MATLAB).

6.7 Using Control-Oriented MATLAB Functions

Some control-oriented MATLAB functions have been already mentioned along the chapter. The purpose of this section is to present simple programs that use those functions, as coding examples for the reader.

Linear systems have a transient behavior, which clearly manifests when, in the absence of any input, the system starts from an initial nonzero state. In stable systems, the transient goes gradually to the zero state. If the system is not stable, the transient could be a permanent oscillation or just go to infinity.

6.7.1 Discrete State Feedback (Pole Placement)

We shall apply the function *place( )*, which facilitates pole placement by means of state feedback.

The case considered is to let the plant start from an initial state and see what happens under the feedback computed by *place(A,B,P)*, where A and B describe the plant and P specifies the desired poles.

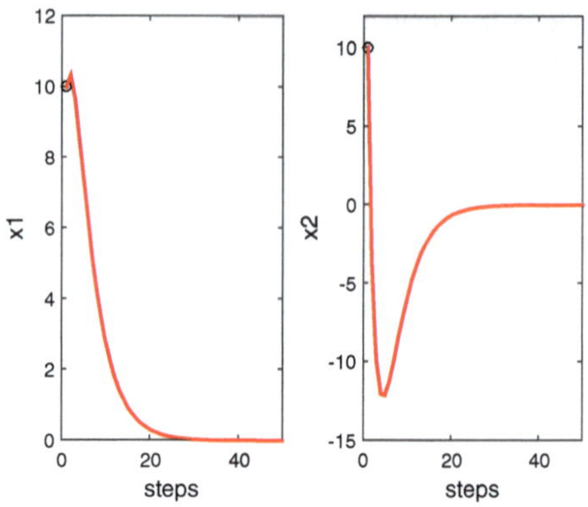

Fig. 6.29 Transient behavior with state feedback (pole placement)

Figure 6.29 shows the result for poles at (0.6, 0.8). As expected, the plant follows a transient toward the state (0, 0). Notice the initial negative oscillation of $x2$; indeed, this is not desired for certain types of plants (for instance, during the take-off of an airplane).

Figure 6.29 has been generated by program 6.14. Notice how simple it is to compute the feedback gain K. The reader is invited to change the values of desired poles and see the result (in discrete systems, poles should be less than 1 for the system to be stable).

Program 6.14 Pole placement example

```
% example of pole placement by state feedback
% using place( ) function
%System matrices
A=[1 0.1; 0 1];
B=[0.005; 0.1];
%Desired poles (eigenvalues)
p1=0.6; p2=0.8;
P=[p1;p2];
%Find feedback
K=place(A,B,P); %feedback gain
%compute system transient under this feedback
x=zeros(2,50); u=zeros(1,50);
x0=[10;10]; %initial state
x(:,1)=x0;
for nn=1:49,
  u(:,nn)=-K*x(:,nn);
  x(:,nn+1)=(A*x(:,nn))+(B*u(:,nn));
end
%display of system transient behaviour
figure(1)
subplot(1,2,1)
plot(x0(1),'ko'); hold on; %initial point
plot(x(1,:),'r-');
```

```
ylabel('x1'); xlabel('steps');
axis([0 50 0 12]);
subplot(1,2,2)
plot(x0(2),'ko'); hold on; %initial point
plot(x(2,:),'r-');
ylabel('x2'); xlabel('steps');
axis([0 50 -15 12]);
%result
K
```

6.7.2 Discrete LQR Feedback Control

We continue with the same plant as before. Now, the feedback is computed with the MATLAB function *lqr( )*.

The case considered is to start from the same initial state as before and see how the transient is under the feedback gain K computed by lqr(A,B,Q,V), where Q and V specify parameters of the optimization objective function.

Figure 6.30 shows the result, being a transient going to state (0, 0). Notice the peak of x1; this could also be not desired (for instance, we do not want the level of water in a reservoir to overflow the wall).

Figure 6.30 has been generated with the program 6.15, which is very similar to program 6.14, except for the use of *dlqr( )* instead of *place( )*. Here, the reader could change the values of Q and V and see the effects.

Program 6.15 Using the LQR MATLAB function

```
% example of linear quadratic regulation
% using dlqr( ) function
%System matrices
A=[1 0.1; 0 1];
B=[0.005; 0.1];
%optimization weights
Q=[1 0;0 1];
V=2;
%Find feedback
[K,S,E]=dlqr(A,B,Q,V); %feedback gain
%compute system transient under this optimal feedback
x=zeros(2,80); u=zeros(1,80);
x0=[10;10]; %initial state
x(:,1)=x0;
for nn=1:79,
  u(:,nn)=-K*x(:,nn);
  x(:,nn+1)=(A*x(:,nn))+(B*u(:,nn));
end
%display of system transient behaviour
```

```
figure(1)
subplot(1,2,1)
plot(x0(1),'ko'); hold on; %initial point
plot(x(1,:),'r-');
ylabel('x1'); xlabel('steps');
axis([0 80 -1 14]);
subplot(1,2,2)
plot(x0(2),'ko'); hold on; %initial point
plot(x(2,:),'r-');
ylabel('x2'); xlabel('steps');
axis([0 80 -10 12]);
%result
K
```

6.7.3 Discrete Kalman State Estimator (Kalman Filter)

We shall estimate the states of a plant being subject to perturbations and noise. No input to the plant is applied, so the plant just follows a transient from the initial state to state (0, 0).

The role of the Kalman filter is to estimate the plant state over time, using a model of the plant and the output $y(n)$ of the plant over time. This output is obtained from noisy measurements.

We shall apply the MATLAB function *kalman( )*, which needs a specification of the plant using the function *ss( )*.

Figure 6.31 shows in lighter gray the transient of the plant (subject to perturbations) and in darker color the state estimation made by the Kalman filter. Notice that we consider that the Kalman filter does not know the initial state of the plant, so the filter starts from (0, 0) instead of (10, 10). After some initial steps, the Kalman filter aligns well with the plant state.

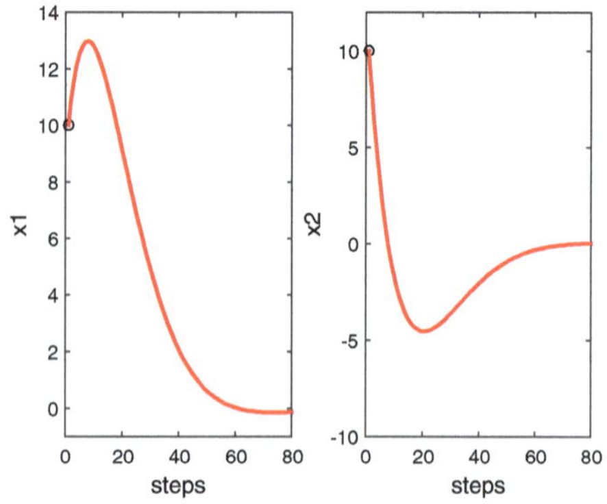

Fig. 6.30 Transient behavior with LQR control

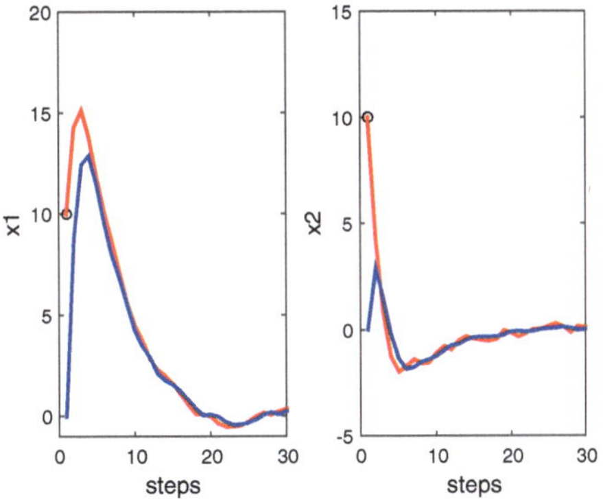

Fig. 6.31 Kalman estimate of autonomous transient behavior

Figure 6.31 has been obtained with program 6.16. If you type on the MATLAB prompt, *help kalman*, you will obtain instructions for using the Kalman estimation as we made in the three lines following *% Kalman estimation*.

The reader is invited to change the values of noises and see the result.

Notice that the plant is not the same as before. The plant used in precedent programs is not itself stable, but it was rendered stable by the feedback control; it is like a bike, which is stable with a driver, but cannot sustain without a driver. Now, for the experiment with Kalman, we preferred a stable plant (the experiment does not use any feedback controller).

Program 6.16 Using the Kalman MATLAB function

```
% example of state estimation
% using kalman( ) function
%System matrices
A=[0.9 0.5; -0.1 0.5]; B=[0.005; 0.1]; G=[0.5;0.5];
C=[1,1]; D=0; H=0.3;
sys=ss(A,[B G],C,[D H],1); %discrete model
%noise covariances
cw=0.1; cv=0.1;
%Apply Kalman
[K,L,P,M,Z]=kalman(sys,cw,cv); %feedback gain
%transient experiment
N=30;
x=zeros(2,N); u=zeros(1,N); y=zeros(1,N);
xe=zeros(2,N); xm=zeros(2,N); ye=zeros(1,N);
x0=[10;10]; %initial state
x(:,1)=x0; xe(:,1)=[0;0]; xm(:,1)=[0;0];
for nn=1:N-1,
  %plant behaviour
  u(:,nn)=0; %no input
  nw=sqrt(cw)*randn(1); nv=sqrt(cv)*rand(1);
  x(:,nn+1)=(A*x(:,nn))+(B*u(nn))+(G*nw);
  y(nn)=(C*x(:,nn))+(D*u(nn))+(H*nv);
```

```
  %Kalman estimation
  xe(:,nn+1)=(A*xe(:,nn))+(B*u(nn))+L*(y(nn)-...
  (C*xe(:,nn))-(D*u(nn)));
  ye(nn)=C*xm(:,nn)+(D*u(nn));
  xm(:,nn)=xe(:,nn)+ M*(y(nn)-(C*xe(:,nn))...
  -(D*u(nn)));
end
%display of system transient behaviour
figure(1)
subplot(1,2,1)
plot(x0(1),'ko'); hold on; %initial point
plot(x(1,:),'r-');
plot(xe(1,:),'b-');
ylabel('x1'); xlabel('steps');
axis([0 N -1 20]);
subplot(1,2,2)
plot(x0(2),'ko'); hold on; %initial point
plot(x(2,:),'r-');
plot(xe(2,:),'b-');
ylabel('x2'); xlabel('steps');
axis([0 N -5 15]);
%result
L
```

6.8 Exercises

1- [6.1] [6.2] A simple Keyne model for a national economy is the following:

$$\begin{pmatrix} C(n+1) \\ I(n+1) \end{pmatrix} = \begin{pmatrix} a & a \\ ab-b & ab \end{pmatrix} \begin{pmatrix} C(n) \\ I(n) \end{pmatrix} + \begin{pmatrix} a \\ ab \end{pmatrix} G(n)$$

$$Y(n) = C(n) + I(n) + G(n)$$

where

C is consumption,

I is investment,

G is government expenditure.

Y is the GNP.

Time increments are in quarter years. Parameters are $a = 0.8$and $b = 1.25$

Design a state feedback for the closed loop to have two eigenvalues at 0.5.

Apply a unit increase in government expenditure. Compare the responses of open-loop and closed-loop systems.

2- [6.3] [6.4] Consider the following discrete-time system:

$$\vec{x}(n+1) = A\,\vec{x}(n) + B\,\vec{u}(n)$$

with

$$A = \begin{pmatrix} 0 & 1 & 0 \\ 0 & 0 & 1 \\ 1 & -1 & 2 \end{pmatrix}; \; B = \begin{pmatrix} 0 \\ 0 \\ 1 \end{pmatrix}, \; Q = \begin{pmatrix} 0 & 0 & 0 \\ 0 & 0 & 0 \\ 0 & 0 & 1 \end{pmatrix}, \; V = 1$$

We want to minimize $J = \sum_n \vec{x}(n)^T \, Q \, \vec{x}(n) + u(n)^T \, V \, u(n)$.

Determine an optimal linear state feedback. Visualize the behavior of the system from an initial nonzero state.

Compare the result with a simple state feedback with a suitable

$$u(n) = -K \, \vec{x}(n)$$

6.8.1 Exercise References

[6.1] https://www.cds.caltech.edu/ murray/books/AM08/pdf/fbs-exercises_27Dec19.pdf
[6.2] https://www.cds.caltech.edu/~murray/books/AM08/pdf/obc08-complete_13Feb08.pdf
[6.3] https://www.dfall.ethz.ch/pandsfiles/exercise04/Quadrotor_PandS_athome_ExerciseSheet04_2020-06-03.pdf
[6.4] https://www.control.lth.se/fileadmin/control/Education/DoctorateProgram/Optimal_Control/2023/ex6_sol.pdf

References

1. Aguirre, L. A., Letellier, C., & Neher Macau, E. E. (2009). Modeling nonlinear dynamics and chaos: A review. *Mathematical Problems in Engineering, 2009*(35). Available from: https://ftp.gwdg.de/pub/EMIS/journals/HOA/MPE/Volume2009/238960.pdf
2. Astakhov, S., Gulai, A., Fujiwara, N., & Kurths, J. (2016). The role of asymmetrical and repulsive coupling in the dynamics of two coupled van der Pol oscillators. *Chaos: An Interdisciplinary Journal of Nonlinear Science, 26*(2), 023102.
3. Åström, K. J., & Kumar, P. R. (2014). Control: A perspective. *Automatica, 50*(1), 3–43. Available from: https://bwang-ccny.github.io/files/ME41100/Control_A_perspective.pdf
4. Auger, F., Hilairet, M., Guerrero, J. M., Monmasson, E., Orlowska-Kowalska, T., & Katsura, S. (2013). Industrial applications of the Kalman filter: A review. *IEEE Transactions on Industrial Electronics, 60*(12), 5458–5471. Available from: https://ieeexplore.ieee.org/stamp/stamp.jsp?arnumber=6400245.
5. Barrios, C., & Motai, Y. (2011). Improving estimation of vehicle's trajectory using the latest global positioning system with Kalman filtering. *IEEE Transactions on Instrumentation and Measurement, 60*(12), 3747–3755. Available from: https://par.nsf.gov/servlets/purl/10053994
6. Beaudry, P., Galizia, D., Portier, F., et al. (2015). Reviving the limit cycle view of macroeconomic fluctuations. Technical report, National Bureau of Economic Research Cam-

bridge. Available from: https://web.archive.org/web/20190502034550id_/https://www.nber.org/papers/w21241.pdf
7. Ben Arosh, L., Cross, M. C., & Lifshitz, R. (2021). Quantum limit cycles and the Rayleigh and van der Pol oscillators. *Physical Review Research, 3*(1), 013130. Available from: https://journals.aps.org/prresearch/pdf/10.1103/PhysRevResearch.3.013130
8. Bernard, P., Andrieu, V., & Astolfi, D. (2022). Observer design for continuous-time dynamical systems. *Annual Reviews in Control, 53*, 224–248. Available from: https://www.sciencedirect.com/science/article/abs/pii/S1367578821000948
9. Chindrus, A., Copot, D., & Caruntu, C. F. (2022). Continuous manufacturing using linear quadratic regulator in the context of cyber-physical systems. In *2022 26th International Conference on System Theory, Control and Computing (ICSTCC)* (pp. 231–236). IEEE.
10. Christopher, C., & Li, C. (2007). *Limit cycles of differential equations*. Springer.
11. Cooper, M., Heidlauf, P., & Sands, T. (2017). Controlling chaos—forced van der Pol equation. *Mathematics, 5*(4), 70. Available from: https://www.mdpi.com/2227-7390/5/4/70
12. Crutchfield, J. P. (2012). Between order and chaos. *Nature Physics, 8*(1), 17–24. Available from: https://fricke.co.uk/Teaching/CS523_2017spring/Readings/Between_order_and_Chaos.pdf
13. Cveticanin, L. (2013). On the Van der Pol oscillator: An overview. *Applied Mechanics and Materials, 430*, 3–13.
14. Dawkins, P. (2022). Phase plane. Available from: http://tutorial-math.wip.lamar.edu/Classes/DE/PhasePlane.aspx
15. Diliberto, S. P. (1967). 8 the Pontryagin maximum principle. In *Mathematics in science and engineering* (Vol. 31, pp. 373–389). Elsevier. Available from: https://www.sciencedirect.com/science/article/abs/pii/S0076539209600448
16. Dorf, R. C., & Bishop, R. H. (2021). *Modern control systems*. Pearson.
17. Driscoll, T. A. (2020). Stability and the phase plane. Available from: https://tobydriscoll.net/udmath305/first_system/phase_plane.html
18. Friedland, B. (2012) *Control system design: An introduction to state-space methods*. Courier Corporation.
19. Ginoux, J.-M., & Letellier, C. (2012). Van der Pol and the history of relaxation oscillations: Toward the emergence of a concept. *Chaos: An Interdisciplinary Journal of Nonlinear Science, 22*(2). Available from: https://arxiv.org/pdf/1408.4890
20. Głazek, S. D., & Wilson, K. G. (2022). Limit cycles in quantum theories. *Physical Review Letters, 89*(23), 230401. Available from: https://arxiv.org/pdf/hep-th/0203088
21. Grewal, M. S., & Andrews, A. P. (2010). Applications of Kalman filtering in aerospace 1960 to the present [historical perspectives]. *IEEE Control Systems Magazine, 30*(3), 69–78. Available from: https://ieeexplore.ieee.org/stamp/stamp.jsp?arnumber=5466132
22. Grewal, M. S., & Andrews, A. P. (2014). *Kalman filtering: Theory and Practice with MATLAB*. John Wiley & Sons.
23. He, J., & Cai, J. (2019). Design of a new chaotic system based on Van der Pol oscillator and its encryption application. *Mathematics, 7*(8), 743. Available from: https://www.mdpi.com/2227-7390/7/8/743
24. Hespanha, J. P. (2005), Lecture notes on LQR/LQG controller design. In *Knowledge creation diffusion utilization*. Available from: https://staff.uz.zgora.pl/wpaszke/materialy/kss/lqrnotes.pdf
25. Hobbelen, D. G. E., & Wisse, M. (2007). *Limit cycle walking*. Citeseer. Available from: https://citeseerx.ist.psu.edu/document?repid=rep1&type=pdf&doi=29e48256175bc1651f5a975313b4f5d9ec0e1e81
26. Jiang, Y., Yin, S., Dong, J., & Kaynak, O. (2020). A review on soft sensors for monitoring, control, and optimization of industrial processes. *IEEE Sensors Journal, 21*(11), 12868–12881. Available from: https://ieeexplore.ieee.org/abstract/document/9235582
27. Kalman, R. E. (1960). A new approach to linear filtering and prediction problems. *Journal Basic Engineering, D 82*(1), 35–44.

28. Kim, W.-J. (2001). Formulating the state-variable form of dynamic systems. Available from: https://wonjongkim.tamu.edu/MEEN364/StateSpace.pdf
29. Kim, Y., Bang, H., et al. (2018). Introduction to Kalman filter and its applications. *Introduction and Implementations of the Kalman Filter, 1*, 1–16. Available from: https://www.intechopen.com/chapters/63164
30. Kirk, D. E. (2004). *Optimal control theory*. Dover.
31. Luenberger, D. (1971). An introduction to observers. *IEEE Transactions on Automatic Control, 16*(6), 596–602. Available from: https://ieeexplore.ieee.org/abstract/document/1099826
32. Mao'an, H. (2017). *Bifurcation theory of limit cycles*. Science Press.
33. May, R. M. (1972). Limit cycles in predator-prey communities. *Science, 177*(4052), 900–902.
34. Nozari, E. (2021). Stability, controllability & state feedback. Available from: https://intra.engr.ucr.edu/~enozari/teaching/ME120_Fall21/Lecture3-Stability,Controllability&StateFeedback.pdf
35. Ogata, K. (2015) *Modern control engineering*. Pearson.
36. Okyere, E., Bousbaine, A., Poyi, G. T., Joseph, A. K., & Andrade, J. M. (2019). LQR controller design for quad-rotor helicopters. *The Journal of Engineering, 2019*(17), 4003–4007. Available from: https://web.archive.org/web/20210715005019id_/https://ietresearch.onlinelibrary.wiley.com/doi/pdfdirect/10.1049/joe.2018.8126
37. Phan, D. C., & Trinh, T. H. (2019). Application of linear quadratic regulator to control directly power for DFIG wind turbine. *Journal of Electrical Systems, 15*(1), 42–52. Available from: https://web.archive.org/web/20200713015517id_/http://journal.esrgroups.org/jes/papers/15_1_4.pdf
38. Plett, G. L. (2016). Linear quadratic regulator. *ECE5530 lecture notes*. Available from: http://mocha-java.uccs.edu/ECE5520/ECE5520-CH08.pdf
39. Poyen, F. B., Mukherjee, D., Banerjee, D., & Guin, S. (2013). Implementation of linear quadratic regulator for CSTR tank. In *Proceedings of Advances in Electronic and Electric Engineering* (pp. 11–15). Available from: https://www.researchgate.net/profile/Faruk-Poyen-Phd/publication/311099537_Implementation_Of_Linear_Quadratic_Regulator_For_CSTR_Tank/links/583d75b208ae2d217554908a/Implementation-Of-Linear-Quadratic-Regulator-For-CSTR-Tank.pdf
40. Raginsky, M., & Liberzon, D. (2018). State estimation and observer. Available from: https://courses.engr.illinois.edu/ece486/sp2018/handbook/lec22.html
41. Reddy, P. B., & Hiskens, I. A. (2005). *Limit-induced stable limit cycles in power systems*. IEEE.
42. Robinett, III R. D., & Wilson, D. G. (2008). What is a limit cycle? *International Journal of Control, 81*(12), 1886–1900. Available from: https://www.tandfonline.com/doi/full/10.1080/00207170801927163
43. Sastry, S., Laine, F., & Claire, T. (2021). Optimal control and the linear quadratic regulator. Available from: https://pages.github.berkeley.edu/ee290-005/sp21-site/assets/lec/LQR_Tomlin.pdf
44. Scrivener, S. L., & Thompson, R. C. (1994). Survey of time-optimal attitude maneuvers. *Journal of Guidance, Control, and Dynamics, 17*(2), 225–233. Available from: https://arc.aiaa.org/doi/abs/10.2514/3.21187
45. Shumafov, M. M. (2019). Stabilization of linear control systems and pole assignment problem: A survey. *Vestnik St. Petersburg University, Mathematics, 52*, 349–367. Available from: https://link.springer.com/article/10.1134/S1063454119040095
46. Sparrow, C. (1982). *The Lorenz equations* (Vol. 269). Springer. Available from: https://link.springer.com/book/10.1007/978-1-4612-5767-7
47. Steindl, A., Edelmann, J., & Plöchl, M. (2020). Limit cycles at oversteer vehicle. *Nonlinear Dynamics, 99*, 313–321. Available from: https://link.springer.com/article/10.1007/s11071-019-05081-8
48. Stone, J. E., Aubert, X. L., Maass, H., Phillips, A. J., Magee, M., Howard, M. E., Lockley, S. W., Rajaratnam, S. M., & Sletten, T. L. (2019). Application of a limit-cycle oscillator model for prediction of circadian phase in rotating night shift workers. *Scientific Reports, 9*(1), 11032. Available from: https://www.nature.com/articles/s41598-019-47290-6

49. Urrea, C., & Agramonte, R. (2021). Kalman filter: historical overview and review of its use in robotics 60 years after its creation. *Journal of Sensors, 2021*(1), 9674015. Available from: https://onlinelibrary.wiley.com/doi/pdf/10.1155/2021/9674015
50. Van Der Pol, B. (1927). VII. forced oscillations in a circuit with non-linear resistance.(reception with reactive triode). *The London, Edinburgh, and Dublin Philosophical Magazine and Journal of Science, 3*(13), 65–80. Available from: https://www.tandfonline.com/doi/pdf/10.1080/14786440108564176
51. Welch, G., Bishop, G., et al. (1995). *An introduction to the Kalman filter*. Chapel Hill. Available from: https://perso.crans.org/club-krobot/doc/kalman.pdf
52. Wong, J. (2019). Phase planes: Definitions, fundamentals. Available from: https://services.math.duke.edu/~jtwong/math356-2019/lectures/Dynamics3_planar_nonlin.pdf
53. Zhu, X., Ur Rehman, K., Wang, B., & Shahzad, M. (2020). Modern soft-sensing modeling methods for fermentation processes. *Sensors, 20*(6), 1771. Available from: https://www.mdpi.com/1424-8220/20/6/1771

Chapter 7
Quadratic Programming

7.1 Introduction

When fitting a function to a cloud of data points, the typical method is to minimize the sum of squared errors. This is a way for avoiding mutual cancellation of positive and negative error values, also emphasizing the impact of large errors.

The sum of squared errors is an important example of quadratic objective function. There are many other application fields, like energy management, or certain economic activities, which require to minimize quadratic objective functions.

The standard general quadratic programming problem can be stated as follows:

$$\min_x \; J(\mathbf{x}) = \frac{1}{2}\,\mathbf{x}^T\, Q\,\mathbf{x} + \mathbf{b}^T\,\mathbf{x}$$

Subject to the next constraints:

$$A\,\mathbf{x} = \mathbf{c}$$

$$C\,\mathbf{x} \leq \mathbf{d}$$

This problem can be seen as an extension of the linear programming problem. But with several peculiarities, which would become clear along this chapter. At first place, notice that the problem is nonlinear (squared variables).

The first sections of this chapter would consider the characteristics of the problem, which depend on the nature of constraints. Some algorithms will be introduced, with special attention to the sequential quadratic programming approach. After that, the chapter will focus on important applications.

In addition to the algorithms that will be introduced next, there are others, like for instance interior point methods, which can be used for many types of problems,

J. M. Giron-Sierra, *Introduction to Optimization with Matlab® Examples*,
https://doi.org/10.1007/978-3-032-05048-9_7

not only quadratic. These more general methods will be treated in other chapters, later on.

7.2 The Simplest Case: No Constraints

In the case of no constraints, the solution can be obtained by equaling to zero the gradient of the objective function:

$$\Delta J^T = Q\mathbf{x} + \mathbf{b} = 0$$

Then

$$\mathbf{x}^* = -Q^{-1}\mathbf{b}$$

This is the first order necessary condition for optimality. We have found a stationary point. The second derivative of J, which is the Hessian, would be just Q and gives us the following information on the stationary point:

- If Q is positive definite, $\mathbf{x}^*$ is a local minimizer
- If Q is negative definite, $\mathbf{x}^*$ is a local maximizer

(for Q definite, the contours of J in 2D problems would be concentric ellipses).

- If Q is positive semidefinite, $\mathbf{x}^*$ is a valley
- If Q is negative semi-definite, $\mathbf{x}^*$ is a ridge
- If Q is indefinite, $\mathbf{x}^*$ is a saddle point

All eigenvalues of Q would be nonzero positive if Q positive definite; or nonzero negative if Q negative definite. If some eigenvalues are nonzero positive and others nonzero negative, Q is indefinite.

Figure 7.1 shows a possible situation, in a two-dimensional quadratic optimization problem. The absolute optimal point is outside the feasible region (determined

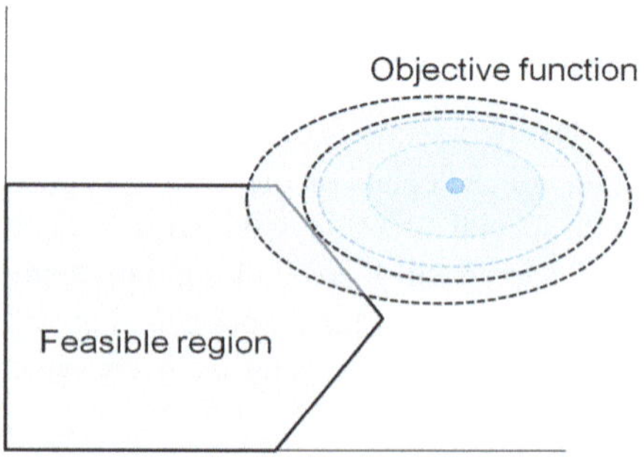

Fig. 7.1 Example of a two-dimensional quadratic optimization problem (the absolute optimal point is outside)

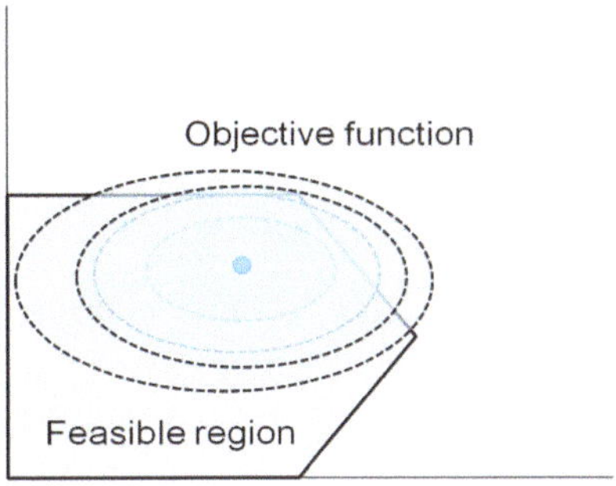

Fig. 7.2 Another example of a two-dimensional quadratic optimization problem (the absolute optimal point is inside)

by constraints). The solution would be at the intersection of the objective function contours and the feasible region.

Figure 7.2 shows another possible case. The absolute optimal point is inside the feasible region and would be the solution of the problem.

7.3 Karush-Kuhn-Tucker (KKT) Conditions in Quadratic Programming

In the general quadratic programming case, with constraints, candidates for the optimal solution can be obtained by equaling derivatives to zero.

Suppose that the optimization problem is stated as follows:

$$\min_{x} J(\mathbf{x}) = \frac{1}{2}\mathbf{x}^T Q\mathbf{x} + \mathbf{b}^T \mathbf{x}$$

Subject to:

$$\mathbf{h}(\mathbf{x}) = C\mathbf{x} - \mathbf{d} = 0$$

$$\mathbf{g}(\mathbf{x}) = \mathbf{x} \geq 0$$

where $\mathbf{h}(\mathbf{x})$ includes equality and inequality constraints.

By using Lagrange multipliers, the problem can be rewritten as

$$\min L(\mathbf{x}, \boldsymbol{\mu}, \boldsymbol{\lambda}) = J(\mathbf{x}) - \boldsymbol{\mu}^{\mathbf{T}}\mathbf{g}(\mathbf{x}) - \boldsymbol{\lambda}^{\mathbf{T}}\mathbf{h}(\mathbf{x})$$

Subject to: $\boldsymbol{\mu} \geq 0$ ($\boldsymbol{\lambda}$ unspecified)

Then, the KKT conditions are

$$\frac{\partial L}{\partial \mathbf{x}} = \nabla_{\mathbf{x}} J - \boldsymbol{\mu}^{\mathbf{T}}\nabla_{\mathbf{x}}\mathbf{g}(\mathbf{x}) - \boldsymbol{\lambda}^{\mathbf{T}}\nabla_{\mathbf{x}}\mathbf{h}(\mathbf{x}) = 0$$

$$\frac{\partial L}{\partial \boldsymbol{\lambda}} = \mathbf{h}(\mathbf{x}) = 0$$

$$\frac{\partial L}{\partial \boldsymbol{\mu}} = \mathbf{g}(\mathbf{x}) - \mathbf{s} = 0 \quad \text{(s, a surplus variable)}$$

$\boldsymbol{\mu} \geq 0$ and $\mu_j \cdot g_j(\mathbf{x}) = 0$ for $j = 1, 2, \ldots N$ (complementary slackness condition)

These conditions could be considered for any of the scenarios that will be treated in the sequel.

7.4 Equality Constraints

Suppose now that we have, specifically, a set of equality constraints to be satisfied. There are some analytical methods for obtaining the solution. Two of them are now introduced.

The linear constraints could be expressed as follows:

$$A\,\mathbf{x} = 0$$

Inequality constraints can be converted to equality constraints by adding artificial variables (this will be considered in the next section).

7.4.1 The Null-Space Approach

Suppose that the constraints are $A\,\mathbf{x} = 0$. Let us apply the following change of variables:

$$Z\,\mathbf{y} = \mathbf{x}$$

where Z is chosen such that $A \cdot Z = 0$. This means that Z is a basis for the null space of matrix A.

When the change of variables is done, constraints disappear. So one has an unconstrained problem to minimize:

$$\frac{1}{2}\mathbf{y}^{\mathbf{T}} Z^T\, Q\, Z\,\mathbf{y} + \mathbf{b}^{\mathbf{T}}\, Z\,\mathbf{y}$$

The solution, after equaling derivatives to zero, would be

$$Z^T\, Q\, Z\,\mathbf{y} = -Z^T\,\mathbf{b}$$

The matrix $Z^T\ Q\ Z$ is called the reduced Hessian. If this matrix is positive definite, the expression above gives the unique solution.

7.4.2 Lagrangian Approach

Assume the constraints have the form:

$$A\ \mathbf{x} = \mathbf{c}$$

By using Lagrangian multipliers, the problem is transformed to minimizing the following function:

$$L(\mathbf{x}, \boldsymbol{\lambda}) = \frac{1}{2}\mathbf{x}^{\mathbf{T}}\ Q\ \mathbf{x} + \mathbf{b}^{\mathbf{T}}\mathbf{x} - \boldsymbol{\lambda}^{\mathbf{T}}\ (A\ \mathbf{x} - \mathbf{c})$$

Using partial derivatives, a set of KKT conditions can be obtained:

$$Q\,\mathbf{x} + \mathbf{b} + A^T\boldsymbol{\lambda} = 0$$

$$A\,\mathbf{x} = \mathbf{c}$$

Or, in matricial format:

$$\begin{pmatrix} Q & A^T \\ A & 0 \end{pmatrix} \begin{bmatrix} \mathbf{x} \\ \boldsymbol{\lambda} \end{bmatrix} = \begin{bmatrix} -\mathbf{b} \\ \mathbf{c} \end{bmatrix}$$

The matrix on the left-hand side is called the *KKT matrix*.

If matrix A is full row rank $m < n$, and the reduced Hessian is positive definite, then the *KKT* matrix is nonsingular, and the equations have a unique solution.

7.5 Inequality Constraints

Consider now the case of inequality constraints:

$$C\,\mathbf{x} \leq \mathbf{d} \text{ with } \mathbf{x} \geq 0$$

This more general quadratic optimization problem has motivated several methods/algorithms for solving it. We shall introduce some of them, starting from a procedure that extends the use of the Linear Programming simplex method.

7.5.1 The Wolfe's Method (Extended Simplex)

Philip Wolfe, University of Princeton, introduced its adaptation of the simplex method for quadratic programming in 1959, [100]. Since then, some improvements and variants have been proposed.

From the Lagrangian function, using derivatives (as we made in the previous section), one obtains a set of equations:

$$Q\mathbf{x} + \mathbf{b} + A^T\boldsymbol{\lambda} \geq 0$$

$$C\mathbf{x} \leq \mathbf{d}$$

Two sets of variables would be added: one set for converting inequalities into equalities and another for generating a linear objective function.

For instance, let us use the example given in [52]. It is the following problem:

Minimize:

$$z = -8\,x_1 - 16\,x_2 + x_1^2 + 4\,x_2^2$$

Subject to:

$$x_1 + x_2 \leq 5$$

$$x_1 \leq 3$$

with all variables ≥ 0.

Figure 7.3 visualizes the optimization problem with contours of the objective function and a colored patch to highlight the feasible region imposed by the constraints.

Figure 7.3 has been generated with Program 7.1 in MATLAB.

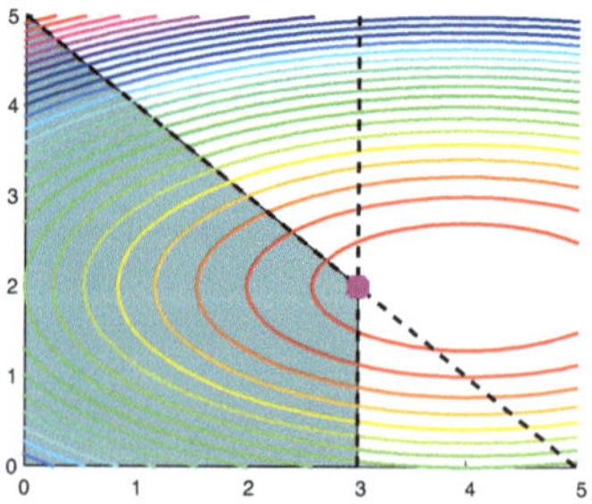

Fig. 7.3 Visualization of the optimization example

Program 7.1 Example of Wolfe's method

```
%Visualization of (Wolfe's method) example
[vx1,vx2] = meshgrid(0:0.1:5, 0:0.1:5);
y = (-8*vx1)-(16*vx2) + (vx1.^2) + 4*(vx2.^2);
figure(1)
colormap('hsv');
patch([0,0,3,3,0],[0,5,2,0,0],[0.5 0.7 0.7]);
hold on;
contour(vx1,vx2,y,25,'LineWidth',1.5);
plot([0,5],[5,0],'k--','LineWidth',2); %constraint 1
plot([3,3],[0,5],'k--','LineWidth',2); %constraint 2
plot(3,2,'m*','LineWidth',5); %solution
```

Let us start the procedure suggested by Wolfe.

First, one builds the Lagrangian function and takes derivatives:

$$2x_1 + \lambda_1 + \lambda_2 - e_1 = 8$$

$$8x_2 + \lambda_1 - e_2 = 16$$

$$x_1 + x_2 + v_1 = 5$$

$$x_1 + v_2 = 3$$

Notice that a first set of artificial variables (e_1, e_2, v_1, v_2) has been added, so constraints are now equalities.

Then, by adding a second set of artificial variables, the problem is transformed to

Minimize:

$$z = a_1 + a_2 + a_3 + a_4$$

Subject to

$$2x_1 + \lambda_1 + \lambda_2 - e_1 + a_1 = 8$$

$$8x_2 + \lambda_1 - e_2 + a_2 = 16$$

$$x_1 + x_2 + v_1 + a_3 = 5$$

$$x_1 + v_2 + a_4 = 3$$

Now it is possible to apply the simplex method of linear programming, taking into account complementary slackness:

x_j and e_j are complementary

λ_j and ν_j are complementary

According to the simplex steps, a maximization criterion is applied for choosing entering variables (when exchanging columns). In our present case, the entering variable should not have its complementary in the basis (or would leave the basis on the same iteration).

The following table summarizes the evolution of the simplex steps in this example:

Iteration	Basis	Solution	Objective value	Entering variable	Leaving variable
1	(a_1, a_2, a_3, a_4)	(8, 16, 5, 3)	32	x_2	a_2
2	(a_1, x_2, a_3, a_4)	(8, 2, 3, 3)	14	x_1	a_3
3	(a_1, x_2, x_1, a_4)	(2, 2, 3, 0)	2	λ_1	a_4
4	$(a_1, x_2, x_1, \lambda_1)$	(2, 2, 3, 0)	2	λ_2	a_1
5	$(\lambda_2, x_2, x_1, \lambda_1)$	(2, 2, 3, 0)	0		

The solution is $(x_1 = 3;\ x_2 = 2)$.

7.5.2 The Active Set Method

When a candidate solution is on the boundary of the feasible region, some of the constraints become equal to zero. The set of constraints equal to zero at a given stage of solution guessing is called the *working set* W_K (also called *active set*).

The active set method iteratively changes the members of W_K. In each step, a constraint is added or removed from W_K in order to make the objective function decrease.

What happens in each step k is the following: the constraints outside W_K are omitted, and then an equality quadratic programming problem (QP) is solved. The solution $\mathbf{x_W}$ gives us a direction to follow:

$$\mathbf{p_k} = \mathbf{x_W} - \mathbf{x(k-1)}$$

where $\mathbf{x(k-1)}$ is the solution found at step $k-1$.

Depending on $\mathbf{p_k}$ we have two alternatives:

- If $\mathbf{p_k} = 0$, the optimum with current W_K. cannot be decreased. If all Lagrange multipliers were non-negative that means the final solution was obtained. Otherwise, a constraint with negative λ is dropped from W_K and a new QP is solved to obtain a new $\mathbf{p_{k+1}}$.
- If $\mathbf{p_k} \neq 0$, a step length must be specified so that $\mathbf{x_{k+1}} + \alpha\, \mathbf{p_k}$ was feasible (with $(0 \leq \alpha \leq 1)$. The parameter α is increased until a constraint not in W_K was reached. This blocking constraint is included in W_K. If it was not possible to reach such a constraint, then α is set to one.

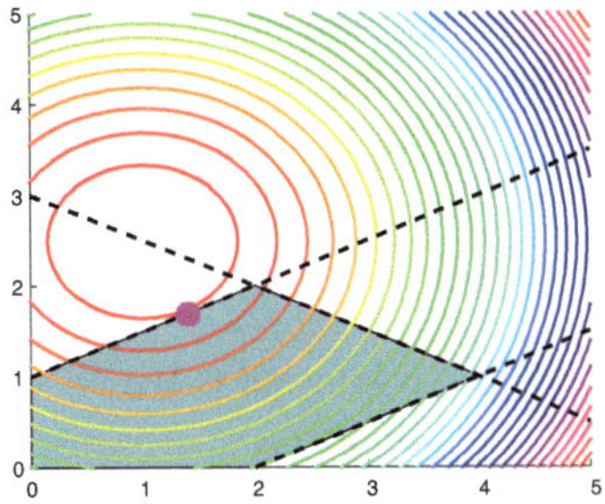

Fig. 7.4 Another optimization example

All this would become clear with the following example, in which we will construct a path towards the solution.

The example is as follows:

Minimize:

$$z = (x_1 - 1)^2 + (x_2 - 2.5)^2$$

Subject to:

$$x_1 - 2x_2 + 2 \geq 0 \ (A)$$

$$-x_1 - 2x_2 + 6 \geq 0 \ (B)$$

$$-x_1 + 2x_2 + 2 \geq 0 \ (C)$$

with: $x_1 \geq 0$ (D), and $x_2 \geq 0$ (E)

Figure 7.4 presents a view of the problem (the feasible region F has been emphasized with a colored patch).

This figure has been generated with Program 7.2.

Program 7.2 Example of active set method

```
%Visualization of Active Set example
[vx1,vx2] = meshgrid(0:0.1:5, 0:0.1:5);
y = ((vx1-1).^2)+((vx2-2.5).^2);
figure(1)
colormap('hsv');
patch([0,2,4,2,0],[1,2,1,0,0],[0.5 0.7 0.7]);
hold on;
contour(vx1,vx2,y,30,'LineWidth',1.5); hold on;
plot([0,5],[1,7/2],'k--','LineWidth',2); %constr 1
plot([0,5],[3,1/2],'k--','LineWidth',2); %constr 2
plot([2,5],[0,3/2],'k--','LineWidth',2); %constr 3
plot(1.4,1.7,'m*','LineWidth',5); %solution
```

Let us detail the steps of the active set method in this example. We could start from the following point (one of the corners of F):

$$\mathbf{x(0)} = \begin{bmatrix} 2 \\ 0 \end{bmatrix}$$

The working set is $W_0 = (C,\ E)$.; and the equality QP would be
Minimize:

$$z = \mathbf{x}^{\mathbf{T}} \begin{pmatrix} 2 & 0 \\ 0 & 2 \end{pmatrix} \mathbf{x} + \begin{pmatrix} -2 \\ -5 \end{pmatrix}^T \mathbf{x} + \frac{29}{4}$$

Subject to:

$$\begin{pmatrix} 1 & 2 \\ 0 & 1 \end{pmatrix} \mathbf{x} = \begin{pmatrix} -2 \\ 0 \end{pmatrix}$$

The *KKT* equations for solving this QP, would be:

$$\begin{pmatrix} 2 & 0 & -1 & 0 \\ 0 & 2 & 2 & 1 \\ -1 & 2 & 2 & 1 \\ 0 & 1 & 0 & 0 \end{pmatrix} \cdot \begin{pmatrix} \mathbf{x} \\ -\boldsymbol{\lambda} \end{pmatrix} = \begin{pmatrix} 2 \\ 5 \\ -2 \\ 0 \end{pmatrix}$$

The solution is:

$$\mathbf{xw} = \begin{pmatrix} 2 \\ 0 \end{pmatrix} \ ; \ \boldsymbol{\lambda}\mathbf{w} = \begin{pmatrix} -2 \\ -1 \end{pmatrix}$$

Therefore $\mathbf{p}_1 = 0$. Since all λ are negative, this is not the optimum.
Then, one of the constraints (C) is removed, so $W_1 = (E)$. A new equality QP should be solved, having the following *KKT* equations:

$$\begin{pmatrix} 2 & 0 & 0 \\ 0 & 2 & 1 \\ 0 & 1 & 0 \end{pmatrix} \cdot \begin{pmatrix} \mathbf{x} \\ -\boldsymbol{\lambda} \end{pmatrix} = \begin{pmatrix} 2 \\ 5 \\ -2 \end{pmatrix}$$

The solution is

$$\mathbf{xw} = \begin{pmatrix} 1 \\ 0 \end{pmatrix} \ ; \ \boldsymbol{\lambda}\mathbf{w} = -5$$

Hence, $\mathbf{p}_2 = \begin{pmatrix} 1 \\ 0 \end{pmatrix} - \begin{pmatrix} 2 \\ 0 \end{pmatrix} = \begin{pmatrix} -1 \\ 0 \end{pmatrix}$ is a search direction. Parameter α is set to one, so the new solution is

$$\mathbf{x}_2 = \begin{pmatrix} 1 \\ 0 \end{pmatrix} \quad ; \quad \lambda_2 = -5$$

Since λ_2 is negative, this is not the optimum. Another constraint is removed, so $W_2 = \mathit{empty}$. Then

$$\begin{pmatrix} 2 & 0 \\ 0 & 2 \end{pmatrix} \mathbf{x} = \begin{pmatrix} 2 \\ 5 \end{pmatrix} \quad \Rightarrow \quad \mathbf{x}_\mathbf{W} = \begin{pmatrix} 1 \\ 2.5 \end{pmatrix}$$

The new search direction would be:

$$\mathbf{p}_3 = \begin{pmatrix} 1 \\ 2.5 \end{pmatrix} - \begin{pmatrix} 1 \\ 0 \end{pmatrix} = \begin{pmatrix} 0 \\ 2.5 \end{pmatrix}$$

From constraint (A), one concludes that $\alpha < 0.6$.

Hence, one moves to $\mathbf{x}_3 = \begin{pmatrix} 1 \\ 1.5 \end{pmatrix}$

Constraint (A) is added to the working set, so $W_3 = (A)$. The *KKT* conditions for the new QP problem would be

$$\begin{pmatrix} 2 & 0 & 1 \\ 0 & 2 & -2 \\ 0 & -2 & 0 \end{pmatrix} \cdot \begin{pmatrix} \mathbf{x} \\ -\boldsymbol{\lambda} \end{pmatrix} = \begin{pmatrix} 2 \\ 5 \\ -2 \end{pmatrix}$$

The solution is

$$\mathbf{x}_\mathbf{W} = \begin{pmatrix} 1.4 \\ 1.7 \end{pmatrix} \quad ; \quad \boldsymbol{\lambda}_\mathbf{W} = 0.8$$

Since the Lagrange multiplier is nonnegative, this is the optimal solution.

Figure 7.5 represents the evolution of the algorithm, with the sequence of steps as a series of arrows starting from (2, 0).

This figure has been generated with Program 7.3, which is a simple modification of the previous program by adding the plot of the steps.

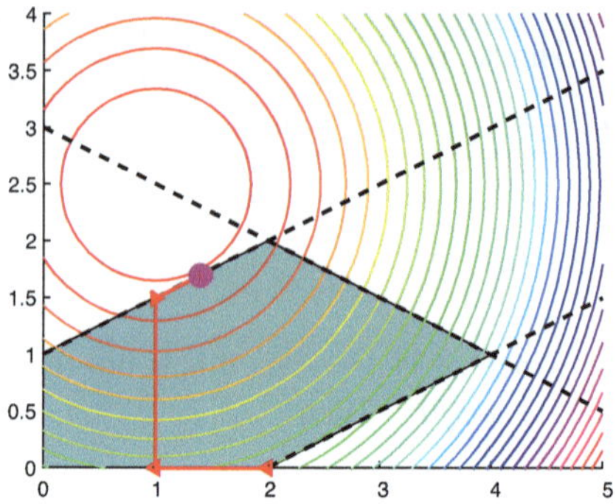

Fig. 7.5 Evolution of active set exploration

Program 7.3 Evolution of active set exploration

```
%Visualization of Active Set example
[vx1,vx2] = meshgrid(0:0.1:5, 0:0.1:4);
y = ((vx1-1).^2)+((vx2-2.5).^2);
figure(1)
colormap('hsv');
patch([0,2,4,2,0],[1,2,1,0,0],[0.5 0.7 0.7]);
hold on;
contour(vx1,vx2,y,30,'LineWidth',1); hold on;
plot([0,5],[1,7/2],'k--','LineWidth',2); %constr 1
plot([0,5],[3,1/2],'k--','LineWidth',2); %constr 2
plot([2,5],[0,3/2],'k--','LineWidth',2); %constr 3
%towards solution ----------------------
plot([2,1],[0,0],'r$<$-','LineWidth',2);
plot([1,1],[0,1.5],'r-','LineWidth',2);
plot([1,1.4],[1.5,1.7],'r$>$-','LineWidth',2);
plot(1.4,1.7,'mo','LineWidth',6); %solution
```

The active set method was introduced in 1971 by Roger Fletcher [36]. The example was taken from [102], which is an important book—a text of reference—on numerical optimization. More mathematical analysis details concerning active set methodology can be found in that book.

7.5.3 Dual Active-Set Methodology

By the way, it can be shown, [24], that the quadratic programming problem can be formulated in two forms:

Primal:

Find a vector **x** and a minimum $\underline{z}$ such that:

$$C\,\mathbf{x} \,\le\, \mathbf{d}, \quad \mathbf{x} \ge 0, \quad \underline{z} = \mathbf{b}^{\mathbf{T}}\,\mathbf{x} + \frac{1}{2}\,\mathbf{x}^{\mathbf{T}} Q\,\mathbf{x}$$

Dual:

Find a vector $\mathbf{y}$ and a maximum $\bar{z}$ such that:

$$-\,Q\mathbf{y}\ \leq \mathbf{b}\quad \mathbf{y}\geq 0,\qquad \bar{z}=-\frac{1}{2}\mathbf{x}^{\mathbf{T}}\,Q\,\mathbf{x}-\mathbf{d}^{\mathbf{T}}\,\mathbf{y}$$

Active-set methods iteratively make steps toward the solution, by solving smaller problems with only the equality constraints in the current active set (that we will denote as A^K, corresponding to the K-step).

In the important contribution of D. Goldfarb and D. Idnani, Bell Laboratories, [43], on a dual method for solving quadratic programming problems, it was argued that one-third to one-half of the computational effort of primal methods was expended in determining a first feasible solution, from which iterations can start. However, based on duality, a first feasible solution for the dual problem is simple to obtain: just the solution for the unconstrained problem $\mathbf{x}^{\circ}\ =\ -Q^{-1}\,\mathbf{b}$.

From the initial feasible solution, a typical approach of dual active set algorithms is to add primal constraints one at a time, choosing among constraints that are violated by current iteration, and resolve the problem to get a new iteration [95].

In a given step K of the algorithm, with an active set W_K, one has to solve:

$$\begin{pmatrix} Q & (A^K)^T \\ A^K & 0 \end{pmatrix} \begin{bmatrix} \mathbf{x}^K \\ \boldsymbol{\lambda}^K \end{bmatrix} = \begin{bmatrix} -\mathbf{b}^K \\ \mathbf{c}^K \end{bmatrix}$$

which corresponds to the pertinent equality constraints at K-step.

The article of Goldfarb and Idnani (GI, for short) suggests also a convenient implementation of its active set method, focusing on good numerical properties. It is well known that matrix inversion, ill-conditioning, and rounding errors may originate practical computational problems. Frequently, those difficulties could be alleviated by factorization mechanisms.

Instead of a direct inversion of the KKT matrix, for solving the equation at K-step, GI proposed a formulation easy to factorize. We will summarize the mathematical reasoning of the GI proposal, based on [21].

The inversion of the KKT matrix could be achieved with

$$(KKT^K)^{-1}=\begin{pmatrix} I & -\,Q^{-1}(A^K)^T \\ 0 & I \end{pmatrix}\begin{pmatrix} Q^{-1} & 0 \\ 0 & S_Q^{-1} \end{pmatrix}\begin{pmatrix} I & 0 \\ -A^K\,Q^{-1} & I \end{pmatrix}$$

where $S_Q =\ -A^K\,Q^{-1}\,(A^K)^T$, is the Schur complement of matrix Q.

In view of the previous equation, we define now two operators (the GI article says they are fundamental for the algorithm).

$$A^{K*}\ =\ (A^K Q^{-1}\,(A^K)^T)^{-1}\,A^K\,Q^{-1}$$

$$\mathrm{H}^K\ =\ Q^{-1}\,(I-\,(A^K)^T\,A^{K*})$$

where A^{K*} is the Moore-Penrose pseudoinverse of A^K under the transformation $\mathbf{y} = Q^{1/2}\mathbf{x}$. And H^K is the inverse Hessian operator in the active set space.

At K step, a new vector $\mathbf{x}^K$ and a set of active constraints are obtained from $\mathbf{x}^{K-1}$, A^{K-1}, and a violated constraint p. A primal-dual update is applied that makes the constraint p active while keeping dual feasibility. This is done as follows:

$$\mathbf{x^K} = \mathbf{x^{K-1}} + \alpha^K \boldsymbol{\Delta x}^\mathbf{K}$$

$$\boldsymbol{\lambda}^\mathbf{K} = \begin{pmatrix} \boldsymbol{\lambda}^{\mathbf{K}-1} \\ \lambda_p^{K-1} \end{pmatrix} + \alpha^K \begin{pmatrix} \boldsymbol{\Delta\lambda}^\mathbf{K} \\ 1 \end{pmatrix}$$

where α^K is the step size, and $\boldsymbol{\Delta x}^\mathbf{K}$, $\boldsymbol{\Delta\lambda}^\mathbf{K}$, are the update directions in the primal and dual space, respectively. These directions are computed as

$$\boldsymbol{\Delta}\,\mathbf{x^K} = -\,\mathrm{H}^{K-1}\,(A_p)^T$$

$$\boldsymbol{\Delta}\,\boldsymbol{\lambda}^\mathbf{K} = -\,A^{(K-1)*}\,(A_p)^T$$

and the step size: $\alpha^K = \min\left\{\alpha_1^K,\, \alpha_2^K\right\}$, with:

$$\alpha_1^K = \min\left\{\min_{j\in(1,2,\ldots\#W_{K-1})}\{-\frac{\lambda_j^{K-1}}{\Delta\lambda_j^K}\,\Big|\,\Delta\lambda_j^K < 0\}\,,\,\infty\right\}$$

$$\alpha_2^K = \frac{c_p - A_p x^{K-1}}{A_p \Delta x^K}$$

is chosen such that $\boldsymbol{\lambda}^\mathbf{K} \geq 0$ and, if possible, the p-th constraint becomes active.

Let us note that the dual step length α_1 is the maximum step that can be taken without violating dual feasibility. In turn, the primal step length α_2 is the step required to make the p constraint active.

Another event at K step is that a blocking constraint m may be dropped from the active set. This constraint is selected according to

$$m = \min\ \arg \min_{j\in(1,2,\ldots\#W_{K-1})}\left\{-\frac{\lambda_j^{K-1}}{\Delta\lambda_j^K}\,\Big|\,\Delta\lambda_j^K < 0\}\,,\,\infty\right\}$$

Three different situations may happen at K step:

- If $\alpha^K \equiv \alpha_1^K$ the m constraint is blocking and must be dropped; only a partial step can be taken.
- If $\alpha^K \equiv \alpha_2^K$ a full step can be taken, and the p constraint is added to the active set.
- If $\alpha^K \equiv \infty$ the original quadratic programming problem is infeasible.

For the computation of the two operators defined above, it would be convenient to employ the Cholesky factorization of Q, in the form: $Q = L\,L^T$; and also the following QR factorization:

$$L^{-1}\,A^K = \begin{bmatrix} Q_1^K\,,\; Q_2^K \end{bmatrix} \begin{bmatrix} R^K \\ 0 \end{bmatrix}$$

Now, define:

$$J^K = \begin{bmatrix} J_1^K\,,\; J_2^K \end{bmatrix} = \begin{bmatrix} L^{-T}\,Q_1^K & L^{-T}\,Q_2^K \end{bmatrix}$$

With these elements, it is possible to compute the two operators as follows:

$$A^{K*} = -\,R^{-1,K}\,(J_1^K)^T$$

$$\mathrm{H}^K = J_2^K\,(J_2^K)^T$$

Hence, the GI algorithm only needs to iteratively update J^K, R^K, departing from initial values $J^0 = L^{-T}$ and $R^0 = M_{0,0}$. The updates can be done through Householder reflections or by Givens rotations.

We refer to the original article [43], for more mathematical details. In fact, the algorithm detailed in the article is the basis of several software packages for solving quadratic programming applications (for instance: [31, 85]).

There is a contribution to the MathWorks File Exchange site, with MATLAB code concerning the Goldfarb and Idnani method [4].

7.5.4 The Lemke's Method and the Linear Complementary Problem

The Lemke's method can be applied to a number of problems, including those related to quadratic programming. It also gives us a good occasion for introducing the linear complementary problem (LCP), which has motivated a lot of research. This part of the chapter will begin with some space devoted to LCP, and then will focus on the complementary pivot algorithm proposed by Lemke in the mid-1960s.

7.5.4.1 The Linear Complementary Problem

In succinct terms, the LCP can be enunciated as follows: given a vector $\boldsymbol{q}$ and a real matrix M, find vectors ω and z which satisfy:

$$\begin{aligned} \boldsymbol{\omega} &= \mathbf{q} + M\,\mathbf{z}, \qquad \boldsymbol{\omega} \geq 0, \quad \mathbf{z} \geq 0 \\ \omega_j \cdot z_j &= 0, \;\forall j \end{aligned}$$

(the second equation, $\omega_j \cdot z_j = 0, \ \forall j$, is the complementary condition, already seen in previous pages. Two vectors complying with this condition are said to be complementary).

How is this related with our subject?

Some responses can be found in the well-known article [25]. The article first examines the primal and dual formulations of the **linear programming** problem:

Primal:

Find a vector **x** and a minimum $\underline{z}$ such that:

$$A\,\mathbf{x} \geq \mathbf{b}, \ \ \mathbf{x} \geq 0, \ \ \underline{z} = \mathbf{c}^T\,\mathbf{x}$$

Dual:

Find a vector **y** and a maximum $\bar{z}$ such that:

$$A^T\mathbf{y} \ \leq \mathbf{c}, \ \ \mathbf{y} \geq 0, \ \ \bar{z} = \mathbf{b}^T\,\mathbf{y}$$

If a solution is found: $\mathbf{b}^T\,\mathbf{y} \ = \mathbf{c}^T\,\mathbf{x}$

By adding slack variables, inequalities are transformed to equalities, and we can write a set of equivalent equations to satisfy:

$$A\,\mathbf{x} \, - \, \mathbf{v} = \mathbf{b}, \ \ \mathbf{x} \geq 0, \ \ \mathbf{v} \geq 0$$

$$A^T\mathbf{y} + \mathbf{u} \, = \mathbf{c}, \ \ \mathbf{y} \geq 0, \ \ \mathbf{u} \geq 0$$

Hence, the linear programming problem is to find vectors **u**, **v**, **x**, **y**, such that:

$$\begin{pmatrix} \mathbf{u} \\ \mathbf{v} \end{pmatrix} = \begin{pmatrix} \mathbf{c} \\ -\mathbf{b} \end{pmatrix} + \begin{pmatrix} 0 & -A^T \\ A & 0 \end{pmatrix} \begin{pmatrix} \mathbf{x} \\ \mathbf{y} \end{pmatrix}$$

with all vectors ≥ 0 and: $\mathbf{u}^\mathbf{T}\,\mathbf{x} + \mathbf{v}^\mathbf{T}\,\mathbf{y} = 0$

Consider now the following definitions:

$$\omega = \begin{pmatrix} \mathbf{u} \\ \mathbf{v} \end{pmatrix}; \ \mathbf{q} = \begin{pmatrix} \mathbf{c} \\ -\mathbf{b} \end{pmatrix}; \ \ M = \begin{pmatrix} 0 & -A^T \\ A & 0 \end{pmatrix}, \ \ \mathbf{z} = \begin{pmatrix} \mathbf{x} \\ \mathbf{y} \end{pmatrix}$$

Compare with the LCP succinct formulation given above. The conclusion is that our linear programming problem is a LCP.

Let us see what happens with **quadratic programming**.

The problem can be formulated as follows: find a vector **x** and minimum $\underline{z}$ such that

$$\underline{z} = \frac{1}{2}\mathbf{x}^T\,Q\,\mathbf{x} + \mathbf{b}^T\,\mathbf{x}$$

$$C\,\mathbf{x} \geq \mathbf{d}, \ \ \mathbf{x} \geq 0$$

(assume that Q is positive semidefinite)

Define **u** and **v** by:

$$\mathbf{u} = Q\,\mathbf{x} - C^T\,\mathbf{y} + \mathbf{b}, \quad \mathbf{v} = C\,\mathbf{x} - \mathbf{d}$$

The KKT conditions mean in this case that the solution $\mathbf{x}^*$ for the minimum should satisfy

$$\mathbf{x}^* \geq 0, \quad \mathbf{u}^* \geq 0, \quad \mathbf{y}^* \geq 0, \quad \mathbf{v}^* \geq 0,$$

$$\mathbf{u}^{\mathbf{T}*} \cdot \mathbf{x}^* = 0, \quad \mathbf{v}^{\mathbf{T}*} \cdot \mathbf{y}^* = 0$$

Therefore, the solution of the quadratic programming problem is to find vectors **u**, **v**, **x**, **y**, such that

$$\mathbf{u} = Q\,\mathbf{x} - C^T\,\mathbf{y} + \mathbf{b}, \quad \mathbf{x} \geq 0, \; \mathbf{y} \geq 0$$

$$\mathbf{v} = C\,\mathbf{x}, - \mathbf{d} \qquad \mathbf{u} \geq 0, \; \mathbf{v} \geq 0$$

$$\mathbf{u}^{\mathbf{T}}\,\mathbf{x} + \mathbf{v}^{\mathbf{T}}\,\mathbf{y} = 0$$

In the same vein as before, if we adopt the following definitions:

$$\omega = \begin{pmatrix} \mathbf{u} \\ \mathbf{v} \end{pmatrix}; \; \mathbf{q} = \begin{pmatrix} \mathbf{b} \\ -\mathbf{d} \end{pmatrix}; \; M = \begin{pmatrix} Q & -C^T \\ C & 0 \end{pmatrix}, \; \mathbf{z} = \begin{pmatrix} \mathbf{x} \\ \mathbf{y} \end{pmatrix}$$

one arrives again to a LCP problem.

The LCP problem also appears in the context of game theory and the Nash equilibrium point. Nonzero-sum two-person games can be treated as *bimatrix games* (one of the matrices corresponds to the strategies of person A, and the other matrix corresponds to the strategies of person B). More information on LCP can be found in [26], a well-recognized book on this topic.

The following simple example [66], would be useful for an intuitive geometrical view of the LCP problem. It is the case of

$$M = \begin{pmatrix} 2 & 1 \\ 1 & 2 \end{pmatrix}, \; \mathbf{q} = \begin{pmatrix} -5 \\ -6 \end{pmatrix}$$

Which leads to the following LCP:

$$\omega_1 - 2\,z_1 - z_2 = -5$$

$$\omega_2 - z_1 - 2\,z_2 = -6$$

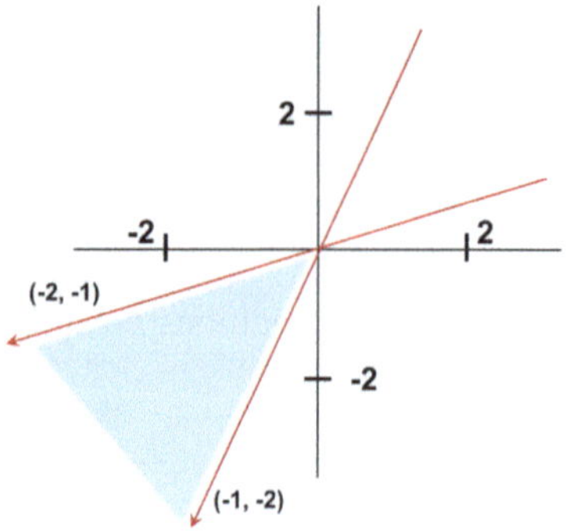

Fig. 7.6 Complementary cone

$$\omega_1,\ \omega_2,\ z_1,\ z_2\ \geq\ 0 \text{ and } \omega_1\, z_1 =\ \omega_2\, z_2\ = 0$$

Or, equivalently:

$$\omega_1 \begin{pmatrix} 1 \\ 0 \end{pmatrix} + \omega_2 \begin{pmatrix} 0 \\ 1 \end{pmatrix} + z_1 \begin{pmatrix} -2 \\ -1 \end{pmatrix} + z_2 \begin{pmatrix} -1 \\ -2 \end{pmatrix} = \begin{pmatrix} -5 \\ -6 \end{pmatrix}$$

In accordance with the constraints, at least one of the variables in each pair $(\omega_j,\ z_j)$ has to be zero. In this situation, a simple approach for solving the problem would be to take ω_1 and ω_2 equal to zero; so, the remaining system is:

$$z_1 \begin{pmatrix} -2 \\ -1 \end{pmatrix} + z_2 \begin{pmatrix} -1 \\ -2 \end{pmatrix} = \begin{pmatrix} -5 \\ -6 \end{pmatrix} = \begin{pmatrix} q_1 \\ q_2 \end{pmatrix} = \mathbf{q}$$

$$z_1,\ z_2\ \geq\ 0$$

This system has a solution iff $\mathbf{q}$ can be expressed as a nonnegative linear combination of the vectors $(-2,\ -1)^T$and $(-1,\ -2)^T$. The set of all nonnegative linear combination of such vectors is the cone shown in Fig. 7.6.

Only if the given vector $\mathbf{q}$ lies in that cone, the LCP has a solution in terms of z_1 and z_2. It easy to confirm that the vector $(-5, -6)^T$ achieves this condition. The solution, in fact, can be easily obtained from the equations: $z_1\ =\ 4/3$; $z_2 =\ 7/3$.

The cone in Fig. 7.6 is a *complementary cone* associated with the LCP. There are other complementary cones corresponding to taking z_1 and ω_2 equal to zero (instead of ω_1 and ω_2), or ω_1 and z_2 equal to zero, or the last possibility: z_1 and z_2 equal to zero. Figure 7.7 visualizes those cones.

In general, for an n-dimensional LCP problem, there would be 2^n complementary cones.

One of the aspects of interaction with robots being investigated from some time ago, is *contacts*. Inside a robotic structure (a robotic arm, a humanoid, etc.) there are contacts and also when a robot touches a man. The article [6] uses a LCP formulation of rigid body dynamics to obtain definitions of contact types. It is shown that contact modes *slide right*, *slide left*, *roll* and *separate*, correspond to

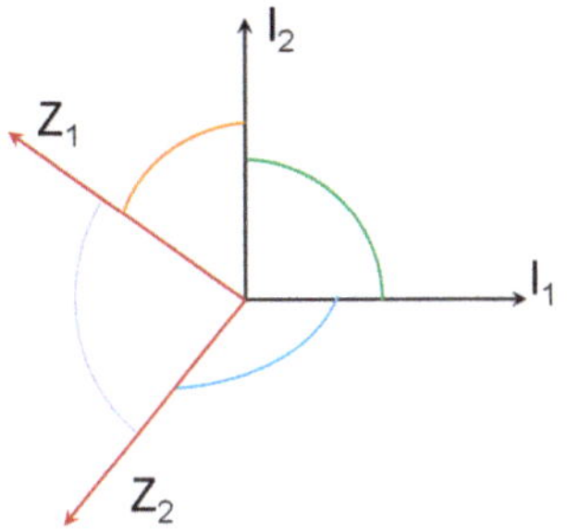

Fig. 7.7 All the complementary cones for a 2D problem

complementary cones of the LCP (see https://scaron.info/robotics/contact-modes.html for a concise introduction to contact modes). The doctoral thesis [30] about LCP solutions includes comments on contacts with some references. Also, the article [40], with an interesting introduction, proposes a LCP-based formulation of contact problems that considers regularized friction.

A modern application of contacts is in video games.

7.5.4.2 The Lemke's Method

Lemke and Howson devised an iterative technique for finding equilibrium points in bimatrix games [58]. Later on, Lemke extended the method for solving LCP problems [57]. It is a pivoting method much in the line of linear programming (actually, it can be put into practice using tableaux).

Some terms and a modification of the problem would be introduced now for the description of the method, [80].

A solution of the LCP is called a "*complementary basic feasible solution* (CBFS)" if it is: (a) a complementary feasible solution and (b) it belongs to a complementary cone whose generators are linearly independent.

Let us add a variable z_0 to the problem, so we have

$$\begin{aligned} &\boldsymbol{\omega} - M\,\mathbf{z} - \mathbf{1}\,z_0 = \mathbf{q} \qquad \boldsymbol{\omega} \geq 0, \quad \mathbf{z} \geq 0 \quad z_0 \geq 0 \\ &\omega_j \cdot z_j = 0, \ \forall j \end{aligned}$$

where $\mathbf{1}$ represents a column matrix of ones: $(1,\ 1,\ \ldots,\ 1)^T$.

Note that $\boldsymbol{\omega} = \mathbf{q} + \mathbf{1}\,z_0$, $\mathbf{z} = 0$ is a complementary solution of the problem.

The added variable facilitates an initial solution, from which iterations will try to drive z_0 out of the basis.

A triplet $(\boldsymbol{\omega},\ \mathbf{z},\ z_0)$ is called an "*almost complementary basic feasible solution* (ACBFS)" if:

- The triplet is a basic feasible solution to:

$$\boldsymbol{\omega} - M\,\mathbf{z} - \mathbf{1}\,z_0 = \mathbf{q} \qquad \boldsymbol{\omega} \geq 0, \quad \mathbf{z} \geq 0 \quad z_0 \geq 0$$

- There is one complementary pair $(\omega_S,\ z_S)$ such both ω_S and z_S are not basic
- And z_0 is basic

The method will pivot from one ACBFS to another (each ACBFS has at most two adjacent ACBFS's). Like in the simplex algorithm, on each iteration, one column of the basis will be removed, and another will be inserted in place.

Here is a succinct description of the method steps:

–Initialization:
Let s be the index for which: $-q = \max\{-q_i\}$.
The initial basis is z_0 and $\omega_j,\ \ j \neq s$.
Set $y = z_S$
–Iterations
Denote as d the basic direction associated with variable y.
If $d \leq 0$ Stop (the feasible region is unbounded)
Otherwise, the variable y enters the basis
Do a minimum ratio test to determine the leaving variable:
If ω_1 just left the basis, set $y = z_1$
If z_1 just left the basis, set $y = \omega_1$
If z_0 left the basis, stop.

Coming back to the quadratic programming problem and its formulation as LCP (based on the KKT conditions), the convergence of Lemke's method is such that, in the absence of degeneracy, the algorithm will find a solution of the KKT equations in a finite number of iterations if (a) Q is positive definite or (b) Q is positive semidefinite and b = 0 or (c) Q has nonnegative elements with positive diagonal entries, [80].

A detailed example of the Lemke's method has been given by Stephen Wright in [101], accessible on the Internet. The case is the following:

Minimize:

$$J = x_1^2 + 1.5\, x_1\, x_2 + 4\, x_2^2 - 3\, x_1 + x_2$$

Subject to:

$$2\, x_1 + 3\, x_2 = 7$$

$$x_1 + x_2 \geq -3$$

Figure 7.8 visualizes the problem. No colored patch has been painted because the feasible region in this case is just the dotted line in black.

Figure 7.8 has been generated by Program 7.4.

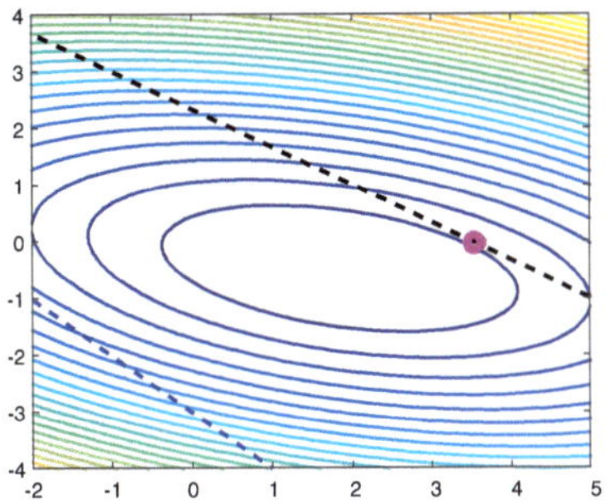

Fig. 7.8 Example for Lemke's method

Program 7.4 Example of Lemke's method

```
%Example of Lemke's method
[vx1,vx2] = meshgrid(-2:0.1:5, -4:0.1:4);
y = (vx1.^2)+ 1.5*(vx1.*vx2) + 4*(vx2.^2) - ...
3*vx1 + vx2;
figure(1)
contour(vx1,vx2,y,25,'LineWidth',1.5); hold on
plot([5,-2],[-1,11/3],'k--','LineWidth',2); %constr 1
plot([-2,1],[-1,-4],'b--','LineWidth',2); %constr 2
plot(3.5469,-0.0312,'mo','LineWidth',5); %solution
```

The KKT conditions for this problem are the following:

$$2x_1 + 1.5x_2 - 2u_1 - u_2 - 3 = 0$$

$$1.5x_1 + 8x_2 - 3u_1 - u_2 + 1 = 0$$

$$2x_1 + 3x_2 - 7 = 0$$

$0 \leq x_1 + x_2 + 3$ and $u_2 \geq 0$ complementary

We can write these equations in LCP format as follows:

$$\mathbf{z} = \begin{pmatrix} x_1 \\ x_2 \\ u_1 \\ u_2 \end{pmatrix} M = \begin{pmatrix} 2 & 1.5 & -2 & -1 \\ 1.5 & 8 & -3 & -1 \\ 2 & 3 & 0 & 0 \\ 1 & 1 & 0 & 0 \end{pmatrix}, \quad \mathbf{q} = \begin{pmatrix} -3 \\ 1 \\ -7 \\ 3 \end{pmatrix}$$

where $\omega_1, = \omega_2 = \omega_3 = 0$ and z_1, z_2, z_3 are free variables.

The book [35] includes several MATLAB programs that can be conveniently used for our example. Those programs are available through the web page [34].

By using *lemketbl()* (one of the cited programs), the following initial tableau for our problem can be created:

	z1	z2	z3	z4	1
w1	2.0000	**1.5000**	−2.0000	−1.0000	−3.0000
w2	1.5000	8.0000	−3.0000	−1.0000	1.0000
w3	2.0000	3.0000	0.0000	0.0000	−7.0000
w4	1.0000	1.0000	0.0000	0.0000	3.0000

We write in bold the pivot. It is desired to get $\omega_1,\ \omega_2,\ \omega_3$ to the top of the tableau, and $z_1,\ z_2,\ z_3$ to the side.

Now, another program, *ljx()*, can be used for pivoting, obtaining the following tableau:

	z1	w1	z3	z4	1
z2	−1.3333	0.6667	1.3333	0.6667	2.0000
w2	**−9.1667**	5.3333	7.6667	4.3333	17.0000
w3	−2.0000	2.0000	4.0000	2.0000	−1.0000
w4	−0.3333	0.6667	1.3333	0.6667	5.0000

Using again *ljx()* for the new pivot, the next tableau was obtained:

	w2	w1	z3	z4	1
z2	0.1455	−0.1091	0.2182	0.0364	−0.4727
z1	−0.1091	0.5818	0.8364	0.4727	1.8545
w3	0.2182	0.8364	**2.3273**	1.0545	−4.7091
w4	0.0364	0.4727	1.0545	0.5091	4.3818

And, with *ljx()* again:

	w2	w1	w3	z4	1
z2	0.1250	−0.1875	0.0938	−0.0625	−0.0312
z1	−0.1875	0.2812	0.3594	0.0937	3.5469
z3	−0.0938	−0.3594	0.4297	−0.4531	2.0234
w4	−0.0625	0.0938	0.4531	0.0312	6.5156

Finally, one could extract the result of the procedure:

	z4	1
z2	−0.0625	−0.0312
z1	0.0937	3.5469
z3	−0.4531	2.0234

The solution **z**, and the Lagrange multipliers **u**, are then:

$$\mathbf{z} = \begin{pmatrix} 3.5469 \\ -0.0312 \end{pmatrix} \text{ and } \mathbf{u} = \begin{pmatrix} 2.0234 \\ 0 \end{pmatrix}$$

7.5.5 Gradient Projection Methodology

Gradient projection methods propose a piecewise search for the optimal solution, using projections of the objective function gradients onto the feasible set boundary. This will be illustrated now with some figures.

Figure 7.9 presents one of the steps of the search, in the context of a minimization problem in a tridimensional space, where the feasible set is a rectangular box. The step departs from a certain point and goes straight toward the minimum, guided by (minus) the gradient. The linear step crosses the feasible set boundary, becoming unfeasible. But we can still use its projection onto that boundary (a face of the box), for approaching the desired solution. Thus, the search trajectory bends and continues. The figure comes from [102]. Notice that the projection of a single line in the gradient direction originates a series of three segments in the example considered.

Fig. 7.9 Example of a gradient projection step

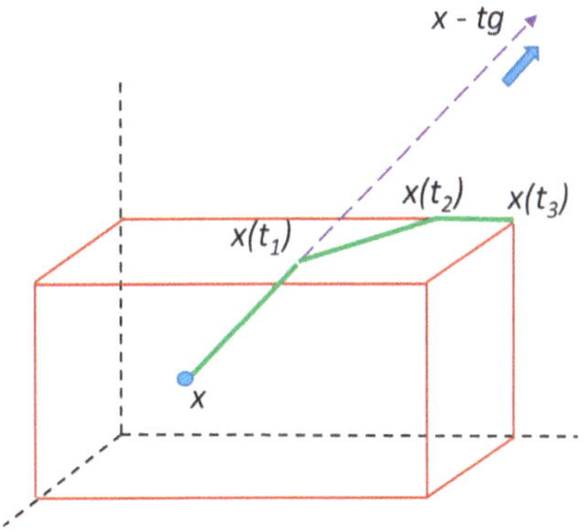

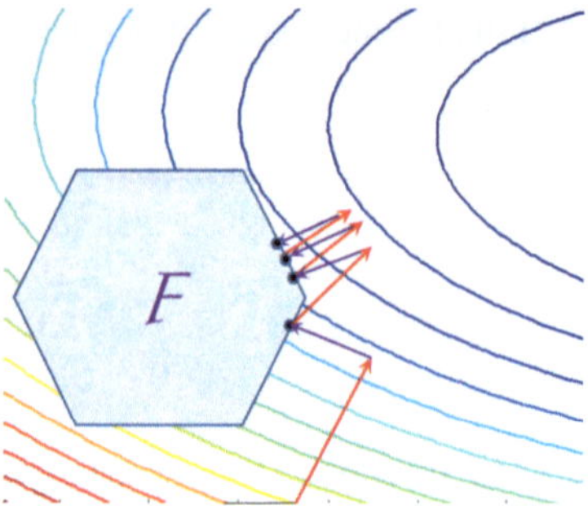

Fig. 7.10 Example of a sequence of steps

Suppose that $\mathbf{x_K}$ is the starting point of a searching step. The next point of the search would be

$$\mathbf{x_{K+1}} = P\ |x_K - \alpha \cdot \nabla \mathbf{J}(\mathbf{x_K})| = P\,(\vec{z}_K)$$

where P is the projection operator and α is the step size.

The operator P refers to a projection of $\vec{z}_K$ onto the feasible set F. The projection of any point q onto F is the closest point $p \in F$ in Euclidean distance. Hence, the projection involves an optimization problem, of distance minimization (distance from a point to a delimited region).

Usually, it is desired to have a simple region F, for easy minimum distance calculation. For this reason, in most cases, the gradient projection method is applied when constraints are simple bounds (like, for instance, a *box* containing variables between $l \leq x \leq h$ limits).

Figure 7.10 shows a typical sequence of steps, starting from an unfeasible point. After a first step approaching F (the feasible set), once we have a projection on the boundary of F, a series of points along the F boundary were obtained, corresponding to the gradient projection steps of the method (a zig-zag of arrows in the figure). The searching along the edge of F would terminate when no significant improvement yields in keeping the zig-zag exploration.

The step size α could be adjusted by the Armijo's rule; see [3]. Besides, the research has proposed several alternatives, including exact steps [46].

In general, the gradient projection procedure is usually faster than the active set method, with less steps.

A basic implementation of the method in MATLAB has been developed, Program 7.5, based on code available from GitHub at [99]. Figure 7.11 has been generated with our program and shows an example of gradient projection application.

As said before, the projection involves a quadratic optimization which has been solved in the program (see function *PG_proj*) by calling the function *quadprog()*, which belongs to the MATLAB Optimization Toolbox.

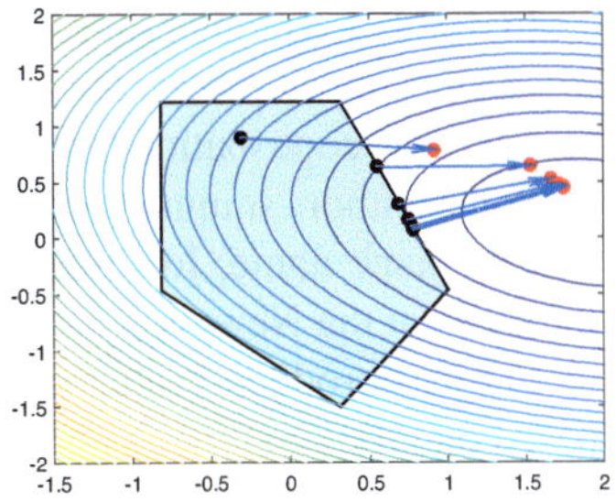

Fig. 7.11 Search done with the gradient projection program

Note (Program 7.5)

The complete listing of the Program 7.5, corresponding to gradient projection has been included in the Appendix for Long Programs.

More details on gradient projection can be found in [7], [84] and [18].

Next subsection deals with an interesting variant, the spectral projected gradient (SPG) method. It also gives us the opportunity for introducing the Barzilai-Borwein steps, considered as an important contribution to the field.

7.5.6 Spectral Projected Gradient Method

The spectrum of a matrix is the set of its eigenvalues. The name "*spectral* gradient" alludes to this concept.

Perhaps it would be opportune to note the importance of eigenvalues for the study of system dynamics (like the control systems to which a chapter has been dedicated before). The eigenvalues govern the time constants and oscillatory (even unstable) behavior of transients.

Eigenvalues are behind the scene (not everybody notice that) when numerical calculus deals with convergence issues of iterative approximation methods. Indeed, the gradient methods we are dealing with in this chapter, belong to this context.

7.5.6.1 Barzilai-Borwein Steps

In several methods (already introduced in this book), the Armijo's rule or other techniques have been employed for establishing sizes of line searching steps. In these techniques, most times a first-order approximation series is applied. Surely, it would be better to use second-order information, but a computational price would be paid.

In 1988, J. Barzilai and J. M. Borwein published a brief article (eight pages), [5], proposing step sizes for gradient methods, by a two-point approximation of the secant. This contribution has received about 3,000 cites.

Some authors summarize the central idea of Barzilai and Borwein (curiously both had the same name, Jonathan), as a cheap approximation of the Hessian, which is the basis of the Newton's method.

Let us introduce in mathematical terms the proposal of Barzilai and Borwein, based on [103].

For brevity, consider the following change of notation:

$\mathbf{g}^{(\mathbf{k})} = \nabla f\left(\mathbf{x}^{(\mathbf{k})}\right)$, and $H^{(k)} = \nabla^2 f\left(\mathbf{x}^{(\mathbf{k})}\right)$

The standard gradient iterative method uses

$$\mathbf{x}^{(\mathbf{k+1})} = \mathbf{x}^{(\mathbf{k})} - \alpha_k \mathbf{g}^{(\mathbf{k})}$$

In turn, the Newton's method uses

$$\mathbf{x}^{(\mathbf{k+1})} = \mathbf{x}^{(\mathbf{k})} - \left(H^{(k)}\right)^{-1} \mathbf{g}^{(\mathbf{k})}$$

The main problem with the Newton's method is that a matrix inversion is needed.

The method of Barzilai and Borwein chooses α_k so that $\alpha_k \mathbf{g}^{(\mathbf{k})}$ approximates $\left(H^{(k)}\right)^{-1} \mathbf{g}^{(\mathbf{k})}$, with no need of computing $\left(H^{(k)}\right)^{-1}$.

For the quadratic programming problem, the gradient is $\mathbf{g}^{(\mathbf{k})} = Q\,\mathbf{x}^{(k)} + \mathbf{b}$ and the Hessian is Q. The Newton step is $-\ Q^{-1}\,\mathbf{g}^{(\mathbf{k})}$.

Then, our target would be to select α_k for a good approximation of $-\ Q^{-1}\,\mathbf{g}^{(\mathbf{k})}$ by $-\,\alpha_k\,\mathbf{g}^{(\mathbf{k})}$.

With this purpose, consider the following definitions:

$$\mathbf{s}^{(\mathbf{k-1})} = \mathbf{x}^{(\mathbf{k})} - \mathbf{x}^{(\mathbf{k-1})} \text{ and } \mathbf{y}^{(\mathbf{k-1})} = \mathbf{g}^{(\mathbf{k})} - \mathbf{g}^{(\mathbf{k-1})}$$

Therefore: $Q\,\mathbf{s}^{(\mathbf{k-1})} = \mathbf{y}^{(\mathbf{k-1})}$.

Based on these terms, our target is choosing α_k so that

$$\left(\alpha_k^{-1}\,I\right)\mathbf{s}^{(\mathbf{k-1})} \approx \mathbf{y}^{(\mathbf{k-1})}$$

The approach of Barzilai and Borwein is to consider a least square minimization of

$$\left\|\left(\alpha_k^{-1}\,I\right)\mathbf{s}^{(\mathbf{k-1})} - \mathbf{y}^{(\mathbf{k-1})}\right\|^2$$

which leads to (*the long step*)

$$\alpha_k = \frac{\left(\mathbf{s}^{(\mathbf{k-1})}\right)^T \mathbf{s}^{(\mathbf{k-1})}}{\left(\mathbf{s}^{(\mathbf{k-1})}\right)^T \mathbf{y}^{(\mathbf{k-1})}}$$

(note that no matrix inversion is needed)

Alternatively, the term to be minimized can be written as

$$\left\| \mathbf{s}^{(\mathbf{k-1})} - \mathbf{y}^{(\mathbf{k-1})} \alpha_k \right\|^2$$

which leads to (*the short step*):

$$\hat{\alpha}_k = \frac{\left(\mathbf{s}^{(\mathbf{k-1})}\right)^T \mathbf{y}^{(\mathbf{k-1})}}{\left(\mathbf{y}^{(\mathbf{k-1})}\right)^T \mathbf{y}^{(\mathbf{k-1})}}$$

These are the Barzilai-Borwein (BB) steps.

At $k = 0$ no information on variables at $(k - 1)$ is available, so one starts the searching with a standard gradient, and then (from $k = 1$) one proceeds with BB steps. Both α_k and $\hat{\alpha}_k$ could be alternated. In general, the use of BB steps drive to fairly good results. Reference [103] reports that for a quadratic case with Q having a condition number of

$$\frac{\lambda_{\max}}{\lambda_{\min}} = 50$$

(a quotient of eigenvalues of Q)

... it happens that the standard steepest descent needs 90 iterations for reaching a distance less than 10^{-8} to the solution, while BB requires only 10 iterations for it. The convergence rate of methods based on BB steps is quite good in quadratic problems, and this motivates a lot of research on the use of BB steps in other types of problems.

In order to give an implementation example, the Program 7.5 has been devised, based on the same example employed in Program 2.15. This is a search for the minimum of a given quadratic function. The result of employing Barzilai-Borwein large steps is visualized in Fig. 7.12.

Figure 7.12 has been obtained with Program 7.5. The process is divided into two phases. During the first phase, a line search is applied along the direction of the first gradient (this is marked with a series of points). In the second phase, a series of four Barzilai-Borwein large steps are used.

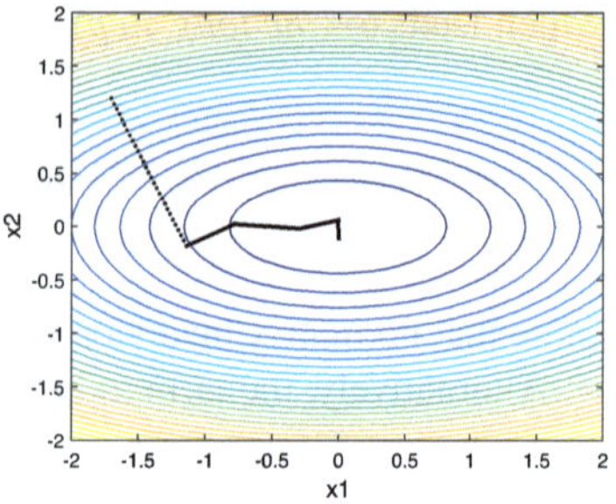

Fig. 7.12 Employing Barzilai-Borwein large steps

Program 7.5 Using Barzilai-Borwein steps

```
% steepest descent example
% using Barzilai-Borwein steps
x1=-2:0.05:2;
x2=-2:0.05:2;
N=length(x1);
y=zeros(N,N); %reserve memory
%
%set of function values
for i=1:N,
  for j=1:N,
    y(j,i)=2*(x1(i)^2)+7*(x2(j)^2);
  end;
end;
%function display--------
figure(1)
contour(x1,x2,y,26); hold on;
xlabel('x1'); ylabel('x2');
%prepare vectors for computations
xi=zeros(2,1); xn=zeros(2,1);
% searching--------------
% initial point
xi=[-1.7; 1.2];
yi=2*(xi(1)^2)+7*(xi(2)^2);
g=[(4*xi(1));(14*xi(2))]; %gradient value
% sign of g along x(1):
sg=1;
if g(1)<0, sg=-1; end;
%
% first search segment
m=g(2)/g(1); %slope
flag=1; yref=yi;
xn(1)=xi(1);
while flag==1,
  xn(2)=m*(xn(1)-xi(1))+xi(2);
  y=2*(xn(1)^2)+7*(xn(2)^2);
  if y>yref,
```

```
    flag=0;
  else
  yref=y;
  plot(xn(1),xn(2),'k.');
  xn(1)=xn(1)-(sg*0.02);
  end;
end;
% now, Barzilai-Borwein steps are used
% more search segments
for M=1:4,
  so=xn-xi;
  xi=xn;
  gro=g;
  g=[(4*xi(1));(14*xi(2))]; %new gradient value
  yo=g-gro;
  %long B-B step
  Lalpha = (so'*so)/(so'*yo);
  xn=xi-(Lalpha*g);
  plot([xi(1) xn(1)],[xi(2) xn(2)],'k-');
end;
```

7.5.6.2 The Spectral Projected Gradient Method

The BB steps can be applied to unconstrained problems, as well as facing constraints. Needless to say, the case of projected gradients belongs to constrained scenarios (on which we project).

The key reference for the SPG method is [9]. In words of this article, *"the Spectral Project Gradient (SPG) method was born from the marriage of the Barzilai-Borwein (spectral) nonmonotone ideas with classical projected gradient strategies"*.

In effect, the SPG method uses BB steps as the step size α in

$$\mathbf{x_{K+1}} = P\,|x_K - \alpha \cdot \nabla \mathbf{J}(\mathbf{x_K})| = P\,(\vec{z}_K)$$

(we repeat here the formula of gradient projection).

Several years later, in 2014, the same authors of [9], published a review of the SPG method and some of its many applications, [10].

As indicated by the title of [9], the sequence of BB steps is nonmonotone. It implies a risk of cycling. This fact was specially highlighted in [28], in relation to box-constrained quadratic programming. The authors of this contribution were Y. H. Dai and R. Fletcher, who frequently added in subsequent publications more insights and improvements to clear the way for BB new uses.

A strategy for circumventing possible drawbacks along BB steps is to substitute the BB steps that may cause problems (for instance, large excursions) by other more suitable steps. See, for example, [59] and [48] (or [49]). The article [59] is also interesting for its links to software implementations.

Not surprisingly, some recent publications do address the possibility of devising a spectral *conjugate* gradient projection method, like in [97].

Many of the contributions about spectral projections deal with the *nonnegative matrix factorization*: a mathematical problem that rises in image processing for *deblurring* and other signal recovery applications.

7.5.7 Conditional Gradient (Frank-Wolfe Algorithm)

The conditional gradient method proposed by Marguerite Frank and Philip Wolfe [37], takes a linear approximation L of the objective function J at point **x(k)** and moves a bit toward the minimizer of L, which is simply based on a first-order Taylor approximation (using the gradient at point **x(k)**).

Figure 7.13 depicts the idea, [16], the dotted line represents the linear approximation of J at **x(k)**. By displacing a parallel copy of this line toward the minimum of J, one intersects a vertex **v** of the feasible set F. This is much like in linear programming (in fact, the simplex algorithm can be employed here). Now, a line connecting **x(k)** and the vertex **v** can be drawn (a dotted line inside the polygon).

The algorithm will take a step on this dotted line towards the vertex line, with no intention to reach **v** but just some advance. So, a new point **x(k + 1)** is established, and there a new linear approximation is made using the gradient at this point, a new vertex is determined, etc. In many practical applications, the resulting trajectory of the steps is in zig-zag.

In mathematical terms, the algorithm starts from some point $\mathbf{x}(0) \in F$ and then applies an iterative search according to

$$\mathbf{x}(\mathbf{k}+\mathbf{1}) = \mathbf{x}(\mathbf{k}) + \delta_k \, (\mathbf{v}(\mathbf{k}) - \mathbf{x}(\mathbf{k}))$$

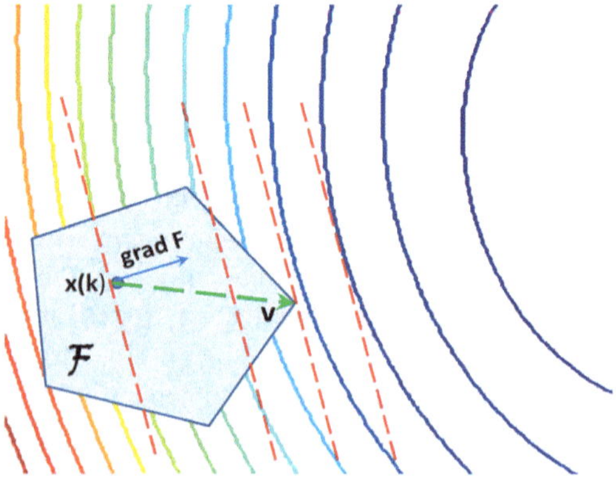

Fig. 7.13 First step of the algorithm

where,

$$\mathbf{v(k)} = \arg\min_{x \in F} \left(J(k) + \nabla \mathbf{J(k)}^{\mathbf{T}} (\mathbf{x} - \mathbf{x(k)}) \right)$$

which simplifies to

$$\mathbf{v(k)} = \arg\min_{x \in F} \left(\nabla \mathbf{J(k)}^{\mathbf{T}} (\mathbf{x} - \mathbf{x(k)}) \right)$$

The step size δ must be $0 < \delta \leq 1$ to guarantee that $\mathbf{x(k+1)} \in F$.

A main advantage of the Frank-Wolfe method is that no projection is made (projection—could entrain computational difficulties). Besides, it has been shown [51, 55] that the method has convergence properties similar to those of projected gradient.

Based on [72] (University of Pisa, Italy), a simple implementation of Frank-Wolfe was made in MATLAB: Program 7.6. Figure 7.14 presents the results for the same case considered in the projected gradient method.

The listing of Program 7.6 is given below. Notice that we employed the function *linprog()* offered by the MATLAB Optimization Toolbox. This function has been already used in the previous chapter on Linear Programming.

Program 7.6 Example of Frank-Wolfe method

```
%Example of Frank-Wolfe method
Q=[0.30 0.08
0.08 0.7];
b=[-1.2 ; -0.5];
x0=[-0.3; 0.9]; %Initial point
%contour data
[vx1,vx2] = meshgrid(-1.5:0.05:2, -2:0.05:2);
y=Q(1,1)*(vx1.^2)+(Q(2,1)+Q(1,2))*vx1.*vx2+...
Q(2,2)*(vx2.^2)+b(1)*vx1+b(2)*vx2;
%-----------------------------------------------
%Define a polygonal feasible set (Ax<=c)
% number of polygon sides
nply = 5;
A = zeros(nply,2);
c = zeros(nply,1);
% find vertices (inside a circle)
theta_ply = (0:(2*pi)/nply:2*pi);
xply = cos(theta_ply);
yply = 1.5*sin(theta_ply-pi/10);
for i = 1:nply
  % compute edge directions p = x2-x1
  p = [xply(i+1)-xply(i)
  yply(i+1)-yply(i)];
  % get the normal direction
```

```
  % rotate 90 degrees counter-clockwise
  a = [p(2),-p(1)];
  a = a/norm(a);
  A(i,:) = a;
  % offset of edges
  c(i) = a*[xply(i); yply(i)];
end
%Algorithm --------------------------------------
% initialization
mingap=0.01;
rx=zeros(50,2);
figure(1)
plot(xply,yply,'k','LineWidth',2); hold on
patch(xply,yply,[0.8 0.95 0.95]);
contour(vx1,vx2,y,35);
% start -------------------------
xk=x0;
niter=1;
rx(1,:)=xk;
while true
  dJk=Q*xk + b; %gradient
  yk = linprog(dJk,A,c,[],[],[],[],[],...
   optimset('Display','off'));
  d= yk - xk;
  gap = -dJk'*d;
  if gap< mingap
    break
  else
    t= 1/(niter+1); %step size
    niter=niter+1;
    xk = xk + (t*d);
    rx(niter,:)=xk;
  end
end
plot(rx(:,1), rx(:,2), 'r-','LineWidth',2);
```

The research has shown considerable interest on the Frank-Wolfe algorithm, and therefore many variants and improvements have been proposed [16]. This has been accompanied by a large variety of applications, [14], including recent initiatives for machine learning.

7.5.8 The Criss-Cross Algorithm and Related Aspects

A number of pivoting methods have been already introduced in the chapter on linear programming and in the present chapter. They are not the only ones, on the contrary: Here lives a fertile research field, adjacent to algebra and numerical calculus with matrices.

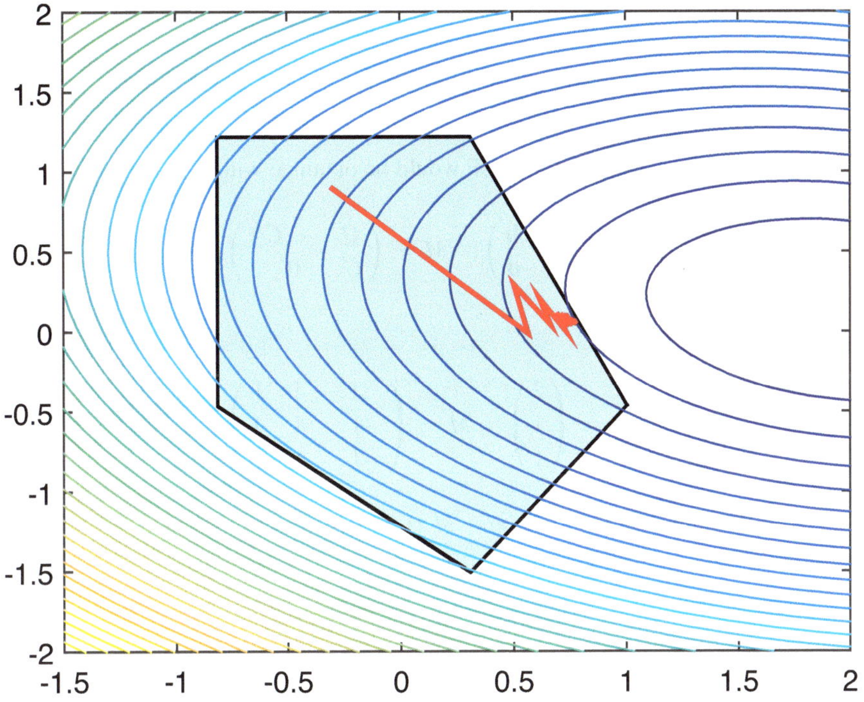

Fig. 7.14 Example of Frank-Wolfe method

An interesting example of other pivoting methods is the criss-cross algorithm, which is linked to certain abstract aspects of combinatorial optimization. We will describe now this algorithm, and then some space will be devoted to a higher level point of view.

The criss-cross algorithm could be seen as an alternative to the simplex method for linear programming. It uses a sequence of tableaux and pivots. The main difference is that only the signs, not the magnitude of the components determine the pivot selection.

It is not necessary to start from any basis, and no work is invested to preserve feasibility along the procedure.

A series of examples would help for introducing the criss-cross algorithm. Let us begin with the following example (from [83]):

Minimize:

$$J = \frac{1}{2}x_1^2 + 2x_2^2 - x_1 x_2 + 2x_1 + x_2$$

Subject to:

$$x_1 + x_2 \leq 1 \text{, and } x_1,\ x_2 \geq 0$$

So we have:

$$Q = \begin{pmatrix} 1 & -1 \\ -1 & 4 \end{pmatrix}; \quad C = (-1 \ \ -1)$$

Recall that the corresponding LCP would be obtained with

$$\boldsymbol{\omega} = \begin{pmatrix} \mathbf{u} \\ \mathbf{v} \end{pmatrix}; \ \mathbf{q} = \begin{pmatrix} \mathbf{b} \\ -\mathbf{d} \end{pmatrix}; \ M = \begin{pmatrix} Q & -C^T \\ C & 0 \end{pmatrix}, \ \mathbf{z} = \begin{pmatrix} \mathbf{x} \\ \mathbf{y} \end{pmatrix}$$

Hence:

$$\boldsymbol{\omega} = \begin{pmatrix} u_1 \\ u_2 \\ v_1 \end{pmatrix}; \ \mathbf{q} = \begin{pmatrix} 2 \\ 1 \\ 1 \end{pmatrix}; \ M = \begin{pmatrix} 1 & -1 & 1 \\ -1 & 4 & 1 \\ -1 & -1 & 0 \end{pmatrix}, \ \mathbf{z} = \begin{pmatrix} x_1 \\ x_2 \\ y \end{pmatrix}$$

Now, all elements of the LCP, $(\boldsymbol{\omega} - M\,\mathbf{z}) = \mathbf{q}$, can be expressed in the following tableau format:

Basis	$\mathbf{z}^T \boldsymbol{\omega}^T$		
ω	$-M$	I	$\mathbf{q}$

After including numbers, the tableau for our LCP, is

Basis	x_1	x_2	y	u_1	u_2	v_1	q
u_1	−1	1	−1	1	0	0	2
u_2	1	−4	−1	0	1	0	1
v_1	1	1	0	0	0	1	1

Notice that the tableau includes primal variables x_1, x_2 and dual variables (in this case, y).

According to the algorithm, terminal tableaux can have one of three possible sign structures. If all components of column **q** are nonnegative, this corresponds to an optimal solution.

If the sign structure was as follows:

basis	$\mathbf{x_1}$	$\mathbf{x_2}$	**y**	$\mathbf{u_1}$	$\mathbf{u_2}$	$\mathbf{v_1}$	**q**
u_1							
u_2							
v_1	+	+	+	0	0	0	

then the case is primal infeasible.

If the sign structure was

Basis	x_1	x_2	y	u_1	u_2	v_1	q
u_1	+	+	+	0	0	0	
u_2	+	+	+	0	0	0	
v_1							

then the case is dual infeasible.

Since the initial tableau for the given example has all components of **q** being nonnegative, then it is already optimal. It says also that the solution is$x_1 = 0,\ x_2 = 0$; which already was evident from a simple inspection of the problem.

Suppose now that we have another optimization problem with an almost identical initial tableau (see column **q**):

Basis	x_1	x_2	y	u_1	u_2	v_1	q
u_1	−1	1	−1	1	0	0	2
u_2	1	−4	−1	0	1	0	−1
v_1	1	1	0	0	0	1	1

This tableau is not terminal. The algorithm dictates that a *diagonal pivot* must be used, in the second row (where column **q** has a minus sign). The pivot should be selected in the nonbasic part (columns $x_1,\ x_2,\ y$). Then, the pivot is −4.

The new tableau would be (notice the change of basis):

Basis	x_1	x_2	y	u_1	u_2	v_1	q
u_1	3/4	0	−5/4	1	1/4	0	7/4
x_2	−1/4	1	1/4	0	−1/4	0	1/4
v_1	5/4	0	−1/4	0	1/4	1	3/4

which is optimal (the solution is $x_1 = 0,\ x_2 = 1/4$).

If the diagonal pivot was 0 instead of −4, this would be useless for pivoting. In that case, an *exchange pivot* would be selected.

Next example (continuing with [83]), give us a case in need of exchange pivot:

Minimize:

$$J = x_1^2 + x_2^2 + x_3^2$$

Subject to:

$$x_1 + 2x_2 - x_3 \leq 4$$

$$x_1 - x_2 + x_3 \leq -2$$

$$\text{and } x_1,\ x_2,\ x_3 \geq 0$$

Then:

$$Q = \begin{pmatrix} 2 & 0 & 0 \\ 0 & 2 & 0 \\ 0 & 0 & 2 \end{pmatrix}; \quad C = \begin{pmatrix} -1 & -2 & 1 \\ -1 & 1 & -1 \end{pmatrix}$$

The initial tableau of the corresponding LCP would be

Basis	y_1	y_2	x_1	x_2	x_3	v_1	v_2	u_1	u_2	u_3	**q**
v_1	0	0	1	2	−1	1	0	0	0	0	4
v_2	0	0	1	−1	1	0	1	0	0	0	-2
u_1	−1	−1	−2	0	0	0	0	1	0	0	0
u_2	−2	1	0	−2	0	0	0	0	1	0	0
u_3	1	−1	0	0	−2	0	0	0	0	1	0

The first negative element in the **q** column is − 2, on the second row. A diagonal pivot in the nonbasic part is not possible (the v_2–y_2 entry is 0). An exchange pivot is then required. We look for the first negative element in row 2 (nonbasic part). It is − 1, the v_2–x_2 entry.

Let us build the next tableau, using the exchange pivot:

Basis	y_1	y_2	x_1	x_2	x_3	v_1	v_2	u_1	u_2	u_3	**q**
v_1	0	0	3	0	1	1	2	0	0	0	0
x_2	0	0	−1	1	−1	0	−1	0	0	0	2
u_1	−1	−1	−2	0	0	0	0	1	0	0	0
u_2	−2	1	−2	0	−2	0	−2	0	1	0	4
u_3	1	−1	0	0	−2	0	0	0	0	1	0

There is a problem with this tableau: the complementary variable of x_2, which is u_2, should not be in the same basis. Therefore, we choose a new pivot, the u_2–y_2 entry, which solves the problem:

basis	y_1	y_2	x_1	x_2	x_3	v_1	v_2	u_1	u_2	u_3	**q**
v_1	0	0	3	0	1	1	2	0	0	0	0
x_2	0	0	−1	1	−1	0	−1	0	0	0	2
u_1	−3	0	−4	0	−2	0	−2	1	1	0	4
y_2	−2	1	−2	0	−2	0	−2	0	1	0	4
u_3	−1	0	−2	0	−4	0	−2	0	1	1	4

The tableau is optimal. The solution is $x_1 = x_3 = 0$, $x_2 = 2$; and the optimal value is 4.

The idea of criss-cross type methods was introduced by Stanley Zionts [107]. After some improvements suggested by the research, a convergent criss-cross method was presented by T. Terlaky [93].

One of the most famous investigators on optimization is R. T. Rockafellar (born 1935), with more than 93,000 cites. Among several important initiatives, he started an abstract combinatorial analysis of linear programming. Following this line, R. G. Bland, M. Las Vergnas, J. Folkman, J. Lawrence, and others established the theory of *oriented matroids* [11]. In 1987, Terlaky generalized the criss-cross method for oriented matroids [94].

A relevant motivation for our interest on this topic is the danger of cycling when applying pivoting methods; a question that can be treated as complications in the exploration of polytopes. In words of [81], the theory of oriented matroids provides a broad setting for the modeling, description, and analysis of combinatorial properties of geometric configurations. It can be applied for arrangements of hyperplanes, directed graphs, linear programming, etc. In particular, it represents directions with signs (+ or −).

The relationship that could be established between the LCP problem and oriented matroids is well illustrated in [38].

There are new applications of the criss-cross method, beyond linear or quadratic programming, for example [50], on hyperbolic programming.

As a matter of curiosity, we could mention that oriented matroids can be found in several scenarios of physics, including loop quantum gravity, [68, 69].

For more details on matroids and linear programming, the doctoral thesis of Z. Zhang [106] would be recommended. It includes a chapter on using machine learning for selecting pivots.

Coming back to the original development of ideas, related to the criss-cross method, we should mention the contributions of R. G. Bland, well known for his proposal of new pivots for the simplex method [12]. This proposal is also based on signs.

A final note: one of the recent heuristic population-based methods was called *criss-cross algorithm* (implemented in MATLAB, available from MathWorks File Exchange). It is not related at all to the method just described here.

7.6 Sequential Quadratic Programming (SQP)

Many optimization problems do not have a quadratic objective function. However, it could be still possible to decompose these problems into quadratic subproblems. In each iterate of the sequential quadratic programming procedure, a Taylor series approximation of the objective and constraints around the current point are considered.

Suppose that the problem at hand is

Minimize $f(\mathbf{x})$

Subject to the next constraints:

$$g_i(\mathbf{x}) = 0, \quad i = 1, 2, \ldots, n_e$$

$$h_i(\mathbf{x}) \geq 0, \quad i = 1, 2, \ldots, n_i$$

The Lagrangian would be:

$$L(\mathbf{x}, \boldsymbol{\lambda}, \boldsymbol{\theta}) = f(\mathbf{x}) - \sum_{i=1}^{ne} \lambda_i \, g_i(\mathbf{x}) - \sum_{i}^{ni} \theta_i \, h_i(\mathbf{x})$$

According to the SQP procedure, at k-th iterate, one considers the following quadratic approximation:

$$\min_{p} \frac{1}{2} \mathbf{p}^{\mathbf{T}} W_k \, \mathbf{p} + \nabla \mathbf{f}(\mathbf{x_k})^{\mathbf{T}} \, \mathbf{p}$$

complying with

$$\nabla \mathbf{g_i}(\mathbf{x_k})^{\mathbf{T}} \, \mathbf{p} + g_i(\mathbf{x}_k) = 0, \quad i = 1, 2, \ldots, n_e$$

$$\nabla \mathbf{h_i}(\mathbf{x_k})^{\mathbf{T}} \, \mathbf{p} + h_i(\mathbf{x}_k) \geq 0, \quad i = 1, 2, \ldots, n_i$$

where W_k is the Hessian of the Lagrangian (that is: $\nabla_x^2 \, L(\mathbf{x_k}, \boldsymbol{\lambda}_\mathbf{k}, \boldsymbol{\theta}_\mathbf{k})$) at the current iterate.

This optimization setting constitutes a subproblem to be solved. Frequently, this is done with an active set procedure.

Once the subproblem was solved, the next iterate is obtained with

$$\mathbf{x_{k+1}} = \mathbf{x_k} + \mathbf{p_k}\ ; \quad \boldsymbol{\lambda}_{\mathbf{k+1}} = \boldsymbol{\lambda}_{\mathbf{k}} + \boldsymbol{\Delta\lambda}_{\mathbf{k}}\ ; \quad \boldsymbol{\theta}_{\mathbf{k+1}} = \boldsymbol{\theta}_{\mathbf{k}} + \boldsymbol{\Delta\theta}_{\mathbf{k}};$$

Several academic publications about SQP, like for instance [85], point out a connection of the SQP algorithm with Newton's method. This can be outlined as follows.

Assume there are only equality constraints. The KKT optimality conditions can be written as

$$\mathbf{F}(\mathbf{x}, \boldsymbol{\lambda}) = \begin{pmatrix} \nabla_{\mathbf{x}} \mathbf{L}(\mathbf{x}, \boldsymbol{\lambda}) \\ \mathbf{g}(\mathbf{x}) \end{pmatrix} = 0$$

Suppose an approximation of the solution $(\mathbf{x_k}, \boldsymbol{\lambda}_{\mathbf{k}})$, the next iterate (following Newton's method) would be

$$\nabla\, \mathbf{F}(\mathbf{x}_k, \boldsymbol{\lambda}_k) \begin{pmatrix} \mathbf{d_k} \\ \mathbf{y_k} \end{pmatrix} + \mathbf{F}(\mathbf{x}_k, \boldsymbol{\lambda}_k) = 0$$

Making the pertinent substitutions:

$$\begin{pmatrix} W_k & -\nabla \mathbf{g}(\mathbf{x_k}) \\ \nabla \mathbf{g}(\mathbf{x_k})^T & 0 \end{pmatrix} \begin{pmatrix} \mathbf{d_k} \\ \mathbf{y_k} \end{pmatrix} + \begin{pmatrix} \nabla \mathbf{f}(\mathbf{x_k}) - \nabla \mathbf{g}(\mathbf{x_k})\, \boldsymbol{\lambda}_{\mathbf{k}} \\ \mathbf{g}(\mathbf{x_k}) \end{pmatrix} = 0$$

Define now: $\mathbf{u_k} = \mathbf{y_k} + \boldsymbol{\lambda}_{\mathbf{k}}$. Then, from above equation, one obtains

$$W_k\, \mathbf{d_k} - \nabla \mathbf{g}(\mathbf{x_k})\, \mathbf{u_k} + \nabla \mathbf{f}(\mathbf{x_k}) = 0$$

$$\nabla \mathbf{g}(\mathbf{x_k})^{\mathrm{T}}\, \mathbf{d_k} + \mathbf{g}(\mathbf{x_k}) = 0$$

which are precisely the equations to solve when adhering to SQP (quadratic optimization subproblems).

The implementation of the SQP approach can be done in several ways. In general, it involves three main tasks to be cycled:

- Updating of W_k
- Solving a quadratic optimization subproblem
- Using line search and a merit function

For the updating of W_k, a quasi-Newton approximation is usually employed, like for instance BFGS (see Chap. 2).

Important functions of the MathWorks Optimization Toolbox, in MATLAB, such as **fmincon**, **fminimax**, **fgoalattain**, and **fseminf**, all use SQP [22]. If the *options* parameter *Display* is set to *iter*, then various informations were given on the march of calculations, including the maximum constraint violation or difficulties with Hessian updating.

For the steps $\mathbf{x_{k+1}} = \mathbf{x_k} + \mathbf{p_k}$; $\boldsymbol{\lambda}_{\mathbf{k+1}} = \boldsymbol{\lambda}_{\mathbf{k}} + \boldsymbol{\Delta\lambda}_{\mathbf{k}}$; $\boldsymbol{\theta}_{\mathbf{k+1}} = \boldsymbol{\theta}_{\mathbf{k}} + \boldsymbol{\Delta\theta}_{\mathbf{k}}$, a line search could be used in order to produce a sufficient decrease in a certain merit function.

The manual [22] extensively gives more SQP calculation details.

It must be emphasized that the SQP method could not converge if the starting point of the search is far from the solution. A lot of research has been devoted to analyze convergence properties, suggesting improvements of SQP in favor of possible globalization or generalizations (see, for example, the surveys [33, 64]).

The idea of solving nonlinear constrained optimization problems by a sequence of quadratic programming subproblems was first suggested in the Doctoral Thesis of R. B. Wilson in 1963. SQP methods were further popularized by M. C. Biggs [8], S. P. Han [47], and M. J. D. Powell [77]. See [41] for references therein. A most cited reference on SQP is [13]. More recently, in 2011, the chapter [42] offers a good treatment of the topic.

In words of [85], sequential programming methods belong to the most powerful optimization algorithms we know today for solving differentiable nonlinear problems (not just quadratic problems).

There are several implementations of SQP in MATLAB that are available from the web, in particular from GitHub. One of these, [SPI] has been chosen as the basis for Program 7.8 since it is simple and illustrative. We added to that code the graphical expression of the problem and the SQP iterations for reaching the solution (Fig. 7.15).

The example considered in Program 7.8 is the minimization of

$$y = x_1^4 - 2x_2 x_1^2 + x_2^2 + x_1^2 - 2x_1 + 5$$

Subject to a single constraint:

$$-(0.25 + x_1)^2 + 0.75 x_2 \leq 0$$

Notice in Program 7.8 that it clearly suggests a possible implementation in symbolic MATLAB, which would be useful for general use.

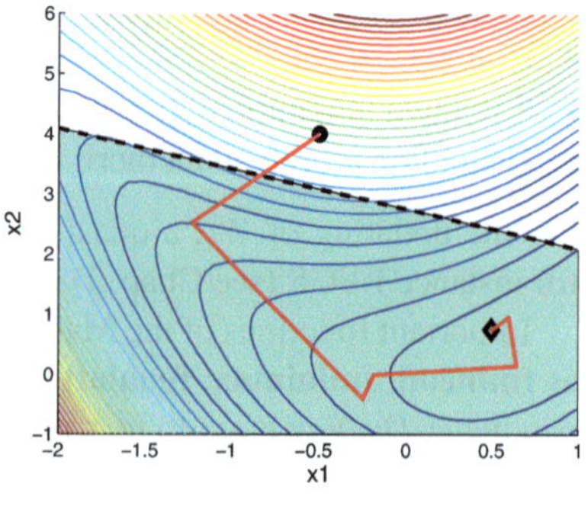

Fig. 7.15 Evolution of SQP iterations for the example considered

Note (Program 7.8)

The complete listing of the Program 7.8, corresponding to the SQP algorithm has been included in the Appendix for Long Programs.

We would like to add some comments about the use of merit functions and certain difficulties that may arise.

The setting considered is a line search, in which:

$$\mathbf{x_{k+1}} = \mathbf{x_k} + \alpha_k \, \mathbf{p_k}$$

and α_k is chosen for a satisfactory descent of a suitable merit function $\Phi()$, so that

$$\Phi(\mathbf{x_k} + \alpha_k \, \mathbf{p_k}) < \Phi(\mathbf{x_k})$$

This should correspond to a better optimization solution and is used for easier calculations.

In case of equality constraints, a simple merit function is the following:

$$\Phi(\mathbf{x}) = f(\mathbf{x}) + \rho \, \|\mathbf{g}(\mathbf{x})\|$$

where $\|\cdot\|$ can be any norm.

In his Doctoral Thesis (1978), N. Maratos [63] pointed out a phenomenon that may occur in certain cases, causing convergence problems. Later on, in 1983 [78], M. J. D. Powell put an example of the so-called *Maratos effect* that, since then, has been frequently cited. It is the following example:

Minimize: $f(\mathbf{x}) = 2\,(x_1^2 + x_2^2 - 1) - x_1$

Subject to: $g(\mathbf{x}) = x_1^2 + x_2^2 - 1 = 0$

The optimal solution is at (1, 0). Notice that the constraint is a circle.

Suppose in this example that we are exploring for the next step, departing from a point:

$$x_k = \begin{pmatrix} \cos\theta \\ \sin\theta \end{pmatrix}$$

at this point, we have $f(\mathbf{x}_k) = -\cos\theta$ and $g(\mathbf{x}_k) = 0$.

Clearly, the simple merit function already mentioned does not recommend the next step to be on the circular constraint.

In turn, the SQP step would be

$$\mathbf{p_k} = \begin{pmatrix} \sin^2\theta \\ -\sin\theta\,\cos\theta \end{pmatrix}$$

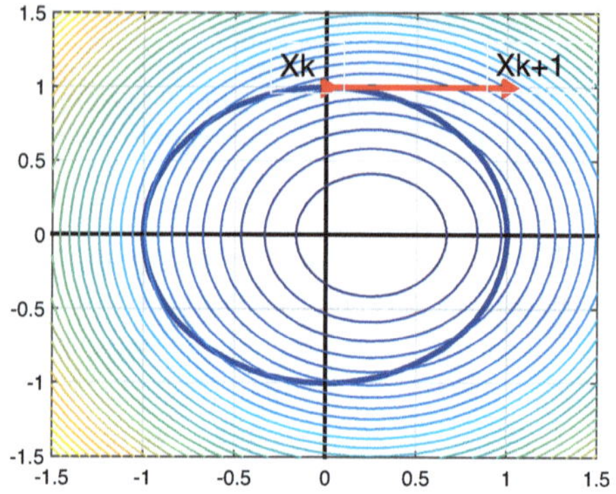

Fig. 7.16 Example for Maratos effect

Figure 7.16 depicts the situation for $\mathbf{x_k} = (0, 1)$. The constraint is shown as non-concentric circle. Using the SQP step (shown as an horizontal arrow) would lead to $\mathbf{x_{k+1}} = (1, 1)$; [96].

Figure 7.16 has been generated with a simple Program 7.7.

Program 7.7 Visualization of the Maratos effect

```
%Example for Maratos effect
figure(1)
%constraint plot
x1o=1; x2o=0;
for theta = 0:0.05: 2*pi
  x1 = cos(theta); x2 = sin(theta);
  plot ([x1o,x1], [x2o, x2], 'b', 'LineWidth',3);
  hold on;
  x1o=x1; x2o= x2;
end;
%axes
plot([-1.5 1.5],[0,0], 'k', 'LineWidth', 2);
plot([0,0], [-1.5 1.5], 'k','LineWidth', 2);
%objective function contours
[vx1, vx2] = meshgrid(-1.5:0.05:1.5, -1.5:0.05:1.5);
y = 2*(vx1.^2 + vx2.^2 -1) - vx1;
contour(vx1, vx2, y, 30, 'LineWidth', 2);
grid;
%SQP step
plot([0, 1], [1, 1],'r->', 'LineWidth', 3);
```

The result of this step would be nearer to the optimal point. However, we would have an increase of the objective function:

$$f(\mathbf{x}_{k+1}) = \sin^2\theta - \cos\theta$$

and a constraint violation:

$$g(\mathbf{x}_{k+1}) = \sin^2 \theta$$

So this step is a candidate that will be rejected, if we use the mentioned merit function. Other merit functions including some curvature modeling might be more appropriate. Having both an increase of the objective function and a constraint violation is a characteristic of the Maratos effect. For more details [44] would be recommended.

7.7 Application to Portfolio Optimization

In 1952, Harry Markowitz proposed a method of Portfolio Optimization, leading to a quadratic programming setting, which meant a kind of revolution in the financial context. Some important authors comment that before this event, quadratic programming did not attract much interest; and—due to the financial use—this quickly changed for high expectations. Nowadays, asking Google about quadratic programming gives 26 million results, while asking about portfolio optimization gives 129 million results. Indeed, after the seminal work of Markowitz, the initiated methodology has evolved, gaining in more comprehensive consideration of important factors.

Figure 7.17 depicts a portrait of professor H. Markowitz, Nobel laureate for Economics in 1990.

This section will focus on the essential approach of Markowitz, which is to propose an adequate framework for optimization. After analyzing the game, he noticed that investors do not like risks: if there were two portfolios offering the same expected return, investors would prefer the less risky one. A way for quantifying risk is to use the *variance* of asset prices.

The optimization problem refers to the portfolio overall risk and return. Diversification may yield a better trade-off between return and risk. It would not be appropriate to include assets positively correlated. The idea is more to have assets compensating each other evolution.

Fig. 7.17 Harry Markowitz

Let us build the mathematical treatment of the problem, as proposed by Markowitz. Denote as x_i the proportions of the capital to form a portfolio. The expected portfolio return would be

$$r_P = \sum_{i=1}^{N} x_i \, r_i$$

The portfolio return variance can be expressed as:

$$\sigma_P^2 = \sum_{i=1}^{N} x_i^2 \sigma_i^2 + \sum_i \sum_{j \neq i} x_i \, x_j \sigma_i \, \sigma_j \, \rho_{ij}$$

where σ_i is standard deviation of returns of asset i, and ρ_{ij} is correlation between the returns of assets i and j.

This last expression can be written as

$$\sigma_P^2 = \sum_i \sum_j x_i \, x_j \sigma_i \, \sigma_j \, \rho_{ij}$$

with $\rho_{ij} = 1$ for $i = j$.

Equivalently, a more simple expression would be:

$$\sigma_P^2 = \sum_i \sum_j x_i \, x_j \sigma_{ij}$$

where $\sigma_{ij} = \sigma_i \, \sigma_j \, \rho_{ij}$ is the covariance of returns of assets i and j. It can be written also as cov(i, j) or cov$_{ij}$.

A frequently used parameter is the portfolio return volatility, which is

$$\sigma_P = \sqrt{\sigma_P^2}$$

Therefore, in short terms, the optimization problem is to

Minimize:

$$\sigma_P^2 = \sum_i \sum_j x_i \, x_j \sigma_{ij}$$

Subject to:

$$r_P = \sum_{i=1}^{N} x_i \, r_i$$

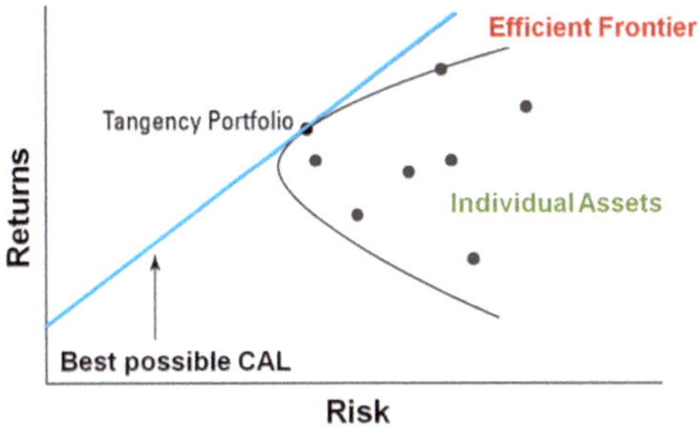

Fig. 7.18 Portfolios and the return vs. risk space

The specialized literature denotes as *mean-variance theory* the contribution of Markowitz. He also introduced a visualization of the investment problem, in terms of return vs. volatility (risk). Figure 7.18 shows an example of this visualization. All possible portfolios can be depicted in this plot and were located inside a region with a parabolic border. The upper part of the parabola is called the *efficient frontier*. The tangent to this frontier is the capital allocation line (CAL).

A basic example of the Markowitz proposal for portfolio optimization is provided below, in the form of the Program 7.8, written in MATLAB. It is a coding based on [79], from GitHub. The program includes some fictional data, taken from [1], which corresponds to six months of economic activity of five companies.

Program 7.8 Using quadprog( ) for Portfolio optimization

```
%Example of Portfolio Optimization with quadprog()
% Assets data ----------------
returns = [0.0229, 0.0359, -0.0058, 0.0335, -0.0043];
R = 0.02; %desired return
covariance = [0.109611 0.048688 0.033886 0.050926...
 0.041998;
0.048688 0.089034 0.031999 0.057688 0.043506;
0.033886 0.031999 0.069938 0.028193 0.030822;
0.050926 0.057688 0.028193 0.211284 0.045765;
0.041998 0.043506 0.030822 0.045765 0.058966;
];
asset_var = diag(covariance); % variance of asssets
asset_std = sqrt(asset_var); % std. dev of asssets
% visualize covariance matrix in colors
figure(1)
imagesc(covariance); colormap('colorcube');
[rows,cols] = size(covariance);
for i = 1:rows
  for j = 1:cols
  textHandles(j,i) = text(j,i,...
    num2str(covariance(i,j)),...
    'horizontalAlignment','center');
  end
end
% Preparing for quadprog() -----------------
```

```
H = covariance;
f = []; %f not used
A1 = returns;
b1 = R;
% portfolio weights (x) equal to 1
A2 = [1,1,1,1,1];
b2 = 1;
% equality constraints (from above specifications)
Aeq = [A1; A2];
beq = [b1; b2];
% inequality constraints (to disallow short-selling)
Aineq = -eye(5);
bineq = zeros(5,1);
% Lower and Upper Bounds on the solution
LB = zeros(5,1);
UB = [1; 1; 1; 1; 1];
% Solving with quadprog ------------------
%(budget allocation)
% adequate option for this case
options=optimset('LargeScale','off','Display','off');
x=quadprog( H, f, Aineq, bineq, Aeq, beq, LB, UB,...
 [],options );
% Results ------
% Portfolio variance (portfolio risk)
port_var = x'*covariance*x;
port_std = sqrt(port_var);
% Portfolio expected return
port_EXPreturn = returns*x;
disp(sprintf('variance = %.3f', port_var));
disp(sprintf('std = %.3f', port_std));
disp(sprintf('expected return = %.2f',...
 port_EXPreturn));
% display
figure(2)
subplot(2,1,1)
bar(returns, 0.3,'c');
title('returns','FontSize',12);
subplot(2,1,2)
bar(x, 0.3, 'r');
title('optimal investment','FontSize',12);
%------------------------------------------------------
%sequence of target returns:
b1 = linspace(0.005,0.1,200);
% Compute a sequence of optimal results
xset = zeros(5,200);
for i=1:200
  beq = [b1(i); b2];
  options=optimset('LargeScale','off',...
   'Display','off');
  xset(: ,i)=quadprog( H, f, Aineq, bineq, Aeq,...
   beq, LB, UB,[], options);
end
optPF_var = zeros(200,1);
```

```
optPF_std = zeros(200,1);
optPF_return= zeros(200,1);
for i=1:length(xset)
  optPF_var(i) = xset(:,i)'*covariance*xset(:,i);
  optPF_std(i) = sqrt(optPF_var(i));
  optPF_return(i) = returns*xset(:,i);
end
figure(3)
plot(optPF_std, optPF_return,'LineWidth',3)
title('Efficient Frontier', 'FontSize',12);hold on;
ylabel('Return');
xlabel('Risk (std)');
scatter(asset_std, returns, 'filled');
axis([0.1 0.5 -0.04 0.08]); hold off;
```

Notice that the heart of Program 7.9 is the use of function "*quadprog*", which belongs to the MATLAB Optimization Toolbox.

The first figure generated by the program (Fig. 7.19) visualizes the matrix of covariances. Notice the symmetries (it is only necessary to store a half of the matrix including the diagonal).

Figure 7.20 depicts the recommendations produced by the Portfolio Optimization. A degree of diversification can be observed. Asset 4 seems to be risky (although highly rewarding).

By a significant computational effort (iterations of *quadprog*), a gross visualization of the efficient frontier can be obtained (Fig. 7.21). Also, the five assets considered in our example were displayed.

For the reader wishing to explore the topic with real financial data, it would be recommended to access the site [90] and the site [87] which shows how to use an important source of information: Yahoo! Finance.

Apart from *quadprog*, there are other software tools that can be used for quadratic programming, like the function *monqp* that belongs to the package FMKL [89] available from GitHub or the professional package CVX [27] for MATLAB.

Fig. 7.19 Visualization of the matrix of covariances

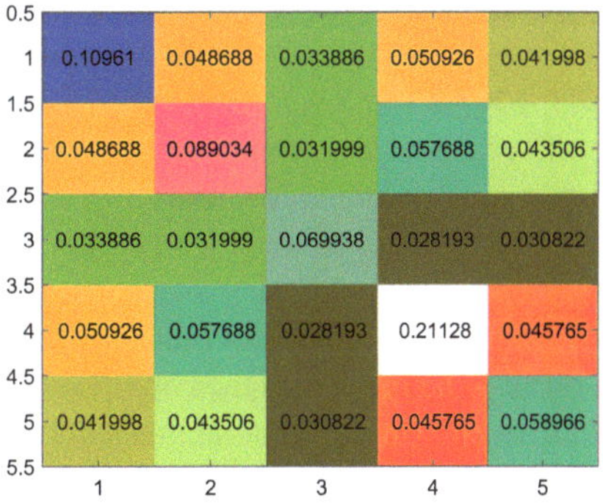

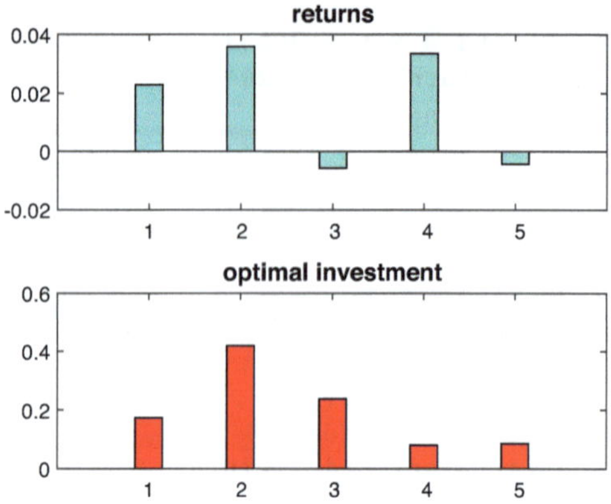

Fig. 7.20 Returns of the assets, and the recommended (optimal) investment proportions

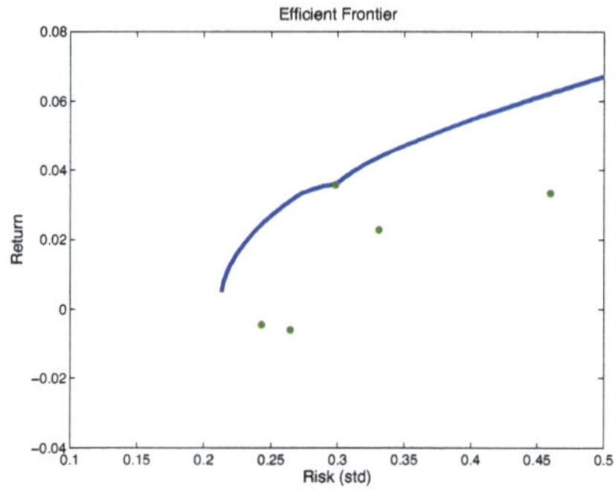

Fig. 7.21 Assets and efficient frontier on the return vs. risk space

Several extensions of the Markowitz method have been proposed, mainly regarding to the constraints. For example cardinality constraints, that handle integer numbers for example, (like, for instance, the number of assets), lower and upper bounds on budget proportions, non-negativity bounds, turnover constraints, etc. See [53] for more details.

Indeed there are criticisms on some aspects of the Markowitz approach, since reality is not so rigid and simple. Other approaches that have been put into practice are the Value-at-Risk model and the Conditional-Value-at-Risk model and more recently a number of alternative perspectives based on machine learning (see [45, 104], for an overview).

For historical accounts on Portfolio Optimization developments, there are several interesting publications, like [56], or [32] more oriented to the financial mentality. In [54], after some history, new trends and challenges were identified.

Academic contributions and deep tutorials are [29, 71, 75].

MathWorks offers a complete Financial Toolbox in MATLAB, including—of course—the mean-variance approach. Likewise, GitHub has 267 public repositories of software (in different languages) for Portfolio Optimization, with some contributions adhering to the machine learning context.

7.8 The Support Vector Machine (SVM)

One of the most important techniques for data classification is the *Support Vector Machine* (SVM). It is frequently used in the context of machine learning.

For a basic example, suppose there are two sets, A and B, of data points, as depicted in Fig. 7.22. We want to determine a line clearly separating both sets, in order to classify data as belonging to A or to B.

To help for analyzing the situation, two polygons were drawn by connecting with lines the points on the borders of A and B. And so, we obtain Fig. 7.23 in which we also named as P and Q the two closest points between those polygons.

Continuing with Fig. 7.23. We added a short segment between P and Q, and we draw another (dotted) line bisecting that segment. This line could serve as a solution: it separates bot data sets. But it looks not quite satisfactory.

By the way, a line (or plane in three-dimensional scenarios) *supports* a class if all points of the class are in one side of the line.

Figure 7.24 shows now the approach proposed by SVM. Two parallel supporting lines were determined such as their mutual distance is maximal. Both supporting lines, V and Z, were tangent to one or another polygon, passing through P and Q.

In this way an optimal corridor is determined, and satisfactory separating line S is obtained in the middle of the corridor.

The points that intersect with the two supporting lines in Fig. 7.24 are called *support vectors* or *support points*. In the case depicted in this figure, P and Q were support vectors; in other cases, more than just two support vectors can exist.

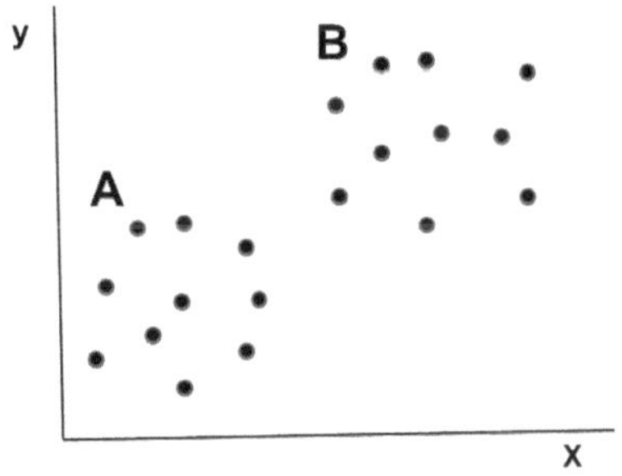

Fig. 7.22 Two sets of data points on the 2D plane

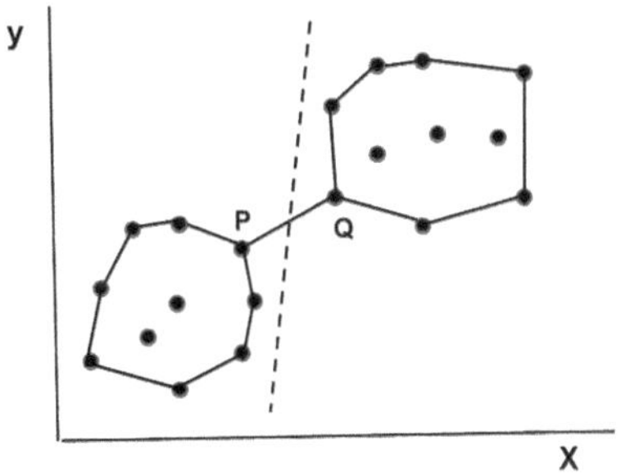

Fig. 7.23 P and Q were the closest points between A and B

Fig. 7.24 Two parallel supporting lines with maximal mutual distance

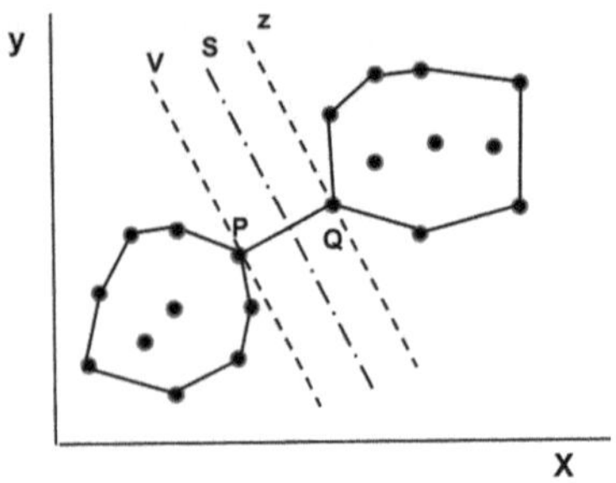

Now, let us translate ideas to a mathematical statement of the problem.

First, note that the equation of a supporting line, $\mathbf{w}^{\mathbf{T}}\mathbf{x} + b = 0$, does not change if $\mathbf{w}$ and b are multiplied by the same constant. It is possible to scale the values of $\mathbf{w}$ and b such that:

$$\mathbf{w}^{\mathbf{T}}\mathbf{x}_i + b \geq +1 \text{ for } \mathbf{x_i} \in A$$

and

$$\mathbf{w}^{\mathbf{T}}\mathbf{x}_i + b \leq -1 \text{ for } \mathbf{x_i} \in B$$

The support vectors are the data points satisfying the equalities.

The distance or margin between the supporting lines corresponding to the two equations above is

$$\gamma = \frac{2}{\|\mathbf{w}\|_2}$$

Hence, the margin maximization problem can be stated as follows:

Minimize

$$\mathbf{w}^{\mathbf{T}}\mathbf{w}$$

Subject to

$$y_i\left(\mathbf{w}^{\mathbf{T}}\mathbf{x_i} + b\right) \geq 1$$

where $y_i = +1$ if $\mathbf{x_i} \in A$, and $y_i = -1$ if $\mathbf{x_i} \in B$.

This is a quadratic optimization problem that can be solved with standard methods.

A typical example for introductory studies on data classification is the case of the Iris flower data set. This set was collected by Edgar Anderson and published in 1935. The data consists of 50 samples from each of three species of Iris flower (*cetosa*, *virginica* and *versicolor*). Each sample includes four measures: the length

Fig. 7.25 The Iris flower

and width of sepals and petals in centimeters. Figure 7.25 shows an Iris flower. The reader can download this data set from [82], which is an impressive site plenty of various data sets.

With the help provided by [61] for a good coding of

$$y_i\,(\mathbf{w}^{\mathbf{T}}\,\mathbf{x_i}\,+\,b)\,\geq 1$$

in the following terms:

$$\begin{pmatrix} y_1 & 0 & 0 & & 0 \\ 0 & y_2 & 0 & & 0 \\ 0 & 0 & \dots & & 0 \\ & & & & \\ 0 & 0 & 0 & \dots & y_L \end{pmatrix} \begin{pmatrix} x_{11} & x_{12} & & x_{1N} & 1 \\ x_{21} & x_{22} & & x_{2N} & 1 \\ \dots & \dots & & & \\ x_{L1} & x_{L2} & & x_{LN} & 1 \end{pmatrix} \begin{pmatrix} w_1 \\ w_2 \\ \vdots \\ w_N \\ b \end{pmatrix} \geq \begin{pmatrix} 1 \\ 1 \\ \vdots \\ 1 \end{pmatrix}$$

(where the right-hand side contains L entries with value 1), it is easy to adequately feed the inputs of *quadprog()*, which is the nucleus of Program 7.9.

This MATLAB program considers the case of two data clusters, depicted in Fig. 7.26. The cluster at the bottom corresponds to *Iris cetosa*. The other, upper, cluster is the union of *versicolor* and *virginica Iris*. The output of *quadprog()* gives the slope of the supporting lines and parameter b.

Program 7.9 SVM classification example

```
% SVM classification example
% IRIS data
% x=sepal width, y=petal length
%read IRIS data
D=dlmread('iris.data'); %columns
%there are 50 points for class A
% and 100 points for class B
X=[D(:,2), D(:,3)]; %two columns (x,y)
Y=ones(150,1); Y(1:50)=-Y(1:50); %labels:1 or -1
```

```
%prepare quadratic optimization
%we multiply with -1 on both sides of constraints
H = eye(3);
H(3,3)= 0;
f = zeros(3,1);
Z = [X ones(150,1)]; %auxiliary matrix
A = -diag(Y)*Z;
b = -ones(150, 1);
% apply quadprog
%no messages...
options=optimset('LargeScale','off','Display','off');
M=quadprog(H, f, A, b, [], [], [], [], [], options);
slop = M(2)/M(1);
%display
figure(1)
%the data
scatter(X(1:50,1),X(1:50,2),32,'k'); hold on;
scatter(X(51:150,1),X(51:150,2),32,'r');
% the lines
vx=[0:0.1:6];
vy=-(M(1)*vx + M(3))/M(2);
%optimal separating line
plot(vx, vy, 'r', 'LineWidth',2);
vyUP = (1-M(1)*vx - M(3))/M(2);
plot(vx, vyUP, 'g', 'LineWidth',2); %margin line
vyDW = (-1-M(1)*vx - M(3))/M(2);
plot(vx, vyDW, 'g', 'LineWidth',2); %margin line
axis([0 6 0 8]);
title('Linear SVM on IRIS data');
xlabel('sepal width'); ylabel('petal length');
```

The result of Program 7.9 is visualized in Fig. 7.26. Notice that there are two support vectors (two tangent points) belonging to the *Iris cetosa* data set. The separating line that was obtained travels the center of the corridor.

According to [62], the seminal ideas giving rise to SVM were first proposed by V. N. Vapnik and A. Ya. Chervonenkis (Institute of Control Sciences of the Russian

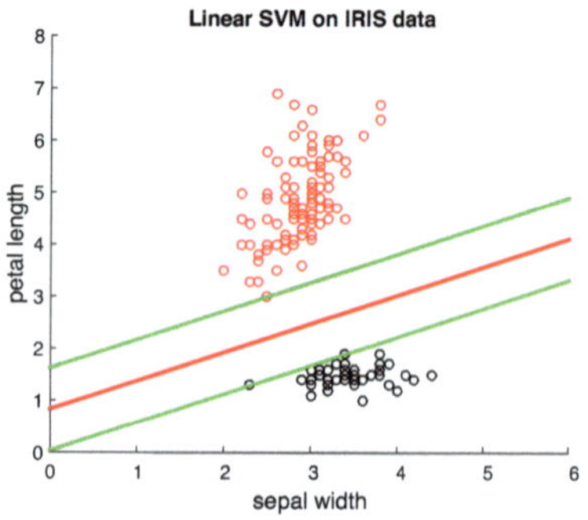

Fig. 7.26 Two parallel supporting lines with maximal mutual distance

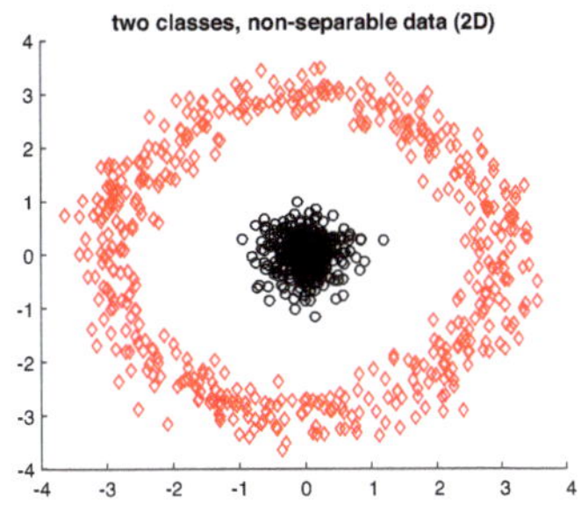

Fig. 7.27 Two non-linearly separable data sets

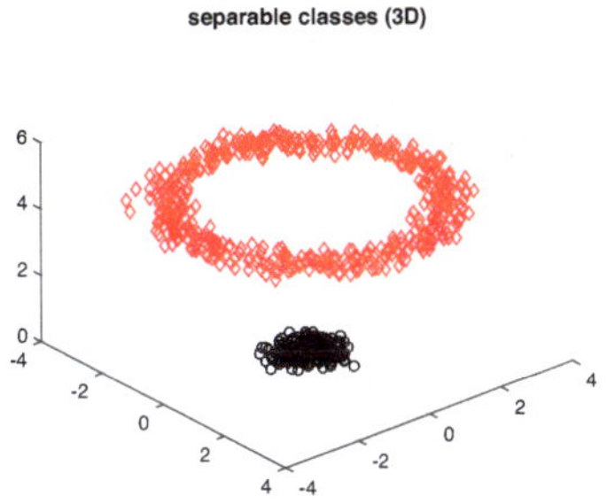

Fig. 7.28 The data sets become separable by a projection to three dimensions

Academy of Sciences, Moscow, Russia). They were published in Russian language in several papers, years 1962, 1963, and 1964.

Later on, a very good idea emerged. It was to apply the so-called "*kernel-trick*".

Let us visualize, in a simple but quite effective form, part of the trick. Figure 7.27 shows two data sets, in concentric circular bands. No straight line could separate both sets (it is a case of nonlinearly separable data).

However, if we go to three dimensional, by adding a third dimension to the data in a way that the outside band would jump, the two data sets could be neatly separated by a plane. This is illustrated in Fig. 7.28 (the reader would easily imagine a separating plane).

In more formal terms, the data $\mathbf{x}$ on the original space were projected to a convenient "*feature space*" through a mapping $\Phi(\mathbf{x})$.

To do justice to the topic we are now visiting, it is worth to mention its relationship with *pattern recognition*, one of the main streams of machine learning.

Consider, in particular, the question of *similarity*. Near my university there is a big rock popularly named as "the cat." Yes, it looks like a cat. Mathematicians would like to express why we see a cat there. So one could extract/measure some peculiar features about the shape we perceive: roundness, vertical-horizontal proportions, whatever... The idea is to project our observations to a n-dimensional "feature space." Something like this happens when medicine assigns a set of symptoms to a disease.

Now, one could study the similarity of two objects by looking at their projections, $\mathbf{x}$ and $\mathbf{y}$, on the feature space. If they are near, one could say these objects are similar. Kernels were introduced as a measure of similarity.

Among the kernel functions that have been proposed, the Gaussian type could be seen as a paradigm of kernel. Its expression is

$$K(\mathbf{x}, \mathbf{y}) = \exp\left(-\frac{1}{2\sigma^2}\|\mathbf{x} - \mathbf{y}\|^2\right)$$

Obviously, the object more alike an object is itself. The Gaussian kernel says: $K(x, x) = 1$. If there was an (Euclidean) distance between **x** and **y**, the more the distance the more $K(x,y)$ approaches zero.

Other kernels would express similarity in other ways, perhaps needing normalization. For the interested reader, more details on kernels can be reached in [74, 86, 88].

It has been found that kernels offer great potential for classification purposes, while being simple to apply. Let us dive in these aspects.

We will use the notation $< \mathbf{x}, \mathbf{y} >$ to indicate the inner product of vectors **x** and **y**. For example, if x = (2,3) and y = (5,8), then $< \mathbf{x}, \mathbf{y} > = (2*5) + (3*8) = 34$. The inner product is also called the "*dot product*".

In mathematical terms, kernels are similarity functions that return inner products between feature mappings:

$$K(\mathbf{x}, \mathbf{y}) = < \Phi(\mathbf{x}),\ \Phi(\mathbf{y}) >$$

The simplest kernel is just $< \mathbf{x}, \mathbf{y} >$. There are polynomial kernels based on $(\mathbf{x}^{\mathbf{T}}\mathbf{y} + b)^P$, also the Gaussian kernel; and many more.

In computer implementations, usually kernels are expressed as positive semidefinite symmetric matrices, such as:

$$K(\mathbf{x}, \mathbf{y}) = \begin{pmatrix} K(x_1, x_1) & K(x_1, x_2) & \ldots & K(x_1, x_n) \\ K(x_2, x_1) & K(x_2, x_2) & \ldots & K(x_2, x_n) \\ \ldots & & & \\ \ldots & & & \\ K(x_n, x_1) & \ldots & K(x_n, x_{n-1}) & K(x_n, x_n) \end{pmatrix}$$

The research has observed that a number of important methods can be expressed in function of inner products $< \mathbf{x}, \mathbf{y} >$, and that they can become more flexible if one directly replace those inner products by kernel functions. This is the "kernel trick." There is no need of knowing $\Phi(\mathbf{x})$ and no need to explicitly compute the feature space mapping.

In 1992, Boser, Guyon, and Vapnik, then at AT&T Bell Laboratories, suggested a way to create nonlinear classifiers by applying the kernel trick [15] (this paper has been cited more than 16,000 times). This was the beginning of a very powerful classification method, able to separate data clusters via curved corridors.

Let us sketch the theory concerning SVM with kernels.

The SVM optimization problem can be re-formulated with Lagrange multipliers $\alpha_i \geq 0$ as follows:

Minimize, with respect to **w** and b:

$$L(\mathbf{w}, b, \boldsymbol{\alpha}) = \|\mathbf{w}\|^2 - \sum_i \alpha_i (y_i (< w, \mathbf{x_i} > +b) - 1)$$

Then, there are two equations:

$$\frac{\partial L}{\partial \mathbf{w}} = 0 \text{ and } \frac{\partial L}{\partial b} = 0$$

From the first equation, one obtains:

$$\mathbf{w} = \sum_i \alpha_i y_i \mathbf{x_i}$$

and from the second:

$$\sum_i \alpha_i y_i = 0$$

The dual optimization problem should have the same solution. After some algebra, this dual can be formulated as follows:

Maximize:

$$L(\boldsymbol{\alpha}) = \sum_i \alpha_i - \sum_i \sum_j \alpha_i \alpha_j \, y_i \, y_j < \mathbf{x_i}, \mathbf{x_j} >$$

Subject to:

$$\alpha_i \geq 0 \text{ and } \sum_i \alpha_i y_i = 0$$

To complete the scene, the separating line can be expressed in two ways:

$$f(\mathbf{x}) = < \mathbf{w}, \mathbf{x} > +b$$

$$f(\mathbf{x}) = \sum_i \alpha_i y_i < \mathbf{x_i}, \mathbf{x} > +b$$

The kernel SVM is obtained by direct application of the kernel trick, implying the following changes:

$$L(\boldsymbol{\alpha}) = \sum_i \alpha_i - \sum_i \sum_j \alpha_i \alpha_j \, y_i \, y_j \, K(\mathbf{x_i}, \mathbf{x_j})$$

$$f(\mathbf{x}) = \sum_i \alpha_i y_i \, K(\mathbf{x_i}, \mathbf{x}) + b$$

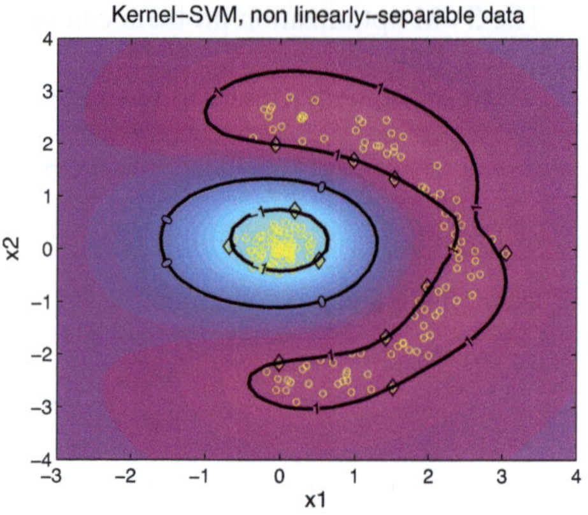

Fig. 7.29 Using Kernel-SVM for classification of nonlinearly separable data

We devised an example, by generating two synthetic datasets that cannot be separated by a straight line. Then a Kernel-SVM, with a Gaussian kernel, has been applied. The result has been visualized in Fig. 7.29, with perhaps too flashy colors.

In the figure, support points (support vectors) have been painted with black diamonds. There are level curves for values −1, 0, and 1; the curve corresponding to zero would be the separating line.

The figure has been generated with the Program 7.10, in MATLAB. Some warnings could appear for certain initial conditions (initial data are generated at random and change from run to run).

Program 7.10 Using kernel-SVM for classification

```
% kernel-SVM classification example
% 2 synthetic data clouds (non linearly separable)
% on 2D
%data (around two circular arcs)
N=200;
M=N/2;
X1=zeros(M,2); X2=zeros(M,2); %two data sets
mu1=0; sigma1=0.35; mu2=2.5; sigma2=0.25;
for nn=1:M,
  R1=mu1+ (sigma1*randn(1));
  phi1=rand(1)*2*pi;
  X1(nn,1)=R1*cos(phi1); X1(nn,2)=R1*sin(phi1);
  R2= mu2+ (sigma2*randn(1));
  phi2=-1.7+(1.1*rand(1)* pi);
  X2(nn,1)=R2*cos(phi2); X2(nn,2)=R2*sin(phi2);
end;
X=[X1;X2];
Y=ones(N,1); Y(1:M,1)=-Y(1:M,1); %labels
%------------------------------------------------
%prepare quadratic optimization
C=10; threshold = 0.05; %for soft margin
```

```
%labels
Dy = diag(Y);
% Gaussian kernel
ps=X*X'; %scalar product
normx=sum(X.^2,2);
[nps,mps]=size(ps);
aux=-2*ps+repmat(normx,1,mps)+repmat(normx',nps,1);
K=exp(-aux/2); %gaussian kernel
H =(Dy * K * Dy);
H = (H+H')/2; %make it symmetric
f = -ones(N,1);
Aeq=Y';
ceq = 0.0;
%bounds
cL= zeros(N,1); cU= repmat(C,N,1);
%optimizer
options = optimset('LargeScale','off','Display',...
'off','MaxIter',1000);
alpha = quadprog(H,f,[],[],Aeq,ceq,cL,cU,[],options);
%index of alphas satisfying conditions
serial_n = (1:size(X,1))';
ser_sv = serial_n((alpha > threshold) & ...
 (alpha < (C -0.01)));
nVectors = size(ser_sv,1);
%support vectors
xsup=X(ser_sv,:);
ysup=Y(ser_sv);
w=alpha(ser_sv).*ysup;
baux=0;
maux=0;
for ii = 1:nVectors
  baux = baux + Y(ser_sv(ii));
  maux= sum(alpha(ser_sv(ii))*Y(ser_sv(ii))*...
   K(ser_sv,ser_sv(ii)));
  baux = baux - maux;
end
b = baux/size(ser_sv,1); %decision function coeff
%----------------------------------------------------
% Prepare for visualization a grid of test points
[xg1,xg2]=meshgrid([-3:0.2:4],[-4:0.2:4]);
[nl,nc]=size(xg1); q=nl*nc;
xt1=reshape(xg1,1,q); xt2=reshape(xg2,1,q);
xt=[xt1;xt2]'; %set of test points
xtest=xt;
%using gaussian kernel
ps=xt*xsup'; %scalar product
normxt=sum(xt.^2,2);
normxsup=sum(xsup.^2,2);
[nps,mps]=size(ps);
yt=zeros(nps,1);
aux=-2*ps+repmat(normxt,1,mps)+...
repmat(normxsup',nps,1);
```

```
Kt=exp(-aux/2); %gaussian kernel
yt(:)=yt(:)+Kt*w(1:mps);
yt=yt+b;
yt2d=reshape(yt,nl,nc);
%-------------------------------------------------
%Display
figure(1)
%background levels
colormap('cool')
contourf(xg1,xg2,yt2d,20);shading flat; hold on
%data
scatter(X(:,1),X(:,2),16,'y')
%separating line and margins
[L,A]=contour(xg1,xg2,yt2d,[-1 0 1],'k');
clabel(L,A); %plot of lines with labels
%support vectors
scatter(xsup(:,1),xsup(:,2),60,'kd');
xlabel('x1');ylabel('x2');
title('Kernel-SVM, non linearly-separable data');
axis([-3 4 -4 4]);
```

Notice that in the program above, there are two new parameters that have not been mentioned yet. These are C and *"threshold"*. It corresponds to an important contribution from C. Cortes and V. Vapnik. They observed that in many scenarios data cannot be neatly separated. For example, symptoms of a disease A that might happen with disease B.

It is now normal—from the comments of Cortes and Vapnik—to distinguish between "*hard-margin*" and "*soft-margin*" situations. The former corresponds to clearly separable data. The second admits some mixing. For this soft-margin scenario, the optimization problem was restated by Cortes and Vapnik in the following terms:

Minimize

$$\mathbf{w}^{\mathbf{T}}\,\mathbf{w} + C\sum_{i=1}^{n}\zeta_i$$

Subject to

$$y_i\left(\mathbf{w}^{\mathbf{T}}\,\mathbf{x_i} + b\right) \geq 1 - \zeta_i\,, \qquad \zeta_i \geq 0$$

The reference publication of Cortes and Vapnik contribution is [23], year 1995. It has received a record number of citations: more than 60,000.

A frequently cited example of intermingled (nonseparable) data is the Wisconsin Breast Cancer Database, available from https://archive.ics.uci.edu/dataset/17/breast+cancer+wisconsin+diagnostic.

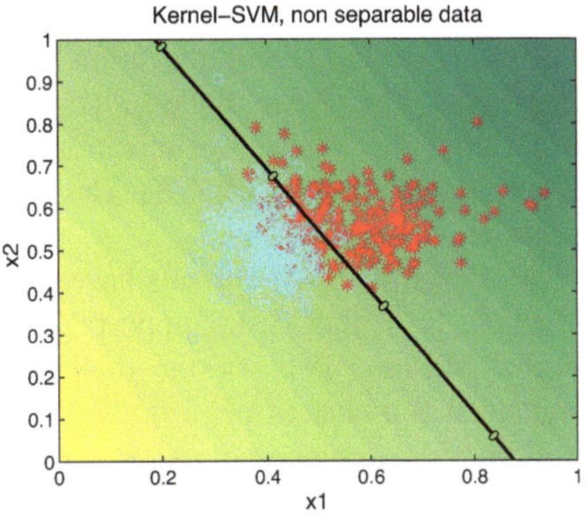

Fig. 7.30 Using Kernel-SVM for classification of intermingled data

We selected this case for SVM application, choosing mean radius (x) 3 <max 30> and mean smoothness (y) 7 <max 0.18> as the variables of interest. We edited the data file, writing − 1 instead of M, and + 1 instead of B.

A program in MATLAB was devised to compute the SVM classifier between malign or benign tumors. The result is shown in Fig. 7.30.

Figure 7.30 was obtained with Program 7.13. It requires some computation time: a little patience would be opportune.

Note (Program 7.13)

The complete listing of the Program 7.13, corresponding to Kernel-SVM has been included in the Appendix for Long Programs.

The Statistics and Machine Learning Toolbox from Mathworks (MATLAB) contains several functions and abundant documentation related to SVM. In turn, GitHub offer 2.783 public software and document repositories about SVM.

In addition to data processing software, there are many repositories of data sets available from the web (see for instance the list of datasets given by Wikipedia). These repositories are actively adding more sets for several applications, including those of artificial intelligence. Nowadays, one could use Google Dataset Search for getting data, being a dataset finder that contains 25 million datasets.

In several cases, these sets have enormous size. Basic SVM codes can be too slow. It is necessary to adapt SVM implementations, to be able to cope with big data. In general, the approach being applied is divide-and-conquer (by partitioning and/or sequencing). Three important alternatives have been proposed: the chunking algorithm [15], the decomposition algorithm [70], and the sequential minimal optimization (SMO) proposed in 1998 by John Platt (Microsoft), [76] (4,150 cites). The SMO method has been widely appreciated and improved in several ways, like the parallel version of [105] (more than 8,300 cites). See [17] for a tutorial on SVM for pattern recognition that discuss some decomposition algorithms (this tutorial received 24,000 cites).

The different SVM versions have some parameters to tune for better behavior. Therefore, we have an optimization need that has been tackled by machine leaning algorithms. The recent article [2] compares the use of Ant Bee Colony, the Genetic Algorithm, the Whale Optimization, and the Particle Swarm Optimization, for hyperparameter tuning of SVM. The references cited in the article cover many other contributions in this sense.

A number of SVM variants have been proposed, like the Fuzzy Support Vector Machine [60], the Twin Support Vector Machine [92], the Distributed Support Vector Machine [91], or the multiclass SVM [98]. Multiclass SVM could be organized as a binary tree [20].

Some SVM comprehensive surveys have been recently published, like [19] and [67].

The Mathworks File Exchange site offers 278 contributions with different variants and applications of SVM.

There are many more applications of quadratic programming, in different activity fields. The article [39] refers to applications on computational geometry (for instance *layout design*), on computational chemistry, on process networks (supply chain, water networks, blending, multicommodity flow, etc.), robotics, telecommunications, and others. As included in the survey [65], several quadratic optimization scenarios are present in electric power systems, as well as in traffic systems (see for instance [73]. In addition digital image restoration and other image processing techniques—including medical contexts—do use quadratic programming.

7.9 Exercises

1- It is desired to minimize the following function:

$$J = x_1^2 + 2x_2^2 - x_1 - 2x_2$$

subject to:

$$x_1 + x_2 = 3$$

$$x_1 > 1$$

$$x_2 \leq 3$$

$$-x_1 + 3x_2 \geq -1$$

$$x_1 + x_2 \leq 5$$

Use Wolfe's method to solve the problem. Visualize in 2D the problem and the result.

2- [7.1] Apply the active set method to problem 1. Visualize in two dimensions the problem and the result. Compare with the Wolfe's method: comments.

3- Minimize with Wolfe's method the following function:

$$J = x_1^2 + x_2^2 - 4x_1 - 4x_2$$

subject to:

$$x_1 + x_2 \leq 2$$

$$x_1 - 2x_2 \leq 2$$

$$-x_1 - x_2 \leq 1$$

$$-2x_1 + x_2 \leq 2$$

Visualize in 2D the problem and the result.

4- [7.1] Apply the active set method to problem 3. Visualize in 2D the problem and the result. Compare with the Wolfe's method: comments.

5- [7.2] Consider the following LCP problem:

$$\vec{\omega} - M\vec{z} = \vec{q}$$

$$\vec{\omega} \geq 0, \vec{z} \geq 0$$

$$\omega_i z_i = 0, \forall i$$

with:

$$q = \begin{pmatrix} 3 \\ 5 \\ -9 \end{pmatrix}, \quad M = \begin{pmatrix} 1 & -1\,. & -1 \\ -1 & 1 & -1 \\ 1 & 1 & 2 \end{pmatrix}$$

Apply the Lemke's method for solving the problem.

6- [7.3] Apply gradient projection to minimize the following function:

$$J = (x_1 - 2)^2 + (x_2 + 1)^2 + (x_3 + 2)^2$$

subject to all variables being non-negative.

Start from (1, 1, 1). Describe the iteration steps.

7- [7.4] Let the quadratic optimization problem be as follows:

Maximize: $J = 32x_1 - x_1^4 + 8x_2 - x_2^2$

Subject to:

$$x_1 - x_2 \leq 1$$

$$3x_1 + x_2 \leq 7$$

$$x_1 \geq 0,\ x_2 \geq 0$$

Apply the Frank-Wolfe method for solving the problem. Visualize in 2D the problem and the iterations of the method.

8- [7.5] Here is the following problem:
Minimize: $J = x_1^4 + x_2^4$
Subject to:

$$x_2 \geq x_1^2 - x_1 + 1$$

$$x_2 \geq x_1^2 - 4x_1 + 6$$

$$x_2 \leq -x_1^2 + 3x_1 + 2$$

Apply sequential quadratic programming. Compute four iterations. Visualize in 2D.

9- [7.6] As it is seen in the previous exercise, sequential quadratic programming can be applied to non-linear problems. For example, let the problem be:
Minimize: $J = e^{-x_1} + (x_2 - 2)^2$
Subject to: $x_1 \cdot x_2 \leq 1$
Clearly, it is a nonlinear problem. Please, compute by hand the matrices for the first three SQP iterations (in each iteration the problem is approximated by a standard QP problem). Visualize in two dimensions the problem and the iteration steps.

10- [7.7] Here is a simple portfolio scenario. We have three assets. The covariance matrix and expected returns are the following:

$$\Sigma = \begin{pmatrix} 2 & 1 & 0 \\ 1 & 2 & 1 \\ 0 & 1 & 2 \end{pmatrix}; \quad m = \begin{pmatrix} 0.4 \\ 0.4 \\ 0.8 \end{pmatrix}$$

Find and visualize the minimum variance portfolio. Visualize the solution in the risk space.

11- Based on diabetes data from:
https://www.csie.ntu.edu.tw/$\sim$cjlin/libsvmtools/datasets/binary.html *#diabetes*
Develop a support vector machine for classification in two classes.

12 Based on ionosphere data from:
https://www.csie.ntu.edu.tw/$\sim$cjlin/libsvmtools/datasets/binary.html *#ionosphere*
Develop a kernel-support vector machine for classification in two classes.

7.9.1 Exercise References

[7.1] https://busoniu.net/teaching/opt18en/ln4_qp.pdf
[7.2] https://riull.ull.es/xmlui/bitstream/handle/915/16279/Problemascomplementarioslineales.pdf?sequence=1&isAllowed=y
[7.3] https://homel.vsb.cz/$\sim$ber95/Methods_of_Optimization/Literature/Ascher/chap9.pdf
[7.4] http://www.aerostudents.com/courses/value-engineering-and-operations/FrankWolfeAlgorithmDemonstration.pdf
[7.5] https://www2.imm.dtu.dk/pubdb/edoc/imm5456.pdf
[7.6] https://ecal.studentorg.berkeley.edu/files/ce191/CH02-Quadratic Programming.pdf
[7.7] https://sites.math.washington.edu/$\sim$burke/crs/408/fin-proj/mark1.pdf

References

1. 112358 Randerson. (2020). Python for finance portfolio optimization. Available from: https://randerson112358.medium.com/python-for-finance-portfolio-optimization-66882498847
2. Ali, Y. A., Awwad, E. M., Al-Razgan, M., & Maarouf, A. (2023). Hyperparameter search for machine learning algorithms for optimizing the computational complexity. *Processes, 11*(2), 349.
3. AN IUSEM. (2003). On the convergence properties of the projected gradient method for convex optimization. *Computational and Applied Mathematics, 22*(1), 37–52.
4. Barraud, A. (2023). A quadratic programming API. Available from: https://es.mathworks.com/matlabcentral/fileexchange/124260-a-quadratic-programming-api
5. Barzilai, J., & Borwein, J. M. (1988). Two-point step size gradient methods. *IMA Journal of Numerical Analysis, 8*(1), 141–148.
6. Berard, S., Egan, K., & Trinkle, J. C. (2004). Contact modes and complementary cones. In *IEEE international conference on robotics and automation, 2004. Proceedings of ICRA'04* (Vol. 5, pp. 5280–5286). IEEE. Available from: https://www.cse.lehigh.edu/~trink/Papers/BETicra04.pdf
7. Bertsekas, D. P. (1976). On the Goldstein-Levitin-Polyak gradient projection method. *IEEE Transactions on Automatic Control, 21*(2), 174–184.
8. Biggs, M. C. (1972). Minimization algorithms making use of non-quadratic properties of the objective function. *IMA Journal of Applied Mathematics, 9*(2), 123–123.
9. Birgin, E. G., Martínez, J. M., & Raydan, M. (2000). Nonmonotone spectral projected gradient methods on convex sets. *SIAM Journal on Optimization, 10*(4), 1196–1211.
10. Birgin, E. G., Martínez, J. M., & Raydan, M. (2014). Spectral projected gradient methods: Review and perspectives. *Journal of Statistical Software, 60*, 1–21.
11. Bjorner, A., Björner, A., Las Vergnas, M., Sturmfels, B., White, N., & Ziegler, G. M. (1999). *Oriented matroids*. Cambridge University Press.
12. Bland, R. G. (1977). New finite pivoting rules for the simplex method. *Mathematics of Operations Research, 2*(2), 103–107.
13. Boggs, P. T., & Tolle, J. W. (1995). Sequential quadratic programming. *Acta Numerica, 4*, 1–51.
14. Bomze, I. M., Rinaldi, F., & Zeffiro, D. (2021). Frank–Wolfe and friends: A journey into projection-free first-order optimization methods. *4OR, 19*, 313–345.

15. Boser, B. E., Guyon, I. M., & Vapnik, V. N. (1992). A training algorithm for optimal margin classifiers. In *Proceedings of the fifth annual workshop on computational learning theory* (pp. 144–152).
16. Braun, G., Carderera, A., Combettes, C. W., Hassani, H., Karbasi, A., Mokhtari, A., & Pokutta, S. (2022). Conditional gradient methods. arXiv preprint arXiv:2211.14103.
17. Burges, C. J. C. (1998). A tutorial on support vector machines for pattern recognition. *Data Mining and Knowledge Discovery, 2*(2), 121–167.
18. Calamai, P. H., & Moré, J. J. (1987). Projected gradient methods for linearly constrained problems. *Mathematical Programming, 39*(1), 93–116.
19. Cervantes, J., Garcia-Lamont, F., Rodríguez-Mazahua, L., & Lopez, A. (2020). A comprehensive survey on support vector machine classification: Applications, challenges and trends. *Neurocomputing, 408*, 189–215.
20. Cheong, S., Oh, S. H., & Lee, S.-Y. (2004). Support vector machines with binary tree architecture for multi-class classification. *Neural Information Processing-Letters and Reviews, 2*(3), 47–51.
21. Cimini, G. (2017). *Complexity certification and efficient implementation of model predictive control for embedded applications*. PhD thesis, Universitá Politechnica delle Marche. Available from: https://iris.univpm.it/retrieve/e18b8790-60ab-d302-e053-1705fe0a27c8/tesi_cimini.pdf
22. Coleman, T., Branch, M. A., & Grace, A. Optimization toolbox for use with matlab. Available from: https://instruct.uwo.ca/engin-sc/391b/downloads/optim_tb.pdf
23. Cortes, C., & Vapnik, V. (1995). Support-vector networks. *Machine Learning, 20*, 273–297.
24. Cottle, R. W. (1963). Symmetric dual quadratic programs. *Quarterly of Applied Mathematics, 21*(3), 237–243.
25. Cottle, R. W., & Dantzig, G. B. (1968). Complementary pivot theory of mathematical programming. *Linear Algebra and Its Applications, 1*(1), 103–125.
26. Cottle, R. W., Pang, J.-S., & Stone, R. E. (2009). *The linear complementarity problem*. SIAM.
27. CVX Research. (2023). CVX, software for disciplined convex programming. Available from: http://cvxr.com/
28. Dai, Y.-H., & Fletcher, R. (2005). Projected Barzilai-Borwein methods for large-scale box-constrained quadratic programming. *Numerische Mathematik, 100*(1), 21–47.
29. Dangi, A. (2013). Financial portfolio optimization: Computationally guided agents to investigate, analyse and invest!? arXiv preprint arXiv:1301.4194
30. DeGuchy, O. (2015). *About solutions of the LCP*. PhD thesis, California State Polytechnic University, Pomona. Available from: https://scholarworks.calstate.edu/downloads/vx021h539?locale=en
31. Di Gaspero, L. (2016). Quadprog++ http://github.com/liuq
32. Elton, E. J., & Gruber, M. J. (1997). Modern portfolio theory, 1950 to date. *Journal of Banking & Finance, 21*(11–12), 1743–1759.
33. Fernández, D., & Solodov, M. (2014). Stabilized sequential quadratic programming: A survey. *Pesquisa Operacional, 34*, 463–479.
34. Ferris, M. C., Mangasarian, O. L., & Wright, S. J. (2007). Linear programming with matlab. Available from: https://research.cs.wisc.edu/math-prog/lpbook/
35. Ferris, M. C., Mangasarian, O. L., & Wright, S. J. (2007). *Linear programming with MATLAB*. SIAM.
36. Fletcher, R. (1971). A general quadratic programming algorithm. *IMA Journal of Applied Mathematics, 7*(1), 76–91.
37. Frank, M., & Wolfe, P. (1956). An algorithm for quadratic programming. *Naval Research Logistics Quarterly, 3*(1–2), 95–110.
38. Fukuda, K., Klaus, L., & Miyata, H. (2013). A new subclass of p-matrix linear complementarity problems. *IEICE Technical Report, 113*(50), 25–32.
39. Furini, F., Traversi, E., Belotti, P., Frangioni, A., Gleixner, A., Gould, N., Liberti, L., Lodi, A., Misener, R., Mittelmann, H., et al. (2019). Qplib: A library of quadratic programming instances. *Mathematical Programming Computation, 11*, 237–265.

40. Gholami, F., Nasri, M., Kövecses, J., & Teichmann, M. (2016). A linear complementarity formulation for contact problems with regularized friction. *Mechanism and Machine Theory, 105*, 568–582. Available from: https://www.sciencedirect.com/science/article/pii/S0094114X16301574
41. Gill, P. E., Murray, W., Saunders, M. A., & Wright, M. H. (1985). Some issues in implementing a sequential quadratic programming algorithm. *ACM Signum Newsletter, 20*(2), 13–19.
42. Gill, P. E., & Wong, E. (2011). Sequential quadratic programming methods. In *Mixed integer nonlinear programming* (pp. 147–224). Springer.
43. Goldfarb, D., & Idnani, A. (1983). A numerically stable dual method for solving strictly convex quadratic programs. *Mathematical Programming, 27*(1), 1–33.
44. Gould, N. (2006) An introduction to algorithms for continuous optimization.
45. Gunjan, A., & Bhattacharyya, S. (2022). A brief review of portfolio optimization techniques. *Artificial Intelligence Review, 56*, 1–40.
46. Hager, W. W., & Park, S. (2004). The gradient projection method with exact line search. *Journal of Global Optimization, 30*, 103–118.
47. Han, S. P. (1976). Superlinearly convergent variable metric algorithms for general nonlinear programming problems. *Mathematical Programming, 11*(1), 263–282.
48. Huang, Y.-K., Dai, Y.-H., & Liu, X.-W. (2021). Equipping the Barzilai–Borwein method with the two dimensional quadratic termination property. *SIAM Journal on Optimization, 31*(4), 3068–3096.
49. Huang, Y., Dai, Y.-H., Liu, X.-W., & Zhang, H. (2022). On the acceleration of the Barzilai–Borwein method. *Computational Optimization and Applications, 81*(3), 717–740.
50. Illés, T., Szirmai, Á., & Terlaky, T. (1999). The finite criss-cross method for hyperbolic programming. *European Journal of Operational Research, 114*(1), 198–214.
51. Jaggi, M. (2013). Revisiting Frank-Wolfe: Projection-free sparse convex optimization. In *International conference on machine learning* (pp. 427–435). PMLR.
52. Jensen, P. A., & Bard, J. F. (2003). Nonlinear programming methods. s2 quadratic programming. In *Operations research models and methods* (Vol. 1). Wiley.
53. Jin, Y., Qu, R., & Atkin, J. (2016). Constrained portfolio optimisation: The state-of-the-art Markowitz models. Available from: https://www.researchgate.net/profile/Rong-Qu-3/publication/301721297_Constrained_Portfolio_Optimisation_The_State-of-the-Art_Markowitz_Models/links/5979e8bd0f7e9b0469ae38ef/Constrained-Portfolio-Optimisation-The-State-of-the-Art-Markowitz-Models.pdf
54. Kolm, P. N., Tütüncü, R., & Fabozzi, F. J. (2014). 60 years of portfolio optimization: Practical challenges and current trends. *European Journal of Operational Research, 234*(2), 356–371.
55. Lacoste-Julien, S., & Jaggi, M. (2015). On the global linear convergence of Frank-Wolfe optimization variants. In *Advances in neural information processing systems* (Vol. 28).
56. Leković, M. M. (2021). Historical development of portfolio theory. *Tehnika, 76*(2), 220–227.
57. Lemke, C. E. (1968). On complementary pivot theory. *Mathematics of the Decision Sciences*, 1, 115–136.
58. Lemke, C. E., & Howson, J. T., Jr. (1964). Equilibrium points of bimatrix games. *Journal of the Society for Industrial and Applied Mathematics, 12*(2), 413–423.
59. Lin, C.-J. (2007). Projected gradient methods for nonnegative matrix factorization. *Neural Computation, 19*(10), 2756–2779.
60. Lin, C.-F., & Wang, S.-D. (2002). Fuzzy support vector machines. *IEEE Transactions on Neural Networks, 13*(2), 464–471.
61. Lindsay, D. (2003). CS392 MATLAB workshop 2 – SVM's. Available from: https://www.robots.ox.ac.uk/~az/lectures/ml/matlab2.pdf
62. London Royal Holloway University. (2020). Support vector machine. Available from: https://cml.rhul.ac.uk/svm.html
63. Maratos, N. (1978). *Exact penalty function algorithms for finite dimensional and control optimization problems*, University of London.

64. Messerer, F., Baumgärtner, K., & Diehl, M. (2021). Survey of sequential convex programming and generalized Gauss-Newton methods. *ESAIM: Proceedings and Surveys, 71*, 64–88.
65. Molzahn, D. K., Dörfler, F., Sandberg, H., Low, S. H., Chakrabarti, S., Baldick, R., & Lavaei, J. (2017). A survey of distributed optimization and control algorithms for electric power systems. *IEEE Transactions on Smart Grid, 8*(6), 2941–2962.
66. Murty, K. G. (1997). Linear complementarity problem, its geometry and applications. In *Linear complementarity, linear and nonlinear programming* (pp. 1–58). Available from: https://public.websites.umich.edu/~murty/books/linear_complementarity_webbook/kat1.pdf
67. Nayak, J., Naik, B., & Behera, H. S. (2015). A comprehensive survey on support vector machine in data mining tasks: Applications & challenges. *International Journal of Database Theory and Application, 8*(1), 169–186.
68. Nieto, J. A. (2004). Chirotope concept in various scenarios of physics. arXiv preprint hep-th/0407093.
69. Nieto, J. A. (2018). Duality, matroids, qubits, twistors, and surreal numbers. *Frontiers in Physics, 6*, 106.
70. Osuna, E., Freund, R., & Girosit, F. (1997). Training support vector machines: An application to face detection. In *Proceedings of IEEE computer society conference on computer vision and pattern recognition* (pp. 130–136). IEEE.
71. Palomar, D. (2021). Portfolio optimization. Available from: https://palomar.home.ece.ust.hk/ELEC5470_lectures/slides_portfolio_optim.pdf
72. Passacantando, M. (2023). frank_wolfe_m. Available from: https://elearning.di.unipi.it/mod/resource/view.php?id=2599&forceview=1
73. Patriksson, M. (2015). *The traffic assignment problem: Models and methods.* Courier Dover Publications.
74. Phillips, J. M., & Venkatasubramanian, S. (2011). A gentle introduction to the kernel distance. arXiv preprint arXiv:1103.16252011.
75. Pichler, A. (2020). Selected topics from portfolio optimization. Available from: https://www.tu-chemnitz.de/mathematik/fima/public/LecturesAndMaterials/PortfolioOptimierung.pdf
76. Platt, J. (1998). Sequential minimal optimization: A fast algorithm for training support vector machines. Available from: https://www.microsoft.com/en-us/research/publication/sequential-minimal-optimization-a-fast-algorithm-for-training-support-vector-machines/
77. Powell, M. J. (1978). The convergence of variable metric methods for nonlinearly constrained optimization calculations. In *Nonlinear programming 3* (pp. 27–63). Academic Press.
78. Powell, M. J. D. (1983). Variable metric methods for constrained optimization. In *Mathematical programming the state of the art: Bonn 1982* (pp. 288–311).
79. R Cope. Constrainedoptimization/assignment1/ex4/markowitz.m. Available from: https://github.com/RamiroCope/ConstrainedOptimization
80. Ralphs, T. Advanced mathematical programming IE417 lecture 24. Available from: https://coral.ise.lehigh.edu/~ted/files/ie417/lectures/Lecture24.pdf
81. Richter-Gebert, J., & Ziegler, G. M. (2017). 6: Oriented matroids. In *Handbook of discrete and computational geometry* (pp. 159–184). Chapman and Hall/CRC.
82. Rong-En-Fan. (2000). Libsvm: Data: Classification, regression, and multi-label. Available from: https://www.csie.ntu.edu.tw/~cjlin/libsvmtools/datasets/
83. Roos, C., Terlaky, T., & de Klerk, E. Nonlinear optimization (WI3 031). Available from: https://www.researchgate.net/publication/228919341_Nonlinear_Optimization_WI3_031
84. Rosen, J. B. (1960). The gradient projection method for nonlinear programming: Part I. linear constraints. *Journal of the Society for Industrial and Applied Mathematics, 8*(1), 181–217.
85. Schittkowski, K. (2005). *QL: A Fortran code for convex quadratic programming-user's guide.* Department of Mathematics, University of Bayreuth, Report, Version, 2.
86. Scholkopf, B., Mika, S., Burges, C. J. C., Knirsch, P., Muller, K.-R., Ratsch, G., & Smola, A. J. (1999). Input space versus feature space in kernel-based methods. *IEEE Transactions on Neural Networks, 10*(5), 1000–1017.
87. Schweitzer, S. (2022). Finance programming – workshop 4. Available from: https://rpubs.com/Stefan240sx/fpw4

88. Scikit-learn. (2023). Pairwise metrics, affinities and kernels. Available from: https://scikit-learn.org/stable/modules/metrics.html#:~:text=Kernels%20are%20measures%20of%20similarity,measure%2C%20such%20as%20a%20kernel
89. Seoane, J. (2013). FSMKL/ monqp.m. Available from: https://github.com/jseoane/FSMKL/blob/master/monqp.m
90. Silicon Cloud Technologies LLC. (2023). Portfolio visualizer. Available from: https://www.portfoliovisualizer.com/asset-correlations
91. Stolpe, M., Bhaduri, K., & Das, K. (2016). Distributed support vector machines: An overview. In *Solving large scale learning tasks. Challenges and algorithms: Essays dedicated to Katharina Morik on the occasion of her 60th birthday* (pp. 109–138).
92. Tanveer, M., Rajani, T., Rastogi, R., Shao, Y.-H., & Ganaie, M. A. (2022). Comprehensive review on twin support vector machines. *Annals of Operations Research, 339*(3), 1–46.
93. Terlaky, T. (1985). A convergent criss-cross method. *Optimization, 16*(5), 683–690.
94. Terlaky, T. (1987). A finite crisscross method for oriented matroids. *Journal of Combinatorial Theory, Series B, 42*(3), 319–327.
95. Turlach, B. A., & Wright, S. J. (2015). Quadratic programming. *Wiley Interdisciplinary Reviews: Computational Statistics, 7*(2), 153–159.
96. Van den Berghen, F. (2004). *CONDOR: A constrained, non-linear, derivative-free parallel optimizer for continuous, high computing load, noisy objective functions*. These de doctorat, Université Libre de Bruxelles, Belgium.
97. Wan, Z., Guo, J., Liu, J., & Liu, W. (2018). A modified spectral conjugate gradient projection method for signal recovery. *Signal, Image and Video Processing, 12*, 1455–1462.
98. Wang, Z., & Xue, X. (2014). Multi-class support vector machine. In *Support vector machines applications* (pp. 23–48). Springer.
99. Wehner, W. (2020). projgrad. Available from: https://github.com/wwehner/projgrad
100. Wolfe, P. (1959). The simplex method for quadratic programming. *Econometrica: Journal of the Econometric Society, 27*, 382–398.
101. Wright, S. J. (2007). Additional notes on solving QPS in nonstandard form via Lemke's method. Available from: https://pages.cs.wisc.edu/~swright/525/handouts/handout_lcp.pdf
102. Wright, S., Nocedal, J., et al. (1999). Numerical optimization. *Springer Science, 35*(67–68), 7.
103. Yin, W. (2015). Math 164: Optimization, Barzilai-Borwein method. Available from: https://www.cmor-faculty.rice.edu/~yzhang/caam565/L1_reg/BB-step.pdf
104. Zanjirdar, M. (2020). Overview of portfolio optimization models. *Advances in Mathematical Finance and Applications, 5*(4), 419–435.
105. Zeng, Z.-Q., Yu, H.-B., Xu, H.-R., Xie, Y.-Q., & Gao, J. (2008). Fast training support vector machines using parallel sequential minimal optimization. In *2008 3rd international conference on intelligent system and knowledge engineering* (Vol. 1, pp. 997–1001). IEEE.
106. Zhang, Z. (2022). *Combinatorial and machine learning problems motivated by the simplex method*. University of California.
107. Zionts, S. (1969). The criss-cross method for solving linear programming problems. *Management Science, 15*(7), 426–445.

Chapter 8
Integer Programming

8.1 Introduction

Integer numbers are connatural to many activities. Cars, bottles, smartphones, etc. are sold in integer numbers; nobody sells three and 1/2 eggs. When designing an industrial facility, the number of needed workers is computed in integer persons.

It is easy to see that there are many cases, where optimization should consider integer or discrete variables.

Most cited books in this field are [107] (12,075 cites) on integer and combinatorial optimization, [93] (11,370 cites) on linear and integer programming, [106] (5,323 cites) on integer programming, and [94] (6,396 cites) on combinatorial optimization. More related to applications are the books [98] and [18].

The chapter starts with theory, involving algorithms, and then includes some important and illustrative applications. There are mixed scenarios, where part of variables are integer and the others not; this will also be considered along the chapter.

8.2 Integer Programming

The optimization problem, for pure integer variables, can be succinctly enunciated as follows:

- The *canonical* form of integer linear programming (ILP) is

  Maximize:

$$z = \vec{c}^{T} \cdot \vec{x}$$

J. M. Giron-Sierra, *Introduction to Optimization with Matlab® Examples*,
https://doi.org/10.1007/978-3-032-05048-9_8

Subject to:

$$A\vec{x} \le \vec{b}\,,\ \vec{x} \ge \vec{0}$$

where all variables included in $\vec{x}$ are integer.

- The *standard* form of ILP is:

  Maximize:

$$z = \vec{c}^T \cdot \vec{x}$$

  Subject to:

$$A\vec{x} = \vec{b}\,,\ \vec{x} \ge \vec{0}$$

where all $x_i \in \vec{x}$, are integer.

A naïve approach for solving this optimization problem is rounding. Conventional linear programming is a first step, and then, in a second step, the result is rounded. Some care is recommended, especially about possible violation of constraints (see, for instance, [84] for a lecture on relaxation).

In general, with sufficient digital processing power and patience, an enumerative approach could be used for not too-large problems. However, many real-life problems are large, and algorithmic methods are welcome.
A first, simple example of integer programming is the following, [89]:

ACME Furniture Company wants to maximize profits. They make \$10 for every table and \$3 for every chair. Employee A can make six tables and seven chairs but does not work more than 40 hours. Employee B can make three tables and one chair but does not work more than 11 hours.

Hence, ACME must solve the following problem:
Maximize:

$$J = 10x_1 + 3x_2$$

Subject to:

$$6x_1 + 7x_2 \le 40$$

$$3x_1 + x_2 \le 11$$

with x_1, x_2 non-negative integers.

Suppose that x_1, x_2 were real numbers and that linear programming was applied. The solution would be

$$x_1 = 3.666;\ x_2 = 0 \text{ and } J = 36.666$$

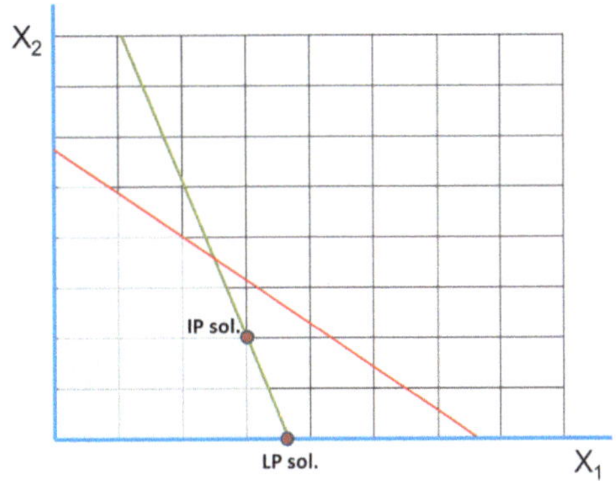

Fig. 8.1 Example of integer programming vs. linear programming

Figure 8.1 depicts the problem with the feasible region, in case of continuous variables. The solution given by linear programming is marked at the bottom right-hand corner.

Assuming that $x_2 = 0$, if we round up x_1 from 3.666 to 4, we get out from the feasible region.

On the opposite direction, if we round down from 3.666 to 3, we get a feasible solution but not necessarily the optimal.

Delving into the problem, the best solution (with integer values) is found to be (3, 2), which is also marked in Fig. 8.1.

The value of J at (3, 0) is 30; while J is 36 at (3, 2).

Since linear programming (LP) is so useful, the research has developed some ways for extending the use of LP in integer programming scenarios. A starting point would be to consider that variables were real, not just integers; this is called *LP relaxation*. Then, we have a feasible set F, like the one shown in A first, simple example of integer Fig. 8.1 (shaded region). The idea then is to explore F (in a divide & conquer manner) or shrink F by adequate cuts, to drive the simplex algorithm toward integer solutions.

Next subsections introduce two main methodologies for solving integer programming problems.

8.2.1 *Cutting Plane Techniques*

Consider another simple example [102]:

Maximize:

$$z = 5\,x_1 + 8\,x_2$$

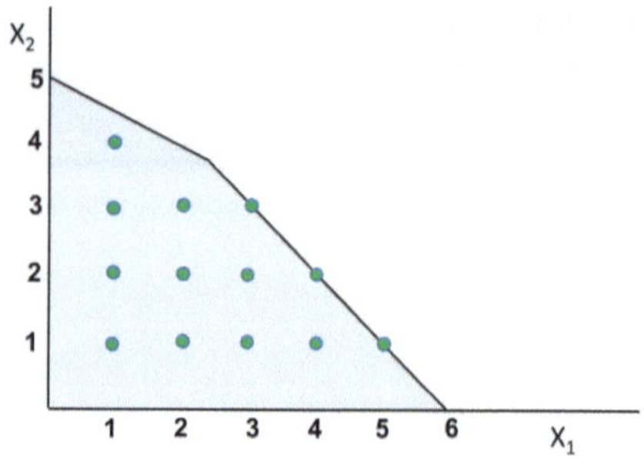

Fig. 8.2 An integer programming example

Subject to:

$$x_1 + x_2 \leq 6$$

$$5x_1 + 9x_2 \leq 45$$

with x_1, x_2 being nonnegative integers.

First, one could try the simplex algorithm, treating the case as a linear programming problem (LP relaxation). The feasible region is represented in Fig. 8.2.

If the solution was an integer, then that is all. But no, the solution takes place at the point $\vec{x}* = (2.25, 3.75)$, so it is not integer.

The reader is invited to run the short Program 8.1, which is based on *linprog( )*, for checking our initial noninteger solution.

Program 8.1 Getting an initial non-integer solution

```
f=[-5, -8];
A=[1 1; 5 9];
b=[6; 45];
lb=[0, 0];
ub=[Inf, Inf];
X=linprog(f,A,b,[],[],lb,ub);
X
```

Let us generate a *cut*, which is a constraint that is satisfied by all feasible integers but not by $\vec{x}*$. For example:

$$2x_1 + 3x_2 \leq 15$$

Figure 8.3 shows this new constraint.

The new constraint is added to the linear programming problem. Now, the solution is $\vec{x}* = (0, 5)$ which is the desired optimal integer solution.

Notice that the cut only removed the noninteger solution; cuts never remove feasible integer solutions.

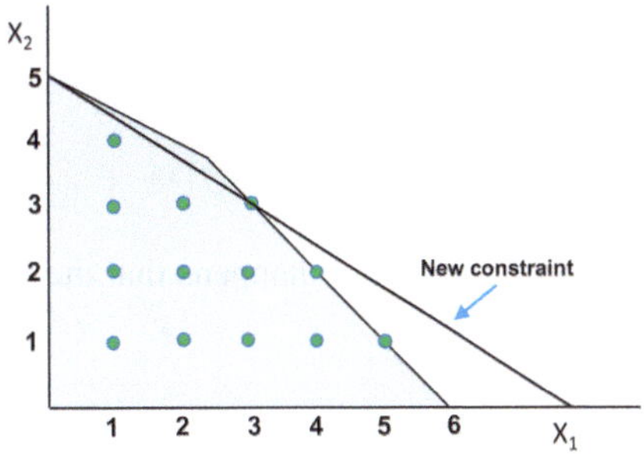

Fig. 8.3 The new constraint (the cut)

Again, the reader can use Program 8.2 for checking the result after adding the new constraint. It is important to notice that *linprog( )* gives us now a result in integer numbers.

Program 8.2 A cut is applied, and integer solution is obtained

```
f=[-5, -8];
A=[1 1; 5 9; 2 3];
b=[6; 45; 15];
lb=[0, 0];
ub=[Inf, Inf];
X=linprog(f,A,b,[],[],lb,ub);
X
```

R.E. Gomory presented in 1958 a method for introducing cuts in the simplex algorithm [44] (2,689 cites). Choose for instance one of the constraints:

$$a_1 x_1 + a_2 x_2 + \ldots + a_n x_n = b$$

where in general, b and the coefficients a_j are not integers.

Denote $floor(\eta)$as $\lfloor \eta \rfloor$. Then, the chosen constraint can be written as

$$[\lfloor a_1 \rfloor + (a_1 - \lfloor a_1 \rfloor)]\, x_1 + \ldots + [\lfloor a_n \rfloor + (a_n - \lfloor a_n \rfloor)]\, x_n = [\lfloor b \rfloor + (b - \lfloor b \rfloor)]$$

Naming as f_i the fractional terms, an equivalent expression would be

$$[\lfloor a_1 \rfloor + f_1)]\, x_1 + \ldots + [\lfloor a_n \rfloor + f_n]\, x_n = [\lfloor b \rfloor + f]$$

Now, we separate fractional and integer terms:

$$f_1 x_1 + \ldots + f_n x_n - f = \lfloor b \rfloor - \lfloor a_1 \rfloor\, x_1 - \ldots - \lfloor a_n \rfloor\, x_n$$

Notice that the right-hand side of this last equation must be an integer. Then, it could be written as

$$f_1x_1 + \ldots + f_nx_n = f + k$$

with k an integer. It happens that the left-hand side of this equation is nonnegative, and therefore,

$$(f + k) \geq 0$$

or:

$$k \geq -f$$

and so, because k is an integer, we can say that $k \geq 0$. Hence, finally, the following Gomory fractional cut can be written:

$$f_1x_1 + \ldots + f_nx_n \geq f$$

An easy way for obtaining Gomory cuts is to write the coefficients in modular way.

For example, suppose the following constraint:

$$3\,x_1 + (1\ 1/4)\,x_2 + (1/3)\,x_3 = 2\ 2/3$$

(notice how coefficients are expressed as integer, space, fraction)

We rewrite the constraint as follows:

$$((1/4)\,x_2 + (1/3)\,x_3) + (3\,x_1 + x_2 - 2) = 2/3$$

or:

$$((1/4)\,x_2 + (1/3)\,x_3) - 2/3 = (3\,x_1 + x_2 - 2)$$

The corresponding cut would be

$$((1/4)\,x_2 + (1/3)\,x_3) \geq 2/3$$

(the recipe is to drop all the integers).

In case of negative coefficients, like in

$$3\,x_1 - (1\ 1/4)\,x_2 + (1/3)\,x_3 = 2\ 2/3$$

one writes instead:

$$3\,x_1 + (-2\ 3/4)\,x_2 + (1/3)\,x_3 = 2\ 2/3$$

Based on these observations, adequate constraints could be added to the simplex algorithm in order to introduce a cut. For example, suppose you arrived to a noninteger optimal in a certain problem, having the following simplified tableau:

Basis	x_1	x_2	s_1	s_2	b
x_1	1	0	0.6	−0.4	3.2
x_2	0	1	−0.4	0.6	3.2
z_j	1.4	1	0	0	
$c_j - z_j$	0	0	0.2	0.2	

Select for instance the x_1 row. The corresponding constraint would be

$$x_1 + 0.6\, s_1 - 0.4\, s_2 = 3.2$$

Therefore, an adequate cut can be

$$0.6\, s_1 - 0.4\, s_2 \geq 0.2$$

This constraint is satisfied by every feasible integer solution but is not feasible for the current optimal solution (which includes $s_1 = 0$; $s_2 = 0$). By adding this constraint to the problem, a different optimal solution would be reached, which might be integer.

Some of the publications that have paid renewed attention to the Gomory cuts developments are [6] and [25]. See [21] and references therein for recent applications.

Figure 8.4 presents a photograph of Ralph E. Gomory. After a while at Princeton University, he joined the Research Division of IBM in 1959, and became Director of Research of that Division in 1970. During those years, he invented the RISC computer architecture. See [24] for the story around Gomory cuts.

Fig. 8.4 Ralph. E. Gomory

Later in this chapter, we will visit again the topic of cuts, with a more formal treatment.

8.2.2 Branch and Bound

It is typical of integer programming scenarios to have decision variables, of the type yes or not (1 or 0). In this case, it is usually convenient to see what happens when one takes one or the other option.

Branch and bound methods divide optimization problems into several subproblems. A simple first example is the following [35]:

Maximize:

$$z = x_1 + x_2$$

Subject to:

$$x_2 - x_1 \leq 2$$

$$8x_1 + 2x_2 \leq 19$$

with x_1, x_2 being nonnegative integers.

Figure 8.5 depicts the situation. If real valued variables were accepted, the solution would take place at *(1.5, 3.5)*; but only integer variables were allowed, and so it is not the desired solution.

Let us decompose the problem into two subproblems, one for $x_1 \leq 1$ and the other with $x_1 \geq 2$. Both subproblems are treated as linear programming cases, with the conventional simplex algorithm. The solutions are

- sub-problem 1: *(1,3), z=4*
- sub-problem 2: *(2,1.5), z=3.5*

When an integer solution is found in a subproblem, no further branching of this subproblem will yield better results. Therefore, our optimal solution has been found, in the subproblem 1.

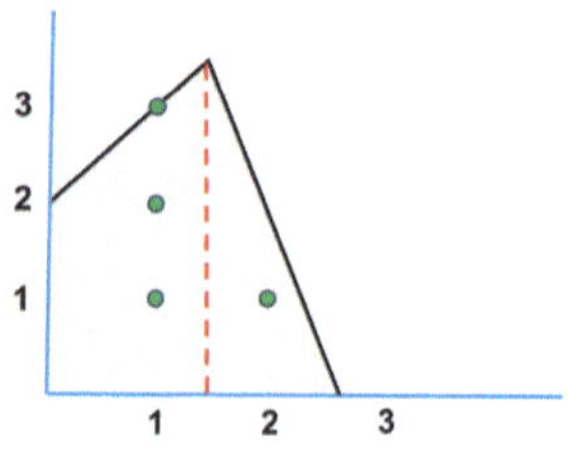

Fig. 8.5 An example of branch and bound

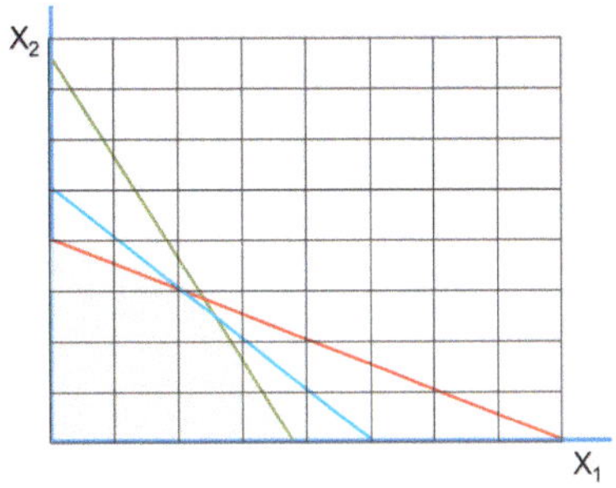

Fig. 8.6 Visualization of the example under study

A more complicated example is the following [22]:
Maximize:

$$z = 3x_1 + 2x_2$$

Subject to:

$$4x_1 + 2x_2 \leq 15$$

$$x_1 + 2x_2 \leq 8$$

$$x_1 + x_2 \leq 5$$

with x_1, x_2 being nonnegative integers.

Figure 8.6 visualizes the feasible set of this problem, corresponding to the constraints (for continuous variables). A grid has been included for indicating the integer values in the region of interest.

By using the simplex algorithm (considering continuous variables), one finds the optimum at *(2.5, 2.5)* with $z = 12$. Let us branch on x_1. Two subproblems are obtained:

- The first, subject to:

$$4x_1 + 2x_2 \leq 15$$

$$x_1 + 2x_2 \leq 8$$

$$x_1 + x_2 \leq 5$$

$$x_1 \geq 3$$

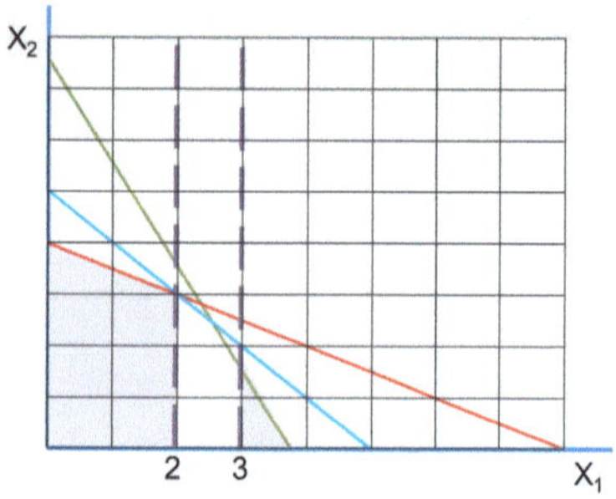

Fig. 8.7 Effect of branching, originating two subproblems

- The second, subject to

$$4x_1 + 2x_2 \leq 15$$

$$x_1 + 2x_2 \leq 8$$

$$x_1 + x_2 \leq 5$$

$$x_1 \leq 2$$

Figure 8.7 shows the two subproblems after branching:

Suppose one chooses the first subproblem, which is situated at the right-hand side of Fig. 8.7 (for x1 ≥3). Using the simplex algorithm for this LP, the solution is (3, 1.5), which is not valid (we need an all-integer solution). Then, let us branch on x_2. Two more subproblems are obtained:

- The third, subject to

$$4x_1 + 2x_2 \leq 15$$

$$x_1 + 2x_2 \leq 8$$

$$x_1 + x_2 \leq 5$$

$$x_1 \geq 3$$

$$x_2 \geq 2$$

- The fourth, subject to

$$4x_1 + 2x_2 \leq 15$$

$$x_1 + 2x_2 \leq 8$$

$$x_1 + x_2 \leq 5$$

$$x_1 \geq 3$$

$$x_2 \leq 1$$

The third subproblem is visualized in Fig. 8.8. No solution can exist in this region.

The fourth subproblem is visualized in Fig. 8.9. Part of the feasible region is captured in this exploration.

The third subproblem is unfeasible. The solution for the fourth subproblem is (3.25, 1), not valid. If one takes the fourth subproblem and branches on x_1, two more subproblems are generated, the 5th with $x_1 \geq 4$, and the 6th with $x_1 \leq 3$. The 5th subproblem is unfeasible, while the 6th gives the solution (3, 1) with $z = 11$. This is a lower bound.

Figure 8.10 visualizes the fifth subproblem. No solution can exist in this region.

Figure 8.11 corresponds to the sixth subproblem. Another part of the feasible region is captured,

Let us come back and see the second subproblem. The optimal solution found with the simplex algorithm is (2, 3) with $z = 12$. This is the optimum.

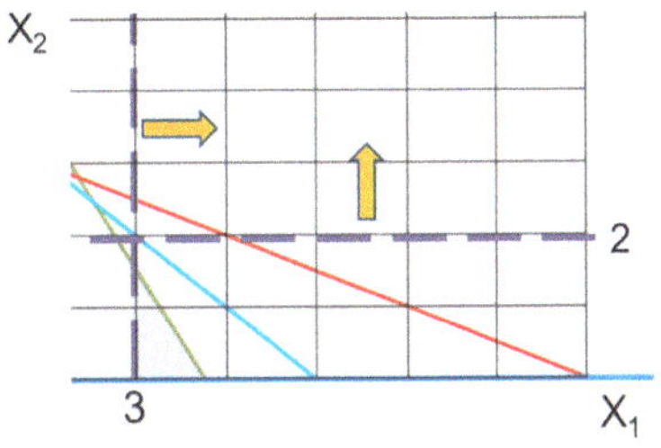

Fig. 8.8 The third sub-problem

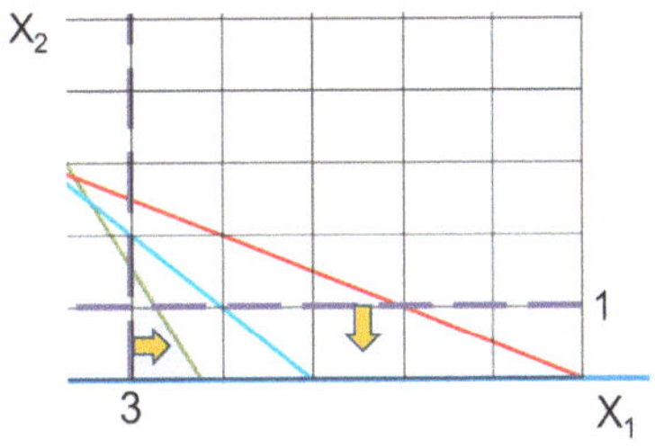

Fig. 8.9 The fourth subproblem

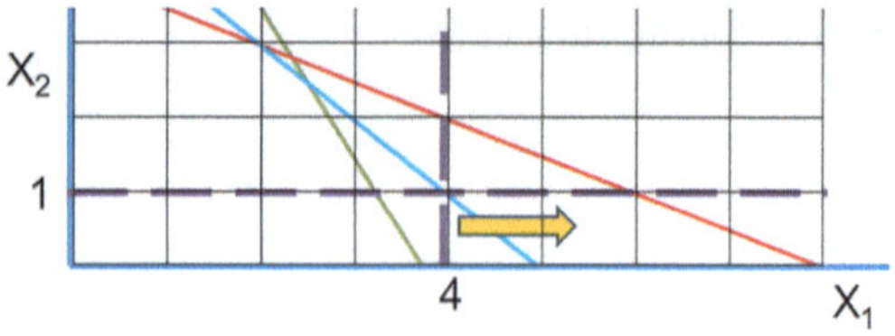

Fig. 8.10 The fifth subproblem

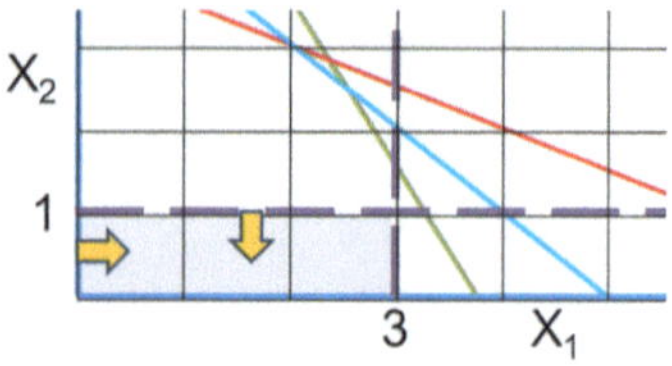

Fig. 8.11 The sixth sub-problem

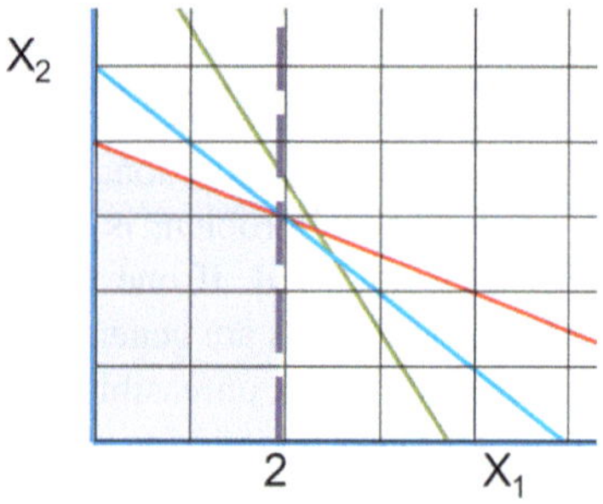

Fig. 8.12 The second subproblem (the optimal is here)

Figure 8.12 visualizes the second subproblem.

It is important to notice that guided by the branching, the conventional simplex algorithm gave us valid all-integer solutions (including the optimal one).

Figure 8.13 summarizes the process that has been followed in this example.

Notice that the process represented in Fig. 8.13 corresponds to a typical decision tree (a major topic in computer science).

In general, there are two exploration alternatives. In the example depicted in Fig. 8.13, *the depth-first* alternative has been selected. The other alternative was *breadth-first* search.

The branch and bound method was proposed in 1960 [62], by A.H. Land and A.G. Doig from the London School of Economics. See [55] (4,450 cites) for an interesting story of what motivated this research and its circumstances.

Software engineers would develop code for branch and bound large applications, using formal methodologies, with data structures, stacks, etc. In our case, we did something more direct, creating Program 8.3 and two auxiliary functions. The capabilities of this program are limited, but it allows to check most of the academic examples of integer programming you can find. The chief intention of the program is to show the importance of guiding the use of *linprog()*, in order to obtain integer solutions, by opportune specification of upper and lower bounds (*lb, ub, lb1, ub1, lb2, ub2*, in the code).

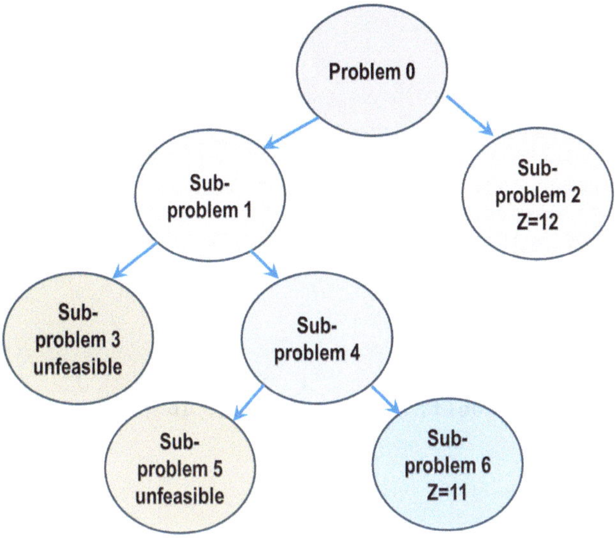

Fig. 8.13 Branch and bound exploration of the example

The structure of the main program corresponds to a tree (with only three levels, as in Fig. 8.12). A key point is to maintain a record of the pivots (where the branching occurs) that appear along the method application; we devised two matrices, *Pval* and *Pix* for this purpose. Indeed, more levels could be (easily) added to the three, increasing the size of the code.

Program 8.3 Branch and bound cxamplc

```
%Branch and bound example
% calculates the minimum solution made of integers
%
% preparation of variables
% global variables
global f; global A; global B;
global Aeq; global Beq;
% Pivots
Pval=zeros(6,12); Pix=zeros(6,12);
nPval=zeros(6,12); nPix=zeros(6,12);
% initialize
niter = 1;
flg_stop=0;
%
%The problem ---------------
% function to be minimized (Winston book)
f = [-8 ; -5];
% constraints
A = [1 1; 9 5];
```

```
B = [6; 45];
Aeq = [];
Beq = [];
% LP bounds:
global lb1; global ub1; global lb2; global ub2;
nvar=2;
lb = zeros(nvar, 1);
ub = (10000)*ones(nvar, 1);
lb1 = lb; lb2 = lb;
ub1 = ub; ub2 = ub;
%
% first application of LP --------------------------
flag_intg = 1;
options = optimoptions('linprog','Display','none');
[X, v, flg] = linprog(f, A, B, Aeq, Beq, lb, ub,...
options);
% solution found or not
if (flg == 1)
  disp('initial LP solution exists:')
  X
  v
else
  disp('no initial LP solution')
  return
end
% see if any variable not integer
% (flag_intg=1, means all-integer)
[flag_intg, var_val, var_index]= fbab_int(X);
% a solution from the very start?
if (flag_intg == 1)
  disp('good initial solution')
  X
  v
  return
end
% continue with branch and bound -------------------
disp('**')
disp('level a) +++++++++++++++++++++++')
Fla=0; Fua=0;
L=1; J=1; %first level, leftmost pivot
Pval(L,J)=var_val; Pix(L,J)=var_index; %first pivot
%generate two branches:
[Fla,Fua,nPval,nPix]=f2bb(Pval,Pix,L,J);
Pval=nPval; Pix=nPix; %new two pivots
% level b ----------------------
disp('**')
disp('level b) +++++++++++++++++++++++')
Flb1=0; Flb2=0; Fub1=0; Fub2=0;
if(Fla==1)
  L=2; J=1; %b-left
  lb2=lb; ub2(Pix(1,1)) = Pval(1,1);
  %generate two branches (level b-left):
  [Flb1,Fub1,nPval,nPix]=f2bb(Pval,Pix,L,J);
```

```
  Pval=nPval; Pix=nPix; %new two pivots
end
if (Fua==1)
  L=2; J=2; %b-right
  ub1=ub; ub2=ub;
  lb1(Pix(1,1)) = Pval(1,1)+1;
  lb2(Pix(1,1)) = Pval(1,1)+1;
  %generate two branches (level b-right)
  [Flb2,Fub2,nPval,nPix]=f2bb(Pval,Pix,L,J);
  Pval=nPval; Pix=nPix; %new two pivots
end
% level c ----------------------
disp('**')
disp('level c) ++++++++++++++++++++++')
% c-left (two branches)
if(Flb1==1)
  L=3; J=1; %c-left-left
  lb1=lb; lb2=lb; ub2=ub;
  ub2(Pix(2,1)) = Pval(2,1);
  %generate two branches:
  [Flc1,Fuc1,nPval,nPix]=f2bb(Pval,Pix,L,J);
end
if(Fub1==1)
  L=3; J=2; %c-left-right
  lb1=lb; lb2=lb; ub2=ub;
  ub2(Pix(2,1)) = Pval(2,1);
  %generate two branches:
  [Flc2,Fuc2,nPval,nPix]=f2bb(Pval,Pix,L,J);
end
% c-right (two branches)
if(Flb2==1)
  L=3; J=3; %c-right-left
  lb1=lb; lb2=lb; ub2=ub;
  ub2(Pix(2,2)) = Pval(2,2);
  %generate two branches:
  [Flc3,Fuc3,nPval,nPix]=f2bb(Pval,Pix,L,J);
end
if(Fub2==1)
  L=3; J=4; %c-right-right
  lb1=lb; lb2=lb; ub2=ub;
  ub2(Pix(2,2)) = Pval(2,2);
  %generate two branches:
  [Flc4,Fuc4,nPval,nPix]=f2bb(Pval,Pix,L,J);
end
```

Notice that the program considers an example of [105] (page 513), which is Maximize:

$$z = 8x_1 + 5x_2$$

Subject to:

$$x_1 + x_2 \leq 6$$

$$9x_1 + 5x_2 \leq 45$$

with x_1, x_2 being non-negative integers.

The function *f2bb()* (code below) uses two times *linprog()* to compute the two branches deriving from a pivot and exploits the information given by *linprog()* about feasibility or not of a solution. Function *f2bb()* also detects that an all-integer solution has been obtained, by using the function *fbab_int()* , to be presented next. The comments in the code uses the term "*fathomed*," typically used in the context of branch and bound to indicate a node with no successors.

Program 8.4 Function for branch and bound

```
%Function
% compute two branches, and analyse them
function [Fgl,Fgu,nPV,nPX] = f2bb(PV,PX,L,J)
global f; global A; global B;
global Aeq; global Beq;
global lb1; global ub1; global lb2; global ub2;
Fgu=1; Fgl=1; % indicate no-phatomed node
K=1+((J-1)*2); %horizontal index for pairs
nPV=PV; nPX=PX;
nPV(L+1,K)=-1; nPX(L+1,K)=-1; %phatomed(default)
nPV(L+1,K+1)=-1; nPX(L+1,K+1)=-1; %phatomed(default)
%
%-------------------------------------------
%take info from pivot input:
var_index = PX(L,J); var_val= PV(L,J);
ub1(var_index) = var_val;
lb2(var_index) = var_val+1;
options = optimoptions('linprog','Display','none');
%
%lower branch:
[X1, v1, flg1] = linprog(f, A, B, Aeq, Beq, lb1,...
 ub1,options);
%upper branch
[X2, v2, flg2] = linprog(f, A, B, Aeq, Beq, lb2,...
 ub2,options);
%
if(flg1==1) %solution exists
  %see if any X1 variable not integer
  [flg_intg1, V1_val, V1_idx]= fbab_int(X1);
  if (flg_intg1 == 1)
    disp('INTEGER SOLUTION -----')
    X1
    v1
    Fgl=0; %fathomed node
```

```
    else
      disp('non-integer solution-----')
      X1
      nPV(L+1,K) = V1_val; %new lower pivot
      nPX(L+1,K) = V1_idx;
    end
  else
    Fgl=0; %fathomed (infeasible)
    disp('$$>> no feasible solution found')
  end
  if(flg2==1) %solution exists
    %see if any X2 variable not integer
    [flg_intg2, V2_val, V2_idx]= fbab_int(X2);
    if (flg_intg2 == 1)
      disp('INTEGER SOLUTION -----')
      X2
      v2
      Fgu=0; %fathomed node
    else
      disp('non-integer solution -----')
      X2
      nPV(L+1,K+1) = V2_val; %new upper pivot
      nPX(L+1,K+1) = V2_idx;
    end
  else
    Fgu=0; %fathomed (infeasible)
    disp('$$>> no feasible solution found')
  end
  return
```

The function *linprog()* could arrive to floating results close to integer values, instead to completely integer values. The function *fbab_int()* (code below) detects this situation and returns the corresponding *floor()* integer value.

Program 8.5 Function that detects non-integer variables

```
%Function
% detect non-integer variables
% the first non-integer is detected,
% and varVal returns its floor value
function [flgInt, varVal, varIx] = fbab_int(X)
flgInt=1; varVal=0; varIx=0; %for all-integer case
k=1; jj=1;
while (k == 1)
  aux1=round(X(jj));
  tol = 0.001;
  if ((X(jj) > aux1+tol) | (X(jj) < aux1-tol))
    flgInt = 0; %non-integer
    varVal = floor(X(jj));
    varIx = jj;
    k=0;
  end
```

```
  if (jj==2)
    k=0;
  else
    jj=jj+1;
  end
end
return
```

The reader could explore with the program academic examples from literature or others of own interest. If any failure was found, please give us some feedback.

Next sections of this chapter will focus on relevant types of integer programming applications.

8.3 Facility Location

The question is to locate facilities in convenient places. It is in reality a large topic, with many branches and related aspects.

Figure 8.14 illustrates the problem.

The following introductory example was taken from [105] (a thick, most cited, useful book):

There are six cities in a certain County. The County has to decide where to build fire stations. At least one fire station should be in within 15 min driving from each city. The question is how many fire stations are needed and where to locate them.

Table 8.1 gives the necessary data (time in minutes):

Let us translate the requirements to mathematical terms. For this purpose, binary variables with values 0 or 1 will be opportune. This kind of variables is typical of decision-making. Variable x1 corresponds to city 1, being 0 if no fire station was to be built there, or 1 if yes. Variables x2, x3... x6 correspond to the other cities.

The target would be to minimize:

$$J = x_1 + x_2 + x_3 + x_4 + x_5 + x_6$$

Fig. 8.14 Metaphor about facility location problem

Table 8.1 Time required for driving between cities

	To					
From	City 1	City 2	City 3	City 4	City 5	City 6
City 1	0	10	20	30	30	20
City 2	10	0	25	35	20	10
City 3	20	25	0	15	30	20
City 4	30	35	15	0	15	25
City 5	30	20	30	15	0	14
City 6	20	10	20	25	14	0

Now, the constraints. To ensure attention to city 1 within 15 min. driving:
$x_1 + x_2 \geq 1$ (see first column of Table 8.1)
(Notice that the solution $x_1 + x_2 = 0$ becomes impossible).
Pertaining to city 2, one has the following constraint:
$x_1 + x_2 + x_6 \geq 1$ (second column of Table 8.1)
And so on with the rest of cities:

$$x_3 + x_4 \geq 1$$

$$x_3 + x_4 \geq 1$$

$$x_4 + x_5 + x_6 \geq 1$$

$$x_2 + x_5 + x_6 \geq 1$$

One optimal solution is $x_2 = x_4 = 1$, $x_1 = x_3 = x_5 = x_6 = 0$. And, therefore, only two fire stations are needed (in cities 2 and 4).

This example belongs to a general class of problems, known as **set-covering problems.** The tutorial [27] briefly describes the topic. Each member of a set A must be covered by a suitable member of set B. For example, set B could be a set of radio antennas and set A a number of smartphones in need of coverage (given by antennas).

It is easy to imagine several scenarios of set coverage, like wi-fi systems, camera surveillance, water distribution, satellite constellations, etc. Actually, there are several variants of the problem and—in parallel—a number of optimization algorithms, exact or approximate. The page https://fastercapital.com/topics/solving-the-set-covering-problem-using-mathematical-optimization-techniques.html offers a sort of compilation of related aspects. Indeed, in the last decades, many bio-inspired algorithms have treated the question, while until the 1990s (*grosso modo*) other more analytical algorithms were proposed (see the survey [101]).

But let us focus now on facility location.

In general, facilities should offer a good service. Commercial facilities want to attract clients and keep the clients satisfied. Distances, costs, and other factors should be considered.

A basic problem is to place a single facility in the presence of several destinations. Each destination has a different associated transportation cost. It could be convenient to determine a location such as to minimize the sum of destination weighted costs. Other possibility is the minimax approach: to find a location such as to minimize the maximum destination weighted cost. The page https://scipbook.readthedocs.io/en/latest/flp.html introduces some variants of the problem.

Most distribution companies use several facilities. The theory has differentiated two cases: uncapacitated and capacitated facilities. We will follow the guide offered by [99] (which includes some code in Julia language) for describing those two scenarios.

8.3.1 Uncapacitated Facility Location

In the uncapacitated facility scenario, we do not have to consider limits on capacity (stocks) of the facility nor on the demand part.

Let us specify the question.

We are given:

- A set $M = \{1, 2, \ldots, m\}$ of clients
- A set $N = \{1, 2, \ldots, n\}$ of sites where a facility can be built

We will consider two types of binary decision variables:

- y_j, indicates whether facility j is built or not
- $x_{i,j}$ indicates whether client i is assigned to facility j

The target of optimization would be to minimize the cost of serving all clients.

This cost includes the fixed cost, $f_j = 1, \forall j$, of building a facility, and the cost of serving clients from the assigned facility ($c_{i,j}$ would be the cost of serving client i from facility j).

The constraints would be that a facility can only serve when it is open and that each client must be served by exactly one facility.

We are now in conditions to write the mathematical statement:

Minimize:

$$J = \sum_{i,j} c_{i,j} x_{i,j} + \sum_{j} f_j y_j$$

Subject to:

$$i \in M; \; j \in N$$

$$\sum_{j} x_{i,j} = 1$$

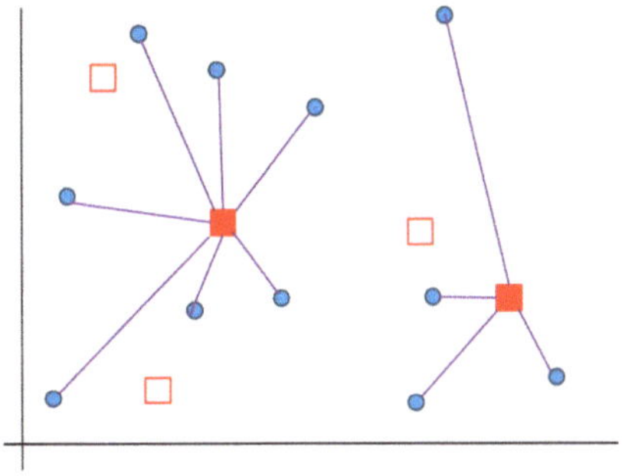

Fig. 8.15 Solution of a facility location problem

$$x_{i,j} \le y_j$$

$$x_{i,j}, y_j \in \{0, 1\}$$

Figure 8.15 visualizes a solution of a certain facility location problem. Facilities are represented by squares and clients by circles situated at given distances from the facilities. A number of facility locations were discarded, as represented by empty squares, while other locations were chosen, as represented by solid squares.

8.3.2 Capacitated Facility Location

In the capacitated facility case, each facility has finite capacity (that cannot be exceeded), and clients have a certain level of demand to be served.

Then,

- Client i has a demand $a_i \ge 0$
- Facility j has a capacity $q_j \ge 0$

Having those limits in consideration, the problem is

Minimize:

$$J = \sum_{i,j} c_{i,j} x_{i,j} + \sum_j f_j y_j$$

Subject to:

$$i \in M; \; j \in N$$

$$\sum_j x_{i,j} = 1$$

$$\sum_i a_i x_{i,j} \leq q_j y_j (\text{capacity constraint})$$

$$x_{i,j}, y_j \in \{0, 1\}$$

In the case of capacitated facilities, the solution can be visualized as in Fig. 8.15, with the following change: facilities with large capacity are represented by large squares, while less capacity is represented by smaller squares. Also, clients demanding more are represented by larger circles, while those with less demand are painted as smaller circles.

The page: https://towardsdatascience.com/optimization-capacitated-facility-location-problem-in-python-57c08f259fe0 presents an example of facility locations for the entire Italy. Many variables, and a detailed treatment (including code in Python language).

The Documentation web site of MATLAB include two specific pages for "Factory, Warehouse, Sales Allocation Model." One of these pages, [79] is devoted to a solver-based approach. The other page, [78] focuses on a problem-based approach. Both pages include detailed commented coding in MATLAB.

One of the packages available from the MathWorks File Exchange site is "FLO-Facility Location Optimizer." It is related to a research project of the Institute for Mathematics of the Martin Luther University Halle-Wittenberg.

Another page with MATLAB code for optimal facility location is [65].

For diving more profoundly in theoretical aspects, see [17] and references therein. Add [60] (1,582 cites) for extensive details.

8.4 Capital Budgeting

This is a classical application that appears in many textbooks, like [105]. For example, consider the case summarized in Table 8.2.

The investor studies four possibilities, having into account that a maximum of $14,000 is available for investment.

As before, binary 0–1 variables will be used for representing decisions.

Then, the net present value (NPV) that the investment would get is

$$J = 16x_1 + 22x_2 + 12x_3 + 8x_4$$

Table 8.2 A basic investment scenario

	Net present value	Cash outflow
1	$16,000	$5,000
2	$22,000	$7,000
3	$12,000	$4,000
4	$8,000	$3,000

Table 8.3 Investment along 3 years

Project	Return	Inv, year 1	Inv. year 2	Inv. year 3
1	0.2	0.5	0.3	0.2
2	0.3	1.0	0.8	0.2
3	0.5	1.5	1.5	0.3
4	0.1	0.1	0.4	0.1
Available capital		3.1	2.5	0.4

The target of the optimization would be to maximize the NPV obtained.

There is a constraint imposed by the $14,000 capital to invest, at most. This constraint can be expressed as follows:

$$5x_1 + 7x_2 + 4x_3 + 3x_4 \leq 14$$

As shown in [105], the optimal solution is $x_1 = 0, x_2 = x_3 = x_4 = 1$, leading to $J = 42,000\$$.

Integer programming problems like the example just posed, with only one constraint, belong to the class of knapsack problems.

The following example is taken from the page [7], which is part of a very interesting set of pages authored by J.E. Beasley. The example contains three columns, as shown in Table 8.3.

The total return to be optimized is given by:

$$J = 0.2x_1 + 0.3x_2 + 0.5x_3 + 0.1x_4$$

The constraints concerning the availability of capital funds per year are the following:

$$0.5x_1 + 1.0x_2 + 1.5x_3 + 0.1x_4 \leq 3.1$$

$$0.3x_1 + 0.8x_2 + 1.5x_3 + 0.4x_4 \leq 2.5$$

$$0.2x_1 + 0.2x_2 + 0.3x_3 + 0.1x_4 \leq 0.4$$

The solution Is:

Projects 3 and 4, yes. Projects 1 and 2, no. Return: 0.6.

In the web page [36], there is a detailed tutorial on how to use integer linear programming for capital budgeting optimization.

Taking as reference the capital budgeting problem paradigm, there is a variety of applications that can be considered in a similar way, like the examples described in [15].

In the financial world, capital budgeting constitutes a complex scenario, being the subject of several treatises on the associated theory and practices, like for example the book [86]. The article [58] presents an impressive review of the research on this topic.

The GitHub page at https://saylordotorg.github.io/text_managerial-accounting/s12-how-is-capital-budgeting-used-.html, explains with some detail how to use capital budgeting to make decisions.

8.5 The Travelling Salesman Problem (TSP)

This section considers an important classical problem, called "**The travelling salesman problem (TSP)**," which also appears in other moments of this book. Tons of (now digital) ink have been used to write in relation with the TSP.

Here is the problem: starting from his home, the salesman wishes to visit several cities and return home at minimal cost. He must visit each city once. A suitable tour should be found.

An intuitive way for describing the problem is through a weighted graph, like the one represented in Fig. 8.16.

In the theory of graphs, a path is a walk in which no vertices are repeated (the path may not visit all vertices). A Hamiltonian path visits—only once—all vertices of the graph, [82].

A cycle is a walk that starts and ends at the same vertex, visiting only once other vertices. A Hamiltonian cycle visits all vertices of the graph.

That is, paths are open; cycles are closed.

If a graph contains a Hamiltonian cycle, then it is a Hamiltonian graph.

In the case of TSP, we are dealing with a weighted graph. One could list all possible Hamiltonian cycles and find the cycle with the least weight. With 4 vertices, we have 6 possible cycles, with 5 vertices, 24 possible cycles.

Using binary variables $x_{i,j}$ that connect (or not), city i with city j, the TSP problem can be described as an integer programming problem where the total

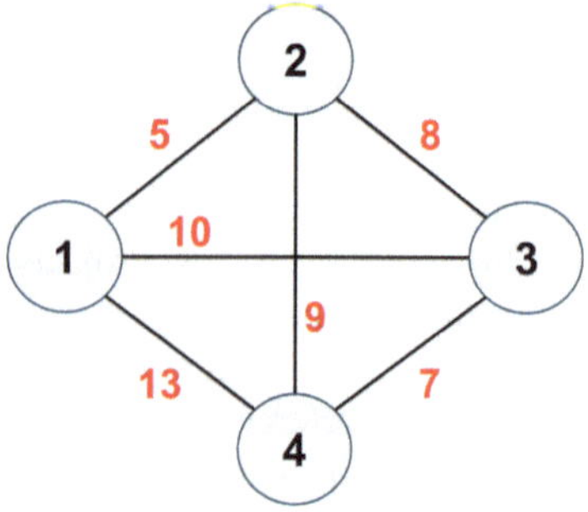

Fig. 8.16 A weighted graph with all connections between four cities

distance of the tour should be minimized. That is:

Minimize:

$$J = \sum_{i} \sum_{j \neq i} c_{i,j} \, x_{i,j}$$

where $c_{i,j}$ is the distance between cities i and j.

A set of constraints should be specified. The most evident are the following:

$$\sum_{i,\, i \neq j} x_{i,j} = 1, \quad \forall j$$

$$\sum_{j,\, j \neq i} x_{i,j} = 1, \quad \forall i$$

since the salesman must enter and leave each city exactly once.

Under these constraints, in the case of six cities, for example, two disjoint sub-tours of three cities each could be found as solution. Other more complicated examples, with more cities and much more sub-tours could happen.

However, only a single tour is what we want. In consequence, additional constraints are required. Actually, several proposals have been suggested along the years.

A little bit of history would be appropriate now. First of all, a relevant fact is that in the 1800s, the famous scientist R.W. Hamilton, and T.P. Kirkman, started to treat a problem similar to the TSP (it was related to a game). In the 1930s, K. Menger, H. Whitney, and M. Flood established more formally the TSP problem.

A naïve strategy for reaching a solution is to go from home to the nearest city; then, once there, repeat the idea, go to the nearest city, etc. It is K. Menger who observes that, in general, this strategy *does not* yield the optimal route.

The term "travelling salesman problem" was first published by Julia Robinson in a 1949 RAND Corporation report. RAND offered prizes for solving the problem, and several scientific contributions came.

In 1954, G. Dantzig, D.R. Fulkerson, and S, Johnson, at that time working with RAND Corporation, developed code for solving the problem, considering it to be a case of integer linear programming and using a cutting plane method [31] (3,589 cites). They were able to tackle tours between 49 cities. A constraint was added for *sub-tour elimination*, establishing the DFJ formulation of the TSP problem.

Notice that behind the TSP, there is a combinatorial explosion scenario. The time to solve a TSP by simple enumeration is proportional to $n!$ (n, the number of cities). Supposing a powerful computer able to do 10^{12} operations per second, a 50 city TSP will take around 3 x 10^{52} seconds. The life of the universe could be estimated as 10^{17} seconds.

The DFJ subtour elimination constraint admits various similar forms in the scientific literature. For example:

$$\sum_{i,j \in S} x_{i,j} \le |S| - 1, \ \ \forall S \subset V, \ \ 2 \le |S| \le n - 1$$

where V is the set of vertices (nodes).

For example, for a subset S of nodes with indices {2, 3, 6}, |S| = 3, this constraint would be

$$x_{3,2} + x_{2,6} + x_{6,3} + x_{2,3} + x_{6,2} + x_{3,6} \le 2$$

A tour among {2, 3, 6} implies that the sum would be at least 3. As the constraint limits this sum to be ≤ 2, it forbids that subtour.

Other possibility to be avoided is subtours of two nodes in S. But this is dealt by constraints where S is {2,3}, {2,6}, and {3,6}.

The number of constraints grows exponentially as the number of cities increases. Still, the problem can be managed if, instead of adding constraints for all possible subsets, only some selected constraints were considered. Departing from a solution of the problem with no subtour elimination constraints, it is not difficult to find the subtour elimination constraints being violated. Then, we add those constraints and solve again the problem. This is iterated until reaching the optimal tour.

A new way for dealing with no subtours was proposed by C.E. Miller, A.W. Tucker, and R.A. Zemlin [80] (2,771 cites). It is called the MTZ formulation.

A dummy variable, u_i, is added for keeping track of the order in which cities are visited.

It is desired to have $u_j \ge u_i + 1, \ if \, x_{i,j} = 1$. One could try to express it with $u_j \ge u_i + x_{i,j}$, but this would say that $u_j \ge u_i$ in case of $x_{i,j} = 0$, which is not right. This is why the MTZ formulation proposes the following constraints:

$$u_j + (n-2) \ge u_i + (n-1)x_{i,j}, \ \text{for all distinct} \ \ i, j \in (2, \ldots, n)$$

Hence, the complete MTZ formulation is

Minimize:

$$J = \sum_{i} \sum_{j \ne i} c_{i,j} \, x_{i,j}$$

Subject to:

$$\sum_{i, i \ne j} x_{i,j} = 1, \ \ \forall j$$

$$\sum_{j, j \ne i} x_{i,j} = 1, \ \ \forall i$$

$$u_i - u_j + (n-1)\,x_{i,j} \le n-2, \quad 2 \le i \ne j \le n$$

$$2 \le u_i \le n$$

The MTZ formulation introduces significantly less constraints than the DFJ formulation. Many publications refer to MTZ as soft and to DFJ as tight. When using MTZ, one could expect a larger number of branch and bound steps to reach a solution, compared to the DFJ approach.

Many mathematicians, and computer science people, have confronted the TSP problem, which is now a kind of benchmark, both for deterministic and stochastic algorithms. The site [23] has registered a lot of information about TSP and the associated history of milestones. For instance, in 1987, it was possible to algorithmically solve a TSP with 2,392 cities. In 2001, the number of cities increased to 15,112. In 2006, it was 85,900 cities (well, in reality the target of the algorithm was no longer cities but the internal architecture of microchips).

Taken from [23], Fig. 8.17 presents a view of the TSP problem solved by Applegate et al. in 2004. The optimal tour included 24,978 cities of Sweden.

Trying not to get the exact solution, but a very good approximation, some stochastic (mostly bioinspired) have been developed and successfully applied to problems with millions of cities, with solutions within 2–3% of the exact optimal tour.

Monster problems, in the form of challenges, have been announced. Like the World TSP, with 1,904,711 cities (see again the site [23]) or the Santa Claus TSP challenge [75]. The Galaxy TSP, [34], is based on an Astronomical data set with 1.69 billion stars.

Fig. 8.17 A view of the 24,978 cities of the Sweden TSP tour (2004)

For more details on the problem, the principal algorithms that have been introduced, and special cases, it would be recommended the excellent page of Wikipedia on the travelling salesman problem. In addition, the scientific articles [63], [87] and [51], offer relevant views of algorithms and the state of the art.

The long article [16] presents an interesting recollection of well-solvable special cases of TSP. Some of those cases are, pyramidal tours, the master tour problem, the bottleneck TSP, the Euclidean TSP (which includes the necklace TSP), the permuted Monge TSP, the TSP with Gilmore-Gomory matrices, etc. In some sense, with this article, one feels the research history that is behind.

Beyond real geographical tour design problems, there are many (sometimes unsuspected) applications of TSP. Books like [88] or [32], devote long sections to such topic. The article [77] complements these descriptions with a broad overview of applications.

An important, industrial application of TSP is related to drilling holes in electronic printed boards (for instance computer motherboards), [70]. Another contemporary application is vehicle routing for drones [59]. And with more proximity to the original TSP problem, there are applications for tourism [54].

As the number of different applications increases, so the variants of the TSP problem and algorithms grow and diversify (see [52] for a survey of variants).

The MathWorks File Exchange site offer some "traditional" programs for TSP, using, for example, branch and bound and lots of stochastic (bio-inspired) algorithms for TSP. The MATLAB Optimization Toolbox deals in particular with the TSP problem.

In GitHub, one finds hundreds of software repositories, with a similar scene: a few traditional programs and many stochastic approaches.

The TSPLIB site http://comopt.ifi.uni-heidelberg.de/software/TSPLIB95/ offers a library of TSP and related problems.

8.6 The Cutting Stock Problem (Column Generation)

As it is said in [48], cutting stock problems occur in a wide variety of industries, and there is a large economic incentive to find better solutions.

This optimization problem that tries to minimize waste provides a good opportunity for introducing the *column generation* methodology.

8.7 The Cutting Stock Problem

Suppose that the ACME company, which produce big rolls of paper in a uniform width of 100 inches, received several orders of rolls of width 14 inches, 31 inches, 36 inches, and 45 inches. See Fig. 8.18 to get an idea of the size of industrial rolls of paper.

Fig. 8.18 Industrial rolls of paper

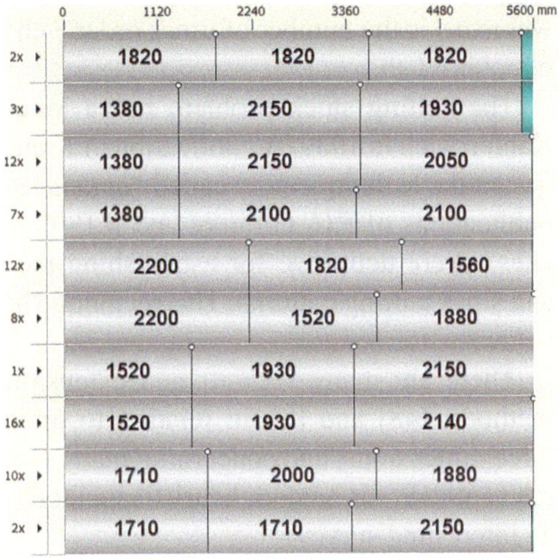

Fig. 8.19 Cutting planification

A single 100 inch roll can be cut into one or more widths; for example, it could be cut into two rolls of 45 inches with a 10 inch roll of scrap.

Here is, for example, a list of current orders [83]:

Order width	Quantity ordered
14	215
31	413
36	567
45	85

The problem is to satisfy all demands with a minimum of scrap.

The common procedure for this application is to establish a list of combinations of cuts, called “*patterns*” or “*configurations*,” and then associate with each pattern an integer variable x_i that represents how many times pattern *i-th* would be used.

Figure 8.19, adapted from [104], visualizes an example of planification of the cuts, with the number of times the different patterns were used.

In mathematical terms, the optimization problem is
Minimize:

$$J = \sum_i c_i \, x_i$$

Subject to:

$$\sum_i a_{i\,j} \, x_i \geq q_j \quad \forall j = 1, \ldots, m$$

where a_{ij} is the number of times order j-th appears in pattern i-th; and c_i is the cost of pattern i-th.

The cutting stock problem has a long story. It was first formulated by L.V. Kantorovich in 1939. A publication of Kantorovich and V.A. Zalgaller, in 1953, suggested the technique now called "*column generation*" for solving the problem; two publications of P.C. Gilmore and R.E. Gomory [41] (2,760 cites) and [42] (1,680 cites), were definitive for establishing the method, based on linear programming and stock cutting patterns. This technique has become popular and will be described below.

There are many variants and extensions of the cutting problem and solution methods, [48] and [8], in relation to optimal cutting in two dimensions or even three dimensions, and concerning many types of solid and liquid materials. This is a relevant field for modern heuristic methods (see, for instance, [108], which includes an interesting survey).

8.7.1 Column Generation

Some optimization problems, when treated with linear programming, may lead to a lot of matrix columns to be considered, so it could be (very) cumbersome to look at all possible columns from the beginning. The method of column generation proceeds column by column. It starts from a subset of the columns and then asks whether any column not in that subset has *negative reduced cost*. After locating the best of these columns with negative reduced cost, that column is added to the subset, and we reoptimize.

Let us briefly introduce the mathematical details.

Recall from the chapter on linear programming the section on duality, in which a primal problem was mentioned in the following form:

Minimize $z = \vec{c}^T \cdot \vec{x}$, subject to $A\,\vec{x} = \vec{b}\,,\; \vec{x} \geq \vec{0},\; \vec{b} \geq \vec{0}$

For which the *dual* would be:

Maximize $p = \vec{b}^T \cdot \vec{y}$, subject to: $A^T \, \vec{y} \leq \vec{c}$, $\vec{y}\; free$

The column generation algorithm starts from a subset of A and calculates the corresponding dual solution $\vec{y}$. This is called the *master problem*.

Then, one determines the column not in A with most negative reduced cost. The reduced cost associated with column A_j is $c_j - \vec{y}A_j$. This is the *column generation subproblem*, also denoted as the *pricing problem* (being, in turn, an optimization problem, similar to the knapsack problem).

The new column is added to A, unless no column was found with negative reduced cost, meaning the end of the iterations.

See [33] for more details on the method. The chapter [73] (1,490 cites) offers a more extended exposition of theoretical and practical aspects involved in column generation. The authors of these publications insist in its real importance when the method is combined with branch and bound, which then becomes *branch and pricing*.

In words of [73], the cutting stock problem is a classic example in the literature on column-generation methodology.

8.8 More About (Algorithmic) Cuts

In the last section, we treated the optimal physical cutting of paper rolls. Now, in this section, we come back to theory, continuing with the topic of shrinking the feasible space exploration by means of cuts.

Along time, several ways of using cuts have been introduced, since this general approach has proved to be effective for large problems.

The purpose of this section is to consider some formalization and variants of the method.

A first point is related to convex regions. This is an important aspect that will be treated in more detail in a next chapter. For the moment, let us express the main concept: in a convex region, any straight segment connection between two arbitrary points of the region is inside the region.

Figure 8.20 shows one example of convex region and one example of nonconvex region.

Some of the figures in previous pages of this chapter, like for instance Fig. 8.2, shows a grid of points (all-integer solutions) inside a feasible region defined by constraints with real-valued variables. Since our target is to determine the best all-integer solution, our exploration space should not go far from the points of interest

Fig. 8.20 Examples of convex and nonconvex regions

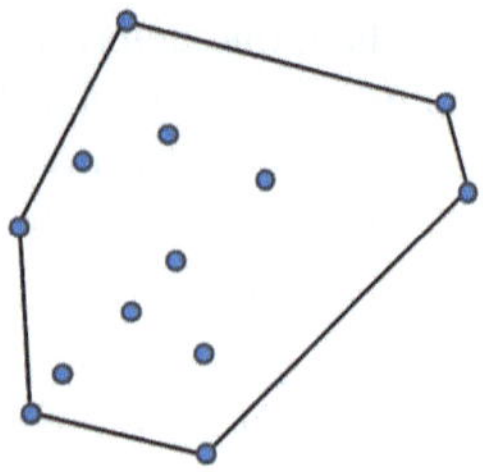

Fig. 8.21 Example of convex hull

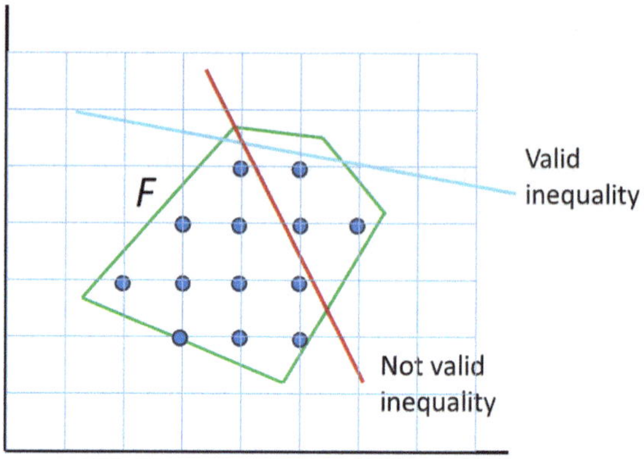

Fig. 8.22 Valid and not valid inequalities

(the grid). Here it is opportune to subscribe to the concept of *convex hull* (also called convex envelope or convex closure).

Given a set of points, the procedure for obtaining the corresponding convex hull is to form a convex polygon by joining frontier points, so all points are contained by the polygon. Figure 8.21 presents an example. Supposing the set is denoted as S, its convex hull can be denoted as *conv(S)*.

It is not necessary to examine one by one the points of S to determine the optimal. The solution would belong to the points forming *conv(S)*.

However, in many cases, the determination of *conv(S)* requires considerable computational effort. It is by using cuts that *conv(S)* could be approximated.

Since cuts are introduced in the problem setting by adding inequalities, let us zoom on inequalities. Some definitions would contribute to a clearer view.

An inequality is said to be *valid inequality* if it keeps all integer solutions in the feasible region F. Figure 8.22 shows a valid inequality and a not valid inequality [19].

A cut is a valid inequality that separates the current fractional optimum from F. See Fig. 8.23.

Sometimes the publications consider cuts being more or less strong. Stronger cuts remove more area of F.

The so-called *Chvátal-Gomory cuts* (CG-cuts) push the inequality into F until it hits an integer solution.

The convex hull can be approximated by possibly infinitely many CG-cuts, forming the Chvátal closure.

Fig. 8.23 A cut

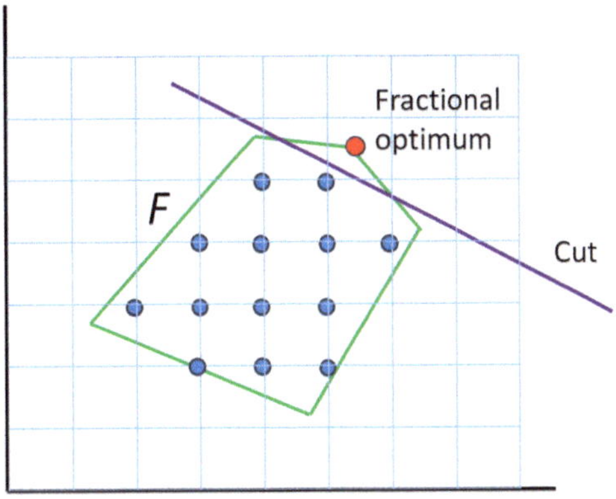

Václav Chvátal, born and educated in Prague, occupied several academic positions in US universities (he had to escape from the Soviet invasion of Czechoslovakia in 1968). He is the author of a famous book on linear programming. In 1973, he published an important article [20] (859 cites) that introduced fundamental concepts on cutting theory, in the line just sketched in this section. This article has been reprinted in 2006. Chvátal did also important contributions to graph theory and combinatorial optimization.

An important observation, in order to create the cuts, is the following: if $x \leq b$ and b is fractional, we can replace the inequality with:

$$x \leq \lfloor b \rfloor$$

Based on this fact, it is possible to apply *integer rounding* as shown in the next example [39]. Suppose the following constraints:

$$x_1 - 2x_2 \geq -5$$

$$2x_1 + 2x_2 \leq 9$$

$$6x_1 + 2x_2 \leq 21$$

(where x_1 , x_2 non-negative integers).

Note that $2x_1 + 2x_2 \leq 9$ can be written as $x_1 + x_2 \leq 4.5$. Then we can replace it with

$$x_1 + x_2 \leq 4$$

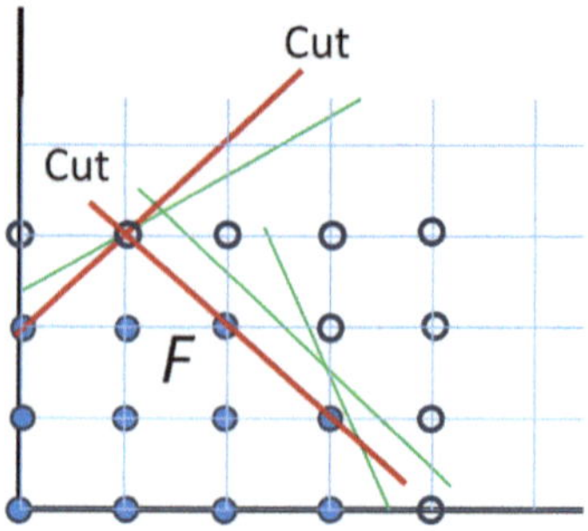

Fig. 8.24 Result of integer rounding: original problem and the new cuts

Likewise, consider $x_1 - 2x_2 \geq -5$, that we re-write as $-x_1 + 2x_2 \leq 5$. Multiply it by 1/2, getting $-1/2x_1 + x_2 \leq 5/2$. Then $-x_1 + x_2 \leq 5/2$ is a valid inequality. Now, apply integer rounding to obtain:

$$-x_1 + x_2 \leq 2$$

In this last case, rounding was used in both sides of the inequality. This is called the *Chvátal-Gomory procedure*.

Figure 8.24 visualizes the feasible region F of the original example and the result of the cuts that have been obtained by integer rounding.

The Chvátal-Gomory procedure can be generalized in several ways. Given the integer set:

$$X = \left\{\mathbf{x} \in Z_+^n \mid \mathbf{A}\,\mathbf{x} \leq \mathbf{b}\right\}$$

where $\mathbf{A}$ is a $m \times n$ matrix and $\mathbf{u} \in R_+^m$; the following inequalities are valid:

$$\sum_{i=1}^{n} \mathbf{u}^{\mathbf{T}} a_i\, x_i \leq \mathbf{u}^{\mathbf{T}}\mathbf{b} \text{ since } \mathbf{u} \geq 0$$

$$\sum_{i=1}^{n} \left\lfloor \mathbf{u}^{\mathbf{T}} a_i \right\rfloor x_i \leq \mathbf{u}^{\mathbf{T}}\mathbf{b} \text{ since } \mathbf{x} \geq 0$$

$$\sum_{i=1}^{n} \left\lfloor \mathbf{u}^{\mathbf{T}} a_i \right\rfloor x_i \leq \left\lfloor \mathbf{u}^{\mathbf{T}}\mathbf{b} \right\rfloor \text{ since } \mathbf{x} \text{ is integer}$$

It is also possible to add the next propositions:

Proposition 1:

Let be the set $X = \{x \in R_+ ;\ y \in Z_+ \mid y \leq b + x\}$, then the inequality

$$y \leq \lfloor b \rfloor + \frac{x}{1 - f}$$

is valid for X.

Proposition 2:
Let be the set $X = \{x \in R_+;\ \mathbf{y} \in Z_+^2 \mid a_1 y_1 + a_2 y_2 \leq b + x\}$, then the inequality

$$\lfloor a_1 \rfloor\, y_1 + \left(\lfloor a_2 \rfloor + \frac{f_2 - f}{1 - f} \right) y_2 \leq \lfloor b \rfloor + \frac{x}{1 - f}$$

is valid for X.
(this proposition is of interest in mixed-integer scenarios).

8.9 Polyhedra, and Unimodular Matrices

In certain types of problems, the simplex algorithm (linear programming) can directly give optimal integer solutions. This section will explore a line of study that sheds light on this aspect.

Recall from the chapter on linear programming that the constraints define a feasible region in the form of a polyhedron (or polytope in N dimensions). Each vertex corresponds to a feasible solution. The simplex algorithm proceeds vertex after vertex, until reaching the optimal one.

Consider now the following definition: any matrix $A \in Z^{m \times n}$ (all entries are integer) of full row rank is said to be ***unimodular***, if every $m \times m$ basis of A (full rank square submatrix) has determinant ± 1.

For that type of matrices, there is a theorem establishing the following: the polyhedron $P = \{\mathbf{x} : A\,\mathbf{x} = \mathbf{b},\ \mathbf{x} \geq 0\}$ is integral for every $\mathbf{b} \in Z^m$, if and only if A is unimodular.

In saying that P is integral, we say that all its vertices correspond to integer solutions.

The reader has surely noticed that the polyhedron P corresponds to the feasible region of a linear programming problem. The important point is that iff matrix A is unimodular, then all vertices correspond to integer feasible solutions.

More definitions: any matrix $A \in Z^{m \times n}$ is ***totally unimodular*** (TUM) if every square submatrix has determinant in $\{0, 1, -1\}$. In particular, all entries of A are in $\{0, 1, -1\}$.

Another theorem says that the polyhedron $P = \{\mathbf{x} : A\,\mathbf{x} \leq \mathbf{b},\ \mathbf{x} \geq 0\}$ is integral for every $\mathbf{b} \in Z^m$, if and only if A is totally unimodular.

See [100], and the book [93] (11,370 cites), for more theoretical details.

There are several important optimization standard problems in connection with TUM matrices.

One of these problems is the ***minimum cost network flow*** (MCNF), which has been already mentioned in the chapter on linear programming. The problem can be treated using a graph like the example represented in Fig. 8.25.

The objective is to find a feasible flow that satisfies all the demands at minimum cost. As represented in Fig. 8.25, each arc connecting two nodes has an associated

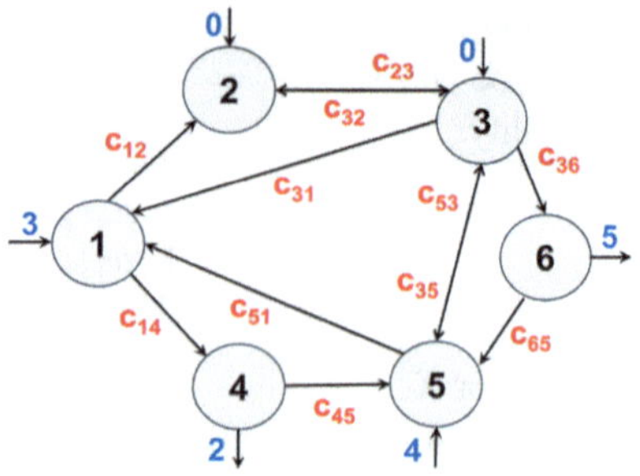

Fig. 8.25 Example of network flow

x12	x14	x23	x31	x32	x35	x36	x45	x51	x53	x65		
1	1	0	-1	0	0	0	0	-1	0	0	=	3
-1	0	1	0	-1	0	0	0	0	0	0	=	0
0	0	-1	1	1	1	1	0	0	-1	0	=	0
0	-1	0	0	0	0	0	-1	0	0	0	=	-2
0	0	0	0	0	-1	0	-1	1	1	-1	=	4
0	0	0	0	0	0	-1	0	0	0	1	=	5

Fig. 8.26 Constraint matrix corresponding to previous figure

cost c_{ij}. Under normalization of magnitudes, we can use variables x_{ij} with values in $\{1, 0, -1\}$ to describe the flow in each arc.

Hence, the optimization problem would be

Minimize:

$$J = \sum_{\text{all arcs}} c_{i,j}\, x_{i,j}$$

Subject to:

$$\sum_j x_{i,j} - \sum_k x_{k,i} = b_i, \quad \forall i \text{ (for each node i in the network)}$$

The table in Fig. 8.26 shows the constraints for the problem described—as a graph—in Fig. 8.25. It should be emphasized that it contains a TUM matrix, with all its entries in $\{0, 1, -1\}$.

Let us add, by the way, that not all matrices with entries in $\{0, 1, -1\}$ are TUM. In general, it is not trivial to determine if a matrix is TUM.

As several publications (for instance, [105], or the specialized [50]) remark, the transportation, assignment, transshipment, shortest path, and critical path problems are all special cases of the minimum-cost network, so their mathematical statements rely on TUM matrices.

8.10 Mixed-Integer Linear Programming

Many applications encounter a mix of continuous and integer variables. A general statement of the optimization problem would be:

Maximize:

$$z = \vec{c}^T \cdot \vec{x} + \vec{h}^T \vec{y}$$

Subject to:

$$A \vec{x} + D \bar{y} \leq \vec{b}$$

where $\vec{x}$ contains the integer variables and $\vec{y}$ the continuous variables.

The methods already introduced for solving linear integer optimization problems can be extended in some way for the mixed-integer field. It is now opportune to consider a simple example, like for instance [105]:

Maximize:

$$z = 2x_1 + x_2$$

Subject to:

$$5x_1 + 2x_2 \leq 8$$

$$x_1 + x_2 \leq 3$$

with x_1 integer, and x_2 continuous; both non-negative.

The geometry of this example is visualized in Fig. 8.27.

By using the linear programming simplex algorithm, one obtains the solution $x_1 = 2/3$, $x_2 = 7/3$, which gives $z = 11/3$. This solution is not valid since we want x_1 to be integer.

Then, let us try branch and bound. Because x_2 is permitted to be fractional, we do not branch on it. We will branch instead on x_1.

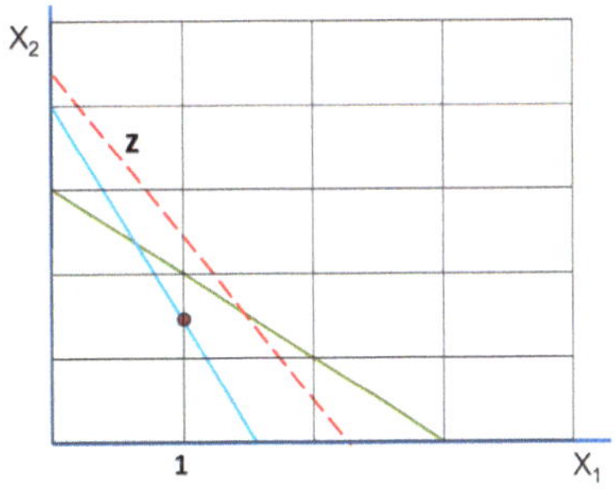

Fig. 8.27 Example of mixed-integer programming

One of the branches would take $x_1 = 0$ and via the simplex algorithm obtains the solution $x_1 = 0, x_2 = 3$, for which $z = 3$. The other branch obtains the solution $x_1 = 1, x_2 = 3/2$, with $z = 7/2$. This last solution is the best.

In some or all the branch and bound nodes, cutting planes could be added to obtain better bounds, so increasing the efficiency and velocity of the tree search. This is called ***branch and cut*** method, which is commonly used.

First initiatives about combining branch and bound, and cutting planes, come from the 1970s. Advancing toward maturity, an important contribution from M. Grötschel, M. Jünger and G. Reinelt appeared in 1984, [45] (477 cites), with application to linear ordering problems. The definitive step was due to M. Padberg and G. Rinaldi, in 1987 and 1991, establishing the full scheme of the method. In particular, the contribution of 1991, [85] (1,758 cites), focused on the travelling salesman problem, being able to cope with 2,392 cities.

When dealing with mixed-integer optimization problems, it is indispensable to look at main professional software efforts. In special, we should cite the CPLEX package. Comparing with nature, big rivers take their water from many affluents, similarly CPLEX embodies main contributions from the optimization community, and it is possible to trace the history of the mixed-integer optimization topic, from the evolution of CPLEX along the years (latest version of CPLEX is 22.1.1). In 1988, R.E. Bixby and J. Lowe started a company for promoting CPLEX, based on classroom code created by Bixby in 1982–1988. CPLEX was acquired by ILOG in 1997; and ILOG was acquired by IBM in 2009. Bixby himself published a short history of CPLEX in [13]. For a rapid and intense overview of what happened along 25 years of CPLEX evolution, we would recommend the presentation [14].

Apart from CPLEX, there are other solvers. For example, Xpress was also founded in the 1980s. Xpress was acquired in 1988 by FICO. When CPLEX was acquired by IBM, Bixby started another development, called Gurobi (now a direct competitor of CPLEX). From Denmark we have MOSEK and from China COPT and MindOpt (Alibaba Research). There are also some more, open-source, packages. See [97] for a commented list of solvers.

A comparison of solvers is given in [53]. Professor Hans Mittelmann, Arizona State University, maintains a benchmark comparing solvers; GitHub offers visualizations of Mittlemann benchmarks.

Professional solvers were frequently used for enormous problems, with perhaps thousands of variables and constraints. Every effort is made for getting satisfactory results in a short time. Based on [2] and other publications, let us survey the most significant ingredients of modern solvers.

8.10.1 Presolving

Nowadays, the common practice is to start with a suitable statement of the problem in hand, applying some **presolving** strategies. Special care should be taken about constraints, to facilitate a fast and straight optimization process.

As said in [40], the basic preprocessing aims at tightening the bounds of the variables, to fix variables that have no influence of the optimal solution, to remove surplus constraints, or to find out inconsistencies leading to infeasibility. More advanced presolving techniques not only remove garbage from specifications but also extract information to be exploited later in the optimization process.

Although it can be technically challenging, some probing could be tried, by fixing binary variables to zero and one, and propagating the effects through the constraints. The analysis of results could detect opportunities for faster processing.

A certain type of component analysis could detect equivalences of binary variables. In this case, such variables could be aggregated (simplifying the problem). Going farther, component analysis may observe a possible decomposition of the constraint matrix into blocks, providing a way for divide-and-conquer.

An *implied integer variable* is a variable declared as continuous but must take integer values in any feasible solution. Presolving could detect such variables and use them as integer when convenient (e.g., for branching).

Many authors cite as relevant source of detailed technical and practical information on presolving (and more aspects of solvers) the contributions of M. Savelsberg, in publications like [92], [56] and [4]. The chapter [76], of A. Martin, is also well appreciated.

8.10.2 Branching

Along the branch and bound process, one must decide about node selection and variable selection. Two things: The review [61] on solvers lists 11 node selection methods and 5 variable selection methods, which have been proposed by the investigators (see references in the review).

Examples of node selection methods are depth-first search, breadth-first search, random search, parallel depth-first search, etc.

In the case of variable selection, the basic proposals are to choose the most fractional variable, to use some branching priorities, to branch on variables with fractional part close to 0.5, etc. Some heuristic rules, based on pseudocosts and estimations, were introduced by [9] (388 cites).

In 1999, J.T. Linderoth and M. Savelsberg made a computational study of different search strategies, [68] (521 cites), with notable results when using the so-called *strong branching* technique (introduced by Applegate et al. in 1996 [3]). The idea is to check the effects of branching for all candidate fractional variables, to decide the actual branching.

Pseudocosts and strong branching can be hybridized. Also, *unreliable* pseudocosts can be considered (*reliability branching*).

The publication [43] sustains that most used strategies nowadays are strong-branching and reliability branching and proposes new branching strategies denoted as "*narrow gauge*" and "*analytical branching*." This article is a good example of the sophistication level of current algorithms for branching.

For a more extended review of the branching advancements, [81] would be recommended.

8.10.3 Cutting Planes

The documentation of CPLEX (*vid.* [29]) contemplates ten (or more, depending on version) types of cuts, including those of Gomory (and Chvátal). For example, there are appropriate cuts for knapsack-like constraints, and *flow path cuts* for constraints defining a path structure in a network, or cuts for generalized upper bound (GUB) constraints for sets of binary variables (where the sum of these variables must be less or equal to one), etc.

For a basic tutorial on the cuts considered in CPLEX, see the presentation [90], which includes examples. It also includes some results based on testing with 80 different optimization problems, for example, the cuts that have better impact on performance were disjunctive cuts (64% improvement), Gomory cuts (22%), and knapsack cuts (18%).

Let us introduce disjunctive cuts by way of a graphical example, [64], which refers to *split cuts*. Recall that our constraints are:

$$A\vec{x} + D\vec{y} \le \vec{b}$$

where $\vec{x}$ contains the integer variables, and $\vec{y}$ the continuous variables.

Figure 8.28 depicts a polyhedron corresponding to the feasible region F (LP relaxation) of a certain MILP problem. Two parallel hyperplanes have been included, by which two polyhedra U and V of interest are obtained. Both U and V are inside F.

The hyperplanes obey to equations $\mathbf{m}^T\mathbf{x} \le h$ and $\mathbf{m}^T\mathbf{x} \le h + 1$, where m is a vector of integer numbers, and h is an integer (Fig. 8.29).

Based on the intersection of the hyperplanes with the boundary of F, we can define a new constraint (the dashed line in Fig. 8.24) that means a split cut. In the

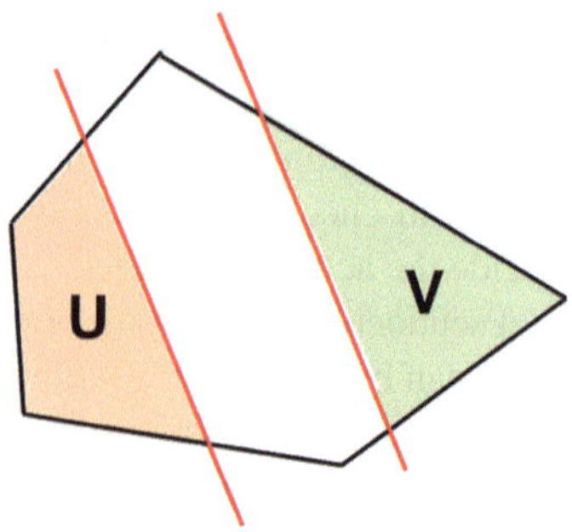

Fig. 8.28 Splitting a polyhedron

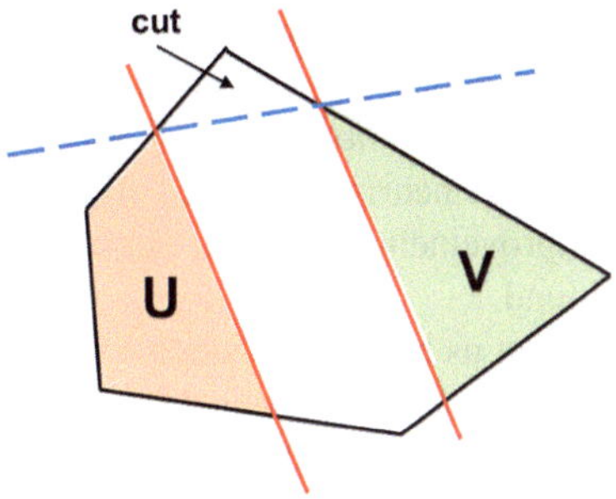

Fig. 8.29 Split cut for polyhedron F

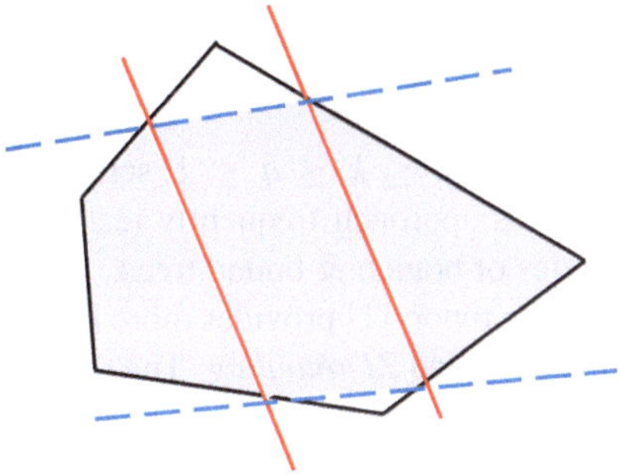

Fig. 8.30 Application of all split cuts

case depicted by Fig. 8.27, we can cut away a triangle, thus shrinking our search space.

Figure 8.30 shows how is the new search space, after applying two split cuts based on the same hyperplanes (notice that in the figure, the two new constraints are not parallel).

The disjunctive technique was introduced in 1979 by Egon Balas [5] (971 cites). Since then, it has been extended and applied for several contexts. The article [66] offers an interesting study of disjunctive cuts, showing its good performances for six types of problems (including the travelling salesman problem).

For a more general view of cuts and closures, it would be suggested to see [26] and [74]. Also, there is a book, [96] on optimization with disjunctive constraints.

8.10.4 Heuristics

Heuristic methods may rapidly provide satisfactory (approximate) solutions for large problems, which could require expensive computation if an exact method was applied. By using opportune "*a priori*" knowledge, heuristic methods can be advantageous for specific problems.

Ample information on how heuristics is applied in mixed-integer optimization, is provided by the thesis of T. Berthold [12], and the article of M. Fischetti and A. Lodi [38]. The thesis classifies the existing heuristics methods into two categories:

start heuristics and *improvement heuristics*. The article prefers to consider two generations of heuristic methods. Some other authors call start heuristics as *construction heuristics*.

Start heuristics aim to find feasible solutions at an early solving stage, while improvement heuristics seek better solutions once a feasible solution has been found.

Let us briefly introduce some of the *start heuristics* methods, departing from the LP relaxation solution ($\vec{x}_R$, $\vec{y}_R$), [39]. For simplicity, integer variables would be binary (0, 1) variables (it is easy to generalize to other ranges of values).

8.10.4.1 Rounding

For some $0 \leq k \leq q \leq 1$, set $x_{Ri} = 0$ if $x_{Ri} \leq k$, or set $x_{Ri} = 1$ if $x_{Ri} \geq q$.

This approach frequently results in infeasibilities. It is mainly used at the deep nodes of branch & bound trees.

The report [1] provides more information on rounding and propagation, including the so-called *ZI-rounding*. The thesis [49] would be also recommended in this area.

8.10.4.2 Diving

Let $x_{Ri} = \min_i(x_{Ri}, 1 - x_{Ri})$. Solve again the LP problem fixing $x_i = \lfloor x_{Ri} \rfloor$

Diving fixes one variable at a time (at each iteration), unless an integral solution was found. It converges rapidly but at a high risk of infeasibility.

The already cited thesis [12] includes an interesting section on some simple divers, including algorithmic and implementation details. Other divers can be found in the thesis [49].

At first sight, diving could be considered as a simplistic idea. However, it raises a rich methodology, as shown in [91] in the context of column generation and the combination of diving with local branching or other heuristics.

8.10.4.3 LP-and Fix

Let $Q = \{i \, / \, x_{Ri} = 0, \text{ or } x_{Ri} = 1\}$, ($Q$ is a set of indexes)

Solve MIP problem fixing $x_i = x_{Ri}\ , \ \forall i \in Q$

This method is related to diving. It is included in most solvers. A description of the algorithm is given in [46], which is a well-documented article on variable fixing heuristics in transport applications.

8.10.4.4 Relax-and Fix

Partition all binary variables into k sets Q_i. Let decisions in set Q_M be more important than decisions in set Q_p, for all pairs $1 \leq M < P \leq k$ for $i = 1$ to k

- solve the MIP problem fixing $x_j \in (0, 1)$, $\forall j \in Q_i$
- Denote as $(\vec{x}_L, \vec{y}_L)$ the solution of this MIP
- Fix $x_j = x_{Lj}$ for all $j \in Q_{i'}$, $i' \leq i$ and $j \notin Q_{i''}$, $i'' > i$

This method is convenient for problems with some priority structure. In a recent article (year 2023), a generalization of the relax-and-fix method has been presented, [57]; it would be recommended to read the introduction of this article, which also mention (with pertinent references) the other heuristic methods we are describing here.

8.10.4.5 Feasibility Pump

Let $\Delta(\cdot)$ be a distance function
Pick a point $(\vec{x}_F, \vec{y}_F)$ belonging to the feasible region F
repeat:
Define the rounding $\tilde{x} = |\vec{x}_F|$
Solve min $\{\Delta(\vec{x}, \tilde{x}) \;/\; (\vec{x}, \vec{y}) \in F\}$, obtain solution $\vec{x}*$
until $\Delta(\vec{x}*, \tilde{x}) = 0$

This method searches in the neighborhood of an LP feasible point. It may have a problem of cycling.

Thc fcasibility pump has been proposed by Fischetti and Lodi. It has attracted a lot of interest, and so many publications, like for instance [49] or [11], describe the method (normally in the light of specific applications, as in [95]).

Regarding to improvement heuristics, here are some examples.

8.10.4.6 Local Branching

Let $(\vec{x}_F, \vec{y}_F)$ be a feasible point for MIP
Add the following cut: $\sum_{i:\, x_{F_i}=0} x_i + \sum_{i:\, x_{F_i}=1} (1 - x_i) \leq k$
Solve again the MIP problem.

This cut defines the k-neighborhood of a feasible point, in order to search around the current solution.

The method was specifically introduced in [37] (1,133 cites), in the year 2003. It is also described in [12] and [38]. A motivating tutorial from A. Lodi is available from [71].

8.10.4.7 RINS: Relaxation Induced Neighborhood Search

Let $(\vec{x}_I, \vec{y}_I)$ be an IP feasible point, and $(\vec{x}_R, \vec{y}_R)$ the relaxation LP solution

Add the following constraints to the MIP:

$$x_i = x_{I\,i}\,, \;\forall i \text{ with } x_{I\,i} = x_{R\,i}$$

Solve again the MIP problem.

This method explores in between the IP feasible point and the LPR solution. It was introduced in 2005, [30].

Before closing this section, we should also emphasize the increasing initiatives for using machine learning in the different facets of mixed-integer programming. The recent survey [109] would be opportune to get an idea of this trend.

8.11 Software

8.11.1 MATLAB Tools

More recent versions (from R2014a) of the MATLAB Optimization Toolbox provide the function *intlinprog( )* for integer or mixed-integer applications. In comparison with the *linprog( )* syntax, it adds an extra argument (denoted as *incon* in the documentation) that indicates which of the variables are integer. It happens that the results for the supposed integers could be not exact (but the user can employ *round( )* for refining the results).

Here is a simple example of using *intlinprog( )*.

Maximize:

$$z = 3x_1 + 2x_2 + x_3$$

Subject to:

$$x_1 + x_2 + x_3 \le 7$$

$$4x_1 + 2x_2 + x_3 = 12$$

$$x_3 \in \{0, 1\}$$

The problem is easily translated to the MATLAB syntax, as shown in Program 8.6:

Program 8.6 Using *intlinprog( )* MATLAB function

```
f=[-3; -2;-1];
intcon=3; %the third variable is integer
A=[1,1,1];
b=7;
Aeq=[4,2,1];
beq=12;
lb=zeros(3,1);
ub=[Inf;Inf;1];
X=intlinprog(f,intcon,A,b,Aeq,beq,lb,ub);
X
```

MATLAB responds to this program as follows:

....

```
LP: Optimal objective value is -12.000000.
Optimal solution found.
Intlinprog stopped at the root node because the
objective value is within a
gap tolerance of the optimal value,
options.AbsoluteGapTolerance = 0.
The intcon variables are integer within tolerance,
options.IntegerTolerance = 1e-05.
X =
0
5.5000
1.0000
```

....

As a second example, we re-visit the case considered for illustration of Gomory cuts:

Maximize:

$$z = 5x_1 + 8x_2$$

Subject to:

$$x_1 + x_2 \leq 6$$

$$5x_1 + 9x_2 \leq 45$$

$$x_1, x_2 \in Z \text{ (integer)}$$

Now, with Program 8.7, we use *intlinprog( )*.

Program 8.7 Another example of using *intlinprog( )* MATLAB function

```
f=[-5, -8];
intcon=1:2;
A=[1 1; 5 9];
b=[6; 45];
lb=[0, 0];
ub=[Inf, Inf];
X=intlinprog(f,intcon,A,b,[],[],lb,ub);
X
```

MATLAB responds to this program as follows:

....

```
LP: Optimal objective value is -41.250000.
Heuristics: Found 1 solution using ZI round.
Upper bound is -39.000000.
Relative gap is 2.50%.
Cut Generation: Applied 1 Gomory cut.
Lower bound is -40.000000.
Relative gap is 0.00%.
Optimal solution found.
Intlinprog stopped at the root node because the
objective value is within a
gap tolerance of the optimal value,
options.AbsoluteGapTolerance = 0.
The intcon variables are integer within tolerance,
options.IntegerTolerance = 1e-05.
X =
0
5
```

....

The value of the objective function for the solution (0, 5) can be easily computed as 40.

Notice in the response of MATLAB that it reports the use of a heuristic method, and a Gomory cut. In other problems, MATLAB would report the use of other type of cut, or another heuristic method.

The MathWorks File Exchange page includes some contributions for linear integer programming and for mixed-integer linear programming.

8.11.2 Other Software Tools

A popular MATLAB toolbox for modeling and optimization is YALMIP, [67, 72] (11,502 cites). It is openly distributed, free. This toolbox relies on external solvers for low-level numerical solution of optimization problems. It can deal with many types of optimization scenarios, not only the linear ones.

The web page [47] from Gurobi comments two free mixed-integer and linear programming solvers, namely GLPK and LP_Solve. The software from Gurobi is compared with these solvers, much in advantage for Gurobi. The package GLPK is part of the GNU Project, and it is reviewed in [103].

J.T. Linderoth and T.K. Ralphs published in 2005 an extensive report on noncommercial software for mixed-integer linear programming, [69]. It mentions many solvers, including GLPK, LP_Solve, and ABACUS, BCP, bonsaiG, CBC, MINTO, and SYMPHONY. After some search from our part, we could add SCIP, [10].

GitHub includes 205 repositories for integer programming (7 in Matlab and 20 in C++) and 194 repositories for mixed integer programming (7 in Matlab and 16 in C++).

Commercial software for mixed integer linear programming, in addition to CPLEX, are COPT, Knitro, Mosek ApS, and Xpress (already cited). See the page GAMS, [28] for more information on integer and mixed-integer programming.

8.12 Exercises

1- [8.1] Solve the problem considered in Sect. 8.2.1, using branch and bound. Visualize and explain the tree.

2- [8.2] Let the integer programming problem be as follows:
Maximize: $J = 2x_1 + 5x_2$
Subject to:

$$4x_1 + x_2 \leq 28$$

$$x_1 + 4x_2 \leq 27$$

$$x_1 - x_2 \leq 1$$

$$x_1, x_2 \geq 0$$

Apply and visualize Gomory cuts to solve the problem.

3- [8.3] Consider the following integer programming problem:
Minimize: $J = x_1 - 2x_2$
Subject to:

$$-4x_1 + 6x_2 \leq 9$$

$$x_1 + x_2 \leq 4$$

$$x_1,\ x_2\ \geq 0$$

Apply and visualize Gomory cuts to solve the problem.

4- [8.3] Let us tackle the following integer programming problem:
   Maximize: $J = 2x_1 + 3x_2$
   Subject to

$$x_1 + 2x_2 \leq 3$$

$$6x_1 + 8x_2 \leq 15$$

$$x_1,\ x_2\ \geq 0$$

Use branch and bound. Visualize and explain the process.

5- [8.4] Consider the following integer programming problem:
   Maximize: $J = x_1 + 4x_2$
   Subject to:

$$2x_1 + 4x_2 \leq 7$$

$$10x_1 + 3x_2 \leq 14$$

$$x_1,\ x_2\ \geq 0$$

Apply cuts for solving the problem. Visualize and explain the process.

6- [8.5] Consider the following integer programming problem:
   Maximize: $J = -x_1 + 4x_2$
   Subject to:

$$-10x_1 + 20x_2 \leq 22$$

$$5x_1 + 10x_2 \leq 49$$

$$x_1,\ x_2\ \geq 0$$

Apply branch and bound for solving the problem. Visualize and explain in detail the process.

7- [8.6] Let the integer programming problem be:
   Maximize: $J = 4x_1 + 5x_2$
   Subject to:

$$x_1 + 4x_2 \leq 10$$

$$3x_1 - 4x_2 \leq 6$$

$$x_1, x_2 \geq 0$$

Apply branch and bound for solving the problem. Visualize and explain in detail the process.

8- [8.7] A certain airport has ten arrival gates, with the following coordinates and number of bags per day:

Gate	A	B	C	D	E	F	G	H	I	J
x coordinate	0	2	5	5	7	10	12	12	15	20
y coordinate	2	4	6	10	15	15	10	6	4	2
bags/day	3,600	2,500	1,800	2,200	1,000	4,500	5,600	1,400	1,800	3,000

Where the passenger pickup point should be placed? (we want to minimize luggage motion).

9- [8.8] Mr. Smith is a ship owner. He has a freighter with 700 tons capacity. Today, Mr. Smith could ship some of the following containers:

Container	c1	c2	c3	c4	c5	c6	c7	c8	c9	c10
Weight	100	155	50	112	70	80	60	118	110	55

Find how to maximize the load.

10- [8.9] Company JMG would consider five capital projects over the next four years. Next table shows the required investments and the corresponding NPV (net present value). The company desires to maximize the benefit.

Project	Capital Required ($ millions)				NPV benefits
	Year 1	Year 2	Year 3	Year 4	($ millions)
A	3	0	0	0	2
B	0	5	1	0	3
C	1	2	1	2	1
D	10	4	2	0	5
E	2	0	5	1	4

Projects D and E are mutually exclusive.
No more than four projects in total can be undertaken.
Projects A and C must be undertaken together.
Project D is dependent of project B (if no B, then no D).
The capital requirements imply four capital requirement constraints, one per year. Those limits are 12, 8, 8, and 4.
Please, find which projects are to be accepted.

11- [8.10] Orders from five destinations are satisfied by a central warehouse. Routes are assigned to different carriers. There are nine feasible routes. Next table contains a short description of the routes and their lengths:

Routes	1	2	3	4	5	6	7	8	9
R. description	1,3,4	1,4,5	1,2,5	2,3,5	2,4,5	3,4,5	1	2,3	4,5
lengths	10	12	12	13	11	9	7	8	8

Please, determine the shortest total distance routing. Each destination must be on at least one of the chosen routes.

12- [8.11] A paper company manufactures rolls of paper in five standard widths, by cutting rolls of 25 m width. Now the company has the following demands:

Widths	5	8	12	15	17
Demand	40	35	30	25	20

Please, find the best cutting planification.

13- [8.1] Consider the following mixed integer problem:
Maximize: $J = -3x_1 - 2x_2 + 10$
Subject to:

$$x_1 - 2x + x_3 = 5/2$$

$$2x_1 + x_2 + x_4 = 3/2$$

$$\forall x_j \geq 0$$

With x_2 and x_3 integer.

Apply branch and bound for solving the problem. Visualize and explain the tree

8.12.1 Exercise References

[8.1] https://web.mit.edu/15.053/www/AMP-Chapter-09.pdf
[8.2] https://coral.ise.lehigh.edu/$\sim$ted/files/computational-mip/lectures/Lecture11.pdf
[8.3] https://www.lix.polytechnique.fr/$\sim$liberti/teaching/isic/isc612-06/exercises-solutions.pdf
[8.4] https://www.doc.ic.ac.uk/$\sim$br/Tutorials/ip2sol.pdf
[8.5] https://www.ie.bilkent.edu.tr/$\sim$mustafap/courses/bb.pdf

[8.6] https://misha.fish/archive/docs/482-spring-2020/lecture33.pdf
[8.7] https://people.brunel.ac.uk/$\sim$mastjjb/jeb/or/facloc.html
[8.8] https://personales.unican.es/castie/chapter2.pdf
[8.9] https://ruby.fgcu.edu/courses/tharring/10183/m8_notes.htm
[8.10] https://www.math.clemson.edu/$\sim$mjs/courses/mthsc.440/integer.pdf
[8.11] https://courses.ie.bilkent.edu.tr/ie400/wp-content/uploads/sites/8/2018/02/IE400_Set5_IP_modelling.pdf

References

1. Achterberg, T., Berthold, T., & Hendel, G. (2012). Rounding and propagation heuristics for mixed integer programming. In *Operations research proceedings 2011: Selected papers of the international conference on operations research (OR 2011)*, August 30–September 2, 2011, Zurich (pp. 71–76). Springer. Available from: https://webdoc.sub.gwdg.de/ebook/serien/ah/ZIB/ZR-11-29.pdf
2. Achterberg, T., & Wunderling, R. (2013). Mixed integer programming: Analyzing 12 years of progress. In *Facets of combinatorial optimization: Festschrift for Martin Grötschel* (pp. 449–481). Springer. Available from: https://link.springer.com/chapter/10.1007/978-3-642-38189-8_18
3. Applegate, D., Bixby, R., Chvátal, V., & Cook, W. (1995). Finding cuts in the TSP (A preliminary report) (No. 95-05). Report.
4. Atamtürk, A., & Savelsbergh, M. W. P. (2005). Integer-programming software systems. *Annals of Operations Research, 140*, 67–124. Available from: https://link.springer.com/content/pdf/10.1007/s10479-005-3968-2.pdf
5. Balas, E. (1979). Disjunctive programming. *Annals of Discrete Mathematics, 5*, 3–51. Available from: https://www.sciencedirect.com/science/article/abs/pii/S016750600870342X
6. Balas, E., Ceria, S., Cornuéjols, G., & Natraj, N. (1996). Gomory cuts revisited. *Operations Research Letters, 19*(1), 1–9. Available from: https://www.sciencedirect.com/science/article/pii/0167637796000077
7. Beasley, J. E. (2013). Or-notes. Available from: https://people.brunel.ac.uk/~mastjjb/jeb/or/ip.html
8. Ben Amor, H., & Valério de Carvalho, J. (2005). Cutting stock problems. In *Column generation* (pp. 131–161). Springer. Available from: https://www.gerad.ca/fr/papers/1137.pdf
9. Bénichou, M., Gauthier, J.-M., Girodet, P., Hentges, G., Ribière, G., & Vincent, O. (1971). Experiments in mixed-integer linear programming. *Mathematical Programming, 1*, 76–94. Available from: https://link.springer.com/article/10.1007/BF01584074
10. Zuse Institute Berlin. (2023). SCIP, solving constraint integer programs. Available from: https://scipopt.org/
11. Bertacco, L. (2006). *Exact and heuristic methods for mixed integer linear programs*. PhD thesis, Universita degli Studi di Padova. Available from: https://www.dei.unipd.it/~fisch/ricop/tesi/2005_Bertacco_PhD_thesis.pdf
12. Berthold, T. (2006). *Primal heuristics for mixed integer programs*. PhD thesis, Zuse Institute Berlin (ZIB). Available from: https://opus4.kobv.de/opus4-zib/frontdoor/index/index/docId/1029
13. Bixby, R. E. (2012). A brief history of linear and mixed-integer programming computation. *Documenta Mathematica, 2012*, 107–121. Available from: https://ems.press/content/book-chapter-files/27357

14. Bixby, R. E. (2014). Business and mathematics: A saga of 25 years of progress in optimization. Available from: https://www.math.uwaterloo.ca/~hwolkowi/henry/teaching/f16/602.f16/602miscfiles/UF_Entrepreneurship_19November2014.pdf
15. BrainKart.com. (2000). Capital budgeting-integer linear programming illustrative applications. Available from: https://www.brainkart.com/article/Capital-Budgeting-Integer-Linear-Programming-Illustrative-Applications_11238/
16. Burkard, R. E., Deineko, V. G., van Dal, R., van der Veen, J. A. A., & Woeginger, G. J. (1998). Well-solvable special cases of the traveling salesman problem: A survey. *SIAM Review, 40*(3), 496–546. Available from: https://pure.tue.nl/ws/portalfiles/portal/2373438/Metis148543.pdf
17. Cantlebary, L., & Li, L. (2020). Facility location problem. Available from: https://optimization.cbe.cornell.edu/index.php?title=Facility_location_problem
18. Chen, D.-S., Batson, R. G., & Dang, Y. (2011). *Applied integer programming: Modeling and solution*. Wiley.
19. Chen, Y. (2011). Am 121: Intro to optimization models and methods. Available from: http://am121.seas.harvard.edu/site/wp-content/uploads/2011/03/lec14-w.pdf
20. Chvátal, V. (1973). Edmonds polytopes and a hierarchy of combinatorial problems. *Discrete Mathematics, 4*(4), 305–337. Available from: https://core.ac.uk/reader/82135850
21. Claussen, G., & Hartmann, A. K. (2022). Cutting-plane algorithms and solution whitening for the vertex-cover problem. *Physical Review E, 106*(3), 035305. Available from: https://journals.aps.org/pre/pdf/10.1103/PhysRevE.106.035305
22. Conitzer, V. (2008). Solving (mixed)integer programs using branch and bound. Available from: https://courses.cs.duke.edu/spring08/cps296.2/branch_and_bound.pdf
23. Cook, W. (2023). Travelling salesman problem. Available from: https://www.math.uwaterloo.ca/tsp/
24. Cornuejols, G. (2012) The ongoing story of Gomory cuts. Available from: https://www.math.uni-bielefeld.de/documenta/vol-ismp/37_cornuejols-gerard.pdf
25. Cornuéjols, G., et al. (2007). Revival of the Gomory cuts in the 1990s. *Annals of Operations Research, 149*(1), 63–66. Available from: https://cgm.cs.mcgill.ca/~avis/courses/567/notes/gomory.pdf
26. Cornuéjols, G., & Li, Y. (2001). Elementary closures for integer programs. *Operations Research Letters, 28*(1), 1–8. Available from: https://www.sciencedirect.com/science/article/abs/pii/S0167637700000675?via%3Dihub
27. Cornuéjols, G., Trick, M. A., & Saltzman, M. J. (2022). A tutorial on integer programming. Available from: https://www.math.clemson.edu/~mjs/courses/mthsc.440/integer.pdf
28. GAMS Development Corp. (2023) Gams documentation center. Available from: https://www.gams.com/latest/docs/index.html
29. ILOG CPLEX. (2001). Cuts. Available from: https://www-eio.upc.es/lceio/manuals/cplex-11/html/usrcplex/solveMIP14.html#:~:text=ILOGCPLEXgeneratesitscuts, constrainttoreflectthiscondition
30. Danna, E., Rothberg, E., & Le Pape, C. (2005). Exploring relaxation induced neighborhoods to improve MIP solutions. *Mathematical Programming, 102*, 71–90. Available from: https://link.springer.com/content/pdf/10.1007/s10107-004-0518-7.pdf
31. Dantzig, G., Fulkerson, R., & Johnson, S. (1954). Solution of a large-scale traveling-salesman problem. *Journal of the Operations Research Society of America, 2*(4), 393–410. Available from: https://web.archive.org/web/20060921030842id_/http://www.cse.wustl.edu/~chen/7102/TSP.pdf
32. Davendra, D. (2010). *Traveling salesman problem: Theory and applications*. BoD–Books on Demand.
33. Desrosiers, J., & Lübbecke, M. E. (2005). A primer in column generation. In *Column generation* (pp. 1–32). Springer. Available from: https://or.rwth-aachen.de/files/research/publications/primer.pdf
34. Drori, I., Kates, B. J., Sickinger, W. R., Kharkar, A. G., Dietrich, B., Shporer, A., & Udell, M. (2020). Galaxytsp: A new billion-node benchmark for TSP. In *Learning meets combinatorial algorithms at NeurIPS2020*. Available from: https://openreview.net/pdf?id=QmKFppOpA7

35. Eisenbrand, F. (2010). Integer programming and branch and bound. Available from: https://www.epfl.ch/labs/disopt/wp-content/uploads/2018/09/scribes11_Boris_Oumow.pdf
36. Eu, Z. (2020). Optimizing capital budgeting using integer linear programming. Available from: https://zhijingeu.medium.com/optimizing-capital-budgeting-using-linear-programming-tools-dfd1dec5a64d
37. Fischetti, M., & Lodi, A. (2003). Local branching. *Mathematical Programming, 98*, 23–47. Available from: https://link.springer.com/content/pdf/10.1007/s10107-003-0395-5.pdf
38. Fischetti, M., & Lodi, A. (2010). Heuristics in mixed integer programming. In *Wiley encyclopedia of operations research and management science*. Available from: https://homepages.cwi.nl/~dadush/workshop/discrepancy-ip/papers/heuristics-survey-fischetti-lodi-11.pdf
39. Gan, H. S., & Akartunalı, K. 620–462 integer programming lecture notes. Available from: https://citeseerx.ist.psu.edu/document?repid=rep1&type=pdf&doi=bcb6f21f353364cdc656c57ff97b2f2dc2cf2ed8
40. Genova, K., & Guliashki, V. (2011). Linear integer programming methods and approaches—a survey. *Journal of Cybernetics and Information Technologies, 11*(1). Available from: https://cit.iict.bas.bg/CIT_2011/v11-1/3-25.pdf
41. Gilmore, P. C., & Gomory, R. E. (1961). A linear programming approach to the cutting-stock problem. *Operations Research, 9*(6), 849–8591961. Available from: https://www.jstor.org/stable/pdf/167051.pdf
42. Gilmore, P. C., & Gomory, R. E. (1963). A linear programming approach to the cutting stock problem—Part II. *Operations Research, 11*(6), 863–888. Available from: https://www.jstor.org/stable/pdf/167827.pdf
43. Glover, F., Shylo, V., & Shylo, O. (2015). Narrow gauge and analytical branching strategies for mixed integer programming. arXiv preprint arXiv:1511.00021. Available from: https://arxiv.org/pdf/1511.00021
44. Gomory, R. E. (2010). Outline of an algorithm for integer solutions to linear programs and an algorithm for the mixed integer problem. In *50 years of integer programming 1958–2008: From the early years to the state-of-the-art* (pp. 77–103). Available from: https://pages.stern.nyu.edu/~rgomory/academic_papers/07_outline.pdf
45. Grötschel, M., Jünger, M., & Reinelt, G. (1984). A cutting plane algorithm for the linear ordering problem. *Operations Research, 32*(6), 1195–1220. Available from: https://pubsonline.informs.org/doi/abs/10.1287/opre.32.6.1195
46. Guimarães, L. R., de Sousa, J. P., & de Athayde Prata, B. (2022). Variable fixing heuristics for the capacitated multicommodity network flow problem with multiple transport lines, a heterogeneous fleet and time windows. *Transportation Letters, 14*(2), 84–93. Available from: https://www.researchgate.net/profile/Bruno-Prata/publication/343689875_Variable_fixing_heuristics_for_the_capacitated_multicommodity_network_flow_problem_with_multiple_transport_lines_heterogeneous_fleet_and_time_window/links/601ed5b245851589398919ec/Variable-fixing-heuristics-for-the-capacitated-multicommodity-network-flow-problem-with-multiple-transport-lines-heterogeneous-fleet-and-time-window.pdf
47. Gurobi. (2023). Open-source mixed-integer and linear programming solvers and software. Available from: https://www.gurobi.com/resources/open-source-mixed-integer-and-linear-programming-solvers/
48. Haessler, R. W., & Sweeney, P. E. (1991). Cutting stock problems and solution procedures. *European Journal of Operational Research, 54*(2), 141–150. Available from: https://deepblue.lib.umich.edu/bitstream/handle/2027.42/29128/0000167.pdf
49. Hendel, G. (2011). *New rounding and propagation heuristics for mixed integer programming*. PhD thesis, Zuse Institute Berlin (ZIB). Available from: https://opus4.kobv.de/opus4-zib/frontdoor/index/index/docId/1332
50. Hochbaum, D. S. (2017). Lecture notes for IEOR 266: Graph algorithms and network flows. Available from: https://hochbaum.ieor.berkeley.edu/files/IEOR266-lecture-notes-F2017.pdf
51. Huerta, I. I., Neira, D. A., Ortega, D. A., Varas, V., Godoy, J., & Asín-Achá, R. (2022). Improving the state-of-the-art in the traveling salesman problem: An anytime automatic algorithm selection. *Expert Systems with Applications, 187*, 115948. Available from: https://www.sciencedirect.com/science/article/pii/S0957417421013014

52. Ilavarasi, K., & Joseph, K. S. (2014). Variants of travelling salesman problem: A survey. In *International conference on information communication and embedded systems (ICICES2014)* (pp. 1–7). IEEE. Available from: https://ieeexplore.ieee.org/stamp/stamp.jsp?arnumber=7033850
53. Jablonskỳ, J., et al. (2015). Benchmarks for current linear and mixed integer optimization solvers. *Acta Universitatis Agriculturae et Silviculturae Mendelianae Brunensis, 63*(6), 1923–1928. Available from: https://acta.mendelu.cz/pdfs/acu/2015/06/12.pdf
54. Jiaoman, D., Lei, L., & Xiang, L. (2018). Travel planning problem considering site selection and itinerary making. In *Proceedings of the 2018 conference on research in adaptive and convergent systems* (pp. 29–36). Available from: https://dl.acm.org/doi/pdf/10.1145/3264746.3264781
55. Jünger, et al. (2010). *50 years of integer programming 1958–2008*. Springer. Available from: https://books.google.es/books?hl=en&lr=&id=bUJc_weiYfkC&oi=fnd&pg=PA104&ots=0zIUOjunNZ&sig=aCx4kLH2ql6fJPiSmnyrC1n1xTw&redir_esc=y#v=onepage&q&f=false
56. Johnson, E. L., Nemhauser, G. L., & Savelsbergh, M. W. P. (2000). Progress in linear programming-based algorithms for integer programming: An exposition. *Informs Journal on Computing, 12*(1), 2–23. Available from: http://cepac.cheme.cmu.edu/pasi2011/library/maravelias/Johnson.2000.JOC.ProgressforIntegerProgramming.pdf
57. Joncour, C., Kritter, J., Michel, S., & Schepler, X. (2023). Generalized relax-and-fix heuristic. *Computers & Operations Research, 149*, 106038. Available from: https://hal.science/hal-04302713/
58. Kengatharan, L. (2016). Capital budgeting theory and practice: A review and agenda for future research. *Research Journal of Finance and Accounting, 7*(1). Available from: https://core.ac.uk/download/pdf/234631219.pdf
59. Khoufi, I., Laouiti, A., & Adjih, C. (2019). A survey of recent extended variants of the traveling salesman and vehicle routing problems for unmanned aerial vehicles. *Drones, 3*(3), 66. Available from: https://www.mdpi.com/2504-446X/3/3/66
60. Klose, A., & Drexl, A. (2005). Facility location models for distribution system design. *European Journal of Operational Research, 162*(1), 4–29. Available from: https://www.sciencedirect.com/science/article/pii/S0377221703008191
61. Kumar, P. H., & Mageshvaran, R. (2020). Methods and solvers used for solving mixed integer linear programming and mixed nonlinear programming problems: A review. *The International Journal of Scientific & Technology Research, 9*(1), 1872–1882. Available from: https://www.ijstr.org/final-print/jan2020/Methods-And-Solvers-Used-For-Solving-Mixed-Integer-Linear-Programming-And-Mixed-Nonlinear-Programming-Problems-A-Review.pdf
62. Land, A. H., & Doig, A. G. (1960). An automatic method for solving discrete programming problems. Available from: https://jmvidal.cse.sc.edu/library/land60a.pdf
63. Laporte, G. (1992). The traveling salesman problem: An overview of exact and approximate algorithms. *European Journal of Operational Research, 59*(2), 231–247. Available from: https://web.ist.utl.pt/~ist11038/CD_Casquilho/TSP1992EJOR_Laporte.pdf
64. Lee, D. (2023). IE631 lecture 17: Split cuts, Gomory's mixed-integer cuts. Available from: https://dabeenl.github.io/IE631_lecture17_note.pdf
65. Lee, S. (2024). Discrete optimization. Available from: https://i-systems.github.io/teaching/ML/iNotes/30_discrete_opt_m.html
66. Letchford, A. N. (2001). On disjunctive cuts for combinatorial optimization. *Journal of Combinatorial Optimization, 5*, 299–315. Available from: https://link.springer.com/content/pdf/10.1023/A:1011493126498.pdf
67. Löfberg, J. (2023). YALMIP toolbox. Available from: https://yalmip.github.io/
68. Linderoth, J. T., & Savelsbergh, M. W.P. (1999). A computational study of search strategies for mixed integer programming. *INFORMS Journal on Computing, 11*(2), 173–187. Available from: https://www2.isye.gatech.edu/people/faculty/Martin_Savelsbergh/publications/ijoc11b.pdf

69. Linderoth, J. T., & Ralphs, T. K. (2005). Noncommercial software for mixed-integer linear programming. *Integer Programming: Theory and Practice, 3*(253–303), 144–189. Available from: https://web.tecnico.ulisboa.pt/mcasquilho/compute/_linpro/2005LinderothR.pdf
70. Litke, J. D. (1984). An improved solution to the traveling salesman problem with thousands of nodes. *Communications of the ACM, 27*(12), 1227–1236. Available from: https://dl.acm.org/doi/pdf/10.1145/2135.2141
71. Lodi, A. (2003). Local branching: A tutorial. *MIC2003, Kyoto*. Available from: http://www.or.deis.unibo.it/research_pages/ORinstances/mic2003-lb.pdf
72. Lofberg, J. (2004). YALMIP: A toolbox for modeling and optimization in MATLAB. In *2004 IEEE international conference on robotics and automation (IEEE Cat. No. 04CH37508)* (pp. 284–289). IEEE. Available from: http://fa.ee.sut.ac.ir/Downloads/AcademicStaff/22/Courses/107/Lecture_01_Paper_01_YALMIPatoolboxformodelingandoptimizationinMATLAB.pdf
73. Lübbecke, M. E., & Desrosiers, J. (2005). Selected topics in column generation. *Operations Research, 53*(6), 1007–1023. Available from: https://api-depositonce.tu-berlin.de/server/api/core/bitstreams/aac69e84-1931-4b33-83b0-f186952aebee/content
74. Marchand, H., Martin, A., Weismantel, R., & Wolsey, L. (2002). Cutting planes in integer and mixed integer programming. *Discrete Applied Mathematics, 123*(1–3), 397–446. Available from: https://www.sciencedirect.com/science/article/pii/S0166218X01003481
75. Mariescu-Istodor, W. R., & Fränti, P. (2020). Santa Claus TSP challenge. Available from: https://cs.uef.fi/sipu/santa/
76. Martin, A. (2001). *General mixed integer programming: Computational issues for branch-and-cut algorithms*. Springer.
77. Matai, R., Singh, S. P., & Mittal, M. L. (2010). Traveling salesman problem: An overview of applications, formulations, and solution approaches. In *Traveling salesman problem, theory and applications* (Vol. 1).
78. MathWorks. (2024) Factory, warehouse, sales allocation model: Problem—based. Available from: https://es.mathworks.com/help/optim/ug/factory-example-problem-based.html
79. MathWorks. (2024). Factory, warehouse, sales allocation model: Solver-based. Available from: https://es.mathworks.com/help/optim/ug/factory-warehouse-sales-allocation-model.html
80. Miller, C. E., Tucker, A. W., & Zemlin, R. A. (1960). Integer programming formulation of traveling salesman problems. *Journal of the ACM (JACM), 7*(4), 326–329. Available from: https://dl.acm.org/doi/pdf/10.1145/321043.321046
81. Morrison, D. R., Jacobson, S. H., Sauppe, J. J., & Sewell, E. C. (2016). Branch-and-bound algorithms: A survey of recent advances in searching, branching, and pruning. *Discrete Optimization, 19*, 79–102. Available from: https://www.sciencedirect.com/science/article/pii/S1572528616000062
82. Naomi. (2023). Travelling salesman problem. Available from: https://www.savemyexams.co.uk/dp/maths_ai-hl/ib/21/revision-notes/3-geometry-trigonometry/3-10-graph-theory/3-10-5-travelling-salesman-problem/
83. NEOS-guide. The cutting stock problem. Available from: https://neos-guide.org/case-studies/sc/mfg/the-cutting-stock-problem/#:~:text=Thecuttingstockproblemisanintegerlinearprogramwith,branchandboundalgorithm
84. Ngo, H. Q. (2005). Approximation algorithms based on lp relaxations. Available from: https://cse.buffalo.edu/~hungngo/classes/2005/594/notes/relaxation-rounding.pdf
85. Padberg, M., & Rinaldi, G. (1991). A branch-and-cut algorithm for the resolution of large-scale symmetric traveling salesman problems. *SIAM Review, 33*(1), 60–100. Available from: https://epubs.siam.org/doi/abs/10.1137/1033004
86. Peterson, P. P., & Fabozzi, F. J. (2002). *Capital budgeting: Theory and practice* (Vol. 10). Wiley.
87. Rego, C., Gamboa, D., Glover, F., & Osterman, C. (2011). Traveling salesman problem heuristics: Leading methods, implementations and latest advances. *European Journal of Operational Research, 211*(3), 427–441. Available from: https://www.sciencedirect.com/science/article/pii/S0377221710006065

88. Reinelt, G. (2003). *The traveling salesman: Computational solutions for TSP applications* (Vol. 840). Springer.
89. Romfo, E., Rende, A., & Redmond, J. (2002). Integer/binary integer programming presentation. Available from: https://web.cortland.edu/matresearch/BinaryInteger.pdf
90. Rothberg, E. (2003). The CPLEX library: Presolve and cutting planes. Available from: https://conferences.mpi-inf.mpg.de/adfocs-03/Slides/Rothberg_4.pdf
91. Sadykov, R., Vanderbeck, F., Pessoa, A., Tahiri, I., & Uchoa, E. (2019). Primal heuristics for branch and price: The assets of diving methods. *INFORMS Journal on Computing, 31*(2), 251–267. Available from: https://inria.hal.science/hal-01237204/document
92. Savelsbergh, M. W. P. (1994). Preprocessing and probing techniques for mixed integer programming problems. *ORSA Journal on Computing, 6*(4), 445–454. Available from: https://pure.tue.nl/ws/portalfiles/portal/1878309/380251.pdf
93. Schrijver, A. (1998). *Theory of linear and integer programming*. Wiley.
94. Schrijver, A., et al. (2003). *Combinatorial optimization: Polyhedra and efficiency* (Vol. 24). Springer.
95. Sharma, S. (2013). Mixed-integer nonlinear programming heuristics applied to a shale gas production optimization problem. Master's thesis, Institutt for teknisk kybernetikk. Available from: https://ntnuopen.ntnu.no/ntnu-xmlui/bitstream/handle/11250/260948/646797_FULLTEXT01.pdf?sequence=2
96. Sherali, H. D., & Shetty, C. M. (2012). *Optimization with disjunctive constraints* (Vol. 181). Springer. Available from: https://link.springer.com/book/10.1007/978-3-642-48794-1
97. SpS. (2020). Solvers for mixed-integer linear programming (MILP)problems. Available from: https://therearchpoint.com/2023/01/19/solvers-for-mixed-integer-linear-programming-milp-problems/
98. Taha, H. A. (2014). *Integer programming: Theory, applications, and computations*. Academic.
99. Tanneau, M., & Montoison, A. (2023). The facility location problem. Available from: https://jump.dev/JuMP.jl/stable/tutorials/linear/facility_location/
100. Thomas, R. R. (2000). Math 583E: Linear and integer polyhedra. Available from: https://sites.math.washington.edu/~thomas/teaching/m583_s2008_web/main.pdf
101. Vemuganti, R. R. (1998). Applications of set covering, set packing and set partitioning models: A survey. In *Handbook of combinatorial optimization: Volume 1–3* (pp. 573–746). Available from: https://link.springer.com/chapter/10.1007/978-1-4613-0303-9_9
102. Wiesemann, W. (2009). Integer programming. Available from: http://worldcolleges.info/sites/default/files/enggnotes/integer.pdf
103. Wikibooks. (2012). GLPK/reviews and benchmarks. Available from: https://en.wikibooks.org/wiki/GLPK/Reviews_and_benchmarks
104. Wikipedia. Cutting stock problem. Available from: https://en.wikipedia.org/wiki/Cutting_stock_problem
105. Winston, W. L. (2022). *Operations research: Applications and algorithms*. Cengage Learning. Available from: https://itslearningakarmazyan.wordpress.com/wp-content/uploads/2015/09/operation-research-aplications-and-algorithms.pdf
106. Wolsey, L. A. (2020). *Integer programming*. Wiley.
107. Wolsey, L. A., & Nemhauser, G. L. (1999). *Integer and combinatorial optimization* (Vol. 55). Wiley.
108. Wu, D., & Yang, G. (2021). A heuristic approach for two-dimensional rectangular cutting stock problem considering balance for material utilization and cutting complexity. *Advances in Materials Science and Engineering, 2021*, 1–15. Available from: https://onlinelibrary.wiley.com/doi/pdf/10.1155/2021/3732720
109. Zhang, J., Liu, C., Li, X., Zhen, H.-L., Yuan, M., Li, Y., & Yan, J. A survey for solving mixed integer programming via machine learning. *Neurocomputing, 519*, 205–217 (2023) Available from: https://arxiv.org/abs/2203.02878

Chapter 9
Convex Programming

9.1 Introduction

In the preceding chapters, a series of optimization methods exploiting linearity have been studied. Once this fundamental theoretical body was mostly established, the research attempts to widen, gradually, the optimization field.

This chapter is devoted to an important stride of the theory expanding effort, taking advantage of *convexity* benefits.

Specifically, this chapter will introduce convex programming (CVX), cone programming (CP), and semidefinite programming (SDP). In addition, two traditional solution methods will be described.

In continuity with the previous chapters, we are considering the presence of constraints.

Figure 9.1 attempts to visualize the areas covered by methods already treated (LP and QP) and the new methods to be introduced. For instance, linear programming (LP) is a particular case of convex programming.

It should be said that reality, as usual, is a little more complex than what Fig. 9.1 shows.

A key property of convex optimization is that any local minimizer is a global minimizer. It does not mean that a convex optimization problem must have a unique minimizer. But if the objective function is strictly convex, then yes: the optimization problem has a unique solution.

As pointed out by A.E. Bryson, it happens that sometimes is difficult to determine that the problem at hand is a convex problem. Therefore, part of the focus of convex optimization was put on convexity itself.

The web page [$REF GHA] would be interesting for more information on basic topics related to this chapter.

We would like to recommend the famous book [19] (71,380 cites) on convex optimization. In particular, page 188 offers a detailed account on the history of the topic, together with references of special relevance.

J. M. Giron-Sierra, *Introduction to Optimization with Matlab® Examples*,
https://doi.org/10.1007/978-3-032-05048-9_9

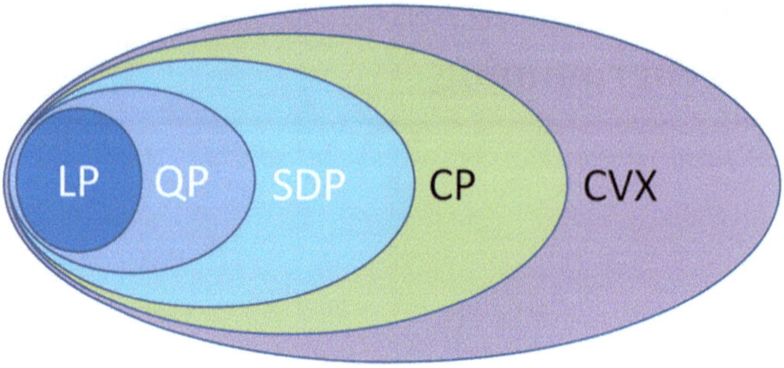

Fig. 9.1 Coverage of Optimization Methods

9.2 Convex Programming

Convex programming is concerned with linear problems and some nonlinear problems.

According to [59], decades ago the optimization research "*thought that the major divide was between linear and nonlinear optimization problems; later people discovered that some nonlinear problems were much harder than others, and the "right" divide was between convex and nonconvex problems*."

Let us introduce the pertinent mathematical definition. Consider a generic optimization problem:

Minimize:

$$f_0(\vec{x})$$

Subject to

$$f_i(\vec{x}) \leq 0, \quad i = 1, 2, \ldots, m$$

$$h_j(\vec{x}) = 0, \quad j = 1, 2, \ldots, r$$

$$\vec{x} \in S$$

The optimization problem is said to be *convex* if all functions $f_k(\vec{x})$, $k = 0, 1, 2, \ldots, m$ are convex and the functions $h_j(\vec{x})$, $j = 1, 2, \ldots, r$ are affine (of the form $A\vec{x} + \vec{b}$), and S is a convex set.

9.2.1 *Convex Sets*

A set S is convex if for any two points $\vec{x}$, $\vec{y} \in S$, the line segment joining $\vec{x}$ and $\vec{y}$ lies in S.

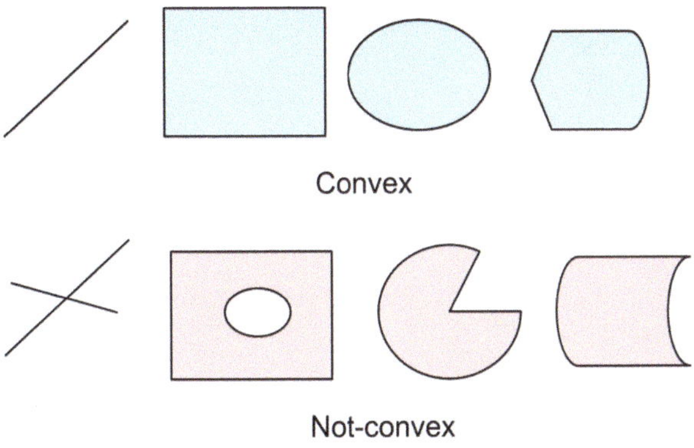

Fig. 9.2 Examples of convex and not-convex sets

In general, a convex set must be solid, with no holes, and with curves outward. Figure 9.2 shows examples of convex and not-convex sets. The intersection of convex sets is a convex set.

Some examples of convex sets:

- Hyperplanes
- Half-spaces
- Norm balls
- Polyhedrons (intersections of a finite number of half-spaces)
- Any bounded polyhedron (called *polytope*)

An important drawback: some common operations on convex sets do not preserve some basic properties (this does not happen with compact and polyhedral sets), [9]. For example, a linearly transformed convex set need not be closed. Also, the vector sum of two closed convex sets needs not to be closed.

Therefore, some analytical difficulties and pathological behaviors can appear.

Some of the operations that preserve convexity are the following:

- The intersection of any number of convex sets is convex
- Given an affine function $f(.)$ from $\Re^n$ to $\Re^m$, a convex set S belonging to $\Re^n$ and another convex set T belonging to $\Re^m$. Then $f(S)$ and $f^{-1}(T)$ are convex sets
- Scaling and translation of convex sets preserve convexity

9.2.2 *Hyperplanes and Epigraph*

The separating hyperplane theorem states that if two convex sets S and T are disjoint ($S \cap T = 0$), then there exists a hyperplane which separates them, (Theorem by Minkowski in 1910) vid. [58]. Figure 9.3 illustrates this theorem.

Given a hyperplane $\vec{a}^T \vec{x}$ and a convex set S, it is said that the hyperplane *supports* S at $\vec{x}_0$ if all points $\vec{x}$ of S satisfy: $\vec{x} \in S \Rightarrow \vec{a}^T \vec{x} \leq \vec{a}^T \vec{x}_0$. The hyperplane is called *supporting hyperplane* and is tangent to S. Figure 9.4 shows an example.

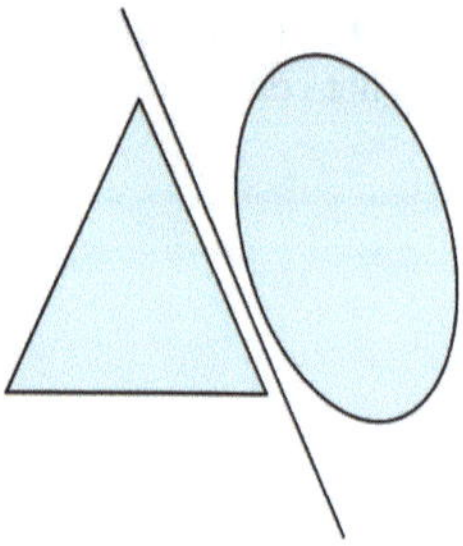

Fig. 9.3 A separating hyperplane

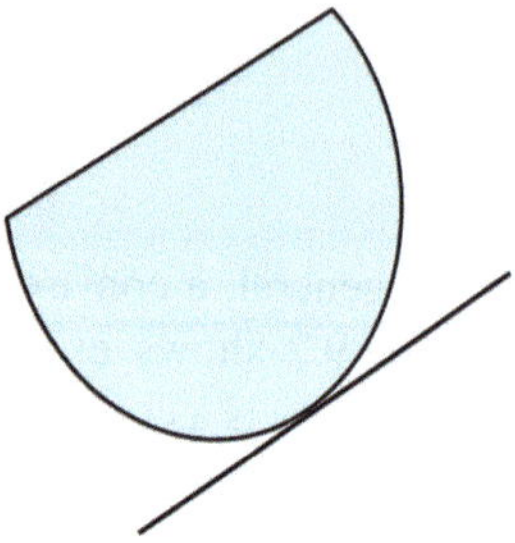

Fig. 9.4 A supporting hyperplane

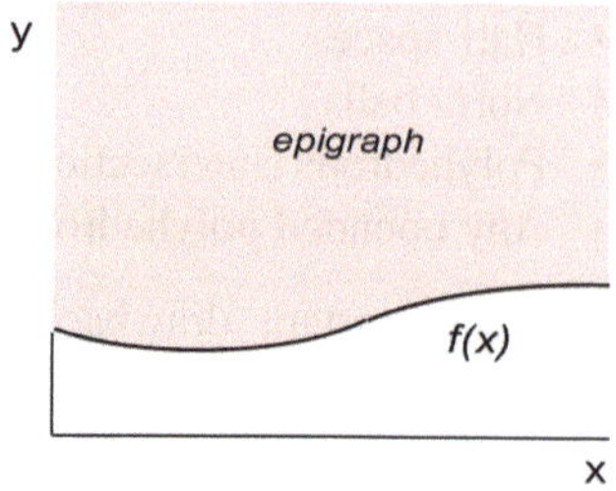

Fig. 9.5 Epigraph of a function f(x)

The supporting hyperplane theorem establishes that there exists a supporting hyperplane at every point of the boundary of a convex set.

The *epigraph* of a function $f(x)$ is the set (x, y) such that $y \geq f(x)$. Figure 9.5 depicts the epigraph of a function $f(x)$; the epigraph being the area above the curve $f(x)$.

Similarly, the *hypograph* of the function $f(x)$ is the area below the curve.

9.2.3 *Convex Functions*

The function $f(x)$ is a convex function if its epigraph is a convex set.

The convexity of a differentiable function can also be characterized by its gradient and Hessian. Based on the first-order Taylor approximation, a first order condition can be enounced as follows, $f(x)$ is convex if

$$f(x) \geq f(x_0) + f'(x_0)(x - x_0)$$

Strict convexity corresponds to having $>$ instead of $\geq$ in this equation.

Likewise, a second-order condition can be enounced, so $f(x)$ is convex if its Hessian is positive semidefinite ($\geq$ 0). Having $>$ instead of $\geq$ implies strict convexity (it is a sufficient, but not necessary condition).

- A function $f(.)$ is concave or strictly concave if $-f(.)$ is convex or strictly convex.
- Affine functions are both convex and concave (the Hessian is zero)
- Quadratic functions $\vec{x}^T P\vec{x}$ are convex if P is positive semidefinite
- The sum of convex functions is also convex

Examples of convex functions:

- Univariate functions like $|x|$, e^x, x^2
- Multivariate functions of the form $a^T x + b$, with a and b being a constant, vector or matrix
- Norms
- x^α(and $|x|^\alpha$) with $\alpha \geq 1$
- $x \log x$ with $x > 0$

A function $f(.)$ is strongly convex with parameter m>0 provided that:

$$f(x) - \frac{m}{2}\|x\|_2^2$$

is a convex function.

($f(.)$ is "as least as convex" as a quadratic function).

As said before, norms are convex. Important examples of norms are the following:

- The l_p norm, defined as

$$\|x\|_p = \left(\sum_{i=1}^{n} |x_i|^p\right)^{1/p}$$

- The trace of a matrix (the sum of its singular values)

The quadratic function,

$$f(\vec{x}) = \frac{1}{2}\vec{x}^T Q\vec{x} + \vec{c}^T\vec{x} + b$$

is convex provided that $Q$is a positive semidefinite matrix.

The quadratic function $f(\vec{x}) = \|A\vec{x} - b\|_2^2$ usually employed for least-squares criteria is hence convex.

The $\max(x_1, \ldots, x_n)$ function is convex.

Since in certain situations it would be difficult to determine the problem at hand is convex, it is recommendable to specify the optimization problems, whenever possible, so they were convex by construction (we shall come back to this point in another section).

9.3 Cone Programming

It can be shown that any convex program can be written as a cone programming [31]. So, cone programming is important, providing at the same time a convenient mathematical framework.

As said in [32], historically, linear programming appeared in 1940s, much earlier than conic programs. Conic programs were introduced in the 1990s as a generalization of linear programming and were shown to enjoy some of the nice theoretical (and sometimes computational) properties of linear programming.

9.3.1 Cones and Cone Programming

A n-dimensional set K is a *cone* if:

$$\vec{x} \in K \;\Rightarrow\; \theta\,\vec{x} \in K\,,\; \forall \theta \geq 0$$

Figure 9.6 shows some examples of 2D cones. Obviously, one of them is not convex.

The intersection of two cones is a cone.

Some definitions:

- A set K is a *convex cone* if K is a cone and K is convex.
- A cone K is a *proper cone* if it is convex, closed, and it is pointed (if $\vec{x} \in K$ and $-\vec{x} \in K$ then $\vec{x} = \vec{0}$).

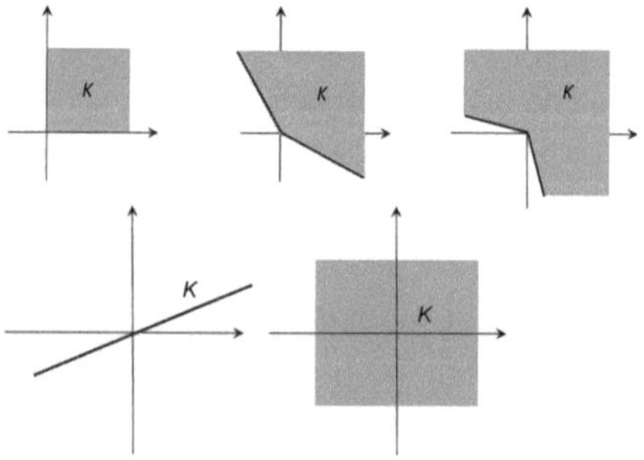

Fig. 9.6 Examples of 2D cones

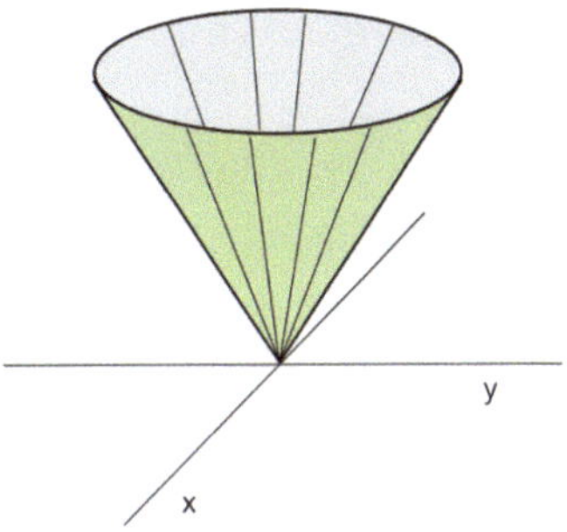

Fig. 9.7 Second-order cone

Important examples of proper cones are the following:

- The non-negative orthant:

$$K = \{\vec{x} \mid x_i \geq 0, \ i = 1 = 1, 2 \ldots, n\}$$

- The cone of positive semidefinite (PSD) matrices:

$$K = \{A \in S^n \mid \vec{u}^T A \, \vec{u} \geq 0, \ \forall \vec{u} \in \Re^n\}$$

with S^n being the set of $n \times n$ symmetric matrices

- Norm cones:

$$K = \{(\vec{x}, t) \mid \ \|\vec{x}\| \leq t\}$$

An example of norm cones is the *second-order cone (SOC)*, which corresponds to $\|\vec{x}\| = \|\vec{x}\|_2$ (that is: $\sqrt{x^2 + y^2} \leq z$). Figure 9.7 depicts a second order cone. This cone is also called the Lorentz cone or, more colloquially, the ice cream cone.

A few lines above, we mentioned the *cone* of PSD matrices. Let us give an explanation. Any matrix could be viewed as a vector of components, from entry-11 to entry-*nn*. For example, a 3×3 matrix could be expressed as a series of nine entries; and this could be represented as a point in a space of nine dimensions. Thus, a set of matrices could form a cone in that space.

9.3.2 Cone Programming

Cone programming is a particular case of convex programming, with the following formulation:

Minimize:

$$f_0(\vec{x})$$

Subject to:

$$f_i(\vec{x}) \leq 0, \quad i = 1, 2, \ldots, m$$

$$h_j(\vec{x}) = 0, \quad j = 1, 2, \ldots, r$$

$$\vec{x} \in K$$

where K is a convex cone.

This problem may adopt more specific forms, depending on the type of cone and on the other aspects of the problem. An important example corresponds to K being a second-order cone; the problem is then called *second-order cone programming*.

9.3.3 Second-Order Cone Programming (SOCP)

Many optimization applications can be formulated as second-order cone programming cases. There are numerous technical and scientific publications taking advantage of this fact in a wide variety of practical examples.

The well-known review articles, [52] (2879 cites) and [3] (1891 cites), on SOCP applications, identify some general categories of problems that can be cast as SOCP:

Quadratic optimization problems with quadratic constraints (QCQP):

- Sums and maxima of norms. Distances
- Optimization problems having hyperbolic constraints
- Robust linear programming
- Robust least-squares

Pierre de Fermat formulated before 1640 the following problem: given three points, A, B, and C (a triangle), locate a point D such the *sum of distances* from D to each of these three points was minimized. This was the beginning of location theory. The problem was geometrically solved by Torricelli in 1645. After a century, in 1750, a generalization of the question was proposed, the Weber problem: find a point in the plane that minimizes the sum of transportation costs from this point to n destinations [48]. It is clearly related to facility location. It was geometrically solved by T. Simpson in 1750. In 1972, L-N. Tellier found a numerical solution. This is a famous example of minimizing a sum of norms. There are versions of the problem that include attraction-repulsion forces.

In the guide to conic optimization provided by [51], an example of hyperbolic constraints was treated, related to inventory control. Another case of hyperbolic constraints was cited in [52], concerning the logarithmic Chebyshev approximation problem.

There is a section in [3] about robust linear programming and robust least squares. Typically, some uncertainty on certain parameters is included in the formulation of the objective function and/or the constraints.

Pertaining applications [52] introduce some examples, about the antenna array weight design, the grasping force optimization (robotics), the design of digital filters, the portfolio optimization with loss risk constraints, the design of truss structures, the optimization of mechanical structures with springs, etc. In turn, [3] cites facility location scenarios, image restoration problems (computer vision), and variations of the portfolio optimization.

Other publications apply SOCP for the optimization of sensor networks, for beamforming in radio communication, minimum surface area problems, AC optimal power flow, limit analysis in load-carrying structures, trajectory planning for hypersonic entry, three-dimensional reconstruction for laparoscopic images, simultaneous localization and mapping (SLAM) in robotics, multiclass SVM, robotic/orthopedic gloves, quantitative finance [34], optimal selection of tree breeding, stability analysis of walking robots, spacecraft attitude control, lens adjustment in semiconductor lithography, etc.

MATLAB provides the solver (a function) *coneprog( )* for second-order cone programming. Another function, *secondordercone( )*is provided for constraint specification.

Program 9.1 gives a simple example of second-order cone programming, extracted from the MATLAB page on *coneprog( )*.

Program 9.1 Example of second order cone programming

```
%example of second order cone programming
% constraint definition -------------
% norm(Asc*x - bsc) <= d'*x - gamma;
Asc = diag([0.9,1/2,0.7]);
bsc = zeros(3,1);
d = [0;0;1];
gamma = 0;
%the constraints:
socCT = secondordercone(Asc,bsc,d,gamma);
%-----------------------------------
f = [-2,-3,1]; %objective function
Aineq = []; bineq = [];
Aeq = []; beq = [];
lb = [-Inf,-Inf,0];
ub = [Inf,Inf,2];
[x,fval]=coneprog(f,socCT,Aineq,bineq,Aeq,beq,lb,ub)
```

As a result, this program prints the following message:

Optimal solution is found.

x =

0.5527

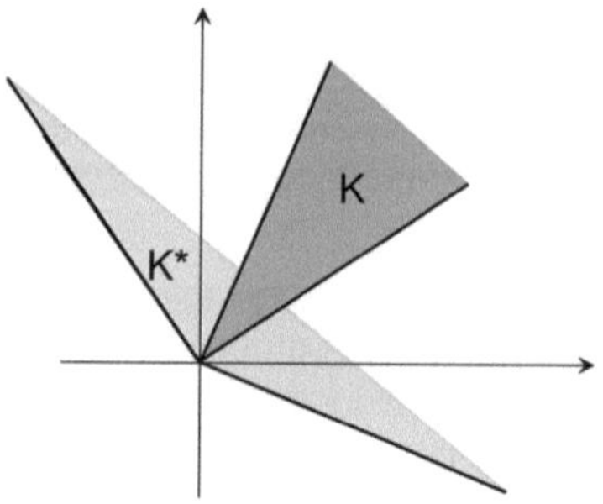

Fig. 9.8 Example of cone K and dual cone K*

2.6777
2.0000
fval =
-7.1386

9.3.4 Dual cones

The theory also considers dual cones. Given a cone K, its *dual cone* $K*$ is the following:

$$K* = \{\vec{y} \mid \vec{x}^T \vec{y} \geq 0, \ \forall \vec{x} \in K\}$$

(recall the equation of the perpendicular to a line).

Figure 9.8 shows an example of two-dimensional cone and dual cone.

The dual cone is always convex, even when K is not. If K is a proper cone, then $K = K**$.

A cone is self-dual if $K* = K$. The positive orthant, the SOC, and the cone of PSD matrices are self-dual cones.

Like in the case of LP, it is possible to work with primal-dual problems based on the Lagrange approach. Concretely, in the following conic programming case [37]:

Inf: $\vec{c}^T \vec{x}$

Subject to: $A\vec{x} = \vec{b}, \ \vec{x} \in K$

If K is a solid, pointed, closed convex cone, the dual problem would be

Sup: $\vec{b}^T \vec{y}$

Subject to: $A^T \vec{y} + \vec{s} = \vec{c}, \ \vec{s} \in K*$

Suppose we have found solutions $\vec{x}*$ and $\vec{y}*$ for the primal and the second problem respectively. Then, one has

$$\vec{b}^T \vec{y}* = (A\,\vec{x}*)^Y \vec{y}* = (\vec{x}*)^T (A^T \vec{y}*) = (\vec{x}*)^T (\vec{c} - \vec{s}*) =$$

$$= \vec{c}^T \vec{x}* - (\vec{s}*)^T \vec{x}* \leq \vec{c}^T \vec{x}*$$

Therefore, if we denote $d* = \vec{b}^T \vec{y}*$ and $p* = \vec{c}^T \vec{x}*$, we find that:

$$\vec{d}* \leq p*$$

This situation is called *weak duality*. It is possible to have a nonzero gap (gap is the difference $p* - d*$).

Zero gap means *strong duality*.

Consider the problem:

Minimize: x_3

Subject to

$$x_1 \geq \sqrt{x_2^2 + x_3^2}$$

$$x_2 \geq x_1 \ ; \ x_3 \geq -1$$

The solutions would be $(x, x, 0)$, with $p* = 0$.

The dual problem is

Maximize: $-y_2$

Subject to:

$$y_1 \geq \sqrt{y_1^2 + (1 - y_2)^2}$$

Now, the solution is $(y, 1)$, with $d* = -1$.

Therefore, there is a nonzero gap $p* - d* = 1$. We have weak, not strong, duality.

It is said that the primal problem is *strictly feasible* if there exists a point $\vec{x} \in int(K)$ in the interior of K, such that $A\vec{x} = \vec{b}$.

The **criterion of Slater** establishes that strong duality holds if the primal problem is strictly feasible and $p*$ is finite. Moreover, if both primal and dual problems are strictly feasible, then $p*$ and $d*$ are attained.

9.3.5 Some Observations

A classical lesson of geometry is the one devoted to the different curves generated by the intersection of a plane and a cone: the conic sections. In the case of conic programming, this is extended to n-dimensions. The feasible region can be the intersection of polyhedrons, ellipsoids, paraboloids, and hyperboloids. Quadratic programming can be easily treated by adding a new variable and considering the quadratic objective function as included in the set of constraints.

There are some peculiarities to have in mind. For instance, the problem:

$$\min\{x_3 \mid x_1 = 1,\ x_2 = 1,\ x_3 \geq \sqrt{x_1^2 + x_2^2}\}$$

has an irrational solution: $x_3 = \sqrt{2}$.

Another case:

$$\inf\{x_3 - x_2 \mid x_1 = 1,\ x_3 \geq \sqrt{x_1^2 + x_2^2}\}$$

Although $x_3 - x_2$ can be made arbitrarily close to 0, it is not possible to actually reach 0. Therefore, in some cases, the infimum (or the supremum of the dual) may not be attained.

9.3.6 *Copositive Programming*

Copositive optimization has been receiving increasing attention because many combinatorial problems can be tackled in this domain.

The concept of copositivity comes from the 1950s. After some contributions from investigators in the 1990s, an article from Bomze et al. [10], year 2000, established a copositive formulation of the quadratic optimization problem (we shall see an example of this), and coined the term "*copositive programming*."

A symmetrical real $n \times n$ matrix M is *copositive* if

$$\vec{x}^T M \vec{x} \geq 0$$

for every nonnegative vector $\vec{x} \geq 0$.

An $n \times n$ matrix A is *completely positive* if it can be factored as $A = B^T B$ where B is an $m \times n$ entrywise nonnegative matrix for some integer m.

The set of copositive matrices form a cone ς. This cone is significantly larger than the semidefinite matrix cone and the nonnegative matrix cone.

We shall denote as $\langle C, X \rangle$the following operation:

$$\langle C, X \rangle = \sum_{i=1}^{n} \sum_{j=1}^{n} c_{ij} x_{ij}$$

A copositive program, [29], is a linear optimization problem in matrix variables of the following form:

Minimize: $\langle C, X \rangle$

Subject to:

$$\langle A_i, X\rangle = b_i \text{ with } i = 1, \ldots, m$$

$$X \in \varsigma$$

The cone ς is not self-dual. Instead, the dual of ς is the cone of completely positive matrices. Both cones, ς and $\varsigma*$ are convex, closed, and pointed, however, not polyhedral. More details and some applications in [1].

As indicated before, there is an illustrative connection between copositive and quadratic programming. Let us quickly explore this aspect.

The standard quadratic optimization problem can be stated as follows:

Minimize: $\vec{x}^T Q \vec{x}$

Subject to:

$$\vec{e}^T \vec{x} = 1$$

$$\vec{x} \geq 0$$

where $\vec{e}$ is a vector of ones.

With some manipulation, the objective function can be written as follows:

$$\vec{x}^T Q \vec{x} = \left\langle Q, \vec{x}\,\vec{x}^T \right\rangle$$

also, $\vec{e}^T \vec{x} = 1$ can be expressed as $\langle E, \vec{x}\,\vec{x}^T\rangle$; with $E = \vec{e}\,\vec{e}^T$.

And so, the quadratic optimization problem can be reformulated as

Minimize: $\langle Q, X\rangle$

Subject to:

$$\langle E, X\rangle = 1$$

$$X \in \varsigma*$$

The interesting point is that all the complexity of the problem has moved into the constraint $X \in \varsigma*$.

For the interested reader it would be recommended to see the thesis [2] and the contributions of Samuel Burer, University of Iowa (for instance, [20] and [21]).

9.4 Semidefinite Programming (SDP)

According to [2], semidefinite programming is one of the most well-studied special cases of cone programming. It is a natural generalization of linear programming, and it is one of the polynomial time solvable classes of cone programming. SDP is frequently used for combinatorial optimization, statistics, structural optimization, etc. Reference publications are the handbook [63] and the article [61] (5800 cites).

The semidefinite optimization problem is a convex optimization problem and is specified in terms of positive semidefinite matrices.

A $n \times n$ real symmetric matrix A is said to be *positive semidefinite* if it satisfies one of the following properties:

- For all $\vec{x} \in \Re^n$, $\vec{x}^T A \vec{x} \geq 0$
- All eigenvalues of A are positive.

(the set of positive semidefinite matrices is denoted as S_+^n)

A matrix P is *positive definite* if for all $\vec{x} \in \Re^n$, $\vec{x}^T P \vec{x} > 0$, $\vec{x} \neq 0$.

(the set of positive definite matrices is denoted as S_{++}^n)

All eigenvalues of a symmetric matrix are real. The corresponding eigenvectors can be chosen so that they are orthogonal. The determinant of the matrix is the product of the eigenvalues; and the sum of all entries is equal to the sum of the eigenvalues (the trace of the matrix).

9.4.1 The SDP Problem

The SDP problem can be seen as a generalization of the LP problem, by using matrices in lieu of variables and parameters. The SDP problem can be stated as follows, [25]:

Minimize:

$$\langle C, X \rangle$$

Subject to

$$\langle A_i, X \rangle = b_i \ , \ i = 1, 2, \ldots, m$$

$$X \geq 0$$

The SDP problem is a conic programming case, since the set of positive semidefinite matrices $X$is a cone.

Recall from the previous section that any matrix could be viewed as a vector of components. So it is possible to define, as in [61], the SDP problem in the following terms:

Minimize:

$$\vec{c}^T \vec{x}$$

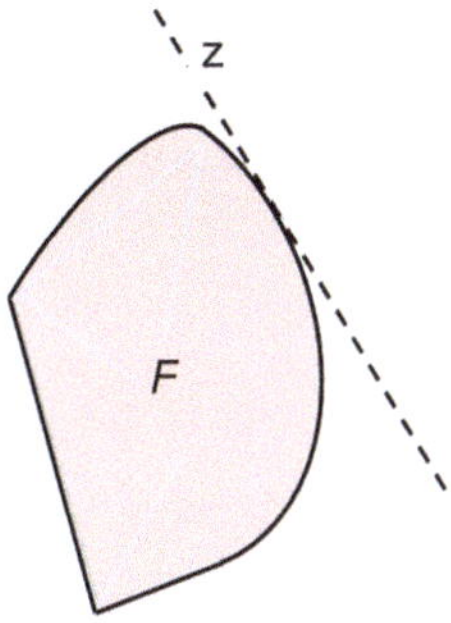

Fig. 9.9 Example of SDP feasibility region

Subject to

$$G(\vec{x}) \geq 0$$

where

$$G(\vec{x}) = G_0 + \sum_{i=1}^{m} x_i G_i$$

and the $m + 1$ matrices $G_0, G_1, \ldots, G_m \in \Re^{n \times n}$ are symmetric.

The expression $G(\vec{x}) \geq 0$ means that G is positive semidefinite. This expression is also called a *linear matrix inequality* (LMI). Some topics of modern control theory are directly linked with the use of LMIs.

Figure 9.9 shows a simple example of the feasible region of a SDP problem. It consists of two line segments and two smooth curved segments. The optimal solution will be at the boundary of the feasible set F. In this sense, the problem has evident similarities with the LP problem.

An Abstract Example

Consider the following simple example of a nonlinear convex optimization problem [61], which can be handled as SDP but not as LP:

Minimize:

$$\frac{(\vec{c}^T \vec{x})^2}{\vec{d}^T \vec{x}} (\text{with } \vec{d}^T \vec{x} > 0)$$

Subject to:

$$A\vec{x} + \vec{b} \geq 0$$

Using a typical trick, it can be written as

Minimize: t

Subject to:

$$A\vec{x} + \vec{b} \geq 0$$

$$\frac{(\vec{c}^T \vec{x})^2}{\vec{d}^T \vec{x}} \leq t$$

In LMI form:

Minimize: t

Subject to:

$$\begin{pmatrix} diag(A\vec{x} + \vec{b}) & 0 & 0 \\ 0 & t & \vec{c}^T \vec{x} \\ 0 & \vec{c}^T \vec{x} & \vec{d}^T \vec{x} \end{pmatrix} \geq 0$$

In this way, a SDP formulation has been obtained. Notice that the matrix has a block-diagonal form. The matrix inequality:

$$\begin{pmatrix} t & \vec{c}^T \vec{x} \\ \vec{c}^T \vec{x} & \vec{d}^T \vec{x} \end{pmatrix} \geq 0 \quad (1)$$

is equivalent to $\vec{d}^T \vec{x} \geq 0$ and $t - \frac{(\vec{c}^T \vec{x})^2}{\vec{d}^T \vec{x}} \geq 0$.

The term, $t - \frac{(\vec{c}^T \vec{x})^2}{\vec{d}^T \vec{x}} \geq 0$, is called the *Schur complement* of $\vec{d}^T \vec{x}$ in the matrix inequality (1).

9.4.2 Duality

(a) **The dual SDP problem**

The dual SDP problem would be:

Maximize:

$$- tr(G_0 Z)$$

Subject to:

$$tr(G_i Z) = c_i \, , \; i = 1, 2, \, \ldots, m$$

$$Z \geq 0$$

where $tr(.)$ is the trace of the matrix, and $Z = Z^T$ is the variable added for the Lagrangian.

The duality gap would be

$$\vec{c}^T\vec{x} + tr(G_0 Z) = tr(G(\vec{x})Z)$$

(b) Another statement of the duality gap
Given a feasible solution X of the primal and a feasible solution (y, S) of the dual, the duality gap is:

$$\langle C, X\rangle - \sum_{i=1}^{m} y_i b_i = \langle S, X\rangle \geq 0$$

If the duality gap was zero, then the solutions X and (y, S) were optimal solutions of the primal and the dual problem respectively.

(c) The dual problem is also SDP
An equivalent expression of the dual SDP problem is:
Maximize:

$$\vec{b}^T\vec{y}$$

Subject to:

$$C - \sum_{i=1}^{m} y_i A_i \geq 0$$

And it happens that this is also a positive semidefinite problem, similar to
Minimize:

$$\vec{c}^T\vec{x}$$

Subject to:

$$G(\vec{x}) \geq 0$$

with:

$$G(\vec{x}) = G_0 + \sum_{i=1}^{m} x_i G_i$$

(d) Primal-dual optimization

In most SDP problems, the optimum of the primal problem is coincident with the optimum of the dual problem, so the duality gap is zero. A combined view of the optimization problem could be to consider:

Minimize:

$$\eta = \vec{c}^T \vec{x} + tr(G_0 Z)$$

Subject to:

$$G(\vec{x}) \geq 0, \;\; Z \geq 0,$$

$$tr(G_i \, Z) = c_i, \;\; i = 1, \, 2, \, \ldots, m$$

This form, which intends to minimize the duality gap η, is called the primal-dual optimization approach.

9.4.3 *Example of SDP problem*

Here is an example of SDP problem [30]:

$$A_1 = \begin{pmatrix} 1 & 0 & 1 \\ 0 & 3 & 7 \\ 1 & 7 & 5 \end{pmatrix} \quad , \quad A_2 = \begin{pmatrix} 0 & 2 & 8 \\ 2 & 6 & 0 \\ 8 & 0 & 4 \end{pmatrix} \quad ; \quad b = \begin{pmatrix} 11 \\ 9 \end{pmatrix}; \quad C = \begin{pmatrix} 1 & 2 & 3 \\ 2 & 9 & 0 \\ 3 & 0 & 7 \end{pmatrix}$$

The variable X:

$$X = \begin{pmatrix} x_{11} & x_{12} & x_{13} \\ x_{21} & x_{22} & x_{23} \\ x_{31} & x_{32} & x_{33} \end{pmatrix}$$

Therefore:

$$\langle C, X \rangle \; = x_{11} + 2x_{12} + 3x_{13} + 2x_{21} + 9x_{22} + 0x_{23} + 3x_{31} + 0x_{32} + 7x_{33}$$

and the constraints $\langle A_i, X \rangle = b_i \;, \; i = 1, \, 2, \, \ldots, m$:

$$x_{11} + 0x_{12} + 1x_{13} + 0x_{21} + 3x_{22} + 7x_{23} + 1x_{31} + 7x_{32} + 5x_{33} = 11$$

$$0x_{11} + 2x_{12} + 8x_{13} + 2x_{21} + 6x_{22} + 0x_{23} + 8x_{31} + 0x_{32} + 4x_{33} = 9$$

All these polynomials can be further simplified taking into account that X is symmetric.

The dual problem of the above primal SDP problem would be
Maximize:

$$\sum_{i=1}^{m} y_i b_i$$

Subject to:

$$\sum_{i=1}^{m} y_i A_i + S = C$$

$$S \geq 0$$

So, in the case of the previous example:
Maximize:

$$11y_1 + 9y_2$$

Subject to:

$$y_1 \begin{pmatrix} 1 & 0 & 1 \\ 0 & 3 & 7 \\ 1 & 7 & 5 \end{pmatrix} + y_2 \begin{pmatrix} 0 & 2 & 8 \\ 2 & 6 & 0 \\ 8 & 0 & 4 \end{pmatrix} + S = \begin{pmatrix} 1 & 2 & 3 \\ 2 & 9 & 0 \\ 3 & 0 & 7 \end{pmatrix}$$

$$S \geq 0$$

These constraints can be also written as:

$$\begin{pmatrix} 1 - 1y_1 - 0y_2 & 2 - 0y_1 - 2y_2 & 3 - 1y_1 - 8y_2 \\ 2 - 0y_1 - 2y_2 & 9 - 3y_1 - 6y_2 & 0 - 7y_1 - 0y_2 \\ 3 - 1y_1 - 8y_2 & 0 - 7y_1 - 0y_2 & 7 - 5y_1 - 4y_2 \end{pmatrix} \geq 0$$

The lecture notes [33] and the web site [41] give more information on SDP.

9.5 Cutting Plane in General Contexts

Cutting planes can be used in other contexts, not only in integer programming problems. Actually, they have earned an important role in continuous variable optimization methods, including nonsmooth scenarios.

This section is based on [16], which contains the mathematical details of the topic. The cutting plane method we are to introduce was proposed by James E Kelley Jr in 1960, [46] (2265 cites).

Fig. 9.10 Example of cutting plane

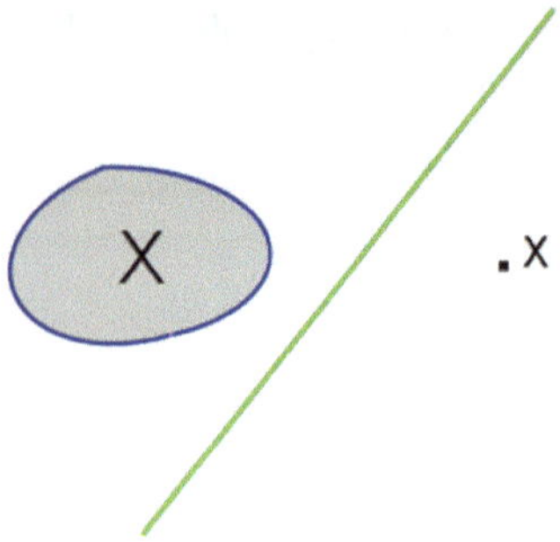

Given a feasible set F, our aim would be to determine a (small) subset X such that the optimal solution belongs to X (called *the target set*). The idea is to apply a sequence of cuts to F, obtaining smaller and smaller regions that contain the optimum.

The cutting plane method uses an *oracle*. When you query the oracle about a certain point $\vec{x} \in \Re^n$, the oracle tells you that $\vec{x} \in X$, so a successful result has been reached, or it gives you the parameters $\vec{a}$, b corresponding to a separating hyperplane between $\vec{x}$ and X :

$$\vec{a}^T\vec{z} \leq b \ for \ z \in X \text{ and } \vec{a}^T\vec{x} \geq b$$

Figure 9.10 illustrates the situation:

The hyperplane is called a *cutting plane* or *cut*. If it contains the point $\vec{x}$, it is called a *neutral cut*. The case depicted in Fig. 9.10 corresponds to a *deep cut*, so the point $\vec{x}$ is inside the half-space on the other side of X.

Suppose the problem was to minimize a convex and differentiable $f_o(\vec{x})$. Given a certain point $\vec{x}$, a cutting plane containing $\vec{x}$ (a neutral cut) is the following:

$$\nabla f_o(\vec{x})^T(\vec{z} - \vec{x}) \leq 0 \quad (2)$$

(supposing that $\vec{x}$ is not the optimal and so $\nabla f_o(\vec{x}) \neq 0$).

Now, let us consider the constraints corresponding to F:

$$f_i(\vec{x}) \leq 0, \ i = 1, 2, \ldots m \text{ (convex and differentiable).}$$

Given a nonfeasible point $\vec{x}$, this means that at least one of the constraints is violated; for instance: $f_j(\vec{x}) > 0$. Then, any feasible $\vec{z}$ satisfies:

$$f_j(\vec{x}) + \nabla f_j(\vec{x})^T(\vec{z} - \vec{x}) \leq 0 \qquad (3)$$

Therefore, this is the equation of a deep cutting plane.

Fig. 9.11 Example of localization polyhedron

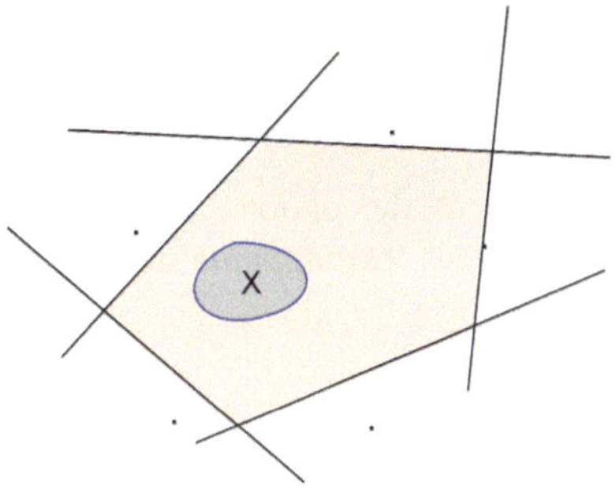

Let us look at the complete problem, for which one could combine the above results:

Minimize $f_o(\vec{x})$

Subject to: $f_i(\vec{x}) \leq 0\,,\ i = 1,\ 2,\ \ldots m$

If the query to the oracle concerns a certain point $\vec{x}$, this point may be not feasible, so a deep cut would be (3). It is called a *feasibility cut*, for one is cutting away, a halfspace of infeasible points.

If the point is feasible and not optimal, then a neutral cut would be (2). It is called an *objective cut*, which cuts out a halfspace of points corresponding to too large objective values.

Basic Algorithm (Localization Polyhedron)

One could start the process choosing a ball large enough to contain X. Then, the oracle is queried at a series of points $\vec{x}_1,\ \vec{x}_2,\ \ldots,\ \vec{x}_k$. The sequence stops if a point belonging to X is found. Otherwise, what is most probable, one has obtained a series of cutting planes that can be written as

$$\vec{a}_i^T \vec{z} \leq b_i\,,\quad i = 1,\ 2,\ \ldots, k \quad (4)$$

These planes would constitute a localization polyhedron P_k, as depicted in Fig. 9.11.

If one adds a new point, in the interior of P_k, not belonging to X, the oracle would return a new cutting plane (a new inequality), and this means a new polyhedron $P_{k+1} \subset P_k$.

The algorithm will iterate, obtaining better and better localization of X. The progress could be measured in several ways, like for instance with a ratio of volumes of P_{k+1} and P_k.

When trying a new query point, one wants to remove as much as possible from the current polyhedron P_k. Given the ignorance about where is X, the best choice on average would be a point near the "center" of P_k and to use a neutral cut.

Several alternatives have been proposed to implement this idea:

- Use the *center of gravity* of the polyhedron P_k

- Obtain the *maximum volume ellipsoid (MVE)* contained in P_k and then use its center
- Choose the *analytic center (ACCPM)* (vid. [18])

One of the options in the *setSolver( )* function of the MATLAB Financial Toolbox, is *'cuttingplane'* for using the Kelley's cutting plane method.

9.6 The Ellipsoid Method

The ellipsoid method is an algorithm for solving general convex optimization problems. It was the first polynomial-time algorithm discovered for linear programs and can be applied for many other scenarios, like combinatorial optimization or machine learning.

The method is a generalization of binary search. Recall that binary search keeps an interval that contains the solution; in each iteration, one tests whether the midpoint of the interval is a solution: if not, the search proceeds with one of two half-intervals. The ellipsoid method generalizes the strategy to n dimensions, using ellipsoids as generalization of intervals.

An ellipsoid consists of all points $\mathbf{x} \in \Re^n$ satisfying the following inequality:

$$(\mathbf{x} - \mathbf{z})^T\ Q\ (\mathbf{x} - \mathbf{z})$$

where $\mathbf{z} \in \Re^n$ is the center of the ellipsoid and Q is any $n \times n$ positive-definite matrix.

The heart of the method is the Löwner-John ellipsoids, [43]. Suppose there is an ellipsoid E_0 with center $\mathbf{z_0}$ that we know contains a convex body P. Suppose also that $\mathbf{z_0} \notin P$ and there is a separating hyperplane such as $\mathbf{a}^\mathbf{T}\mathbf{x} \geq \mathbf{a}^\mathbf{T}\mathbf{z}_0$ for all $\mathbf{x} \in P$. Then we can find a smaller ellipsoid that still contains P. This process can be iterated, from one ellipsoid to another. Figure 9.12 illustrates the construction from an ellipsoid E_t to the next one E_{t+1}. The hyperplane is labelled as H_t.

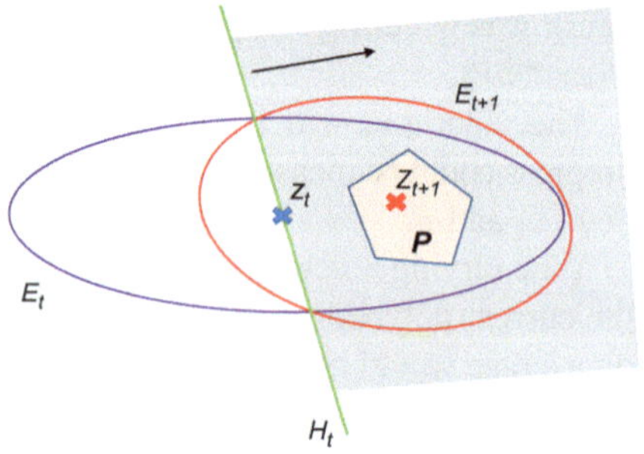

Fig. 9.12 Löwner-John ellipsoids

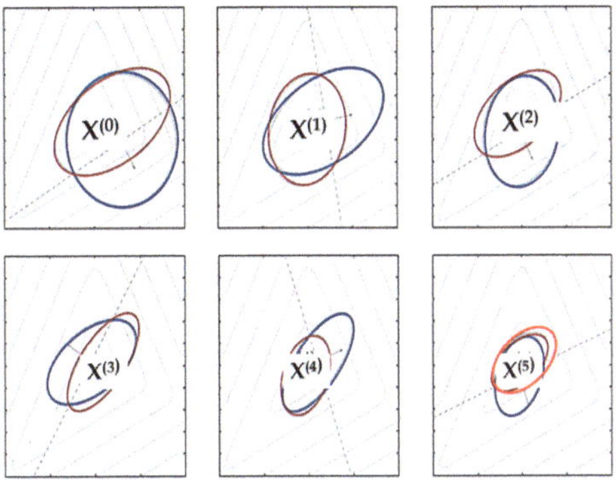

Fig. 9.13 Evolution of ellipsoids

The ellipsoid method takes *P* as the feasible region defined by the constraints, applies iterations of the Löwner-John ellipsoids, and uses a separation oracle (like in the case of cutting planes).

Figure 9.13 depicts a sequence of the algorithm iterations. The volume of ellipsoids decreases from one to another iteration. It may happen that the final ellipsoid, sufficiently small in volume, has to be very long and flat.

The origins of the ellipsoid method could be situated in the Soviet Union, during the 1970s, with contributions from N.Z. Shor, D.B. Yudin, and A. Nemirovski. In 1979, L.G. Khachiyan published his seminal work, showing that linear programming problems can be solved in polynomial time, by using the ellipsoid method. This feature is of great theoretical interest.

Notice that the ellipsoid method does not rely upon the linearity of both the constraints and the objective function (which can be nondifferentiable).

The algorithm is easy to implement, requiring not much storage. It is not fast however and may suffer from numerical instability.

The hyperplane acts as a cut. In fact, the original algorithm applies "*neutral cuts*." For improvement purposes, the research has proposed deep cuts, surrogate cuts, shallow cuts, etc. (see for instance [6, 40] and the document [14] of S.P. Boyd on the ellipsoid method).

One of the contributions available from the MathWorks File Exchange site (packages in MATLAB) is devoted to the Ellipsoid Method, which includes visualization features. The link is

https://es.mathworks.com/matlabcentral/fileexchange/47837-ellipsoid-method?s_tid=srchtitle

9.7 Other Related Optimization Methods

Let us introduce some other optimization methods, giving opportune references for the interested reader, and some indications on typical uses.

Like in the case of mixed-integer quadratic programming, there are methods that mix integers and convex programming, or cone programming, or semidefinite programming.

The article [8] presents a survey of mixed-integer second-order cone programming. It includes descriptions of a number of algorithms for solving this problem. For example, the rounding cuts introduced in [5] for a branch and bound framework. Some applications are also described in [8], like portfolio optimization, option pricing, network design and operations, cellular networks, power distribution systems, scheduling and logistics, etc. A contemporary application example is given by [36], about optimal generation scheduling in microgrids (including electric vehicle charging).

A generic mixed-integer semidefinite programming solver is explored in [35], including branch and bound rules, a dual fixing technique for relaxation strengthening, presolving, cutting planes, etc. One of the test scenarios considered is truss topology design. Two years after, [47] proposed a faster procedure for branch and cut, finding better results for the truss topology design problem. There are also application examples of high current impact, like the research by [26] on navigation planning for unmanned aerial vehicles in cluttered environments.

This said, we should emphasize the rapidly expanding interest of mixed integer nonlinear programming (MINLP), which could be or not convex. The amplitude of the topic is being reflected now by a good number of books dealing with general o particular aspects. A popular, long article (almost a book) that we could mention is [7] (800 cites, year 2013). This article covers convex and nonconvex MINLP, paying attention to solving methods, software, applications in control systems, etc.

Likewise, there is an interesting survey article (year 2012), [22], on nonconvex MINLP, that looks in detail types of branch and bound techniques, some heuristics, and the special quadratic case.

More specific treatments are the Thesis [49], on polyhedral outer approximations for convex MINLP (the dissertation contains a good, concise review of convex MINLP), and the article [53] on the polyhedral approximation for mixed integer convex optimization. The recent Thesis [55], University of Malaga, put the focus on decomposition methods for MINLP, and has an application section devoted to energy systems planning (by heuristic column generation).

There are numerous applications of MINLP in aerospace systems, like the recent example of [56], or [23] about avoidance of collisions between aircrafts. Another expanding field is removable energy systems; for example, [42] includes MINLP in the analysis of a home energy system with aerogenerator and batteries.

The MathWorks File Exchange site includes the software *MINLP*, for solving optimization problems. In addition, the MATLAB Toolbox for finance includes the function *setSolverMINLP( )* to choose a MINLP solver in portfolio management scenarios.

9.8 Sequential Versions

There are sequential versions of the methods described in this chapter. Concerning sequential convex programming, the article [54] surveys contributions about three methods: sequential convex programming (SCP), sequential convex quadratic programming (SCQP), and sequential quadratically constrained quadratic programming (SQCQP). The lecture notes [28] provides a basic tutorial on SCP. The Thesis [27] contains some applications of SCP, for optimal control. Time-optimal control is also the subject of [12].

Aerospace optimization problems are considering sequential versions. This is the case, for example, of [45] that uses real-time sequential conic optimization for the landing guidance of a rocket. Likewise, [62] employs sequential semidefinite programming for hypersonic trajectory optimization.

The article [11] introduces GuSTO, an algorithmic framework for robot trajectory optimization, based on sequential convex programming. This framework, written in the Julia language, is available from GitHub. Also in GitHub, one can access the *SCP Toolbox* (sequential convex programming) oriented to trajectory optimization. Another GitHub repository of interest is *scvx_matlab*, a general purpose sequential convex programming in MATLAB. There are many other GitHub repositories with similar contents. In fact, asking Google for *github sequential convex programming*, gives 1,640,000 results. It would be recommended to see the entry *semidefinite-programming* in GitHub, which gives 54 public repositories.

9.9 Disciplined Convex Programming and CVX

By using the CVX tool, one can be sure that a given optimization problem is convex and that it is formulated in adequate way. If that problem is not convex, CVX rejects it.

CVX imposes some conventions, called the disciplined convex programming ruleset (DCP ruleset), that the user must adopt when specifying a problem. These conventions enable CVX to automatically verify convexity, and to perform a transformation of the specification to the target language. See [24] for a detailed description of DCP.

There are versions of CVX for MATLAB, Octave, Python, R, Julia, etc. Instructions for easy installation of CVX in MATLAB can be found with the following link:

https://es.mathworks.com/matlabcentral/answers/511063-installing-cvx-in-matlab

Here is an example of problem specification with CVX:

Program 9.2 Using CVX

```
m = 16 ; n = 8;
A=randn(m,n);
b=randn(m,1);
cvx_begin
variable x(n);
minimize(norm(A*x-b) );
cvx_end
```

This program will minimize $\|A\,\mathbf{x} - b\|_2$. Notice that, once variables were initialized, the main code is inside cvx_begin … cvx_end.

Another example, which deals with a second order cone programming problem [60], is the following:

Program 9.3 Another example of CVX use

```
c = [1; 1; -1; 1; -1];
A = [1 1 1 1 1; 1 2 3 4 5];
b = [1; 10];
cvx_begin
variable x(5);
minimize( c'*x );
subject to
A*x == b;
{x(1:2), x(3)} == lorentz(2);
{x(4), x(5)} == lorentz(1);
x(5) >= 1;
cvx_end
```

Disciplined convex programming was the subject of the doctoral thesis, [39] (Stanford University, 2004), by Michael C. Grant, with Stephen P. Boyd as principal advisor.

Professor S.P. Boyd, Stanford University, has a record number of cites about his research in optimization, near 240,000.

The article [38] (2965 cites, 56 pages) gives a complete description of disciplined convex programming, with abundant bibliographic references. A brief tutorial can be found in the book [15]. In addition, the document [17] offers a large collection of examples, academic exercises, and practical applications of convex programming (with some computational solutions in CVX).

9.10 Software

One of the Wikipedia pages offers a list of optimization software. It includes around 50 proprietary software products and then 12 examples of open-source libraries. Some (17) proprietary products can be free for academic use.

The article [50] offers a good review and comparison of solvers for convex MINLP. More recently, [4] compares five MINLP solvers.

Putting the focus on products with MATLAB compatibility, we could mention CVX, YALMIP, MOSEK, UNLocBox, OPTI Toolbox, and TOMLAB. UNLocBox is explicitly developed for convex optimization. OPTI Toolbox has been discontinued although it persists on GitHub. TOMLAB has a wide range of optimization toolboxes, including a solver for MINLP.

Stephen P. Boy has a website, [13] with about 25 software developments, in several languages (some in MATLAB).

Alfonso, a MATLAB package for nonsymmetric conic optimization was introduced in the year 2021, [57].

A helping hand for using/programming MATLAB in optimization problems is the web page [44].

9.11 Exercises

1- [9.1] Here is the following conic programming problem:

Minimize: $J = x_1 + x_2 - x_3 + x_4 - x_5$

Subject to:

$$x_1 + x_2 + x_3 + x_4 + x_5 = 1$$

$$x_1 + 2x_2 + 3x_3 + 4x_4 + 5x_5 = 10$$

$$\sqrt{x_1^2 + x_2^2} \leq x_3$$

$$|x_4| \leq x_5 \ ; \ x_5 \geq 1$$

Find the solution by using MATLAB functions. Add comments on coding aspects.

2- [9.2] Consider the following conic programming problem:

Minimize: $J = x_4 + x_5 + x_6$

Subject to:

$$x_1 + x_2 + 2x_3 = 1$$

$$2x_5 x_6 \geq x_3^2$$

$$\sqrt{x_1^2 + x_2^2} \leq x_4$$

$$x_1,\ x_2,\ x_3 \geq 0$$

Find the solution with MATLAB functions. Add comments on coding aspects.

3- [9.3] Consider the following conic programming problem:

Maximize: $J = x_1^{1/3} + (x_1 + x_2 + 0.1)^{1/4}$

Subject to:

$$(x_1 - 0.5)^2 + (x_2 - 0.6)^2 \leq 1$$

$$(x_1 - x_2) \leq 1$$

Find the solution with MATLAB functions. Visualize in two dimensions the problem and the result obtained.

4- [9.4] Let the problem be

Minimize: $J = x_1 + 2x_2$

Subject to:

$$\begin{pmatrix} (1 - x_2) & x_2 \\ x_2 & (1 + x_1) \end{pmatrix} \quad semidef+$$

Find the solution with CVX and MATLAB. Visualize in two dimensions the problem and the result obtained.

5- [9.5] Consider the following problem:

Minimize: $J = 2x_1 + 3x_2$

Subject to:

$$\begin{aligned} -x_1 - x_2 &\leq -2 \\ 3x_1 &\leq 4 \\ -4x_1 + 4x_2 &\leq 6 \end{aligned}$$

Based on the code offered by [9.5], apply the ellipsoid method for solving the problem. Visualize in two dimensions the evolution of the ellipsoids. Try some changes in the constraints, adding comments on the results.

6- [9.6] Using CVX, develop a MATLAB program for the application of SVM (support vector machine) in the Iris Flower classification (as in Chap. 7). Visualize the results. Add comments on the programming approach and its efficiency.

7- [9.7] Define a portfolio problem as in Chap. 7 (probably you could get data of your interest from the Internet), and apply CVX and MATLAB for solving it. Visualize the result.

9.11.1 Exercise References

[9.1] https://quoctd.web.unc.edu/wp-content/uploads/sites/10081/2018/09/lecture2-disciplined-convex-programming.pdf
[9.2] https://docs.mosek.com/latest/toolbox/tutorial-cqo-shared.html
[9.3] https://docs.mosek.com/latest/toolbox/tutorial-affine-conic.html
[9.4] https://www.damtp.cam.ac.uk/user/hf323/lent2017_topics_convex_optimisation/lecture5.pdf
[9.5] https://es.mathworks.com/matlabcentral/fileexchange/47837-ellipsoid-method
[9.6] https://oceandatamining.sciencesconf.org/conference/oceandatamining/TP_Linear_SVM.pdf
[9.7] https://www.maths.ed.ac.uk/hall/NATCOR_2014/Intro_to_cvx.pdf

References

1. Berman, A., Bomze, I. M., Dür, M., & Shaked-Monderer, N. (2018). Copositivity and complete positivity. Technical report, Mathematisches Forschungsinstitut Oberwolfach. Available from: https://publications.mfo.de/bitstream/handle/mfo/3616/OWR_2017_52.pdf?sequence=1
2. Ahmed, F. (2010). Relations between semidefinite, copositive, semi-infinite and integer programming. Master's thesis, University of Twente. Available from: http://essay.utwente.nl/59751/1/MA_thesis_F_Ahmed.pdf
3. Alizadeh, F., Goldfarb, D. (2003). Second-order cone programming. *Mathematical Programming, 95*(1), 3–51. Available from: https://citeseerx.ist.psu.edu/document?repid=rep1&type=pdf&doi=67ae3a751b1a41682b3e780951d0c4d1fc5af484
4. Alkhalifa, L. (2022). Comparison between five minlp solvers and new results related to trigonometric functions. *Fractals, 30*(10), 2240249. Available from: https://www.worldscientific.com/doi/pdf/10.1142/S0218348X22402496
5. Atamtürk, A., Narayanan, V. (2010). Conic mixed-integer rounding cuts. *Mathematical Programming, 122*(1), 1–20. Available from: https://link.springer.com/content/pdf/10.1007/s10107-008-0239-4.pdf
6. Beck, A., Sabach, S. (2012). An improved ellipsoid method for solving convex differentiable optimization problems. *Operations Research Letters, 40*(6), 541–545. Available from: https://num.math.uni-goettingen.de/~ssabach/BS2012.pdf
7. Belotti, P., Kirches, C., Leyffer, S., Linderoth, J., Luedtke, J., & Mahajan, A. (2013). Mixed-integer nonlinear optimization. *Acta Numerica, 22*, 1–131. Available from: https://www.cambridge.org/core/services/aop-cambridge-core/content/view/2D0CE8CDA53363A31ADE8689565517BD/S0962492913000032a.pdf/mixedinteger_nonlinear_optimization.pdf
8. Benson, H. Y., & Sağlam, Ü. (2013). Mixed-integer second-order cone programming: A survey. In *Theory driven by influential applications* (pp. 13–36). INFORMS. Available from: https://pubsonline.informs.org/doi/pdf/10.1287/educ.2013.0115
9. Bertsekas, D. (2010). Lecture slides on convex analysis and optimization. Available from: https://dspace.mit.edu/bitstream/handle/1721.1/76254/6-253-spring-2010/contents/lecture-notes/MIT6_235S10_lec01.pdf
10. Bomze, I. M., Dür, M., De Klerk, E., Roos, C., Quist, A. J., & Terlaky, T. (2000). On copositive programming and standard quadratic optimization problems. *Journal of Global Optimization, 18*, 301–320. Available from: https://pure.uvt.nl/ws/portalfiles/portal/844622/oncopositive.pdf

11. Bonalli, R., Cauligi, A., Bylard, A., & Pavone, M. (2019). Gusto: Guaranteed sequential trajectory optimization via sequential convex programming. In *2019 International Conference on Robotics and Automation (ICRA)* (pp. 6741–6747). IEEE. Available from: https://arxiv.org/pdf/1903.00155.pdf
12. Bonalli, R., Lew, T., & Pavone, M. (2022). Analysis of theoretical and numerical properties of sequential convex programming for continuous-time optimal control. *IEEE Transactions on Automatic Control, 68*(8), 4570–4585. Available from: https://ieeexplore.ieee.org/stamp/stamp.jsp?arnumber=9895309
13. Boyd, S. (2022). Software. Available from: https://web.stanford.edu/~boyd/software.html
14. Boyd, S., & Barratt, C. (2008). Ellipsoid method. *Notes for EE364B*. Stanford University, 2008. Available from: https://stanford.edu/class/ee364b/lectures/ellipsoid_method_notes.pdf
15. Boyd, S.,& Grant, M. (2008). Disciplined convex programming and cvx. Available from: https://see.stanford.edu/materials/lsocoee364a/review3_single.pdf
16. Boyd, S., & Vandenberghe, L. (2007). Localization and cutting-plane methods. *From Stanford EE 364b lecture notes*, 386. Available from: https://see.stanford.edu/materials/lsocoee364b/05-localization_methods_notes.pdf
17. Boyd, S., & Vandenberghe, L. (2010). Additional exercises for convex optimization. Available from: http://seas.ucla.edu/~vandenbe/ee236b/homework/bv_cvxbook_extra_exercises.pdf
18. Boyd, S., Vandenberghe, L., & Skaf, J. (2008) Analytic center cutting-plane method. Available from: https://see.stanford.edu/materials/lsocoee364b/06-accpm_notes.pdf
19. Boyd, S. P., & Vandenberghe, L. (2004). *Convex optimization*. Cambridge University Press.
20. Burer, S. (2009). On the copositive representation of binary and continuous nonconvex quadratic programs. *Mathematical Programming, 120*(2), 479–495. Available from: https://people.duke.edu/~psun/Burer.2009.pdf
21. Burer, S. (2015). A gentle, geometric introduction to copositive optimization. *Mathematical Programming, 151*, 89–116. Available from: http://sburer.github.io/papers/046-ismp2015.pdf
22. Burer, S., & Letchford, A. N. (2012). Non-convex mixed-integer nonlinear programming: A survey. *Surveys in Operations Research and Management Science, 17*(2), 97–106. Available from: https://core.ac.uk/download/pdf/9660837.pdf
23. Cafieri, S., Conn, A. R., & Mongeau, M. (2022). An MINLP and a continuous optimization formulations for aircraft conflict avoidance via heading and speed deviations. *HAL Open Science*. Available from: https://hal.science/hal-03629088/document
24. CVX Research, Inc. (2013). The DCP ruleset. Available from: http://cvxr.com/cvx/doc/dcp.html
25. De Klerk, E. (2006). *Aspects of semidefinite programming: interior point algorithms and selected applications* (Vol. 65). Springer. Available from: http://lib.ysu.am/disciplines_bk/4453e27f733389456514ef1208ebfe5f.pdf
26. Deits, R., & Tedrake, R. (2015). Efficient mixed-integer planning for UAVs in cluttered environments. In *2015 IEEE International Conference on Robotics and Automation (ICRA)* (pp. 42–49). IEEE. Available from: https://ieeexplore.ieee.org/stamp/stamp.jsp?arnumber=7138978&casa_token=G3YXu3nViJoAAAAA:Szzfxdnocd-pt8D06BQIIq1jutSBAxrlqiO-NBlDya4OqucRwAWJbTTgbGFvX79cx9abWkvQdw
27. Dinh, Q. T. (2012). *Sequential convex programming and decomposition approaches for nonlinear optimization*. PhD thesis, KU Leuven. Available from: https://quoctd.web.unc.edu/wp-content/uploads/sites/10081/2016/09/Tran_PhD_Thesis_onlinever.pdf
28. Duchi, J., Boyd, S., & Mattingley, J. (2018). Sequential convex programming. In *Lecture notes EE364b*. Stanford University. Available from: https://web.stanford.edu/class/ee364b/lectures/seq_notes.pdf
29. Dür, M. (2010). Copositive programming—a survey. In *Recent Advances in Optimization and its Applications in Engineering: The 14th Belgian-French-German Conference on Optimization* (pp. 3–20). Springer. Available from: https://pure.rug.nl/ws/portalfiles/portal/2573836/2010RecAdvOptimApplEngDur.pdf
30. Epelman, M. A. (2007). IOE 511/math 652: Continuous optimization methods, section 1. Available from: http://www-personal.umich.edu/~mepelman/teaching/511notesFA07.pdf

31. Fairbrother, J., Letchford, A. (2011). Mathematical programming through cones. Available from: https://www.lancaster.ac.uk/pg/fairbrot/files/master.pdf
32. Fawzi, H. (2018). Topics in convex optimisation (michaelmas 2018). *Conic programming, Lecture notes 3*, Cambridge University. Available from: http://www.damtp.cam.ac.uk/user/hf323/M18-OPT/lecture3.pdf
33. Freund, R. M. (2004). *Introduction to semidefinite programming (SDP)*. Massachusetts Institute of Technology (pp. 8–11). Available from: https://ise.ncsu.edu/wp-content/uploads/sites/9/2015/07/sdp094_digest.pdf
34. Fylstra, D. (2005). Introducing convex and conic optimization for the quantitative finance professional. *Wilmott Magazine* 18–22. Available from: https://www.solver.com/files/_document/ConvexConicWilmott.pdf
35. Gally, T., Pfetsch, M. E., & Ulbrich, S. (2018). A framework for solving mixed-integer semidefinite programs. *Optimization Methods and Software, 33*(3), 594–632. Available from: https://wwwopt.mathematik.tu-darmstadt.de/preprints/ulbrich/Sulb2017_6.pdf
36. Shaghaghi-shahr, G., Sedighizadeh, M., Aghamohammadi, M., & Esmaili, M. (2020). Optimal generation scheduling in microgrids using mixed-integer second-order cone programming. *Engineering Optimization, 52*(12), 2164–2192. Available from: https://doi.org/10.1080/0305215X.2019.1695790
37. Glineur, F. (2001). *Topics in convex optimization: Interior-point methods, conic duality and approximations*. PhD Thesis, Polytechnic College of Mons. Available from: https://theses.hal.science/tel-00006861/document
38. Grant, M., Boyd, S., & Ye, Y. (2006) Disciplined convex programming. In: *Global optimization: From theory to implementation* (pp. 155–210). Available from: https://web.stanford.edu/~boyd/papers/pdf/disc_cvx_prog.pdf
39. Grant, M. C. (2004). *Disciplined convex programming*. PhD Thesis, Stanford University. Available from: https://web.stanford.edu/~boyd/papers/pdf/mcg_thesis.pdf
40. Grötschel, M., Lovász, L., Schrijver, A., Grötschel, M., Lovász, L., Schrijver, A. (1993). The ellipsoid method. In: *Geometric Algorithms and Combinatorial Optimization* (pp. 64–101). Available from: https://www.mpi-inf.mpg.de/fileadmin/inf/d1/ellipsoid-lovasz.pdf
41. Helmberg, C. (2013). Semidefinite programming. Available from: https://www-user.tu-chemnitz.de/~helmberg/semidef.html
42. Hemmati, R. (2017). Technical and economic analysis of home energy management system incorporating small-scale wind turbine and battery energy storage system. *Journal of Cleaner Production, 159*, 106–118. Available from: https://ui.adsabs.harvard.edu/abs/2017JCPro.159..106H/abstract
43. Henk, M. (2012). löwner-john ellipsoids. *Documenta Mathematica, 95*, 106. Available from: http://ftp.gwdg.de/pub/misc/EMIS/journals/DMJDMV/vol-ismp/24_henk-martin.pdf
44. Hussain, S. Z. (2023). Optimization using MATLAB. Available from: https://www.zakirtechblog.com/post/matlab-optimization/#example-on-how-to-model-socp-with-_coneprog_-of-matlab
45. Kamath, A. G., Elango, P., Yu, Y., Mceowen, S., Carson, J. M. III, & Açıkmeşe, B. (2022). Real-time sequential conic optimization for multi-phase rocket landing guidance. arXiv preprint arXiv:2212.00375. Available from: https://arxiv.org/pdf/2212.00375.pdf
46. Kelley, J. E., Jr. (1960). The cutting-plane method for solving convex programs. *Journal of the Society for Industrial and Applied Mathematics, 8*(4), 703–712. Available from: https://epubs.siam.org/doi/abs/10.1137/0108053?journalCode=smjmap.1
47. Kobayashi, K., Takano, Y. (2020). A branch-and-cut algorithm for solving mixed-integer semidefinite optimization problems. *Computational Optimization and Applications, 75*, 493–513. Available from: https://link.springer.com/article/10.1007/s10589-019-00153-2
48. Krarup, J., Vajda, S. (1997). On Torricelli's geometrical solution to a problem of Fermat. *IMA Journal of Management Mathematics, 8*(3), 215–224. Available from: https://ieeexplore.ieee.org/abstract/document/8132576

49. Kronqvist, J. (2018). *Polyhedral outer approximations in convex mixed-integer nonlinear programming*. PhD Thesis, Åbo Akademi University. Available from: https://www.doria.fi/bitstream/handle/10024/160918/kronqvist_jan.pdf?sequence=2
50. Kronqvist, J., Bernal, D. E., Lundell, A., & Grossmann, I. E. (2019). A review and comparison of solvers for convex MINLP. *Optimization and Engineering, 20*, 397–455. Available from: https://link.springer.com/article/10.1007/s11081-018-9411-8
51. Letchford, A. N., & Parkes, A. J. (2018). A guide to conic optimisation and its applications. *RAIRO-Operations Research, 52*(4–5), 1087–1106. Available from: http://www.numdam.org/item/10.1051/ro/2018034.pdf
52. Lobo, M. S., Vandenberghe, L., Boyd, S., & Lebret, H. (1998). Applications of second-order cone programming. *Linear Algebra and Its Applications, 284*(1–3), 193–228 (1998) Available from: https://www.sciencedirect.com/science/article/pii/S0024379598100320
53. Lubin, M., Yamangil, E., Bent, R., & Vielma, J. P. (2018). Polyhedral approximation in mixed-integer convex optimization. *Mathematical Programming, 172*, 139–168. Available from: https://juan-pablo-vielma.github.io/publications/Polyhedral-Approximation-in-Mixed-Integer-Convex.pdf
54. Messerer, F., Baumgärtner, K., & Diehl, M. (2021). Survey of sequential convex programming and generalized Gauss-Newton methods. *ESAIM: Proceedings and Surveys, 71*, 64–88. Available from: https://www.esaim-proc.org/articles/proc/pdf/2021/02/proc2107107.pdf
55. Muts, P., et al. (2021). *Decomposition methods for mixed-integer nonlinear programming*. PhD Thesis, Universidad de Málaga. Available from: https://riuma.uma.es/xmlui/bitstream/handle/10630/23145/TD_MUTS_Pavlo.pdf?sequence=1
56. Norheim, J., & de Weck, O. (2021). Co-optimizing spacecraft component selection, design, and operation with minlp. In *2021 IEEE Aerospace Conference (50100)* (pp. 1–10). IEEE. Available from: https://ieeexplore.ieee.org/stamp/stamp.jsp?arnumber=9438148
57. Papp, D., & Yıldız, S. (2022). alfonso: Matlab package for nonsymmetric conic optimization. *INFORMS Journal on Computing, 34*(1), 11–19. Available from: https://arxiv.org/pdf/2101.04274.pdf
58. Soltan, V. (2021). Support and separation properties of convex sets infinite dimension. *Extracta Mathematicae*. Available from: https://revista-em.unex.es/index.php/EM/article/view/2605-5686.36.2.241
59. Tibshirani, R. (2015). Convexity and optimization. Available from: https://www.stat.cmu.edu/~larry/=sml/convexopt.pdf
60. Tran-Dinh, Q. (2018). A brief introduction to conic programming. Available from: https://quoctd.web.unc.edu/wp-content/uploads/sites/10081/2018/09/lecture2-disciplined-convex-programming.pdf
61. Vandenberghe, L., & Boyd, S. (1996). Semidefinite programming. *SIAM Review, 38*(1), 49–95. Available from: https://www.ise.ncsu.edu/fuzzy-neural/wp-content/uploads/sites/9/2015/07/SDP-Review-1.pdf
62. Wang, Z., & Grant, M. J. (2017). Hypersonic trajectory optimization by sequential semidefinite programming. In *AIAA Atmospheric Flight Mechanics Conference* (p. 0248). Available from: https://www.researchgate.net/profile/Zhenbo-Wang-4/publication/308202891_Hypersonic_Trajectory_Optimization_by_Sequential_Semidefinite_Programming/links/58922455a6fdcc1b4146bf1a/Hypersonic-Trajectory-Optimization-by-Sequential-Semidefinite-Programming.pdf
63. Wolkowicz, H., Saigal, R., & Vandenberghe, L. (2012). *Handbook of semidefinite programming: Theory, algorithms, and applications* (Vol. 27). Springer.

Chapter 10
Interior Point Methodology

10.1 Introduction

Interior point methods (IPM) became quite popular after 1984, when Karmarkar, with the AT&T Bell Laboratories, announced in [30] (8,467 cites) a fast polynomial-time method for linear programming. In fact, the comment of specialists was that an *interior-point revolution* emerged from then. It also came to headlines of newspapers, like The New York Times or The Wall Street Journal.

The basic IPM method is iterative; it starts from an interior point of F and then it uses a modified Newton's method to approximate the solution.

Let us sketch the basic idea by means of Fig. 10.1, [43]. The situation depicted on the left hand corresponds to a gradient descent algorithm, departing from a point inside F (an interior point). Once the boundary of F is reached, many zig-zags may appear. This problem disappears when using the IPM methodology, as shown on the right hand side of the figure. By means of a barrier function, IPM avoids the intersection with the perimeter.

This chapter shall introduce IPM focusing in a first step on the linear programming scenario, so important. Then, the chapter will continue with other optimization areas where interior point methods can be applied, like QP, SDP, etc.

Nowadays, most optimization solvers (software) include interior point procedures. Besides, a fruitful field of research has emerged around IPM, about understanding of its behavior, the introduction of improvements, and extending its applications.

Main references for the matters covered in this chapter are the book of J. Nocedal and S.J. Wright [49] (39,640 cites), the book of A.V. Fiacco and G.P. McCormick [23] (5,425 cites), and the seminal monograph of Y Nesterov and A. Nemirovski [47] (7,039 cites). Also, both Nesterov and Nemirovski, and collaborators, published important articles that will be referenced when opportune along the chapter.

J. M. Giron-Sierra, *Introduction to Optimization with Matlab® Examples*,
https://doi.org/10.1007/978-3-032-05048-9_10

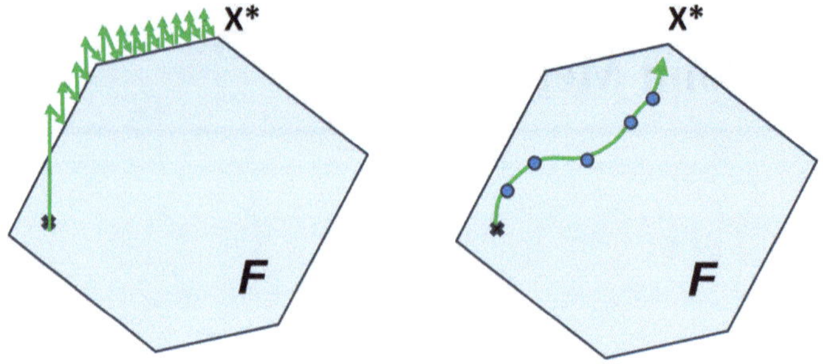

Fig. 10.1 Iterations to the solution: (left) gradient descent, (right) interior point method

Other relevant books are [12] (72,045 cites) by S.P. Boyd and L. Vanderberghe and more recently [58] (a collection of chapters by different authors) or the new edition of [69]. The book [51] includes coding examples in MATLAB.

10.2 Preliminaries

For an easier reading of the next mathematical materials, a short remembering of fundamental methodological aspects, already treated in previous chapters, is now included.

The Newton's method

The iterative Newton's method is used to solve:

$$f(x) = 0$$

The method is based on a Taylor series approximation:

$$f(x + \Delta x) \approx f(x) + f'(x)\, \Delta x$$

The Newton step is given by

$$\Delta x = -\frac{f(x)}{f'(x)}$$

And the iteration is

$$x^{(k+1)} = x^{(k)} + \Delta x^{(k)} = x^{(k)} - \frac{f\left(x^{(k)}\right)}{f'\left(x^{(k)}\right)}$$

When the method converges, it does so quadratically. Of course, the method can be effortlessly generalized for n dimensions, using gradients.

Suppose you want to find the minimum of a function $g(x)$ using the Newton's method. Then, you take $f(x) = g'(x)$. In case of n variables, with $g(x_1, x_2, \ldots, x_n)$, the Newton step would be

$$\Delta\vec{x} = -\frac{\nabla g(\vec{x})}{\nabla^2 g(\vec{x})}$$

so there is a gradient in the numerator and a Hessian in the denominator.

Lagrange's multipliers

Suppose you have the following optimization problem:

Minimize:

$$f(\vec{x})$$

Subject to:

$$g_i(\vec{x}) = 0, \ i = 1, 2, \ldots, m$$

The idea of Lagrange for solving this problem is to use a set of multipliers λ_i to build the following function:

$$L(\vec{x}, \vec{\lambda}) = f(\vec{x}) - \sum_{i=1}^{m} \lambda_i \, g_i(\vec{x})$$

And then minimize this function, which is called the Lagrangian function, by solving the following equation system:

$$\frac{\partial L}{\partial x_j} = \frac{\partial f}{\partial x_j} - \sum_{i=1}^{m} \lambda_i \cdot \frac{\partial g_i}{\partial x_j} = 0, \ j = 1, 2, \ldots, n$$

$$\frac{\partial L}{\partial \lambda_i} = -g_i(\vec{x}) = 0, \ i = 1, 2, \ldots, m$$

These equations can be solved by Newton's method.

Inequalities. Duality

Let us take into account inequalities in the constraints, so the problem becomes

Minimize:

$$f(\vec{x})$$

Subject to:

$$h_i(\vec{x}) \le 0, \; i = 1, 2, \ldots, m$$

$$l_i(\vec{x}) = 0, \; i = 1, 2, \ldots, r$$

This problem will be denoted as the "*primal problem*".
The following Lagrangian is now formed:

$$L(\vec{x}, \vec{u}, \vec{v}) = f(\vec{x}) + \sum_{i=1}^{m} u_i \, h_i(\vec{x}) + \sum_{i=1}^{r} v_i \, l_i(\vec{x})$$

A Lagrangian *dual function* is considered:

$$q(\vec{u}, \vec{v}) = \min_{\vec{x}} L(\vec{x}, \vec{u}, \vec{v})$$

(some specialists prefer to use:

$$q(\vec{u}, \vec{v}) = \inf_{\vec{x}} L(\vec{x}, \vec{u}, \vec{v})$$

The infimum is used instead of the minimum, because the Lagrangian might not have a minimum on F).

An important fact is the following: denote as $f*$ the solution of the primal problem; then $q(\vec{u}, \vec{v})$ is always a lower bound of $f*$. This property is called '*weak duality*'.

Then, it would make sense to look for a set of values $(\vec{u}*, \vec{v}*)$ such that the Lagrangian dual function was maximized. This leads to the Lagrange "*dual problem*":

Maximize:

$$q(\vec{u}, \vec{v})$$

Subject to:

$$\vec{u} \ge \vec{0}$$

The dual problem is always convex. The primal and dual optimal values, $f*$ and $q*$, always satisfy weak duality: $f* \ge q*$.

The difference:

$$\eta\,(\vec{x}, \vec{u}, \vec{v}) = f(\vec{x}) - q(\vec{u}, \vec{v})$$

is called the '*duality gap*' between $\vec{x}$ and $\vec{u}, \vec{v}$.

Under certain conditions, it is possible to have $f* = q*$; in this case we have '*strong duality*', and the duality gap is zero. For instance, the Slater's condition for convex primal problems establishes that if there is an $\vec{x}$ such that

$$h_1(\vec{x}) < 0,\ h_2(\vec{x}) < 0,\ \ldots,\ h_m(\vec{x}) < 0,\ \text{and}$$

$$l_1(\vec{x}) = 0,\ l_2(\vec{x}) = 0,\ \ldots,\ l_r(\vec{x}) = 0$$

then we have strong duality.
KKT conditions: In case of strong duality, one finds that

$$u_i^* h_i(\vec{x}*) = 0, \quad i = 1, 2, \ldots, m$$

which is called the '*complementary slackness*' condition.

Joining together the facts associated with strong duality, one obtains the *Karush-Kuhn-Tucker (KKT)* optimality conditions:

$$h_i(\vec{x}*) \leq 0\,,\ i = 1, 2,\ \ldots, m$$

$$l_i(\vec{x}*) = 0\,,\ i = 1, 2,\ \ldots, r$$

$$u_i^* \geq 0\,,\ i = 1, 2,\ \ldots, m$$

$$u_i^* h_i(\vec{x}*) = 0\,,\ i = 1, 2,\ \ldots, m$$

$$\nabla f(\vec{x}*) + \sum_{i=1}^{m} u_i^* \nabla h_i(\vec{x}*) + \sum_{i=1}^{r} v_i^* \nabla l_i(\vec{x}*)) = 0$$

Conversely, if the problem is convex and $\vec{x}*,\ \vec{u}*,\ \vec{v}*$ satisfy the KKT conditions, then these values are (primal, dual) optimal.
In the particular case of a linear programming problem,

Minimize: $\vec{c}^T \cdot \vec{x}$
Subject to: $A\,\vec{x} = \vec{b}\ ,\ \vec{x} \geq \vec{0}$

the dual problem is:

Maximize: $\vec{b}^T \cdot \vec{\lambda}$
Subject to: $A^T\,\vec{\lambda} + \vec{s} = \vec{c}\ ,\ \vec{s} \geq \vec{0}$

Then, the Lagrangian can be written as follows:

$$L\,(\vec{x},\,\vec{\lambda},\,\vec{s}) = \vec{c}^T\vec{x} - \vec{\lambda}^T(A\,\vec{x} - \vec{b}) - \vec{s}^T\vec{x}$$

And the KKT conditions would be

$$A\vec{\lambda} + \vec{s} = \vec{c}$$

$$A\vec{x} = \vec{b}$$

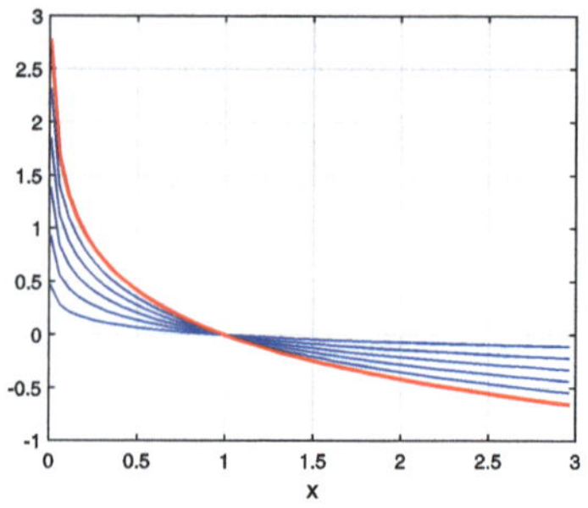

Fig. 10.2 Log-barrier function

$$\vec{x} \geq 0$$

$$\vec{s} \geq 0$$

$$x_i s_i = 0 \text{ for } 0 \leq i \leq n$$

(this system of equations will be mentioned several times in the next pages).

10.2.1 Barrier Functions: Central Path

Usually some of the constraints are of the type $x_i \geq 0$ (non-negativity of variables). The idea of the barrier function is to keep solutions at the interior of F building a barrier that prevents variables from reaching the boundary $x_i = 0$. For instance, in the case of:

Minimize $f(\vec{x})$ subject to $\vec{x} \geq 0$

an equivalent unconstrained problem is written as:

Minimize $B(\vec{x}|\mu) = f(\vec{x}) - \mu \sum_{i=1}^{m} \log(x_i)$

Figure 10.2 shows a series of curves corresponding to the function $-\mu \log(x)$, for $\mu = 0.1, 0.2, \ldots, 0.6$ and $x = 0.01, 0.06, \ldots, 3$. As x approaches zero, the value of the function approaches to infinity.

Figure 10.2 has been generated with Program 10.1. The curve starting from $(0, 0.5)$ corresponds to $\mu = 0.1$ while the curve starting from $(0, 2.7)$ corresponds to $\mu = 0.6$

Program 10.1 Displaying barrier functions

```
%plots of log barrier function
%
x=0.01:0.05:3;
figure(1);
for nn=1:6,
```

```
  mhu = nn*0.1;
  z = -mhu * log(x);
  plot(x,z,'b'); hold on;
end
xlabel('x'); grid;
```

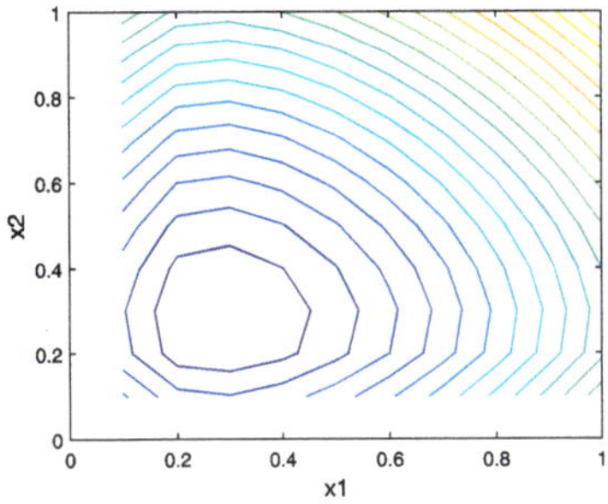

Fig. 10.3 Contours of B for $\mu = 0.7$

A simple and interesting example is given in [38]:

Minimize $(x_1 + 1)^2 + (x_2 + 1)^2$ subject to $x_1 \geq 0,\ x_2 \geq 0$

The unconstrained minimum is at $(-1, -1)$, but the constrained minimum is at $(0, 0)$.

The barrier function is introduced as follows:

$$B = (x_1 + 1)^2 + (x_2 + 1)^2 - \mu \log(x_1) - \mu \log(x_2)$$

Figure 10.3 depicts the contours of the barrier function for $\mu = 0.7$.

Figure 10.3 has been obtained with Program 10.2.

Program 10.2 Contours of a barrier

```
% example of barrier
% 2D contour of barrier function
v=0:0.1:1;
[vx1,vx2]=meshgrid(v);
y=((vx1+1).^2)+((vx2+1).^2);
figure(1)
mhu=0.7;
z=((vx1+1).^2)+((vx2+1).^2) - (mhu*log(vx1)) - ...
  (mhu*log(vx2));
contour(vx1,vx2,z,20); hold on;
xlabel('x1'); ylabel('x2');
```

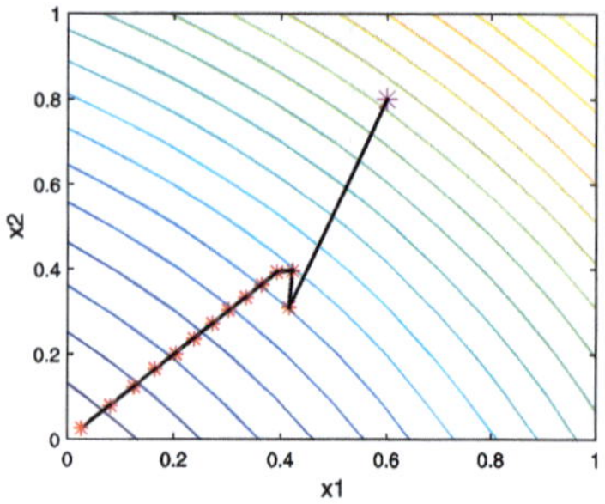

Fig. 10.4 Barrier based iterations

The gradient components of the barrier function would be

$$\frac{\partial B}{\partial x_1} = 2(x_1 + 1) - \frac{\mu}{x_1}$$

$$\frac{\partial B}{\partial x_2} = 2(x_2 + 1) - \frac{\mu}{x_2}$$

and the Hessian entries:

$$H_{11} = 2 + \frac{\mu}{x_1^2}; \; H_{12} = 0; \; H_{21} = 0; \; H_{22} = 2 + \frac{\mu}{x_2^2}$$

Given a value of $\mu = 0.7$, the corresponding minimum of B would be found with Newton steps:

$$\Delta\vec{x} = -\frac{\nabla B(\vec{x})}{\nabla^2 B(\vec{x})}$$

Program 10.3 runs a loop of μ values from 1.3 to 0.1 (decreasing in 0.1 steps). For each value of μ, a single Newton step is computed to find the corresponding minimum of B. The result is the trajectory depicted in Fig. 10.4.

Program 10.3 Example of barrier based Newton iterations

```
% example of barrier
% Newton iterations in each step
%
v=0:0.1:1;
[vx1,vx2]=meshgrid(v);
y=((vx1+1).^2)+((vx2+1).^2);
%
% function display-------------
figure(1)
contour(vx1,vx2,y,20); hold on;
```

```
xlabel('x1'); ylabel('x2');
%
% trajectory to the solution
% initial point
x1o=0.6; x2o=0.8;
plot(x1o,x2o,'m*','MarkerSize',12);
for mhu=1.3:-0.1:0.1,
  %gradient
  gr1=(2*(x1o+1))-(mhu/x1o);
  gr2=(2*(x2o+1))-(mhu/x2o);
  g=[gr1;gr2];
  %Hessian
  h11=2+(mhu/(x1o^2));
  h22=2+(mhu/(x2o^2));
  H=[h11 0; 0 h22];
  delta=H\-g; %Newton step
  x1n=x1o+delta(1);
  x2n=x2o+delta(2);
  plot([x1o x1n],[x2o x2n],'k','LineWidt',2);
  plot(x1n,x2n,'r*','MarkerSize',8)
  x1o=x1n; x2o=x2n;
end
xlabel('x1'); ylabel('x2');
```

In the simple example we selected, it is easy to find the analytical minimization of B, given a value of μ. From $\partial B/\partial x_i = 0, \ i = 1, 2$, one obtains the solution:

$$x_1(\mu) = x_2(\mu) = -\frac{1}{2} + \frac{1}{2}\sqrt{1 + 2\mu}$$

Notice that the solution tends to (0,0) as $\mu \to 0$. This is a general result of the barrier function method [23]. In iterative schemes, there would be a sequence of solutions $\vec{x} * (\mu_1), \vec{x} * (\mu_2), \ldots, \vec{x} * (\mu_n)$ as $\mu \to 0$. This sequence is called the *central trajectory* or the *central path.*

Figure 10.5 depicts the sequence of analytical solutions for μ values from 2 to 0.05 (decreasing in 0.05 steps).

Figure 10.5 has been obtained with Program 10.4.

Program 10.4 Sequence of analytical solutions

```
% example of barrier
% analytic solution
%
v=0:0.1:1;
```

```
[vx1,vx2]=meshgrid(v);
y=((vx1+1).^2)+((vx2+1).^2);
%
% function display-------------
figure(1)
contour(vx1,vx2,y,20); hold on;
xlabel('x1'); ylabel('x2');
%
% trajectory to the solution
for mhu= 2:-0.05:0.05;
  x1 = (-0.5 + (0.5 * sqrt((1+ 2*mhu))));
  x2 = x1;
  plot(x1,x2,'r*','MarkerSize',10);
end
xlabel('x1'); ylabel('x2'); grid;
```

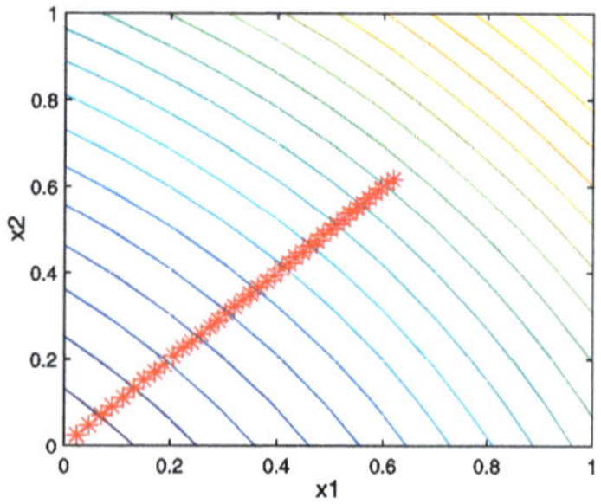

Fig. 10.5 Evolution of barrier minimum with $\mu \to 0$

In the case of linear programming, the optimal solution will be located at the boundary of F. The central path would be a series of interior points, tending to this optimal solution.

In the general case of a $B(\vec{x}|\mu)$ to be minimized, the set of equations $\partial B/\partial x_i = 0,\ i = 1, \ldots, m$ would be written, and then one could use the Newton's method (or any other) to solve them.

The barrier function technique for linear programming was introduced in 1967, in [23]. The subtitle of this book is *Sequential Unconstrained Minimization Techniques* (also known as SUMT).

It is convenient to use *self-concordant* barriers (we will introduce the self-concordant property some few lines below).

Examples of self-concordant-barrier functions are the following:

- $-\log(\det(A))$ for the positive definite cone
- $-\log(\vec{x}^T A\,\vec{x} + \vec{b}^T\vec{x} + \vec{c})$ for the convex set $\vec{x}^T A\,\vec{x} + \vec{b}^T\vec{x} + \vec{c} > 0$ when A is positive semidefinite
- $-\log(y^2 - \|\vec{x}\|^2)$ for the second order cone $\|\vec{x}\| \leq y$

The lecture notes [35] includes many more examples of self-concordant barriers.

Self-concordant functions

After some initial years of applying new software tools based on interior points, it was recognized that problems of convergence and numerical error sensitivity may appear. A deeper analysis of the convergence of the Newton's method was done, and the theory of *self-concordant* functions was developed along several papers. The book of [47] in 1994 helped definitively to establish this new methodology.

The intuition is that locally, in self-concordant functions, the Hessian does not change too fast. It can be shown that this is linked to guaranteeing a suitable (quadratic) convergence of the Newton steps, [21].

A self-concordant (convex) function f(t) obeys:

$$|f'''(t)| \leq 2\, f''(t)^{3/2} \tag{10.1}$$

What it is important in this expression (10.1) is that the third derivative is bounded at every point by the second derivative.

By using self-concordant barriers, one can define closeness to the central path and the policy to update μ in order that the Newton steps stay close to the central path, with not much fluctuation (this is implied by the derivative bounds). The result is a better convergence of the interior-point method, [47].

All linear and quadratic functions are self-concordant. The sum of two self-concordant functions is also self-concordant.

An interesting property of self-concordant barriers is that they let you know how close to an optimal point, the current algorithm step is; and you can obtain a upper bound on the number of iterations required to approach the optimum within a certain ε.

See [28] for a detailed exposition on self-concordant barriers.

10.3 Initial Overview

It is time to introduce the usual formalism employed by interior point methods, together with the main procedural aspects. The guide for this section is [49].

A linear programming problem (LP) would be used as a paradigmatic example. Recall that the associated KKT conditions for LP are the following:

$$A\vec{\lambda} + \vec{s} = \vec{c}$$

$$A\vec{x} = \vec{b}$$

$$x_i s_i = 0 \text{ for } 0 \leq i \leq n$$

$$\vec{x} \geq 0, \quad \vec{s} \geq 0$$

Newton-based methods can be iteratively applied to the equalities, employing search directions and step lengths for which $\vec{x}$ and $\vec{s}$ remain non-negative (this is the difficult aspect).

During the search for a next point, in each iteration, a measure of the point suitability can be the average of $x_i s_i = 0$ for $0 \leq i \leq n$. This quantity is called "*the duality measure*" and is

$$\mu = \sum_{i=1}^{n} x_i s_i = (\vec{x}\,\vec{s})/n$$

For convenience, let us express the KKT conditions in matrix format:

$$F(\vec{x}, \vec{\lambda}, \vec{s}) = \begin{pmatrix} A^T \vec{\lambda} + \vec{s} - \vec{c} \\ A\vec{x} - \vec{b} \\ X S \vec{e} \end{pmatrix} = 0$$

$$\vec{x} \geq 0, \quad \vec{s} \geq 0$$

where:

$$X = \text{diag}(x_1, \ldots, x_n), \quad S = \text{diag}(s_1, \ldots, s_n) \text{ and } \vec{e} = (1, \ldots, 1)^T.$$

The idea of Newton for estimating a search direction is to form a linear model of $F()$ around the current point (this model is a Jacobian) and then solve the following system:

$$J(\vec{x}, \vec{\lambda}, \vec{s}) \begin{pmatrix} \Delta\vec{x} \\ \Delta\vec{\lambda} \\ \Delta\vec{s} \end{pmatrix} = -F(\vec{x}, \vec{\lambda}, \vec{s})$$

It is customary to consider the following residuals:

$$\Delta_P = A\vec{x} - \vec{b}, \quad \Delta_D = A^T \vec{\lambda} + \vec{s} - \vec{c}$$

Hence, the Newton equations can be written as follows:

$$\begin{pmatrix} 0 & A^T & I \\ A & 0 & 0 \\ S & 0 & X \end{pmatrix} \begin{pmatrix} \Delta\vec{x} \\ \Delta\vec{\lambda} \\ \Delta\vec{s} \end{pmatrix} = \begin{pmatrix} -\Delta_P \\ -\Delta_D \\ -X S \vec{e} \end{pmatrix}$$

Probably, a full step in the Newton direction would get out from our feasible region. Therefore, a line search in that direction is opportune, so the new iterate would be

$$(\vec{x}, \vec{\lambda}, \vec{s}) + \alpha \left(\Delta\vec{x}, \Delta\vec{\lambda}, \Delta\vec{s}\right)$$

with $\alpha \in (0, 1]$.

The Newton direction is also called the *affine direction*.

It is normally preferred to use other directions, not so directly aimed to the solution. A centering parameter $\sigma \in (0, 1]$ is included so that the iterations steps move toward a point such as $x_i s_i = \sigma \mu$.

Therefore, the Newton equation is modified as follows:

$$\begin{pmatrix} 0 & A^T & I \\ A & 0 & 0 \\ S & 0 & X \end{pmatrix} \begin{pmatrix} \Delta\vec{x} \\ \Delta\vec{\lambda} \\ \Delta\vec{s} \end{pmatrix} = \begin{pmatrix} -\Delta_P \\ -\Delta_D \\ -X S \vec{e} + \sigma \mu \vec{e} \end{pmatrix}$$

The central path is an arc of points inside the feasible region, each point of the central path satisfies the following system:

$$A \vec{\lambda} + \vec{s} = \vec{c}$$

$$A \vec{x} = \vec{b}$$

$$x_i s_i = \tau \text{ for } 0 \leq i \leq n$$

$$\vec{x} > 0, \quad \vec{s} > 0$$

Notice the presence of the scalar parameter τ.

Another, equivalent formulation of the central path is

$$F(\vec{x}_\tau, \vec{\lambda}_\tau, \vec{s}_\tau) = \begin{pmatrix} 0 \\ 0 \\ \tau \vec{e} \end{pmatrix}, \quad (\vec{x}_\tau, \vec{s}_\tau) > 0$$

In general, the central path converges to the solution as $\tau \to 0$.

Many interior point algorithms take the iteration steps toward points of the central path, instead of affine direction steps. In this way, longer steps are normally possible keeping inside $(\vec{x}_\tau, \vec{s}_\tau) > 0$.

If the point $(\vec{x}, \vec{\lambda}, \vec{s})$ is feasible, then $\Delta_P = 0$ and $\Delta_D = 0$, and so the search direction is:

$$\begin{pmatrix} 0 & A^T & I \\ A & 0 & 0 \\ S & 0 & X \end{pmatrix} \begin{pmatrix} \Delta\vec{x} \\ \Delta\vec{\lambda} \\ \Delta\vec{s} \end{pmatrix} = \begin{pmatrix} 0 \\ 0 \\ -X S \vec{e} + \sigma \mu \vec{e} \end{pmatrix}$$

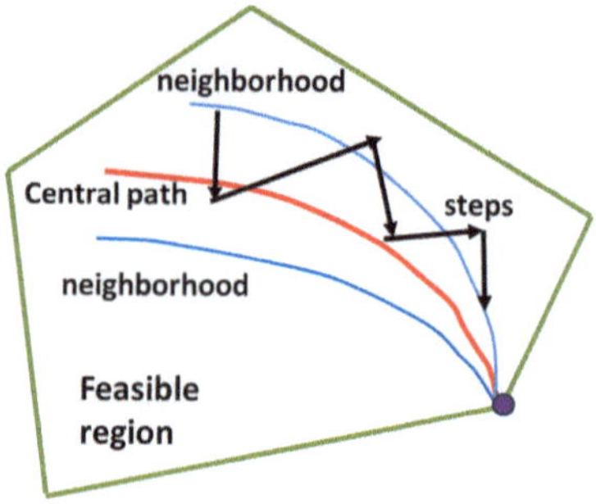

Fig. 10.6 Sketch of predictor-corrector steps toward the optimal solution

The searching direction corresponding to $\sigma = 1$ is called the *centering direction*. It leads to the point $(\vec{x}_\mu, \vec{\lambda}_\mu, \vec{s}_\mu)$ of the central path where $x_i s_i$ are identical to the current μ. Indeed, the direction corresponding to $\sigma = 0$ is the affine direction.

The research has proposed a good number of directions. Most of them can be cast into three categories: path-following methods, affine-scaling algorithms, and potential reduction algorithms.

Path-following

Path-following algorithms follow to a certain extent the central path, according to values assigned to the centering parameter. Several ways of doing so can be classified as short-step, long-step, or predictor-corrector algorithms:

- Short-step: the centering parameter is set close to 1; then the iterations stay near the central path.
- Long-step: values of the centering parameter move away from 1, toward 0. Iterations are farther from the central path.
- Predictor-corrector methods alternate steps with $\sigma = 1$ and steps with $\sigma = 0$.

Figure 10.6 depicts in a simplified manner the evolution of predictor-corrector steps, toward the optimal solution. The neighborhood around the central path is usually described as a horn.

Affine scaling

After the impacting contribution of Karmarkar, the research tried to simplify some aspects of his algorithm. One of the results was the affine-scaling algorithms. As years passed by, it was discovered that this methodology was already proposed in 1967 by Dikin in Russia, long before Karmarkar. Affine-scaling algorithms do not consider the central path.

The polyhedral shape of the feasible region F would cause the zig-zag bad behavior represented in Fig. 10.1. In order to avoid this, affine-scaling algorithms consider ellipsoids inside F. For iteration $\vec{x}^k$ an ellipsoid centered at $\vec{x}^k$ is built; and then one optimizes the problem on this ellipsoid, obtaining $\vec{x}^{k+1}$. Then, a new ellipsoid centered at $\vec{x}^{k+1}$ is built and so on…

Let us construct an ellipsoid inside F and centered at $\vec{x}^k$, by means of scaling, [26].

The original LP problem is

Minimize $\vec{c}^T\vec{x}$ subject to $A\vec{x} = \vec{b}$ with $\vec{x} \geq 0$

It can be restated as follows:

Minimize $\left(D^k\vec{c}\right)^T\vec{w}$ subject to $\left(A\,D^k\right)\vec{w} = \vec{b}$ with $\vec{w} \geq 0$

where D^k is a positive diagonal matrix: $D^k = \mathrm{diag}\left(x_1^k, x_2^k, \ldots, x_n^k\right)$

Hence, the variable $\vec{x}$ has been scaled by $\vec{x} = D^k\,\vec{w}$ (this is why the method is called affine-scaling).

Notice that $\vec{w} = (1, 1, \ldots, 1) = \vec{e}$ is feasible.

We shall restrict the minimization problem to a unit ball centered at $\vec{e}$:

Minimize $\left(D^k\vec{c}\right)^T\vec{w}$ subject to $\left(A\,D^k\right)\vec{w} = \vec{b}$ with $\|\vec{w} - \vec{e}\| \leq 1$

Going back to the original space, this is equivalent to

Minimize $\vec{c}^T\vec{x}$ subject to $A\vec{x} = \vec{b}$ with $\left\|\left(D^k\right)^{-1}\vec{x} - \vec{e}\right\| \leq 1$

The expression $\left\|(D^k)^{-1}\vec{x} - \vec{e}\right\| \leq 1$ corresponds to an ellipsoid centered at $\vec{x}^k$, called the *Dikin ellipsoid*. The minimum over this ellipsoid is obtained at $\vec{x}^k + \Delta\vec{x}^k$, in which:

$$\Delta\vec{x}^k = -\frac{D^k P_Q D^k\vec{c}}{\left\|P_Q D^k\,\vec{c}\right\|}$$

where $Q = A\,D^k$ and $P_Q = I - \left[Q^T\left(Q\,Q^T\right)^{-1}Q\right]$

(P_Q is the projection matrix onto the null space of Q)

See the web page [41] for a numerical example of the affine-scaling algorithm.

Potential Reduction

Potential reduction algorithms use a potential function that is reduced in each iteration, [26]. A popular potential function, proposed by Tanabe, Todd, and Ye, [59], is the following:

$$\Phi\left(\vec{x}, \vec{s}\right) = \rho\,\log\vec{x}^T\vec{s} - \sum_{i=1}^{n}\log\,x_i s_i$$

with $\rho > n$.

This potential can be rewritten as

$$\Phi\left(\vec{x}, \vec{s}\right) = (\rho - n)\,\log\left(\vec{x}^T\vec{s}\right) - \sum_{i=1}^{n}\log\frac{x_i s_i}{\vec{x}^T\vec{s}/n} + n\,\log\,n$$

Here, the first term tends to minus infinity when $(\vec{x}, \vec{s})$ tends to the optimal solution; and the second term is zero for centered iterates, tending to infinity when

Fig. 10.7 Narendra Karmarkar in 1984

iterates deviate and approach the boundary of F. It is similar to the path-following behavior, except for one thing: instead of following a predefined sequence, μ has to be recomputed at each step.

The basic algorithm computes the Newton step $\left(\Delta\vec{x}^k, \Delta\vec{\lambda}^k, \Delta\vec{s}^k\right)$ and then determines the next state:

$$\left(\vec{x}^{k+1}, \vec{\lambda}^{k+1}, \vec{s}^{k+1}\right) = \left(\vec{x}^{k}, \vec{\lambda}^{k}, \vec{s}^{k}\right) + \alpha^k\left(\Delta\vec{x}^k, \Delta\vec{\lambda}^k, \Delta\vec{s}^k\right)$$

where α^k minimizes the potential at the $(k+1)$ point (μ must be re-evaluated at this point). The new point must be inside F.

Potential reduction methods have been extended for several types of optimization problems, including primal-dual and/or affine versions of the algorithm, [6].

Karmarkar's algorithm is a potential-reduction algorithm, based on the following potential function:

$$\Phi(\vec{x}) = \rho \log\left(\vec{c}^T\vec{x} - Z\right) - \sum_{i=1}^{n} \log x_i$$

with $\rho = n + 1$ and Z a lower bound of the optimal objective value.

Figure 10.7 shows one of the few photographs of N. Karmarkar you can find, corresponding to the time when he presented his algorithm. He then was a professor at M.I.T. (1991) and Princeton (1996). Afterward he came to India, Mumbai, and later on Puna, where he is investigating on supercomputing architectures.

The algorithm of Karmarkar is an iterative algorithm that uses projective transformations and the potential function introduced above. In the extensive article of M.J. Todd on potential reduction methods [59], there is a detailed description of this algorithm (also in [6]).

Some other historical landmarks relative to the interior point methodology are the following: in 1967, the introduction of barrier functions by A.V. Fiacco and G. McCormick; in 1988, J. Renegar [53] proposed a method that forms a succession of smaller subsets of F, each containing the solution. In 1987 (published in 1989, [39]), a celebrated contribution by N. Megiddo took place, introducing a primal-dual framework for path-following algorithms: in words of F.A. Potra and S.J. Wright [52], a new era was inaugurated with this contribution.

We would recommend [36] and [55] for interesting details, with opportune references, about the chronological evolution and expansion of interior-point theory.

10.4 Interior Point Method and Linear Programming

The first target of the interior point methodology, from 1984, was linear programming (LP). It is clearly an important target, with large impact in financial and economy fields.

This section is divided into two parts, devoted first to primal and then to primal-dual methodologies. The second one is the most important in practical terms. This said, the primal alternative is nevertheless interesting for introducing fundamental concepts.

Our main references in this section are [38, 54] and the motivating historical account given by [67] (435 cites) about the interior point revolution.

We will insist in several points already introduced in the previous section.

10.4.1 Primal Methodology

Consider a standard LP problem, with the following formulation:

Minimize $\vec{c}^T \vec{x}$ subject to $A\vec{x} = \vec{b}$ with $\vec{x} \geq 0$

The application of KKT conditions to this problem conducts to the following system of equations:

$$A\vec{\lambda} + \vec{s} = \vec{c}$$

$$A\vec{x} = \vec{b}$$

$$\vec{x} \geq 0, \quad \vec{s} \geq 0$$

$$x_i s_i = 0 \text{ for } 0 \leq i \leq 1$$

(this last equation can be written as $X S \vec{e} = 0$)

Let us associate to our problem the following barrier function:

$$B(\vec{x}, \mu) = \vec{c}^T \cdot \vec{x} - \mu \sum_{i=1}^{n} \log(x_i)$$

Now, let us change our focus. Instead of solving the original LP problem, we will consider the next problem:

Minimize $B(\vec{x}, \mu)$ subject to $A\vec{x} = \vec{b}$ (we can drop the constraint $\vec{x} \geq \vec{0}$)

The Lagrangian for this problem is

$$L(\vec{x}, \vec{\lambda}, \vec{s}) = \vec{c}^T \cdot \vec{x} - \mu \sum_{i=1}^{n} \log(x_i) - \vec{\lambda}^T (A\vec{x} - \vec{b})$$

Hence:

$$\nabla_x L(\cdot) = \vec{c} - \mu X^{-1} \vec{e} - A^T \vec{\lambda}$$

Therefore, according to the KKT conditions, there exists a vector $\vec{\lambda}$ such that:

$$\mu X^{-1} \vec{e} + A^T \vec{\lambda} = c$$

$$A\vec{x} = \vec{b}$$

If we introduce the variable $\vec{s} = \mu X^{-1} \vec{e}$, an equivalent system would be

$$A^T \vec{\lambda} + \vec{s} = \vec{c}$$

$$A\vec{x} = \vec{b}$$

$$XS\vec{e} = \mu \vec{e}$$

Notice that the system is quite similar to the KKT conditions for the original LP problem. The only difference concerns the complementary condition: instead of 0, now we have $\mu \vec{e}$. Then, driving μ to zero makes the solution of the barrier minimization problem to be the solution of the LP.

The same iterative procedure we applied in Sect. 10.2.1 can be used now for solving the barrier minimization problem. We would alternate between taking a step to get a sufficient decrease of $B(\vec{x}, \mu)$, and then diminish μ. Steps are as follows: $\vec{x}_{k+1} = \vec{x}_k + \alpha \vec{p}$, where $\vec{p}$ gives a search direction, and α is a parameter (the step size.)

The particular search direction can be provided by the Newton's method (or any other method), and the step size would be modulated for maintaining the search inside $A\vec{x} = \vec{b}$ (this is the *feasible point descent method*). For the Newton's method, one needs the gradient and the Hessian, which are

$$\vec{g} = \nabla_x B(\vec{x}, \mu) = \vec{c} - \mu X^{-1}\vec{e}$$

$$H = \nabla_x^2 B(\vec{x}, \mu) = \mu X^{-2}$$

Thus, it is desired to find a Newton step such as
Minimize $\min_p \frac{1}{2} \vec{p}^T H \vec{p} + \vec{g}^T \vec{p}$
Subject to $A(\vec{x} + \vec{p}) = \vec{b}$
By considering the KKT conditions for this sub-problem, one obtains:

$$\begin{pmatrix} H & A^T \\ A & 0 \end{pmatrix} \begin{pmatrix} -\vec{p} \\ \vec{\lambda} \end{pmatrix} = \begin{pmatrix} \vec{c} - \mu X^{-1}\vec{e} \\ 0 \end{pmatrix}$$

Let us elaborate on the first associated equation:

$$- H\vec{p} + A^T\vec{\lambda} = \vec{c} - \mu X^{-1}\vec{e}$$

Multiply both sides by $A X^2$ and rearrange:

$$A X^2 A^T \vec{\lambda} = A X^2 \vec{c} - \mu A X \vec{e} + A X^2 H \vec{p}$$

Supposing $\vec{p}$ is a solution, then it should satisfy $A(\vec{x} + \vec{p}) = \vec{b}$, while also $A\vec{x} = \vec{b}$. Therefore $A\vec{p} = 0$. Besides, $H = \mu X^{-2}$. Using these facts, the above equation can be simplified to give

$$A X^2 A^T \vec{\lambda} = A X^2 \vec{c} - \mu A X \vec{e}$$

Since the matrix X is positive definite, ($\vec{x} \geq 0$), and A is full rank, this equation has a unique solution for $\vec{\lambda}^*$. Hence, we can use the following search direction:

$$\vec{p} = \vec{x} + \frac{1}{\mu} X^2 \left(A^T \vec{\lambda}^* - \vec{c}\right) \tag{10.2}$$

(this comes from
$- H\vec{p} + A^T\vec{\lambda} = \vec{c} - \mu X^{-1}\vec{e} \quad \rightarrow \vec{p} = \frac{1}{H}\left(\mu X^{-1}\vec{e} + A^T\vec{\lambda} - \vec{c}\right) \quad \rightarrow$
$\rightarrow \vec{p} = \frac{1}{\mu X^{-2}}\left(\mu X^{-1}\vec{e} + A^T\vec{\lambda} - \vec{c}\right) \rightarrow \quad \vec{p} = \frac{1}{\mu} X^2\left(A^T\vec{\lambda} - \vec{c}\right) + X\vec{e})$
In order to formulate an algorithm, we choose $\rho \in (0, 1)$ and set $\mu_{k+1} = \rho\,\mu_K$. Here is a primal interior point algorithm:

Algorithm 1

choose $\rho \in (0, 1)$ and $\mu_0 > 0$ sufficiently large
choose a point $\vec{x}_0$ so that $A\,\vec{x} = \vec{b}$ and $\vec{x} \geq 0$

$$k = 0$$

For k = 0, 1, 2..., do

$$\mu_k = \rho\,\mu_{k-1}$$

Compute the Newton's direction $\vec{p}$ using (10.2).
Find α for the minimum of $B(\vec{x}_k, \mu_k)$, where $\vec{x}_k = \vec{x}_{k-1} + \alpha\,\vec{p}$
subject to $A\,\vec{x}_k = \vec{b}$

$$\vec{x}_k \leftarrow \vec{x}_{k-1} + \alpha\,\vec{p}$$

$$k \leftarrow k + 1$$

end For

The use of the primal method is not without some problems that may occur. According to the comments of [67], the Hessian may become ill-conditioned at some trajectory points; and poor scaling of the search direction in early iterations could exist (some precautions should be taken).

10.4.2 Primal-Dual Methodology

After the initial route based on the primal approach, a second alternative supported by the primal-dual methodology appeared, being now generally preferred since it enjoys better performances.
Let us apply the primal-dual approach to the linear programming problem.

The primal problem is

Minimize $\vec{c}^T \cdot \vec{x}$ subject to $A\,\vec{x} = \vec{b}$, $\vec{x} \geq \vec{0}$.

The dual problem is

Maximize $\vec{b}^T \cdot \vec{\lambda}$ subject to $A^T\,\vec{\lambda} + \vec{s} = \vec{c}$, $\vec{s} \geq \vec{0}$.

(the vector $\vec{s}$ is an artificial variable called the *dual slack*).

The duality gap would be $\eta = \vec{c}^T\vec{x} - \vec{b}^T\,\vec{\lambda}$.

Now, barrier functions are introduced and two Lagrangians are built, [38], corresponding to the primal and the dual problem:

$$L_P(\vec{x}, \vec{\lambda}) = \vec{c}^T\vec{x} - \mu \sum_{i=1}^{m} \log(x_i) - \vec{\lambda}^T\left(A\vec{x} - \vec{b}\right)$$

$$L_D(\vec{x}, \vec{\lambda}, \vec{s}) = \vec{b}^T \vec{\lambda} + \mu \sum_{i=1}^{n} \log(s_i) - \vec{x}^T \left(A^T \vec{\lambda} + \vec{s} - \vec{c}\right)$$

After setting the derivatives of the Lagrangians to zero, one obtains just three equations:

$$A\,\vec{x} = \vec{b}$$

$$A^T\,\vec{\lambda} + \vec{s} = \vec{c}$$

$$x_i\, s_i = \mu,\ \ i = 1, \ldots, n$$

(with $\vec{x} \geq \vec{0},\ \ \vec{s} \geq \vec{0}$)

Hence, one could apply the Newton's method to solve the following equations:

$$A\,\vec{x} - \vec{b} = \vec{0}$$

$$A^T\,\vec{\lambda} + \vec{s} - \vec{c} = \vec{0}$$

$$X\,S\,\vec{e} - \mu\,\vec{e} = \vec{0}$$

The Newton step can be computed from

$$J\left(\vec{x}^{(k)}\right) \Delta\vec{x} = -\vec{f}\left(\vec{x}^{(k)}\right)$$

where J (.) is the Jacobian.

Let:

$$\Delta_P = A\,\vec{x}^0 - b$$

$$\Delta_D = A^T\vec{\lambda}^0 + \vec{s}^0 - \vec{c}$$

Then, taking the Jacobian of the three equations above, and the residuals Δ_P, Δ_D, the Newton step equation can be expressed as follows:

$$\begin{pmatrix} A & 0 & 0 \\ 0 & A^T & I \\ S & 0 & X \end{pmatrix} \begin{bmatrix} \Delta\,\vec{x} \\ \Delta\,\vec{\lambda} \\ \Delta\,\vec{s} \end{bmatrix} = \begin{bmatrix} -\Delta_P \\ -\Delta_D \\ \mu\,\vec{e} - X S\,\vec{e} \end{bmatrix} \tag{10.3}$$

In order to solve this equation, the procedure would be to solve first:

$$\left(A\,S^{-1} X\,A^T\right) \Delta\vec{\lambda} = \vec{b} - \mu\,A\,S^{-1}\vec{e} - A\,S^{-1} X\,\Delta_D \tag{10.4}$$

obtaining $\Delta\vec{\lambda}$; and then calculate:

$$\Delta\vec{s} = -\Delta_D - A^T \Delta\vec{\lambda}$$

$$\Delta\vec{x} = S^{-1}\left(\mu\,\vec{e} - XS\vec{e} - X\,\Delta\vec{s}\right)$$

This equation gives a Newton step. However, the actual next iterate to be applied would be

$$\left(\vec{x}, \vec{\lambda}, \vec{s}\right) + \alpha\left(\Delta\vec{x}, \Delta\vec{\lambda}, \Delta\vec{s}\right)$$

where $\alpha \in (0, 1]$ is chosen so the next solution is feasible.

In the experimental work described in [38], the sequence of μ values are calculated in function of the gap:

$$\mu_{k+1} = \frac{\vec{c}^T\vec{x}_k - \vec{b}^T\vec{\lambda}_k}{\theta(n)}, \quad (k = 0, 1, 2, \ldots)$$

where:

$$\theta(n) = \begin{cases} n^2 \text{ if } n \leq 5{,}000 \\ n\sqrt{n} \text{ if } n > 5{,}000 \end{cases}$$

In summary, here is an algorithm for the primal-dual interior point method.

Algorithm 2

choose $\rho \in (0, 1)$ and $\mu_0 > 0$ sufficiently large
choose a point $(\vec{x}_0, \vec{\lambda}_0, \vec{s}_0)$ so that $\vec{x}_0$ is feasible for the primal problem
and $(\vec{\lambda}_0, \vec{s}_0)$ is feasible for the dual
For k = 0, 1, 2..., do

$$\mu_k = \rho\,\mu_{k-1}$$

Compute the primal-dual Newton's direction $\vec{p}$ using (10.3)
Find α for the minimum of $B(\vec{x}_k, \mu_k)$, where $\vec{x}_k = \vec{x}_{k-1} + \alpha\,\vec{p}$
subject to $A\,\vec{x}_k = \vec{b}$

$$\vec{x}_k \leftarrow \vec{x}_{k-1} + \alpha\,\vec{p}$$

$$k \leftarrow k + 1$$

end For

There are some implementation aspects that deserve special attention. It happens that about 90% of the computational time when using the primal-dual method is consumed for solving equation (10.4), [38]. This equation can be expressed in a very short form:

$$\left(A\,D\,A^T\right)\vec{d} = \vec{r}$$

Five percent of computation time is devoted to form the matrix $A\,D\,A^T$, which is symmetric and positive definite.

The Cholesky Factorization $A\,D\,A^T = L\,L'$ is the preferred way for solving the equation (matrix L is lower triangular). In many cases, the matrix $A\,D\,A^T$ becomes increasingly ill-conditioned along iterations, causing difficulties for the Cholesky factorization. Some heuristics to facilitate the factorization would be applied when forming the matrix (reordering rows and columns). Likewise, it is opportune to apply some preconditioning [52].

Starting point

In practice, it is important to start the iterations from a suitable point. A repertory of methods for determining a good starting point has been proposed. We shall focus on the procedure described in [49] (pg. 410), which reasonably satisfies the primal and dual equality constraints. The idea is to solve two problems:

Minimize $\frac{1}{2}\vec{x}^T\vec{x}$ subject to $A\,\vec{x} = \vec{b}$

and,

Minimize $\frac{1}{2}\vec{s}^T\vec{s}$ subject to $A^T\vec{\lambda} + \vec{s} = \vec{c}$

The solution to both problems is given by:

$$\vec{x} = A^T\,\Omega\,\vec{b}, \quad \vec{\lambda} = \Omega\,A\,\vec{c}, \quad \vec{s} = \vec{c} - A^T\,\vec{\lambda}$$

where, $\Omega = \left(A\,A^T\right)^{-1}$

Normally, $\vec{x}$ and $\vec{s}$ have non-positive components. To circumvent this difficulty, define:

$$\delta_x = \max(-(3/2)\min_i(x_i)\,,\,0) \quad , \quad \delta_s = \max(-(3/2)\min_i(s_i)\,,\,0)$$

and adjust: $\widehat{x} = \vec{x} + \delta_x\vec{e}, \widehat{s} = \vec{s} + \delta_s\vec{e}$

To make sure that $\widehat{x}$ and $\widehat{s}$ components are not too dissimilar and not near zero, two more scalars are defined:

$$\varphi_x = \frac{1}{2}\frac{\widehat{x}^T\widehat{s}}{\vec{e}^T\widehat{s}} \quad \text{and} \quad \varphi_s = \frac{1}{2}\frac{\widehat{x}^T\widehat{s}}{\vec{e}^T\widehat{x}}$$

And finally, we arrive to the desired starting point:

$$x_O = \widehat{x} + \varphi_x \vec{e}, \quad s_O = \widehat{s} + \varphi_s \vec{e}, \quad \lambda_O = \vec{\lambda}$$

Mehrotra's predictor-corrector

In 1989, S. Mehrotra proposed a *predictor-corrector* method that now is widely used (see [40] (2,540 cites) for his article in 1992). The method proceeds in two steps, introducing some changes in Eq. (10.3). The first step considers Eq. (10.3) with $(X S \vec{e})$ instead of $(\mu \vec{e} - X S \vec{e})$; so a Newton step (an affine-scaling step) is done:

$$\begin{pmatrix} A & 0 & 0 \\ 0 & A^T & I \\ S & 0 & X \end{pmatrix} \begin{bmatrix} \Delta \vec{x}^{aff} \\ \Delta \vec{\lambda}^{aff} \\ \Delta \vec{s}^{aff} \end{bmatrix} = \begin{bmatrix} -\Delta_P \\ -\Delta_D \\ X S \vec{e} \end{bmatrix} \tag{10.5}$$

It happens that a full step in the affine direction results in the complementary condition not being satisfied:

$$\left(x_i + \Delta x_i^{aff}\right)\left(s_i + \Delta s_i^{aff}\right) = x_i s_i + x_i \Delta s_i^{aff} + s_i \Delta x_i^{aff} + \Delta x_i \Delta s_i^{aff} =$$

$$= \Delta x_i \Delta s_i^{aff} \neq 0$$

In order to correct for this error (this is the second step), we can solve the following equation:

$$\begin{pmatrix} A & 0 & 0 \\ 0 & A^T & I \\ S & 0 & X \end{pmatrix} \begin{bmatrix} \Delta \vec{x}^{\text{cor}} \\ \Delta \vec{\lambda}^{\text{cor}} \\ \Delta \vec{s}^{\text{cor}} \end{bmatrix} = \begin{bmatrix} 0 \\ 0 \\ -\Delta X^{aff} \Delta S^{aff} \vec{e} \end{bmatrix} \tag{10.6}$$

According to [49], the combined step:

$$\left(\Delta x_i^{aff}, \Delta \lambda_i^{aff}, \Delta s_i^{aff}\right) + \left(\Delta x_i^{\text{cor}}, \Delta \lambda_i^{\text{cor}}, \Delta s_i^{\text{cor}}\right)$$

gives usually better results than the pure Newton step.

Still, a centering parameter would continue to be used. Practical algorithms consider the maximum allowable step lengths in the affine direction, being:

$$\alpha_{aff}^{pri} = \min\left(1, \min_{\beta} - \frac{x_i}{\Delta x_i^{aff}}\right), \quad \text{where } \beta \text{ is } i : \Delta x_i^{aff} < 0$$

$$\alpha_{aff}^{dual} = \min\left(1, \min_{\beta} - \frac{s_i}{\Delta s_i^{aff}}\right), \quad \text{where } \beta \text{ is } i : \Delta s_i^{aff} < 0$$

and a parameter μ_{aff} is introduced:

$$\mu_{aff} = (x + \alpha_{aff}^{pri} \Delta x^{aff})^T \ (s + \alpha_{aff}^{dual} \Delta s^{aff})/n$$

(n is the space dimension)

Then, the centering parameter σ is chosen with the following heuristic:

$$\sigma = \left(\frac{\mu_{aff}}{\mu} \right)^3$$

The contributions of predictor, corrector, and centering can be aggregated into a single system:

$$\begin{pmatrix} A & 0 & 0 \\ 0 & A^T & I \\ S & 0 & X \end{pmatrix} \begin{bmatrix} \Delta \vec{x} \\ \Delta \vec{\lambda} \\ \Delta \vec{s} \end{bmatrix} = \begin{bmatrix} -\Delta_P \\ -\Delta_D \\ -XS\,\vec{e} - \Delta X^{aff} \Delta S^{aff}\, \vec{e} + \sigma\, \vec{e} \end{bmatrix}$$

The solution of this system gives the search direction. Since the Jacobian is the same for all the calculations needed (equations AA, BB, and computation of σ), not much computer effort is added to the interior point procedure.

Program 10.5 gives an example of Mehrotra's method implementation in MATLAB. It is based on [34] with some minor modifications. It includes a short *ad-hoc* function, called *Meh_dist( )*.

Program 10.5 Example of Mehrotra predictor-corrector algorithm

```
% example of Mehrotra predictor-corrector algorithm
% the problem
c=[-2; -3];
A=[3 7;
5 2];
b=[21; 10];
% using linprog to get a reference solution
[X,Fval]=linprog(c,A,b,[],[],zeros(1,2));
disp('using linprog:-------------------')
X
Fval
% using the Mehrotra algorithm
% the problem with artificial variables for Ax = b
c=[-2; -3; 0; 0];
A=[3 7 1 0;
5 2 0 1];
b=[21; 10];
```

```
m=size(A,1) ; n=size(A,2);
% find an initial point
R = builtin('_cholinf', A*A');
y = R$R'$A*c)); %this is Lambda
s0 = c-(A'*y); x0 = A' * (R$R'\b));
dx = max(1.5*min(x0),0); ds = max(-1.5*min(s0),0);
x = x0 + dx + 0.5*((x0+dx)' * (s0+ds))/sum(s0+ds);
s = s0 + ds + 0.5*((x0+dx)' * (s0+ds))/sum(x0+dx);
xi=x; si=s;
nb = mean(abs(b));
mu=[];
% loop
for nn=1:30 %(edit max iter)
  mu(end+1) = sum(x.*s)/n;
  %right-hand terms:
  Dp = A*x - b; Dd = A'*y + s -c; rxs = x.*s;
  %break point
  aux = mean(abs(Dp));
  if mu(end)<1e-6 && aux<1e-6*nb,
    disp('using Mehrotra:----------------------')
    disp('num of iterations:');
    nn
    x
    MeH_Fval =c'*x
    break;
  end
  R = builtin('_cholinf', A*diag(sparse(x./s))*A');
  % affine step direction
  r = -Dp + A* ((-x.*Dd + rxs)./s);
  dy_a = R$R'\r); ds_a = -Dd - A'*dy_a;
  dx_a = -(rxs + x.*ds_a)./s;
  xaff_step = Meh_dist(x,dx_a);
  saff_step = Meh_dist(s, ds_a);
  muaff = sum((x+(xaff_step*dx_a)).*...
  (s+(saff_step*ds_a)))/n;
  % centering parameter:
  sigma = min((muaff/mu(end))^3,1);
  % corrector
  rxs = -(sigma * mu(end) - dx_a.*ds_a);
  r = A* (rxs./s);
  dy_c = R$R'\r); ds_c = - A'*dy_c; dx_c
   = -(rxs + x.*ds_c)./s;
  dx = dx_a + dx_c; dy = dy_a + dy_c;
  ds = ds_a + ds_c;
  x_step = min(0.99*Meh_dist(x,dx),1);
  s_step = min(0.99*Meh_dist(s,ds),1);
```

```
  x = x + x_step*dx; y = y + s_step*dy;
  s = s + s_step*ds;
end
if nn == 30,
  disp ('no convergence');
end
%Function
% function used by Mehrotra program
function [stp] = Meh_dist(x, dx)
st = -x./dx;
st(dx>=0) = Inf;
stp = min(st);
if stp>1,
  stp=1;
end
end
```

Notice the use of the MATLAB function *builtin( )* with the option *'_cholinf'*, in order to apply a Cholesky factorization of the form R' R (upper triangular matrices).

The first lines of Program 1.1 are devoted to get a reference solution by calling *linprog( )*. In continuation, the program computes another solution using the Mehrotra's algorithm. The outputs of the program are the two solutions (that should be similar) and the number of iterations invested by Mehrotra's method.

It is opportune to remark that the MATLAB function *linprog( )* has an option called *'interior-point'*, by which *linprog( )* is commanded to use an interior point method.

Observations

An important reason for using a parameter μ is the following: if you set $\mu = 0$ then some of the components of $\vec{x}$ and $\vec{s}$ become zero (on the boundary of the problem) and then it cannot be guaranteed that the Jacobian is non-singular. In such a case, the algorithm gets stuck. The consequence is that it is important to always use interior points.

The direct application of KKT conditions leads to a formulation equivalent to having $\mu = 0$, but the central path approach prefers to use a perturbed version of KKT by using the parameter μ.

It has been noticed in many applications that interior-point algorithms only need some tens of iterations to obtain good approximations of the optimum, being sufficient to terminate the iteration process.

10.5 Application to Quadratic Programming

The interior point method was extended for quadratic programming in 1989 by R.D.C. Monteiro and I. Adler [1]. There were some precedents, as referenced in this article.

In line with the treatment of [49], let us focus on convex quadratic problems stated as follows:

Minimize:

$$q(x) = \frac{1}{2}\vec{x}^T Q\vec{x} + \vec{c}^T \vec{x}$$

Subject to:

$$A\vec{x} \geq \vec{b}$$

where Q is symmetric positive semidefinite.

The KKT conditions would be

$$Q\vec{x} - A^T\vec{\lambda} + \vec{c} = 0$$

$$A\vec{x} - \vec{b} \geq 0$$

$$(A\vec{x} - \vec{b})_i\,\lambda_i = 0, \quad i = 1, 2, \ldots, m$$

$$\vec{\lambda} \geq 0$$

Introducing slack variables (vector $\vec{s}$), we rewrite the KKT conditions as follows:

$$Q\vec{x} - A^T\vec{\lambda} + \vec{c} = 0$$

$$A\vec{x} - \vec{b} \geq 0$$

$$s_i\,\lambda_i = 0, \quad i = 1, 2, \ldots, m$$

$$(\vec{s}, \vec{\lambda}) \geq 0$$

A complementary measure μ can be assigned to iterations:

$$\mu = \frac{\vec{s}^T\vec{\lambda}}{m}$$

Analogously to the previous section, denote:

$$\Lambda = \mathrm{diag}\{\lambda_1, \lambda_2, \ldots, \lambda_n\}$$

$$S = \mathrm{diag}\{s_1,\, s_2, \ldots, s_n\}$$

Also, like in the section before, one applies the Newton's method and obtains:

$$\begin{pmatrix} A & 0 & -I \\ Q & -A^T & 0 \\ 0 & S & \Lambda \end{pmatrix} \begin{bmatrix} \Delta\vec{x} \\ \Delta\vec{\lambda} \\ \Delta\vec{s} \end{bmatrix} = \begin{bmatrix} -\Delta_P \\ -\Delta_D \\ \sigma\,\mu\,\vec{e} - \Lambda S\,\vec{e} \end{bmatrix} \tag{10.7}$$

In order to solve this equation, a first step is to solve:

$$(Q + A^T S^{-1} \Lambda\, A)\, \Delta\vec{x} = -\Delta_D + A^T\, S^{-1} \Lambda\, (-\Delta_P - \vec{s} + \sigma\,\mu\,\Lambda^{-1}\,\vec{e})$$

The main computational effort of the interior point method would be invested in this step. Usually, a modified Cholesky factorization would be applied to the matrix on the left-hand side.

Generally, the Mehrotra's algorithm would be employed, with the affine predictor step and the corrector step, arriving to the following aggregated system:

$$\begin{pmatrix} A & 0 & -I \\ Q & -A^T & 0 \\ 0 & S & \Lambda \end{pmatrix} \begin{bmatrix} \Delta\vec{x} \\ \Delta\vec{\lambda} \\ \Delta\vec{s} \end{bmatrix} = \begin{bmatrix} -\Delta_P \\ -\Delta_D \\ \sigma\,\mu\,\vec{e} - \Lambda S\,\vec{e} \,-\, \Delta\Lambda^{aff}\, \Delta S^{aff}\,\vec{e} \end{bmatrix} \tag{10.8}$$

There are some details concerning the selection of step length that deserve special attention, [49]. Suppose one defines the new iterate as follows:

$$(x^{k+1}, y^{k+1}) = (x^k, y^k) + \alpha^{\mathrm{Pr}}(\Delta x, \Delta y) \quad , \quad \lambda^{k+1} = \lambda^k + \alpha^{Dual}\,\Delta\lambda$$

From equations above, it can be deduced that:

$$\Delta_P^{k+1} = (1 - \alpha^{\mathrm{Pr}})\,\Delta_P^K$$

$$\Delta_D^{k+1} = (1 - \alpha^{Dual})\,\Delta_D^K + (\alpha^{\mathrm{Pr}} - \alpha^{Dual})\,Q\,\Delta\vec{x}$$

Hence, an increase of Δ_D along iterations would be possible, causing divergence problems. An obvious solution is to use equal values for primal and dual step lengths. However, user's experience is in favor of using unequal values (see [49] and [15] for more details).

Here is the predictor-corrector algorithm proposed in [49] for practical quadratic programming:

A basic, direct implementation of Algorithm 3 in MATLAB has been devised, Program 10.6. According to comments of investigators, the selection of the initial point has some influence on the satisfactory behavior of the process. In our case, the initial values for position, slacks, and Lagrange multipliers came after some trial

Algorithm 3

From a starting point with $(\vec{s}_0, \vec{\lambda}_0) > 0$
For $k = 0, 1, 2\ldots$
  Set $(\vec{x}, \vec{s}, \vec{\lambda}) = (\vec{x}_k, \vec{s}_k, \vec{\lambda}_k)$ and solve (10.7) with $\sigma = 0$, to determine $(\Delta\vec{x}^{aff}, \Delta\vec{s}^{aff}, \Delta\vec{\lambda}^{aff})$
  Compute $\mu = \vec{s}^T\vec{\lambda} / m$
  Find $\hat{\alpha}_{aff} = \max\left[\alpha \in (0, 1] / (\vec{s}, \vec{\lambda}) + \alpha(\Delta\vec{s}^{aff}, \Delta\vec{\lambda}^{aff}) \geq 0\right]$
  Compute $\mu_{aff} = (\vec{s} + \hat{\alpha}_{aff}\Delta\vec{s}^{aff})^T(\vec{\lambda} + \hat{\alpha}_{aff}\Delta\vec{\lambda}^{aff}) / m$
  Set $\sigma = (\mu_{aff} / \mu)^3$
  Solve (10.8) to find $(\Delta\vec{x}, \Delta\vec{s}, \Delta\vec{\lambda})$
  Choose $\tau_k \in (0, 1)$ and set $\hat{\alpha} = \min(\alpha_k^{pri}, \alpha_k^{dual})$
  Set $(\vec{x}_{k+1}, \vec{s}_{k+1}, \vec{\lambda}_{k+1}) = (\vec{x}_k, \vec{s}_k, \vec{\lambda}_k) + \hat{\alpha}(\Delta\vec{x}, \Delta\vec{s}, \Delta\vec{\lambda})$
end For

and error: there are procedures for the initialization subproblem—as included in the precedent program—but we preferred here a more simple code.

Program 10.6 Example of interior-point quadratic programming

```
%Example of interior-point quadratic programming
%problem specification ---------
Q=[2 0;
0 2];
c=[-2 ; -5];
%constraints polygon
A=[1 -2;
-1 -2];
b= [-2 ; -6];
x0=[3; 0.2]; %Initial point -------
n=2; %dim
E=[1;1]; %1's column vector
%reference solution using quadprog(),
% using -A and -b -------------------------
solRef = quadprog(Q,c,-A,-b,[],[],[],[],x0,...
optimset('Display','off'));
%preparing for IPM ------------
% for steps recording
Nit=100;
ux=zeros(2,Nit); us=zeros(2,Nit); uy=zeros(2,Nit);
umalpha=zeros(1,Nit); usigma= zeros(1,Nit);
mu=[];
s= [1.15;1]; % tentative initial values
y= [40;40]; %these are the lambdas
x=x0;
```

```
%loop -------------------------------------------
for nn=1:Nit, %(edit max iter)
  mu(end+1) = (sum(y.*s))/n;
  %affine step ---------------------
  Dp=A*x-b-s; Dd=Q*x-(A'*y)+c;
  rxs=-s.*y;
  xaux = (s.* ((A'\Q)*E)) + (y.*(A*E));
  raux = rxs - (s.*(A'\Dd)) - (y.*Dp);
  dx_a = raux ./ xaux;
  ds_a = Dp + (A*dx_a);
  dy_a = A'$Dd + (Q*dx_a));
  % determine step length alpha_aff
  % (simple line search)
  paux = 1; alpha=1;
  while paux > 0
    yaux= min(y+(alpha*dy_a));
    saux= min(s+(alpha*ds_a));
    if (yaux<0 || saux<0),
      alpha=alpha-0.001;
      if alpha<=0,
        alpha =0; alpha_aff=0;
        paux=0;
      end
    else
      alpha_aff=alpha;
      paux=0;
    end
  end
  muaff = sum((y+(alpha_aff*dy_a).*...
  (s+(alpha_aff*ds_a))))/n;
  % centering parameter
  sigma = min((muaff/mu(end))^3,1);
  % aggregated step -----------------
  rxs = (sigma * mu(end) -y.*s -dy_a.*ds_a);
  raux = rxs - (s.*(A'\Dd)) - (y.*Dp);
  dx = raux ./ xaux;
  ds = Dp + (A*dx);
  dy = A'$Dd + (Q*dx));
  % step length
  tau=0.1;
  my = -tau*y./dy; my(dy>0)=Inf; laux1=min(my);
  ms = -tau*s./ds; ms(ds>0)=Inf; laux2=min(ms);
  malpha = min(laux1,laux2);
  x = x + malpha*dx; y = y + malpha*dy; s = ...
```

```
    s + malpha*ds;
  %save data:
  ux(:,nn)=x; us(:,nn)=s; uy(:,nn)=y;
  umalpha(nn)=malpha; usigma(nn)=sigma;
end
% figure with contours
%contour data ----------
[vx1,vx2] = meshgrid(0:0.05:6, 0:0.05:4);
z=0.5*Q(1,1)*(vx1.^2)+(Q(2,1)+Q(1,2))*vx1.*vx2+...
0.5*Q(2,2)*(vx2.^2)+c(1)*vx1+c(2)*vx2;
figure(1)
colormap('hsv');
patch([0,2,6,0,0],[1,2,0,0,1],[0.3 0.8 0.7]);
hold on;
contour(vx1,vx2,z,30,'LineWidth',1.5);
plot([0,5],[1,7/2],'k--','LineWidth',2); %constr 1
plot([0,5],[3,1/2],'k--','LineWidth',2); %constr 2
plot(solRef(1), solRef(2),'bo','MarkerSize',10,...
'LineWidth',3); %solution
plot(ux(1,:),ux(2,:),'or');
x
J=(0.5*x'*Q*x)+(c'*x)
figure(2)
subplot(1,2,1)
plot(us(1,:),'r'); hold on;
plot(us(2,:),'b');
subplot(1,2,2)
plot(uy(1,:),'r'); hold on;
plot(uy(2,:),'b');
figure(3)
subplot(1,2,1)
plot(umalpha(1,:),'r');
subplot(1,2,2)
plot(usigma(1,:),'m');
figure(4)
plot(mu,'k');
```

Indeed this program could be improved in many ways, to prevent numerical difficulties (there are matrix inversions), to overcome convergence traps, to get a better final approach, etc.

Figure 10.8 shows the interior-point trajectory toward the optimal solution, departing from point (3, 0.2). This is one of the figures generated by Program 10.6.

Next two plots (Fig. 10.9) depict the evolution of slack variables $\vec{s}$ and Lagrange multipliers $\vec{y}$ (i.e. $\vec{\lambda}$), along the trajectory depicted in Fig. 10.8.

Figure 10.10 shows the evolution of $\hat{\alpha}$ and σ along the trajectory in Fig. 10.8.

Finally, Fig. 10.11 depicts the evolution of the μ parameter.

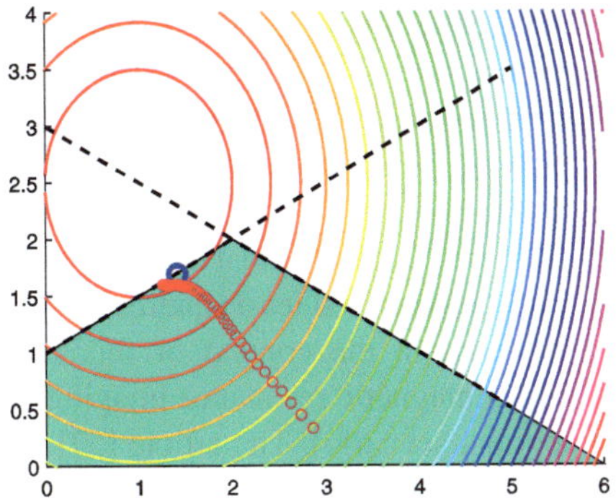

Fig. 10.8 Interior-point trajectory from (3, 0.2)

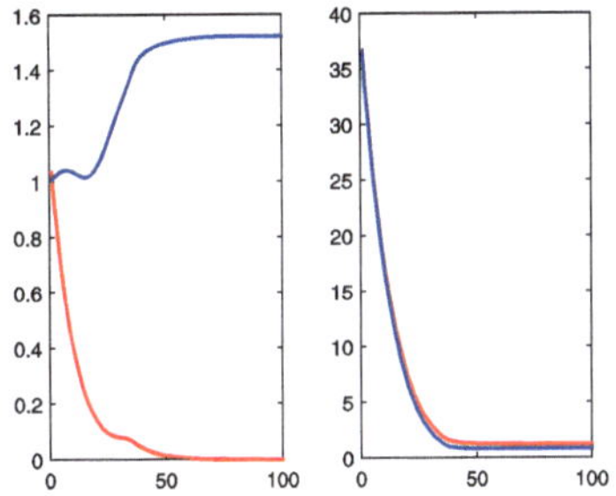

Fig. 10.9 Evolution of $\vec{s}$ (left) and $\vec{y}$ (right) along the trajectory from (3, 0.2)

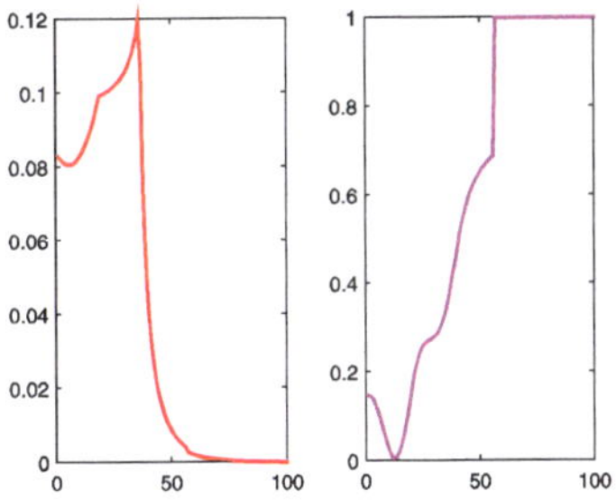

Fig. 10.10 Evolution of $\hat{\alpha}$ (left) and σ (right) along the trajectory from (3, 0.2)

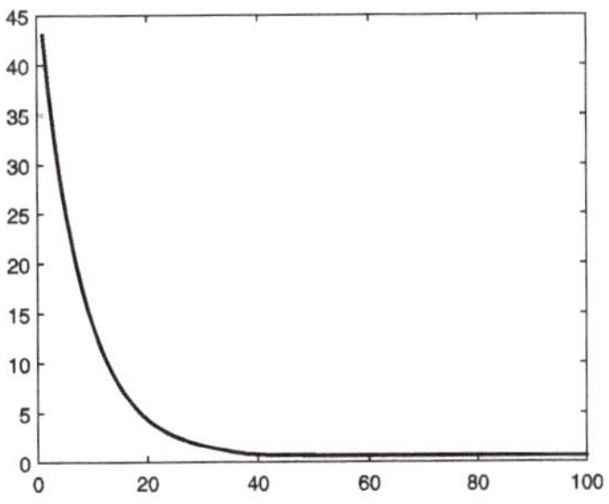

Fig. 10.11 Evolution of μ along the trajectory from (3, 0.2)

Among the many applications of quadratic programming by interior-point methods, we wish to cite some illustrative and promising areas.

There is a lot of research on Support Vector Machines (SVMs) training in special circumstances. In [22], it was proposed to use interior-point methods for massive SVMs and massive meaning quadratic programs with between 10 and 100 million variables. Other, more recent publications on SVMs and interior-point methods are [20] and [14].

It seems surprising that the article [32], titled "An interior-point method for large-scale l_1-regularized least squares," had received 2,391 cites. One of the applications considered in the context of this contribution is inversion problems related to medical magnetic resonance imaging (this makes our surprise to disappear: it is an important scenario). The article [57] gives more information on this medical facet. Actually, a basic search on this topic, interior-point methods and medical imaging, gives quite a number of highly cited contributions.

Speaking of images. Some of the methods and techniques being applied for digital processing of pictures and movies, with restoration or inpainting purposes for instance, come to the optimization field in the form of quadratic objective functions and lots of (space-time) data. For example, applications that use a *Total Variation* criterion for image quality evaluation: some old roots are here, in 1996 we have [37] using interior-point methods for image restoration (many more contributions of research in this line followed).

From a more general point of view, it would be recommended to see [18], about sparse approximations with interior point methods. This is a nuclear topic for disparate applications in finance, medical imaging, data classification, etc.

In a different atmosphere, which is related to energy, a quadratic optimization source of problems comes from electric power generation and management. So, again, it is not strange to have research contributions based on employing interior-point methods. Article [29] is a kind of programmatic insight (year 1999) welcoming interior-point methods in power engineering. Asking Google today about *electric power interior point methods* yields 3,930,000,000 results, an incredible number.

10.6 Application to Convex Programming

The main applications that will be considered in some detail next are cone programming and semidefinite programming. Considerable research efforts have been devoted to these convex programming categories. The high interest of this field can be somewhat gauged by asking Google, where you find *millions* of responses about interior point methods for conic or semidefinite programming.

10.6.1 Cone Programming

Several interior point approaches have been proposed for the cone programming scenario. A well-known solution approach is the algorithm introduced by E.D. Andersen, C. Roos, and T. Terlaky, [5]. In fact, it has been selected by MATLAB for the function *coneprog( ).* We will now describe in concise way the algorithm, which we will call the ART algorithm.

At this point, it should be highlighted the important contributions of. Y.E. Nesterov and M.J. Todd to the solid progress of the interior point methodology. For instance, as reported in [5], the ART algorithm we are going to introduce is based on a publication of Nesterov and Todd, [48], which shows that the primal-dual approach keeps its efficiency when applied to cones, provided that these cones were homogeneous and self-dual (that is, *self-scaled*, in terminology of Nesterov and Todd).

A principal reference for our description would be, logically, [5]. It would be complemented by [33].

10.6.1.1 Preliminaries

One of the authors referred in [5] points out that the only self-scaled cones of interest are the first orthant cone, the quadratic cone, and the cone of positive semidefinite matrices.

The ART algorithm focuses on second-order cones.

The second-order cone K is defined as $K = K^1 \times K^2 \times \ldots \times K^k$. That is: the direct product of k cones, where K^j is one of the three cones below:

A- $\Re_+ = \{\vec{x}^i \in \Re \,/\, \vec{x} \geq 0\}$

B- quadratic cone $K^q = \{\vec{x}^{\,i} \in \Re^{n_i} \,/\, (x_1^i)^2 \geq \left\|\vec{x}^{\,i}_{2:n_i}\right\|^2,\ x_1^i \geq 0\}$

C- rotated quadratic cone:

$$K^r = \{\vec{x}^{\,i} \in \Re^{n_i} \,/\, 2\,x_1^i x_2^i \geq \left\|\vec{x}^{\,i}_{3:n_i}\right\|^2,\ x_1^i,\ x_2^i \geq 0\}$$

These three cones are self-dual, $K = K^*$.

The variable $\vec{x}$ will be partitioned according to the cones:

$$\vec{x} = \begin{pmatrix} \vec{x}^{\,1} \\ \vec{x}^{\,2} \\ \vdots \\ \vec{x}^{\,k} \end{pmatrix}$$

We shall consider two matrices associated with each cone, Q^i, T^i with dimension $n^i \times n^i$. Depending on the type of cone (A, B, or C) these matrices are defined as

A- $Q^i = 1, \;\; T^i = 1$
B- $Q^i = \text{diag}\,(1, \; -1, \; \ldots, -1)$ and $T^i = \text{diag}\,(1, \; 1, \; \ldots 1)$
C-

$$Q^i = \begin{pmatrix} 0 & 1 & 0 & \ldots & 0 \\ 1 & 0 & 0 & \ldots & 0 \\ 0 & 0 & -1 & \ldots & 0 \\ \vdots & \vdots & & \ddots & \vdots \\ 0 & 0 & 0 & \ldots & -1 \end{pmatrix} \text{ and } T^i = \begin{pmatrix} \beta & \beta & 0 & & \ldots & 0 \\ \beta & \beta & 0 & & \ldots & 0 \\ 0 & 0 & 1 & 0 & & \ldots 0 \\ \vdots & \vdots & & \ddots & & \\ 0 & 0 & 0 & & \ldots & 1 \end{pmatrix} \tag{10.9}$$

where $\beta = 1/\sqrt{2}$

This definition of matrices Q^i allow an equivalent way of stating B and C cone types:

B- quadratic cone $\text{K}^q = \{\vec{x}^{\,i} \in \Re^{n_i} \,/\, (\vec{x}^{\,i})^T Q^i \vec{x}^{\,i}, \; x_1^i \geq 0\}$
C- rotated quadratic cone $\text{K}^r = \{\vec{x}^{\,i} \in \Re^{n_i} \,/\, (\vec{x}^{\,i})^T Q^i \vec{x}^{\,i}, \; x_1, \; x_2 \geq 0\}$

During the mathematical development of the algorithm, some "arrow head" matrices will be employed, with the following formalism:

$$V = \text{mat}\,(\vec{v}) = \begin{pmatrix} v_1 & \vec{v}_{2:n}^{\,T} \\ \vec{v}_{2:\,n} & v_1 I \end{pmatrix}$$

10.6.1.2 Adaptation of the Interior-Point Method

First, let us state the optimization problem and then proceed with the solution approach.

The conic optimization problem can be expressed as

Minimize: $\vec{c}^T \vec{x}$

Subject to:

$$A\vec{x} = \vec{b}$$

$$\vec{x} \in \text{K}$$

where K is a pointed closed convex cone.

The dual problem is
Maximize: $\vec{b}^T \vec{\lambda}$
Subject to:

$$A^T \vec{\lambda} + \vec{s} = \vec{c}$$

$$\vec{s} \in \mathrm{K}^*$$

where K* is the dual cone.

The primal problem is feasible if it has at least one feasible solution. It is strictly feasible if it has a feasible solution belonging to the interior of K. *Mutatis mutandis* with respect to the dual problem.

The ART algorithm employs a generalization of the Goldman-Tucker homogeneous model (a combined primal-dual statement of the optimization problem), in the following simplified terms:

$$A\vec{x} - \vec{b}\tau = 0$$

$$A^T \vec{\lambda} + \vec{s} - \vec{c}\tau = 0$$

$$-\vec{c}^T \vec{x} + \vec{b}^T \vec{\lambda} - \kappa = 0$$

where two nonnegative variables τ and κ have been added.

A solution to this system of equations is said to be *complementary*, if the complementary gap $\vec{x}^T \vec{s} + \tau\kappa$ is zero.

(See [4] and references therein for more details on the homogeneous model)

Part of the KKT conditions are the complementary slackness conditions. In order to consider them, let us define X^i and S^i for each type of cone:

A- $x^i \in \Re_+$: $X^i = x^i$ and $S^i = s^i$
B- $\vec{x}^i \in K^q$: $X^i = \mathrm{mat}(\vec{x}^i)$ and $S^i = \mathrm{mat}(\vec{s}^i)$
C- $\vec{x}^i \in K^r$: $X^i = \mathrm{mat}(T\,\vec{x}^i)$ and $S^i = \mathrm{mat}(T\,\vec{s}^i)$

Now, it can be shown [33] that, given $\vec{x}, \vec{s} \in K$, then $\vec{x}$ and $\vec{s}$ are complementary ($\vec{x}^T \vec{s} = 0$) if and only if:

$$X^i S^i e^i = S^i X^i e^i = 0 \text{ for } i = 1, \ldots, k$$

Let us come to a key point of the algorithm. We can define the **central path** as follows, [5]:

$$A\vec{x} - \vec{b}\tau = \gamma\,(A\,\vec{x}^{(0)} - \vec{b}\tau^{(0)})$$

$$A\vec{\lambda} + \vec{s} - \vec{c}\tau = \gamma\,(A\,\vec{\lambda}^{(0)} + \vec{s}^{(0)} - \vec{c}\tau^{(0)})$$

$$-\vec{c}^T \vec{x} + \vec{b}^T \vec{\lambda} - \kappa = \gamma\,(-\vec{c}^T \vec{x}^{(0)} + \vec{b}^T \vec{\lambda}^{(0)} - \kappa^{(0)})$$

$$X S \vec{e} = \gamma \mu^{(0)} \bar{e}$$

$$\tau \kappa = \gamma \mu^{(0)}$$

As can be seen, the central path is parameterized by the centering parameter γ (between 0 and 1).

Besides,

$$\mu^{(0)} = \frac{(\vec{x}^{(0)})^T s^{(0)} + \tau^{(0)} \kappa^{(0)}}{k+1}$$

And,

$$X = \text{diag}\{X^1, \ X^2, \ldots, X^k\} \text{ and } S = \text{diag}\{S^1, \ S^2, \ldots, S^k\}$$

The equations of the central path just written could become a not well-defined scenario. A well-defined search direction can be ensured by using scaling before applying Newton's method and then scaling back to the original space [33].

Therefore:

$W^i \in \Re^{n_i \times n_i}$ is a *scaling matrix* if:

(a) W^i is a symmetric positive definite matrix, and
(b) $W^i Q^I W^i = Q^i$

The scaled point $\widehat{x}, \ \widehat{s}$ is obtained with the transformation:
$\widehat{x} = \Theta W \vec{x}$ and $\widehat{s} = (\Theta W)^{-1} \vec{s}$
where,

$$W = \begin{pmatrix} W^1 & 0 & \ldots 0 \\ 0 & W^2 & \ldots 0 \\ \vdots & \ddots & \ldots \vdots \\ 0 & 0 & W^k \end{pmatrix} \quad \Theta = \begin{pmatrix} \Theta^1 I_{n_1} & 0 & 0 \\ 0 & \Theta^2 I_{n_2} & \ldots 0 \\ \vdots & \ddots & \ldots \vdots \\ 0 & 0 & \Theta^k I_{n_k} \end{pmatrix}$$

Hence, the scaled search direction would be

$$A \Delta_x - \vec{b} \Delta_\tau = (\gamma - 1) \left(A \vec{x}^{(0)} - \vec{b} \tau^{(0)} \right)$$

$$A \Delta_\lambda + \Delta_s - \vec{c} \Delta_\tau = (\gamma - 1) \left(A \vec{\lambda}^{(0)} + \vec{s}^{(0)} - \vec{c} \tau^{(0)} \right)$$

$$- \vec{c}^T \Delta_x + \vec{b}^T \Delta_\lambda - \Delta_\kappa = (\gamma - 1) \left(- \vec{c}^T \vec{x}^{(0)} + \vec{b}^T \vec{\lambda}^{(0)} - \kappa^{(0)} \right)$$

$$\widehat{X}^{(0)} T (\Theta W)^{-1} \Delta_s + \widehat{S}^{(0)} T \Theta W \Delta_x = - \widehat{X}^{(0)} \widehat{S}^{(0)} \vec{e} + \gamma \mu^{(0)} \bar{e}$$

$$\tau^{(0)} \Delta_\kappa + \kappa^{(0)} \Delta_\tau = - \tau^{(0)} \kappa^{(0)} + \gamma \mu^{(0)}$$

where,

$$T = \begin{pmatrix} T^1 & 0 & \dots 0 \\ 0 & T^2 & \dots 0 \\ \vdots & & \ddots \dots \vdots \\ 0 & 0 & T^k \end{pmatrix}$$

The interior point iterations will come from one point to the next as follows:

$$\begin{pmatrix} \vec{x}^{(1)} \\ \tau^{(1)} \\ \vec{\lambda}^{(1)} \\ \vec{s}^{(1)} \\ \kappa^{(1)} \end{pmatrix} = \begin{pmatrix} \vec{x}^{(0)} \\ \tau^{(0)} \\ \vec{\lambda}^{(0)} \\ \vec{s}^{(0)} \\ \kappa^{(0)} \end{pmatrix} + \alpha \begin{pmatrix} \Delta\vec{x} \\ \Delta\tau \\ \Delta\vec{\lambda} \\ \Delta\vec{s} \\ \Delta\kappa \end{pmatrix}$$

The scaling suggested by [48] is chosen such that:

$$\Theta\, W\, \vec{x} = \widehat{x} = \widehat{s} = (\Theta\, W)^{-1}\, \vec{s}$$

Implying that:

$$\vec{s} = \left(W\, \Theta^2\, W\right) \vec{x}$$

With this option, the computation of involved parameters and matrices is not expensive, with the following expressions:

$$\theta_i^2 = \sqrt{\frac{(\vec{s}^i)^T\, Q^i \vec{s}^i}{(\vec{x}^i)^T\, Q^i \vec{x}^i}}$$

(cones)

A- $W^i = \frac{1}{\theta_i}((X^i)^{-1}\, S^i)^{\frac{1}{2}} = 1$

B- $W^i = \begin{pmatrix} w_1^i & \left(\vec{w}^i_{2:n^i}\right)^T \\ \vec{w}^1_{2:n^i} & I + \frac{\vec{w}^i_{2:n^i}\,(\vec{w}^i_{2:n^i})^T}{1+w_1^i} \end{pmatrix} = -Q^i + \frac{(\vec{e}^i+\vec{w}^i)(\vec{e}^i+\vec{w}^i)^T}{1+(\vec{e}^i)^T\vec{w}^i}$

where

$$\vec{w}^i = \frac{\theta_i^{-1}\vec{s}^i + \theta_i\, Q^i\, \vec{x}^i}{\sqrt{2}\sqrt{(\vec{x}^i)^T\vec{s}^i + \sqrt{(\vec{x}^i)^T\, Q^i\vec{x}^i\,(\vec{s}^i)^T\, Q^i\vec{s}^i}}}$$

Furthermore:

$$(W^i)^2 = -Q^i + 2\vec{w}^i(\vec{w}^i)^T$$

C- $W^i = -Q^i + \frac{(T^i\vec{e}^i+\vec{w}^i)(T^i\vec{e}^i+\vec{w}^i)^T}{1+(\vec{e}^i)^T T^i\vec{w}^i}$

10.6.1.3 Predictor-Corrector Method

Like in previous sections, the Mehrotra's algorithm can be adapted as follows.

First, the predictor direction can be found by solving:

$$\begin{pmatrix} A & -\vec{b} & 0 & 0 & 0 \\ 0 & -\vec{c} & A^T & I & 0 \\ -\vec{c}^T & 0 & \vec{b}^T & 0 & -1 \\ \widehat{S}\,T\,\Theta\,W & 0 & 0 & \widehat{X}\,T\,(\Theta W)^{-1} & 0 \\ 0 & \kappa & 0 & 0 & \tau \end{pmatrix}\begin{pmatrix} \Delta\vec{x}^{aff} \\ \Delta\tau^{aff} \\ \Delta\vec{\lambda}^{aff} \\ \Delta\vec{s}^{aff} \\ \Delta\kappa^{aff} \end{pmatrix} = \begin{bmatrix} -\Delta_P \\ -\Delta_D \\ -\Delta_G \\ -\widehat{X}\widehat{S}\,\vec{e} \\ -\tau\,\kappa \end{bmatrix}$$

Next, the step length along this direction can be determined:

$$\alpha_P = \arg\max\Big(\alpha \in [0,1] \,/\, (\vec{x},\tau) + \alpha\left(\Delta\vec{x}^{aff}, \Delta\tau^{aff}\right),\ (\vec{s},\kappa) + \alpha\left(\Delta\vec{s}^{aff}, \Delta\kappa^{aff}\right) \in \mathrm{K}\Big)$$

define also:

$$\mu_P = \Big(\left(\vec{x} + \alpha_P\Delta\vec{x}^{aff}\right)^T\left(\vec{s} + \alpha_P\Delta\vec{s}^{aff}\right) + \\ + \left(\tau + \alpha_P\Delta\tau^{aff}\right)\left(\kappa + \alpha_P\Delta\kappa^{aff}\right) / (k+1)\Big)$$

After some approximations, [33], the equation to solve for corrector-centering direction is established as

$$\begin{pmatrix} A & -\vec{b} & 0 & 0 & 0 \\ 0 & -\vec{c} & A^T & I & 0 \\ -\vec{c}^T & 0 & \vec{b}^T & 0 & -1 \\ \widehat{S}\,T\,\Theta\,W & 0 & 0 & \widehat{X}\,T\,(\Theta W)^{-1} & 0 \\ 0 & \kappa & 0 & 0 & \tau \end{pmatrix}\begin{pmatrix} \Delta\vec{x}^{\mathrm{cor}} \\ \Delta\tau^{\mathrm{cor}} \\ \Delta\vec{\lambda}^{\mathrm{cor}} \\ \Delta\vec{s}^{\mathrm{cor}} \\ \Delta\kappa^{\mathrm{cor}} \end{pmatrix} = \begin{bmatrix} \sigma\,\Delta_P \\ \sigma\,\Delta_D \\ \sigma\,\Delta_G \\ \sigma\,\mu_P\,\vec{e} - \Delta\widehat{X}_P\,\Delta_P\widehat{S}\,\vec{e} \\ \sigma\mu_P - \Delta\tau_P\,\Delta\kappa_P \end{bmatrix}$$

10.6.2 Semidefinite Programming

An important specific branch of convex programming is SDP. The topic was introduced in 1963, by Bellman and Fan [8]. The generalization of interior-point methods from LP to SDP was introduced in [2].

The mathematical description of the path-following interior point methodology for SDP would be based on [27], with background from [63] (which was the main reference for SDP in the previous chapter).

Recall that the SDP problem is

$\min \langle C,\ X\rangle$, subject to $A\,X\ =\ \vec{b}$ with X positive semi-definite

The following **barrier function** is frequently used for SDP problems:

$$B(\vec{x})\ =\ -\log\left(\det\left(X\right)\right)$$

(other barrier functions can be used)

This barrier function belongs to the class of strongly self-concordant functions. The gradient of $B(\vec{x})$ is $-\ X^{-1}$.

The function $B(\cdot)$ can be regarded as a potential function corresponding to a repelling force from each constraint hyperplane. Note that, in the case of linear inequalities $\vec{a}_i^T\ \vec{x}\ +\ b_i\ \geq\ 0,\ \ i=1,\ldots,n:$

$$\nabla B(\vec{x})\ =\ \sum_{i=1}^{n}\frac{1}{\vec{a}_i^T\,\vec{x}+b_i}\,\vec{a}_i\ =\ \sum_{i=1}^{n}\frac{-1}{r_i\,\|\vec{a}_i\|}\,\vec{a}_i$$

where r_i is the distance from $\vec{x}$ to the i-th constraint hyperplane.

Figure 10.12 presents photographs of Nemirovski and Nesterov, who substantially contributed to solve semidefinite optimization problems with interior point algorithms. Arkadi Nemirovski is a professor at the Georgia Institute of Technology. Yurii Nesterov is a professor at the University of Louvain. Both Nemirovski and Nesterov received the John von Neumann theory and the George B. Dantzig, prizes, among other recognitions.

Path-following interior-point algorithm

Let us develop an interior point algorithm based on the barrier function.

First, one establishes the following auxiliary problem:

$\min \langle C,\ X\rangle\ -\ \mu\ \log\left(\det(X)\right)$, subject to $A\,X\ =\ \vec{b}$

with X definite positive

This problem is transformed into an unconstrained problem by building the next Lagrangian:

$$L(X,\ \lambda)\ =\ \langle C, X\rangle\ -\ \mu\ \log\left(\det(X)\right)\ +\ \left\langle\vec{\lambda},\ \vec{b}\ -\ A\,X\right\rangle.$$

Fig. 10.12 A. Nemirovski (left) and Y. Nesterov (right)

Gradients are now applied to find the optimum:

$$\nabla_x L(X, \lambda) = C - \mu X^{-1} - A^T \vec{\lambda} = 0$$

$$\nabla_\lambda L(X, \lambda) = \vec{b} - A\vec{x} = 0$$

In preparation for a primal-dual formulation, one sets: $S = \mu X^{-1}$
And so, one obtains the following KKT conditions:

$$A X = \vec{b}$$

$$A^T \vec{\lambda} + S = C$$

$$X Z = \mu I$$

with X, C definite positive

For a sequence of $\mu_k > 0$, $k = 1, \ldots N$ with $\mu_k \rightarrow 0$, the solutions of the KKT equations give us the central path, which converges to the optimum.

In view of using the Newton method, one considers:

$$F(X, \vec{\lambda}, S) = \begin{pmatrix} AX - \vec{b} \\ A^T \vec{\lambda} + S - C \\ XZ - \mu I \end{pmatrix} = 0$$

then one forms

$$J(X, \vec{\lambda}, S) \begin{pmatrix} \Delta X \\ \Delta \vec{\lambda} \\ \Delta S \end{pmatrix} = -F(X, \vec{\lambda}, S)$$

obtaining the following system:

$$\begin{pmatrix} 0 & A^T & I \\ A & 0 & 0 \\ S & 0 & X \end{pmatrix} \begin{pmatrix} \Delta X \\ \Delta \vec{\lambda} \\ \Delta S \end{pmatrix} = \begin{pmatrix} -\Delta_P \\ -\Delta_D \\ \mu I - X S \end{pmatrix} \tag{10.10}$$

which yields the Newton direction for the next step.

There are serious difficulties associated with the equations above. First: X and S do not commute ($X S \neq S X$). Also, in general no solution with both ΔX and ΔS symmetric could be found. In particular, the solution for square matrices ΔX, ΔS, would yield a symmetric ΔS but (in general) a nonsymmetric ΔX. This is a problem, because one needs the next step $X + \alpha \Delta X$ to be a symmetric positive definite matrix.

Some alternatives for circumventing the difficulty have been proposed. Let us mention three of them (see [27] and references therein):

A first approach tolerates ΔX to be unsymmetric; but the skew-symmetric part of ΔX is ignored and the symmetric part is taken as step direction:

$$\widehat{\Delta X} S + X \Delta Z = \mu I - X Z, \quad \Delta X = \frac{\widehat{\Delta X} + \widehat{\Delta X}^T}{2}$$

The second approach is based on self-scaled barrier functions. The following scaling matrix is used:

$$W = X^{1/2} (X^{1/2} S X^{1/2})^{-1/2} X^{1/2}$$

(this matrix satisfies that: $W^{-1/2} X W^{-1/2} = W^{1/2} S W^{1/2}$)

Matrix W permits to reformulate the complementary condition as:

$$W^{-1} \Delta X W^{-1} + \Delta S = \mu X^{-1} - S$$

Any solution to this equation would be symmetric.

The third approach modifies the complementary condition to allow for only symmetric updates. One considers the linearization of $X S + S X - \mu I$ as follows:

$$\Delta X S + X \Delta S + S \Delta X + \Delta S X = 2\mu I - X S - S X$$

Then, solutions of ΔX and ΔS would be symmetric if X and S were close to the central path.

All three search directions already introduced belong to a general scenario obtained with the next symmetrization operator:

$$H_P(M) = \frac{1}{2}\left(P\,M\,P^{-1} + (P\,M\,P^{-1})^T\right)$$

where P is a nonsingular matrix, $P \in S_n$.

Applying the operator to the complementary condition, gives

$$H_P(X\,S + \Delta X\,S + X\,\Delta S) = \mu\,I$$

The first approach corresponds to $P = S^{1/2}$. The second approach to $P^T P = W$. And the third corresponds to $P = I$.

Many other search directions have been investigated. In [60], M.J. Todd describes up to 20 techniques for deriving search directions.

In summary, here is an algorithm for solving a SDP problem, [27]:

Algorithm 4

Repeat:
  choose μ
  Compute $(\Delta X,\ \Delta\vec{\lambda},\ \Delta\, S)$ by solving the system (10.10)
  choose some $\alpha \in (0,\ 1]$ so that $X + \alpha\, X$ and $S + \alpha\, S$ remain
    positive definite
  Set $(X,\ \vec{\lambda},\ S) := (X + \alpha\,\Delta X,\ \vec{\lambda} + \alpha\,\Delta\vec{\lambda},\ S + \alpha\,\Delta S)$
until a satisfactory approximation to the optimal is reached

In practical terms, the procedure for solving the system (10.10) begins by expressing ΔX in terms of ΔS in the third equation. Then, the second equation is used to express ΔS in terms of $\Delta\vec{\lambda}$. And finally, one obtains $\Delta\vec{\lambda}$ from the first equation, which now reads as

$$A\left(X\,A^T(\Delta\vec{\lambda})\,S^{-1}\right) = \vec{b} - A\left(\mu\,S^{-1} + X\left(C - A^T\vec{\lambda} - S\right)S^{-1}\right)$$

The matrix $M = A\,(X\,A^T(\cdot)\,S^{-1})$ is positive definite if both X and S are definite positive and A has full row rank. The factorization of matrix M is usually the most expensive part of the computations (in certain cases, the line search for some $\alpha \in (0,\ 1]$ could be also cumbersome). More numerical details in [27] and [52].

10.6.2.1 Affine Scaling

Two primal-dual variants of affine scaling have been proposed by Etienne de Klerk [17], Delft University. Both variants are based on the following scaling matrix:

$$D := S^{1/2}\left(S^{1/2}\, X\, S^{1/2}\right)^{1/2} S^{-1/2}$$

(related to the contributions of Nesterov, Ye and Todd in [48])

The matrix D can be used to scale the variables X and S to the same symmetric positive definite matrix V, since:

$$D^{-1/2}\, X\, D^{-1/2} = D^{1/2}\, S\, D^{1/2} = V$$

It happens that

$$V^2 = D^{-1/2}\, X\, S\, D^{1/2} \propto XS$$

that is: matrix V^2 is symmetric positive definite and has the same eigenvalues of XS. This symmetrization of XS is called NT scaling.

The scaled search directions are defined as

$$D_X = D^{-1/2}\, \Delta X\, D^{-1/2}$$

$$D_S = D^{1/2}\, \Delta S\, D^{1/2}$$

and the scaled primal-dual direction is defined as

$$D_V = D_X + D_S$$

After a feasible primal-dual step, the duality gap becomes, [17]:

$$tr((X + \Delta X)\,(S + \Delta S)) = tr((V + D_X)\,(V + D_S)) = tr(V^2 + V\, D_V)$$

Now, let us introduce the two variants: the Dikin-type direction (so named by de Klerk) and the primal-dual affine direction.

The *Dikin-type direction* is derived by minimization of the duality gap over the primal-dual Dikin ellipsoid, as follows:

$$D_V^* = \arg\min_{D_V} \left\{ tr(V^2 + V\, D_V) \,/\, \left\| V^{-1/2}\, D_V\, V^{-1/2} \right\| \leq 1 \right\}$$

This problem has the following solution:

$$D_V^* = D_X^* + D_S^* = -\frac{V^2}{\|V^2\|}$$

To transform back to the un-scaled space, one pre&post multiplies by $D^{1/2}$, obtaining:

$$\Delta X + D\,\Delta S\,D = -\frac{-X\,S\,X}{(tr(X\,S)^2)^{1/2}}$$

The Daikin direction is found by solving this equation, subject to the next conditions:

$$tr(A_i\,\Delta X) = 0, \qquad i = 1, \ldots, m$$

$$\sum_{i=1}^{m} \Delta\,\lambda_i\,A_i + \Delta S = 0$$

The *primal-dual affine-scaling direction* is defined by a simple change in the norm to be obeyed:

$$D_V^* = \arg\min_{D_V} \left\{ tr(V^2 + V\,D_V) \,/\, \left\| V^{-1/2}\,D_V\,V^{-1/2} \right\|_2 \leq 1 \right\}$$

It is deduced that the solution of this problem is $D_V^* = -V$
As before, getting back to the un-scaled space, one obtains:

$$\Delta X + D\,\Delta S\,D = -X$$

The direction is found by solving this equation, subject to the same conditions as before.

The two short-step algorithms can both be described as follows:

Algorithm 5

From a strictly feasible $(X^0,\ S^0)$
While $tr(X,\ S) > \varepsilon$, do:
  Compute $(\Delta X,\ \Delta S)$, being a Dikin-type step,
    or a primal-dual affine-scaling step
  Set $X := X + \alpha\,\Delta X$, $S := S + \alpha\,\Delta S$
end

where α is

$\alpha = \frac{1}{\sqrt{n}\,\tau_0}$ in the case of Dikins type steps

or,

$\alpha = \frac{1}{n\,L\,\tau_0}$ in the case of primal-dual affine-scaling

and where: $L = \log \frac{tr(X^0 S^0)}{\varepsilon}$

10.6.2.2 Primal-Dual Path Following Using Nesterov and Todd Scaling

According to the system (10.10), we need to compute $(\Delta X,\ \Delta S)$ such that approximately:

$$(X + \Delta X)\,(S + \Delta S) = \mu\, I$$

Using the same scaling matrices that we employed for affine-scaling, the previous expression can be rewritten as follows:

$$(V + D_X)\,(V + D_S) = \mu\, I$$

Let us take a weakened version of this condition, replacing the left-hand side with its symmetric part:

$$\frac{1}{2}\left[(V + D_X)(V + D_S) + ((V + D_X)\,(V + D_S))^T\right] = \mu\, I$$

By neglecting cross terms, this equation can be linearized as follows:

$$\frac{1}{2}\,[(D_X + D_S)\,V + V(D_X + D_S)] = \mu\, I - V^2$$

There is unique symmetric solution for this equation, which is

$$D_V = D_X + D_S = \mu\, V^{-1} - V$$

By pre & post multiplying by $D^{1/2}$, one obtains the Nesterov-Todd (NT) equations:

$$\Delta X + D\,\Delta S\,D = \mu\, S^{-1} - X \tag{10.11}$$

Subject to:

$$tr(A_i \Delta X) = 0, \quad i = 1, \ldots, m$$

$$\Delta S = \sum_{i=1}^{m} \Delta\lambda_i A_i$$

It is possible to get ΔS from the last equation and then obtain ΔX from (10.11). The result $(\Delta X,\ \Delta S)$ is the *NT direction*.

Primal-dual potential reduction

In his introduction of potential reduction methods, de Klerk comments that these methods were the first interior-point methods to be extended from LP to SDP;

mostly due to the work of Nesterov and Nemirovski [47] and Alizadeh [2]. See [3], or [62], for convenient surveys of potential reduction methods.

At each iteration, the potential should achieve a certain decrease. Commonly, a potential of the type Tanabe-Todd-Ye is chosen, with the following expression:

$$\Phi(X, S) = (n + \nu\sqrt{n}) \log (tr(X\,S) - \log \det(X\,S) - n \log n$$

where $\nu \geq 1$.

The algorithm proposed by Nesterov and Todd is the following:

Algorithm 6

From a strictly feasible starting pair $(X^0,\ S^0)$
While $tr(X,\ S) > \varepsilon$, do:
  Compute $\mu = tr(X,\ S)\,/\,(n + \nu\sqrt{n})$
  Compute $(\Delta X,\ \Delta\,S)$ from (10.11)
  Find $(\alpha,\ \beta) = \arg\min\ \Phi(X + \alpha\,\Delta X,\ S + \beta\,\Delta S)$
  Set $X := X + \alpha\,\Delta X,\ S := S + \beta\,\Delta S$
end

10.7 Extending the Field

A large variety of research works are extending the interior-point method (IPM) field, both in the type of problems that can be tackled and in algorithms and applications. Nonlinear programming using IPM is an important example of what is being done. In addition, several established techniques are revisited under IPM new optics. And the possibility of hybrids combining IPM and heuristics is becoming an interesting trend.

10.7.1 Nonlinear Programming

A great deal of research on interior point methods is devoted now to nonlinear applications, which is a much broader field than linear cases.

R.J. Vanderbei, D.F. Shanno, and collaborators, developed a software tool (a solver) named LOQO for quadratic programming, based on interior-point methods. Taking advantage of the success attained by this solver, it was proposed in [64] to extend the solver methodology for nonconvex nonlinear programming. Let us describe the substance of their way of attack.

The following optimization problem was considered:

Minimize: $f(\vec{x})$
Subject to: $h_i(\vec{x}) \geq 0, \quad i = 1, \ldots m,$
where $\dim(\vec{x}) = n$; and $f(\vec{x})$ and $h_i(\vec{x})$ are twice continuously differentiable.
By adding slack variables w_i, the problem can be re-formulated as follows:
Minimize: $f(\vec{x})$
Subject to: $\vec{h}(\vec{x}) - \vec{w} = 0, \vec{w} \geq 0$
On the basis of a barrier function, this problem can be transformed to
Minimize: $f(\vec{x}) - \mu \sum\limits_{i=1}^{m} \log(w_i)$
Subject to: $\vec{h}(\vec{x}) - \vec{w} = 0$
The corresponding Lagrangian would be:

$$L(\vec{x}, \vec{w}\, \vec{\lambda}, \mu) = f(\vec{x}) - \mu \sum_{i=1}^{m} \log(w_i) - \vec{\lambda}^T (\vec{h}(\vec{x}) - \vec{w})$$

In order to find the minimum, we equal derivatives to zero obtaining the following conditions:

$$\nabla_x L(\cdot) = \nabla f(\vec{x}) - \nabla \vec{h}(\vec{x})^T \vec{\lambda} = 0$$

$$\nabla_w L(\cdot) = -\mu W^{-1} \vec{e} + \vec{\lambda} = 0$$

$$\nabla_\lambda L(\cdot) = \vec{h}(\vec{x}) - \vec{w} = 0$$

where:
$W = \mathrm{diag}(w_1, w_2, \ldots, w_m)$, and $\nabla \vec{h}(\vec{x})$ is the Jacobian matrix of $\vec{h}(\vec{x})$
We multiply now the second equation by W, obtaining the standard primal-dual system:

$$\nabla f(\vec{x}) - \nabla \vec{h}(\vec{x})^t \vec{\lambda} = 0$$

$$-\mu \vec{e} + W \Lambda \vec{e} = 0$$

$$\vec{h}(\vec{x}) - \vec{w} = 0$$

with: $\Lambda = \mathrm{diag}(\lambda_1, \lambda_2, \ldots)$.

Having in mind the established formulations of linear and quadratic programming, we shall simplify the notation by denoting:

$$H(\vec{x}, \vec{\lambda}) = \nabla^2 f(\vec{x}) - \sum_{i=1}^{m} \lambda_i \nabla^2 h_i(\vec{x})$$

$$A(\vec{x}) = \nabla \vec{h}(\vec{x})$$

Hence, the Newton system for our problem can be expressed as follows:

$$\begin{pmatrix} H(\vec{x},\vec{\lambda}) & -A(\vec{x})^T & 0 \\ A(\vec{x}) & 0 & -I \\ 0 & W & \Lambda \end{pmatrix} \begin{bmatrix} \Delta\vec{x} \\ \Delta\vec{\lambda} \\ \Delta\vec{w} \end{bmatrix} = \begin{bmatrix} -\nabla f(\vec{x}) + A(\vec{x})^T\vec{\lambda} \\ -\vec{h}(\vec{x}) + \vec{w} \\ \mu\vec{e} - W\Lambda\vec{e} \end{bmatrix}$$

This system is nonsymmetric. It can be transformed to a symmetric one multiplying by -1 the first equation and by $-W^{-1}$the third equation:

$$\begin{pmatrix} H(\vec{x},\vec{\lambda}) & -A(\vec{x})^T & 0 \\ A(\vec{x}) & 0 & -I \\ 0 & -I & -W^{-1}\Lambda \end{pmatrix} \begin{bmatrix} \Delta\vec{x} \\ \Delta\vec{\lambda} \\ \Delta\vec{w} \end{bmatrix} = \begin{bmatrix} \sigma \\ \rho \\ -\gamma \end{bmatrix}$$

where the vectors σ, γ and ρ correspond to the following equations:

$$\sigma = -\nabla f(\vec{x}) - A(\vec{x})^T\vec{\lambda}$$

$$\rho = \vec{w} - \vec{h}(\vec{x})$$

$$\gamma = \mu W^{-1}\vec{e} - \vec{\lambda}$$

Here, ρ measures primal infeasibility and σ dual infeasibility.

At the hour of solving the system, the usual first step is to eliminate $\Delta\vec{w}$ based on the third equation:

$$\Delta\vec{w} = W\Lambda^{-1}(\gamma - \Delta\vec{\lambda})$$

Then, one obtains the following *reduced KKT system* (this is a typical approach in most interior-point contexts):

$$\begin{pmatrix} -H(\vec{x},\vec{y} & A(\vec{x})^T \\ A(\vec{x}) & W\Lambda^{-1} \end{pmatrix} \begin{pmatrix} \Delta\vec{x} \\ \Delta\vec{\lambda} \end{pmatrix} = \begin{pmatrix} \sigma \\ \rho + W\Lambda^{-1}\gamma \end{pmatrix}$$

The corresponding iterative algorithm would solve the system and use the resulting directions $\Delta\vec{x}$, $\Delta\vec{\lambda}$, $\Delta\vec{w}$ for computing the next state:

$$\vec{x}^{(k+1)} = \vec{x}^{(k)} + \alpha_k\Delta\vec{x}^{(k)}$$

$$\vec{\lambda}^{(k+1)} = \vec{\lambda}^{(k)} + \alpha_k\Delta\vec{\lambda}^{(k)}$$

$$\vec{w}^{(k+1)} = \vec{w}^{(k)} + \alpha_k\Delta\vec{w}^{(k)}$$

The article [64] continues with implementation details, mainly concerning the computation of the step lengths (that should be short enough for guaranteed convergence), and the use of merit functions for helping in this task. A year later, Shanno and Vanderbei published in [56] an extension of their work, improving LOQO computational features.

Mathematicians know that nonlinear problems could be well compared to a jungle, with traps and surprises. It is not strange to detect difficulties when putting algorithms into practice. For example, concerning the present case, in year 2000 Wächter and Biegler [65] provided an example for which primal-dual methods fail to converge. This fact motivated several reactions. In the same year 2000, H.Y. Benson, Vanderbei, and Shanno published an article, [9], considering the case put by [65] and offering three ways of resolution (one of these already in use by LOQO) and not only this: the article dives into the issue and identifies six distinct types of problems that can cause troubles for LOQO (and for other solvers).

In [10], Benson, Venderbei, and Shanno considered the use of the so-called *filter methods*, introduced by Fletcher and Leyffer [24] (1,329 cites), instead of merit functions, inside LOQO. They conclude that filter-based algorithms are superior in terms of efficiency to ones with only merit functions.

Many other investigators participated in the development of interior-point concepts and algorithms for nonlinear programming. A popular survey of those efforts up to year 2002 is [25].

In this context, a celebrated contribution (with 9,892 cites) has been [66], by Wätcher and Biegler. They present a primal-dual algorithm (described in detail) with a filter line search method. The filter method is a tool to guarantee global convergence; the underlying concept being that trial points (during the progress toward the solution) are accepted if they improve the objective function **or** improve the constraint violation.

The state of the art in interior methods by 2008 is well described by Nemirovski and Todd in [46]. Concerning the application to nonlinear programming, they recognize two main approaches based on line searches (an example is the Vanderbei et al.'s way) or on trust regions. Some of the software tools they mention, able to deal with nonlinear problems, are LOQO (Vanderbei and Shanno and KNITRO/DIRECT (Waltz, Morales, Nocedal).

10.7.2 Other Variants and Extensions

Interior-point methods can be combined with other methods, as it has been proposed by many investigators. Let us briefly give a sketch of the ideas and contributions put on scene.

Active set
The article [61] introduces an active-set-like strategy to deal with dual slacks, when applying interior-point methods for linear programming.

Cutting planes
In his doctoral thesis [44], J. Naoum-Sawaya introduces novel approaches that use interior-point cutting planes methods for integer and mixed-integer programming. The thesis also presents an interior-point branch and cut algorithm.

Branch and bound
As described in the precedent chapter, the original branch and bound algorithm uses several times the simplex method for linear programming. In 1991, reference [11] suggested to use an interior-point algorithm instead of the simplex method inside the steps of the branch and bound algorithm.

The idea has been further developed in several publications, including [42] which proposes a branch-price-and-cut algorithm and [45] with a branch-and-cut algorithm for mixed integer programming, both published in 2013.

Sequential quadratic programming (SQP)
The article [19] focuses on an important class of minimization problems that arises from the discretization of optimal control systems and that are formulated as a specially structured nonlinear programming problem. The optimization by SQP solves a sequence of QP subproblems that the article treats with trust region interior-point techniques (in particular, the article uses an affine-scaling interior-point method).

With a more general aim, in 1999, R.H. Byrd, M.E. Hribar, and J. Nocedal [13] (2,179 cites) proposed an algorithm for large-scale nonlinear programming, based on a barrier approach and trust regions.

In 2008, [7] presented in detail an interior-point SQP algorithm for nonlinear programming. The algorithm results in a more flexible choice of initial points. This publication includes an interesting comparison with standard SQP for a set of benchmark problems (for instance, it requires much less iterations in most cases).

Mixing with heuristic approaches
More recently, several publications have considered the hybridization of heuristic approaches and interior-point methods. We just mention a few of them.

The article [68] proposed a hybrid genetic algorithm-interior point method for optimal reactive power flow (a typical problem in electricity distribution). Something in this line is the approach of [31], coupling an evolutionary algorithm and a local search with interior-point methods.

In [16], the fusion of particle swarm optimization and primal-dual interior-point algorithms was proposed.

See [50] for a peculiar combination of interior-point methods and a metaheuristics called *path relinking search*.

10.8 Interior-Point Methods and MATLAB

This short section is devoted to signal the presence of interior-point methods (IPM) in the optimization functions of MATLAB and to mention some software tools in MATLAB that use IPM.

It has been already remarked that IPM is one of the options in the important MATLAB function *linprog( )*; the same happens with *quadprog( )*.

The MATLAB function *mpcInteriorPointSolver( )* is devoted to quadratic programming problems using an IPM with Mehrotra's procedure.

The MATLAB general purpose optimization function *fmincon( )* has a repertory of five algorithms to choose; one of these is IPM.

As commented before, *coneprog( )* is also based on IPM.

MATLAB offers several toolboxes that use IPM: the Optimization Toolbox, the Financial Toolbox, the Global Optimization Toolbox, etc. In particular, the Financial Toolbox includes functions for mixed-integer nonlinear programming.

Some of the programs/packages that can be obtained from the MathWorks File Exchange site are related to or do use IPM. Like for instance one about "Interior Penalty Method," or the "Optimization in MATLAB: An introduction to Quadratic Program" (associated to a webinar), or "Constraint-reduced predictor corrector IPM for semidefinite programming," etc. Some other entries can be localized asking for Mehrotra.

The GitHub site includes 55 repositories of software related to interior-point methods.

Note that Wikipedia gives an impressive list of optimization software, with dozens of products written in FORTRAN, C, Java, MATLAB, etc. Some of the tools using interior-point methods, like PCx, BPMPD, and HOPDM, are free. Many solvers, like CPLEX, Gurobi, LINDO, BARON, and LIPSOL, offer ways to be combined with MATLAB (e.g., routines that can be called from MATLAB). Interior-point methods are included in most solvers.

Examples of tools already mentioned in this chapter are LOQO, IPOPT, and KNITRO. Let us add two more:

The TOMLAB/KNITRO Toolbox integrates the sparse nonlinear IPM and active-set solver KNITRO with TOMLAB. In turn, TOMLAB has a wide range of optimization toolboxes in MATLAB.

The MOSEK Optimization Toolbox for MATLAB includes IPM for many types of problems.

10.9 Exercises

1- [10.1] One of the typical benchmark non-linear functions is the Rosenbrock function, which reads as follows:

$$J = 100\left(x_2 - x_1^2\right)^2 + (1 - x_1)^2$$

This is also a non-convex problem.

Try to find the minimum of this function by an interior point method. First, use a primal approach and then a primal-dual approach. Visualize and compare results.

2- [10.2] Let the optimization problem be as follows:

Maximize: $J = 3x_1 + 5x_2$

Subject to:

$$3x_1 + 2x_2 \leq 18$$

$$2x_2 \leq 12$$

$$x_1 \leq 3$$

$$x_1, x_2 \geq 0$$

Solve the problem with a primal-dual interior point method. Visualize and comment the result.

3- [10.3] Let the optimization problem be as follows:

Maximize: $J = x_1 + 2x_2$

Subject to:

$$2x_1 + 2x_2 \leq 10$$

$$4x_1 + x_2 \leq 10$$

$$4x_1 + 2x_2 \leq 12$$

$$x_1 + 2.2x_2 \leq 10$$

$$x_1 \leq 2.3$$

$$x_1, x_2 \geq 0$$

Solve the problem with a primal-dual interior point method. Visualize and comment the result.

4- [10.1] Consider the following optimization problem:
Minimize: $J = -x_1 x_2 - x_2 x_3$
Subject to:

$$x_1^2 - x_2^2 + x_3^2 - 2 \leq 0$$

$$x_1^2 + x_2^2 + x_3^2 - 10 \leq 0$$

Solve the problem with a primal-dual interior point method. Visualize and comment the result.

5- [10.4] Let us tackle the following optimization problem:
Minimize: $J = x_2(5 + x_1)$
Subject to:

$$x_1 x_2 \geq 5$$

$$x_1^2 + x_2^2 \leq 20$$

Employ a primal-dual interior point method to solve the problem. Visualize and comment details of the iterations.

6- [10.5] Consider the following optimization problem:
Minimize: $J = \frac{1}{2}x_1^2 + x_2^2 - x_1 x_2 - 2x_1 - 6x_2$
Subject to:

$$-x_1 - x_2 \geq -2$$

$$x_1 - 2x_2 \geq -2$$

$$-2x_1 - x_2 \geq -3$$

$$x_1, x_2 \geq 0$$

Solve the problem with a primal-dual interior point method. Visualize and comment the result.

10.9.1 Exercise References

[10.1] https://matpower.org/docs/MIPS-manual-1.5.pdf
[10.2] https://people.math.sc.edu/kellerlv/Primal_Dual_Interior_Point_Method.pdf
[10.3] https://benthamopen.com/contents/pdf/TOORJ/TOORJ-3-1.pdf
[10.4] https://optimization.cbe.cornell.edu/index.php?title=Interior-point_method_for_NLP

[10.5] https://acme.byu.edu/00000179-afb2-d74f-a3ff-bfbb15880000/interiorpoint-quad19-pdf

References

1. Adler, L., & RDC Monteiro. (1989). Interior path following primal-dual algorithms, Part II: Convex quadratic programming. *Mathematical Programming, 44*(1), 43–66. Available from: https://adler.ieor.berkeley.edu/ilans_pubs/path_following_2_1989.pdf
2. Alizadeh, F. (1992). Optimization over the positive-definite cone: Interior point methods and combinatorial applications. In *Advances in optimization and parallel computing*. North-Holland.
3. Alizadeh, F., & Schmieta, S. (2000). Symmetric cones, potential reduction methods and word-by-word extensions. In *Handbook of semidefinite programming: Theory, algorithms, and applications* (pp. 195–233). Springer. Available from: https://link.springer.com/chapter/10.1007/978-1-4615-4381-7_8
4. Andersen, E. D. (2009). The homogeneous and self-dual model and algorithm for linear optimization. Technical report, TR-1-2009, MOSEK ApS. http://docs.mosek.com. Available from: https://docs.mosek.com/whitepapers/homolo.pdf
5. Andersen, E. D., Roos, C., & Terlaky, T. (2003). On implementing a primal-dual interior-point method for conic quadratic optimization. *Mathematical Programming, 95*, 249–277. Available from: https://link.springer.com/content/pdf/10.1007/s10107-002-0349-3.pdf
6. Anstreicher, K. M. (1996). Potential reduction algorithms. In *Interior point methods of mathematical programming* (pp. 125–158). Springer. Available from: https://citeseerx.ist.psu.edu/document?repid=rep1&type=pdf&doi=7bd935cbe5e15be3669ad8285c4092864033240d
7. Bashir, H. A., Liang, X., & Li, S. (2008). Applying infeasible interior point method to SQP for constrained nonlinear programming. In *2008 international conference on computer science and software engineering* (Vol. 1, pp. 399–402). IEEE.
8. Bellman, R., & Fan, K. (1963). On systems of linear inequalities in hermitian matrix variables. *Convexity, 7*, 1–11.
9. Benson, H. Y., Shanno, D. F., & Vanderbei, R. J. (2000). *Interior-point methods for nonconvex nonlinear programming: Jamming and comparative numerical testing*. Operations Research and Financial Engineering, Princeton University, ORFE-00-02, p. 59. Available from: https://citeseerx.ist.psu.edu/document?repid=rep1&type=pdf&doi=487069c40f30c6c7321ef8ec29d2879995b8e261
10. Benson, H. Y., Vanderbei, R. J., & Shanno, D. F. (2002). Interior-point methods for nonconvex nonlinear programming: Filter methods and merit functions. *Computational Optimization and Applications, 23*, 257–272. Available from: https://link.springer.com/content/pdf/10.1023/A:1020533003783.pdf
11. Borchers, B., & Mitchell, J. E. (1991). Using an interior point method in a branch and bound algorithm for integer programming. Technical report, Mathematical Sciences, Rensselaer Polytechnic. Available from: http://euler.nmt.edu/~brian/intptbb.pdf
12. Boyd, S. P., & Vandenberghe, L. (2004). *Convex optimization*. Cambridge University Press. Available from: https://ds.amu.edu.et/xmlui/bitstream/handle/123456789/10327/convex%20optimization%20%282004%29.pdf?sequence=2&isAllowed=y
13. Byrd, R. H., Hribar, M. E., & Nocedal, J. (1999). An interior point algorithm for large-scale nonlinear programming. *SIAM Journal on Optimization, 9*(4), 877–900. Available from: https://citeseerx.ist.psu.edu/document?repid=rep1&type=pdf&doi=0c1c4bbdd7467c5ba1818b2e7a360e768b067d2c
14. Castro, J. (2022). New interior-point approach for one-and two-class linear support vector machines using multiple variable splitting. *Journal of Optimization Theory and Applications*, 1–34. Available from: https://link.springer.com/article/10.1007/s10957-022-02103-1

15. Curtis, F., & Nocedal, J. (2007). Steplength selection in interior-point methods for quadratic programming. *Applied Mathematics Letters, 20*(5), 516–523. Available from: https://core.ac.uk/reader/82482440
16. Dada, E. G., & Ramlan, E. I. (2018). PDPSO: The fusion of primal-dual interior point method and particle swarm optimization algorithm. *Malaysian Journal of Computer Science, 31*(1), 17–34. Available from: https://mjlis.um.edu.my/index.php/MJCS/article/view/10557
17. De Klerk, E. (2006). *Aspects of semidefinite programming: Interior point algorithms and selected applications* (Vol. 65). Springer. Available from: http://lib.ysu.am/disciplines_bk/4453e27f73338945651 4ef1208ebfe5f.pdf
18. De Simone, V., di Serafino, D., Gondzio, J., Pougkakiotis, S., & Viola, M. Sparse approximations with interior point methods. *SIAM Review, 64*(4), 954–988 (2022) Available from: https://optimization-online.org/wp-content/uploads/2021/03/8274-1.pdf
19. Dennis, J. E., Heinkenschloss, M., & Vicente, L. N. (1998). Trust-region interior-point sqp algorithms for a class of nonlinear programming problems. *SIAM Journal on Control and Optimization, 36*(5), 1750–1794. Available from: https://repository.rice.edu/server/api/core/bitstreams/8188f2f3-354d-45ee-9bf4-ee66324a5d18/content
20. Djemai, S., Brahmi, B., & Bibi, M. O. (2016). A primal–dual method for SVM training. *Neurocomputing, 211*, 34–40. Available from: https://dl.acm.org/doi/abs/10.1016/j.neucom.2016.01.103
21. Fawzi, H. (2019). Lecture 14: Self-concordant functions. Available from: https://www.damtp.cam.ac.uk/user/hf323/M19-OPT/lecture14.pdf
22. Ferris, M. C., & Munson, T. S. (2002). Interior-point methods for massive support vector machines. *SIAM Journal on Optimization, 13*(3), 783–804. Available from: https://epubs.siam.org/doi/abs/10.1137/S1052623400374379
23. Fiacco, A. V., & McCormick, G. P. (1990). *Nonlinear programming: Sequential unconstrained minimization techniques*. SIAM.
24. Fletcher, R., & Leyffer, S. (2002). Nonlinear programming without a penalty function. *Mathematical Programming, 91*, 239–269. Available from: https://link.springer.com/content/pdf/10.1007/s101070100244.pdf
25. Forsgren, A., Gill, P. E., & Wright, M. H. (2002). Interior methods for nonlinear optimization. *SIAM Review, 44*(4), 525–597. Available from: https://epubs.siam.org/doi/pdf/10.1137/S0036144502414942
26. Glineur, F. (2001). *Topics in convex optimization: Interior-point methods, conic duality and approximations*. PhD thesis, Polytechnic College of Mons. Available from: https://theses.hal.science/tel-00006861/document
27. Helmberg, C. (2002) Semidefinite programming. *European Journal of Operational Research, 137*(3), 461–482. Available from: https://edocs.tib.eu/files/e001/322654831.pdf
28. Hildebrand, R. (2022). Lecture 14: Self-concordant barriers. Available from: https://membres-ljk.imag.fr/Roland.Hildebrand/emo/interior_point_methods.pdf
29. Kai, X., & Song, Y. H. (1999). Interior point methods and applications in power systems. In *Modern Optimisation Techniques in Power Systems* (pp. 205–245). Available from: https://link.springer.com/chapter/10.1007/978-94-015-9189-8_9
30. Karmarkar, N. (1984). A new polynomial-time algorithm for linear programming. In *Proceedings of the sixteenth annual ACM symposium on theory of computing* (pp. 302–311). Available from: https://dl.acm.org/doi/pdf/10.1145/800057.808695
31. Kelner, V., Capitanescu, F., Léonard, O., & Wehenkel, L. (2008). A hybrid optimization technique coupling an evolutionary and a local search algorithm. *Journal of Computational and Applied Mathematics, 215*(2), 448–456. Available from: http://www.turbo.ulg.ac.be/research/paper/ACOMEN2005.pdf
32. Kim, S.-J., Koh, K., Lustig, M., Boyd, S., & Gorinevsky, D. (2007). An interior-point method for large-scale l_1 regularized least squares. *IEEE Journal of Selected Topics in Signal Processing, 1*(4), 606–617. Available from: https://web.stanford.edu/~boyd/papers/pdf/l1_ls.pdf

33. Kuo, Y.-J., & Mittelmann, H. D. (2004). Interior point methods for second-order cone programming and or applications. *Computational Optimization and Applications, 28*, 255–285. Available from: https://link.springer.com/content/pdf/10.1023/B:COAP.0000033964.95511.23.pdf
34. Lee, Y. T. (2018). Lecture 14: Primal-dual interior-point method. Available from: https://yintat.com/teaching/cse599-winter18/14.pdf
35. Lee, Y. T. (2018). Lecture 15: Self-concordant functions. Available from: https://yintat.com/teaching/cse599-winter18/15.pdf
36. Lesaja, G. (2009). Introducing interior-point methods for introductory operations research courses and/or linear programming courses. *Open Operational Research Journal, 3*, 1. Available from: https://digitalcommons.georgiasouthern.edu/cgi/viewcontent.cgi?article=1116&context=math-sci-facpubs
37. Li, Y., & Santosa, F. (1996). A computational algorithm for minimizing total variation in image restoration. *IEEE Transactions on Image Processing, 5*(6), 987–995. Available from: https://www.researchgate.net/profile/Yuying-Li-11/publication/3326483_Santosa_F_A_computational_algorithm_for_minimizing_total_variation_in_image_restoration_IEEE_Trans_Image_Process_5_987-995/links/555e19aa08ae6f4dcc8dd149/Santosa-F-A-computational-algorithm-for-minimizing-total-variation-in-image-restoration-IEEE-Trans-Image-Process-5-987-995.pdf
38. Marsten, R., Subramanian, R., Saltzman, M., Lustig, I., & Shanno, D. (1990). Interior point methods for linear programming: Just call Newton, Lagrange, and Fiacco and McCormick! *Interfaces, 20*(4), 105–116. Available from: https://pubsonline.informs.org/doi/abs/10.1287/inte.20.4.105
39. Megiddo, N. (1989). Pathways to the optimal set in linear programming. In *Progress in mathematical programming: Interior-point and related methods* (pp. 131–158). Springer. Available from: https://theory.stanford.edu/~megiddo/pdf/pathways.pdf
40. Mehrotra, S. (1992). On the implementation of a primal-dual interior point method. *SIAM Journal on Optimization, 2*(4), 575–601. Available from: https://epubs.siam.org/doi/abs/10.1137/0802028
41. Mitchell, J. (2010). Interior point methods. Available from: https://homepages.rpi.edu/~mitchj/handouts/interior_html/interior.html
42. Munari, P., & Gondzio, J. (2013). Using the primal-dual interior point algorithm within the branch-price-and-cut method. *Computers & Operations Research, 40*(8), 2026–2036. Available from: https://www.maths.ed.ac.uk/~gondzio/reports/ipdcgm.pdf
43. Musco, C. (2016). Lecture 17: Interior point methods. Available from: https://www.cs.princeton.edu/courses/archive/fall18/cos521/Lectures/lec17.pdf
44. Naoum-Sawaya, J. (2011). *Interior point cutting plane methods in integer programming*. PhD thesis, University of Waterloo. Available from: https://uwspace.uwaterloo.ca/bitstream/handle/10012/6105/Naoum-Sawaya_Joe.pdf?sequence=1&isAllowed=y
45. Naoum-Sawaya, J., & Elhedhli, S. (2013). An interior-point benders based branch-and-cut algorithm for mixed integer programs. *Annals of Operations Research, 210*, 33–55. Available from: https://link.springer.com/content/pdf/10.1007/s10479-010-0806-y.pdf
46. Nemirovski, A. S., & Todd, M. J. (2008). Interior-point methods for optimization. *Acta Numerica, 17*, 191–234. Available from: https://www2.isye.gatech.edu/~nemirovs/Published.pdf
47. Nesterov, Y., & Nemirovskii, A. (1994). *Interior-point polynomial algorithms in convex programming*. SIAM.
48. Nesterov, Y. E., & Todd, M. J. (1997). Self-scaled barriers and interior-point methods for convex programming. *Mathematics of Operations Research, 22*(1), 1–42. Available from: https://web.archive.org/web/20120712175458id_/http://people.orie.cornell.edu/~miketodd/orie6327/nestonn2.pdf
49. Nocedal, J., & Wright, S. J. (1999). *Numerical optimization*. Springer. Available from: https://sharif.ir/~pishvaie/Articles/OptimizationArticles/Semester_9293_2/SomeUsefulRefs/NumericalOptimization%20-%20J.%20Nocedal, Stephen J. Wright (Springer,1999).pdf.

50. Plateau, A., Tachat, D., & Tolla, P. (2002). A hybrid search combining interior point methods and metaheuristics for 0–1 programming. *International Transactions in Operational Research, 9*(6), 731–746. Available from: https://www.angelfire.com/ak6/invo_escom/lectura31_programacion_0_1_entera.pdf
51. Ploskas, N., Samaras, N., et al. (2017). *Linear programming using MATLAB®* (Vol. 127). Springer. Available from: https://link.springer.com/content/pdf/10.1007/978-3-319-65919-0.pdf
52. Potra, F. A., & Wright, S. J. (2000). Interior-point methods. *Journal of Computational and Applied Mathematics, 124*(1–2), 281–302. Available from: https://pages.cs.wisc.edu/~swright/papers/potra-wright.pdf
53. Renegar, J. (1988). A polynomial-time algorithm, based on Newton's method, for linear programming. *Mathematical Programming, 40*(1–3), 59–93. Available from: https://link.springer.com/article/10.1007/BF01580724
54. Robere, R. (2012). Interior point methods and linear programming. *University of Toronto*, 1–15. Available from: https://faculty.ksu.edu.sa/sites/default/files/Interior%20Point%20Methods%20and%20Linear%20Programming.pdf
55. Shanno, D. (2012). Who invented the interior-point method? *Optimization Stories, Documenta Mathematica, Journal der Deutschen Mathematiker-Vereinigung, S*, 55–64. Available from: https://www.kurims.kyoto-u.ac.jp/EMIS/journals/DMJDMV/vol-ismp/20_shanno-david.pdf
56. Shanno, D. F., & Vanderbei, R. J. (2000). Interior-point methods for nonconvex nonlinear programming: Orderings and higher-order methods. *Mathematical Programming, 87*, 303–316. Available from: https://link.springer.com/content/pdf/10.1007/s101070050116.pdf
57. Su, G., Wang, L., Zhou, X., & Nie, S. (2019). Inversion method of two-dimensional low-field nuclear magnetic resonance relaxation spectrum using primal–dual interior method for convex objectives. *Journal of Medical Imaging and Health Informatics, 9*(5), 993–999. Available from: https://www.ingentaconnect.com/contentone/asp/jmihi/2019/00000009/00000005/art00023
58. Terlaky, T. (2013). *Interior point methods of mathematical programming* (Vol. 5). Springer. Available from: https://link.springer.com/book/10.1007/978-1-4613-3449-1
59. Todd, M. J. (1997). Potential-reduction methods in mathematical programming. *Mathematical Programming, 76*(1), 3–45. Available from: https://link.springer.com/content/pdf/10.1007/BF02614377.pdf
60. Todd, M. J. (1999). A study of search directions in primal-dual interior-point methods for semidefinite programming. *Optimization Methods and Software, 11*(1–4), 1–46. Available from: https://ecommons.cornell.edu/server/api/core/bitstreams/ddfe8d52-ca36-4e30-a619-3d031e0e1398/content
61. Tone, K. (1993). An active-set strategy in an interior point method for linear programming. *Mathematical Programming, 59*(1–3), 345–360. Available from: https://link.springer.com/article/10.1007/BF01581252
62. Tunçel, L. (2000). Potential reduction and primal-dual methods. In *Handbook of semidefinite programming: Theory, algorithms, and applications* (pp. 235–265). Available from: https://link.springer.com/chapter/10.1007/978-1-4615-4381-7_9
63. Vandenberghe, L., & Boyd, S. (1996). Semidefinite programming. *SIAM Review, 38*(1), 49–95. Available from: http://luthuli.cs.uiuc.edu/~daf/courses/Opt-2019/Combinatorialpapers/0.pdf
64. Vanderbei, R. J., & Shanno, D. F. (1999). An interior-point algorithm for nonconvex nonlinear programming. *Computational Optimization and Applications, 13*, 231–252. Available from: https://link.springer.com/content/pdf/10.1023/A:1008677427361.pdf
65. Wächter, A., & Biegler, L. T. (2000). Failure of global convergence for a class of interior point methods for nonlinear programming. *Mathematical Programming, 88*, 565–574. Available from: https://www.scholars.northwestern.edu/en/publications/failure-of-global-convergence-for-a-class-of-interior-point-metho
66. Wächter, A., & Biegler, L. T. (2006). On the implementation of an interior-point filter line-search algorithm for large-scale nonlinear programming. *Mathematical Programming, 106*, 25–57. Available from: https://link.springer.com/content/pdf/10.1007/s10107-004-0559-y.pdf

67. Wright, M. (2005). The interior-point revolution in optimization: History, recent developments, and lasting consequences. *Bulletin of the American Mathematical Society, 42*(1), 39–56. Available from: https://www.ams.org/journals/bull/2005-42-01/S0273-0979-04-01040-7/S0273-0979-04-01040-7.pdf
68. Yan, W., Liu, F., Chung, C. Y., & Wong, K. P. (2006). A hybrid genetic algorithm-interior point method for optimal reactive power flow. *IEEE Transactions on Power Systems, 21*(3), 1163–1169. Available from: https://www.ecs.csun.edu/~bruno/IEEEpapers/01664951.pdf
69. Ye, Y. (2011). *Interior point algorithms: Theory and analysis*. Wiley.

Chapter 11
Evolutionary Optimization: Introduction

11.1 Introduction

In many real-life optimization scenarios, analytical methods are not easy to apply. It is still possible to try systematic exploration, but this brute-force approach could imply combinatorial explosions. A standard example is the traveling salesman problem (TSP), which, in the case of just $n = 20$ cities to visit, leads to 20! combinations (an enormous number with 19 digits).

However, a much lower computational effort could be chosen and successfully exerted, for certain problems, by using dices: random searching.

Indeed, the term "*dice*" suggests gambling and casinos. In fact, we find in the recent history of science the emergence of Monte-Carlo methods. It was E. Fermi who started to use them in 1930. Later, by the end of the 1940s, S. Ulam and John von Neumann gave relevance to these methods, for the research on neutron diffusion processes. Nicholas Metropolis, colleague of von Neumann and Ulam, suggested the name "Monte-Carlo" for the algorithms and methods they were developing.

By the way, soon in this chapter, we will deal with "*genetic algorithms*," in which typically one uses a roulette-wheel. This algorithm is an archetypical example of evolutionary algorithm.

Being so popular, genetic algorithms are not the only bio-inspired methods. Many other exist. A representative part of them will be introduced in this chapter, after considering with some detail the genetic algorithms.

There are other evolutionary methods not directly inspired by biology but related to mathematics, physics, or other areas. The chapter will start with an optimization method called "*Simulated Annealing*," which has proved good results in many fields of application and that it is inspired by a physical process.

In general, the stochastic methods (evolutionary methods are a subset of stochastic ones) would not find the exact optimum but a useful approximation in less time, which could be opportune for tentative studies.

J. M. Giron-Sierra, *Introduction to Optimization with Matlab® Examples*,
https://doi.org/10.1007/978-3-032-05048-9_11

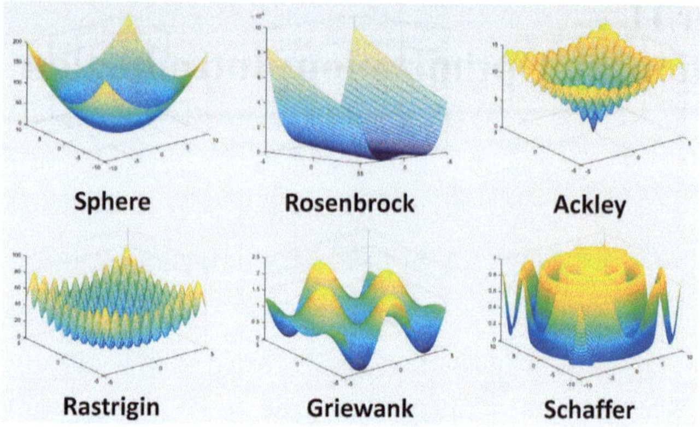

Fig. 11.1 Examples of benchmark functions

Recall from previous chapters that **local** optima may happen in certain cases. This represents a risk for exploration methods that might be trapped in a local optimum. In general, stochastic searching are suitable for escaping from traps.

There is an overwhelming quantity of evolutionary methods. This might cause, at first, a sense of disorientation. For this reason, we wish to give gradually an idea of the present situation, attacking the topic by parts. Actually, instead of a very long chapter, it was preferred to divide the matter into three smaller, more manageable chapters. Now, in the first one, we initiate the study by a selection of a few paradigmatic algorithms, which contain most of the fundaments of the rest of stochastic algorithms.

A series of programs in MATLAB will be provided for demonstration and for personal exploration by the reader, who is invited to play with the programs modifying parameters and seeing what may happen.

It is customary to use some peculiar functions as benchmarks, in order to evaluate and compare optimization methods. The page [26], for example, offers a large set of these functions, which usually have a lot of local minima.

Figure 11.1 shows in three dimensions a few examples of benchmark functions used in optimization. In this chapter, we will instead use projections to two dimensions with contours, for clearer visualization of where are the candidate solutions.

11.2 Simulated Annealing

At high temperature, the atoms of metals, or other materials, have greater mobility and freely move. If temperature is lowered, the mobility decreases. With a suitable policy of temperature lowering, the atoms could form a crystal structure, which corresponds to a global minimum energy configuration.

One could imitate this process, in order to reach a minimum.

The *simulated annealing* algorithm considers small random displacements of an atom, causing a change in energy. If this change is negative, the new configuration is accepted. If the change in energy is positive, the new configuration could still be accepted according to the Boltzmann probability factor:

$$P = \exp\left(\frac{-\Delta E}{T}\right)$$

where ΔE is the energy change, and T is the temperature.

Notice that P gets smaller as temperature T decreases. Also, P would diminish as (positive) ΔE increases.

In case of positive ΔE, the corresponding value of P is compared to a random number drawn from a uniform distribution between 0 and 1. If the random number is smaller than P, the configuration is accepted [1].

When applied to optimization problems, an analogy is considered between energy and the objective function.

Typically, for a given temperature, the algorithm checks around 1000 close neighbor candidates, and then temperature is decreased by a certain step. There are examples that start from $T = 1000$ and that apply a cooling rate of 0.9.

We will now consider the following objective function, as a simple and interesting case

$$y = 0.2 + x_1^2 + x_2^2 - 0.1\,\cos(6\pi\, x_1) - 0.1\,\cos(6\pi\, x_2)$$

This function is similar to the Rastrigin function, one of the benchmarks listed in the mentioned page. This example has a global minimum at the origin of coordinates.

Figure 11.2 presents, with circles, the final positions of candidates, after running the Program 11.1. For clarity purposes, we represented the objective function using contours. The best candidate is very close to the exact solution. Notice that candidates tend to cover the whole search space, as would be expected from Monte-Carlo-based explorations. This coverage helps to escape from local traps.

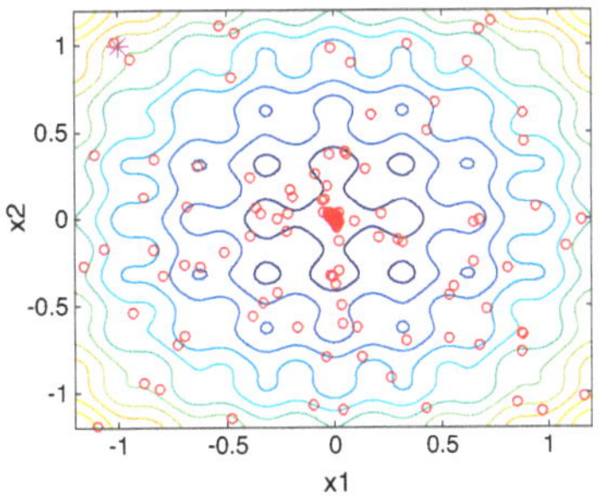

Fig. 11.2 Simulated annealing searching for minimum

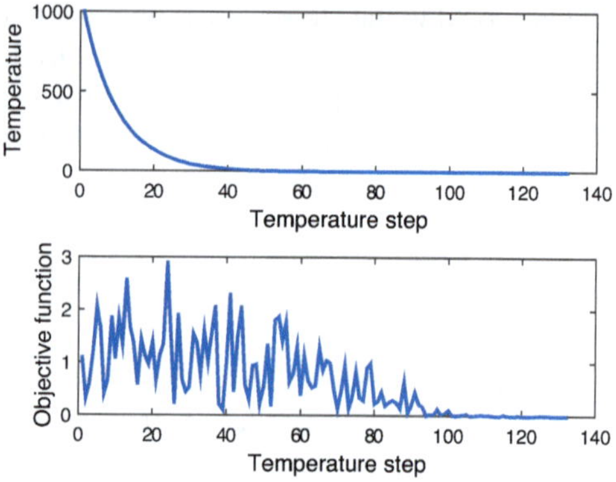

Fig. 11.3 Evolution of temperature and "energy"

Figure 11.3 depicts the evolution of temperature and "*energy*" during the simulated annealing process. This energy corresponds to the best objective function value.

Both Figs. 11.2 and 11.3 have been obtained with the program 11.1, which provides a simple example of MATLAB implementation of simulated annealing.

Program 11.1 Simulated annealing example

```
%Simulated Annealing example
% objective function display
[vx1,vx2]=meshgrid(-1.2:0.02:1.2, -1.2:0.02:1.2);
y= 0.2+(vx1.^2)+(vx2.^2)-(0.1 * cos(6*pi*vx1))- ...
  (0.1 * cos(6*pi*vx2));
%
% function display-------------
figure(1)
contour(vx1,vx2,y,11); hold on;
xlabel('x1'); ylabel('x2');
%
% searching parameters -------------------
T=1000; %initial temperature
Tmin=0.001; %final temperature
%maximum number o iterations, per temp. stage
Nmax=1000;
Ct=0.9; %temp. lowering rate
%
x1min=-1.2; x1max=1.2;
x2min=-1.2; x2max=1.2;
sigma1=(x1max-x1min)/6;
sigma2=(x2max-x2min)/6;
x1n=0; x2n=0; dE=0; r=0; P=0; %auxiliary variables
vT=zeros(1,200); vy=zeros(1,200);
% initial point
x1=-1; x2=1; plot(x1,x2,'m*','MarkerSize',12);
xlabel('x1'); ylabel('x2');
```

```
%start of annealing process
jj=0; %counter of temperature steps
while (T>Tmin)
  for nn=1:Nmax
    %new candidate using normal distribution
    x1n=x1+(sigma1^2)*randn(1);
    x2n=x2+(sigma2^2)*randn(1);
    if ((x1n<=x1max) & (x1n>=x1min) & (x2n<=x2max) &
     (x2n>=x2min))
      y=0.2+(x1^2)+(x2^2)-(0.1*cos(6*pi*x1))-...
        (0.1*cos(6*pi*x2));
      yn=0.2+(x1n^2)+(x2n^2)-(0.1*cos(6*pi*x1n))-...
      (0.1*cos(6*pi*x2n));
      dE=yn-y;
      %deciding acceptance
      if dE<0
        x1=x1n; x2=x2n; %new point is accepted
      else
        r=rand(1);
        P=exp(-dE/T); %Boltzmann prob.
        if (r<=P)
          x1=x1n; x2=x2n; %new point is accepted
        end
      end
    end
  end
  jj=jj+1; plot(x1,x2,'ro');
  y= 0.2+(x1^2)+(x2^2)-(0.1*cos(6*pi*x1))- ...
  (0.1*cos(6*pi*x2));
  vT(jj)=T; vy(jj)=y; %save evolution data
  T=T*Ct; %temperature lowering
end;
x1
x2
y= 0.2+(x1^2)+(x2^2)-(0.1*cos(6*pi*x1))- ...
(0.1*cos(6*pi*x2))
figure(2)
subplot(2,1,1);
plot(vT(1:jj));
xlabel('Temperature step');
ylabel('Temperature');
subplot(2,1,2);
plot(vy(1:jj));
xlabel('Temperature step');
ylabel('Objective function');
```

The simulated annealing algorithm was introduced by Kirpatrick et al., [$REF KPK], in 1983. Since then, it has been cited by around 54.500 publications. Clearly, it can be considered as a most important algorithm.

The article [$REF SMN], year 2006, provides an extensive survey of simulated annealing variants and applications. Many other publications give account of interesting and relevant applications in a large variety of fields.

11.3 Tabu Search

Tales about been lost on a desert say that after some walking you discover your own footprints on the sand, walking in circles and no good for a searching method, if it strikes again a place already visited.

Tabu search tries to avoid cycles in short-term explorations, although it permits revisits in the long term. It uses a tabu list of prohibited moves, which is updated in each searching step.

The tabu search can be used to guide processes based on moves from one candidate solution to others.

After starting from an initial solution $Z(n)$, a set of nontabu moves is explored. The best new solution $Z(n+1)$ (better than the initial solution) is chosen, and the tabu list is updated as follows: the set of moves that would lead back to $Z(n)$ is added to the list, and sufficiently old entries of the list are eliminated.

Iterations continue until no move could improve the solution reached so far.

The tabu search was created by Fred W. Glover in 1986 and formalized three years later [$REF GLO1] [$REF GLO2]. In addition to the tabu list, the method includes intensification and diversification of the searching.

Let us give more details by considering the classical Traveling Salesman Problem (TSP). The salesman should visit N places, being only once in each place, and minimizing the total distance covered. No subtours are allowed.

Suppose N = 6. An initial solution can be built at random, like for instance:

3	6	2	5	4	1

(It is understood that the initial place is 1, which would be also the final one)

By the way, a probably better initial solution—although non-optimal—could be obtained by going from 1 to the nearest one; once there, choose the nonvisited nearest one and so on.

Now, let us explore *neighborhood* solutions, which are obtained by a pair wise exchange of any two nodes (except for the last), like the following:

3	**4**	2	5	**6**	1

(This was a swapping of positions 2 and 5, in the order of visits)

Once a best move was determined (with any suitable algorithm), the tabu table is updated. Let us use a diagram to represent this process (Fig. 11.4):

In Fig. 11.4, the two columns at the right-hand side register the best four moves and their corresponding effect on the objective function value (which is total distance). Once the 2, 3 swapping was done, we include a 3 in the tabu table (this number is used as counter).

Fig. 11.4 Updating the tabu list

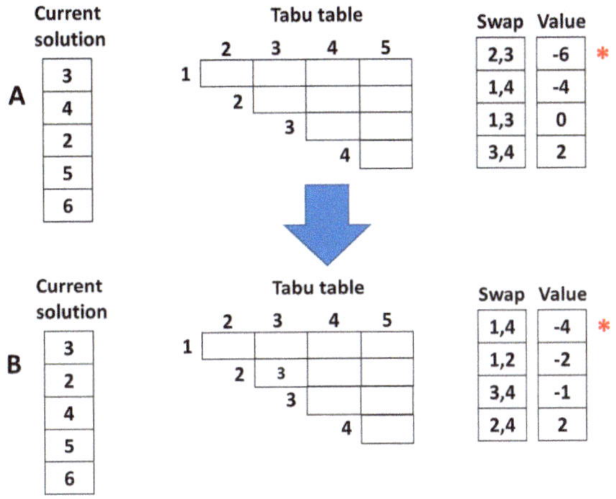

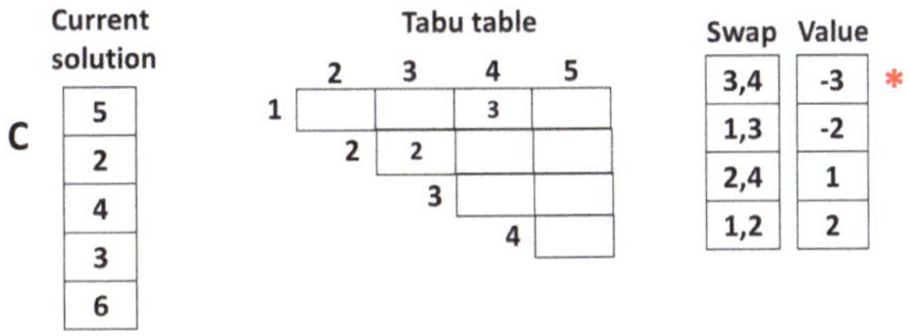

Fig. 11.5 Updating the tabu list

Figure 11.5 depicts the next step of the procedure. Notice that the entry corresponding to the 2, 3 swap, is decreased (now is 2). Being a tabu option, this move has not been considered as candidate.

There is an "*aspiration*" criterion, by which a tabu move is allowed if, compared to the better of possible swaps, is still better (leading to a better value improvement).

To escape from local traps, it is convenient to diversify the searching process, driving it to new regions. Another triangular table could be added, to annotate the frequency of certain swaps. This information would be used to penalize more frequent swaps and to promote infrequent swaps.

Based on a table of distances between main cities in the United States, we developed a MATLAB program in order to apply Tabu search for a TSP problem with ten cities. Figure 11.6 presents the scenario we chose:

New York, Los Angeles, Chicago, Phoenix, Denver, Dallas, Seattle, St. Louis, Minneapolis, Houston.

By the way, Google Map has found an optimal routing for ground transportation; our program, instead, consider distances by air.

Program 11.2 just covers the basic aspects of Tabu search and does not include aspiration, intensification, or diversification. For larger TSP problems (much more than ten cities), it would be opportune to add these features. The program, as is,

Fig. 11.6 An experimental scenario for TSP on US

would be good for inspection of the execution parts of interest; like for instance, visualize what happens with matrices along main steps.

The first part of the program applies a nearest neighbor algorithm, to get an initial solution, which is the following tour:

3 8 4 2 5 6 10 97

The total distance corresponding to this tour is 8376 miles.

Once obtained a solution of reference, the second part of the program applies pair swapping and a tabu table. There were four successive tour optimizations:

3 8 4 2 5 6 10 7 9

3 8 4 2 7 6 10 5 9

3 8 4 2 7 5 10 6 9

3 8 4 2 7 5 6 10 9

No more progress could be obtained by pair swapping, so this last tour was the proposed optimized solution.

The total distance corresponding to this optimized tour is 6740 miles. A substantial saving.

Looking at the tabu table, it included three nonzero entries: 3, 2, and 1. In this example, it was not necessary to apply any tabu.

An important observation is that, if all cities were forming a kind of "circle," the initial solution provided by nearest neighbor algorithm would be the best (tabu search not needed).

More in general, optimization problems may exhibit certain structural properties, which could help for efficient searching. This means that the searching algorithm, in turn, should be able to benefit from that specific help. This is one of the reasons for having so many searching algorithms.

Program 11.2 Example of tabu search

```
%Tabu search example
%New York - (2)Los Angeles - (3)Chicago - (4)Phoenix
% - (5)Denver - (6)Dallas - (7)Seattle - (8)St. Louis
% - (9)Minneapolis - (10) Houston
%intercities distance (in miles) symmetric matrix:
D = zeros(10,10);
D(1,:)= [0, 2451, 713, 2145, 1631, 1374, 2408,...
 875, 1018, 1420];
D(2,:)= [2451, 0, 1745, 357, 831, 1240, 959, 1589,...
 1524, 1374];
D(3,:)= [713, 1745, 0, 1453, 920, 803, 1737, 262,...
 355, 940];
D(4,:)= [2145, 357, 1453, 0, 586, 887, 1114, 466,...
 1280, 1017];
D(5,:)= [1631, 831, 920, 586, 0, 663, 1021, 796,...
 700, 879];
D(6,:)= [1374, 1240, 803, 887, 663, 0, 1681, 547,...
 862, 225];
D(7,:)= [2408, 959, 1737, 1114, 1021, 1681, 0,...
 1724, 1395, 1891];
D(8,:)= [875, 1589, 262, 466, 796, 547, 1724, 0,...
 466, 679];
D(9,:)= [1018, 1524, 355, 1280, 700, 862, 1395,...
 466, 0, 1056];
D(10,:)= [1420, 1374, 940, 1017, 879, 225, 1891,...
 679, 1056, 0];
for i=1:10
D(i,i)=100000; %diagonal penalty
end;
Daux=D;
%initialization
tour= zeros(1,8);
%--------------------------------------------------
%build initial tour, by nearest neighbors
k=1; Tdist=0;
Daux(:,1)=100000; %preserve origin
for n=1:9
  [mindist,ix]=min(Daux(k,:));
  Tdist=Tdist+mindist; %add to total distance
  tour(n)=ix; %new destination
  Daux(:,ix)=100000; %no re-visits
  k=ix;
end
Tdist=Tdist+D(ix,1);
disp('initial tour solution, between 1 and 1');
tour
disp('with total distance:');
Tdist
%--------------------------------------------------
%try modifications of the tour by pair exhanging
```

```
%while updating a Tabu table
%Tabu table (only upper triangle for tenure entries)
Tbt=zeros(9,9);
%Swap evaluation table (only upper triangle is used)
SE=zeros(9,9);
%start iterations
ff=1; %flag for termination
while (ff==1)
  TdistREF=Tdist;
  % evaluate swaps in the current tour
  for i=1:9
    for j=(i+1):9
      ixo=1; ntour=tour; Tdist=0;
      ntour(i)=tour(j); ntour(j)=tour(i); %swapping
      for n=1:9
        ixf=ntour(n);
        Tdist=Tdist+D(ixo,ixf);
        ixo=ixf;
      end
      Tdist=Tdist+D(ixf,1);
      SE(i,j)=Tdist-TdistREF;
      %penalty for tabu move
      if(Tbt(i,j)>0)
        SE(i,j)=100000;
      end
    end
  end
  %find best non-tabu move --------------------
  Vaux=min(min(SE)); %best distance change
  if(Vaux==0)
    Tbt %print last Tabu table
    ff=0 %termination
  else
    %update tabu table --------------
    for i=1:9
      for j=1:9
        if Tbt(i,j)>0
          Tbt(i,j)=Tbt(i,j)-1;
        end
      end
    end
    %position of best move in SE
    [ii,jj]=find(SE==Vaux);
    ntour=tour;
    %best swapping:
    ntour(ii)=tour(jj); ntour(jj)=tour(ii);
    tour=ntour; %the new tour is accepted
    Tbt(ii,jj)=3; %new tabu mark
    ntour %print solution
    Tdist=TdistREF+Vaux %Vaux is negative
  end
end
```

11.4 Genetic Algorithms

Imagine a wild territory, with mountains and valleys, and not explored yet. You could send there one explorer, to determine where the highest mountain is, or the deepest valley. But it would be more effective to parachute at random many explorers. They would be able to compare results, so the team could avoid being trapped by local minima or maxima.

This is a main idea of many methods based on a *population* of candidate solutions.

Genetic algorithms (GA) are an archetypical type of bio-inspired methods based on populations. GAs mimic the evolution of biological organisms. There is population of individuals or chromosomes, each representing a candidate solution of our optimization problem. The next generation is obtained by cross breeding of selected parents. Some mutations may be allowed. Generations are repeated until a certain criterion was met.

Figure 11.7 presents a basic block diagram of a genetic algorithm.

Note that we speak in plural: genetic algorithms. This is so because there exist many variants in the evolutionary spirit of the algorithm.

A key aspect is how to represent problems in terms of chromosomes. In general, a chromosome is a series of numbers: binary, integers, and real. Historically, the first GA versions used binary numbers.

Some creativity may be needed for a suitable coding. Our particular experience is related to beer fermentation. During this fermentation, temperature is controlled, starting from cold wort, adding yeast, letting temperature increase (but not too much), and then, after some days, cooling the result (hopefully good beer). Therefore, there is a temperature profile that should be followed by the process. We want an optimal process, fast, cheap, and with good resulting quality. To represent

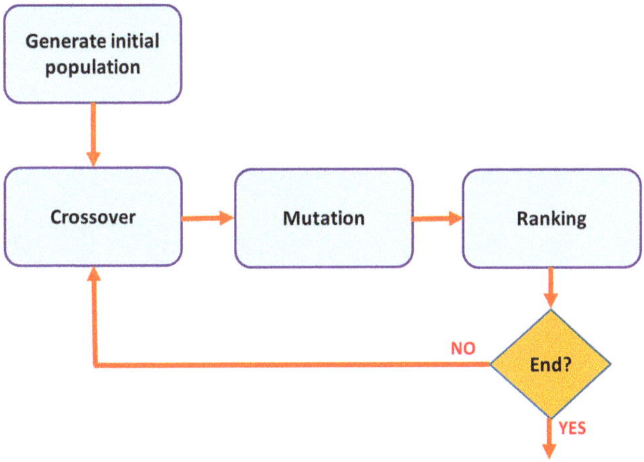

Fig. 11.7 Block diagram of a Genetic Algorithm

the problem, we divided the temperature profile into linear segments of 1 hour duration. Temperatures at segment corners, real numbers, were the "genes" forming a chromosome (each chromosome represents a complete temperature profile). After some computational activity with GA, an approximation to the optimal profile was obtained and successfully checked on the pilot fermentation plant (80 liters) we built [$REF JMG].

With some alike precedents in the 1960s, the genetic algorithms acquired popularity when the book *Adaptation in Natural and Artificial Systems* (1975) was published [13]. The author was John Holland, University of Michigan. It has been cited more than 70,000 times by investigators.

We made a simple implementation of GA, in order to solve the same problem as in the section on simulated annealing. The chromosomes were just pairs (x1, x2), coded as real numbers. The initial population was generated at random. Each chromosome leads to a certain value of the objective function. A ranking of the chromosomes was computed, based on how good the corresponding objective function value is.

The crossover operation between two "parents" (chromosomes), (xa1, xa2), (xb1, xb2), generates a "child" (xa1, xb2). Both parents were chosen at random (roulette-wheel). In our program, only one child was generated in each generation step; and this child substitutes the worst chromosome.

Mutation was certainly radical, by choosing at random one of the chromosomes and substituting it by a newly generated chromosome. This has the risk of "killing" the best chromosome thus far.

For code simplification, two MATLAB functions were built, *fit()* and *The_problem()*.

Figure 11.8 depicts the initial population of chromosomes (small black circles). Recall that each chromosome is a candidate solution of the minimization problem. It is also clear that the solution is (0, 0), with the objective function being 0 at this point.

By letting the execution of the program going on, a series of best chromosomes were obtained. Frequently a best chromosome stays on top along several genera-

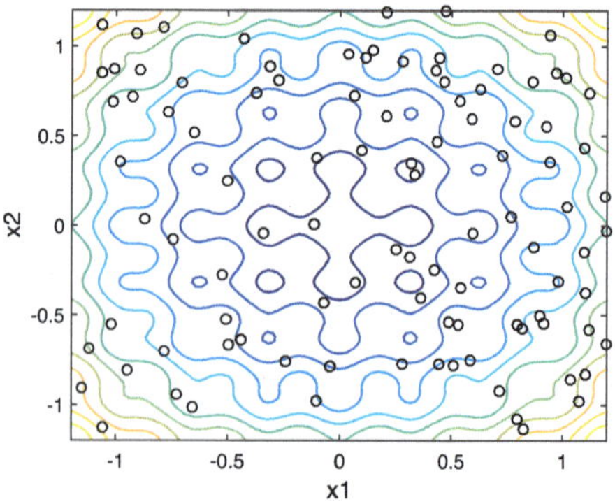

Fig. 11.8 Initial population of chromosomes

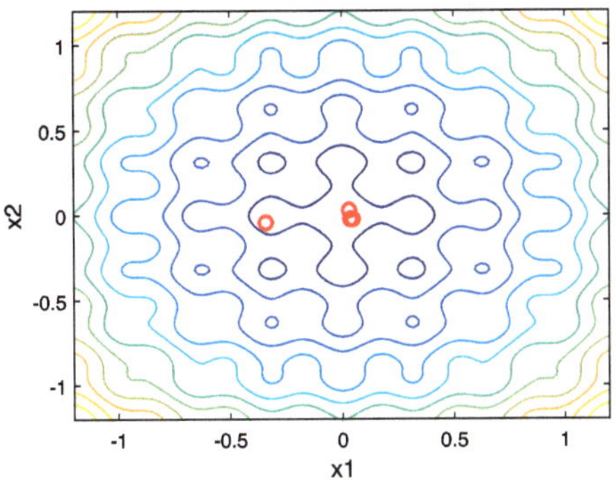

Fig. 11.9 History of best chromosomes along generations

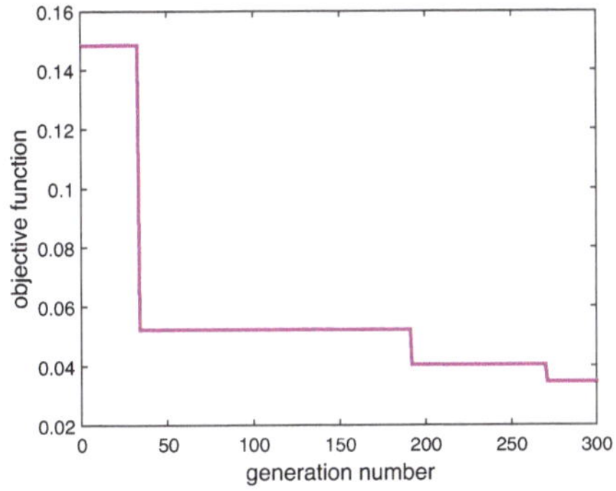

Fig. 11.10 History of the best value of the objective function along generations

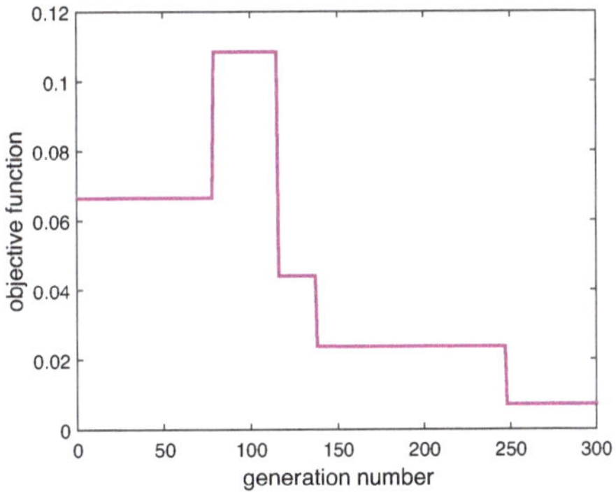

Fig. 11.11 Another history of the best value of the objective function along generations

tions, until the algorithm finds a better substitute. Figure 11.9 depicts the history of best chromosomes (small circles near origin), along the number of generations we specified.

Figure 11.10 presents how the best value of the objective function evolves along generations. In this example, the objective function value remains constant or decreases.

Figures 11.8, 11.9 and 11.10 have been generated with the program 11.3. Notice that the program can be run several times, obtaining different results in each time. Figure 11.11 shows, for example, a case where, due to a mutation, the trajectory of the objective function value increases in a certain moment of the evolution.

The reader is invited to run several times program 11.3, to see how different the evolutions could be; including, for instance, the case when an initial chromosome falls on the minimum (or very near), and so the evolution almost does not happen.

Program 11.3 Genetic algorithm example

```
%Genetic Algorithm example
% objective function display
[vx1,vx2]=meshgrid(-1.2:0.02:1.2, -1.2:0.02:1.2);
figure(1)
y=The_problem(vx1,vx2); %using a function for display
contour(vx1,vx2,y,11); hold on;
xlabel('x1'); ylabel('x2');
%algorithm parameters -------------------
Ncr=100; %number of chromosomes
Ngen=300; %number of generations
x1min=-1.2; x1max=1.2;
x2min=-1.2; x2max=1.2;
%start the algorithm--------------------
chr=zeros(2,Ncr); %chromosomes
child=zeros(2,1);
z=zeros(1,Ncr); %fitness value
hcr=zeros(2,Ngen); %for evolution history
hz=zeros(1,Ngen); % " "
%producing an initial population
%and evaluate fitness of each cromosome
for nn=1:Ncr
  chr(1,nn)=x1min+((x1max-x1min)*rand(1));
  chr(2,nn)=x2min+((x2max-x2min)*rand(1));
  %use of function fit():
  z(nn)=fit(chr(1,nn),chr(2,nn));
end
%plot initial population
plot(chr(1,:),chr(2,:),'ko');
zaux=z;
%generations start ------------------------------
for gg=1:Ngen
  %ranking of chromosomes
  [zaux,indx]=sort(z);
  ib=indx(1);
  hcr(:,gg)=chr(:,ib); %best chromosome
  hz(gg)=fit(chr(1,ib),chr(2,ib)); %minimum of fit()
  %1 child by crossover
  % parents random selection
  naux=Ncr-1;
  p1=1+round(naux*rand(1)); p2=1+round(naux*rand(1));
  child(1,1)=chr(1,p1); child(2,1)=chr(2,p2);
  %the new child substitutes worst individual
  iw=indx(Ncr);
  chr(:,iw)=child(:);
  %mutation of 1 chromosome
```

```
    %from time to time
    if rand(1)<0.2
      pm=1+round(naux*rand(1));
      chr(1,pm)=x1min+((x1max-x1min)*rand(1));
      chr(2,pm)=x2min+((x2max-x2min)*rand(1));
    end
    %evaluation of new population
    for nn=1:Ncr
      %use of function fit():
      z(nn)=fit(chr(1,nn),chr(2,nn));
    end
  end
hz(Ngen)
hcr(:,Ngen)
%history of best chromosomes
figure(2)
y=The_problem(vx1,vx2); %using a function for display
contour(vx1,vx2,y,11); hold on;
xlabel('x1'); ylabel('x2');
plot(hcr(1,:),hcr(2,:),'ro');
figure(3)
plot(hz,'m');
xlabel('generation number');
ylabel('objective function');
%For objective function display as contours
function y = The_problem(vx1,vx2)
y= 0.2+(vx1.^2)+(vx2.^2)-(0.1 * cos(6*pi*vx1))- ...
(0.1 * cos(6*pi*vx2));
%end
%evaluation function
function objv = fit(x1,x2)
objv= 0.2+(x1^2)+(x2^2)-(0.1 * cos(6*pi*x1))-...
(0.1 * cos(6*pi*x2));
%end
```

In case of using a binary representation, chromosomes are series of 0 or 1 numbers. The crossover operation may choose at random the position to cut parent chromosomes, in order to exchange segments. Likewise, the mutation may choose at random a position and change the 1 by 0 or vice versa.

In any case, binary, or real, etc., one could protect best chromosomes from dying out. Some investigators introduce adaptation of parameters (mutation rate, number of selected parents, etc.) along generations. It is not rare that, after many iterations of the genetic operators, the population of chromosomes loses variety; some supervision could be devised to enrich this variety from time to time.

We could say that a general drawback to avoid is premature convergence to a solution, which could mean a trap.

In most cases, the best solution found with GA would be a reasonable approximation to the optimum. Once the GA final approximation was obtained, it is usual to apply then a local search in order to get nearer to the exact solution.

11.5 About Targets and Approaches

The previous sections on simulated annealing, tabu search and genetic algorithms, have given us a good set of important ideas, opening the doors to the new optimization field we are now considering. It is a good moment for discussing some generalizations and to acquire a basic perspective of stochastic approaches.

Concerning the target one pursuits, many times it is a matter of several objectives and concerns. From our research on beer fermentation, which has been mentioned before, we could cite as example that the main target was to get a good product in less time, in order to earn more. This means, among other economic factors, not to incur in higher costs. Our specific problem was to devise a temperature profile, driving the fermentation process to suitable results as fast as possible. But it happens that some undesired by-products could develop, depending on the shape of the temperature profile. What we mean with "good product" includes certain limits on byproduct concentrations.

Most aspects touched during the basic description we made about the fermentation example could be read as constraints. A more modern view is to see it as a multiobjective problem [28]. This topic will be the subject of a next chapter.

Recall that the classical "trick" when there are constraints is to form a unique objective function, by using Lagrange multipliers (which add the constraints).

Suppose our problem has two main objectives, J_1 and J_2. One could combine them, to form a single objective as follows:

$$J = J_1 + \alpha J_2$$

This seems good, but α should be specified, doing justice to the nature of the specific problem in hand.

Notice also, that a minimum value of J, say for instance 4, may equally correspond to $J_1 = 2$ and $\alpha J_2 = 2$, or $J_1 = 1$ and $\alpha J_2 = 3$, etc. This could lead to unsatisfactory proposed solutions.

The Pareto front helps to understand the situation. Figure 11.12 depicts a case with two objectives and a set of possible combinations with same global result. It is usually a good compromise, to choose a point paying equal attention to J_1 and J_2 (the arrow in the figure signals this point).

In the fermentation case, J_1 may correspond to time and J_2 to by-products; both J_1 and J_2 should be minimized.

Notice that the good compromise can be expressed as a minimax criterion. The maximum of (J_1, J_2) must be minimized; therefore, instead of a 1+3 solution, one chooses 2+2.

The discussion on how to deal with desires and constraints does not end here. In more general cases, the fitness evaluation step of any algorithm could be accomplished with a decision subprocess, where factors are considered with some reasoning (for instance, with fuzzy logic, or a set of rules).

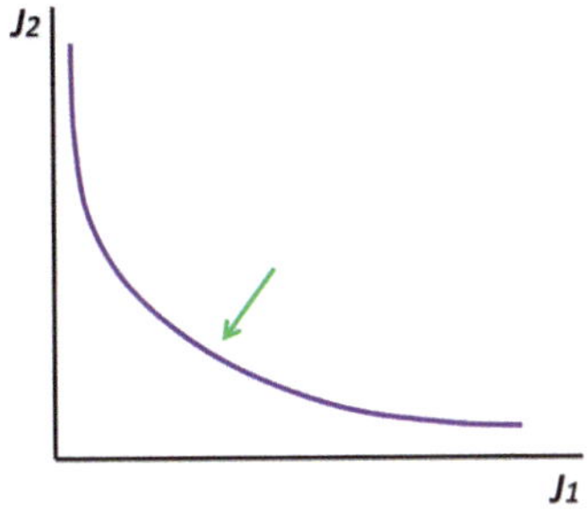

Fig. 11.12 Sketch of a Pareto front

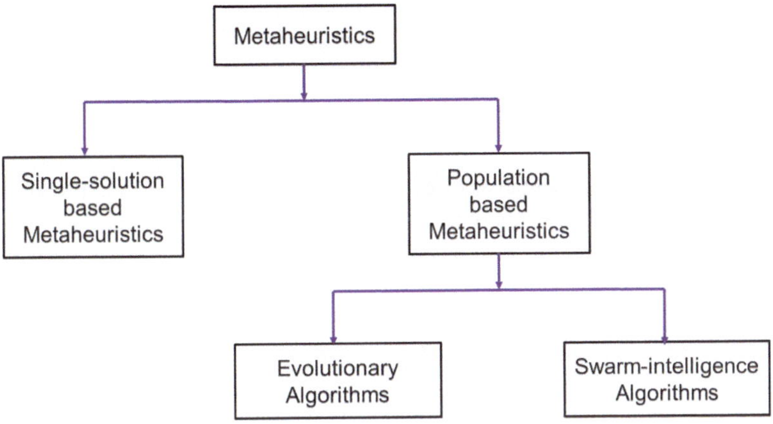

Fig. 11.13 Basic taxonomy of metaheuristic searching methods

An important and most convenient feature of intelligent searching methods is that they usually provide the opportunity of benefit from our "a priori" knowledge. For example, one knows that variables should be of such and such value range and precision, and this drives to a suitable problem representation. Or one owns some hints of where the solution should be, or what constraints are decisive, and this leads to a proper fitness evaluation.

The increasing technological power makes possible to employ many searching processes in parallel. For example, parallel GAs with several diverse populations at the same time. Some schemes include interactions between these populations. It is also possible to use several (many) digital processors, for sharing the searching work. Of course, intelligent supervision levels could be there for riding the parallel searching.

Let us take some height for a perspective of our present topic. Figure 11.13, which we reproduce from [15], shows a simple diagram that would be helpful. Both simulated annealing and tabu search, as already described, bring examples of single-solution-based metaheuristics. Genetic algorithms correspond to population-based metaheuristics.

In addition to Gas, there are other evolutionary algorithms. The next section will briefly introduce them.

The other block, related to swarm based approaches, requires much more space, which will be distributed in several sections. The first one in this series will be devoted to *ant colonies*, which is a relevant archetype of swarm optimization.

Swarms have attracted a lot of attention from several scientific fields. There are mysteries about how living organism swarms react so quickly to dangers, how they organize internally, etc. Moreover, some of swarm behaviors seem to be optimal for certain tasks (mostly related to survival) and are a source of inspiration for new algorithms.

11.6 Evolutionary Algorithms

The most popular representative of evolutionary algorithms is GAs, already introduced. This section focuses on other types of evolution-related algorithms.

11.6.1 Genetic Programming

Computer sentences and programs can be represented as trees. For instance, the equation $x * ((x/y) - sin(x)) + exp(x)$ can be represented as shown in Fig. 11.14.

Now, suppose you generate at random a large set of these trees. Then, you can apply crossover or mutation operations on the trees. Examples were given in Fig. 11.15, about crossover; and Fig. 11.16, about mutation. Crossover is done by exchanging subtrees.

For mutation, a random node in the tree is chosen and replaced with a different random generated subtree [29].

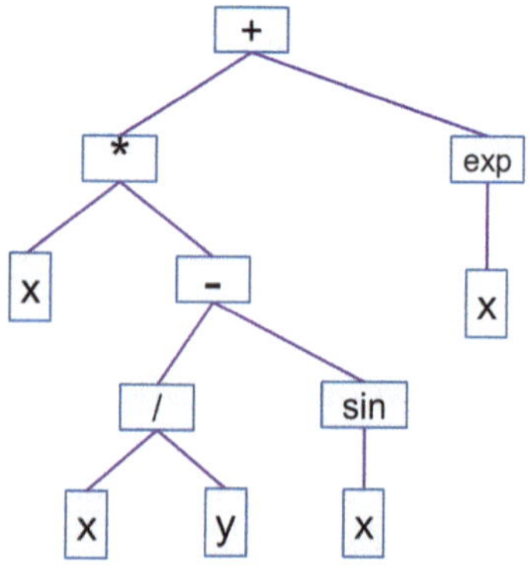

Fig. 11.14 Tree representation of the equation example

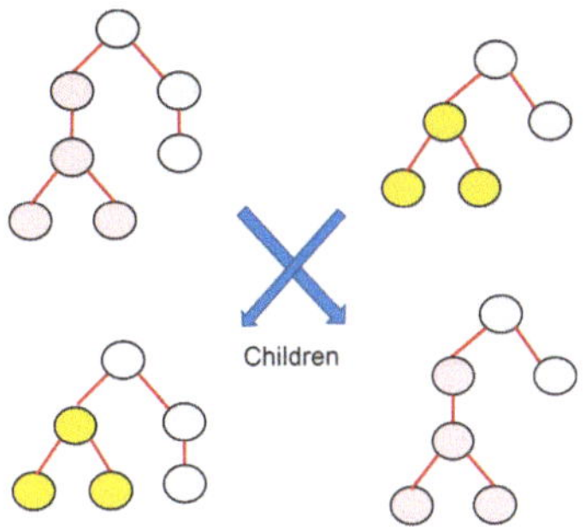

Fig. 11.15 Crossover operator on trees

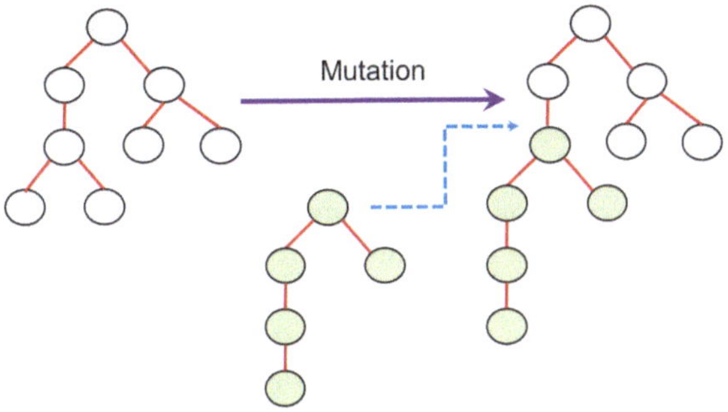

Fig. 11.16 Mutation operator on a tree

The central idea of *Genetic Programming* (GP) is to generate a suitable program for the desired optimization, based on a population of programs and the use of genetic operators.

Some commentators say (in a futuristic pitch) that, with GP, we are in the presence of programs writing programs.

An application example may well be the fitting of a certain set of data, by a function to be established with GP. This relates to the familiar fitting of data by polynomials but goes further by opening the opportunity of using nonpolynomial functions, which could be more in concert with the data at hand.

In fact, one of the successful applications of GP is for process modeling.

The fundamental concepts of GP were introduced by John Koza in 1988. After this year, J. Koza and collaborators have produced more than 200 scientific publications on GP.

The procedure suggested by Koza includes four steps. During step 1, one chooses a possible set of operations and variables that could be useful for the programs to be built. Then, in step 2, an initial population of random trees (programs) is generated. The trees should be syntactically correct. Now, step 3 is devoted to calculate the fitness of each tree, using a set of reference cases for which the correct behavior is known. And step 4 focuses on creating a new population, by selection of best individuals, crossover, and some mutations (if any). It is recommended to keep around 10% of the old population and replace the rest (90%).

Steps 3 and 4 are repeated until a termination criterion was met.

A convenient introductory reading about GP is [17], written in 2007 by J. Koza. Also, coauthored by Koza, there are two extensive tutorials [18, 19], which include brief descriptions of a good number of applications. For more theoretical and practical details, the book [21] would be recommended.

As resources for practicing and programming, we could mention the MATLAB Genetic Programming Toolbox (GPLAB), [24]; and the 388 repositories related to genetic psrogramming in GitHub [11], with abundant implementation examples, mainly in language Python.

11.6.2 Evolution Strategies

To capture the general idea of *Evolution Strategies* (ES), we could use the metaphor of a "black-box" optimization [10], box with inputs and outputs, and a set of knobs. We move the knobs until some progress was achieved and continue with the knobs until reaching a best solution. These knobs correspond to parameters (for instance, the weights in a neural network).

Perturbations are applied to the knobs at random, and the results are evaluated. Once a set of perturbations has been tested, a gradient toward better solutions can be estimated. Then, new, updated perturbations in the gradient direction are applied and evaluated. And so on, until reaching a satisfactory state. Figure 11.17 illustrates the evolutionary concept.

The evolution strategies were introduced in 1964 by three students—Bienert, Rechenberg and Schwefel—at the Technical University of Berlin, which were dealing with an aerodynamic design problem.

In Fig. 11.12, the clouds of points correspond to Gaussian random distributions. This is the usual scenario with ES. Each cloud has a center (the mean, $\mathbf{m}$) and shape determined by a covariance matrix. This shape would be circular or elliptical. There is a scale factor σ that modifies the size of the cloud.

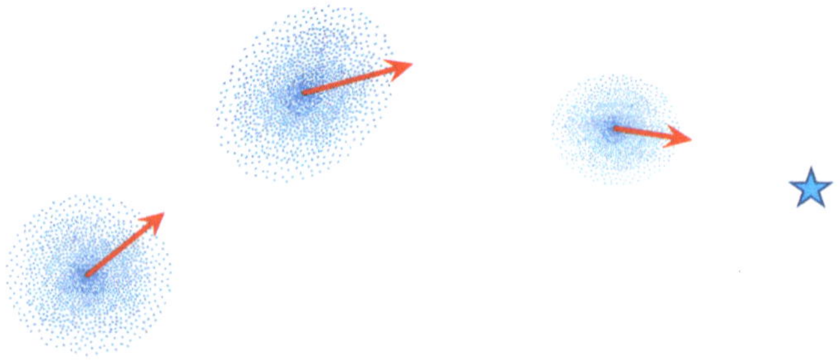

Fig. 11.17 Example of ES behavior towards optimum

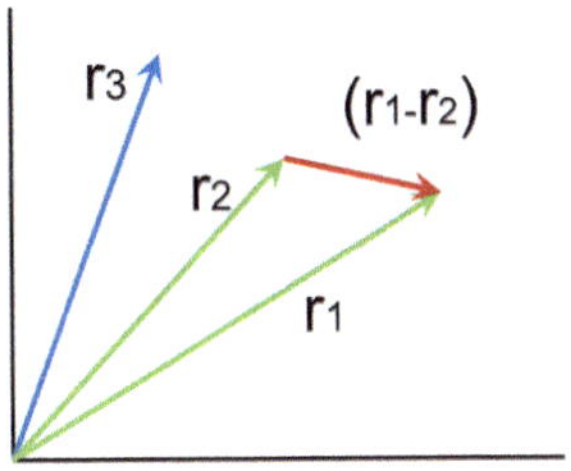

Fig. 11.18 Example of vectors composing the perturbation vector ***u***

There is a mutation operator that generates new individuals **x**, according to **x** = **m**+ σ y, where ***y*** belongs to a zero-mean Gaussian distribution with covariance matrix **C**.

After generation and evaluation of the initial population of candidates, mutation takes place for producing new candidates.

Once a new generation was created, it is evaluated, and selection mechanisms are applied. Then, the distribution parameters are updated (**m, C,** σ); and we loop again from the mutation step (until a termination criterion is satisfied). See [20] for much more details on the ES methodology and its variants. An important version is CMA-ES, [12], where CMA stands for *Covariance Matrix Adaptation*.

In principle, ES provides an easy method for a black-box approach to optimization, including continuous time contexts.

There are several ES implementations examples in the MATLAB File Exchange web site.

11.6.3 Differential Evolution

Frequently the optimization solution is a series of values (parameters and/or variables). Let us consider these series as vectors.

The *Differential Evolution* (DE) starts with a population of vectors, representing candidate solutions. The next step is to perturb each vector, so from each vector $\mathbf{x}_j$ one obtains a new vector $\mathbf{y}_j$.

Perturbation of a vector involves two actions.

First, three vectors, $\mathbf{r}_1$, $\mathbf{r}_2$, $\mathbf{r}_3$, are generated at random. Then, a perturbation vector **u** is computed as u = $\mathbf{r}_3 + \mathbf{F}(\mathbf{r}_1 - \mathbf{r}_2)$. Figure 11.18 depicts an example of the mentioned vectors.

Second, use crossover of **x** and **u**, as follows:

```
j = rand(1,N) //find an index j, at random
For i = 1 to N
```

(continued)

```
    if (rand(0,1)<CR or i==j)
      y[i] = u[i]
    else
      y[i] = x[i]
    end
  end For
  return y
```

where CR is the crossover rate.

After perturbation of each **x** to obtain a certain **y**, change **x** to **y** if fitness(**y**) > fitness(**x**). This is the selection step.

Perturbation and selection are repeated until the termination criterion is satisfied.

The DE method was introduced by K. Price and R. Storn in 1994. Since then, several investigators have found DE to be better, in some aspects, than other stochastic searching methods. See [5, 25], for more details and current state of the art.

There are several DE implementation examples in the MATLAB File Exchange website. In addition, GitHub offers 157 repositories of DE code, in various programming languages.

11.7 Particle Swarm Optimization

The origin of the particle swarm concept was the (graphical) simulation of a simplified social system. It initially focused on bird flocks (Fig. 11.19). A bird flock flies in a certain environment looking for the best place to rest, considering safety, water and food access, sufficient space, etc. The simulation had to include nearest-neighbor velocity matching, multidimensional search, and accelerations by distance. Birds were considered as particles.

At some point of the research with the simulation, it was realized that, actually, it could be used for optimization purposes.

The particle swarm optimization (PSO) algorithm was introduced in 1995 by J. Kennedy and R. Eberhart [16] (87,647 cites), and soon various improvements were proposed by Y. Shi and others. See [23] for a history of precedents and further developments and variants. J. Kennedy is a social psychologist; R. Eberhart is an electrical engineer (professor at Purdue University).

PSO is an important source of inspiration for social-based optimization algorithms. Compared to GA, there are no mutations, crossover, etc. Only visual information exchange between birds and their neighbors.

Fig. 11.19 Bird flock

The original PSO algorithm considers that all the particles share information and that any particle knows what is the best position ever visited by any particle in the swarm.

In general, PSO is applied to multidimensional optimization problems.

Each particle has a position and a velocity. These variables are updated at time steps. Then

$$\mathbf{x}_{i,d}(t+1) = \mathbf{x}_{i,d}(t) + \mathbf{v}_{i,d}(t+1)$$

where d corresponds to the dimension being considered.

The velocities are updated with

$$\mathbf{v}_{i,d}(t+1) = \mathbf{v}_{i,d}(t) + C_1 * rnd(0,1) * [\mathbf{pb}_{i,d}(t) - \mathbf{x}_{i,d}(t)] +$$

$$+ C_2 * rnd(0,1) * [\mathbf{gb}_{i,d}(t) - \mathbf{x}_{i,d}(t)]$$

where ***gb*** (global best) is the best of all the visited locations of all the particles, and ***pb*** (personal best) is the best of all locations visited by the *i-th* particle.

The term beginning with C_1 expresses a cognitive component (concerning the *i-th* particle experience). The third term, beginning with C_2, corresponds to the social component (the bird belongs to a flock).

The PSO procedure can be expressed as follows:

Generate a population of particles at random locations and velocities in D dimensions
Repeat:
  Compute the value of the objective function for each *i-th* particle, *vobj(i)*
    (this value depends on particle's location)
  For each particle, compare *vobj(i)* with *pbest(i)*
    (the best value already obtained by the particle):

(continued)

If *vobj(i)* is better than *pbest(i)*, then set *pbest(i)* = *vobj(i)*.
And change ***pb*** to store the new best particle location

Identify the particle with best value of the objective function.
Compare this value with *gbest* (global best):

If the new global best value is better than *gbest*, substitute accordingly and change ***gb*** to store the new global best location.

Update position and velocity of each particle

Until a final condition is met

There are several MATLAB implementations that can be accessed from the Internet. For reasons of simplicity and clear structure, the example provided by [$REF KHA] in the MATLAB File exchange site has been preferred as guide for developing our program 11.4, for PSO illustration.

Program 11.4 PSO algorithm example

```
%PSO algorithm
%simple example
Np=50; %number of particles
Niter=40; %maximum number of iterations
C1=2.03; %cognitive coefficient
C2=2.03; %social coefficient
W=0.9; %initial inertia
minW=0.4; %last inertia
alpha=((W-minW)/Niter); %for W decrease monitoring
Lb=-5; %lower bound search space
Ub=5; %upper bound search space
init_Vel= 0; % initial particles Velocity
var_W= W; % for plotting
%Generation of particle swarm ---------------------
Aux.pos=[]; Aux.val=[]; Aux.velocity=[];
prt=repmat(Aux, Np, 1); %space for Np particles
%particles initialization
for i=1:Np
  %random prt 2D positions inside search space
  prt(i).pos= Lb + (Ub-Lb)*rand(1,2);
  %evaluate objective function at prt positions
  %(use function Fitness)
  prt(i).val=Fitness(prt(i).pos);
  %set initial prt velocity
  prt(i).velocity=init_Vel.*ones(1,2);
end
%save personal best of each particle,
%which is itself at initial iteration
bprt=prt;
```

```
% determine the global best among the bparticles
[aux, idx]= min([bprt.val]);
% save the Global optimum
gprt= prt(idx);
%prepare for main loop
% average of objective function values:
avrg= mean([prt.val]);
best= gprt.val; %best ojective function value
iter= 1; % number of iteration
W_vector= W;
%main loop ------------------------------------------
while iter<Niter
  for i=1:Np
    %update velocities
    prt(i).velocity= W.*prt(i).velocity...
     + C1.*rand(1, 2).*(bprt(i).pos - prt(i).pos)...
     +C2.*rand(1, 2).*(gprt(iter).pos - prt(i).pos);
    %update position
    prt(i).pos= prt(i).pos + prt(i).velocity;
    %%use of boundaries after update:
    %when crossing up
    crup= prt(i).pos >= Ub;
    prt(i).pos(crup)= Ub;
    %when crossing down
    crdn= prt(i).pos <= Lb;
    prt(i).pos(crdn)= Lb;
    %evaluate vobj at prt position
    prt(i).val= Fitness(prt(i).pos);
    %the optimum is a minimum
    if prt(i).val<bprt(i).val
      bprt(i)=prt(i);
    end
  end
  if W>minW
    W= W - alpha;
  end
  iter= iter+1;
  [aux, idx]= min([bprt.val]);
  gprt(iter)= bprt(idx);
  best(iter)= gprt(iter).val;
  %avrg(iter)= mean([prt.val]);
  W_vector= [W_vector W];
end
%best position
gprt(end).pos
%best value
best(end)
% results ----------------------------------
figure(1)
[X, Y]= meshgrid(Lb:0.1:Ub, Lb:0.1:Ub);
Z= X.^2 + Y.^2 + (25 * (sin(X).^2 + sin(Y).^2));
contour(X,Y,Z,20); hold on;
```

```
for i=1:Np
  plot(prt(i).pos(1), prt(i).pos(2),'or');
end;
xlabel('X'); ylabel('Y');
figure(2)
subplot(1,2,1)
plot(1:Niter, best, 'r');
ylabel('best value'); xlabel('iter');
subplot(1,2,2)
plot(1:Niter, W_vector, 'm');
ylabel('Inertia W'); xlabel('iter')
% Function
function [v]= Fitness(S)
v= S(1).^2 + S(2).^2 + (25 * (sin(S(1)).^2 + ...
 sin(S(2)).^2));
```

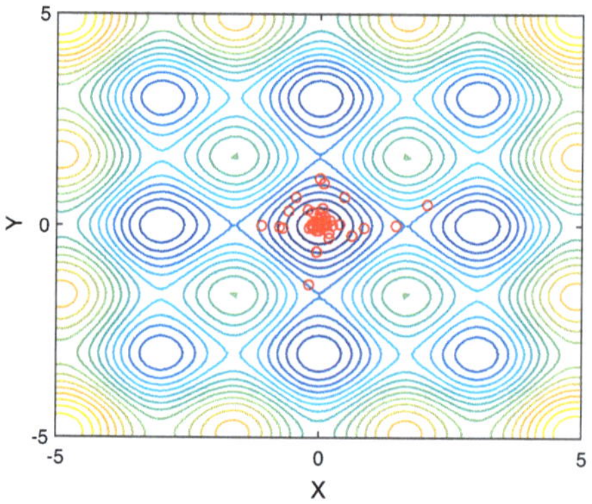

Fig. 11.20 Final position of particles

As the algorithm starts, the generated particles randomly cover all the search space. Gradually, from iteration to iteration, the particles move towards minima.

The two last lines of the listing above correspond to the objective function that has been chosen as example, which is called the *egg-crate function*:

$$f(x) = x_1^2 + x_2^2 + 25\,(\sin^2(x_1) + \sin^2(x_2))$$

This bidimensional function has a zero global minimum at (0,0) and several nonzero local minima.

Figure 11.20 depicts a contour plot of the objective function, together with the final position reached by the PSO particles after 40 iterations of the algorithm. The best location found by the particles is very close to (0,0).

Figure 11.21 depicts on the left how the best value found in each PSO iteration, progress toward an almost zero minimum. On the right, we see the linear descent of the inertia linked to the algorithm march.

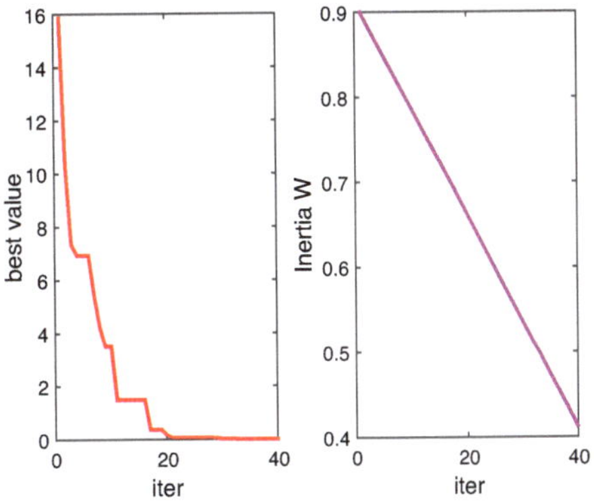

Fig. 11.21 Evolution of best found value (left-hand), and inertia (right hand) along iterations of the algorithm

In 2015, a comprehensive review of PSO was published [30]. It includes a description of relevant modifications, variations, and hybridization of the algorithm. In addition, the article comments various applications in electrical and electronic engineering, automation and control, communications, operations research, mechanical engineering, fuel and energy, medicine, chemistry, and biology.

The GitHub site for *particle-swarm-optimization* offers 380 repositories of documentation and software.

11.8 Ant Colonies

During the 1990s, Marco Dorigo and collaborators developed an optimization metaheuristic based on the behavior of ant colonies [6]. It has been shown that this methodology can be successfully applied in important areas of optimization.

Figure 11.22 presents a photograph of Marco Dorigo, who has been cited about 110,000 times by scientific publications.

Ants are able to find shortest paths from their nest to food locations. The key point is a chemical mechanism that ants use for information exchange. Ants walking from

Fig. 11.22 Marco Dorigo

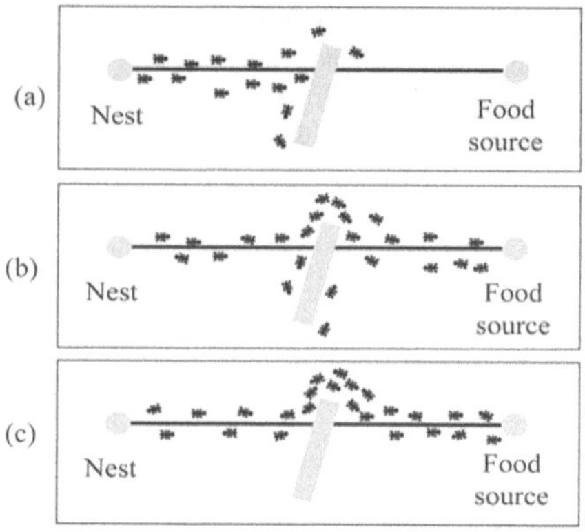

Fig. 11.23 Ants are able to find shorter paths

or to a food source, deposit on the ground a substance called pheromone. It can be smelled by ants. With time the pheromone evaporates.

The ants follow the trail indicated by pheromone. Suppose there are two paths leading to the same source of food. If one of the paths is longer than the other, this would mean less pheromone density and less probability that ants prefer this path. The consequence after some time would be the predominance of only one path, the shortest one. Figure 11.23 illustrates the ant colony behavior [27].

The pheromone-based mechanism provides an example of what is called "*stimergy*": indirect communication among self-organizing emergent systems, through individuals modifying their environment. Notice also that while ants continue to use a path, they contribute to keep and increase the pheromone concentration, meaning a *self-reinforcement* process.

The first ant algorithm, by Dorigo and collaborators, was called "*Ant System*" (AS). After few years, they generalized the algorithm and introduced the "*Ant Colony Optimization*" (ACO) metaheuristic. ACO is structured around three main functions [3]:

- *AntSolutionsConstruct*
  A set of candidate solutions are built by the ants, moving through states of the problem by using state-transition rules.
- *Pheromone Update*
  The pheromone associated with solutions is updated, including evaporation effects.
- *DeamonActions*
  This is optional. Some extra updates could be decided, that, for instance, could be to increase the pheromone of the best solution.

These three functions were iteratively applied, until a termination criterion was satisfied.

Following the coding example given in [14], which can be obtained from the MATLAB File Exchange site, we have developed a simple implementation

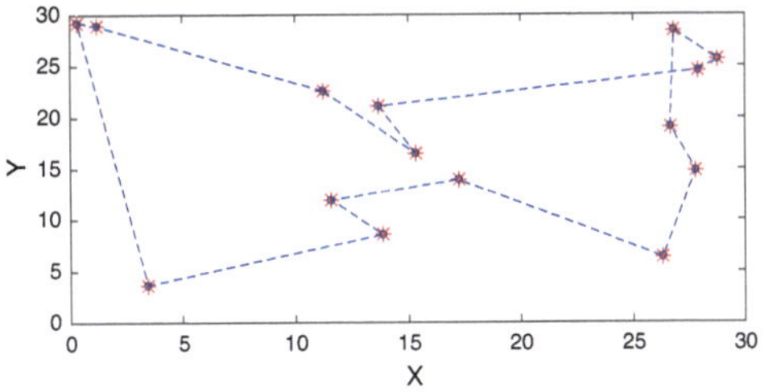

Fig. 11.24 The best tour found by ants for a TSP example

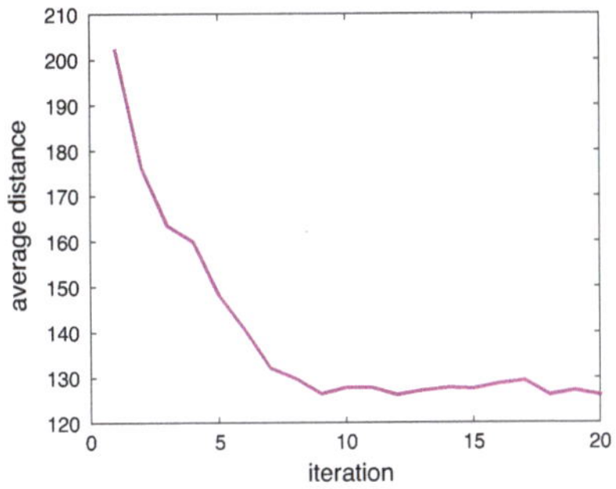

Fig. 11.25 Evolution of the search: best distance vs. iteration

(Program 11.5) for TSP problems. No daemon actions were considered. The results are satisfactory, not requiring much computing time.

The implementation is structured as a main program and three auxiliary functions.

Note (Program 11.5)

The complete listing of the program 11.5, about Ant Colonies, has been included in the Appendix for Long Programs.

Figure 11.24 presents a result for a TSP problem with 15 cities. The locations of these cities are generated at random, being different each time you run the program (this is interesting for experimenting with various possible scenarios).

Figure 11.25 depicts how the best (minimum) distance found by the algorithm, evolves along several iterations of the search. Although the general trend is to descend, there are scenarios where this curve would have intervals of increasing-then-decreasing (like the low hill you can notice near iteration 12 in this example).

More information on ant colony optimization can be found in the scientific articles [4, 7]. There are also books on this theme, like [2] or [22].

The MATLAB website for file exchange includes several implementations of ant colony algorithms, for different applications. Likewise, GitHub offer 242 code repositories on ant-colony-optimization.

Typically, bioinspired algorithms are modified by investigators, which propose interesting variants. Ant colonies are no exception: there are several versions, according with types of possible applications, or related to aspects deserving more

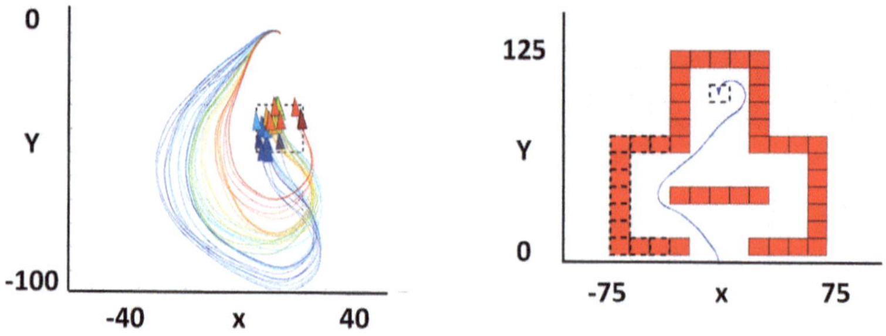

Fig. 11.26 Using Ant Colony for optimization of ship paths

research. In our case, we can mention one of our publications, which apply ant colony optimization to derive suitable trajectories of ships, in the presence of obstacles and in difficult maneuvering scenarios, like navigation and docking in harbors [8]. In this publication, the peculiar dynamics of ships (only rudder, large inertia) has been embedded in the optimization algorithm.

Figure 11.26 presents two views of the optimization process related to trajectories of ships. By using ants, a set of trajectory candidates for ships arriving to port are found. These trajectories are shown on the left-hand of the figure. The end condition for the ship is to stop pointing to North. The curvature of trajectories are in accordance with the turning radius of the ship. On the right-hand of the figure, we plot a schematic harbor, with a mouth, an obstacle to avoid, and a final place for docking. The ship should stop pointing to the mouth. The trajectory has been selected as the best proposed by the algorithm. Of course, further conditions may be added, such as minimum allowable distances to walls.

One of the standard optimization problems, which in fact constitute a benchmark for comparison of algorithms, is the Travelling Salesman Problem (TSP). We already considered this problem when we introduced tabu search and also in the ant colony program recently listed. The case is that ant algorithms are quite successful for TSP.

It seems opportune to mention that we proposed a variant of the ant colony algorithm, in which there were ants in charge of exploration, and other in charge of carrying food [9]. This proposal contemplates a typical dilemma of searching methods: to explore or to exploit. We found that the proportion of explorers or workers should be adaptive along the course of events, since a particular food resource could become exhausted or other problems could appear. Our algorithm reached a state-of-the-art level for TSP, being able to deal with problems including thousands of cities.

11.9 Exercises

1- [11.1] Consider the following function, which corresponds to a potential well:

$$J = x^4 - 16x^2 + 5x$$

Apply simulated annealing to approximate the solution. Visualize the evolutionary process.
2- [11.2] Let the optimization problem be as follows:
Minimize: $J = \exp(x_1)\,(4x_1^2 + 2x_2^2 + 4x_1x_2 + 2x_2 + 1)$
Subject to:

$$x_1x_2 \geq -10$$

$$x_1x_2 - x_1 - x_2 \leq -1.5$$

$$-100 \leq x_1 \leq 10$$

$$0 \leq x_2 \leq 100$$

Please, use tabu search to approximate the solution. Describe the evolutionary process.
3- Using information from the Internet, create yourself a TSP problem with 20 cities in Belgium. You are planning a touristic tour on bike. Use tabu search for selecting the shortest path.
4- [11.3] Let us choose the following function:

$$f = 1 + \exp(-x^2)\cos(36x)$$

If you plot this function, you will notice it has many oscillations, with a series of local maxima and minima.

Please, find with genetic algorithms the absolute maximum. Visualize the evolution of the process.
5- Place eight queens on a chessboard, so none of them can attack the other (this problem would be revisited in a next chapter).

Use genetic algorithms. Describe how you solve the problem, adding some comments on the evolution you observe in several runs.
6- [11.4] [11.5] The application of evolutionary algorithms implies sometimes intelligence and patience for adapting the algorithm to the problem in hand. One example of this is the Facility Layout problem, which can involve complex conditions to meet. The essence of the problem is to assign a number of machines to locations within a plant, trying to minimize material handling costs. Next image gives an idea of the problem in modern industry.

We shall follow the guide provided by the publication [11.4]. Please apply genetic algorithms to solve the problem with 10 machines and 25 locations (a 5×5 grid). Visualize the behavior of the algorithm and some results (several runs of the algorithm). Add comments.

It is suggested to extend the exercise by considering [11.5], where the size of working cells and other details are taken into account.

7- [11.6] Annie Hu runs a rental car company with sites at five cities. At the end of this week, the stocks are the following:

Location	Stock
Alphaville	3 cars
Beantown	1 car
Catcity	7 cars
Dogburg	2 cars
Elmoland	1 car

These fiction towns are on the same long road. Driving a car from one town to the next costs $100.

Annie wants to move the cars, so next week (to satisfy the demands), she will have:

Location	Stock
Alphaville	2 cars
Beantown	4 cars
Catcity	1 car
Dogburg	1 cars
Elmoland	6 cars

Please, find with genetic Algorithms the cheapest car moving. Explain your approach and the result found after several runs. Describe what happens with the process evolution.

8- [11.7] Here is a simple portfolio problem. The covariance matrix is

	IBM	WMT	SEHI
IBM	0.017087987	0.003298885	0.001224849
WMT	0.003298885	0.005900944	0.004488271
SEHI	0.001224849	0.004488271	0.063000818

And the average returns are:

	IBM	WMT	SEHI
Average return	0.026	0.008	0.074

Please find an optimal solution employing particle swarm optimization. Visualize the process evolution.

9- [11.8] Let us visit an example of vehicle routing problems. It is desired to service a number of customers with a fleet of vehicles. There could be one or several depots. Every route starts in a depot, taking several customers in a given sequence along the route, and then comes back to the depot. Costs should be minimized (shortest path). The vehicles can carry a limited number of passengers.

For our particular problem, there is one depot and nine stops for taking passengers. Next table give the mutual distances (the first row/column refer to the depot):

	D	1	2	3	4	5	6	7	8	9
D		12	11	7	10	10	9	8	6	12
1	12		8	5	9	12	14	16	17	22
2	11	8		9	15	17	8	18	14	22
3	7	5	9		7	9	11	12	12	17
4	10	9	15	7		3	17	7	15	18
5	10	12	17	9	3		18	6	15	15
6	9	14	8	11	17	18		16	8	16
7	8	16	18	12	7	6	16		11	11
8	6	17	14	12	15	15	8	11		10
9	12	22	22	17	18	15	16	11	10	

Number of customers in each stop:

The maximum capacity of vehicles is 40 passengers.

i	1	2	3	4	5	6	7	8	9
n	10	15	18	17	3	5	9	4	6

Please, use ant colony method for determining the plan of routes (hint: the solution is a set of three routes).

10- [11.9] [11.10] [11.11] Probably in the near future, humans shall step on Mars. Next figure (from page [11.10], depicts the trajectory to follow:

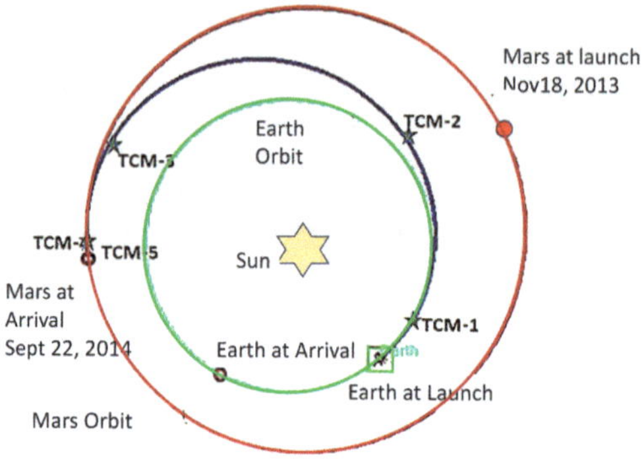

Indeed, cost minimization is desired, and this depends on the launching date. Please, apply genetic algorithms to obtain suitable trajectories. The publication [11.9] provides an opportune guidance. Visualize the behavior of the algorithm (several runs), adding comments.

See page [11.11] for more details on the Mars mission.

11- [11.12] [11.13] Robert H. Goddard had important contributions to our Space Age, in particular about rockets and satellite launching. He patented 214 inventions. Part of the heritage he left for us was—year 1919—a problem to solve: how to reach extreme altitudes? (how to spend the rocket's available fuel?) It is, till now, a famous problem (26 million answers on Goggle).

A peculiarity of the optimization problem is the influence of atmospheric drag, which depends on rocket's velocity, altitude, and other factors. The page [11.12] offers an animated simulation of the rocket motion and associated variables (in this simulation the rocket is capable of landing at the end, sounds SpaceX?).

Please, apply differential evolution to solve the problem. The publication [11.13] would help. Describe how the optimization method was applied, and visualize the results.

12- [11.14] We have not finished with the lunar mission yet. Before the final landing phase (considered in a previous exercise), there is a sequence of orbital maneuvers for approaching this last phase. Specifically, when the space vehicle is near the Moon, it enters into a lunar parking orbit; and from this orbit it descends using horizontal braking and then vertical braking.

Please, apply the two approaches described in [11.14] to find the optimal trajectory after parking. These approaches are particle swarm optimization and differential evolution.

11.9.1 Exercise References

[11.1] https://dialnet.unirioja.es/descarga/articulo/7233334.pdf
[11.2] https://dialnet.unirioja.es/descarga/articulo/5815968.pdf
[11.3] https://www.flll.jku.at/div/teaching/Ga/GA-Notes.pdf
[11.4] https://www.sciencedirect.com/science/article/pii/0360835294900493
[11.5] Solving the Facility Layout Problem with Genetic Algorithm
[11.6] https://andrejb.net/csi2101/notes/25L01.pdf
[11.7] https://www2.isye.gatech.edu/~sahmed/isye6669/notes/portfolio.pdf
[11.8] https://personal.fd.cvut.cz/rybicale/TGA_06_VRP.pdf
[11.9] https://www.engr.psu.edu/CE/Divisions/Hydro/Reed/Education/CE563Projects/Weeks.pdf
[11.10] https://www.planetary.org/space-images/maven-trajectory-to-mars
[11.11] https://marspedia.org/Earth-Mars_Transfer_Trajectory
[11.12] https://www.yoptimization.com/goddardLanding
[11.13] *Solution of singular optimal control problems using the improved differential evolution algorithm*
[11.14] *Fuel optimum lunar soft landing trajectory design using different solution schemes*

References

1. APMONITOR. (2020). Simulated annealing tutorial. Available from: http://apmonitor.com/me575/index.php/Main/SimulatedAnnealing
2. Barbosa, H. (2013) *Ant colony optimization: Techniques and applications*. BoD–Books on Demand.
3. Binitha, S., & Siva Sathya, S., et al. (2012). A survey of bio inspired optimization algorithms. *International Journal of Soft Computing and Engineering, 2*(2), 137–151. Available from: https://citeseerx.ist.psu.edu/document?repid=rep1&type=pdf&doi=45618bef1e4be5cfd2c999d42398166621fd7649
4. Blum, C. (2005). Ant colony optimization: Introduction and recent trends. *Physics of Life reviews, 2*(4), 353–373. Available from: https://www.iiia.csic.es/~christian.blum/downloads/blum_aco_PLR2005.pdf
5. Das, S., & Suganthan, P. N. (2010). Differential evolution: A survey of the state-of-the-art. *IEEE Transactions on Evolutionary Computation, 15*(1), 4–31. Available from: https://www.cp.eng.chula.ac.th/~prabhas/teaching/ec/ec2012/das-de-sota-2011.pdf
6. Dorigo, M., Birattari, M., & Stutzle, T. (2007). Ant colony optimization. *IEEE Computational Intelligence Magazine, 1*(4), 28–39. Available from: https://tiezhongyu2005.github.io/resources/popularization/ACO_2006.pdf

7. Dorigo, M., & Stützle, T. (2019). Ant colony optimization: overview and recent advances. In *Handbook of metaheuristics* (pp. 311–351). Available from: https://www.cheps-kmutt.com/_files/ugd/2ab5f6_e7c2bb16e24243b6ab44367b60a260bf.pdf#page=248
8. Escario, J. B., Jimenez, J. F., & Giron-Sierra, J. M. (2010). Optimization of autonomous ship maneuvers applying swarm intelligence. In *2010 IEEE International Conference on Systems, Man and Cybernetics* (pp. 2603–2610). IEEE. Available from: https://ieeexplore.ieee.org/abstract/document/5641909
9. Escario, J. B., Jimenez, J. F., & Giron-Sierra, J. M. (2015). Ant colony extended: experiments on the travelling salesman problem. *Expert Systems with Applications, 42*(1), 390–410. Available from: https://www.sciencedirect.com/science/article/abs/pii/S0957417414004709?via%3Dihub
10. Frigeni, F. (2018). Evolution strategies. Available from: https://medium.com/swlh/evolution-strategies-844e2694e632
11. GitHub. (2022). Genetic-programming. Available from: https://github.com/topics/genetic-programming
12. Hansen, N. (2016). The CMA evolution strategy: A tutorial. Available from: https://arxiv.org/pdf/1604.00772
13. Holland, J. H. (1992). *Adaptation in natural and artificial systems: An introductory analysis with applications to biology, control, and artificial intelligence*. MIT Press.
14. Ibrahim, S. (2022). Ant colony optimization (ACO) to solve traveling salesman problem (TSP). Available from: https://es.mathworks.com/matlabcentral/fileexchange/51113-ant-colony-optimization-aco-to-solve-traveling-salesman-problem-tsp?tab=reviews
15. Katoch, S., Chauhan, S. S., & Kumar, V. (2021). A review on genetic algorithm: past, present, and future. *Multimedia Tools and Applications, 80*(5), 8091–8126. Available from: https://link.springer.com/content/pdf/10.1007/s11042-020-10139-6.pdf
16. Kennedy, J., & Eberhart, R. (1995). Particle swarm optimization. In *Proceedings of ICNN'95-International Conference on Neural Networks* (Vol. 4, pp. 1942–1948). IEEE. Available from: https://ieeexplore.ieee.org/stamp/stamp.jsp?arnumber=488968
17. Koza, J. R. (2007). Introduction to genetic programming. In *Proceedings of the 9th Annual Conference Companion on Genetic and Evolutionary Computation* (pp. 3323–3365). Available from: https://dl.acm.org/doi/pdf/10.1145/1274000.1274116
18. Koza, J. R., & Poli, R. (2003). A genetic programming tutorial. Available from: https://www.ibug.doc.ic.ac.uk/media/uploads/documents/courses/GeneticProgramming-tutorial-2003.pdf
19. Langdon, W. B., Poli, R., McPhee, N. F., & Koza, J. R. (2008). Genetic programming: An introduction and tutorial, with a survey of techniques and applications. In *Computational intelligence: A compendium* (pp. 927–1028). Springer. Available from: https://wiki.eecs.yorku.ca/course_archive/2010-11/W/4403/_media/gp1.pdf
20. Li, Z., Lin, X., Zhang, Q., & Liu, H. (2020). Evolution strategies for continuous optimization: A survey of the state-of-the-art. *Swarm and Evolutionary Computation, 56*, 100694. Available from: https://www.sciencedirect.com/science/article/pii/S221065021930584X
21. O'Neill, M., Poli, R., Langdon, W. B., & McPhee, N. F. (2009). *A field guide to genetic programming*. Springer. Available from: https://www.zemris.fer.hr/~yeti/studenti/izvori/A_Field_Guide_to_Genetic_Programming.pdf
22. Ostfeld, A. (2011). *Ant Colony Optimization: Methods and applications*. BoD–Books on Demand.
23. Poli, R., Kennedy, J., & Blackwell, T. (2007). Particle swarm optimization: an overview. *Swarm Intelligence, 1*(1), 33–57. Available from: https://link.springer.com/content/pdf/10.1007/s11721-007-0002-0.pdf
24. Silva, S., & Almeida, J. (2003). GPLAB-a genetic programming toolbox for matlab. In *Proceedings of the Nordic MATLAB Conference* (pp. 273–278). Citeseer. Available from: https://eden.dei.uc.pt/~sara/gplab.nmc-2003.pdf

25. Storn, R. (1996). On the usage of differential evolution for function optimization. In *Proceedings of North American Fuzzy Information Processing* (pp. 519–523). IEEE. Available from: https://citeseerx.ist.psu.edu/document?repid=rep1&type=pdf&doi=77fb358c060e0fc377cbda912d01b4e1a186557b
26. Surjanovic, S., & Bingham, D. (2013). Virtual library of simulation experiments. Available from: https://www.sfu.ca/~ssurjano/optimization.html
27. Tang, Z., Sonntag, M., & Gross, H. (2019). Ant colony optimization in lens design. *Applied Optics, 58*(23), 6357–6364. Available from: https://opg.optica.org/ao/abstract.cfm?uri=ao-58-23-6357
28. Van Veldhuizen, D. A., & Lamont, G. B. (2000). Multiobjective evolutionary algorithms: Analyzing the state-of-the-art. *Evolutionary Computation, 8*(2), 125–147. Available from: https://ieeexplore.ieee.org/abstract/document/6788123
29. Verma, A. (2020). An introduction to genetic programming. Available from: https://medium.datadriveninvestor.com/an-introduction-to-genetic-programming-94ad22adbf35
30. Zhang, Y., Wang, S., & Ji, G. (2015). A comprehensive survey on particle swarm optimization algorithm and its applications. *Mathematical Problems in Engineering, 2015*. Available from: https://onlinelibrary.wiley.com/doi/pdf/10.1155/2015/931256

Chapter 12
Evolutionary Optimization: Bio-inspired Methods

12.1 Bio-Inspired Methods Related to Zoology

Some scientific journals on biology are considering bio-inspired optimization methods. One would notice the mentality of biologists in how they classify these methods, although different biologist' points of view exist. In any case, the fact is that there are articles in these journals, classifying the bio-inspired optimization methods as corresponding to invertebrates, or vertebrates; considering also solitaire animals, or those acting in groups, or even diverse types of swarms.

This perspective gives an initial idea of how complicated is to introduce bio-inspired methods.

It is opportune to cite here the prolific work of Dr. Xin-She Yang, at several UK universities, who has proposed various bio-inspired optimization algorithms. Examples of these proposals are the *firefly* algorithm (2008), the *cuckoo search* algorithm (2009), the *bat* algorithm (2012), the *flower pollination* algorithm (2012), etc. Dr. Yang has published 572 contributions and has been cited more than 70,000 times by other researchers.

Another creator of many bio-inspired algorithms is Gai-Ge Wang, Ocean University of China, who has proposed (with co-workers) the *Monarch butterfly optimization* algorithm (2015), the *earthworm optimization* algorithm (2015), the *elephant herding optimization* (2015), the *moth search* algorithm (2016), etc. He also contributed to variants of other bio-inspired algorithms.

In this chapter, we will try to go from invertebrates to vertebrates, and in the case of vertebrates, we will look at water, land, and then air. Recall that we have already included the case of ant colonies.

J. M. Giron-Sierra, *Introduction to Optimization with Matlab® Examples*,
https://doi.org/10.1007/978-3-032-05048-9_12

12.2 Artificial Bee Colony

Karl von Frisch received the Nobel Prize in 1973 for his discoveries about the "*dancing of honey bees*." He proposed that this dance is a language, and in fact, he deciphered aspects of it. This was a revolutionary view, which awakened controversy until being gradually accepted.

He observed that once a bee finds food at a certain location, many others soon appear there. Later, with patience, he found that when the "*scout*" bees discover a food source, they come back to the hive and execute a dance before other bees. There are two types of dance: the round dance causes bees to look for food at short distance (up to 50m); the waggle dance (in the shape of an 8) informs bees about the direction and distance (and quantity) of a food source. After a number of bees have visited candidate sources, there is a voting process that chooses the source having more bees dancing for it.

Figure 12.1, extracted from [9], depicts the two types of bee dance.

The Artificial Bee Colony (ABC) meta-heuristic algorithm was introduced by D. Karaboga in 2005 [33] (8949 cites). There are three groups of bees: employed bees, onlookers, and scouts. The food sources represent the solutions of an optimization problem, and the quantity of nectar corresponds to the objective function value.

The employed bees search food around the food source in their memory, and—by dancing—they share the information on new discovered food sources to the onlooker bees. Food sources with higher quality (nectar amount) have a large chance to be selected by onlooker bees, to go there. Once there, they look around for good neighbors, evaluating nectar amounts.

Some employed bees abandon exhausted food sources and become scouts in search of new sources.

In the ABC algorithm, the first half of the bee swarm consists of employed bees, the other half being onlookers. The number of bees is denoted as SN. The algorithm

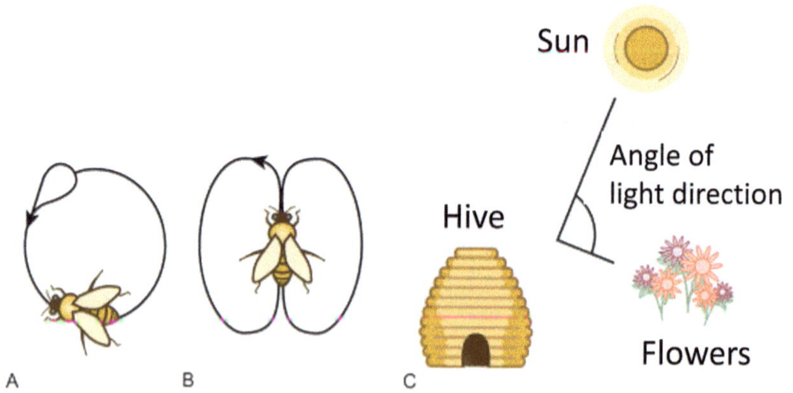

Fig. 12.1 Dance of bees: (**a**) round dance, (**b**) waggle dance

begins with the generation at random of SN solutions (food locations, in a space of D dimensions). D corresponds to the number of parameters to be optimized.

Denote as X_i the position of food source number i. This is one of the candidate solutions. The coordinates of X_i would be $(x_{i,1}, x_{i,2}, \ldots, x_{i,D})$. These coordinates correspond to parameters to be optimized.

When the ABC algorithm starts, it generates at random a population of food sources, according to:

$$x_{i,j} = l_j + rand(0, 1) * (u_j - l_j)$$

where l_j is the lowest value that the *j-th* coordinate can have. And u_j is the highest value *j-th* coordinate can have.

Then, the ABC algorithm repeats three phases:

- *Employed bees phase:*

Employed bees keep in memory a known food source and search in the vicinity for new food sources having more nectar. If they find a better source, it memorizes the new position and forgets the old one.

The search could be guided by the following expression:

$$v_{i,j} = x_{i,j} + \Phi_{i,j}(x_{i,j} - x_{k,j})$$

where coordinate $x_{i,j}$ is changed to $v_{i,j}$; and $\Phi_{i,j}$ is a random number between [-1, 1]. The index k is also randomly chosen, but it must be different from i.

- *Onlooker bees phase:*

Employed bees give food source information to onlooker bees waiting in the hive. Then onlookers choose a food source depending on the probability value associated with that source, p_i:

$$p_i = \frac{fit_i}{\sum_{n=1}^{SN} fit_n}$$

where fit_i is the fitness value of the *i-th* solution.

Once an onlooker comes to the chosen food source, it searches around like the employed bees for a better source.

The probabilistic mechanism tends to recruit more onlookers toward richer sources; which means a positive reinforcement.

- *Scout bees phase:*

If a position cannot be improved over a predefined number of iterations, then the food source is abandoned. Scout bees find replacements, using the same formula that was employed for initial generation of solutions.

Let us choose as example the same objective function considered in previous sections (the function similar to the Rastrigin benchmark). Figure 12.2 offers a 3D

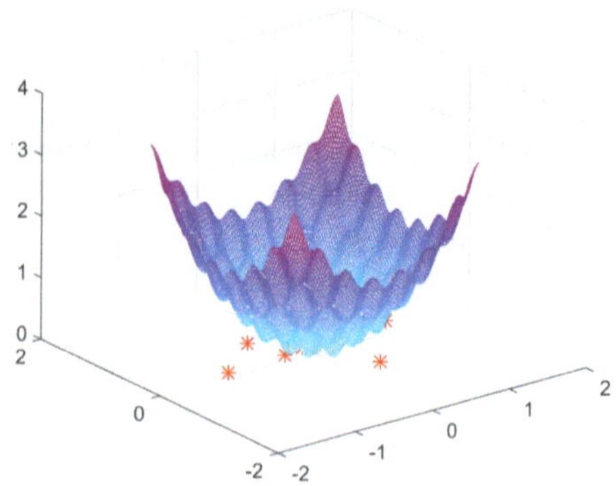

Fig. 12.2 Example of objective function. Initial ABC source population

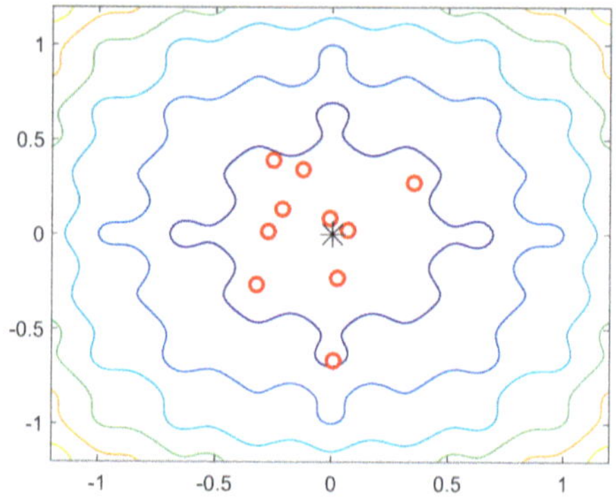

Fig. 12.3 Final result of ABC

view of the function, adding an initial population of solutions (a set of asterisks) as generated by the ABC algorithm.

Figure 12.3 presents—using function contours—the result of the ABC algorithm, with final source candidates. The best candidate marked in black. This candidate is the final solution determined by the algorithm; solution that is quite close to the exact one (0,0). Computation was very fast.

Following, with some changes, the MATLAB implementation example given in [38], a simple version has been devised -program 12.1- for the objective function we selected.

Program 12.1 Artificial Bee Colony example

```
%Artificial Bee Colony example
%one auxiliary function is employed
%algorithm parameters
SN=10; %population size
niter=50; %number of iterations
flim=3; %maximum number of failures
%objective function limits
l=[-1.2 -1.2]; %lower limits
u=[1.2,1.2]; %upper limits
%preparation of vectors
```

```
D=length(u); %dimension of the problem space
OFV=zeros(SN,1); %objective function values
fit=zeros(SN,1); %fitness values
trial=zeros(SN,1); %trial vector
%generate initial population
PP=repmat(l,SN,1)+ repmat((u-l),SN,1).*rand(SN,D);
%display of objective function,
%and initial population
[vx1,vx2]=meshgrid(-1.2:0.02:1.2, -1.2:0.02:1.2);
y=0.2+(vx1.^2)+(vx2.^2)-(0.1*cos(6*pi*vx1))- ...
 (0.1*cos(6*pi*vx2));
figure(1)
colormap('cool')
mesh(vx1,vx2,y); hold on;
plot(PP(:,1),PP(:,2),'r*');
%evaluate initial population
for i=1:SN
  x1=PP(i,1); x2=PP(i,2);
  OFV(i)=0.2+(x1^2)+(x2^2)-(0.1*cos(6*pi*x1))- ...
  (0.1*cos(6*pi*x2));
  aux=OFV(i);
  if aux>=0
    fit(i)=1/(1+aux);
  else
    fit(i)=1+abs(aux);
  end
end
%memorize best initial solution,
%and corresponding objective function value
[best_obj,ix]= min(OFV);
best_loc= PP(ix,:);
%start the algorithm--------------------------------
for g = 1: niter
  %employed bee phase ------
  %uses a function
  for jj=1:SN
    [trial,PP,fit,OFV]=genNewsolution(l,u,SN,jj,...
     PP,fit,trial,OFV,D);
  end
  %onlooker bee phase ------
  %vector of probabilities:
  pblty = 0.9 * (fit/max(fit)) + 0.1;
  mm = 0; nn = 1;
  while (mm<SN)
    if (rand(1) < pblty(nn))
      [trial,PP,fit,OFV]=genNewsolution(l,u,SN,nn,...
       PP,fit,trial,OFV,D);
      mm = mm+1;
    end
    nn = 1+ mod(nn,SN);
  end
  [best_obj,ix]= min([OFV;best_obj]);
  combined_sol = [PP;best_loc];
  best_loc= combined_sol(ix,:);
```

```
    %scout bee phase ------
    [val,ix] = max(trial);
    if (val> flim)
      trial(ix) = 0;
      %generate random solution:
      PP(ix,:) = l + (u-l).*rand(1,D);
      %evaluate new solution
      x1=PP(ix,1); x2=PP(ix,2);
      OFV(ix)=0.2+(x1^2)+(x2^2)-(0.1*cos(6*pi*x1))-...
      (0.1*cos(6*pi*x2));
      aux=OFV(ix);
      if aux>=0
        fit(ix)=1/(1+aux);
      else
        fit(ix)=1+abs(aux);
      end
    end
end
[best_fitness,ix] = min([OFV;best_obj]);
combined_sol = [PP;best_loc];
best_loc= combined_sol(ix,:);
best_loc
best_fitness
figure(2)
contour(vx1,vx2,y); hold on;
plot(PP(:,1),PP(:,2),'ro');
plot(best_loc(1), best_loc(2), 'k*',...
 'MarkerSize', 14);
%Function
%generate new solution
function [trial,PP,fit,OFV]=genNewsolution(l,u,SN,...
n,PP,fit,trial,OFV,D)
%select the variable to change:
k = 1+round((D-1)*rand(1));
q = 1+round((SN-1)*rand(1)); %select a neighbour
%neighbour should not be the current solution
while (q==n)
  q = 1+round((SN-1)*rand(1));
end
Xnew = PP(n,:); %reserve for new solution
%random number between -1 and +1:
Phi = -1 + (2*rand(1));
%new solution:
Xnew(k) = PP(n,k) + Phi*(PP(n,k)- PP(q,k));
%put new solution inside limits
Xnew(k)=min(Xnew(k),u(k));
Xnew(k)=max(Xnew(k),l(k));
%value of objective function, and fitness,
%for new variable
x1=Xnew(1); x2=Xnew(2);
obv=0.2+(x1^2)+(x2^2)-(0.1*cos(6*pi*x1)) -...
 (0.1*cos(6*pi*x2));
if obv>=0
  fitNewsol=1/(1+obv);
else
```

```
  fitNewsol=1+abs(obv);
end
if (fitNewsol > fit(n)) %updating after new solution
  PP(n,:) = Xnew;
  fit(n) = fitNewsol;
  OFV(n) = obv;
  trial(n) = 0;
else
  trial(n) = trial(n) + 1;
end
```

A number of improvements and versions of the Artificial Bee Colony algorithm have been proposed, like [24, 52]. An interesting survey of variants and applications is [34].

The MathWorks MATLAB File Exchange web site offers several implementations of the ABC algorithm. Likewise, GitHub contains 33 repositories of software, related to artificial-bee-colony (of special interest is SWARMLIB, which attacks most of the bio-inspired methods considered in this chapter).

There are frequent publications on the medical use of the Artificial Bee Colony algorithm, like is the case of [25] and [26] for brain tumor detection (image processing).

12.3 Firefly Algorithm

Fireflies are winged beetles or insects that produce light and blinking at night. The flash light is used for several purposes, like attracting preys, or for communication with other fireflies (Fig. 12.4).

The Firefly Algorithm (FA), proposed in year 2008, is based on flashing patterns and behavior of fireflies [53]. The algorithm considers three idealized rules:

- Fireflies are unisex, so that one firefly will be attracted to other fireflies regardless of their sex.

Fig. 12.4 The flash light of a firefly

- The attractiveness is proportional to the brightness, and they both decrease as their distance increases.

Therefore, for any two fireflies, the less bright one will move toward the brighter one.

If there is no brighter one that a particular firefly, it will move at random

- The brightness of a firefly corresponds to the value of the objective function (at the position of the firefly)

The variation of attractiveness β with distance r, is governed by:

$$\beta = \beta_0 \cdot \exp(-\gamma\, r^2)$$

where β_0 is the attractiveness at $r = 0$.

The *i-th* firefly will move to another brighter *j-th* firefly according to:

$$x_i(g+1) = x_i(g) + \beta_0 \cdot \exp(-\gamma\, r^2) \cdot (x_j(g) - x_i(g)) + \alpha_g \cdot \varepsilon_i(g)$$

The second term on the right-hand side of the equation corresponds to attraction, and the third are random numbers. Variable g is the generation number.

The parameter α can be modulated along generations:

$$\alpha_g = \alpha_0 \cdot \delta(g)\ with \quad 0 < \delta < 1$$

The authors of the algorithm recommend δ to be between 0.95 and 0.97.

Taking the elements just introduced, the procedure corresponding to FA can be summarized as follows:

```
Generate at random a population of m fireflies
Evaluate f(x[i]), the value of the objective function (the brightness),
for each firefly
Repeat:
      for i=1 to m
         for j=1 to m
            if (f(x[j] > f(x[i]) then:
               Calculate distance r(j,i)
               Update attractiveness βi = β0 exp(−γr(j, i)^2)
               Generate random vector u[i]
               Evaluate new solution x[i] = x[i] + βi (x[j]-x[i]) + u[i]
            end
         end
      end
```

(continued)

```
    Generate random vector u[k]
    Best firefly should move at random: x[k] = x[k] + u[k]
until a final condition was met
```

Many successful applications of the firefly algorithm have been reported (see [61]), like digital image compression, design of mechanical structures (e.g., antennas), task scheduling, dispatch problems, etc. It has been noticed in comparative studies (like again in [61]) that the firefly algorithm is more efficient than genetic algorithms, requiring much less computational effort.

An implementation of the algorithm is available from the MATLAB File Exchange site, [57]. In addition, GitHub offers 38 repositories of software and documentation on firefly algorithm.

12.4 Glowworm Swarm Optimization Algorithm

Glowworms are firefly larvae, capable of light emission, which serves for several purposes. Inspired by the behavior of these worms, K.N. Krishnanand and D. Ghose proposed the Glowworm Swarm Optimization (GSO) algorithm, in 2005 [36]. The algorithm was originally created for collective robotic studies: it is supposed that each worm has a sensor of light with a certain radial range.

In the GSO algorithm, glow intensity corresponds to the value of the objective function at the worm location. Initially, a population of worms is generated at random locations covering the area of interest. Glowworms of brighter intensities attract other glowworms with lower intensities. The decision to move discard distant glowworms when there is a sufficient number of neighbors, or when they are out of sensor range.

Light emission is due to a substance called *luciferin*. In GSO, initially, all glowworms have the same quantity of luciferin. The iterative procedure of the algorithm includes three phases:

- Luciferin update. This is done with:

$$L_i(g) = (1 - \rho) L_i(g - 1) + \gamma \cdot J(x_i(g))$$

where g is generation number, $L_i(g)$ is the luciferin quantity belonging to *i-th* worm, parameter ρ corresponds to decay of luciferin quantity, parameter γ is the luciferin enhancement constant, and $J(x_i(g))$ is the value of the objective function at the worm location.

- Movement, using a probabilistic transition rule. The probability of movement toward a neighbor is given by:

$$p_{jz}(g) = \frac{L_z(g) - L_j(g)}{\sum_{k \in N_j(g)} L_k(g) - L_j(g)}$$

where Nj is the set of neighbors of *j-th* worm.

The discrete model of worm motion is given by:

$$x_j(g+1) = x_j(g) + s\,\frac{x_z(g) - x_j(g)}{Dist(jz)}$$

where s is the step size

- Neighborhood range update rule.

$$rd_j(g) = \min\{rs,\ \max[0,\ rd_j(g-1) + \beta(nt - |N_j(g-1)|)]\}$$

where $rd_j(g-1)$ is the range of the radial sensor.

Based on these elements, the procedure corresponding to GSO can be summarized as follows:

```
Deploy n glowworms at random
Repeat:
    Update luciferin for each i-th glowworm
    Move each i-th glowworm:
    for j=1 to n
       Compute probability of movement p_ji
       Use the discrete motion model to compute x_i(g+1)
       Update neighborhood range rd_i
    end
until a final condition was met
```

A consequence of considering sensor range limits is that glowworms can split into subgroups, allowing for the identification of several peaks in multimodal optimization. This would be an advantage compared to FA.

Several variants of the GSO algorithms have been proposed, being applied to robot collectives, image processing, dispatch problems, etc. (see the reviews [32, 63]. More information can be found in [37].

The GitHub page [13] offers an implementation of GSO in MATLAB.

Note that another algorithm, called *Galactic Swarm Optimization*, exist, with the same acronym GSO. The MATLAB File Exchange site includes an implementation of this algorithm. A Google search on the galactic algorithm finds more than 1 million responses.

Other bio-inspired algorithms based on invertebrates are the *Monarch butterfly optimization algorithm* and the *moth search algorithm* (both already cited); the *ant lion optimizer*, introduced in 2015 by S. Mirjalili [40], with more than 2300 cites. The *Dragonfly algorithm*, also by Mirjalili in 2016, [41] (more than 1900 cites). In addition, the *grasshopper optimization algorithm*, by S. Saremi et al, in 2017 [45], with 1800 cites. The *fruit fly optimization algorithm*, introduced in 2012 by W.T. Pan [47], with 1600 cites. And several algorithms inspired by butterflies.

12.5 Krill Herd Algorithm

Krills are quite important for maintaining life in the world. They are small crustaceans being found in all oceans. They feed on phytoplankton and zooplankton and in turn serve as main food for many larger animals. The Antarctic krill in the Southern Ocean makes up an estimated biomass of 379,000,000 tones.

Figure 12.5 presents a photograph of a few krill individuals (image taken from the web).

The algorithm we introduce now is inspired by the krill herding behavior for getting their food. This algorithm was proposed in 2012 by Gandomi and Alavi [23], as a meta-heuristic swarm intelligence optimization method. The authors took note of observations from Hofmann et al. in 2004 [29] about krill motion, which seems to result from three factors: movements induced by other krill individuals, foraging activity, and physical diffusion. A Lagrangian model of the change in individual krill position was formulated [29] as follows:

$$\frac{d X_i}{dt} = N_i + F_i + D_i$$

The expression refers to the *i-th* krill individual and the terms on the right-hand side are the three motion factors just mentioned.

Fig. 12.5 Krills (food for many sea inhabitants)

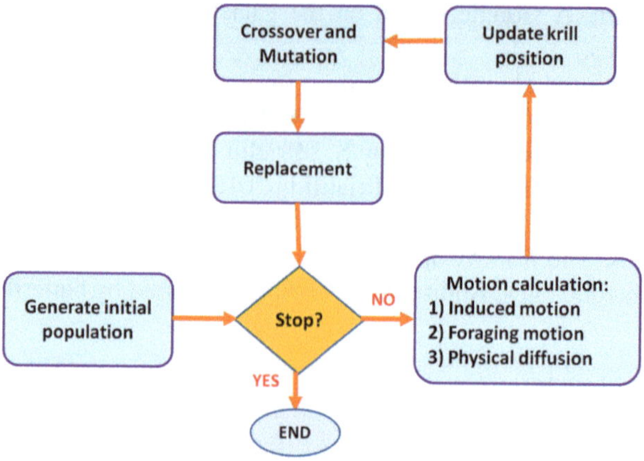

Fig. 12.6 Block diagram of the algorithm

Figure 12.6 provides a block diagram that visually describes the main steps of the krill herd (KH) algorithm. Notice that crossover and mutation, clearly related to genetic algorithms, are included. Krill population search for food, which position is the optimal point of interest for our practical purposes.

Now, let us focus on mathematical models for the three factors, as introduced in [23].

(a) Motion induced by other krill individuals

For the *i-th* krill individual, the induced motion could be described with:

$$N_i^{new} = N^{\max} \alpha_i + \omega_n N_i^{old}$$

and,

$$\alpha_i = \alpha_i^{local} + \alpha_i^{target}$$

where $N^{\max}$ is the maximum induced speed (around 0.01 m/s), α_i is the direction of motion, ω_n is the inertia weight. Also, N_i^{old} is the last induced motion, α_i^{local} is the local effect due to the neighbors, and α_i^{target} is the target direction effect (given by the best krill individual).

The effect of neighbors in the motion of an individual could be estimated as:

$$\alpha_i^{local} = \sum_{j=1}^{NN} \hat{K}_{ij} \cdot \hat{X}_{ij}$$

with:

$$\hat{X}_{ij} = \frac{X_j - X_i}{\|X_j - X_i\| + \varepsilon}$$

$$\hat{K}_{ij} = \frac{K_i - K_j}{K^{worst} - K^{best}}$$

where K^{worst} and K^{best} are the worst and the best fitness values of the krill individuals, while K_i is the fitness value of the *i-th* krill individual, K_j is the fitness of the *j-th* neighbor, and $j \in \{1, 2, \ldots, NN\}$. Vector X corresponds to the position of krill individuals. The parameter ε is quite small (e.g., 10^{-6}).

A total number N of krill individuals is considered. Taking this population, a sensing distance (a constant) is defined as follows:

$$d_{s,i} = \frac{1}{5N}\sum_{j=1}^{N}\|X_i - X_j\|$$

where $i \in \{1, 2, \ldots, N\}$.

If the distance of any two krill individuals is less than the sensing distance, they are neighbors (the *i-th* individual has NN neighbors).

The effect of the krill individual with best fitness, into the *i-th* individual is described with,

$$\alpha_i^{target} = C^{best} \cdot \hat{K}_{i,best} \cdot \hat{X}_{i,best}$$

where C^{best} is an empirical coefficient given by:

$$C^{best} = 2\left(rand + \frac{I}{I_{\max}}\right)$$

where I and $I_{\max}$ are the actual and maximum number of iterations.

(b) Foraging motion

There are two main terms of relevance, the food attraction and the previous krill experience:

$$F_i = V_f\,\beta_i + \omega_f\,F_i^{old}$$

and,

$$\beta_i = \beta_i^{food} + \beta_i^{best}$$

where V_f is the foraging speed, ω_f is the inertia weight. Also, F_i^{old} is the last foraging motion, β_i^{food} is the food attraction, and β_i^{best} is the effect of the best fitness of the *i-th* individual during its history. Foraging speed is taken as 0.02 m/s.

The center of food in each iteration can be computed as a center of mass:

$$X^{food} = \frac{\sum_{i=1}^{N} \frac{1}{K_i} X_i}{\sum_{i=1}^{N} \frac{1}{K_i}}$$

And the food attraction for the *i-th* individual could be described with,

$$\beta_i^{food} = C^{food} \cdot \hat{K}_{i,food} \cdot \hat{X}_{i,food}$$

where C^{food} is empirically defined as:

$$C^{food} = 2\left(1 - \frac{I}{I_{\max}}\right)$$

The effect of the best fitness of the *i-th* individual during the history is defined as:

$$\beta_i^{best} = \hat{K}_{i,best} \cdot \hat{X}_{i,best}$$

where $K_{i,best}$ is the best previously visited position by the *i-th* individual.

(c) Physical diffusion

This is a random process that could be modeled as:

$$D_i = D^{\max}\left(1 - \frac{I}{I_{\max}}\right)\delta$$

where $D^{\max}$ is the maximum diffusion speed and δ is the uniform random directional vector between -1 and 1.

Coming back to the Lagrangian model, the three phenomena just described in mathematical terms can be added for computing dX_i/dt, and therefore, we can express the motion process of the KH algorithm as follows:

$$X_i(t + \Delta t) = X_i(t) + \Delta t \frac{dX_i}{dt}$$

where Δt can be obtained from:

$$\Delta t = C_t \sum_{j=1}^{NV} (UB_j - LB_j)$$

where NV is the number of variables, UB_j and LB_j are upper and lower bounds of *j-th* variables, and C_t is a constant (it could be 0.5).

For crossover, the *m-th* component of X_i is exchanged as follows:

$$x_{i,m} = \begin{cases} x_{r,m} \; rand_{i,m} < Cr \\ x_{i,m} \; else \end{cases}$$

with,

$$Cr_i = 0.2 \, \hat{K}_{i,best}$$

where Cr_i is the crossover probability of the *i-th* krill individual.

Another mutation probability (with a value around 0.7) is applied to components of the position of krill individuals.

Program 12.2 is based, with minor modifications, in the MATLAB implementation of the Krill Herd algorithm given in [2], available from the MathWorks File Exchange site. Notice that the program does not include mutation.

Note (Program 12.2)

The complete listing of the program 12.2, about Krill Herd algorithm, has been included in the Appendix for Long Programs.

Notice that, for capability demonstration purposes, the program considers an optimization search space of ten dimensions. The objective function we selected is the Ackley function for ten variables. Figure 12.7 shows the evolution of the best result along iterations.

According to [8], the krill herd algorithm is a powerful search technique, since it contains both exploration and exploitation strategies. It also enjoys simplicity, flexibility, and computational advantages. Several variants and hybridizations have been proposed (see the review [50]), having attracted a lot of interest.

The optimization algorithm has been applied to electric and power systems, wireless and network systems, neural network training, document clustering, job-shop scheduling, etc.

Fig. 12.7 Evolution of best result

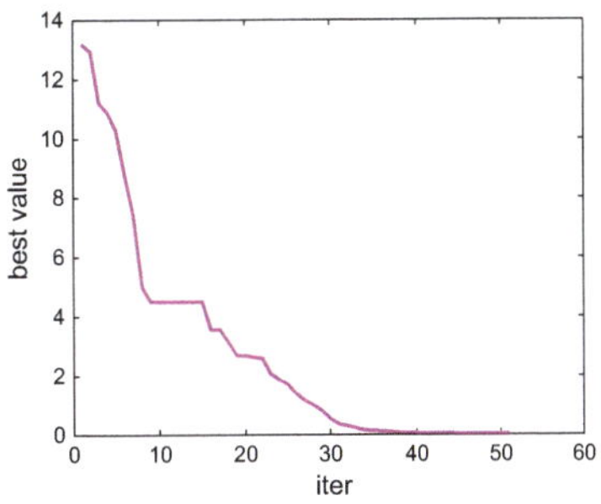

A number of other optimization algorithms, based on the aquatic medium, have been introduced. For example, *fish swarm optimization* proposed in 2002 by Li et al. [46], or *dolphin echolocation* proposed by Kaveh and Farhoudi in 2013 [35]. In addition, there are at least two algorithms named as *dolphin swarm optimization*, [62], or [51].

The *whale optimization algorithm*, proposed by Mirjalili and Lewis in 2016 [43], is now quite popular (6.649 citations in scientific publications). Another more recent (2017) proposal from S. Mirjalili is the *salp swarm algorithm* [42].

Other algorithms, not so popular, are the *sailfish optimizer*, the *shark smell optimization algorithm*, the *multi-objective seagull optimization algorithm*, the *jellyfish search algorithm*, the *coral reef optimization algorithm*, etc.

12.6 Shuffled Frog Leaping Algorithm

As cited in [3], a famous biologist, naturalist, E.O. Wilson, wrote:

"In theory, at least, individual members of the school (the author refers to fish schools) can profit from discoveries and previous experience of all other members of the school during the search for food. This advantage can become decisive, outweighting the disadvantages of competition for food items, whenever the resource is unpredictably distributed in patches."

It is so suggested that social sharing of information offers an evolutionary advantage. This is the basis of a number of bio-inspired algorithms providing the share of information, like in the case of ant colonies and also in the shuffled frog leaping algorithm (SFLA).

The SFLA algorithm is based on evolution of memes carried by the interactive individuals and a global exchange of information. It combines the benefits of local search by PSO and the idea of mixing information from parallel local searches.

Figure 12.8 represents the jumping of a frog that possibly is abandoning a certain local entourage, to visit another place of interest.

Fig. 12.8 Frog jumping

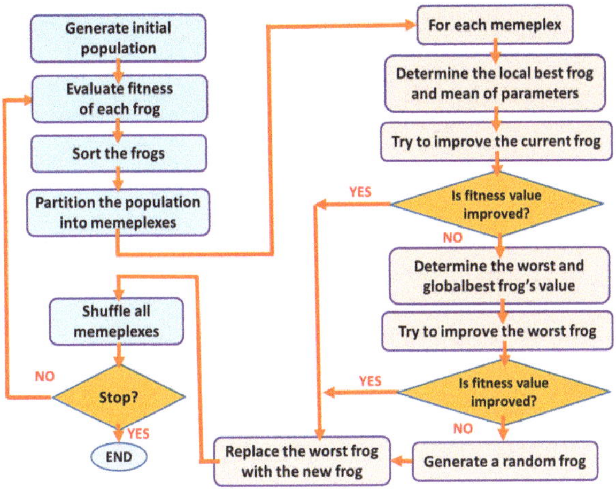

Fig. 12.9 Flowchart of the SFLA algorithm

In SFLA, frogs (solutions) participate in a memetic evolution. In fact, individual frogs are not so important. They are hosts for memes (SFLA uses a memetic vector).

The population of frogs is partitioned into sub-sets, which are called "*memplexes*." The different memeplexes are considered as different cultures of frogs. Within each memeplex, the individual frogs hold ideas, which can be influenced by the ideas of other frogs, being subject to a memetic evolution. After a defined number of evolution steps, ideas are passed among memeplexes in a shuffling process (memeplexes are further reorganized).

The SFLA is a meta-heuristic introduced in 2006 by Eusuff and Lansey [19]. The algorithm conducts an iterative global search, which includes local searching in each memeplex. Figure 12.9 depicts a flowchart of the algorithm. The blocks on the right-hand side correspond to the local searching.

More details of the SFLA algorithm can be found in [3, 12].

Program 12.2 provides an implementation of SFLA, based with some modifications on the code offered by [21] in the MathWorks File Exchange site.

Program 12.2 Example of shuffled frog leaping algorithm

```
%Simple coding of shuffled frog leaping algorithm
m=4; % Number of Memeplexes
n=4;% Number of frogs in each memeplex
F=m*n; % Number of frogs
Nvar=10; % Number of variables
MaxIter=1000; % Maximum number of Iteration
%for cost tracking along iterations:
bestc=zeros(MaxIter,1);
w=1;
C1=1; C2=2;
% initial population of frogs
population=cell(1,F);
```

```
for i=1:F
  population{i}=frogmaker(Nvar);
end
% optimization loop:
for iter=1:MaxIter
  for i=1:F
    cost(i)=objfcn(population{i});
  end
  [vals,idxs]=sort(cost);
  best_cost=vals(1);
  Xg=population{idxs(1)}; %best global frog
  bestc(iter)=best_cost; %for plotting
  %sort population
  aux=prod(size(population));
  for nn=1:aux
    U2{nn}=population{idxs(nn)};
  end
  population=U2;
  % making memeplexes:
  for i=1:m
    X={};
    for j=(i-1)*n+1:i*n
      X=[X,population{j}];
    end
    Xb=X{1}; % best local
    Xw=X{end}; % Worst local
    aux=prod(size(X));
    for j=2:aux-1
      X2=w*X{j}+C1*rand*(Xb-Xw);
      if objfcn(X2)<objfcn(X{j})
        X{j}=X2;
      else
        X2=w*X{j}+C2*rand*(Xg-Xw);
        if objfcn(X2)<objfcn(X{j})
          X{j}=X2;
        end
      end
    end
    X{end}=frogmaker(Nvar); % Replace the worst frog
    for nx=1:numel(X)
      Xcost(nx)=objfcn(X{nx});
    end
    [nnx,Xidx]=sort(Xcost);
    aux=prod(size(X));
    for nn=1:aux % Sort population
      Y2{nn}=X{Xidx(nn)};
    end
    X=Y2;
    memeplex{i}=X;
  end
  newpop=[];
  for i=1:m
    newpop=[newpop,memeplex{i}];
```

```
  end
  population=newpop;
end
best_solution=Xg
best_cost
figure(1)
ni=1:MaxIter;
plot(ni,bestc(ni),'m');
xlabel('iter'); ylabel('best value');
%Function
%objective function
function cost=objfcn(X)
cost=0;
aux=prod(size(X));
for i=1:aux
cost=cost+X(i)^2;
end
%Function
%make a frog
function frog=frogmaker(NF)
minv=-5;
maxv=5;
frog=minv+(maxv-minv)*rand(1,NF);
```

The objective function considered in the code above is simply the sum of squares, for an optimization problem in ten dimensions. Figure 12.10 shows the evolution of the best result along generations (each time the program is run, other—always convergent—evolutions happen).

The review article [48] presents many applications of SFLA in water distribution systems, job-shop scheduling, distribution networks, drone flight control, knapsack problem, flood control, etc. This same article comments a good number of modifications and improvements of the basic SFLA algorithm.

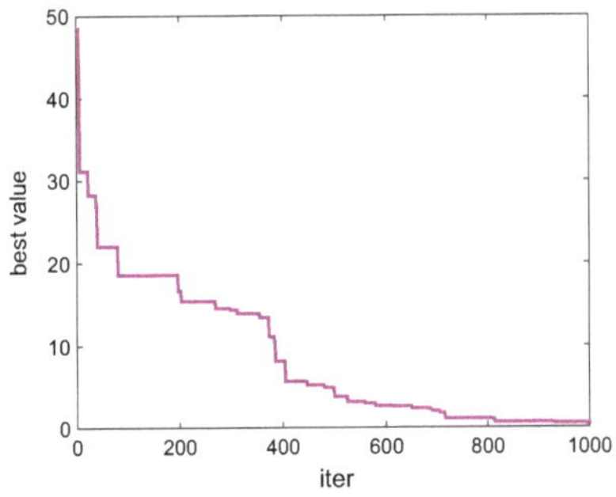

Fig. 12.10 Evolution of best result

12.7 Grey Wolf Algorithm

The grey wolf optimizer (GWO) algorithm is a swarm based meta-heuristic inspired by the hunting process found in grey wolves . Wolves prefer to live in a group (a pack) of 12–15 members on average. There is a social hierarchy, with alpha, beta, delta, and omega members. Grey wolves hunt in groups. The way they hunt corresponds to the hierarchy, as we will detail when describing the algorithm. The main phases of hunting are:

1. Tracking, chasing and approaching the prey
2. Pursuing, encircling and harassing the prey until it stops moving
3. Attack the prey

Figure 12.11 shows an illustrative scene.

The GWO algorithm was proposed by Mirjalili et al. in 2014 [44]. In this algorithm, the hunting is guided by the alpha (α), beta (β), and delta (δ) wolves. The omega (ω) wolves follow those other three.

The mathematical description of how the wolves proceed is fairly simple.

(a) Encircling prey

The encircling behavior could be modeled as follows:

$$X(t + 1) = X_p(t) - A \cdot D$$

where all terms are vectors, Xbeing the position of a wolf, X_p the position of the prey, D is distance, and A a coefficient. This distance is computed as:

$$D = \left| C \cdot X_p - X \right|$$

Coefficients A and C were calculated as:

$$A = 2 a\, r_1 - a$$

Fig. 12.11 Wolves hunting a bison

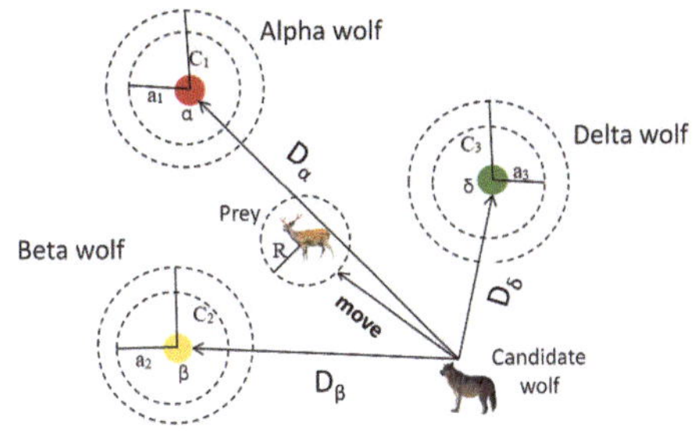

Fig. 12.12 Update of wolf position

$$C = 2\,r_2$$

where the components of vector a are linearly decreased from 2 to 0 over the course of iterations, and r_1, r_2 are random vectors in [0, 1].

(b) Hunting

We suppose that alpha, beta, and delta wolves have better knowledge of the potential location of the prey, so we require the other wolves to update their positions according to the three best search agents. In mathematical terms, this updating would be as follows:

$$D_\alpha = |C_1 \cdot X_\alpha - X|; \quad D_\beta = |C_2 \cdot X_\beta - X|; \quad D_\delta = |C_3 \cdot X_\delta - X|$$

$$X_1 = X_\alpha - A_1 \cdot D_\alpha; \quad X_2 = X_\beta - A_2 \cdot D_\beta; \quad X_3 = X_\delta - A_3 \cdot D_\delta$$

$$X(t+1) = \frac{X_1 + X_2 + X_3}{3}$$

Figure 12.12 depicts how a wolf updates its position, taking into account where alpha, beta and delta wolves are.

One of the items offered by the MathWorks File Exchange site is a collection of optimization programs from [7], which corresponds to some of the algorithms we are introducing. Program 12.3 provides an example of MATLAB implementation of the Grey Wolf Optimizer, based in one of these programs.

Program 12.3 Example of Grey Wolf Optimization

```
%Example of Grey Wolf Optimization
Npar = 2;
VarLow=[-5 -5];
VarHigh = [5 5];
% parameters
WolfSize = 50;
MaxIter = 30;
% initialize a random value as best value
alpha.Position=rand(1,Npar).*(VarHigh-VarLow)+VarLow;
```

```
alpha.Fit = objFit(alpha.Position);
GB=alpha.Fit;
% Reserve memory space
X = repmat(struct('Position',zeros(1,Npar),'Fit',...
zeros(1,Npar)),WolfSize,1);
%Prepare for graphic 2D maps
[xx, yy]= meshgrid(VarLow(1):0.1:VarHigh(1),...
 VarLow(2):0.1:VarHigh(2));
zz= xx.^2 + yy.^2 + (25 * (sin(xx).^2 + sin(yy).^2));
%Initialization-------------------------
X(1) = alpha; % first wolf is alpha
for nn = 2:WolfSize
  %initial position of wolves
  X(nn).Position=rand(1,Npar).*(VarHigh-VarLow)+...
    VarLow;
  % fitness of the solutions
  X(nn).Fit=objFit(X(nn).Position);
  % evaluate alpha
  if (X(nn).Fit < alpha.Fit)
    alpha.Position = X(nn).Position;
  alpha.Fit = X(nn).Fit;
  end
end
%plot of initial position of wolves
figure(1)
contour(xx,yy,zz,20); hold on;
for i=1:WolfSize
  plot(X(i).Position(1), X(i).Position(2),'or');
end;
xlabel('X'); ylabel('Y');
% sort wolves according to fitness
[nnx, sortind] = sort([X.Fit]);
X = X(sortind);
%initialize alpha beta and delta
alpha = X(1); Beta = X(2); delta = X(3);
%initialize A and C vectors
a=2*ones(1,Npar);
A1=2*rand(1,Npar).*a -a;
A2=2*rand(1,Npar).*a -a;
A3=2*rand(1,Npar).*a -a;
C1=2*rand(1,Npar); C2=2*rand(1,Npar);
C3=2*rand(1,Npar);
% Main Loop----------------------------
for jj = 1:MaxIter
  for ii = 1:WolfSize
    %compute encircling vectors
    Da=abs(C1.*alpha.Position-X(ii).Position);
    Db=abs(C2.*Beta.Position-X(ii).Position);
    Dd=abs(C3.*delta.Position-X(ii).Position);
    %update wolf position
    X1=alpha.Position - A1.*Da;
    X2=Beta.Position - A2.*Db;
```

```
    X3=delta.Position - A3.*Dd;
    X(ii).Position = (X1 + X2 + X3)/3;
    %observe constraints
    X(ii).Position=limiter(X(ii).Position,...
     VarHigh,VarLow);
    %update fitness
    X(ii).Fit=objFit(X(ii).Position);
    % update alpha if better
    if X(ii).Fit < alpha.Fit
      alpha = X(ii);
    end
  end
  % sort wolves according to fitness
  [nnx, sortind] = sort([X.Fit]);
  X = X(sortind);
  %update beta and delta
  Beta = X(2);
  delta = X(3);
  %update A and C vectors
  a=2*(1-jj/MaxIter);
  A1=2*rand(1,Npar).*a -a;
  A2=2*rand(1,Npar).*a -a;
  A3=2*rand(1,Npar).*a -a;
  C1=2*rand(1,Npar); C2=2*rand(1,Npar);
  C3=2*rand(1,Npar);
  % store the best value in each iteration
  GB = [GB alpha.Fit];
end
fprintf('The best value is :');
alpha.Position
alpha.Fit
% final position of wolfves
figure(2)
[xx, yy]= meshgrid(VarLow(1):0.1:VarHigh(1),...
 VarLow(2):0.1:VarHigh(2));
zz= xx.^2 + yy.^2 + (25 * (sin(xx).^2 + sin(yy).^2));
contour(xx,yy,zz,20); hold on;
for i=1:WolfSize
  plot(X(i).Position(1), X(i).Position(2),'or');
end;
xlabel('X'); ylabel('Y');
% best result evolution
figure(3)
plot(0:MaxIter,GB,'m');
xlabel('Iter');
ylabel('Objective Function');
%Function
%objective function
function fitness = objFit(x)
fitness= x(1)^2 + x(2)^2 + (25 * (sin(x(1))^2 +...
 sin(x(2))^2));
```

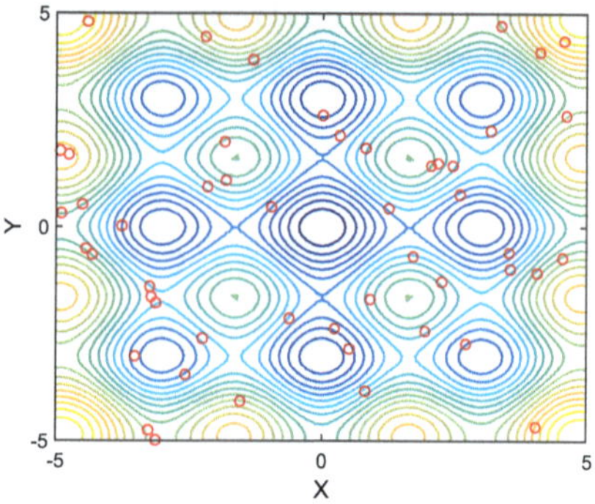

Fig. 12.13 Initial position of wolves

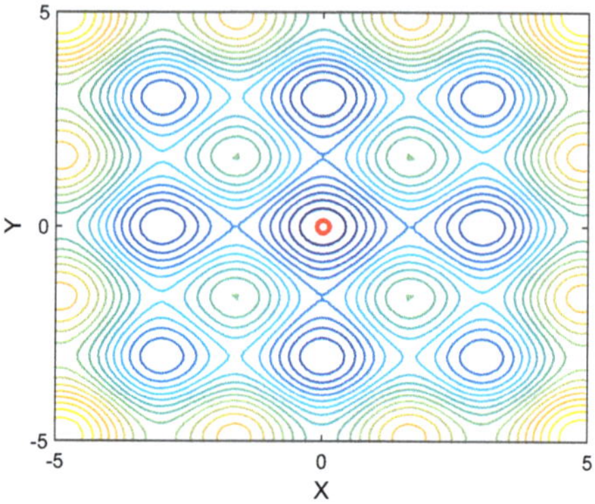

Fig. 12.14 Final position of wolves

We have selected the same objective function that we used in the Particle Swarm optimization.

Every time the program was executed, different results were obtained. Next three figures show for example what happened in one of the runs.

Figure 12.13 visualizes on a contour map of the objective function, the initial position of the 50 wolves. They were covering, at random, the searching space.

After 30 iterations of the algorithm, the wolves moved toward the optimal position. Figure 12.14 shows the result: the pack of wolves thicken in a small space. Our experience with many runs of the program is that sometimes they got not exactly in the optimal position (even, a few times, on a local minimum).

As shown in Fig. 12.15, the evolution of best results along iterations of the algorithm, begins with an abrupt descent. The slope of this descent is controlled by parameter a.

The original article where the Grey Wolf Optimizer was proposed, [44], has attracted a lot of interest, with about 10,000 cites from other publications. Several improvements and hybridizations of the algorithm have been proposed, as reviewed in [20]. Article that also briefly comments many applications, ranging from machine learning and Big Data, to electronic and civil engineering, and medical contexts.

Some other algorithms based on vertebrates running on ground have been introduced. Like those focusing on monkeys; for example, the *spider monkey optimization* proposed in 2014 by J.C. Bansai et al. [6], the *monkey search*

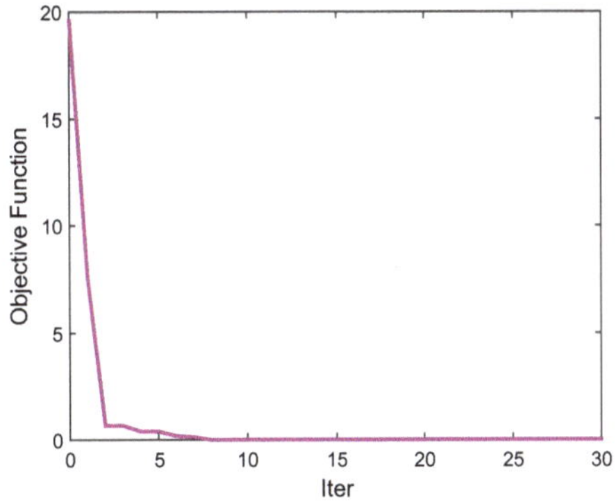

Fig. 12.15 Evolution of best result

metaheuristic, the *climb monkey algorithm*, the *monkey swarm algorithm*, the *monkey King evolution*, the *blue monkey algorithm*, etc. Article [15] gave in 2017 a survey of this approach.

Other swarm-based algorithms are the *cat swarm optimization algorithm* proposed by S-C Chu, et al. [11], in 2006. A review of this algorithm, improvements, and applications is [1] (year 2020).

Recently, in 2021, a *rat swarm optimizer* has been proposed [16].

As mentioned before, the *elephant herding optimization algorithm* has been introduced in 2015. Also, the *spotted hyena optimizer* has been proposed [17] in 2017. With some similarities with the grey wolf algorithm, not for hunting but for protecting, the *border collie optimization* has been introduced by [18] in 2020.

With a more general view, the *group search optimizer* was proposed in 2009 by [27]. It refers to many animal species. Likewise, a *collective animal behavior algorithm* was introduced in 2012, and some optimization algorithms based on *predator-prey* frameworks have been proposed.

Of course, this is not an exhaustive account of algorithms inspired by terrestrial animals.

12.8 Bats

The bat algorithm was introduced in 2012, in the article [60]. This paper has a long part devoted to echolocation of microbats, which is a sort of sonar that these mammals use to detect prey (insects), avoid obstacles, etc. They emit short pulses of sound, in the range 25–150 KHz, so its wavelength is in the range 2–14 mm (same order of prey sizes). Normally, microbats emit about 10–20 sound bursts per second. However, when hunting for prey, this rate increases to about 200 pulses per second.

Figure 12.16, taken from [14] depicts the bio-inspired idea of the bat algorithm. The position of each bat corresponds to a solution in the optimization search space.

Fig. 12.16 Microbats in search of preys

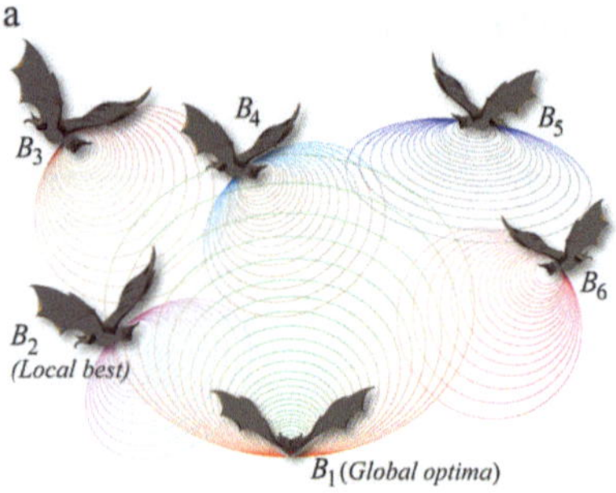

With some idealization of the microbat behavior, the bat algorithm uses the following three rules:

- Bats use echolocation to sense distance, and they also "know" the difference between food/prey and backgrounds barriers
- Bats fly randomly with velocity v_i and position x_i while emitting sound pulses. They can automatically adjust the wavelength of these pulses and the rate of pulse emission r in the range 0-1, depending on the proximity of their target.
- The loudness of sound can vary from a large A_0 to a minimum A_{min}.

The bat algorithm repeats at time steps g, the following updates:

$$f_i = f_{\min} + (f_{\max} - f_{\min}) \cdot \beta$$

$$\mathbf{v}_i(g) = \mathbf{v}_i(g-1) + [\mathbf{x}_i(g) - \mathbf{x}*]\, f_i$$

$$\mathbf{x}_i(g) = \mathbf{x}_i(g-1) + \mathbf{v}_i(g)$$

where β is a random vector drawn from a uniform distribution. The frequencies of sound pulses can vary from *fmin* = *0*, to *fmax* = *100*. The current best solution (position) is denoted as **x***.

After the mentioned updating, some changes in position and parameters may be allowed, as indicated in the bat algorithm pseudo-code:

```
Generate at random a bat population, with certain positions, velocities,
sound frequencies, pulse rates, and loudness.
Repeat:
    Update of frequencies, velocities and positions
    if (rand > r_i)
       Select a solution among the best solutions
       Generate a new solution by random walk around the selected solution
    end
    Generate a new solution by flying randomly
    if(rand <A_i & f(x_i) <f(x*))
```

(continued)

```
        Accept new solution
        Increase r_i and reduce A_i
      end
      Rank the bats and find the current best solution
  Loop end when final condition is met
```

The random walk is done as follows:

$$\mathbf{x}_{new} = \mathbf{x}_{old} + \varepsilon \, A^g$$

where ε is a random number between -1 and 1, and A^g is the average loudness of all bats.

The increase of r_i obeys to:

$$r_j^{g+1} = r_j^0 \, [1 - \exp(-\gamma \, g)]$$

And the loudness decrease is:

$$A_j^{g+1} = \alpha \, A_j^g$$

See the original publication [60] for more details.

Below we include a simple implementation of the bat algorithm ($Program 12.4), following the example given by Yang in the MATLAB File Exchange site [55].

Program 12.4 Bat algorithm example

```
%BAT algorithm
%simple example
Np=20; % Population size, typically 20 to 40
it_max=500; % Maximum number of iterations
A=1; % Initial loudness (constant or decreasing)
% The initial pulse rate (constant or decreasing)
r0=1;
alpha=0.97; % Parameter alpha
gamma=0.1; % Parameter gamma
% Frequency range
Freq_min=0; % Frequency minimum
Freq_max=2; % Frequency maximum
iter=0; % iteration counter
d=10; %dimension of search space
% Initialization of all the arrays
Freq=zeros(Np,1); % Frequency-tuning array
v=zeros(Np,d); % Equivalnet velocities or increments
Lb=-5*ones(1,d); % Lower bounds
```

```
Ub=5*ones(1,d); % Upper bounds
%for fmin tracking along evolution:
vfmin=zeros(1,it_max/50);
nvfx=1; %index for vfmin
% Initialize the population/solutions
for i=1:Np,
  Sol(i,:)=Lb+(Ub-Lb).*rand(1,d);
  Fitness(i)=Fun(Sol(i,:));
end
% Find the best solution of the initial population
[fmin,ix]=min(Fitness);
best=Sol(ix,:);
%main loop --------------------------------
while (iter<it_max)
  % Varying loudness (A) and pulse emission rate (r)
  r=r0*(1-exp(-gamma*iter));
  A=alpha*A;
  % Loop over all bats/solutions
  for i=1:Np,
    Freq(i)=Freq_min+(Freq_max-Freq_min)*rand;
    v(i,:)=v(i,:)+(Sol(i,:)-best)*Freq(i);
    S(i,:)=Sol(i,:)+v(i,:);
    % Check a switching condition
    if rand<r,
      S(i,:)=best+0.1*randn(1,d)*A;
    end
    % apply bounds
    S(i,:)=simplebounds(S(i,:),Lb,Ub);
    % Evaluate new solutions
    Fnew=Fun(S(i,:));
    % If the solution improves or not too loudness
    if ((Fnew<=Fitness(i)) & (rand>A)),
      Sol(i,:)=S(i,:);
      Fitness(i)=Fnew;
    end
    % Update the current best solution
    if Fnew<=fmin,
      best=S(i,:);
      fmin=Fnew;
    end
  end
  iter=iter+1; % Update iteration counter
  % Save the result every 50 iterations
  if ~mod(iter,50),
    vfmin(nvfx)=fmin;
    nvfx=nvfx+1;
  end
end % End of the main loop
% Output the best solution
disp('final result-------------');
fmin
best
```

```
figure(1)
nn=1:(it_max/50);
plot(nn*50, log10(vfmin(nn)),'m');
xlabel('iter'); ylabel('log10(fmin)');
%Function
function s=simplebounds(s,Lb,Ub)
% Apply the lower bound
ns_tmp=s;
I=ns_tmp<Lb;
ns_tmp(I)=Lb(I);
% Apply the upper bounds
J=ns_tmp>Ub;
ns_tmp(J)=Ub(J);
% Update this new move
s=ns_tmp;
%Function
%objective function
function z=Fun(x)
% Optimal solution fmin=0 at (2,2,...,2):
z=sum((x-2).^2);
```

We use again the sphere function (a sum of squares) as the objective function. Program 12.5 considers ten dimensions.

Figure 12.17 depicts the evolution of log10 of the best result, along iterations. Although we let the program run for 500 iterations, the execution is fairly fast.

Applications of the bat algorithm are abundant, like in mechanical designs, combinatorial optimization, scheduling, data clustering, image processing, etc. [54]. See the recent review [64], year 2020, of bat algorithm applications and variants.

12.9 Crow Search

Crows watch other birds, observe where the other birds hide their food, and steal it once the owner leaves. In turn, crows hide the stolen food taking precautions (Fig. 12.18).

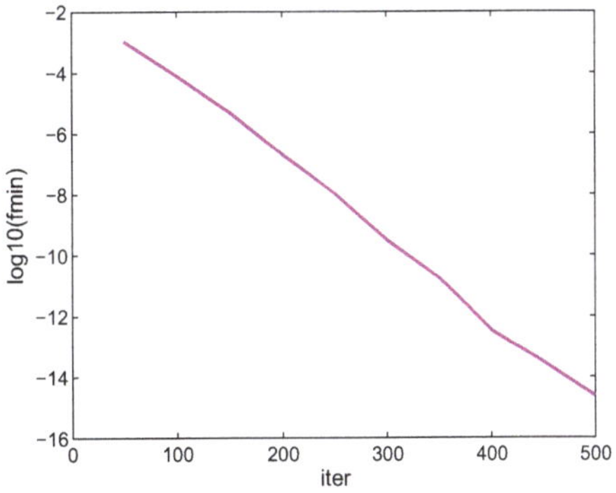

Fig. 12.17 Evolution of the best solution (logarithm) along generations

Fig. 12.18 Crow carrying food

The crow search algorithm was proposed in 2016 by A. Askarzadeh, in the article [5] that has been cited by more than 1500 publications.

The algorithm considers N crows in a d-dimensional space. The position of crows is handled as a matrix:

$$Crows = \begin{bmatrix} x_1^1 \; x_2^1 \; \ldots x_d^1 \\ x_1^2 \; x_2^2 \; \ldots x_d^2 \\ \vdots \;\; \vdots \, \vdots \\ x_1^N \; x_2^N \; \ldots x_d^N \end{bmatrix}$$

Each crow has a memory in which the position of its hiding place is memorized. These places are handled with another matrix:

$$Memory = \begin{bmatrix} m_1^1 \; m_2^1 \; \ldots m_d^1 \\ m_1^2 \; m_2^2 \; \ldots m_d^2 \\ \vdots \;\; \vdots \, \vdots \\ m_1^N \; m_2^N \; \ldots m_d^N \end{bmatrix}$$

Suppose that at a certain iteration of the algorithm, crow j wants to visit his hiding place, and crow i decides to follow crow j, for discovering that hiding place. If crow j does not know that it is being followed, crow i will approach the hiding place. But if crow j detects the danger, then it would deceive the follower by going to any other place. These two possibilities are described as follows:

$$X^{i,iter+1} = \begin{cases} X^{i,iter} + r_i \times fl^{i,iter} \times (M^{j,iter} - X^{i,iter}) r_i \geq AP^{i,iter} \\ a\ random \quad position \quad otherwise \end{cases}$$

where *fl* denotes flight length, r_i is a random number (uniform distribution), and *AP* denotes awareness probability of crow j. Large values of *AP* increase diversification of search, while small values increase intensification.

Figure 12.19 shows the effect of *fl* on the search features. Small values of *fl* lead to local search, near $X^{i,iter}$, while larger values result in global search.

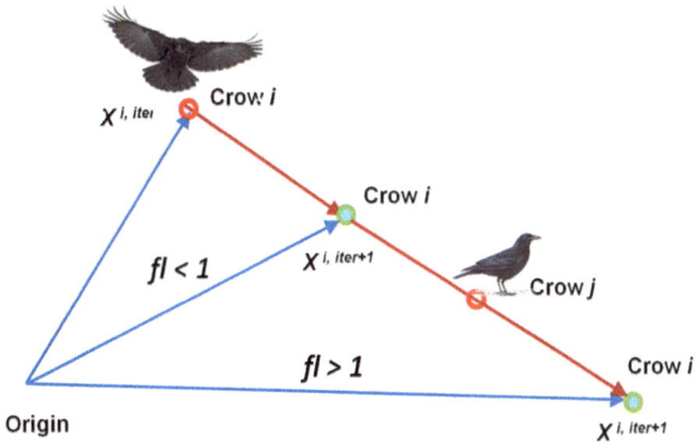

Fig. 12.19 Influence of fl in the process

The crows update their memories as follows:

$$M^{i,iter+1} = \begin{cases} X^{i,iter+1} & if \ \ f(X^{i,iter+1}) \ is \ better \ than \ f(M^{i,iter}) \\ M^{i,iter} & otherwise \end{cases}$$

where $f()$ is the objective function.

The target of the search algorithm is to find places where the objective function is optimal.

The crow search procedure can be summarized as follows:

Generate at random a population of *N* crows (matrix of crow positions)
Initialize *fl* and *AP*
Evaluate the value of the objective function for each crow
Initialize the memory matrix
Repeat a number of iterations:
For i=1 to N
    Randomly choose a crow to be followed
    Give *AP* a value
    if (r_j > AP) then,
        Update position of crow i (formula)
    else:
        Position of crow i = a random position
end For
Check the feasibility of new positions
Evaluate fitness of new positions
Update the memory of crows

Program 12.5 gives an example of MATLAB implementation of the crow search algorithm, based with a few changes on the code of [4], available from the MathWorks File Exchange site.

Program 12.5 Crow search example

```
%Example of Crow Search algorithm
format long;
pd=10; % Problem dimension
N=20; % Flock (population) size
AP=0.1; % Awareness probability
fl=2; % Flight length (fl)
[x l u]=init(N,pd); % initial position of crows
xn=x;
ft=Cfitness(xn,N,pd); % Fitness evaluation
mem=x; % Memory of positions
fit_mem=ft; % Memory of fitness
%main loop-----------------------------
tmax=200; % Maximum number of iterations
for t=1:tmax
  % Generation candidate crows for chasing:
  num=ceil(N*rand(1,N));
  for i=1:N
    if rand>AP
   %Generation of a new position for crow i (state 1)
      xnew(i,:)= x(i,:)+fl*rand*(mem(num(i),:)-...
       x(i,:));
    else
      for j=1:pd
   %Generation of a new position for crow i (state 2)
        xnew(i,j)=l-(l-u)*rand;
      end
    end
  end
  xn=xnew;
  % Fitness evaluation of new solutions:
  ft=Cfitness(xn,N,pd);
  for i=1:N % Update position and memory
    if xnew(i,:)>=l & xnew(i,:)<=u
      x(i,:)=xnew(i,:); % Update position
      if ft(i)<fit_mem(i)
        mem(i,:)=xnew(i,:); % Update memory
        fit_mem(i)=ft(i);
      end
    end
  end
  % Best found value until iteration t:
  ffit(t)=min(fit_mem);
end
ngbest=find(fit_mem== min(fit_mem));
g_best=mem(ngbest(1),:)
%plot of best result evolution
figure(1)
```

```
plot(log10(ffit),'m');
%Function
%initialize position of crows
% Function for initialization:
function [x, l, u]=init(N,pd)
l=-10; u=10; % Lower and upper bounds
%Generation of initial solutions (crows position):
for i=1:N
  for j=1:pd
    % Position of the crows in the space:
    x(i,j)=l-(l-u)*rand;
  end
end
%Function
%Fitness evaluation
% Function for fitness evaluation:
function ft=Cfitness(xn,N,pd)
for i=1:N
  ft(i)=sum(xn(i,:).^2); % Sphere function
end
```

The program considers a search space in ten dimensions. Once again, the sphere function (a sum of squares) has been chosen as objective function.

Figure 12.20 shows the evolution of the log10 of the best solution, along generations.

The code of this algorithm is not complicated, and its execution is fast.

As it is shown in review articles, such [30] or [39], the research has proposed a number of improvements, variants, and hybridizations of the algorithm. These reviews consider also several applications, in power engineering, computation and artificial intelligence, image processing, chemical engineering, water management, wireless networks, etc.

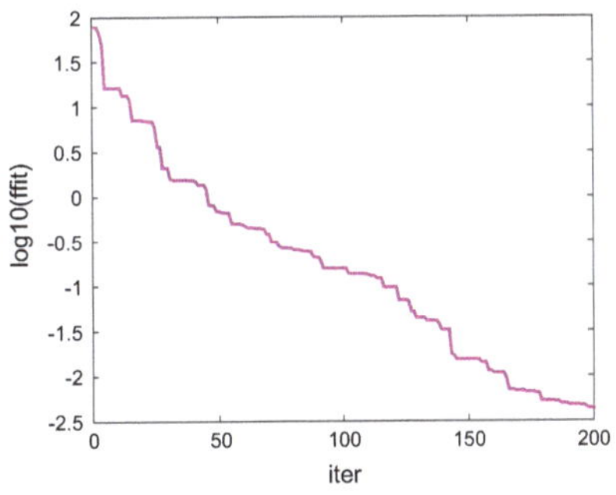

Fig. 12.20 Evolution of the best solution (logarithm) along generations

12.10 Cuckoo Search

The cuckoo search (CS) algorithm is inspired by the brooding parasitism of some cuckoo species. They lay their eggs in the nest of other host birds (often other species). A hidden cuckoo female awaits nearby a target nest, for a chance to lay there its eggs (these females usually lay 16 to 24 eggs). There are cuckoo species that produce eggs similar to those of a few selected host species. If a host discovers alien eggs, it can throw them out or simply abandon the nest.

Cuckoo females prefer nests where eggs from the host are recent. In general, cuckoo eggs hatch slightly earlier than the host eggs. Newborn cuckoo chicks push the host eggs out of the nest and even imitate the calls of host chicks asking for food (Fig. 12.21).

In 2009, X-S. Yang and S. Deb proposed an algorithm for optimization, based on the behavior of cuckoos [58].

The CS algorithm considers a population of nests. Each nest has a set of coordinates in a space of d dimensions. The idealized behavior of cuckoos obeys to three rules:

- Each cuckoo lays one egg at a time, dumping it in a nest chosen at random.
- The best nests with high-quality eggs are kept over to the next generation.
- The egg laid by a cuckoo may be discovered by the host with a certain probability. In case of discovery, the host may get rid of the alien egg, or abandon the nest to build a new one.

Then, each egg in a nest represents a candidate solution, which is evaluated according to an objective function. The cuckoo egg represents a new coming solution, which could replace worse solutions.

The cuckoo exploration is not based on a random walk but instead on Lévy flights [58]. It has been observed that the flight of many creatures use a series of

Fig. 12.21 Cuckoos and parasitic breeding on other's nest

straight intervals and sudden perpendicular turns. Compared to the random walk, Lévy flights allow larger changes from current position in the search space [22].

When the *i-th* cuckoo generates a new solution, a Lévy flight is performed:

$$\mathbf{x}_i(g+1) = \mathbf{x}_i(g) + \eta \otimes levy(\lambda)$$

where $\otimes$ indicates entrywise multiplication, η is the stepsize, g is the generation number, and the Lévy distribution corresponds to:

$$levy(\lambda) \sim g^{-\lambda}$$

The generation of random numbers with Lévy flights includes two steps. First, a random direction is selected, based on a uniform distribution. Second, a step is computed. This could be done in several ways, like, for instance, using the so-called Mantegna algorithm. This algorithm starts by obtaining a factor σ:

$$\sigma = \left(\frac{\Gamma(1+\beta) \cdot \sin((\pi\,\beta)/2)}{\Gamma((1+\beta)/2 \cdot \beta \cdot 2^{(\beta-1)/2})} \right)^{1/\beta}$$

where Γ is the gamma function, and $\beta = 3/2$.

Then, based on a normal random distribution to generate a value r, an auxiliary value u is obtained:

$$u = \sigma\, r$$

Now, a step length is computed with:

$$\xi = \frac{u}{|\nu|^{(1/\beta)}}$$

where ν is a value generated from a normal random distribution, with zero-mean and standard deviation 1.

Finally, a stepsize is computed with:

$$\eta = 0.01\, \xi\, (x - x_{best})$$

Following one of the implementations offered by the MATLAB File Exchange site [56], a program has been written (program 12.7). For this case, which belongs to the atmosphere of birds, we selected again the egg-crate function ([31]) as the objective function.

Figure 12.22 displays in 3D this objective function. It has a global minimum at (0,0) and many other local minima.

An important parameter in program 12.7 is *tol* (tolerance). It sets how far from exact solution you will accept as result. In the example of egg-crate, the obvious exact result concerning the objective function is 0 (at point (0,0)); if you set

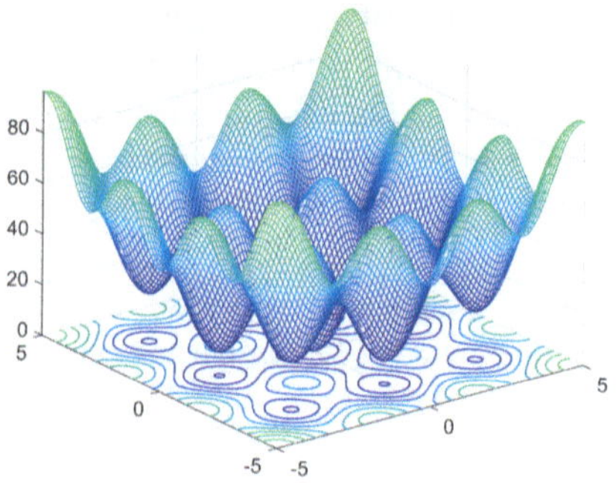

Fig. 12.22 Egg-crate objective function

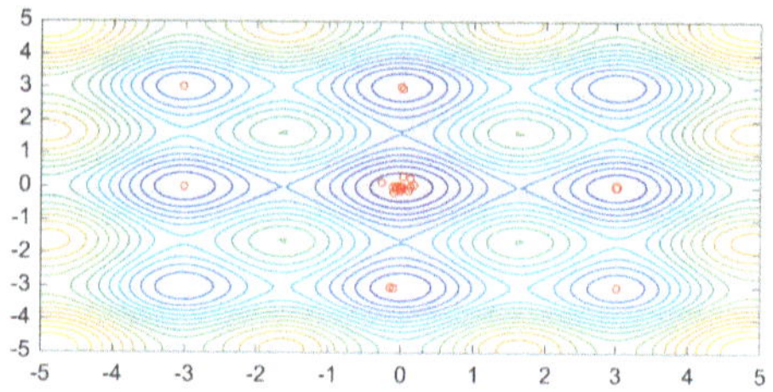

Fig. 12.23 Final set of solutions

tol=0.001, the execution of the program will stop as soon the result of iterations becomes lower than 0.001. Our experience is that this happens after some seconds of computer work.

Note (Program 12.7)

The complete listing of the program 12.7, about Cuckoo Search algorithm, has been included in the Appendix for Long Programs.

The program has been run many times, obtaining in each run a different result. Figure 12.23 presents an example of final result, which clearly shows some crows on local minima; anyway, the algorithm selects the best solution (in this case, the nearest to origin).

Although the example considered in program 12.7 is bi-dimensional, the cuckoo search is usually applied to problems in more dimensions. In fact, the example treated in [56] has 15 dimensions.

The publication in which the Cuckoo search was introduced, [58], compares this method against Genetic Algorithms, for several benchmark problems. In every case, cuckoo search gets satisfactory solutions with much less computational effort.

There are several variants and hybrids of CS, as reviewed in [22] and [49]. Many successful applications of CS have been reported, [59], like path planning of mobile robots, task scheduling in cloud computing, wireless sensor localization, data clustering, neural network training, curve data fitting, etc.

GitHub offers 18 repositories on cuckoo-search, including software and documentation.

We have to introduce a word of caution. There are many publications of cuckoo search (over 7000), so it is clearly very popular. But there are doubts, as expressly analyzed in [10], about the extent this algorithm contributes with new ideas to optimization. It is not only an issue with cuckoo search, the mentioned article does criticize other metaphor-based algorithms.

Other algorithms related to birds are the *Harris hawks optimization algorithm*, introduced in 2019 by A.A. Heidari [28], with more than 4465 cites from other publications (there are several modifications and hybridizations of this algorithm). Also, the *bald eagle search optimization algorithm*, introduced in 2020, and the *bird mating optimizer*, proposed in 2014.

From the side of swarms, we could mention the *bird swarm algorithm*, proposed in 2016, and the *chicken swarm optimization*, introduced in 2014.

12.11 Exercises

1- [12.1] Here is a simple instance of a standard problem, which is to schedule University events to resources such as lecturers, classrooms, and time slots while considering some constraints. For our simple case, we have a data set with the following entries:

#ROOMS: RoomName RoomCapacity RoomType

#COURSES: CourseName NumOfLecturers NumOfLessons NumOfLabs

#LECTURERS: LectureName CourseA CourseB ...

#Student groupS: groupName NumOfStudents CourseA CourseB ...

The data set is the following:

ROOMS	COURSES	LECTURERS	Student groupS
D1 200 0	CALC 2 1 0	SVEN CALC MULT	COMP 1 200 CALC JAVA
D45 40 1	JAVA 1 0 1	BERT JAVA SCON OOPC	COMP 2 120 MULT CTEC
E35 40 1	MULT 2 0 1	KARL CSEC	COMP 3 70 CSEC SCON
SPEL 40 2	CTEC 1 2 0	GUNN CTEC	
SPOR 30 2	CTEC 1 2 0		
	SCON 1 1 1		

A course week has 5 days. The days is divided into four time slots, two hours each. Each course has one to three events per week.

There are three types of events. Lab events have a maximum size of 25 students. Lesson events have a maximum size of 40. Lecture events can afford a complete group.

Student groups form several events for one lesson or lab.

The constraints are as follows:

- All courses need to have their required hours scheduled.
- An invalid timeslot may not be used.
- A student or an instructor cannot be at two places at the same time.
- A room cannot hold more students or instructors than its capacity.
- A room cannot be used for more than one lecture at the same time.
- A lecture with special equipment requirements needs to be scheduled to a room that accommodates those requirements.

The problem is to assign events to resources.

Please find a solution using Artificial Bee Colony (other methods could be explored and compared).

2- [12.2] Recall the materials covered in section on SVM in Chap. 7.

Please, determine with the Firefly algorithm the SVM for the Wisconsin Breast Cancer dataset. Visualize the result. Describe the evolution of the process.

3- [12.3] A network of mobile robots can be used for multiple odor source localization. The 2D effect of those sources can be modeled as follows:

$$J(x, y) = \sum_{i=1}^{Q} a_i \exp(-b_i[(x - x_i)^2 + (y - y_i)^2])$$

where Q represents the number of odor sources and $(x_i, y_i)$the location of each peak. Take the following values for our case:

	1	2	3
a_i	3	2.5	2.5
b_i	0.01	0.04	0.02
x_i	– 10	0	12
y_i	– 10	7	– 5

Suppose that 20 robots are governed by a glowworm algorithm. Suppose the robots are points. There are three odor sources. Run the algorithm and show in simulation the behavior of the robots. Notice that a similar problem is the use of robots for military mine localization and removal.

4- [12.4] This is a simplified example of optimal energy dispatch. There is a power plant with six turbines for electricity generation. The production cost of *i-th* generator is:

$$F_i(P_i) = a_i\, P_i^2 + b_i\, P_i + c_i$$

where P_i is the power generated by the unit.

The production cost should be minimized, provided the electricity demand P_D was satisfied.

Transmission losses, P_L, must be taken into account. Therefore, we have:

$$\sum_{i=1}^{6} P_i = P_D + P_L$$

The transmission losses can be modeled as follows:

$$P_L = \sum_{I=1}^{6} \sum_{J=1}^{6} P_i B_{ij} P_j + \sum_{i=1}^{6} B_{0i} P_i + B_{00}$$

The generators cannot give more than a maximum power limit, $P_i^{\max}$, nor less than a minimum power limit $P_i^{\min}$.

Next table gives the specifications of the generation system:

Unit	$P_i^{\min}$	$P_i^{\max}$	a_i	b_i	c_i
1	100	500	240	7	0.007
2	50	200	200	10	0.0095
3	80	300	220	8.5	0.009
4	50	150	200	11	0.009
5	50	200	220	10.5	0.008
6	50	120	190	12	0.0075

For a simple scenario, the *B* coefficients would be taken as:

$$B_{ij} = 0.002; \;\; B_{0i} = -0.0006; \;\; B_{00} = 0.056$$

(a more complete view would take 36 coefficient values for B_{ij} and 6 for B_{0i})

Please, use the Krill Herd algorithm to solve the problem with adequate production of each generator. Visualize the evolution of the optimization process.

5- [12.5] Water distribution is a critical service for humans and plants. Optimization problems in this context are quite varied. Next, a simple municipal project is considered, as depicted in the following diagram:

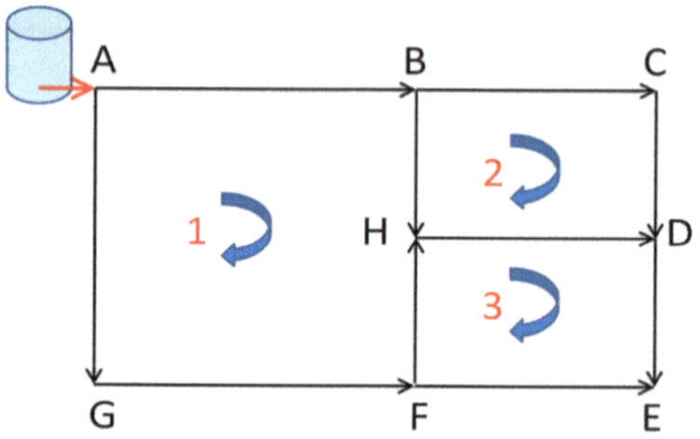

It is desired to minimize the cost of building the system. For simplification, the diameter of the pipe was chosen as the parameter to be modified, in order to carry the water to each node maintaining a pressure above a certain minimum.

Supposing linear conditions, the cost to be minimized is given by:

$$J = \sum_{j=1}^{NP} L_i C_{pi}(D_i) + \sum_{k=1}^{NR} C_{rk}\,(H_k)$$

where the first sum is the cost of pipes, which depends on length and a coefficient associated to pipe diameter, and the second sum is the cost of pressure generators (e.g., a tank with a certain water level).

The pressure at a node must be between a minimum and a maximum. In this example, between 60 and 100.

One can select pipes with 150 mm diameter (cost $40 a meter), or pipes with 200 mm diameter (cost $50 a meter). The cost of the tank is $4000.

Get from Internet reasonable impact of pipe diameter and length on pressure loss. The tank contains water at 40 m height (tank on top of a tower).

Next table gives the lengths (m) of pipes:

Segment	AB	BC	BH	CD	DH	DE	EF	FG	HF	GA
Length	400	300	150	150	300	150	300	400	150	300

The demands (L/s) are the following:

Node	A	B	C	D	E	F	G	H
Demand	− 230	50	20	20	40	30	40	30

Please, use the Suffled Frog Leaping algorithm to find the minimum cost solution. Visualize the evolution of the optimization process.

6- [12.6] This is an exercise related to a famous algorithm for data clustering, called k-means.

Recall from the section on SVM, in Chap. 7, that a dataset concerning the Iris Flower was employed. The data were plotted in a figure, to visualize how a separating line can be determined between two groups (clusters) of data.

The main target of the k-means algorithm is to determine a series of centroids. In the case of the two clusters that we can discern at first sight, we are speaking of two centroids, one for each cluster. The idea is similar to locating the center of gravity of an object. Data points belong to one or another cluster according with their Euclidean distance to the centroids.

Next figure shows an example of four data clusters, which can be clearly identified. Centroids were added (star symbols).

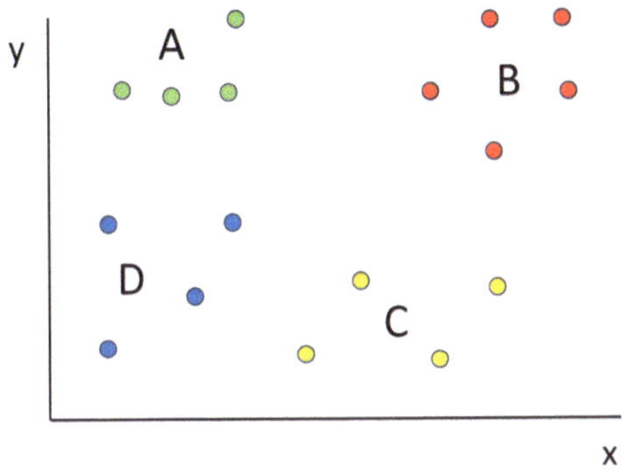

In reality, the Iris Flower dataset considers three species of the flower, which means three clusters (but two of these are not so clearly separated).

Please, use the Grey Wolf algorithm to find three centroids, and so the three clusters of Iris Flower data. Visualize the evolution of the process, adding comments on the data classification quality.

7- [12.7] [12.8] Heuristic optimization algorithms can be also applied for mechanical engineering. An example that is proposed in this exercise is the design of a three-bar truss structure, as described in the next figure:

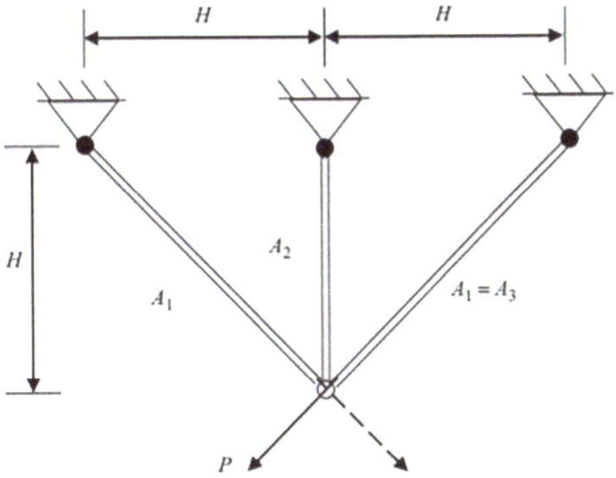

The engineer wants to specify cross sectional areas, $A_1 = A_3 = x_1$ and $A_2 = x_2$ such that the volume of the truss structure is minimized. Therefore, the criterion to be minimized is:

$$J = (2\sqrt{2}x_1 + x_2)\,H$$

There are three constraints:

$$g_1 = \frac{\sqrt{2}x_1 + x_2}{\sqrt{2}x_1^2 + 2x_1x_2}P - \sigma \le 0$$

$$g_2 = \frac{x_2}{\sqrt{2}x_1^2 + 2x_1x_2}P - \sigma \le 0$$

$$g_3 = \frac{1}{x_1 + \sqrt{2}x_2}P - \sigma \le 0$$

where $0 \le x_1 \le 1$ and $0 \le x_2 \le 1$. The parameters are:

H	100 cm
P	2 KN/cm^2
σ	2 KN/cm^2

Please, apply the Bat algorithm to find the optimal solution.

8- [12.9] This is an exercise related to the training of artificial neural networks (ANN), one of the main branches of artificial intelligence. Normally, ANNs are trained by a gradient-based method, but it is also possible to employ heuristics.

Next figure depicts a basic model (diagram) of one neuron:

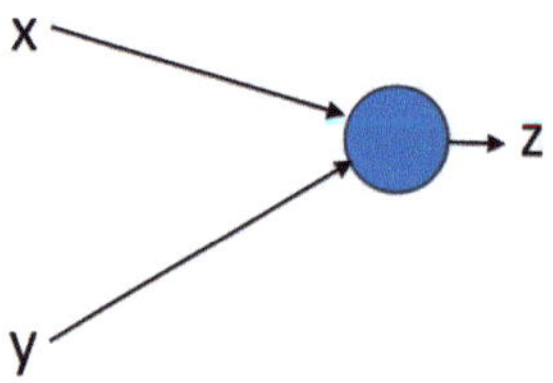

The neuron in this diagram has two inputs, (x, y), and one output z. The output depends on the inputs as follows:

$$z = f\Big(\sum_i \omega_i x_i + b\Big)$$

where $0 \leq \omega_i \leq 1$ are weights, $x_i$are the inputs (x and y in the figure), b is bias, and $f(q)$is the activation function. The research has considered many types of activation function; the most popular is the sigmoid function, which is:

$$f(q) = \frac{1}{1 + \exp(-q)}$$

Neurons can be interconnected, forming several types of structures. Next figure shows a simple case, which will be used for our exercise. There are three layers. We should find the values of weights for the six connections between the first and second layer and another six weights for the connections between second and third layer. Add also the bias for each neuron in the second and third layer. Suppose unity weight and no bias for the input neurons.

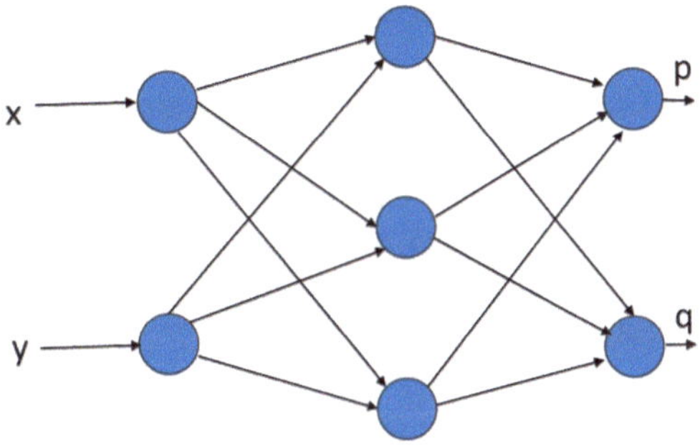

Now, let us use the network for data classification. The next figure shows an example, with four clusters of data.

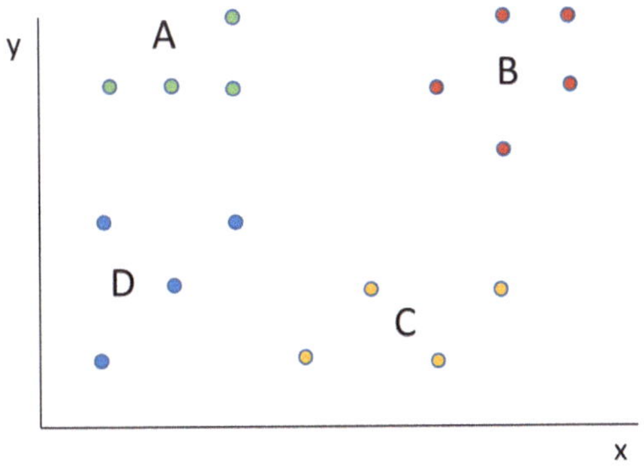

Another way of expressing the data is the following table:

Data points (x,y)	Class
(1,5), (2,5), (3,5), (3,6)	A
(6,5), (7,4), (7,6), (8,6), (8,7)	B
(4,1), (5,2), (6,1), (7,2)	C
(1,1), (1,3), (2,2), (3,3)	D

So, there are 17 cases to be classified by the ANN. The two outputs of the ANN can provide four states (0,0), (0,1), (1,0), and (1,1) corresponding to classes A, B, C, or D. Since real numbers are used for weights and outputs, do not expect exactly 0 or 1, but something like 0.12, 0.87, or so. The idea is to apply the inputs, see the corresponding outputs and modify the weights to get closer to a correct classification (several iterations).

Please, apply the Cuckoo search algorithm for training the ANN. Visualize the evolution of the process. Comment the results.

Once the ANN was trained, you can do the following experiment: take a new datum (a new point in the 2D data figure), and see if the ANN gets a correct classification. This is why ANN are interesting; in principle, they can tackle new (unexpected) classification demands.

12.11.1 Exercise References

[12.1] https://www.diva-portal.org/smash/get/diva2:771121/FULLTEXT01.pdf
[12.2] https://onlinelibrary.wiley.com/doi/pdf/10.1155/2015/212719
[12.3] A Network Robot System for Multiple Odor Source Localization using Glowworm Swarm Optimization Algorithm (ResearchGate).
[12.4] https://www.sciencedirect.com/science/article/pii/S0142061513004742
[12.5] https://www.revistaespacios.com/a20v41n32/a20v41n32p07.pdf
[12.6] https://ictactjournals.in/paper/IJSC_Vol_11_Iss_3_Paper_6_2353_2358.pdf
[12.7] https://staff.fmi.uvt.ro/ daniela.zaharie/am2016/proiecte/tehnici/Bat Algorithm/Bat_algorithm_EngineeringComputations.pdf
[12.8] https://neorl.readthedocs.io/en/latest/examples/ex6.html#:~:text= *The Three bar truss design, for stress (Îf) constraints.*
[12.9] https://www.nature.com/articles/s41598-021-93108-9

References

1. Ahmed, A. M., Rashid, T. A., & Saeed, S. A. M. (2020). Cat swarm optimization algorithm: A survey and performance evaluation. *Computational Intelligence and Neuroscience, 2020*(1), 4854895. Available from: https://onlinelibrary.wiley.com/doi/pdf/10.1155/2020/4854895
2. Amir. (2022). Krill Herd algorithm. Available from: https://es.mathworks.com/matlabcentral/fileexchange/55486-krill-herd-algorithm?s_tid=srchtitle
3. Amiri, B., Fathian, M., & Maroosi, A. (2009). Application of shuffled frog-leaping algorithm on clustering. *The International Journal of Advanced Manufacturing Technology, 45*(1), 199–209. Available from: https://link.springer.com/content/pdf/10.1007/s00170-009-1958-2.pdf
4. Askarsadeh, A. (2016). Crow search algorithm. Available from: https://www.mathworks.com/matlabcentral/fileexchange/56127-crow-search-algorithm?s_tid=ta_fx_results
5. Askarzadeh, A. (2016). A novel metaheuristic method for solving constrained engineering optimization problems: Crow search algorithm. *Computers & Structures, 169*, 1–12. Available from: https://www.matlabi.ir/wp-content/uploads/bank_papers/Matlabi.ir_636693349929032253.pdf
6. Bansal, J. C., Sharma, H., Jadon, S. S., & Clerc, M. (2014). Spider monkey optimization algorithm for numerical optimization. *Memetic Computing, 6*, 31–47. Available from: https://fardapaper.ir/mohavaha/uploads/2020/03/Fardapaper-Spider-Monkey-Optimization-algorithm-for-numerical-optimization.pdf
7. Benedict, A. (2018). Collection of optimization algorithms. Available from: https://www.mathworks.com/matlabcentral/fileexchange/69468-collection_of_optimization_algorithms?s_tid=srchtitle
8. Bolaji, A. L., Al-Betar, M. A., Awadallah, M. A., Khader, A. T., & Abualigah, L. M. (2016). A comprehensive review: Krill herd algorithm (KH) and its applications. *Applied Soft Computing, 49*, 437–446. Available from: https://dl.acm.org/doi/abs/10.1016/j.asoc.2016.08.041
9. Breed, M. D., & Moore, J. (2021). *Animal behavior*. Academic Press.
10. Camacho-Villal6n, C. L., Dorigo, M., & Stützle, T. (2022). An analysis of why cuckoo search does not bring any novel ideas to optimization. *Computers & Operations Research, 142*, 105747. Available from: https://iridia.ulb.ac.be/~ccamacho/publications/COR-CuckooSearch.pdf
11. Chu, S.-C., Tsai, P.-W., & Pan, J.-S. (2006). Cat swarm optimization. In *PRICAI 2006: Trends in Artificial Intelligence: 9th Pacific Rim International Conference on Artificial Intelligence* (pp. 854–858). Springer. Available from: https://link.springer.com/chapter/10.1007/978-3-540-36668-3_94
12. Chu, S.-C., Tsai, P.-W., & Pan, J.-S. (2006). Application of the shuffled frog leaping algorithm for the optimization of a general large-scale water supply system. *Water Resources Management, 23*(4), 797–823. Available from: https://link.springer.com/content/pdf/10.1007/s11269-008-9300-6.pdf
13. Daniel, A. (2020). GSO-matlab. Available from: https://github.com/danielalienyi/gso-matlab/blob/master/GSO.m
14. Deng, B., & Shi, Y. (2019). Modeling and optimizing the composite prepreg tape winding process based on grey relational analysis coupled with BP neural network and bat algorithm. *Nanoscale Research Letters, 14*(1), 1–14. Available from: https://link.springer.com/article/10.1186/s11671-019-3118-4
15. Devi, R. V., & Sathya, S. S. (2017). Monkey behavior based algorithms-a survey. *International Journal of Intelligent Systems and Applications, 9*(12), 67. Available from: https://www.mecs-press.org/ijisa/ijisa-v9-n12/IJISA-V9-N12-7.pdf
16. Dhiman, G., Garg, M., Nagar, A., Kumar, V., & Dehghani, M. (2021). A novel algorithm for global optimization: Rat swarm optimizer. *Journal of Ambient Intelligence and Humanized Computing, 12*, 8457–8482. Available from: https://link.springer.com/article/10.1007/s12652-020-02580-0

17. Dhiman, G., & Kumar, V. (2017). Spotted hyena optimizer: A novel bio-inspired based metaheuristic technique for engineering applications. *Advances in Engineering Software, 114*, 48–70. Available from: https://www.infona.pl/resource/bwmeta1.element.elsevier-51c05983-cf2f-3f8a-a294-612a091176b1
18. Dutta, T., Bhattacharyya, S., Dey, S., & Platos, J. (2020). Border collie optimization. *IEEE Access, 8*, 109177–109197. Available from: https://web.archive.org/web/20201108114638id_/https://ieeexplore.ieee.org/ielx7/6287639/6514899/09106341.pdf
19. Eusuff, M., Lansey, K., & Pasha, F. (2006). Shuffled frog-leaping algorithm: A memetic meta-heuristic for discrete optimization. *Engineering Optimization, 38*(2), 129–154. Available from: https://www.tandfonline.com/doi/abs/10.1080/03052150500384759
20. Faris, H., Aljarah, I., Al-Betar, M. A., & Mirjalili, S. (2018). Grey wolf optimizer: A review of recent variants and applications. *Neural Computing and Applications, 30*, 413–435. Available from: https://link.springer.com/article/10.1007/s00521-017-3272-5
21. Farzaneh, M. (2020). A new shuffled frog leaping optimization algorithm. Available from: https://es.mathworks.com/matlabcentral/fileexchange/74109-a-new-shuffled-frog-leaping-optimization-algorithm-1?s_tid=prof_contriblnk
22. Fister Jr, I., Fister, D., & Fister, I. (2013). A comprehensive review of cuckoo search: Variants and hybrids. *International Journal of Mathematical Modelling and Numerical Optimisation, 4*(4), 387–409. Available from: https://www.iztok-jr-fister.eu/static/publications/22.pdf
23. Gandomi, A. H., & Alavi, A. H. (2012). Krill herd: A new bio-inspired optimization algorithm. *Communications in Nonlinear Science and Numerical Simulation, 17*(12), 4831–4845. Available from: https://ui.adsabs.harvard.edu/abs/2012CNSNS..17.4831G/abstract
24. Gao, W.-F., & Liu, S.-y. (2012). A modified artificial bee colony algorithm. *Computers & Operations Research, 39*(3), 687–697. Available from: https://dl.acm.org/doi/abs/10.1016/j.cor.2011.06.007
25. Hancer, E., Ozturk, C., & Karaboga, D. (2013). Extraction of brain tumors from MRI images with artificial bee colony based segmentation methodology. In *2013 8th International Conference on Electrical and Electronics Engineering (ELECO)* (pp. 516–520). IEEE. Available from: https://www.ttcenter.ir/ArticleFiles/ENARTICLE/3138.pdf
26. Hassen, O. A., Abter, S. O., Abdulhussein, A. A., Darwish, S. M., Ibrahim, Y. M., & Sheta, W. (2021). Nature-inspired level set segmentation model for 3D-MRI brain tumor detection. *CMC-Computers Materials & Continua, 68*(1), 961–981. Available from: https://cdn.techscience.cn/ueditor/files/cmc/TSP_CMC_68-1/TSP_CMC_14404/TSP_CMC_14404.pdf
27. He, S., Wu, Q. H., & Saunders, J. R. (2009). Group search optimizer: An optimization algorithm inspired by animal searching behavior. *IEEE Transactions on Evolutionary Computation, 13*(5), 973–990. Available from: https://ieeexplore.ieee.org/abstract/document/5196714
28. Heidari, A. A., Mirjalili, S., Faris, H., Aljarah, I., Mafarja, M., & Chen, H. (2019). Harris hawks optimization: Algorithm and applications. *Future Generation Computer Systems, 97*, 849–872. Available from: https://dl0.programstore.ir/files/Uploades/Lib/PDF/HarrisHawksOptimization.pdf
29. Hofmann, E. E., Haskell, A. E., Klinck, J. M., & Lascara, C. M. (2004). Lagrangian modelling studies of antarctic krill (euphausia superba) swarm formation. *ICES Journal of Marine Science, 61*(4), 617–631. Available from: https://academic.oup.com/icesjms/article/61/4/617/606038
30. Hussien, A. G., Amin, M., Wang, M., Liang, G., Alsanad, A., Gumaei, A., and Chen, H. (2020). Crow search algorithm: Theory, recent advances, and applications. *IEEE Access, 8*, 173548–173565. Available from: https://ieeexplore.ieee.org/abstract/document/9195808/
31. Bossek, J. (2020). Egg crate function. Available from: http://jakobbossek.github.io/smoof/reference/makeEggCrateFunction.html#egg-crate-function
32. Kalaiselvi, T., Nagaraja, P., & Basith, Z. A. (2017). A review on glowworm swarm optimization. *International Journal of Information Technology, 3*(2), 49–56. Available from: https://www.americaspg.com/articleinfo/3/show/2043
33. Karaboga, D., et al. (2005). An idea based on honey bee swarm for numerical optimization. Technical report, Erciyes University, Engineering Faculty, Computer. Available from: https://abc.erciyes.edu.tr/pub/tr06_2005.pdf

34. Karaboga, D., Gorkemli, B., Ozturk, C., & Karaboga, N. (2014). A comprehensive survey: Artificial bee colony (abc) algorithm and applications. *Artificial Intelligence Review, 42*(1), 21–57. Available from: https://i2pc.es/coss/Docencia/SignalProcessingReviews/Karaboga2012.pdf
35. Kaveh, A., & Farhoudi, N. (2013). A new optimization method: Dolphin echolocation. *Advances in Engineering Software, 59*, 53–70. Available from: https://tarjomefa.com/wp-content/uploads/2016/04/4588-English.pdf
36. Krishnanand, K., & Ghose, D. (2009). Detection of multiple source locations using a glowworm metaphor with applications to collective robotics. In *Proceedings 2005 IEEE Swarm Intelligence Symposium, 2005. SIS 2005* (pp. 84–91). IEEE. Available from: http://eprints.iisc.ac.in/27755/
37. Krishnanand, K., & Ghose, D. (2009). Glowworm swarm optimization for simultaneous capture of multiple local optima of multimodal functions. *Swarm Intelligence, 3*(2), 87–124. Available from: https://link.springer.com/content/pdf/10.1007/s11721-008-0021-5.pdf
38. SKS Labs. (2022). Artificial bee colony optimization. Available from: https://ch.mathworks.com/matlabcentral/fileexchange/74122-artificial-bee-colony-optimization
39. Meraihi, Y., Gabis, A. B., Ramdane-Cherif, A., & Acheli, D. (2021). A comprehensive survey of crow search algorithm and its applications. *Artificial Intelligence Review, 54*(4), 2669–2716. Available from: https://su.edu.ye/wp-content/uploads/sites/42/2024/01/A-comprehensive-survey-of-Crow-Search-Algorithm-and-its.pdf
40. Mirjalili, S. (2015). The ant lion optimizer. *Advances in Engineering Software, 83*, 80–98. Available from: https://dl0.programstore.ir/files/Uploades/Lib/PDF/The%20Ant%20Lion%20Optimizer.pdf
41. Mirjalili, S. (2016). Dragonfly algorithm: A new meta-heuristic optimization technique for solving single-objective, discrete, and multi-objective problems. *Neural Computing and Applications, 27*, 1053–1073. Available from: https://e-tarjome.com/storage/uploaded_media/2022-09-12/1662953417_F1914-English-e-tarjome.pdf
42. Mirjalili, S., Gandomi, A. H., Mirjalili, S. Z., Saremi, S., Faris, H., & Mirjalili, S. M. (2017). Salp swarm algorithm: A bio-inspired optimizer for engineering design problems. *Advances in Engineering Software, 114*, 163–191. Available from: https://zeus.inf.ucv.cl/~bcrawford/DOCTORADO-METAHEURISTICAS/Papers/Salp-Swarm-Algorithm--A-bio-inspired-optimizer-for_2017_Advances-in-Engineer.pdf
43. Mirjalili, S., & Lewis, A. (2016). The whale optimization algorithm. *Advances in Engineering Software, 95*, 51–67. Available from: https://programstore.ir/wp-content/uploads/2019/07/WOA.pdf
44. Mirjalili, S., Mirjalili, S. M., & Lewis, A. (2014). Grey wolf optimizer. *Advances in Engineering Software, 69*, 46–61. Available from: https://zeus.inf.ucv.cl/~bcrawford/DOCTORADO-METAHEURISTICAS/Papers/Grey%20Wolf%20Optimizer.pdf
45. Mirjalili, S. Z., Mirjalili, S., Saremi, S., Faris, H., & Aljarah, I. (2018). Grasshopper optimization algorithm for multi-objective optimization problems. *Applied Intelligence, 48*, 805–820. Available from: https://link.springer.com/article/10.1007/s10489-017-1019-8
46. Neshat, M., Sepidnam, G., Sargolzaei, M., & Toosi, A. N. (2014). Artificial fish swarm algorithm: A survey of the state-of-the-art, hybridization, combinatorial and indicative applications. *Artificial Intelligence Review, 42*(4), 965–997. Available from: https://link.springer.com/content/pdf/10.1007/s10462-012-9342-2.pdf
47. Pan, W.-T. (2012). A new fruit fly optimization algorithm: Taking the financial distress model as an example. *Knowledge-Based Systems, 26*, 69–74. Available from: https://dl.acm.org/doi/abs/10.1016/j.knosys.2011.07.001
48. Sarkheyli, A., Zain, A. M., & Sharif, S. (2015). The role of basic, modified and hybrid shuffled frog leaping algorithm on optimization problems: A review. *Soft Computing*, 19(7):2011–2038. Available from: https://link.springer.com/article/10.1007/s00500-014-1388-4
49. Shehab, M., Khader, A. T., & Al-Betar, M. A. (2017). A survey on applications and variants of the cuckoo search algorithm. *Applied Soft Computing, 61*, 1041–1059. Available from: https://www.iztok-jr-fister.eu/static/publications/22.pdf

50. Wang, G.-G., Gandomi, A. H., Alavi, A. H., and & Gong, D. (2019). A comprehensive review of krill herd algorithm: Variants, hybrids and applications. *Artificial Intelligence Review, 51*(1), 119–148. Available from: https://link.springer.com/article/10.1007/s10462-017-9559-1
51. Wu, T.-Q., Yao, M., & Yang, J.-H. (2016). Dolphin swarm algorithm. *Frontiers of Information Technology & Electronic Engineering, 17*(8), 717–729. Available from: https://link.springer.com/article/10.1631/FITEE.1500287
52. Xu, Y., Fan, P., & Yuan, L. (2013). A simple and efficient artificial bee colony algorithm. *Mathematical Problems in Engineering, 2013*. Available from: https://onlinelibrary.wiley.com/doi/pdf/10.1155/2013/526315
53. Yang, X.-S. (2009). Firefly algorithms for multimodal optimization. In *International Symposium on Stochastic Algorithms* (pp. 169–178). Springer. Available from: https://arxiv.org/pdf/1003.1466
54. Yang, X.-S. (2013). Bat algorithm: Literature review and applications. arXiv preprint. Available from: https://arxiv.org/pdf/1308.3900
55. Yang, X.-S. (2020). The standard bat algorithm. Available from: https://ch.mathworks.com/matlabcentral/fileexchange/74768-the-standard-bat-algorithm-ba
56. Yang, X.-S. (2022). Cuckoo search (CS) algorithm. Available from: https://ch.mathworks.com/matlabcentral/fileexchange/29809-cuckoo-search-cs-algorithm
57. Yang, X.-S. (2022). Firefly algorithm. Available from: https://www.mathworks.com/matlabcentral/fileexchange/29693-firefly-algorithm
58. Yang, X.-S., & Deb, S. (2009). Cuckoo search via Lévy flights. In *2009 World Congress on Nature & Biologically Inspired Computing (NaBIC)* (pp. 210–214). IEEE. Available from: https://staff.fmi.uvt.ro/~daniela.zaharie/am2016/proiecte/tehnici/CuckooSearch/CuckooSearch1.pdf
59. Yang, X.-S., & and Deb, S. (2014). Cuckoo search: Recent advances and applications. *Neural Computing and Applications, 24*(1), 169–174. Available from: https://link.springer.com/article/10.1007/s00521-013-1367-1
60. Yang, X.-S., & Gandomi, A. H. (2012). Bat algorithm: A novel approach for global engineering optimization. *Engineering Computations*. Available from: https://arxiv.org/pdf/1211.6663
61. Yang, X.-S., & He, X. (2013). Firefly algorithm: Recent advances and applications. arXiv preprint arXiv:1308.3898. Available from: https://arxiv.org/pdf/1308.3898
62. Yong, W., Tao, W., Cheng-Zhi, Z., & Hua-Juan, H. (2016). A new stochastic optimization approach—dolphin swarm optimization algorithm. *International Journal of Computational Intelligence and Applications, 15*(2), 1650011. Available from: https://text2fa.ir/wp-content/uploads/Text2fa.ir-A-New-Stochastic-Optimization-Appro_compressed.pdf
63. Zainal, N., Zain, A. M., Radzi, N. H. M., & Udin, A. (2013). Glowworm swarm optimization (GSO) algorithm for optimization problems: A state-of-the-art review. *Applied Mechanics and Materials, 421*, 507–511. Available from: https://www.scientific.net/amm.421.507
64. Zebari, A. Y., Almufti, S. M., & Abdulrahman, C. M. (2020). Bat algorithm (BA): Review, applications and modifications. *International Journal of Scientific World, 8*(1), 1–7.

Chapter 13
Evolutionary Optimization: Other Metaphors

13.1 Introduction

This chapter continues with the topic of stochastic optimization, considering now other varied sources of inspiration, related to microorganisms, physics, mathematics, biogeography, etc.

Most probably the reader, after the previous chapters, has been tempted by the following question: what is the best algorithm, anyway?

Of course, this question could be linked to another one: should I delve with still more algorithms?

By the middle of the 1990s, a series of theorems were published under the umbrella of a (not) surprising consideration: **there is no free lunch**. It should be said, at this stage, that such no-free-lunch (NFL) theorems focus on searching and stochastic algorithms, employed for optimization or other topics. The publications of D. Wolpert and W.C. Macready, [76, 78], mark the departure of an interesting, philosophical/scientific debate on what can be expected from our algorithms.

The main observation can be inferred from the table in Fig. 13.1. A set of benchmark problems was selected, and another set of well-established stochastic methods was applied to these problems. The table shows the "optimal" results obtained by each algorithm.

It was found that all these methods obtained—on column averages—the *same performance*. It happened also that for each of the problems, some of the methods were good, while others not so much. *No method was an absolute winner for all problems.*

In words of Wolpert and Macready: "any two algorithms are equivalent when their performance is averaged across all possible problems." A consequence is that if a method is superior on some problems, then it will be inferior in other problems.

The NFL theorem assumes that the search space in the problems considered do have no underlying structure (where a priori knowledge or just gradient methods offer advantages).

J. M. Giron-Sierra, *Introduction to Optimization with Matlab® Examples*,
https://doi.org/10.1007/978-3-032-05048-9_13

	Algorithm 1	Algorithm 2	Algorithm 3	Algorithm 4	Algorithm 5	Algorithm 6	Algorithm 7	Algorithm 8	...
Problem 1	78.90158096	38.18696053	83.9788141	3.128185533	93.71767489	3.612131384	38.02555482	46.02033283	...
Problem 2	63.63661246	51.21726878	6.915100117	92.46504485	20.63056606	90.15194724	6.628150576	88.92628997	...
Problem 3	5.467817525	78.82129795	19.01963224	16.18471759	59.57316925	26.61430506	41.45446652	62.38540108	...
Problem 4	40.96337067	55.59045049	25.47959077	77.75563723	90.98183523	42.23275523	92.4381591	80.17316672	...
Problem 5	17.32640301	80.17604054	48.01380213	9.378352179	13.25844413	66.24497877	17.39991202	46.86218446	...
Problem 6	2.90117365	14.18732284	88.12091607	28.32526953	88.17950692	43.16349405	78.48956349	76.09121009	...
Problem 7	74.22339559	71.35440724	46.26625983	69.9710712	66.9510279	68.97533166	14.29350951	56.8139594	...
Problem 8	69.06790479	89.53420767	17.7105817	71.3419208	48.8622438	3.348772613	70.81053152	3.855765825	...
Problem 9	19.94675498	3.137513385	10.68373549	4.011603637	49.49135388	37.92530089	99.49914362	54.10622766	...
Problem 10	7.510870987	58.55534993	57.60647147	80.17271882	80.41639739	25.77488384	55.59960103	94.67596268	...
Problem 11	98.30840803	40.16271408	15.063453	80.71102508	67.38435353	2.092705478	54.93369837	34.34560747	...
Problem 12	56.35291015	99.47783881	73.23060569	79.11112105	58.89165367	51.21548188	72.3854659	54.63516655	...
Problem 13	42.95441914	5.055088383	20.45995021	60.02150262	2.129162205	0.03549031414	90.26590811	1.821852475	...
Problem 14	44.26664262	55.68963431	33.72502344	56.30721179	88.24480947	42.89040502	29.76489645	6.234549423	...
Problem 15	91.00330356	24.51201295	90.63002494	53.41813975	93.87696033	28.00711639	23.69333881	40.15298867	...
...	...	...	...	...	...	...	...	...	...
Average	100	100	100	100	100	100	100	100	

Fig. 13.1 Results of popular algorithms for a set of benchmark problems

Fig. 13.2 A humorous glimpse to NFL theory

The denomination as *no-free-lunch* is suggestive enough to trigger a lot of discussion, comments, and even illustrations with a touch of humor, like the one displayed in Fig. 13.2.

When asking Google about no-free-lunch theorem, one obtains 3,500,000 results. It is an evident sign of interest. In confirmation of this, the already mentioned publication [76] has been cited more than 12,800 times. The scientific article [4], year 2019, offers an extensive review of the research on NFL theory. Also, as in many other topics, Wikipedia would be recommended for more details and links.

13.2 Microorganisms

The intriguing strategies of microorganisms, like bacteria, virus, etc., have motivated considerable interest of the research for good reason: our survival. Part of what we know at present—much more rest to be discovered—is also serving as source of inspiration for our optimization algorithms.

This section specifically takes into consideration the artificial immune system algorithm, which is highly representative of the algorithms inspired by microorganisms.

13.2.1 Artificial Immune System

Our immune system, so important to repel microbial attacks, is a fruitful source of inspiration for algorithmic developments. Some key features of this system are distributed memory, robustness, self-organizing, and decentralized control mechanisms.

An influential, early overview of "*Artificial Immune Systems*," was due to D. Dasgupta [21], in 1993. Nowadays, we would speak of a large area of research, which follows a variety of theories and methods, like immune network algorithms (INA), negative selection algorithms (NSA), danger theory algorithms (DTA), clonal selection algorithms (CSA), etc.

We shall focus on clonal selection algorithms, which probably are the most popular. According to the Brunel's theory (year 1959), only cells (for instance, B cells) that can recognize an antigen will proliferate. Recall that antigens are our enemy.

When an antigen binds with a B cell (which becomes a selected cell), rapid proliferation of the cell is activated (clonal expansion and production of antibodies). During the duplication of cells, somatic hypermutations happen that result in plasma cells having higher affinities or memory cells that retain the antigenic pattern for future infections. Figure 13.3, taken from [82], represents this mechanism. The whole process of selection and mutation is called the maturation of the immune response. From our optimization perspective, higher affinity means a better solution.

In 2002, De Castro and Timmis [23] proposed the CLONALG algorithm, to solve pattern recognition problems. The main steps of this algorithm, which is frequently used for optimization purposes, are the following ([37]):

- *Initialization*: generate at random a population of Nindividuals (B-cells)
- *Evaluation*: for each pattern (or antigen) determine its affinity (match) with each individual of the population
- *Selection and cloning*: select a number (cN) of the best highest affinity individuals, and generate copies of them proportionally to their affinity
- *Hypermutation*: mutate all copies with a rate proportional to their affinity
- *Receptor editing*: add the mutated individuals to the population and re-select a number (d) of the maturated (optimized) individuals as memory

These steps are to be repeated until a certain final condition was reached. Usually, individuals are encoded as bit strings.

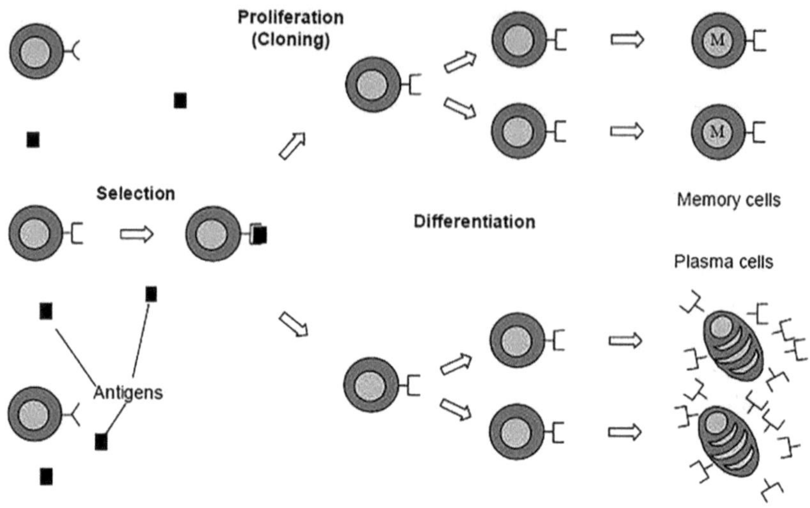

Fig. 13.3 Clonal selection and expansion

In the selection and cloning step, a ranking of the selected individuals is done. Then, the proliferation rate for the individual in the i-th position of the rank is given by:

$$N_c = round\left(\frac{\beta \cdot N}{i}\right)$$

where β is the maximal number of copies parameter.

Simpler versions of the algorithm, just use:

$$N_c = round\,(\beta \cdot N)$$

Concerning hypermutation, the probability of mutation can be chosen as,

$$p = \exp(-\rho \cdot f)$$

where f is the affinity normalized scoring.

More details of the algorithm can be found in [41].

The objective function chosen for our program is the Ackley function, widely used for testing optimization algorithms [70]. Its expression for bi-dimensional scenarios is the following:

$$f(x) = -20\exp\left(-0.2\sqrt{0.5(x^2+y^2}\right) - \exp\left(0.5\left(\cos 2\pi\, x + \cos 2\pi\, y\right)\right) + e + 20$$

We include now a simple implementation (program 13.1) of the algorithm, in accord with the listing offered by [30] in the MATLAB File Exchange site. The program considers an objective function (corresponding to affinity) is two dimensions. The program creates an initial population of 50 antibodies, with values encoded as 44 bit chains (22 bits for x coordinate, 22 bits for y coordinate). The evaluation of the objective function for each individual implies a decoding task. The program uses two auxiliary functions.

Notice how hypermutation of an individual is achieved, by flipping some bits of this individual, bits being selected at random.

Program 13.1 Example of clonal selection algorithm

```
%Clonal selection algorithm
%simple example
Np = 50; % population size
gen = 30; % number of generations
pm = 0.1; % mutation probability
d = 0.1; % population suffering random reshuffle
beta = 0.1; % proportion of clones
%generate antibody population (binary chains)
AB=2.*rand(Np,44)-1;
for i=1:Np
  for j=1:44
    if AB(i,j)>0
      AB(i,j)=1;
    else
      AB(i,j)=0;
    end
  end
end
%initial scenario
Lb = -32; % lower bound
Ub = 32; % upper bound
fbest = 0; % global f best (minimum)
%for objective function plot (use function)
x = meshgrid(linspace(Lb, Ub, 61));
y = meshgrid(linspace(Lb, Ub, 61))';
vxp = x;
vyp = y;
vzp = Fobj([x(:),y(:)]);
vzp = reshape(vzp,size(x));
figure(1)
contour(vxp, vyp, vzp,20); hold on;
%decoding (use function)
x = Dcod(AB(:,1:22),Lb,Ub);
y = Dcod(AB(:,23:end),Lb,Ub);
fit = Fobj([x(:),y(:)]);
%plot initial population
plot3(x,y,fit','k*');
%hypermutation parameters
pma = pm; %initial value
```

```
pmr = 0.8; %control rate
itpm = gen; %restoring iterations
%loop preparation
vfx=zeros(gen,1);
figure(2)
contour(vxp, vyp, vzp,20); hold on;
%main loop -----------------------------------------
for iter= 1:gen
  %decoding
  x = Dcod(AB(:,1:22),Lb,Ub);
  y = Dcod(AB(:,23:end),Lb,Ub);
  fit = Fobj([x(:),y(:)]);
  %plot of solutions
  plot3(x,y,fit','b.');
  %selection
  [a, ix]=sort(fit);
  valx=x(ix);
  valy=y(ix);
  %clone
  T=[];
  cs=zeros(Np,1);
  pcs=zeros(Np,1);
  for i = 1:Np
    cs(i) = round(beta*Np);
    pcs(i) = sum(cs);
    T = [T; ones(cs(i),1) * AB(ix(end-i+1),:)];
  end
  %hypermutation
  M = rand(size(T,1),44) <= pm;
  T = T - 2 .* (T.*M) + M;
  T(pcs,:) = AB(fliplr(ix),:);
  %decoding
  x = Dcod(T(:,1:22),Lb,Ub);
  y = Dcod(T(:,23:end),Lb,Ub);
  fit = Fobj([x(:),y(:)]);
  %re-selection
  aux = [0 pcs'];
  pcs = aux';
  for i = 1:Np
    [aux,index] = min(fit(pcs(i)+1:pcs(i+1)));
    bcs(i)=index;
    bcs(i) = bcs(i) + pcs(i);
  end
  %insertion
  AB(fliplr(ix),:) = T(bcs,:);
  %repertoire shift
  nedit = round(d*Np);
  %replace
  ZB=2.*rand(nedit,44)-1;
  for i=1:nedit
    for j=1:44
```

```
      if ZB(i,j)>0
        ZB(i,j)=1;
      else
        ZB(i,j)=0;
      end
    end
  end
  AB(ix(end-(nedit-1):end),:) = ZB;
  if rem(iter,itpm)==0
    pm = pm*pmr;
    if rem(iter, 10*itpm)==0
      pm = pma;
    end
  end
  vfx(iter) = a(1);
%loop end
end
% results display ------------------
figure(3)
Lb=-5; Ub=5;
nx = meshgrid(linspace(Lb, Ub, 61));
ny = meshgrid(linspace(Lb, Ub, 61))';
vxp = nx; vyp = ny; vzp = Fobj([nx(:),ny(:)]);
vzp = reshape(vzp,size(nx));
colormap('vga');
contour(vxp, vyp, vzp,30); hold on;
plot3(x,y,fit','r*');
figure(4)
plot(vfx);
xlabel('iter')
ylabel('Best f(x,y)')
% numerical result --------------------------------
x = valx(1)
y = valy(1)
fx = vfx(end)
% Function
function z = Dcod(AB,Lb,Ub)
AB = fliplr(AB);
s = size(AB);
aux = 0:1:21;
aux = ones(s(1),1)*aux;
x1 = sum((AB.*2.^aux),2);
% values between bounds
z = Lb + x1' .* (Ub - Lb)/(2^22 - 1);
% Function
function z= Fobj(xx)
d = size(xx,2);
z = -20*exp(-0.2*sqrt(1/d*sum(xx.^2,2)))-...
 exp(1/d*sum(cos(2*pi*xx),2)) + 20 + exp(1);
```

Figure 13.4 shows an initial population of solutions over 2D contours of the objective function. These solutions explore with no special preference the region

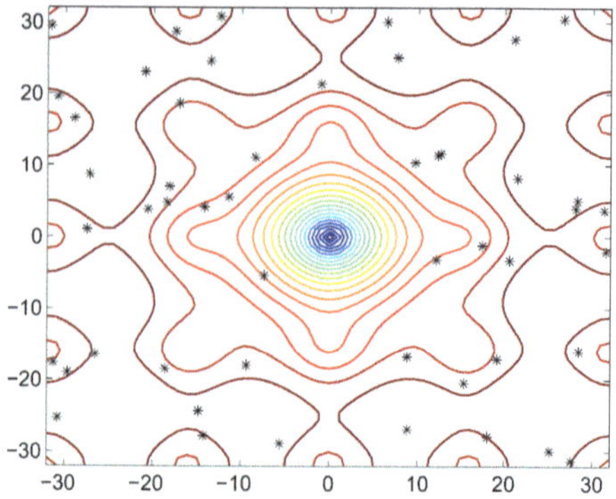

Fig. 13.4 Initial population of solutions

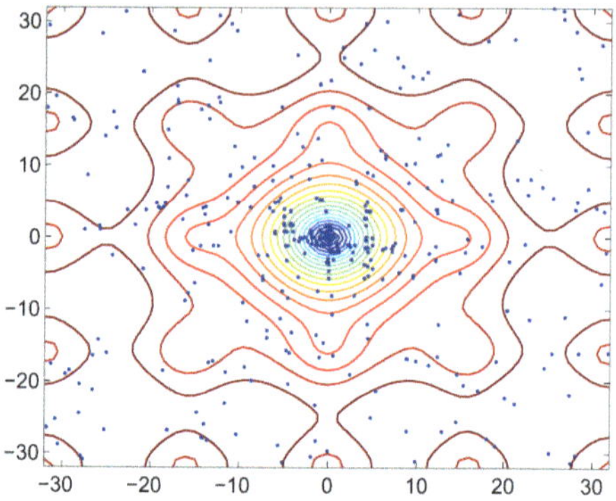

Fig. 13.5 Solutions along evolution

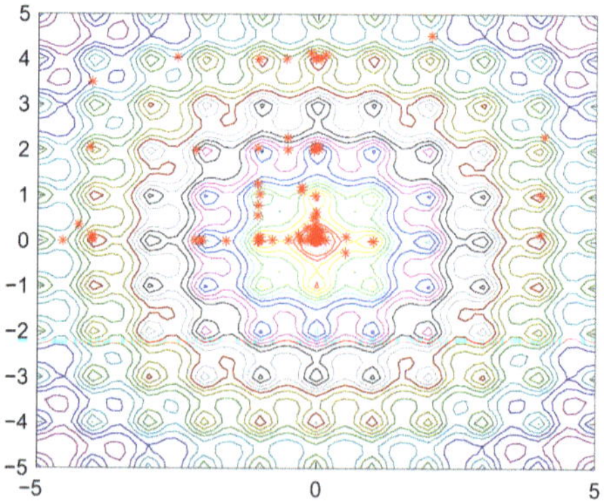

Fig. 13.6 Final population of solutions

to be studied. Each time the program is executed, a different initial population is generated.

Once the program starts to run, we depict in Fig. 13.5 the successive solutions being explored along the 30 generations specified in the program. As evolution advances, the solutions tend to get closer to the origin (which is the exact optimum).

Figure 13.6 zooms on the final situation, after the 30 generations. Notice the shrinking of scale in the figure. Most solutions were near the origin.

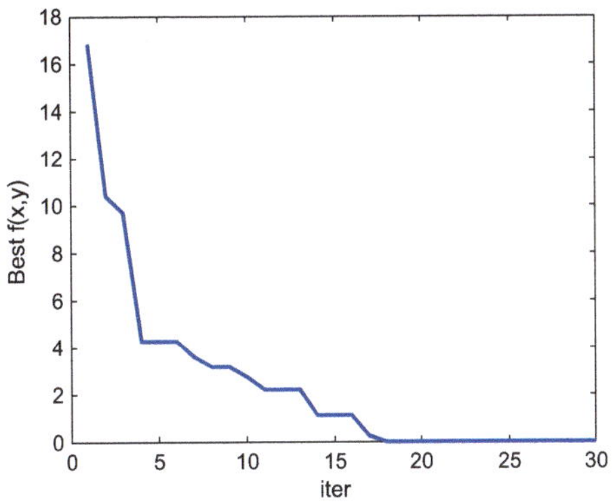

Fig. 13.7 Evolution of best result

Finally, Fig. 13.7 depicts the evolution of the best solution along the 30 generations.

The research activity triggered by the artificial immune system approach is intense, deepening in conceptual and modeling aspects, proposing improvements and variants, and exploring different applications. An interesting overview of this is provided by [22].

Another important stochastic algorithm inspired by microorganisms is the *bacterial foraging* optimization algorithm, proposed in 2002 by K.M. Passino, [56] (3770 cites). The MathWorks File Exchange site offers several MATLAB implementations of the algorithm. Also, GitHub offer several repositories of programs related to bacterial foraging; one of these repositories, [48], is in MATLAB, and there is a publication with the scientific basis of this implementation [40].

The article [33] presents an impressive review of variants and applications of the bacterial foraging algorithm.

Other algorithms we could mention are: the *bacterial colony* optimization (2012, 133 cites), the *virus optimization* algorithm (2016, 65 cites) (there are other algorithms inspired by coronavirus), and the *artificial algae* algorithm (2015, 277 cites). There are many more, less cited, based on epidemic processes or specific virus and diseases.

13.3 Human-Related

Humans boast a lot of activities and phenomena that can inspire optimization algorithms, for instance, education, social matters, arts, work, sports, etc.

For this section, two examples of human-inspired algorithms were selected. One is related to music and the other to geopolitics.

Fig. 13.8 Robotic music

13.3.1 Harmony Search

Frequently, music composers depart from a certain melody and start to improve it by checking changes in some of the notes. We could imitate this habit.

In our case, a melody is a series of values (a vector).

With some imagination, one could see for the next future humanoids improvising (optimal) music in real time while playing an instrument. This is represented in Fig. 13.8.

Of course, the music-inspired algorithm targets at achieving next-to-optimal solutions for N-dimensional problems, based on "melodies" with Nnotes.

The harmony search is a population-based algorithm. Initially, a number of melodies were generated with notes at random. These melodies were stored in a Harmony Memory (HM) matrix. Then, new melodies were iteratively obtained from this initial set, by improvements (improvisations) being controlled with two parameters: the harmony memory considering rate (HMCR) and the pitch adjusting rate.

Each melody is a vector of N components. The number of melodies saved in HM is denoted as HMS (harmony memory size).

The *i-th* note of a melody can only have values in a specified range: $L_i \leq x_i \leq H_i$.

After the initialization step, the next step of the algorithm is devoted to improvisation. A new melody is generated. There are two alternatives for its first note:

- With probability given by HMCR (a value between 0 and 1), this note is chosen from any value belonging to the first column of matrix HM
- With probability 1-HMCR, this note is chosen at random, in the range $L_1 \leq x_1 \leq H_1$

For the second, third, … N, notes, the procedure is the same *mutatis mutandis* (using the second, third, etc., columns of HM).

Now, every component of the new melody is examined to determine whether it should be pitch adjusted. The first new note would be adjusted with probability PAR (a value between 0 and 1), or not adjusted with probability 1-PAR. The same with the rest of the notes.

In case of being adjusted, a value α is added to the note pitch. This value is computed with: bw x $u(-1, 1)$. Where bw is a certain distance bandwidth, and $u(-1, 1)$ is a uniform random distribution between -1 and 1.

Next, if the new melody is better than the worst melody in HM, the new melody is included in HM and the worst melody excluded. Matrix HM is sorted according with the objective function.

The algorithm decides now to stop, or to come back for repeating the procedure.

Regarding historical aspects of the algorithm, the Harmony Search was proposed in year 2001, by Z. W. Geem et al. [32]. It is a well-received contribution, which has been cited by more than 6700 scientific publications.

In a recent review about the algorithm, [5], it is said that the algorithm attempts to balance exploitation and exploration. Exploitation is dominated by PAR and bw parameters, whereas exploration is controlled by HMCR. Also, the harmony search algorithm involves aspects in common with other algorithms. For instance, it keeps memory of visited vectors in the HM matrix, similar to was made in Tabu search. Likewise, it combines solutions to create new ones, like in Genetic Algorithms (GA), only that GA employs two parents, while harmony search uses all the solutions contained in HM. In both GA and harmony search, there are mutation mechanisms.

Many improvements and variants of the original harmony search algorithm have been proposed. One of these, published in [50], gives the substance of a MATLAB implementation that can be accessed through the Mathworks File Exchange site.

Program 13.2 is based, with some modifications, on that implementation, [28]].

The optimization problem we chose for program testing is one of the cases considered in [50], denoted as "constrained function V." We should find the minimum of:

$$f(x_1, x_2) = (x_1^2 + x_2 - 11)^2 + (x_1 + x_2^2 - 7)^2$$

Subject to the following constraints:

$$g_1(x_1, x_2) = 4.84 - (x_1 - 0.05)^2 - (x_2 - 2.5)^2 \geq 0$$

$$g_2(x_1, x_2) = x_1^2 + (x_2 - 2.5)^2 - 4.84 \geq 0$$

$$0 \leq x_1 \leq 6 \; ; \; 0 \leq x_2 \leq 6$$

Figure 13.9 shows the evolution of the best solution, along 1000 iterations of the algorithm. Not having any constraint, the minimum of the objective function would take place at (3,2), with a zero value. However, the constraints (the thick arc you can see in the figure) oblige for a different best solution, (2.246..., 2.381...), giving a value of 13.59....

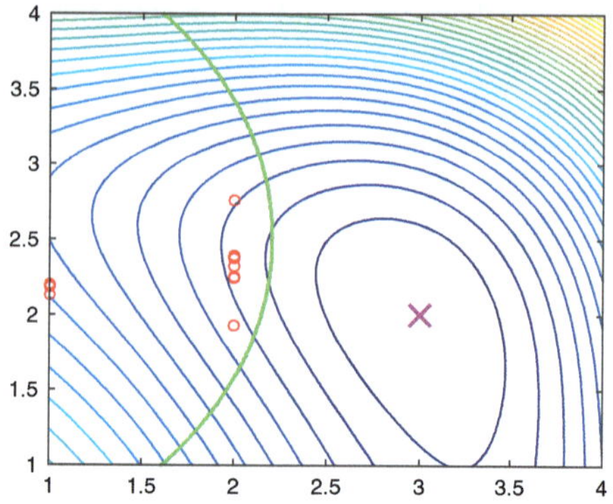

Fig. 13.9 Solutions along every 100 iterations

The solutions depicted in the figure (small circles) correspond to one of the trials we ran. The best solution at the end of this run was (2.2016, 2.3795), with a value of 14.9838. Other executions of the program would give different results.

Program 13.2. uses 4 brief functions.

Program 13.2 Harmony search example

```
%Harmony search example
Nvar=2; %number of variables
Nq=6; %number of inequality constraints
Ne=0; %number of equality constraints
MaxItr=1000; % maximum number of iterations
HMS=8; % harmony memory size
HMCR=0.9; % harmony consideration rate 0< HMCR <1
PARmin=0.4; % minimum pitch adjusting rate
PARmax=0.9; % maximum pitch adjusting rate
bwmin=0.01; % minimum bandwidth
bwmax=0.5; % maximum bandwidth
PVB=[1.0 4.0;1.0 4.0]; % range of variables
%Reserve memory for matrices
HM=zeros(HMS,Nvar); %harmony matrix
NCHV=zeros(1,Nvar);
BestGen=zeros(1,Nvar);
Hfitness=zeros(1,HMS);
BW=zeros(1,Nvar);
gx=zeros(1,Nq);
BestObjVal=zeros(1,MaxItr/100);
%Initialization
% random values for HM entries
for i=1:HMS
  for j=1:Nvar
    HM(i,j)= randval(PVB(j,1),PVB(j,2));
  end
  Hfitness(i)= HObjVal(HM(i,:),Nq);
end
% ---------------------------------------
% visualization of objective function
```

```
[vxp,vyp] = meshgrid(linspace(1.0, 4.0, 120));
for i=1:120
  for j=1:120
    vzp(i,j)= (((vxp(i,j)^2)+vyp(i,j)-11)^2)+
     ((vxp(i,j)+(vyp(i,j)^2)-7)^2);
  end
end
figure(1)
contour(vxp,vyp,vzp,30); hold on;
%plot costrained region
for x=1.5:0.005:2.5
  for y=1:0.01:4
    gz1=4.84 - (x-0.005)^2 -(y-2.5)^2;
    gz2=x^2 +(y-2.5)^2 - 4.84;
    if((gz1>0) & (gz2>0))
      plot(x,y,'g.');
    end
  end
end
%plot of known best place (3,2)
plot(3,2,'mx','MarkerSize',14);
%main loop ------------------------------
Iter = 0; ncent=0; CentCount=0;
while(Iter<MaxItr)
  PAR=(PARmax-PARmin)/(MaxItr)*Iter+PARmin;
  coef=log(bwmin/bwmax)/MaxItr;
  for pp =1:Nvar
    BW(pp)=bwmax*exp(coef*Iter);
  end
  % improvise a new harmony vector
  for i =1:Nvar
    ran = rand(1);
    if( ran < HMCR ) % memory consideration
      index = randint(1,HMS);
      NCHV(i) = HM(index,i);
      pvbRan = rand(1);
      if( pvbRan < PAR) % pitch adjusting
        pvbRan1 = rand(1);
        result = NCHV(i);
        if( pvbRan1 < 0.5)
          result =result+ rand(1) * BW(i);
          if( result < PVB(i,2))
            NCHV(i) = result;
          end
        else
          result =result- rand(1) * BW(i);
          if( result > PVB(i,1))
            NCHV(i) = result;
          end
        end
      end
    else
      % random selection:
```

```
      NCHV(i) = randval( PVB(i,1), PVB(i,2) );
    end
  end
  NewFit = HObjVal(NCHV, Nq);
  %Update-----------------------------
  if(Iter==0)
    BestFit=Hfitness(1);
    for i = 1:HMS
      if( Hfitness(i) < BestFit )
        BestFit = Hfitness(i);
        BestIndex =i;
      end
    end
    BestObjVal(1)=BestFit;
    WorstFit=Hfitness(1);
    for i = 1:HMS
      if( Hfitness(i) > WorstFit )
        WorstFit = Hfitness(i);
        WorstIndex =i;
      end
    end
  end
  if (NewFit< WorstFit)
    if( NewFit < BestFit )
      HM(WorstIndex,:)=NCHV;
      BestGen=NCHV;
      Hfitness(WorstIndex)=NewFit;
      BestIndex=WorstIndex;
    else
      HM(WorstIndex,:)=NCHV;
      Hfitness(WorstIndex)=NewFit;
    end
    WorstFit=Hfitness(1);
    WorstIndex =1;
    for i = 1:HMS
      if( Hfitness(i) > WorstFit )
        WorstFit = Hfitness(i);
        WorstIndex =i;
      end
    end
  end
  %---------------
  Iter=Iter+1;
  [BestFitness, aix] = min(Hfitness);
  %plot of best solution thus far
  %every 100 iterations
  ncent=ncent+1;
  if ncent==100
    ncent=0;
    CentCount=CentCount+1;
    BestObjVal(1+CentCount)=BestFitness;
    plot(HM(aix,:),'ro');
  end
end %end of loop
```

```
%plot of result evolution
figure(2)
aux=BestObjVal(:);
nc=0:(MaxItr/100);
plot(nc,aux,'m');
xlabel('Iter'); ylabel('Best f(x,y)');
% Function
function val1=randval(Maxv,Minv)
val1=rand(1)*(Maxv-Minv)+Minv;
% Function
function val2=randint(Maxv,Minv)
val2=round(rand(1)*(Maxv-Minv)+Minv);
% Function
function S =HObjVal(sol,Nq)
%F(x) = f(x) + penalty:
S = ((sol(1)^2+sol(2)-11)^2)+
((sol(1)+sol(2)^2-7)^2)+Heg(sol,Nq);
% Function
function Q=Heg(sol,Nq)
% constraints g(x) > 0
gx(1)=4.84 -(sol(1)-0.005)^2-(sol(2)-2.5)^2;
gx(2)=sol(1)^2+(sol(2)-2.5)^2-4.84;
gx(3)=6-sol(1);
gx(4)=sol(1);
gx(5)=6-sol(2);
gx(6)=sol(2);
% static penalty function to handle constraints
Q = 0;
for i=1:Nq
  if(gx(i)<0)
    Q = Q - 1000 * gx(i);
  end
end
```

Figure 13.10 depicts the evolution of the best solution along the 1000 iterations, taking points every 100 iterations. In most trials of the program, it decreases very rapidly at the beginning.

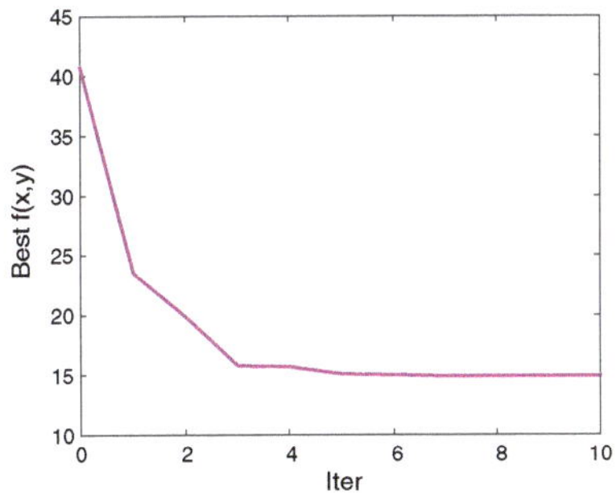

Fig. 13.10 Evolution (every 100 iterations) of best result

The already mentioned review, [5], gives an account of a good number of applications of Harmony Search algorithm variants, in the real world. Of course, it includes not only music composition but also many others: in civil and mechanical engineering, computer science, robotics, logistics, electrical engineering, energy, medicine, etc.

13.3.2 Imperialist Competitive Algorithm

Consider a world with a certain number of countries. A few of these countries are imperialist countries: they own significant power. Others could be considered as colonies: they are influenced by one or another imperialist country. As a result of influence, colonies tend to equal their pole of attraction. There exists competition between empires.

The imperialist competitive algorithm was introduced in 2007 by E. Atashpaz-Gargari and Caro Lucas (both University of Tehran) in 2007, [10]. It has been cited by 3100 publications.

A country is defined by:

$$country = [p_1, p_2, \ldots, p_{N_{var}}]$$

where N_{var} is the number of variables (which could represent level of culture, economic data, religious interest, etc.).

The algorithm initialization includes the generation at random of a number N_{pop} of countries (*pop* denotes population). An objective function (a cost) would be defined, in order to evaluate countries. Some of the best countries, having lowest cost, would be selected as imperialist countries. The number of these imperialist countries would be N_{imp} and the number of colonies N_{col}.

We have:

$$N_{col} = N_{pop} - N_{imp}$$

Part of the initialization is devoted to form empires. Colonies are divided among empires. The initial number of colonies of an empire should be proportional to its power.

Let us specify *power* in mathematical terms.

Denote as c_n the cost of the *n-th* imperialist country. The normalized cost C_n of that country is computed as:

$$C_n = \max_j(c_j) - c_n$$

where $j \in (1, \ldots, N_{imp})$.

Denote as *csum* the sum of imperial countries costs, $\sum_{j}^{N_{imp}} c_j$. Then, the *power* of each imperialist country is calculated with:

$$p_n = \frac{c_n}{csum}$$

The normalized power of the *n-th* imperialist country is the number of colonies owned by that country, which is:

$$NC_n = round(p_n \cdot N_{col})$$

As a colony benefits from the empire it belongs, the colony moves toward the imperialist country. This corresponds to an *assimilation process*. Let d be the direction from colony to imperialist country. The colony moves a distance x in this direction, where x is a random number (uniform distribution) in the interval $(0, \beta \cdot d)$. Parameter β is a value between 1 and 2.

In order to consider some possible deviation respect to d, an angle of deviation θ is possible, where θ is a random number (uniform distribution) in the interval $(-\varphi, \varphi)$. A convenient value for φ would be $\pi/4$ of radian.

Some colonies might resist to the march of events, by means of a *revolution*. In this case, sudden random changes in the position of a colony do happen. This is equivalent to mutation in genetic algorithms.

After any assimilation and revolution step, costs of the colonies are compared with those of imperialist countries. Should a colony be better (less cost) than an imperialist country, then we swap the imperialist with that colony.

As said before, there is competition. Let us start to look at this aspect.

The total cost, *TC*, of an empire would be computed as:

$$TC_n = Cost(imperialist_n) + \xi \cdot mean(Cost(colonies\ of\ empire_n))$$

where parameter ξ (between 0 and 1) takes into account the importance of colonies.

Weaker empires would have higher costs.

All empires try to take possession of colonies of other empires. The imperialist competition would produce a decrease in the power of weaker empires while increasing the power of stronger empires. The algorithm models the competition by choosing any of the weakest colonies of weakest empires and giving the opportunity to all empires of taking possession of these colonies.

Figure 13.11 (taken from [10]) visualizes the competitive situation, with a weakest country at the door of owner-changing.

In this competition, each of the empires will have a likelihood of effectively take that possession. Let express it with mathematics.

Fig. 13.11 Stronger empires tend to grow

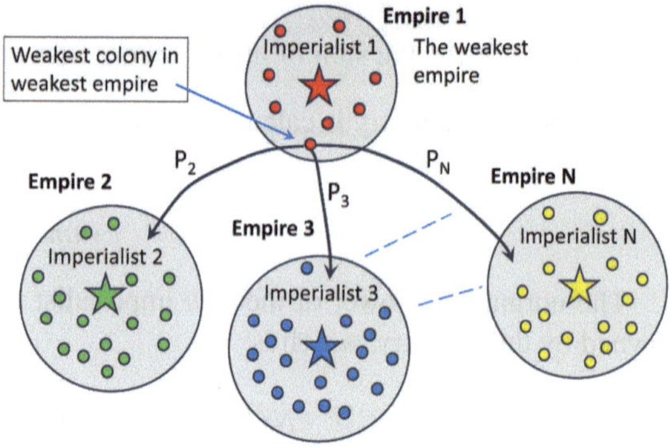

The normalized cost of an empire would be:

$$NTC_n = \max_j(TC_j) - TC_n$$

where $j \in (1, \ldots, N_{imp})$.

And, if we denote as *NTCsum*, the sum of normalized costs of empires, $\sum_j^{N_{imp}} NTC_j$, then the possession probability of an empire would be:

$$p_n = \frac{NTC_n}{NTCsum}$$

Now, we form the vector P, which contains the possession probabilities of all empires:

$$P = [p_1, p_2, \ldots, p_{N_{imp}}]$$

And we build another vector R, also with N_{imp} components:

$$R = [r_1, r_2, \ldots, r_{N_{imp}}]$$

in which, each of the components, r_j, is generated at random (uniform distribution), with values between 0 and 1.

Next, we compute the vector $D = P - R$.

The weakest colony is then assigned to the empire with the largest value in D.

During iterations of the algorithm, powerless empires will collapse, losing all their colonies.

Needless to say, each country (a vector of values) means a candidate solution for the optimization problem at hand.

To summarize, here is a description of the algorithm in pseudo-code terms:

```
Generate at random a population of N_pop countries
Initialize the world of empires:
Compute the evaluation cost of each country
Sort population in accordance with costs
Select a few (N_imp) imperialist countries
Normalize the cost of each imperialist country
Compute the normalized power of each imperialist country
Assign colonies to empires
Loop:
   (Assimilation) move colonies to pertinent imperialist countries
   Compute costs of assimilated countries
   Perform revolutions
   if the cost of a new colony is less than cost of imperialist, then
   Exchange colony and imperialist
   Select the weakest colony (colonies) from weakest empire and
   Assign it (them) to the empire with most likelihood to possess it
   (Elimination) If there is an imperialist with no colonies, eliminate
   this imperialist country
until final condition was met
```

Program 13.3 gives an example of MATLAB implementation of the algorithm. It required a number of fixes and modifications of a program available at GitHub [9]. There are some more programs related to the imperialist algorithm at the MathWorks File Exchange site.

Note (Program 13.3)

The complete listing of the program 13.3, about Imperialist Competitive algorithm, has been included in the Appendix for Long Programs.

For the objective function, we selected one of the cases studied in [10] (the case G1), which is:

$$f = x \cdot \sin(4x) + 1.1 \cdot y \cdot \sin(2y)$$

It is known that the minimum of this function takes place at (9.039, 0.8668), with a value of − 18.5547-

As in the other stochastic algorithms already introduced, every time you run the program, you obtain different (sub-optimal) results. Figure 13.12 corresponds to one of the trials and shows the initial distribution of imperialist countries. Notice how they tend to be at local minima (they are selected countries).

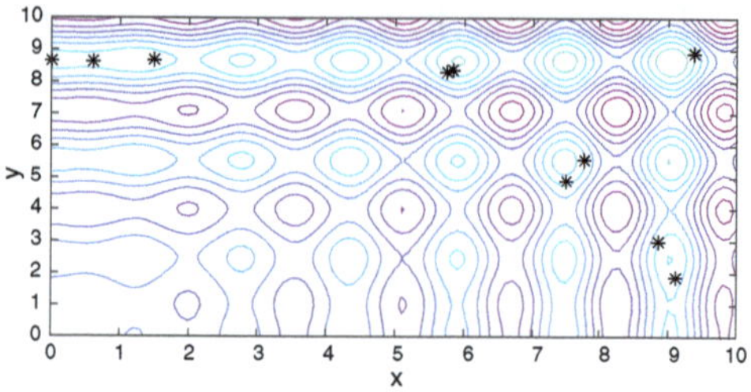

Fig. 13.12 Initial distribution of imperialist countries

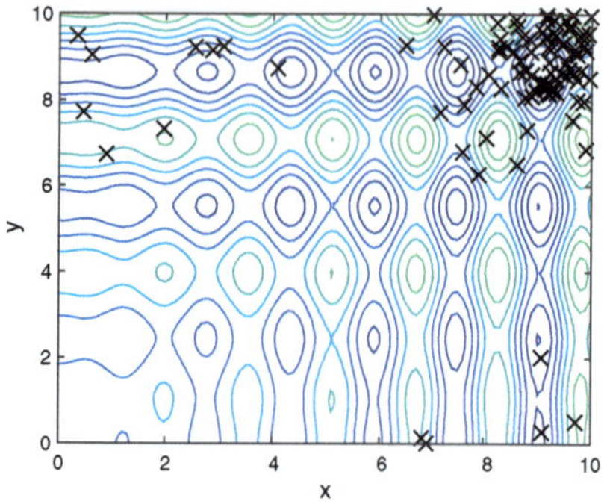

Fig. 13.13 Final distribution of colonies

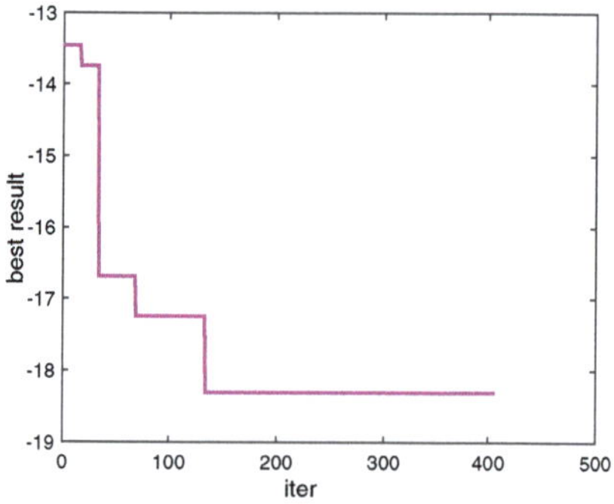

Fig. 13.14 Evolution of best result along iterations

After 400 iterations (this is when *nfe* becomes >40,000), there exists just one empire, and its colonies were distributed as shown in Fig. 13.13. Most of them were close to the optimal solution (near the right-hand upper corner)

Figure 13.14 depicts the evolution of the best result as the iterations proceed. Our experience, after many trials, is that the algorithm exhibits an abrupt descent at the beginning.

The result corresponding to Fig. 13.14 was (9.0512, 8.3489), with a value of −18.3049.

Some variants and hybridizations of the algorithm have been proposed, as reviewed in [43]. According to this article, the most popular application domain of the algorithm is engineering (some of this type of applications were summarized in the article).

We had curiosity for possible societal and political applications of the algorithm. With respect to it, we found researches on team formation in social networks, also about spam detection, optimization of well placement or energy use and environmental emissions in agriculture, the suitable location of emergency medical services, node placement of wireless sensors, biomarker discovery for cancer prediction, etc.

Other human-related stochastic algorithms we could mention are: the *teaching-learning* optimization algorithm (year 2011, 3700 cites), the *brain-storm* optimization algorithm (2011, 632 cites), the *gaining-sharing* knowledge optimization (2020, 173 cites), the *volleyball premier league* algorithm (2022, 144 cites), the *social-based* algorithm (2013, 117 cites), the *social emotional* optimization algorithm (2010, 75 cites), the *human-inspired* algorithm (2022, 57 cites), the *political optimizer* (2020, 193 cites), *the student psychology based* optimization algorithm (2020, 93 cites), etc.

We would like to recommend the article [27] published in 2021. The authors propose a *cooperation search* algorithm, imitating the structure and relationships of company teams (staff, supervisors, directors, etc.). Special merits of this publication are the experimental testing, and the methodical analysis of results.

There is a recent, extensive review of human-inspired optimization algorithms, [61]. The review conveys 51 algorithms. Also, there is another review, [45], focusing on socio-inspired optimization, covering algorithms like the *anarchic society* optimization (aside from imperialist entities, there is room for others), the *social group* optimization, the *election* algorithm, the *ideology* algorithm, etc.

Teaching and learning in particular give fertile *humus* for several algorithms. This is the specific subject of [64], a review published in 2022, in which a table of 43 algorithms is included, and two of them are discussed in detail.

When asking Google for human inspired optimization algorithms, one obtains 20,700,000 results.

13.4 Physics

Natural phenomena studied by Physics is a source of inspiration for stochastic optimization algorithms (more than 9,600,000 results for a Google search on it). Back in 2013, the review [12] on Physics-inspired optimization algorithms, covered around 28 of these algorithms. More recently, the brief overview [19], comments with descriptive illustrations a selection of physics-inspired metaheuristic techniques.

In order to gain perspective on where optimization algorithms come from, it would be interesting to explore what was coming from Iran academic institutions. A most cited researcher is A. Kaveh, Iran University of Science and Technology at Tehran. Together with his collaborators, he has introduced the *charged system search* optimization, the *ray optimization*, the *colliding bodies* optimization method, the *dolphin echolocation* method, the *thermal exchange* optimization, the *water evaporation* optimization, the *magnetic charged system* search, etc. In addition, he contributed to improvements and applications of the *Big-Bang crunch* algorithm, *ant colonies*, *imperialist competitive* algorithm, *whale* optimization, *cuckoo* search, *harmony search*, and many more. Another important researcher is Caro Lucas, University of Tehran; together with his collaborators, he introduced the *imperialist competitive* algorithm, the *weed colonization* algorithm, etc.

The province of Kerman, Iran, is the place where the Shahid Bahonar University was founded, decades ago. There, we find Prof. Hossein Nezamabadi-pour. Together with collaborators, he introduced the *gravitational* search algorithm (this will be the subject of the next sub-section); and contributed to improvements and applications of *ant colonies*, harmony search, *particle swarm* optimization, *artificial immune system* optimization, etc. Along years, the mentioned university has been the origin of several branches, like the Graduate University of Advanced Technology. There we find A.Askarzadeh, the author of the *crow search* algorithm, and the *bird mating* optimizer, and contributor to variants and applications of *artificial bee swarm* optimization, *harmony* search, *free* search, *bat* algorithm, *ant colonies*, etc.

We will consider now the gravitational search algorithm.

13.4.1 Gravitational Search Algorithm

The gravitational search algorithm was introduced in 2009 [62]. It has been cited by more than 6400 publications.

The algorithm chooses as metaphor celestial objects being subject to gravitational interaction. It is not difficult to see similarities with the imperialist competitive and other algorithms. Objects with larger mass would exert greater attraction.

Initially, a population of N objects is created, where each object has a certain position in a d-dimensional space. There is an objective function associated to that space. According to its position, a value of the function is assigned to each object. When the case is one of maximization, the larger the value the heavier the object, as represented in Fig. 13.15.

Of course, in case of minimization is the opposite of the figure: the less the function value at object position, the heavier this object.

The total force on the *i-th* object, caused by the rest of forces would be computed as follows (F_{ii} is zero):

$$F_i(t) = \sum_{j \in (1..N)} rand_j \cdot F_{i,j}(t)$$

The attraction force between *i-th* and *j-th* objects would be:

$$F_{i,j}(t) = G(t) \frac{M_i(t) \cdot M_j(t)}{R_{i,j}(t)^P + \varepsilon} \left(x_j(t) - x_i(t)\right)$$

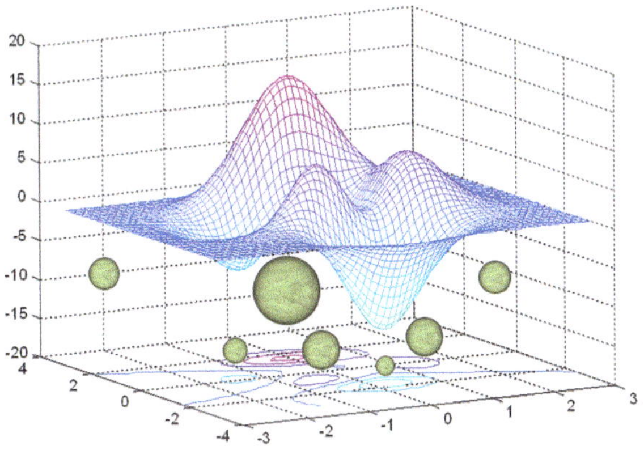

Fig. 13.15 Object mass corresponding to positions and objective function

where $R_{i,j}$ is the Euclidean distance between the objects. Masses were computed according to the position of objects and the objective function. Force and position are vectors in d-dimensions.

The original article makes a distinction of three types of mass: active, passive, and inertial mass. However, when translating to algorithm, it equals all three to the same value. The gravitational constant decreases with age; this could be considered with:

$$G(t) = G(t_0)\left(\frac{t_0}{t}\right)^{\beta}, \quad \beta < 1$$

(or any other adequate expression)

The motion of objects (acceleration, velocity, change of position) would be computed as follows:

$$a_i(t) = \frac{F_i(t)}{M_i(t)}$$

$$v_i(t+1) = rand_i \cdot v_i(t) + a_i(t)$$

$$x_i(t+1) = x_i(t) + v_i(t+1)$$

(in the expressions above, $rand_i$ and $rand_j$ are random vectors with uniform distribution in the interval 0..1)

The updating of masses would be computed with:

$$m_i(t) = \frac{fit_i(t) - worst(t)}{best(t) - worst(t)}$$

And then,

$$M_i(t) = \frac{m_i(t)}{\sum_{j=1}^{N} m_j(t)}$$

In case of minimization,

$$best(t) = \min_{j\varepsilon(1..N)} fit_j(t)$$

$$worst(t) = \max_{j\varepsilon(1..N)} fit_j(t)$$

Or, in case of maximization,

$$best(t) = \max_{j\varepsilon(1..N)} fit_j(t)$$

$$worst(t) = \min_{j\varepsilon(1..N)} fit_j(t)$$

In certain variants of the algorithm, a number K of objects with largest mass (objects with the best objective values) were selected. We will denote as *kbest* the set of these objects. K will decrease with time.

Given any of the N objects, the total gravitational force exerted on it by the K heaviest objects would be:

$$F_i(t) = \sum_{j \epsilon kbest} rand_j \cdot F_{i,j}(t)$$

(the algorithm only pays attention to the gravitational influence of the K heaviest objects).

Program 13.3. provides an example of implementation of the algorithm in MATLAB. It has been produced with non-minor modifications of the code available from GitHub [8]. It would be certainly possible to create faster versions of the program, by using vectorization. The objective function that was selected is the Ackley function, already used in previous programs.

Program 13.3 Example of gravitational search algorithm

```
%Example of Gravitatioal Search algorithm
clear all
%Number of Population &  Dimensions
pop_size = 60;
dim = 2;
%Iteration Condition
max_iter = 30;
```

```
%Domain of search space
from = -30;
to = -1*from;
%Nfe Condition
nfe = 1; max_nfe = 20000;
%Initializing Random Population
X=from+((to-from)*rand(pop_size,dim));
%Calculating Fitness
F_result = zeros(1, pop_size);
for i = 1:pop_size
  F_result(1,i) = GRobj(X(i,:));
  nfe = nfe + 1;
end
%Initializing Best & Worst Fitness
%(minimization problem)
fit_best = min(F_result(1,:));
fit_worst = max(F_result(1,:));
%gravitation
G = zeros(1, max_iter);
G0 = 0.98; beta=0.99;
empty = zeros(1,pop_size);
emptyX = zeros(pop_size,dim);
m = empty;
M = empty;
F_ij = emptyX;
F = emptyX;
acceleration = emptyX;
velocity = emptyX;
BestFit = zeros(1,max_iter);
%Main Loop -------------------------------------------
for iter = 1:max_iter
  %Gravitational Coefficient
  %tentative:
  G(1,iter) = G0 * ((max_iter/(iter+max_iter))^beta);
  %Calculating Mass for each Object
  m(1,:)=(F_result(1,:)-...
  fit_worst)/(fit_best-fit_worst);
  M(1,:) = (m(1,:)/sum(m));
  %Forces acting on Objects
  for nn=1:pop_size
    F(nn,:)=zeros(1,dim);
    for jj=1:pop_size
      dd=sum((X(nn,:)-X(jj,:)).^2);
      if (dd~=0)
        den = sqrt(dd) + eps;
        F_ij(nn,:)= G(1,iter)*...
        ((M(1,nn)*M(1,jj))/den)*(X(jj,:)-X(nn,:));
        F(nn,:)=F(nn,:)+(rand(1,dim).*F_ij(nn,:));
      end
    end
  end
  %Dynamics
  for nn = 1:pop_size
```

```
        acceleration(nn,:) = F(nn,:) / (M(1,nn) + eps);
        velocity(nn,:)=rand(1)*velocity(nn,:) +...
         acceleration(nn,:);
        X(nn,:)=rand(1)*X(nn,:)+velocity(nn,:);
    end
    %Computing the Fitness
    for nn = 1:pop_size
        F_result(1,nn) = GRobj(X(nn,:));
        nfe = nfe + 1;
        if (nfe == max_nfe)%Termination Condition
            break;
        end
    end
    %Best and worst fitness
    [fit_best,index] = min(F_result(1,:));
    fit_worst = max(F_result(1,:));
    BestFit(iter)=fit_best;
end %end of main loop
%Results----------------
X(index,:)
fit_best
figure(1)
L=-4; U=4;
nx = meshgrid(linspace(L, U, 31));
ny = meshgrid(linspace(L, U, 31))';
for i=1:31
    for j=1:31
        nz(i,j)=GRobj([nx(1,i),ny(j,1)]);
    end
end
colormap('vga');
contour(nx, ny, nz,8); hold on;
plot(X(:,1),X(:,2),'k*');
figure(2)
plot(0:max_iter-1,BestFit,'m');
xlabel('iter'); ylabel('Best result');
% Function
%Ackley
function [Out] = GRobj(X)
T1 = -20*exp(-0.2*(sqrt(sum(X.^2,2)*1./size(X,2))));
T2 = exp(sum(cos(X*2*pi),2)*1./size(X,2));
Out = T1-T2+20+exp(1);
```

As usual in our context, every run of the program gives different results. Figure 13.16 shows one example of this (the figure includes contours corresponding to the objective function). In all trials, the final result would have all candidates close to the optimal solution, which is (0,0) with the objective function value equal to zero.

Figure 13.17 shows the evolution of the best result along iterations, corresponding to one of the program executions. In general, the shape is not smooth (enough jumping).

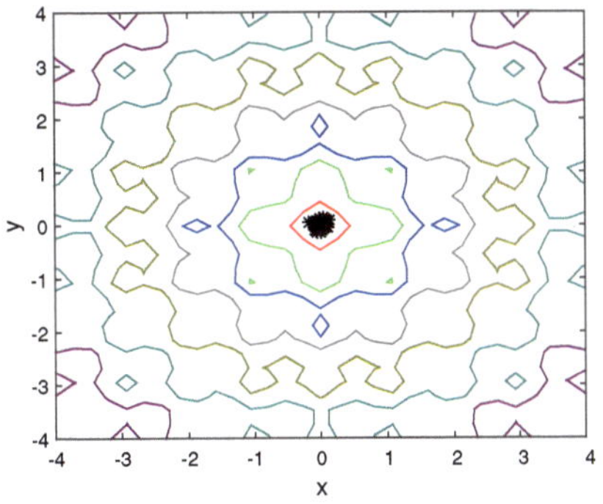

Fig. 13.16 One of the final results of the algorithm

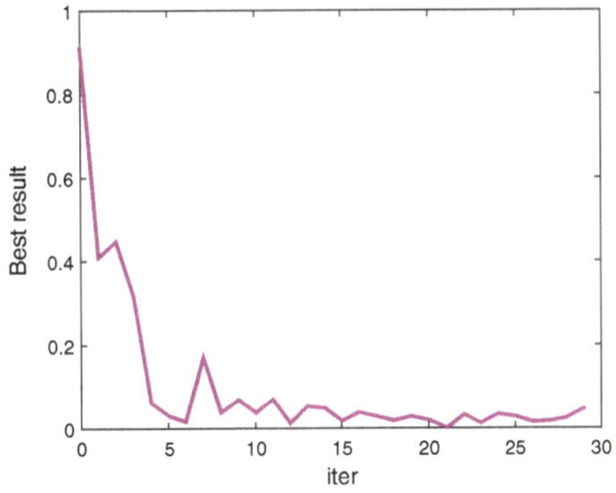

Fig. 13.17 Evolution of best result along iterations

The solution reached in the example just shown was (0.0133, 0.0059), with a value of 0.0469.

According to the article [63], which is a survey of the topic in the year 2018, around eight major variants of the gravitational search algorithm have been proposed. The main fields of application are computer science (for instance, pattern recognition), engineering (power systems, communication, control, mechanical, etc.), and mathematics.

Other algorithms related to physics are: the *ray* optimization (rays of light) (2012, 702 cites), the *Henry gas solubility* optimization (2019, 474 cites), the *thermal exchange* optimization (2017, 344 cites), the *Archimedes optimization* algorithm (2021, 324 cites), the *nuclear reaction* optimization algorithm (2019, 47 cites), the *galaxy gravity optimization* (2017, 9 cites), etc.

The detection of exoplanets, a hot topic for the public in general, is based on transit detection in time-series photometry. Recently, in 2022, the *transit search* has been proposed as optimization algorithm (there is an available MATLAB implementation in the MathWorks File Exchange site).

Going beyond traditional cosmology, a *multi-verse optimization* algorithm has been proposed by [54] (year 2016). The algorithm considers a set of universes, with black holes, white holes, and wormholes. It is very suggestive, and so it has been cited more than 1700 times by scientific publications.

There is a new breed of stochastic algorithms, related to quantum mechanics. Most of them are reviewed in the article [36], about recent advances in *quantum-inspired* metaheuristics. More related to quantum computers, I am glad to cite [6], developed in my Department (University Complutense of Madrid, Spain).

13.5 Math

Random search started to be applied in the 1950s, when digital computers appeared on practical scene. Also, stochastic optimization approaches started to be investigated. A significative event was the proposal in 1955, by G.B. Dantzig (the inventor of Linear Programming), of a stochastic programming framework for "*Linear Programming under Uncertainty*". Genetic algorithms, which triggered the bio-inspired evolutionary algorithms, came later on, in the 1970s (precedents in the 1960s). Nowadays, evolutionary algorithms are part of the Mathematics body: stochastic methods.

Then, one could say that Mathematics is the mother of evolutionary methods, and every evolutionary algorithm has to do with Mathematics. In any case, some evolutionary algorithms are more directly inspired by Mathematics (in particular, geometry, game theory, numerical optimization, and approximation theory): this section is devoted to this kind of algorithms.

13.5.1 Sine Cosine Algorithm

Trigonometric functions are fundamental for several fields of science and engineering. Important examples of it are the Fourier transform, or the three-phase electric power systems. As we will see now, they could be also employed for optimal solutions searching.

The sine cosine algorithm was proposed by S. Mirjalili in 2016 [53] (2900 cites). It is a population-based heuristic algorithm, in which the positions of individuals iteratively move toward the optimal solution, based on the following equations:

$$X_i(t+1) = X_i(t) + r_1 \times \sin(r_2) \times |r_3\, P_i(t) - X_i(t)|$$

$$X_i(t+1) = X_i(t) + r_1 \times \cos(r_2) \times |r_3\, P_i(t) - X_i(t)|$$

In general, these equations are combined to be used as follows:

$$X_i(t+1) = \begin{cases} X_i(t) + r_1 \times \sin(r_2) \times |r_3\, P_i(t) - X_i(t)| & r_4 < 0.5 \\ X_i(t) + r_1 \times \cos(r_2) \times |r_3\, P_i(t) - X_i(t)| & r_4 \geq 0.5 \end{cases}$$

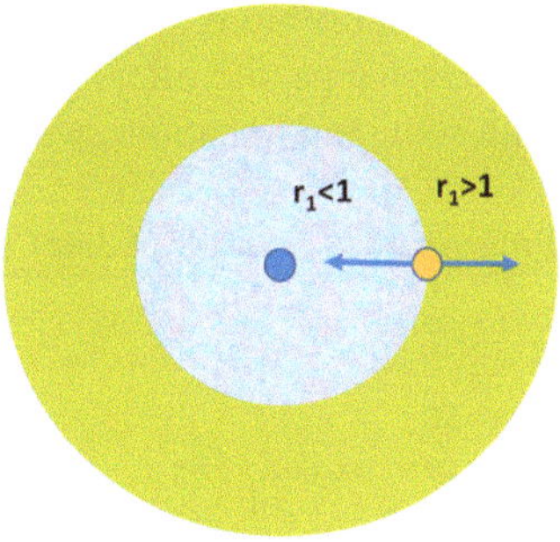

Fig. 13.18 Position updating, according to r_1

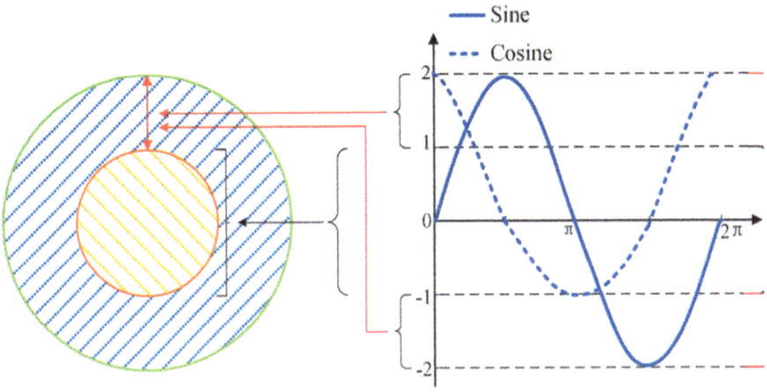

Fig. 13.19 How sine and cosine were combined for motion updating

where X_i represents the position of current individual (in a d-dimensional space), and P_i is the best individual's position. Parameters r_1, r_2, r_3, are random (uniform distribution).

Depending on the value of r_1, the next position would move toward the best solution, or outward it (for exploration). Figure 13.18 illustrates this feature. The optimal solution is represented by a central circle, while the current individual position corresponds to the circle at the frontier of two regions.

The parameter r_1 would linearly decrease from a certain value a, to 0. It is updated according to:

$$r_1 = a - t\frac{a}{T_{\max}}$$

where $T_{\max}$ is the maximum number of algorithm iterations.

Parameter r_2 , which is in the range $[0 \ldots 2\pi]$, determines how big are the movements, while parameter r_3 emphasizes ($r_3 > 1$) or de-emphasizes ($r_3 < 1$) the influence of destination on position updating, r_3 is in the range $[0 \ldots 2]$. In turn, the parameter r_4 is used as a switch between sine or cosine functions (this parameter is in the range $[0 \ldots 1]$.

Figure 13.19 illustrates how sine and cosine functions were combined, as determined by the random parameters.

The algorithm is simple, easy to be implemented as a program. The MathWorks File Exchange site offers an example of coding it in MATLAB [52]. Based on it, with some modifications, we obtained the program 13.4, listed below.

Program 13.4 Example of Sine Cosine algorithm

```
% Example of Sine Cosine algorithm
%
Nsa=50; %number of search agents
max_iteration=20;
dim=2; %search space dimension
Lb=-5; Ub=5; %variable bounds
X=Lb+(rand(Nsa,dim).*(Ub-Lb)); %agent initialization
Dest_pos=zeros(1,dim);
Dest_fit=inf;
BestVal=zeros(1,max_iteration);
objV = zeros(1,size(X,1));
% Calculate the fitness of the first set
% and find the best one
for i=1:size(X,1)
  objV(1,i)=SCobj(X(i,:));
  if i==1
    Dest_pos=X(i,:);
    Dest_fit=objV(1,i);
  elseif objV(1,i)<Dest_fit
    Dest_pos=X(i,:);
    Dest_fit=objV(1,i);
  end
end
BestVal(1)=Dest_fit;
%Main loop --------------------------------
iter=2; % (first iteration was already dedicated
% to initialization)
while iter<=max_iteration
  a = 2;
  % r1 decreases linearly from a to 0:
  r1=a-iter*((a)/max_iteration);
  % Update the position of solutions
  % with respect to destination
  for i=1:size(X,1) % in i-th solution
    for j=1:size(X,2) % in j-th dimension
      % Update r2, r3, and r4
      r2=(2*pi)*rand(1);
      r3=2*rand(1);
      r4=rand(1);
      if r4<0.5
        X(i,j)= X(i,j)+...
        (r1*sin(r2)*abs(r3*Dest_pos(j)-X(i,j)));
      else
        X(i,j)= X(i,j)+...
       (r1*cos(r2)*abs(r3*Dest_pos(j)-X(i,j)));
      end
    end
  end
```

```
  for i=1:size(X,1)
    % If solutions out limits, bring back
    Flag4ub=X(i,:)>Ub;
    Flag4lb=X(i,:)<Lb;
    X(i,:)=(X(i,:).*(~(Flag4ub+Flag4lb)))+...
     Ub.*Flag4ub+Lb.*Flag4lb;
    % Calculate the objective values
    objV(1,i)=SCobj(X(i,:));
    % Update the destination if there is
    % a better solution
    if objV(1,i)<Dest_fit
      Dest_pos=X(i,:);
      Dest_fit=objV(1,i);
    end
  end
  BestVal(iter)=Dest_fit;
  iter=iter+1;
end
Dest_fit
Dest_pos
figure(1)
colormap('vga');
[xx, yy]= meshgrid(Lb:0.2:Ub, Lb:0.2:Ub);
zz= xx.^2 + yy.^2 + (25 * (sin(xx).^2 + sin(yy).^2));
contour(xx,yy,zz,26); hold on;
for i=1:Nsa
  plot(X(i,1), X(i,2),'ko');
end;
xlabel('X'); ylabel('Y');
figure(2)
plot(0:1:max_iteration-1, BestVal(:),'m');
xlabel('iter'); ylabel('Best result');
```

The objective function was the same as the one employed for Particle Swarm optimization.

Program 13.5 Fitness function used in the example of Sine Cosine algorithm

```
% Function
function fitness = objFit(x)
fitness= x(1)^2 + x(2)^2 + (25 * (sin(x(1))^2 +...
 sin(x(2))^2));
```

Figure 13.20 depicts, on a contour plot of the objective function, the final situation after 20 iterations. Most individuals got close to the optimum (the exact optimal solution is (0,0) with a value of zero). A few other individuals came near local optima.

The algorithm exhibits fast convergence, as shown in Fig. 13.21.

Both Figs. 13.20 and 13.21 correspond to one of the program execution trials, which gave as best result (0.0024, −0.0015) with a value of 0.000204.

There are two review articles, [3] and [31], both year 2021, that give an optimistic judgment of the algorithm, in terms of popularity and applications.

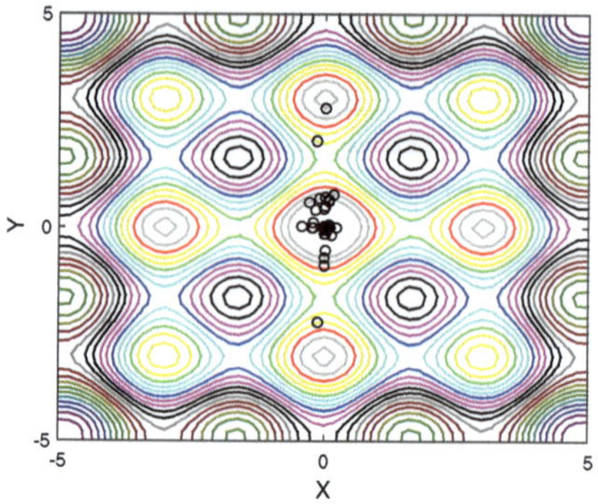

Fig. 13.20 Final solutions after 20 iterations

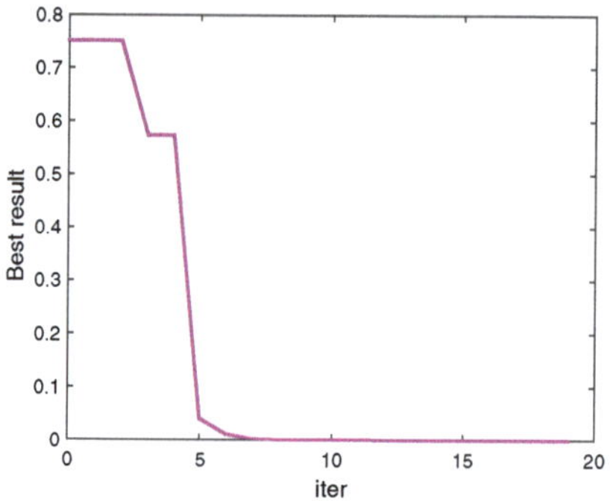

Fig. 13.21 Evolution of best result along iteration

As a consequence of the interest it awakened, more than nine variants and six hybridizations of the algorithm have already been proposed.

There is another reason for recommending [31]. It is that the article includes an extensive and ordered account of stochastic algorithms, indicating bibliographical references.

In addition to the sine cosine algorithm, we should mention other math-inspired algorithms, like the classic *population-based incremental learning* (1994, 1755 cites), the *arithmetic optimization* algorithm (2021, 904 cites), the *interior search* algorithm (2014, 390 cites), the *Gaussian swarm* (2004, 242 cites), the *spiral optimization* algorithm (2014, 108 cites), the *Golden sine* algorithm (2017, 93 cites), the *game theory-inspired* evolutionary algorithm (2017, 19 cites), the *Shell game optimization* (2020, 30 cites), the *team game optimization* algorithm (2018, 25 cites), the *circle search* algorithm (2022), the *geometric mean optimizer* (2022), etc.

13.6 Nature

Nature provides a considerable variety of evolutionary processes, which serve as inspiration for the design of many possible optimization algorithms. This section contains two illustrative examples of these algorithms.

Fig. 13.22 Archetype scenario for biogeography

13.6.1 Biogeography-Based Optimization

In the article where the biogeography-based optimization (BBO) was introduced, [67], its author, Dan Simon (Cleveland State University), devotes large space to describe the interesting background being considered. He refers in particular to the classic book of Robert McArthur and Edward Wilson, entitled *The Theory of Island Biogeography* (1967). This work focused on mathematical models for the extinction and migration of species and ignited a new field of research (as D. Simon remarks, only in the year 2005, more than 25,000 research papers were published on this topic) (Fig. 13.22).

As being said in the survey article [34], in BBO an "island" is considered as a habitat isolated from other habitats, instead of an area surrounded by water. Given a good island, many kinds of species would live in it. Probably, the island could be near saturation, and so there would be a high emigration rate and almost no immigration. On the contrary, a poor island would not source much emigration and would be open for high immigration rate. It can be easily guessed that this basic explanation does stir up many different opinions and models.

Concerning the algorithm proposed in the foundational article, [67] (near 4000 cites), it is a population-based metaheuristics, involving migration and mutation processes. An index, denoted HIS (*habitat suitability index*) is defined in order to evaluate the quality of an island (i.e. habitat).

In our context of optimization purposes, islands are solutions, and migration means interaction between solutions.

The search space would have d dimensions. A habitat H is a vector of d numbers. There is a population of habitats. For **migration**, a habitat H_i will be selected with probability proportional to its immigration rate. Then, one applies the next procedure:

```
For j=1 to N
      Select Hj with probability proportional to its emigration rate
```

(continued)

```
    if Hj is selected:
        Select at random a component ck from Hj
        Replace with ck a component of Hi randomly chosen
    end
End For
```

Therefore, for the selected H_i, there would be probably several immigrations from several H_j.

Each habitat has a probability of existence P_i as solution to a given problem. This probability, and a mutation rate m_i, can be computed from immigration and emigration rates (see [67]).

For **mutation** one takes one by one the habitats (for i=1..N), computes their mutation rates, and applies the following procedure (for each habitat H_i):

```
For j=1 to d
    Generate a random value R
    if R <  m_i, then
        Generate a feasible value F in searching space
        Equal j-th component of H_i to F
    end
End For
```

Like with many other algorithms, there are MATLAB implementations of the BBO algorithm, available from the MathWorks File Exchange site or from GitHub. Interestingly, there is also an application of BBO in MATLAB for the N-Queens problem, and this is an opportunity to include an example of discrete optimization.

If you ask Google for the **N-Queens problem**, you will find around 38 million answers. The problem appeared in a German journal in 1848, for an 8 × 8 chessboard. You should place eight queens so none attack each other. No two queens on the same diagonal. This problem, as it was found, has 92 different solutions. But what happens for, say, a 1000 × 1000 board? As reported in the web page [69], the N-queens general problem has been recently targeted by M. Simkin, Harvard University and other scientists. It is an open story…

Figure 13.23, taken from [69], illustrates the N-Queens problem.

What we could try is to use BBO for this problem. Precisely, the article [35] proposes this approach. And, as we just emphasized, there is a MATLAB implementation in this sense, available from GitHub [55]. Based on this code, with some changes, a program (program 13.6) was obtained. Note that three functions

Fig. 13.23 The N-queens problem

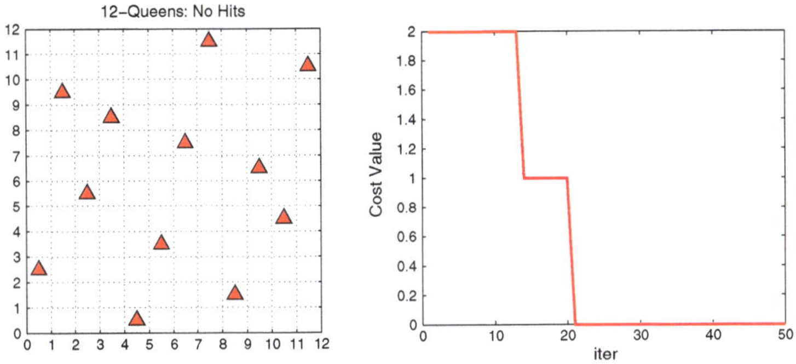

Fig. 13.24 (left) An optimal solution; (right) Evolution of criterion along iterations

are used. In this listing, a problem with 12 queens has been specified, but equally other number of queens can be considered.

Note (Program 13.6)

The complete listing of the program 13.6, about the use of Biogeography-based algorithm to the N-queens problem, has been included in the Appendix for Long Programs.

Most trials of the program give optimal results where no queen can attack the others. Figure 13.24 shows one of the results obtained (each time the program is run, different results are obtained). As shown on the right hand side of the figure, the algorithm converges in a few iterations.

Less frequently, program trials get suboptimal results where diagonal moves exist (allowing a queen to attack another queen). Figure 13.25 shows one of these cases. The program draws lines indicating diagonal undesired moves. In the case of Fig. 13.25, there is only one of these lines.

The program is not a perfect being, it would require some improvements.

The aforementioned survey, [34] reports as many as six mathematical models of immigration and emigration, which have been explored with the BBO algorithm. It also informs about several variants and hybridizations of the algorithm and briefly comments a number of applications in power engineering, communications, combinatorial problems, transportation, medicine, etc.

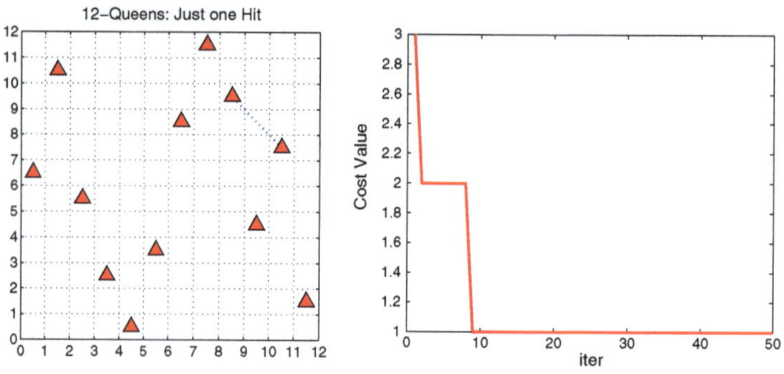

Fig. 13.25 (left) A sub-optimal solution; (right) Evolution of criterion along iterations

Fig. 13.26 Pollination in a real natural scenario

13.6.2 Flower Pollination Algorithm

It is estimated that about 80% of all plant species are flowering species. The main purpose of flowers is reproduction via pollination. Biotic agents, like bees and other insects, and non-biotic agents, like wind, etc., help to transfer pollen between flowers of the same plant (a local process), or farther (other plants, global process).

Figure 13.26 visualizes a typical pollination scene.

The flower pollination algorithm was proposed in [80], year 2012. This publication has been cited more than 2200 times. The algorithm is an abstraction of pollination in nature, which leads to the following pseudo-code:

```
Generate at random a population of flowers (problem solutions)
Evaluate flowers and find the best solution
Define a switch probability p (between 0 and 1)
While the stop condition is not met:
        For each flower
           if rand() <p
              Global pollination based on Lévy distribution
```

(continued)

```
    else
        Select at random two flowers
        Do local pollination
    end
    Evaluate new solutions
    Update solutions with better new ones
    end
  Keep the current best solution
End
```

From a searching perspective, global pollination corresponds to exploration, and local pollination corresponds to exploitation.

For simplicity, the algorithm assumes that each flower only produces one pollen gamete. A solution vector is equivalent to a flower and/or pollen gamete.

Global pollination is represented by the following equation:

$$\mathbf{x_i}(t+1) = \mathbf{x_i}(t) + L\,(\mathbf{x}_i(t) - \mathbf{g}_*)$$

where $\mathbf{x_i}$ is the i-th pollen or solution vector, and $\mathbf{g}_*$ is the current best solution. The value of parameter L is drawn from a Lévy distribution:

$$L \sim \frac{\lambda\,\Gamma(\lambda)\,\sin(\pi\lambda/2)}{\pi}\,\frac{1}{s^{1+\lambda}}$$

Local pollination is represented by the next equation:

$$\mathbf{x_i}(t+1) = \mathbf{x_i}(t) + \varepsilon\,(\mathbf{x}_j(t) - \mathbf{x}_k(t))$$

where $\mathbf{x}_j$ and $\mathbf{x}_k$ are pollens from different flowers of the same plant. The value of parameter ε is drawn from a uniform distribution between 0 and 1.

Based on a program provided by Xin-She Yang (the author of the algorithm), which is available from MatWorks File Exchange site, [81], an implementation example in MATLAB has been developed, program 13.6:

Program 13.6 Flower pollination example

```
%Example of Flower Pollination
n=20; % Population size, typically 10 to 25
p=0.8; % probability switch
N_iter=200; % Total number of iterations
d=2; % Dimension of the search space
Lb=-5*ones(1,d);
Ub=5*ones(1,d);
Rfbest=zeros(1,round(N_iter/20));
Sol=zeros(n,d);
```

```
% Initialize the population/solutions
for i=1:n,
  Sol(i,:)=((Ub-Lb).*rand(1,d))+Lb;
  Fitness(i)=F_Fun(Sol(i,:));
end
%for objective function plot
[vx1,vx2]=meshgrid(-5:0.2:5, -5:0.2:5);
y=(vx1.^2) + (vx2.^2)+ (25*((sin(vx1)).^2)) +
(25*((sin(vx2)).^2));
%initial population plot
figure(1)
colormap('vga');
contour(vx1,vx2,y,15); hold on;
plot(Sol(:,1),Sol(:,2),'k*');
% Find the current best
[fmin,I]=min(Fitness);
best=Sol(I,:);
S=Sol;
Rfbest(1)= fmin;
Ntw=0; Mc=1;
% main loop --------------------------------
for iter=1:N_iter,
  for i=1:n,
    if rand>p,
      L=F_Levy(d);
      dS=L.*(Sol(i,:)-best);
      S(i,:)=Sol(i,:)+dS;
      % Check limits/bounds
      S(i,:)=F_simplebounds(S(i,:),Lb,Ub);
      % If not, then local pollenation of
      % neighbor flowers
    else
      epsilon=rand;
      % Find random flowers in the neighborhood
      JK=randperm(n);
      % If the flowers the same or similar species,
      % then they can be pollenated,
      % otherwise, no action.
      S(i,:)=S(i,:)+epsilon*(Sol(JK(1),:)-...
        Sol(JK(2),:));
      % Check if the simple limits/bounds are OK
      S(i,:)=F_simplebounds(S(i,:),Lb,Ub);
    end
    % Evaluate new solutions
    Fnew=F_Fun(S(i,:));
    % If fitness improves (better solutions found),
    % update them
    if (Fnew<=Fitness(i)),
      Sol(i,:)=S(i,:);
      Fitness(i)=Fnew;
    end
    % Update the current global best
    if Fnew<=fmin,
      best=S(i,:);
```

```
      fmin=Fnew;
    end
  end
  % Save best value every 20 iterations
  Ntw=Ntw+1;
  if Ntw==20
    Ntw=0; Mc=Mc+1;
    Rfbest(Mc)=fmin;
  end
end
% Output/display
disp(['Total number of evaluations:...
 ',num2str(N_iter*n)]);
disp(['Best solution=',num2str(best),' fmin=',...
num2str(fmin)]);
%final population plot
figure(2)
colormap('vga');
contour(vx1,vx2,y,15); hold on;
plot(Sol(:,1),Sol(:,2),'k*');
figure(3)
nn=0:1:Mc-1;
plot(nn, log10(Rfbest(nn+1)), 'm');
xlabel('iter/20'); ylabel('log10(best result)')
%function
function L=F_Levy(d)
% Draw n Levy flight sample
% Levy exponent and coefficient
beta=3/2;
sigma=(gamma(1+beta)*sin(pi*beta/2)/...
(gamma((1+beta)/2)*beta*2^((beta-1)/2)))^(1/beta);
u=randn(1,d)*sigma;
v=randn(1,d);
step=u./abs(v).^(1/beta);
L=0.01*step;
%function
function s=F_simplebounds(s,Lb,Ub)
% Application of simple constraints
% Apply the lower bound
ns_tmp=s;
I=ns_tmp<Lb;
ns_tmp(I)=Lb(I);
% Apply the upper bounds
J=ns_tmp>Ub;
ns_tmp(J)=Ub(J);
% Update this new move
s=ns_tmp;
%function
function z=Fun(u)
% egg-crate
z=(u(1)^2) + (u(2)^2)+ (25*(sin(u(1))^2)) +...
 (25*(sin(u(2))^2));
```

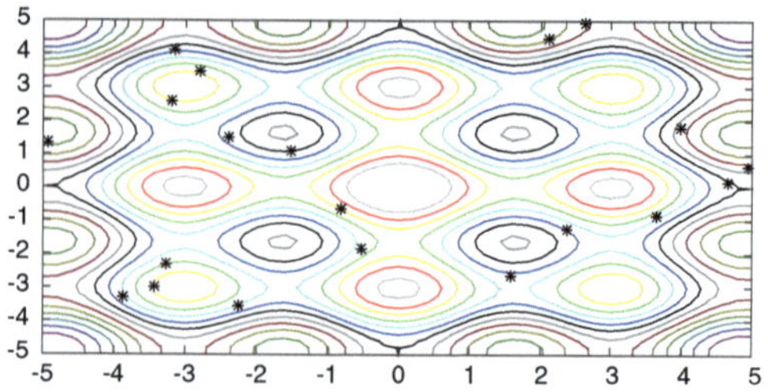

Fig. 13.27 An initial population of candidate solutions

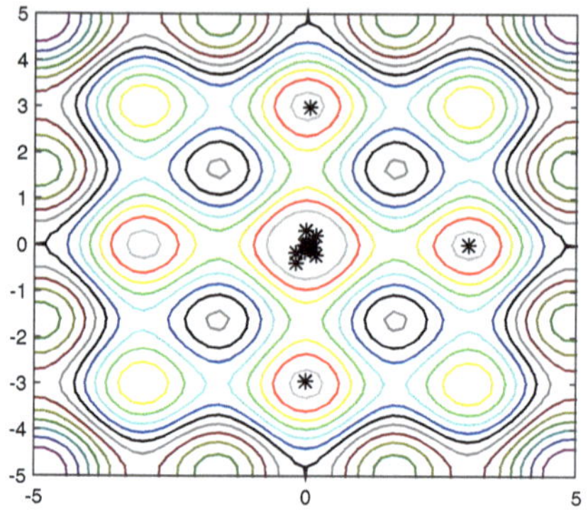

Fig. 13.28 Final population of candidate solutions

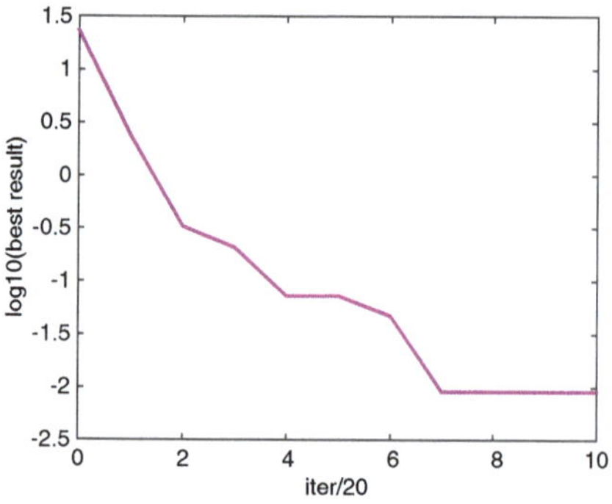

Fig. 13.29 Evolution of best solution along iterations

As it can be noticed, from the listing above, we selected the *egg-crate function* as the objective function to be optimized.

Let us run the program (as usual, each run has its own history and gives different results). Figure 13.27 shows in 2D contours of the objective function and an initial population of candidate solutions (asterisks).

After 200 iterations, we stop the program execution. Figure 13.28 shows where the candidate solutions are now. Most of them are close to the optimal solution. Others (a few) were trapped by local minima.

Figure 13.29 depicts the evolution of the best solution along iterations. Note that ticks on the horizontal axis are numbered every 20 iterations. Likewise, the vertical axis represents log10 of the best solution.

The article [1], published in 2019, offers a comprehensive review of the flower pollination algorithm, with literature references about its applications in computer science, engineering, meteorology, medicine, etc. It also reports five proposed variants and several hybridizations. The article chooses a particular problem (the design of a spring) and compares flower pollination with other well-known heuristic algorithms. The results are clearly in favor of flower pollination. However, it should be mentioned that the qualitative and quantitative analysis of flower pollination, reported in [25] (year 2015), says that the performances of the algorithm are less than average, compared to state-of-the-art algorithms (the study is based on a series of benchmark problems).

Other nature-inspired algorithms we could mention are: the *water cycle* algorithm (2012, 1262 cites), the *lightning search* algorithm (2015, 319 cites), the *water wave* optimization (2015, 497 cites), the *water evaporation* optimization (2016, 252 cites), the *river formation dynamics* algorithm (2007, 172 cites), the *turbulent flow of water-based optimization* algorithm (2020, 60 cites), the *rain-fall* optimization algorithm (2017, 130 cites), the *coral reefs* optimization algorithm (2014, 197 cites), etc.

Coincident with the complaints expressed by [15] about the cuckoo search algorithm, another publication by the same authors [16] seed doubts on the so-called *intelligent water drops* algorithm (2009, 508 cites), in the sense that it could be considered similar to the ant colonies algorithm.

13.7 Other Algorithms

Inevitably, there exist other stochastic algorithms that do not fit in the categories we guessed. This section intends to introduce somehow that open field.

13.7.1 Fireworks Algorithm

As a firework explodes, a shower of sparks appears in the adjacent area. Those sparks will explode again and generate new showers of sparks in smaller areas. Gradually, the sparks will search the whole solution space [71] (Fig. 13.30).

The fireworks optimization algorithm was introduced in 2010 by Y. Tan and Y. Zhu, [72] (1076 cites). As a first step, the algorithm chooses at random N locations in the search space, locations where N fireworks are set off. Then after explosions, the locations of sparks are obtained and evaluated. If a termination condition is met, the algorithm stops. If not, N locations are selected from current sparks and fireworks for the next generation of explosions.

There are four main operators in the algorithm, [46]: the explosion operator, the mutation operator, the selection operator, and the mapping rule. The explosion operator is the basic responsible of searching. If the fitness of an explosion is relatively good, then more sparks will be generated within a smaller region around, for exploitation purposes. Otherwise (bad explosion), a few sparks will be generated

Fig. 13.30 Fireworks and sparks

within a larger region, for global exploration. Aside from explosion sparks, a number M of mutation sparks are also generated.

The design of explosions determines two parameters: the number of sparks generated by each firework and the amplitude of explosion for each firework.

Denote as X_i the location of the *i-th* firework, in a d-dimensional space. The number of explosion sparks would be computed as follows:

$$\lambda_i = \lambda \cdot \frac{\max_K(f(X_k)) - f(X_i)}{\sum_j (\max_K(f(X_k)) - f(X_j))}$$

where $\lambda = \sum_i \lambda_i$and $f(\cdot)$ is the objective function. Parameter λ controls the total number of explosion sparks in each generation. The smaller $f(X_i)$ is, the more explosion sparks are generated. Upper and lower limits were imposed to λ_i:

$$\lambda_i = \begin{cases} round(a\lambda) \ if \ \lambda_i < a\lambda \\ round(b\lambda) \ if \ \lambda_i > b\lambda \\ round(\lambda_i) \ otherwise \end{cases}$$

with $(a < b < 1)$.

The explosion amplitude of each firework obeys to the following equation:

$$A_i = A \cdot \frac{f(X_i) - \min_K(f(X_k))}{\sum_j (f(X_j) - \min_K(f(X_k)))}$$

Parameter A controls the total explosion amplitude. The smaller $f(X_i)$, the smaller the amplitude.

It is assumed that the *j-th* spark generated by the *i-th* firework undergoes explosion effects in z dimensions, where:

$$z = round\,(d \cdot rand(0, 1))$$

The algorithm computes the location of the *j-th* spark adding random displacements in the z dimensions. The procedure for it is:

Compute the displacement:

$$h = A_i \cdot rand(-1, 1)$$

For each dimension k belonging to pre-selected z dimensions of X_i, do:

$$X_{i,k} \leftarrow X_{i,k} + h$$

end For
if new position is out of bounds, map it to another position within bounds:

$$X_{i,k} \leftarrow lb_k + \left|X_{i,k}\right| \cdot \mathrm{mod}(ub_k - lb_k)$$

As said before, some mutation sparks are generated. Each mutation spark is produced by choosing at random a firework X_i and a number z of dimensions and then replacing $X_{i,k}$ with $(eX_{i,k})$, where e is drawn from a Gaussian distribution with mean 1 and standard deviation 1.

There is a GitHub repository of firework algorithm variants, [79], in language Python. Likewise, there are libraries of swarm optimization algorithms that include the firework algorithm, like in the *SwarmPackagePy* (GitHub), [68] (also in language Python).

Many improvements and variants of the algorithm have been proposed, as reviewed in [46] (year 2019). Most of them focus on the algorithm operators. In addition, several hybridizations have been introduced. Also, the algorithm has been adapted for multi-objective optimization.

Applications range from machine learning, scheduling/routing problems, engineering, and automatic control, to image processing, robotics, teams of satellites, etc.

13.7.2 Mine Blast Algorithm

Consider a difficult and risky task: to clear a mine field. Mine bombs of different sizes and explosive powers are planted under the ground. Explosion of a mine throws pieces of shrapnel, which may collide with other bombs resulting in their explosion. Our objective is to locate the most dangerous mine; this would be the optimal point (Fig. 13.31).

Fig. 13.31 Clearing a mine field

The mine blast algorithm starts with a first explosion at the "first shot point." The explosion produces a number N_s of shrapnel pieces. Locations of mines, in a field of m dimensions, are represented with X_i.

The location of the first shot would be:

$$X_0 = LB + rand() \times (UB - LB)$$

where *LB* and *UB* are lower and upper bounds of search space.

Now, explosion X_n is responsible for the consequent explosion at X_{n+1}:

$$X_{n+1} = \Delta X_n + \left(\exp(-\sqrt{\frac{m_{n+1}}{d_{n+1}}}\right) X_n$$

where ΔX_n is the position of the shrapnel, d is distance from original explosion, and m is direction of motion from that explosion:

$$d_{n+1} = \sqrt{(X_{n+1} - X_n)^2 + (F_{n+1} - F_n)^2}$$

$$m_{n+1} = \frac{F_{n+1} - F_n}{X_{n+1} - X_n}$$

where F is the objective function value. The initial value d_0 is computed as *UB-LB*.

Two phases of the algorithm iterations are considered: an initial exploration phase and then an exploitation phase. In order to control the transition, a parameter μ (exploration factor) is introduced. While the iteration number is less than μ, we have:

$$d_{n+1} = d_n \times (randn(\cdot))^2$$

$$\Delta X_{n+1} = d_{n+1} \times \cos(k\theta)$$

where θ is $360°/N_s$.

Once the number of iterations k surpass μ:

$$\Delta X_{n+1} = d_n \times rand(\cdot) \times \cos(k\theta)$$

$$d_{n+1} = d_n \,/\, \exp(k/\alpha)$$

where α is a constant for gradual reduction of distances.

The mine blast algorithm was introduced in 2012, [66] (762 cites). Several engineering applications have been published. See [7] for a flowchart of the algorithm steps.

13.7.3 Joining Forces

There are ways for combining forces, in search of an optimal solution.

Genetic algorithms become popular in the 1990s, more or less at the same time when supercomputers, based on parallel processing, started to proliferate. It was natural to apply supercomputers to genetic algorithms. For instance, by running several genetic algorithms in parallel. The research proposed some mechanisms for introducing beneficial interactions among those algorithms (eg., exchanging better individuals). A most cited survey of parallel genetic algorithms is [18], and a more recent panoramic view is [38]. Note that nowadays parallel processing can be done with personal computers and that not only genetic algorithms but many others can be executed in parallel.

Moreover, since the no free lunch theorem notices that for a given problem, one algorithm would render good results, while other would be worse: it seems a convenient idea to run several different algorithms (perhaps at the same time) for the problem at hand and then choose the final better result. In the case of program GODLIKE, available at the MathWorks File Exchange site, this optimizer simultaneously run genetic algorithm, particle swarm optimization, adaptive simulated annealing, and differential evolution, in a linked manner, occasionally exchanging populations. This can be seen as cooperation.

A relevant aspect of evolution of living organisms is *co-evolution*. For example, plants and their pollinators have evolved in mutual benefit. This is also the case of some symbiotic relationships.

Figure 13.32 presents a typical symbiosis example

Note that co-evolution can be also be present in antagonist organisms.

Extending the bio-inspiration approach, the research on optimization algorithms is considering also co-evolution. Most ideas of this approach were presented in the article [58], on *coevolutionary principles*.

In coevolutionary algorithms the fitness of an individual is *subjective*: the individuals are evaluated based on the interactions with other individuals. We could have several convenient solutions (the Doctoral Thesis [29] (180 cites), deals in depth with this topic).

Fig. 13.32 Symbiosis example

A remarkable observation of the mentioned article is that frequently two types of coevolution have been distinguished: *cooperative*, and *competitive*, and that such distinction is connected to game theory.

Let us highlight that a significant event in this context was the publication by Wolpert and Macready, in 2005, of an article entitled: "*Coevolutionary Free Lunches*", [77]. Recall that these were also the authors of the key article on no free lunch theorems [76]. According to the new article, co-evolution opens the door to more successful algorithms or procedures (for certain scenarios). The article contains an interesting introduction, with a special mention of game theory aspects.

Co-evolution is also being considered under the perspective of multi-agent systems. A good introduction of this field is [13]. More recently, [14] offers interesting views on evolutionary and coevolutionary multi-agent systems, considering optimization applications (co-evolution adds diversity to the population, thus improving exploration capabilities).

Let us draw attention to some revealing steps of the history of co-evolutionary methods. The first significant work to make use of co-evolution and subjective fitness was due to Robert Axelrod, in his famous book "*The Evolution of Cooperation*", [11] (more than 38,000 cites). Axelrod investigated on coevolving automata to play the iterated Prisoner's Dilemma.

It is usually considered that the article of W.D. Hillis, [42] (1560 cites), in a scientific journal of Physics, was the first *mise en scène* of coevolutionary algorithms. This contribution demonstrated, for the first time, the superiority of a coevolutionary algorithm compared to an evolving algorithm for the same problem (of sorting networks). It was also the first work to use a predator-prey model (in particular, for a host-parasite case) for co-evolution.

The first coevolutionary algorithms were competitive (based on predator-prey models or others). It was in 1994 when Potter and De Jong proposed a cooperative co-evolution, [59] (1800 cites). The cooperative scheme was to divide a problem into components, and then a subpopulation (sub-species) was assigned to each component. Each subpopulation evolves independently. A solution is created by assembling of component solutions. The number of components can also evolve.

After 1994 a plethora of research contributions appeared. We could pick out some of these contributions, in order to indicate significant directions the research has taken.

For instance, [75] and other publications of R. Paul Wiegand and collaborators, which work on the analysis of cooperative co-evolution, based on evolutionary games. Precedent research identified some problems and difficulties of co-evolution, like cycling behaviors: so analysis efforts were needed to see why and to find remedies.

In [44], Jansen and Wiegand analyzed the advantages of the cooperative coevolutionary approach over traditional evolutionary algorithms, for optimization application. The main advantage they found is the increased exploratory possibilities of co-evolution.

Concerning games, the research has extensively explored *coevolutionary games*, as reviewed in [57] (1700 cites).

An academic contribution—already cited—of special interest is the Doctoral Thesis of S.G. Ficici [29]. Its initial chapters describe the main branches of the coevolutionary research topic: competition, cooperation, games, optimization, etc. In another chapter, the notion of *solution concept* was introduced (this is the main topic of the dissertation). The connection of co-evolution with multi-objective optimization was highlighted. Examples of solution concepts, as summarized in [24], are:

- Simultaneous maximization of outcomes against all opponents
- The Nash equilibrium
- The Pareto optimal equivalence set
- Maximization of expected utility

The criterion used to select individuals depends on the solution concept the experimenter wishes to approximate.

As noticed in [24], the individuals participating in co-evolution have also the role of evaluators. Hence, it is not strange that some publications explicitly consider the evolution of evaluators. A rough example, in familiar terms, might be what happens with soccer: players evolve, and rules/referees evolve, and the public and the media evolve, and there is interaction (and, perhaps, we could say that contemporary soccer is better than before, it is not?).

An example of contribution that leads to practical algorithms is [20]. It proposes an optimization search based on symbiotic multi-species coevolution. Species extinction and speciation events are also considered. The new algorithm is named as PS2O and clearly outperforms PSO (particle swarm optimization). The article includes pseudo-code of the algorithm.

By the way, there is a lot of interest on symbiotic-inspired algorithms for optimization, (see [47] and the survey [2]). There is a MATLAB program with a symbiotic optimization procedure available from the MathWorks File Exchange site, [60].

The intention of the article, [39], is to prove that there are potential free lunches in coevolutionary optimization. On this purpose, the authors propose a framework for the fusion of *many* evolutionary and swarm intelligence algorithms. Armed with this framework, the authors obtain notably competitive results facing 50 benchmark functions and comparing with six well-established evolutionary methods.

Much more could be said, but for space reasons, it would be better to recommend some recent reviews, like {$REF ATO] (2018) on multi-objective optimization with coevolutionary algorithms, or [49] on cooperative co-evolutionary algorithms. GitHub offers several repositories of coevolutionary programs (for example, [26], or [65]).

Before closing the series of chapters on evolutionary optimization, we wish to add references to general sources on information and programs.

S. Mirjalili provided a new MATLAB optimization toolbox, [51], available from the MathWorks File Exchange site. This toolbox includes Grey Wolf, Ant Lion, Dragonfly, Moth-Flame, Multi-Verse, Sine Cosine, and Whale optimization algorithms. Also, this author offers many other algorithm implementations through the mentioned web site.

Another provider on many algorithm implementations is Jingwei Too through the MatWorks File Exchange site (see [73], on *wrapper feature selection*).

We would recommend the article [74], with a critical survey of nature-inspired algorithms. It covers by groups many algorithms, with pertinent comments.

For a compilation of algorithms, see the *bestiary* [17].

13.8 Exercises

1- [13.1] [13.2] The Job Shop Scheduling problem is of crucial interest in industrial production. A number n of jobs should be adequately assigned to a number m of machines. Next figure shows an example of scheduling:

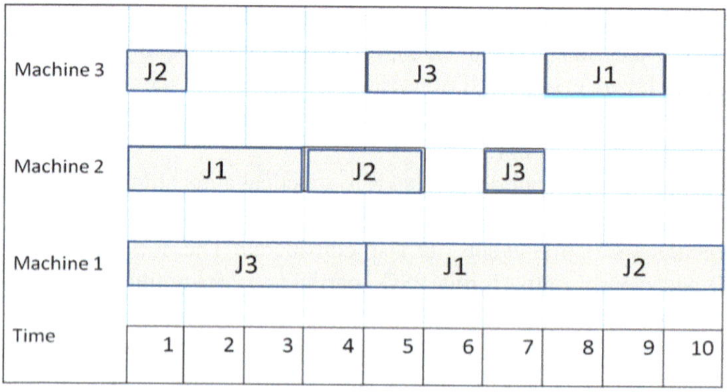

Notice that job J3 at machine 3 waits for J3 to finish at machine 1, etc.

The goal to optimize is minimization of makespan. Makespan is the total time needed to finish all jobs.

There are constraints. One constraint is that two operations of a job cannot be processed at the same time. Another constraint no more than one job can be handled by a machine at the same time.

Here is the specification of a scheduling problem:

Jobs	Machine-time units
1	(1,3), (2,2), (3,2)
2	(1,2), (3,1), (2, 4)
3	(2,4), (3,3)

For instance, (1, 3) means machine 1 during 3 time units.

Please, find minimum makespan using Artificial Immune System.

2- [13.3] [13.4] The nurse rostering problem has NP-hard complexity. Hospital nurses should be assigned to shifts. You can find in the scientific articles large size problems, with hospitals having 1700 nurses or even more. This exercise will be far more modest, although several relevant details will surface.

There are hard constraints and soft constraints. Typically, there are three shifts: morning, afternoon, and night. Hard constraints would be that a nurse should work no more that one shift per day and that all patients should have nursing coverage, etc. Soft constraints, desirable to achieve, could be minimum or maximum number of shifts per week; a nurse does not work a late-night shift followed by a morning shift the next day, etc.

Our problem will be the following: it is the case of a hospital department with 12 nurses. We need to plan next week. Next table gives the preference (soft constraints) of the nurses:

Nurse	A	B	C	D	E	F	G	H	I	J	K	L
Shifts	S1,S3	S2	S3	S1,S2,S3	S2,S3	S1,S3	S1,S2,S3	S3	S2	S1,S2	S2,S3	S1

Hospital rules (hard constraints) are:

- For each shift, one or more nurses must be assigned
- A nurse is not allowed to work two consecutive shifts
- A nurse is not allowed to work more than five shifts a week

Please, apply the Harmony Search algorithm to solve the problem. Describe how you apply the algorithm and the evolution of the iterations.

In case your curiosity demands data for larger problems, consult the page [13.5].

3- [13.6] [13.7] Wireless sensor networks (WSN) are used in the open field, like parking lots, air quality monitoring systems, precision agriculture, forests, etc. They are also used in large buildings, health monitoring, machinery/process maintenance in industrial plants, etc. WSN can include hundreds of sensors, and are usually organized as several clusters. Each cluster can have a selected node, a head. The heads transmit condensed data to one sink, which could be a radio-device for remote monitoring of the WSN. Most times, the sensor are cheap, running on limited energy, with short range communication capabilities.

There are several aspects of WSNs that demand optimization, motivating a lot of research publications [13.8]. In this exercise, we shall focus on cluster head selection.

Please, generate 120 simulated sensors, deploying them at random on a 300×300 m square, for (uniform) covering of the area. In order to apply the Imperialist Competitive algorithm, ten sensors would be randomly chosen as heads. Those heads are empires. The rest of sensors are colonies. Each head is given a number of colonies considering distances. See [13.5] for more details of the proposed method application.

Visualize the evolution of the process in order to select best heads. Explore what minimum communication range is appropriate for the sensors.

4- [13.9] Let us consider the aircraft landing planning (ALP), which is an aspect of air traffic control. The optimal ALP, at certain time slots, on a certain runway (if there were several runways) is a NP-hard, nonlinear and non-convex problem.

The real situation is that, although the airport has a set of planned landing times (PLT), some circumstances could induce deviations, and we get a set of allocated landing times (ALT) that differs from PLT. The main goal is minimizing flight delays, so ALT approached PLT.

The delay of aircraft i in the runway r, is:

$$\Delta_{ir} = \begin{cases} ALT_{ir} - PLT_{ir} \; if \; ALT_{ir} > PLT_{ir} \\ 0, \qquad\qquad\qquad otherwise \end{cases}$$

A parameter is introduced for the representation of the order of aircrafts landing.

$$Seq_{ij} = \begin{cases} 1 \;\; \textit{if aircraft j lands after aircraft i} \\ 0, \; otherwise \end{cases}$$

Another parameter says whether aircraft i and aircraft j land on the same runway, or not:

$$SL_{ij} = \begin{cases} 1 \;\; \textit{if aircraft i and aircraft j land on the same runway} \\ 0, \; otherwise \end{cases}$$

And the fact that aircraft i lands on runway r is indicated as follows:

$$RL_{ir} = \begin{cases} 1 \;\; \textit{if aircraft i lands on runway r} \\ 0, \; otherwise \end{cases}$$

A minimum safe separation time between consecutive landings can be specified as follows:

$$SP_{ij} = \begin{pmatrix} 1.0 & 1.0 & 1.0 \\ 1.5 & 1.5 & 1.0 \\ 2.0 & 1.5 & 1.0 \end{pmatrix}$$

The rows represent leading aircrafts and the columns trailing ones. Aircrafts can be small, large, or huge. The numbers are delays.

Here is a scenario with PLT values (in time units):

Flight	Airplane type	Runway 1	Runway 2	Runway 3
A	H	12	11	10
B	L	15	17	19
C	H	7	9	8
D	H	6	7	8
E	S	10	13	15
F	H	7	6	5
G	L	15	17	19
H	H	7	8	9
I	H	6	7	8
J	S	9	12	15
K	H	6	5	4
L	L	9	7	6
M	S	7	8	9
N	L	6	7	8
P	H	9	8	7

Please use the Gravitational Search algorithm for landing planning, so as to minimize the sum of all delays. Visualize the process.

5- [13.10] One of the preferred uses of the sine cosine algorithm is for mechanical engineering. The scientific literature includes several optimization examples, which are employed for comparison of optimization algorithms. Among those examples, we selected for this exercise the case of pressure vessel design. Next figure depicts the device to be well designed and built:

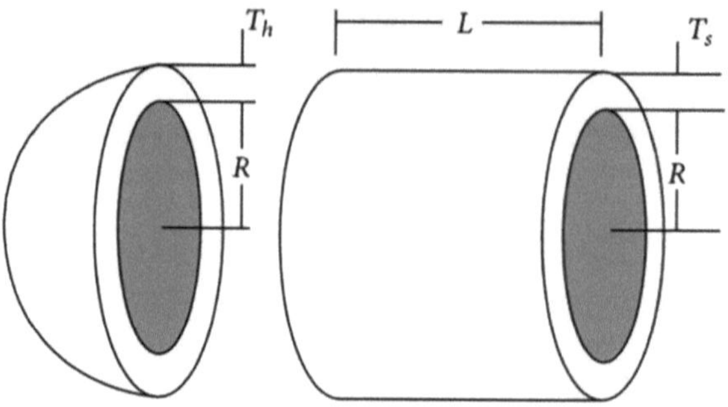

Four variables have to be computed: the thickness of the shell, the thickness of the heads, the internal radius (x_3), and the length (x_4) of the cylindrical section. The cost of the device should be minimized, under certain mechanical (stress, yield) constraints. Specifically, our problem is formulated in the following mathematical terms:

Minimize: $J = 0.6224\, x_1\, x_3\, x_4 + 1.7781\, x_2\, x_3^2 + 3.1611\, x_1^2\, x_4 + 19.84\, x_1^2\, x_3$

Subject to:

$g_1 = -x_1 + 0.0193\, x_3 \leq 0$ (circumferential stress)

$g_2 = -x_2 + 0.00954\, x_3 \leq 0$ (longitudinal stress)

$g_3 = -\pi\, x_3^2 x_4 - \frac{4}{3}\pi\, x_3^3 + 129600 \leq 0$ (volume)

$g_4 = x_4 - 240 \leq 0$ (length)

$1 \times 0.0625 \leq x_1,\ x_2 \leq 9 \times 0.0625$ (thickness, discrete steps)

$$10 \leq x_3,\ x_4 \leq 200$$

Please, apply the sine cosine algorithm to solve the problem. Describe how the method was applied, and the evolution of iterations.

6- [13.11] [13.12] [13.13] This exercise is an invitation to explore sociological studies with modern tools. The particular problem we visit now is community detection. This is the case of a karate club in a certain university. The data concerning this club can be obtained from [13.11] (a page with several interesting details). The fact is that the club splits into two clubs due to internal divisions. Next figure (taken from [13.12]) shows the topological structure of the club (34 members):

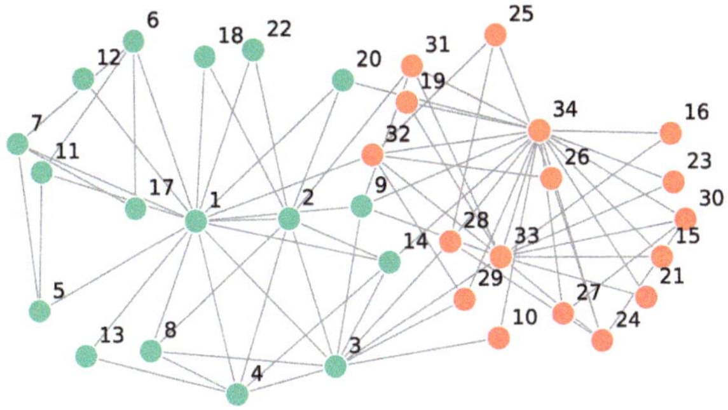

Following the guide provided by [13.13], please apply the biogeography-based optimization algorithm for community detection in this network.

(note) This topic is of evident interest for the analysis of social media.

7- [13.14] [13.15] Recall the kernel-SVM contents in Chap. 7. Now, we want to train a kernel-SVM for a customer-purchases dataset that you can obtain from [13.14]. Follow the guide provided by [13.15].

Please, apply Flower Pollination algorithm for data classification with a kernel SVM, with a gaussian kernel. Visualize the results and add comments on the algorithm behavior.

8- [13.16] Let us insist in kernel-SVM. The mentioned publication [13.15] gets farther by considering a multi-class classification, using a dataset related to wines [13.16].

Please, apply the Fireworks algorithm for training a kernel-SVM, with gaussian kernel, dividing the wine data into three classes. Visualize results and add comments (perhaps you could compare with previous algorithm).

9- [13.17] [13.18] Recall the portfolio optimization section in Chap. 7. A more ambitious dataset can be obtained from [13.17], taking real-world data.

Most heuristics methods have been applied for portfolio optimization, as can be seen in [13.18].

Please, use the Mine Blast algorithm for portfolio optimization concerning any of the datasets from [13.17]. Visualize the results and add comments (perhaps some other algorithm would be better?).

13.8.1 Exercise References

[13.1] https://link.springer.com/content/pdf/10.1007/s00170-005-0226-3.pdf
[13.2] https://www.jiii.org/uploadfile/2021/0611/20210611112336890.pdf
[13.3] https://nottingham-repository.worktribe.com/preview/714701/INS13hsa.pdf
[13.4] Solving Nurse Scheduling/Rostering Problems in Python (Medium)
[13.5] https://www.schedulingbenchmarks.org/nrp/

[13.6] https://www.academia.edu/download/60177740/01_Paper20190801-86618-1vp01ve.pdf
[13.7] https://journals.utm.my/jurnalteknologi/article/view/17685/7955
[13.8] https://pmc.ncbi.nlm.nih.gov/articles/PMC4541950/
[13.9] https://www.ias.ac.in/article/fulltext/sadh/041/02/0183-0191
[13.10] https://ijisrt.com/assets/upload/files/IJISRT22DEC721_(1)_(2).pdf
[13.11] http://konect.cc/networks/ucidata-zachary/
[13.12] Semi-Supervised Learning Guided by the Modularity Measure in Complex Networks.
[13.13] A Biogeography-Based Optimization Algorithm for Community Detection In Complex Networks (IEEE).
[13.14] https://hands-on.cloud/svm-python-tutorial/#h-binary-data-set-for-svm-algorithm
[13.15] https://hands-on.cloud/wp-content/uploads/2024/01/customer_purchases.zip
[13.16] https://scikit-learn.org/stable/modules/generated/sklearn.datasets.load_wine.html
[13.17] https://repod.icm.edu.pl/dataset.xhtml?persistentId=doi:10.18150/ CZYLOV
[13.18] https://link.springer.com/article/10.1007/s00500-023-08177-x

References

1. Abdel-Basset, M., & Shawky, L. A. (2019). Flower pollination algorithm: a comprehensive review. *Artificial Intelligence Review, 52*(4), 2533–2557. Available from: https://link.springer.com/article/10.1007/s10462-018-9624-4
2. Abdullahi, M., Ngadi, M. A., Dishing, S. I., Abdulhamid, S. M., & Usman, M. J. (2020). A survey of symbiotic organisms search algorithms and applications. *Neural Computing and Applications, 32*(2), 547–566. Available from: https://link.springer.com/article/10.1007/s00521-019-04170-4
3. Abualigah, L., & Diabat, A. (2021). Advances in sine cosine algorithm: A comprehensive survey. *Artificial Intelligence Review, 54*(4), 2567–2608. Available from: https://zeus.inf.ucv.cl/~bcrawford/Tesistas2022/Abualigah-Diabat2021_Article_AdvancesInSineCosineAlgorithmA.pdf
4. Adam, S. P., Alexandropoulos, S.-A. N., Pardalos, P. M., & Vrahatis, M. N. (2019). No free lunch theorem: A review. In *Approximation and optimization* (pp. 57–82). Available from: https://link.springer.com/chapter/10.1007/978-3-030-12767-1_5
5. Ala'a, A., Alsewari, A. A., Alamri, H. S., & Zamli, K. Z. (2019). Comprehensive review of the development of the harmony search algorithm and its applications. *IEEE Access, 7*, 14233–14245. Available from: https://ieeexplore.ieee.org/abstract/document/8616762
6. Alarcón Moreno, R. (2022). Study and application of quantum optimization algorithms. Bachelor Thesis, Universidad Complutense de Madrid. Available from: https://docta.ucm.es/rest/api/core/bitstreams/a34cd08f-bb66-4f87-af7a-fb2fd72504e7/content
7. Alattar, A., Selem, S., Metwally, H., Ibrahim, A., Aboelsaud, R., Tolba, M., & El-Rifaie, A. (2019). Performance enhancement of micro grid system with SMES storage system based on mine blast optimization algorithm. *Energies, 12*. Available from: https://www.mdpi.com/1996-1073/12/16/3110

8. Anonymous. (2021). Gravitational search algorithm implementation. Available from: https://github.com/swarm-intelligence/GS-Algorithm
9. Anonymous. (2021). Imperialist competitive algorithm implementation. Available from: https://github.com/swarm-intelligence/IC-Algorithm
10. Atashpaz-Gargari, E., & Lucas, C. (2007). Imperialist competitive algorithm: an algorithm for optimization inspired by imperialistic competition. In *2007 IEEE congress on evolutionary computation* (pp. 4661–4667). IEEE. Available from: https://homepages.ecs.vuw.ac.nz/~yimei/resources/ICA_Atashpaz.pdf
11. Axelrod, R. (1984). *The evolution of cooperation*. Basic Books.
12. Biswas, A., Mishra, K., Tiwari, S., & Misra, A. (2013). Physics-inspired optimization algorithms: A survey. *Journal of Optimization, 2013*. Available from: https://onlinelibrary.wiley.com/doi/pdf/10.1155/2013/438152
13. Bull, L. (1998). Coevolutionary computation: an introduction. Available from: https://citeseerx.ist.psu.edu/document?repid=rep1&type=pdf&doi=d42960eee2417f352acf041d2ef716c9e5037d58
14. Byrski, A., Dreżewski, R., Siwik, L., & Kisiel-Dorohinicki, M. (2015). Evolutionary multi-agent systems. *The Knowledge Engineering Review, 30*(2), 171–186. Available from: https://home.agh.edu.pl/~drezew/papers/drezewski2015evolutionary.pdf
15. Camacho-Villalón, C. L., Dorigo, M., & Stützle, T. (2022). An analysis of why cuckoo search does not bring any novel ideas to optimization. *Computers & Operations Research, 142*, 105747. Available from: https://iridia.ulb.ac.be/~ccamacho/publications/COR-CuckooSearch.pdf
16. Camacho-Villalón, C. L., Dorigo, M., & Stützle, T. (2019). The intelligent water drops algorithm: why it cannot be considered a novel algorithm. *Swarm Intelligence, 13*(3), 173–192. Available from: https://drive.google.com/file/d/1sShLl1Mknl3Mt-0W-JhavKQi87tUbydV/view
17. Campelo, F. (2021). Evolutionary computation bestiary. Available from: http://fcampelo.github.io/EC-Bestiary/.
18. Cantú-Paz, E., et al. (1998). A survey of parallel genetic algorithms. *Calculateurs Paralleles, Reseaux et Systems Repartis, 10*(2), 141–171. Available from: https://www.cs.uoi.gr/~itsoulos/genetic/parallel.pdf
19. Chattopadhyay, S., Marik, A., & Pramanik, R. (2022). A brief overview of physics-inspired metaheuristic optimization techniques. arXiv preprint. Available from: https://arxiv.org/pdf/2201.12810
20. Chen, H., & Zhu, Y. (2008). Optimization based on symbiotic multi-species coevolution. *Applied Mathematics and Computation, 205*(1), 47–60. Available from: https://core.ac.uk/download/pdf/189273433.pdf
21. Dasgupta, D. (1993). An overview of artificial immune systems and their applications. In *Artificial immune systems and their applications* (pp. 3–21). Available from: https://link.springer.com/chapter/10.1007/978-3-642-59901-9_1
22. Dasgupta, D., Yu, S., & Nino, F. (2011). Recent advances in artificial immune systems: Models and applications. *Applied Soft Computing, 11*(2), 1574–1587. Available from: https://staff.fmi.uvt.ro/~daniela.zaharie/am2016/proiecte/tehnici/AIS/AIS_advances.pdf
23. De Castro, L. N., Castro, L. N., & Timmis, J. (2002). *Artificial immune systems: A new computational intelligence approach*. Springer. Available from: https://link.springer.com/book/9781852335946
24. De Jong, E. D., & Bucci, A. (2006). DECA: Dimension extracting coevolutionary algorithm. In *Proceedings of the 8th Annual Conference on Genetic and Evolutionary Computation* (pp. 313–320). Available from: http://bobby.cs-i.brandeis.edu/papers/dejong_bucci_gecco2006.pdf
25. Draa, A. (2015). On the performances of the flower pollination algorithm–qualitative and quantitative analyses. *Applied Soft Computing, 34*, 349–371. Available from: https://dl.acm.org/doi/abs/10.1016/j.asoc.2015.05.015
26. EMI Group. (2020). Incremental optimization. Available from: https://github.com/EMI-Group/Incremental-Optimization

27. Feng, Z.-K., Niu, W.-J., & Liu, S. (2021). Cooperation search algorithm: A novel metaheuristic evolutionary intelligence algorithm for numerical optimization and engineering optimization problems. *Applied Soft Computing, 98*, 106734.
28. Fesanghary, M. (2010). Harmony search algorithm. Available from: https://es.mathworks.com/matlabcentral/fileexchange/28850-harmony-search-algorithm
29. Ficici, S. G. (2004). *Solution concepts in coevolutionary algorithms*. PhD thesis, Citeseer. Available from: http://bobby.cs-i.brandeis.edu/papers/ficici_thesis_04.pdf
30. Franco, D. (2018). CLONALG—clonal selection algorithm for optimization problems. Available from: https://ch.mathworks.com/matlabcentral/fileexchange/67010-clonalg-clonal-selection-algorithm-for-optimization-problems
31. Gabis, A. B., Meraihi, Y., Mirjalili, S., & Ramdane-Cherif, A. (2021). A comprehensive survey of sine cosine algorithm: variants and applications. *Artificial Intelligence Review, 54*(7), 5469–5540. Available from: https://link.springer.com/article/10.1007/s10462-021-10026-y
32. Geem, Z. W., Kim, J. H., & Loganathan, G. V. (2001). A new heuristic optimization algorithm: Harmony search. *Simulation, 76*(2), 60–68. Available from: https://journals.sagepub.com/doi/abs/10.1177/003754970107600201
33. Guo, C., Tang, H., Niu, B., & Lee, C. B. P. (2021). A survey of bacterial foraging optimization. *Neurocomputing, 452*, 728–746. Available from: https://repository.um.edu.mo/handle/10692/40526
34. Guo, W., Chen, M., Wang, L., Mao, Y., & Wu, Q. (2017). A survey of biogeography-based optimization. *Neural Computing and Applications, 28*(8), 1909–1926. Available from: https://link.springer.com/article/10.1007/s00521-016-2179-x
35. Habiboghli, A., & Jalali, T. (2017). A solution to the n-queens problem using biogeography-based optimization. *International Journal of Interactive Multimedia and Artificial Intelligence*. Available from: https://reunir.unir.net/bitstream/handle/123456789/11744/ijimai20174_4_3_pdf_22761.pdf?sequence=1
36. Hakemi, S., Houshmand, M., KheirKhah, E., & Hosseini, S. A. (2022). A review of recent advances in quantum-inspired metaheuristics. *Evolutionary Intelligence*. Available from: https://link.springer.com/article/10.1007/s12065-022-00783-2
37. Haktanirlar Ulutas, B., & Kulturel-Konak, S. (2011). A review of clonal selection algorithm and its applications. *Artificial Intelligence Review, 36*(2), 117–138. Available from: https://link.springer.com/content/pdf/10.1007/s10462-011-9206-1.pdf
38. Harada, T., & Alba, E. (2020). Parallel genetic algorithms: a useful survey. *ACM Computing Surveys (CSUR), 53*(4), 1–39. Available from: https://dl.acm.org/doi/abs/10.1145/3400031
39. He, M., Hu, Y., Chen, H., Sun, L., Wang, X., Su, W., Liu, F., Liang, X., & Ma, L. (2019). Lifecycle coevolution framework for many evolutionary and swarm intelligence algorithms fusion in solving complex optimization problems. *Swarm and Evolutionary Computation, 47*, 3–20.
40. Hernández-Ocaña, B., García-López, A., Hernández-Torruco, J., & Chávez-Bosquez, O. (2022). Bacterial foraging based algorithm front-end to solve global optimization problems. *Intelligent Automation & Soft Computing, 32*(3). Available from: https://cdn.techscience.cn/ueditor/files/iasc/TSP_IASC-32-3/TSP_IASC_23570/TSP_IASC_23570.pdf
41. Hetmaniok, E., Zielonka, A., & Słota, D. (2012). Application of the clonal selection algorithm for reconstruction of the third kind boundary condition. *Zeszyty Naukowe. Matematyka Stosowana/Politechnika Śląska*. Available from: https://www.infona.pl/resource/bwmeta1.element.baztech-dd1e0d4d-1001-44ad-8d7f-9a9950276acb
42. Hillis, W. D. (1990). Co-evolving parasites improve simulated evolution as an optimization procedure. *Physica D: Nonlinear Phenomena, 42*(1–3), 228–234. Available from: https://www.cse.unr.edu/~sushil/class/gas/papers/DannyHillisCoevolution.pdf
43. Hosseini, S., & Al Khaled, A. (2014). A survey on the imperialist competitive algorithm metaheuristic: Implementation in engineering domain and directions for future research. *Applied Soft Computing, 24*, 1078–1094. Available from: https://www.infona.pl/resource/bwmeta1.element.elsevier-c1cb3d2c-263c-3a4f-923a-07ccc384f896

44. Jansen, T., & Wiegand, R. P. (2004). The cooperative coevolutionary (1+ 1) ea. *Evolutionary Computation, 12*(4), 405–434. Available from: https://www.eecs.ucf.edu/~wiegand/papers/JansenWiegand2004ecj.pdf
45. Kumar, M., & Kulkarni, A. J. (2019). Socio-inspired optimization metaheuristics: A review. In *Socio-cultural inspired metaheuristics* (pp. 241–265). Available from: https://link.springer.com/chapter/10.1007/978-981-13-6569-0_12
46. Li, J.. & Tan, Y. (2019). A comprehensive review of the fireworks algorithm. *ACM Computing Surveys (CSUR), 52*(6), 1–28.
47. Lin, N., Jing, S., Liang, X., Yuan, W., & Chen, H. (2015). Biomimicry of symbiotic multi-species coevolution for global optimization. *Metallurgical & Mining Industry, 7*. Available from: https://www.metaljournal.com.ua/assets/Journal/english-edition/MMI_2015_7/029Na%20Lin187-194.pdf
48. Lopez, G. (2023). TS-MBFOA_GUI. Available from: https://github.com/garcialopez/TS-MBFOA-GUI
49. Ma, X., Li, X., Zhang, Q., Tang, K., Liang, Z., Xie, W., & Zhu, Z. (2019). A survey on cooperative co-evolutionary algorithms. *IEEE Transactions on Evolutionary Computation, 23*(3), 421–441. taff.fmi.uvt.ro/ daniela.zaharie/ma2019/Projects/ResearchPapers/CooperativeCoevolution/CooperativeCoevolution_Survey_2019.pdf
50. Mahdavi, M., Fesanghary, M., & Damangir, E. (2007). An improved harmony search algorithm for solving optimization problems. *Applied Mathematics and Computation, 188*(2), 1567–1579. Available from: https://www.infona.pl/resource/bwmeta1.element.elsevier-554cf7d4-cc8e-3151-9328-f994689910d8
51. Mirjalili, S. (2018). a new MATLAB optimization toolbox. Available from: https://es.mathworks.com/matlabcentral/fileexchange/47465-sos-m?s_tid=srchtitle
52. Mirjalili, S. (2018). SCA:a sine cosine algorithm. Available from: https://es.mathworks.com/matlabcentral/fileexchange/54948-sca-a-sine-cosine-algorithm?s_tid=ta_fx_results
53. Mirjalili, S. (2016). SCA: a sine cosine algorithm for solving optimization problems. *Knowledge-based Systems, 96*, 120–133. Available from: https://dl0.programstore.ir/files/Uploades/Lib/PDF/SCA.pdf
54. Mirjalili, S., Mirjalili, S. M., & Hatamlou, A. (2016). Multi-verse optimizer: A nature-inspired algorithm for global optimization. *Neural Computing and Applications, 27*(2), 495–513. Available from: https://dl0.programstore.ir/files/Uploades/Lib/PDF/Multi-Verse%20Optimizer.pdf
55. Mousavi, S. M. H. (2022). N-queen BBO. Available from: https://github.com/SeyedMuhammadHosseinMousavi/n-Queen-BBO#readme
56. Passino, K. M. (2002). Biomimicry of bacterial foraging for distributed optimization and control. *IEEE Control Systems Magazine, 22*(3), 52–67. Available from: https://ieeexplore.ieee.org/abstract/document/1004010
57. Perc, M., & Szolnoki, A. (2010). Coevolutionary games—a mini review. *BioSystems, 99*(2), 109–125. Available from: http://www.matjazperc.com/publications/BioSystems_99_109.pdf
58. Popovici, E., Bucci, A., Wiegand, R. P., & De Jong, E. D. (2012). Coevolutionary principles. Available from: https://www.cs.tufts.edu/comp/150GA/handouts/nchb-main.pdf
59. Potter, M. A., & De Jong, K. A. (1994). A cooperative coevolutionary approach to function optimization. In *International Conference on Parallel Problem Solving from Nature* (pp. 249–257). Springer. Available from: https://cs.gmu.edu/~mpotter/pubs/ppsn94.pdf
60. Prayogo, D. (2015). SOS.m. Available from: https://es.mathworks.com/matlabcentral/fileexchange/47465-sos-m?s_tid=srchtitle
61. Rai, R., Das, A., Ray, S., & Dhal, K. G. (2022). Human-inspired optimization algorithms: Theoretical foundations, algorithms, open-research issues and application for multi-level thresholding. In *Archives Of Computational Methods In Engineering* (pp. 1–40). Available from: https://link.springer.com/article/10.1007/s11831-022-09766-z
62. Rashedi, E., Nezamabadi-Pour, H., & Saryazdi, S. (2009). GSA: A gravitational search algorithm. *Information Sciences, 179*(13), 2232–2248. Available from: https://dl0.programstore.ir/files/Uploades/Lib/PDF/Gravitational%20Search%20Algorithm.pdf

63. Rashedi, E., Rashedi, E., & Nezamabadi-Pour, H. (2018). A comprehensive survey on gravitational search algorithm. *Swarm and Evolutionary Computation, 41*, 141–158.
64. Ristic, O., Milunović Koprivica, S., &S Milošević, M. (2022). Teaching and learning inspired optimization algorithms: A review. In *Proceedings TIE 2022 9th International Scientific Conference Technics and Informatics in Education*. University of Kragujevac, Faculty od Technical Sciences, Čačak. Available from: http://www.ftn.kg.ac.rs/konferencije/TIE2022/docs/papers/S413_25.pdf
65. Robitash-Chandra. (2020). Cooperative-coevolution-CMAES. Available from: https://github.com/EMI-Group/Incremental-Optimization
66. Sadollah, A., Bahreininejad, A., Eskandar, H., &S Hamdi, M. (2013). Mine blast algorithm: A new population based algorithm for solving constrained engineering optimization problems. *Applied Soft Computing, 13*(5), 2592–2612. Available from: https://tarjomefa.com/wp-content/uploads/2016/08/4846-English.pdf
67. Simon, D. (2008). Biogeography-based optimization. *IEEE Transactions on Evolutionary Computation, 12*(6), 702–713. Available from: https://staff.fmi.uvt.ro/~daniela.zaharie/ma2017/projects/techniques/BBO/BigeographyOptimiz.pdf
68. SISDevelop. (2017). Swarmpackagepy 1.0. Available from: https://pypi.org/project/SwarmPackagePy/
69. Sloman, L. (2021). Mathematician answers chess problem about attacking queens. *Quanta Magazine*. Available from: https://www.quantamagazine.org/mathematician-answers-chess-problem-about-attacking-queens-20210921/#:~:text=Then%2Dqueensproblemis, nowallbutsolvedit
70. Surjanovic, S., & Bingham, D. (2013). Ackley function. Available from: https://www.sfu.ca/~ssurjano/ackley.html
71. Tan, Y., Yu, C., Zheng, S., & Ding, K. (2013). Introduction to fireworks algorithm. *International Journal of Swarm Intelligence Research (IJSIR), 4*(4), 39–70. Available from: https://www.igi-global.com/article/introduction-to-fireworks-algorithm/105168
72. Tan, Y., & Zhu, Y. (2010). Fireworks algorithm for optimization. In *International Conference in Swarm Intelligence* (pp. 355–364). Springer. Available from: https://fardapaper.ir/mohavaha/uploads/2018/12/Fardapaper-Fireworks-Algorithm-for-Optimization.pdf
73. Too, J. (2021). a new matlab optimization toolbox. Available from: https://es.mathworks.com/matlabcentral/fileexchange/84139-wrapper-feature-selection-toolbox
74. Tzanetos, A., & Dounias, G. (2021). Nature inspired optimization algorithms or simply variations of metaheuristics? *Artificial Intelligence Review, 54*(3), 1841–1862. Available from: https://link.springer.com/article/10.1007/s10462-020-09893-8
75. Wiegand, R. P., Liles, W. C., & De Jong, K. A. (2002). Analyzing cooperative coevolution with evolutionary game theory. In *Proceedings of the 2002 Congress on Evolutionary Computation. CEC'02 (Cat. No. 02TH8600)* (Vol. 2, pp. 1600–1605). IEEE. Available from: https://www.cse.unr.edu/~sushil/class/gas/papers/cea/cec02-egt.pdf
76. Wolpert, D. H., & Macready, W. G. (1997). No free lunch theorems for optimization. *IEEE Transactions on Evolutionary Computation, 1*(1), 67–82. Available from: https://marksmannet.com/RobertMarks/Classes/ELC5370/Files/IEEE_NoFreeLunch.pdf
77. Wolpert, D. H., & Macready, W. G. (2005). Coevolutionary free lunches. *IEEE Transactions on Evolutionary Computation, 9*(6), 721–735. Available from: https://core.ac.uk/download/pdf/10516539.pdf
78. Wolpert, D. H., Macready, W. G., et al. (1996). No free lunch theorems for search. Technical report, Technical Report SFI-TR-95-02-010, Santa Fe Institute. Available from: http://www.no-free-lunch.org/WoMa96b.pdf
79. Xu, W., Li, Y. (2018). firework-algorithm. Available from: https://github.com/cilatpku/firework-algorithm
80. Yang, X.-S. (2012). Flower pollination algorithm for global optimization. In *International Conference on Unconventional Computing and Natural Computation* (pp. 240–249). Springer. Available from: https://arxiv.org/pdf/1312.5673

81. Yang, X.-S. (2014). Flower pollination algorithm. Available from: https://es.mathworks.com/matlabcentral/fileexchange/45112-flower-pollination-algorithm
82. Zukhri, Z., et al. (2020). Hybridization of genetic algorithm and artificial immune system for assignment problem. In *IOP Conference Series: Materials Science and Engineering* (Vol. 803, pp. 012024). IOP Publishing.

Chapter 14
Multi-Objective Optimization

14.1 Introduction

Many daily-life decisions imply a satisfactory compromise between certain extremes, for example, using the car to move from home to another town. If you go very fast, be prepared to pay a high fuel consumption; if you go very slow, the trip would take a long-long time. Probably, you would prefer a certain balance: not much cost, not much time. This means two objectives.

This was an introductory example, which could transmit the sensation that multi-objective optimization is a simple topic, with perhaps some difficulties about (personal) preferences. As it will be seen in this chapter, this is not completely the case.

A motivating benefit of the chapter is that the concepts and methods to be introduced would be helpful for a clearer judgment of situations and alternatives, when making decisions.

One of the issues to be considered is ordering. An example would show what could it mean [138]. Suppose you want to buy a car. Figure 14.1 shows some options to consider on a 2D plane. The horizontal axis corresponds to longer mileage, the vertical axis to cheaper cost.

Now, if you start by seeking longer mileage (horizontal move) and then the cheapest car (vertical direction), the answer is car E. However, if you start in the vertical direction—cheaper cars—and then you follow the horizontal direction, the answer is car G. Same problem, two different answers.

There are many professional situations in which multi-objective optimization/decisions appear such as finance, medicine, engineering, drug design, political responsibilities, family, etc.

Classical methods for multi-objective optimization have two roads to choose: one wears the name of *Pareto*, and the other is scalarization. Both roads are not so different in the end. More options exist, including recent ones, and there are several related topics, belonging to decision-making, games, etc.

J. M. Giron-Sierra, *Introduction to Optimization with Matlab® Examples*,
https://doi.org/10.1007/978-3-032-05048-9_14

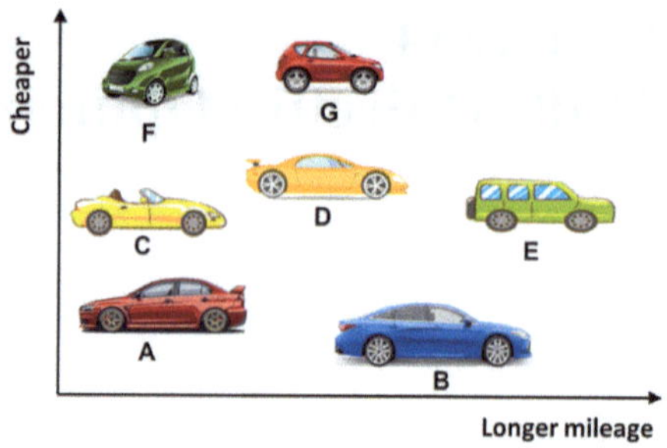

Fig. 14.1 Choosing a car

There has been an explosion of articles, books, etc., devoted to evolutionary methods, like those inspired in biology. Some chapters of our book have been devoted to that topic. Since those methods prove to be useful for multi-objective optimization purposes, they will be considered under this aspect by the end of present chapter. In any case, the main accent we shall adopt in the next pages belongs to the analytical perspective. One of the books in the analytical line is [83].

14.2 Fundamental Concepts

Let us formalize the problem. In general, there will be a set (a vector) of objectives (J_1, J_2, ..., J_m). Also, there will be a set (a vector) of variables (x_1, x_2, ..., x_n). Each objective would be a function of this set of variables, so there is set (a vector) of functions (f_1, f_2, ..., f_m).

Figure 14.2 depicts a simple example with three objectives and two variables.

The space on the left in Fig. 14.2 is called the *decision variable space*, and the space on the right is called the *objective function space*. Variables are confined into a feasible region F (not necessarily an ellipse, of course). Notice that the arrowed curve represents the set of three functions. The image of F under the set of functions would be denoted as Z (not necessarily a cube).

Now a most important concept: the ***Pareto dominance***.

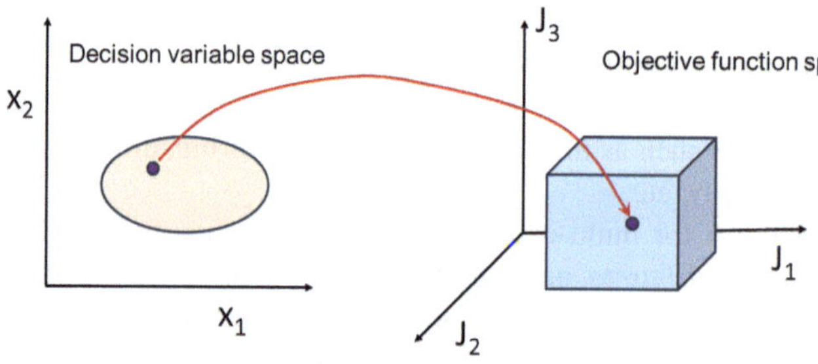

Fig. 14.2 A simple example, with three objectives and two variables

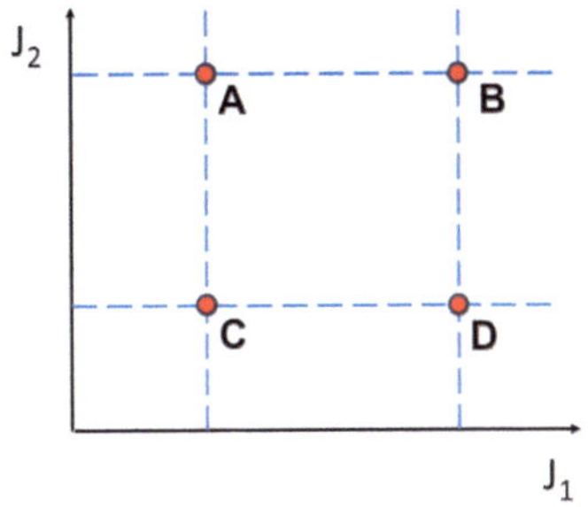

Fig. 14.3 Four vectors in the objective functional space

Suppose two m-dimensional points A and B (vectors) on the objective function space. It is said that A *dominates* B, if and only if:

$$\forall i \in \{1, \ldots, m\} : a_i \leq b_i$$

and,

$$\exists i \in \{1, \ldots, m\} : a_i < b_i$$

where a_i are the components of vector A, and b_i the components of vector B.

By using dominance, it is possible to compare (to an extent) solutions in the objective function space. If a vector in this space is not dominated by any other vector, we say that it is a *nondominated* vector.

Figure 14.3 depicts four vectors in a bidimensional objective function space. The purpose of the figure is to discuss some mutual-relation aspects.

Vector C is strictly less than B in both objectives (then, C dominates B). Also, vector C dominates A, since with respect to J_1 both vectors are equal, but in J_2 vector C is strictly less than A. By similar reasons (*mutatis mutandis*), C also dominates D. Hence, C is a nondominated vector.

Some vectors in the figure can be *incomparable*. For example, A and D, because in J_1 vector A is less than D, but in J_2 ,vector D is less than A.

On the basis of these concepts, Pareto defined optimality as follows:

A solution **x** belonging to F is optimal if there does not exist another solution **y** in F such that **f(y)** dominates **f(x)**.

This is called *Pareto-optimality*. Clearly, the image of an optimal solution is a nondominated vector.

Before continuing with more theoretical substance, let us contemplate some history of the topic.

In 1881, Prof. F.Y. Edgeworth was the first to define an optimum for multicriteria economic decision making. His academic work took place at King's College, London, and later at Oxford Economics.

Vilfredo Pareto was born in Paris, 1848. After his doctorate at University of Turin, he worked as engineer in Florence. In 1898, he becomes the Chair of

Fig. 14.4 Francis Ysidro Edgeworth, and Vilfredo Pareto

Political Economy at the University of Lausanne, where he developed his theory contributions, including the Pareto optimum.

In the 1970s, Prof. Wolfram Stadler, San Francisco State University, began to apply Pareto's concepts to engineering and science (see [146, 147], both by W. Stadler, with historical notes on the topic development from roots).

Figure 14.4 presents photographs of F.Y. Edgeworth (on the left), and V. Pareto (on the right).

Two heavily cited books are [150] (5855 cites) and [44] (5752 cites). Both texts refer in the title to multiple criteria or multi-criteria optimization; this fact reflects the original interest for decision-making. The expression multi-objective took precedence afterward, motivated by mathematical and engineering trends.

14.3 The Pareto Front and Related Aspects

In general, for a given multi-objective problem, several Pareto-optimal solutions would exist. These solutions constitute the Pareto optimal set.

The image of the Pareto-optimal solutions constitutes the so-called *Pareto front*.

Figure 14.5 illustrates the scenario: points in dark gray correspond to Pareto-optimal solutions.

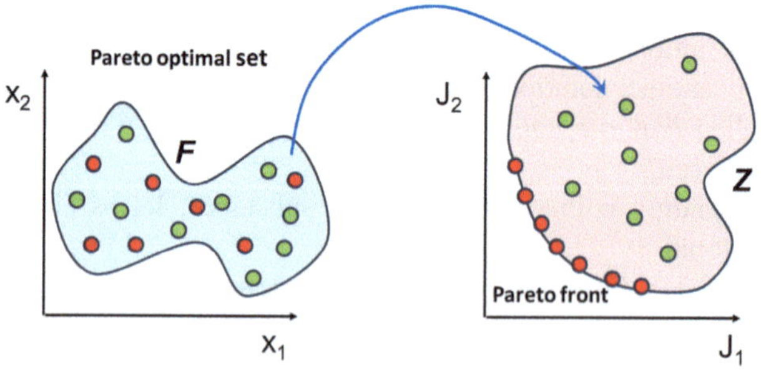

Fig. 14.5 Pareto optimal set, and Pareto front

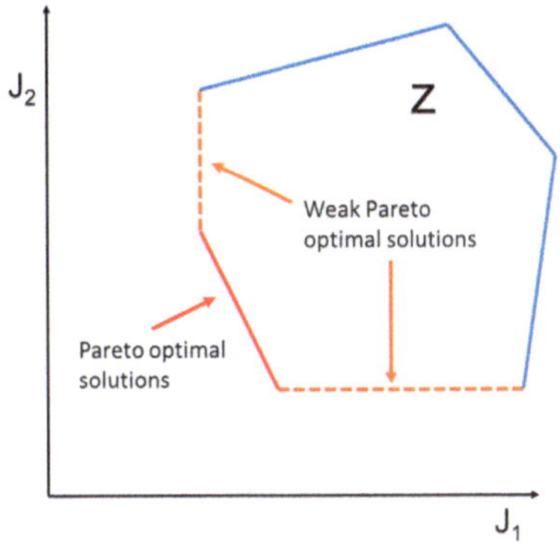

Fig. 14.6 Example of Pareto optimal solutions, and weak Pareto optimal solutions

In familiar terms, looking at the bidimensional example shown in Fig. 14.5, it can be said that each nondominated point forming the Pareto front have no neighbor on the west, nor on south (in the objective function space).

Another aspect of interest is the *weak Pareto optimality*. It is said that a solution **x** belonging to F is weakly Pareto optimal if and only if there does not exist another point **y** in F such that $\mathbf{f}(\mathbf{y}) < \mathbf{f}(\mathbf{x})$. That is, there is no point **y** that improves all of the objective functions simultaneously.

Figure 14.6 exemplifies the concept of weak Pareto optimal solutions, compared to the Pareto optimal solutions.

We will include below a program that illustrates the use of Pareto fronts. By slight modifications of [148], the following two objective functions have been devised to be treated in that program:

$$J_1 = x_1^4 + x_2^4 + x_1 x_2 - x_1^2 x_2^2 - 9\, x_1^2 + 25$$

$$J_2 = x_1^4 + x_2^4 + x_1 x_2 - x_1^2 x_2^2 - 3\, x_2^2 + 5$$

Figure 14.7 presents contour plots of J1, on the left, and J2, on the right. The minima of these functions are at a certain mutual distance. The plots are useful for choosing a pertinent range of exploration for the decision variables.

The Fig. 14.7 has been generated with the Program 14.1, in MATLAB.

Program 14.1 Visualization of J1 and J2

```
% visualization of J1 and J2
b = 0:0.05:3;
[vx1,vx2] = meshgrid(b);
J1= vx1.^4+vx2.^4+vx1.*vx2-(vx1.^2).*(vx2.^2) -...
9*vx1.^2 + 25;
J2= vx1.^4+vx2.^4+vx1.*vx2-(vx1.^2).*(vx2.^2) -...
3*vx2.^2 + 5;
figure(1)
subplot(1,2,1)
```

```
contour(vx1,vx2,J1,30,'LineWidth',1.5);
xlabel('x1'); ylabel('x2');
subplot(1,2,2)
contour(vx1,vx2,J2,30,'LineWidth',1.5);
xlabel('x1'); ylabel('x2');
```

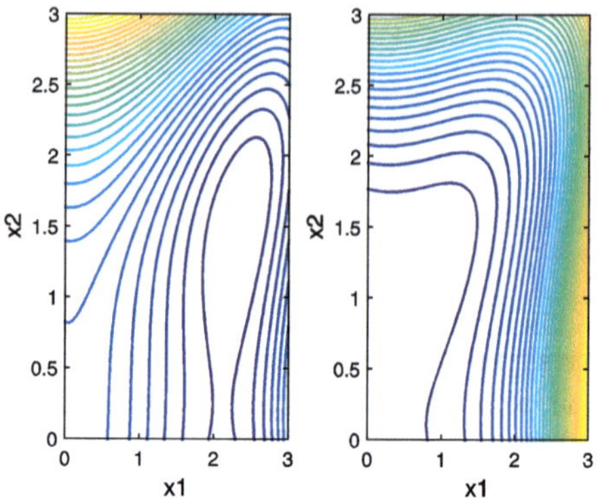

Fig. 14.7 Example functions

To show how the Pareto front could be obtained by computer, a program in MATLAB has been devised, Program 14.2. It is based on [97]. The program expects that the objective functions were non-negative (as usual) in the region of decision variables to be explored.

In the line of Monte Carlo methods (see for instance [REF GIR]), the idea is to generate at random possible solutions in the decision variable space and plot their images in the objective function space. Part of the code is devoted to detect which of the image points are nondominant, so they belong to the Pareto front.

Here is the program.

Program 14.2 Example of Pareto analysis using random data

```
% example of Pareto analysis on two spaces
% generation at random of decision variable
% space points
N = 500; % number of points
Var = 3; % variance
x= Var.*rand(N,2);
%image of those points due to the next functions:
% memory reserve:
J1=zeros(N,1); J2=zeros(N,1); J = zeros(N,2);
for i =1:N
   J1(i)=x(i,1)^4 + x(i,2)^4 + x(i,1)*x(i,2) -...
   (x(i,1)^2)*(x(i,2)^2) - 9*x(i,1)^2+ 25;
   J2(i)=x(i,1)^4 + x(i,2)^4 + x(i,1)*x(i,2) -...
   (x(i,1)^2)*(x(i,2)^2) - 3*x(i,2)^2+ 5;
   J(i,1)=J1(i); J(i,2)=J2(i); % to form a matrix
end
% nondominated points detection in objective
```

```
% function space
ndmt= J; % nondominated points (initialization)
ndmt_ix = 1:N;
other_ix = [];
% (no other row can be better in all elements)
i = 1; j = 1;
while i <= size(ndmt,1)
  r = size(ndmt,1);
  bb = ones(r,1)*ndmt(i,:)-ndmt;
  bb(i,:) = '';
  if any(all(bb'>=0))
    ndmt(i,:)=[];
    other_ix(j) = ndmt_ix(i);
    ndmt_ix(i) = '';
    j=j+1;
    i=i-1;
  end
  i = i+1;
end
if isempty(ndmt_ix)
  disp('No Pareto points found');
end
% minimax nondominated point
[m, ix]=min(max(J(ndmt_ix,:)'));
px = ndmt_ix(ix); % index of the point
J(px,:)
% nondominant point closest to origin
D = []; np = length(ndmt_ix);
for i=1:np,
  D(i) = ndmt(i,1)^2 + ndmt(i,2)^2;
end
[d, jx] = sort (D);
ix = jx(1);
pd=ndmt_ix(ix); %index of the point
J(pd,:)
% display ----------------------------------
figure(1)
plot(J1(:),J2(:),'bo'); hold on
plot(J1(ndmt_ix), J2(ndmt_ix), 'r*');
% minimax point:
plot(J1(px), J2(px), 'k+', 'MarkerSize', 16,...
 'LineWidth',2);
plot([0 J1(px)],[0 J2(px)],'k--');
% closest-to-0 point:
plot(J1(pd), J2(pd), 'mx', 'MarkerSize', 16,...
 'LineWidth',2);
plot([0 J1(pd)],[0 J2(pd)],'m--');
axis([0 45 0 45]);
xlabel('J1'); ylabel('J2');
figure(2)
plot(x(:,1),x(:,2),'bo'); hold on
plot(x(ndmt_ix,1), x(ndmt_ix,2), 'r*');
% minimax point:
```

```
plot(x(px,1), x(px,2),'k+', 'MarkerSize', 16,...
  'LineWidth',2);
%closest-to-0 point:
plot(x(pd,1), x(pd,2),'mx', 'MarkerSize', 16,...
  'LineWidth',2);
axis([0 3 0 3]);
xlabel('x1'); ylabel('x2');
```

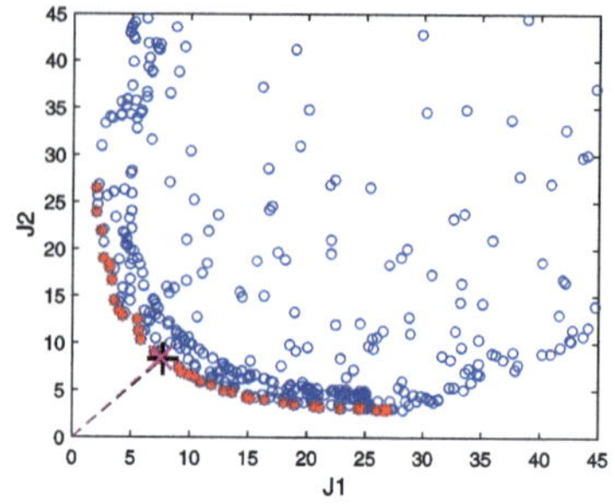

Fig. 14.8 Pareto front of the example

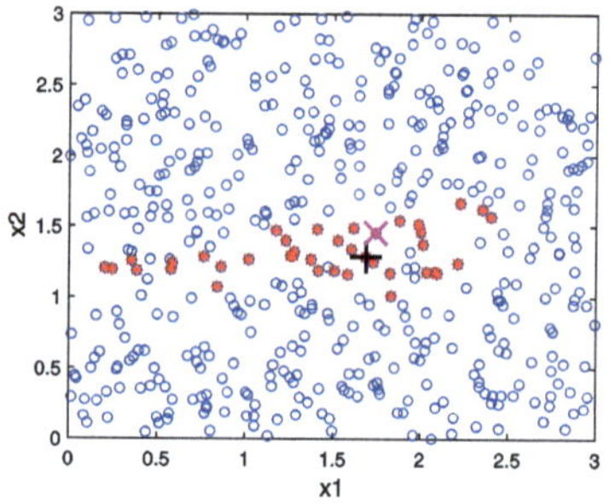

Fig. 14.9 Solutions (decision variables space) of the example

The result of the program are two figures. Figure 14.8 shows in the objective function space, the image of the solutions been tried. The program detects the nondominated points and paint them in dark gray: this is the Pareto front.

Figure 14.9 shows in the decision variable space the solutions that have been tried. Since they are generated at random, each time you run the program you obtain a different set of solutions (points).

Using the indexes of the Pareto front points, it is possible to go backward to identify the corresponding Pareto-optimal solutions in the decision variable space. Those points are also painted in dark gray.

At the time of making decisions, it would be interesting to determine which of the Pareto front points complies with a minimax criterion. For each of the Pareto front points the maximum of their two components is selected, and then the minimum of these maxima is determined, identifying to which point it belongs. This point is marked in Figs. 14.8 and 14.9 with a black cross.

Another Pareto front point of interest is the point with minimum distance to the origin. This point is marked with a dark X. Normally, the closest-to-origin point is near the minimax point. The solutions corresponding to those points have been backwards identified and marked in the decision variables space.

Experts in the use of Pareto fronts take into consideration the so-called nadir and ideal objective vectors (in the objective function space). The components of the *nadir vector* are ($sup\, f_1(x^p)$, $sup\, f_2(x^p)$,... $sup\, f_m(x^p)$), where x^p denote the vectors belonging to the Pareto front. The components of the *ideal vector* are ($inf\, f_1(x^p)$, $inf\, f_2(x^p)$,... $inf\, f_m(x^p)$). The nadir and ideal objective vectors define the upper and lower bounds of the Pareto front.

Another vector of interest is the *utopian objective vector*, which components are:

$$z_i^{utop} = z_i^{ideal} - \varepsilon, \ \forall i \in (1, \ldots, m)$$

where ε is a small constant.

One of the examples provided by MATLAB on Internet shows how to generate Pareto fronts in 3D [149] for problems having three objectives. An interesting publication about visualization of 3D Pareto fronts is [74] (see also [75]).

The MATLAB Optimization Toolbox offers several functions being useful for multi-objective purposes. Concerning Pareto in particular, the functions of interest are *paretosearch( )* and *paretoplot( )*. Other more or less related functions are *gamultiobj( )* (based on Genetic Algorithms), *fminimax( )*, and *fgoalattain( )*.

In the statistics context, another popular analysis tool is the Pareto chart, which combines a bar chart with a line graph. Some software routines related to the Pareto chart from Wolfram, Mathematica, and other companies, are called *paretoplot( )* or something similar, which should not be confused with the *paretoplot( )* of MATLAB. The function in MATLAB for generation of Pareto charts is *pareto( )*.

The famous 80/20 principle belongs to the statistical context.

In addition, there exist a kind of random numbers, which obeys to a Pareto distribution. This aspect is considered in the MATLAB Financial Toolbox.

The use of Pareto fronts is being extended for integer and mixed integer optimization scenarios, like in [26]. It would give Pareto fronts separated into several segments or patches.

14.4 Weighted Sum Approach

The weighted sum method allows the multi-objective problem to be transformed into a single-objective problem, as follows:

$$J = \sum_i w_i J_i$$

where the sum of weights w_i is usually chosen to be 1.

Once the decision-maker has fixed a set of weights, the question is to find the values of the decision variables that minimize J.

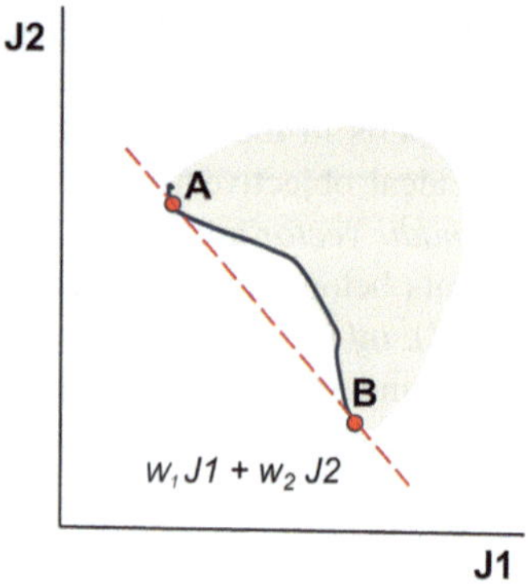

Fig. 14.10 Concave part of a Pareto front

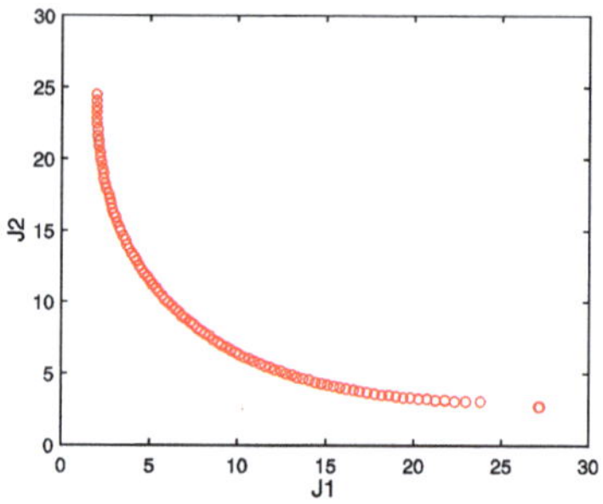

Fig. 14.11 Pareto front of the example

It is an intuitive method, in which the preferences of the decision-maker are easily translated to mathematics, giving higher weight to more important objectives (to be minimized).

In many cases, the Pareto front can be obtained by exploring values of weights. However, if the Pareto front has some concave sections, these cannot be reached with sum of weights, [50] (617 cites).

Figure 14.10 depicts a Pareto front with some part being concave. The weighted sum is unable to get to any of Pareto optimal solutions between A and B, since the line resulting from any choice of weights will hit A or B before any of the points between them [65].

We considered again the example employed in the precedent section and devised a program in MATLAB, Program 14.3, which tests the use of a weighted sum for solving the multi-objective problem. The program generates three figures. Since there are two weights and they sum 1, only one parameter L (range 0..1) suffices for exploring our weighting alternatives.

Figure 14.11 shows that we are able to obtain the Pareto front, by moving L step by step, from 0 to 1, and minimizing the weighted sum in each step.

By using indexes, we backtrack the values of decision variables corresponding to the Pareto front. This is shown in Fig. 14.12.

Figure 14.13 presents an interesting visualization of what happens as we explore the values of L from 0 to 1 and minimize the weighted sum at each step. The values of J1 (starting from (0, 27)) and J2 (curve starting from (0, 3)) are represented. The curve from (0, 3) to (1, 3) is the weighted sum.

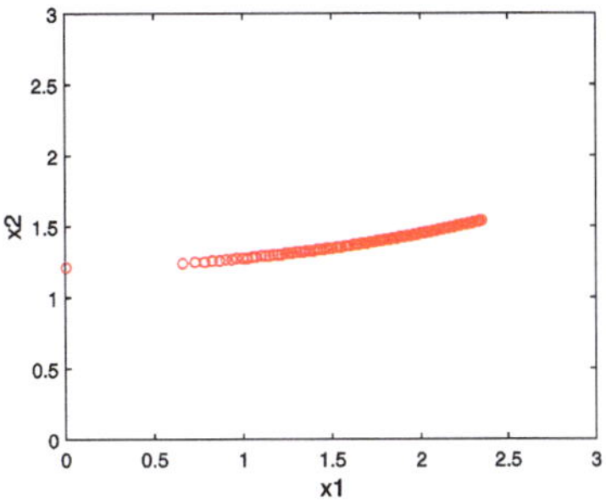

Fig. 14.12 Solutions (decision variables space) of the example

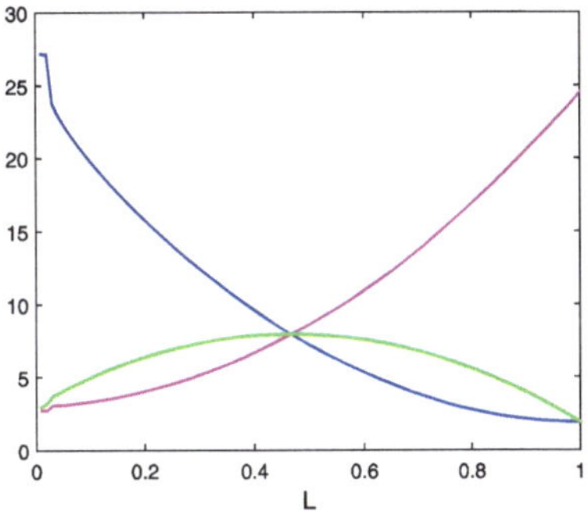

Fig. 14.13 J1 (curve starting from (0, 27)), J2 (starting from (0, 3)), and weighted sum (from (0, 3) to (1, 3)), of the example

The crossing of J1 and J2 curves take place at the minimax point.

In general, the weighted sum approach should be careful with normalization, [98]. It makes no sense to compare objectives of different orders of magnitude without normalization. The extremes of the J1 and J2 curves in Fig. 14.13 show that both J1 and J2 are reasonably aligned in the order of values.

As said before, the three figures above have been generated by a program in MATLAB, which is listed below. Notice the use of the MATLAB function *fmincon( )*.

Program 14.3 Weighted sum approach: an example

```
% example of sum of weights
N=100;
% space reserve
xopt=zeros(N,2);
J1=zeros(N,1); J2=zeros(N,1); J=zeros(N,1);
% prepare the objective function
%(weighted sum, L is the 0..1 weight)
f=@(x,L)((L*(x(1)^4 + x(2)^4 + x(1)*x(2) -...
  (x(1)^2)*(x(2)^2) - 9*x(1)^2+ 25 )) +...
  ((1-L)*((x(1)^4 + x(2)^4 + x(1)*x(2) -...
   (x(1)^2)*(x(2)^2) - 3*x(2)^2+ 5))));
% calculate optimal values for a series of L values
x0= [1,1]; %initial values of x
for ix=1:N,
```

```
    L = ix/N;
    fun = @(x)f(x,L);
    options = optimoptions('fmincon','Display','off');
    % Matlab function for minimization:
    [x, fval] = fmincon(fun,x0,[],[],[],[],[0 0],...
    [3 3],[],options);
    xopt(ix,:)= x;
    J(ix) = fval;
    J1(ix)=x(1)^4+x(2)^4+x(1)*x(2)-...
    (x(1)^2)*(x(2)^2)- 9*x(1)^2+ 25;
    J2(ix)=x(1)^4+x(2)^4+x(1)*x(2)-...
     (x(1)^2)*(x(2)^2)- 3*x(2)^2+ 5;
end
% Objective space (Pareto front)
figure(1)
for ix=1:N,
   plot(J1(ix), J2(ix), 'ro'); hold on;
end;
axis([0 30 0 30]);
xlabel('J1'); ylabel('J2');
% decision variable space
figure(2)
for ix=1:N,
   plot(xopt(ix,1), xopt(ix,2), 'ro'); hold on;
end;
axis([0 3 0 3]);
xlabel('x1'); ylabel('x2');
% combined plot
figure(3)
jx=(1:N)*(1/N); %values of L
plot(jx,J1,'b-','LineWidth',1.5); hold on;
plot(jx,J2, 'm-','LineWidth',1.5);
plot(jx,J, 'g-', 'LineWidth',2);
xlabel('L');
% minimax J value
disp('minimax point');
[val,kx]=min(max(J1,J2));
xopt(kx,:)
disp('J1 and J2 values');
J1(kx)
J2(kx)
```

It was in 1963 when L. Zadeh showed for the first time that the KKT conditions imply that nondominated solutions can be obtained by a weighted sum of objective functions, [167] (1657 cites). There are precedents of this approach from 1955, for two-objective problems [61].

14.5 Other Scalarization Approaches

The term *scalarization* refers to reducing a multi-objective optimization problem to a parameter-dependent single objective optimization problem.

Just in the section above, we have seen an important scalarization method, frequently put into practice: the weighted sum approach.

During last decades, the research has contributed with improvements and new proposals regarding scalarization methods (see for instance [27]). This section would concisely introduce some of these scalarization methods that enjoy a certain degree of maturity and popularity.

Convenient references for this section would be the introduction [77], and the thesis [155].

14.5.1 The ε-Constraint Method

The ε-constraint method is well known, and frequently used, to solve multi-objective problems. One of the objectives is minimized, while the other objectives are considered as constraints bound by a set of constants $\vec{\varepsilon}$.

Then the multi-objective problem is transformed into a classical problem:

Minimize J_k
Subject to $J_i(\vec{x}) \ \leq \ \varepsilon_i$ for all $i \neq k$

The minimization process starts by fixing lower and upper bounds for objective J_k. Then the initial values of the $\vec{\varepsilon}$ constants are made equal to the upper bound, and the minimization problem is solved, providing a first point of the Pareto front. Now, the $\vec{\varepsilon}$ constants are decreased by a certain δ, and the minimization is repeated, obtaining a second point. This is repeated while the $\vec{\varepsilon}$ constants were $\geq$ the lower bound.

This method was introduced in 1971 by Y.Y. Haimes, et al., in 1971, [67] (1953 cites). A more recent article, [31] (year 2013) gives an interesting survey of the modern status and improvements of the method.

The article [120] (year 2016) provides a convenient computational procedure to apply this method. It also includes theoretical aspects.

Based on [4], available from the MathWorks File Exchange site, a MATLAB program (Program 14.4) has been developed to illustrate the ε-constraint method. Here is the code. The same example of precedent sections is again considered.

Program 14.4 Example of epsilon-constraint method

```
% example of epsilon-constraint method
% objective function (J1):
fun = @(x)x(1)^4 + x(2)^4 + x(1)*x(2) -...
```

```
 (x(1)^2)*(x(2)^2) - 9*x(1)^2 + 25;
% process preparation
x0 = [1; 1]; % Starting point
UB = [3; 3]; % Upper bound
LB = [0; 0]; % Lower bound
options = optimoptions('fmincon','Display','off');
N = 50; % Number of Pareto points
% N values of epsilon
epsi_min = 0; epsi_max = 25;
epsi = epsi_min:(epsi_max - epsi_min)/(N-1):epsi_max;
% reserve memory space
xopt = zeros(N,length(x0));
J1 = zeros(N,1);
J2 = zeros(N,1);
% Solve scalarized problem
% for each epsilon value: -----------
for i=1:N
  limt=epsi(i);
[xopt(i,:),J1(i)]=fmincon(fun, x0, [], [], [], [],...
 LB, UB,@(x)nonlcon_epsi(x,limt), options);
  x=xopt(i,:);
  J2(i)=x(1)^4+x(2)^4+x(1)*x(2)-(x(1)^2)*(x(2)^2)-...
   3*x(2)^2 + 5;
end
%display -------------------
figure(1);
plot(xopt(:,1), xopt(:,2), 'rs'); hold on
xlabel('x1')
ylabel('x2')
figure(2);
plot(J1(:), J2(:), 'ro'); hold on
xlabel('J1')
ylabel('J2')
The program uses the following function.
function [C,Ceq] = nonlcon_epsi(x, limt)
% constraint function (based on J2)
C=x(1)^4+x(2)^4+x(1)*x(2)-(x(1)^2)*(x(2)^2)-...
3*x(2)^2+ 5 - limt;
Ceq = [];
```

Notice that the program employs the MATLAB function *fmincon( )*.

By using the program, one obtains two figures with the results. Figure 14.14 shows the optimal solutions on the decision variables space.

And Fig. 14.15 shows the Pareto front corresponding to the example. Some patience would be suitable until getting the results.

The ε-constraint method can work for non-convex parts of Pareto fronts.

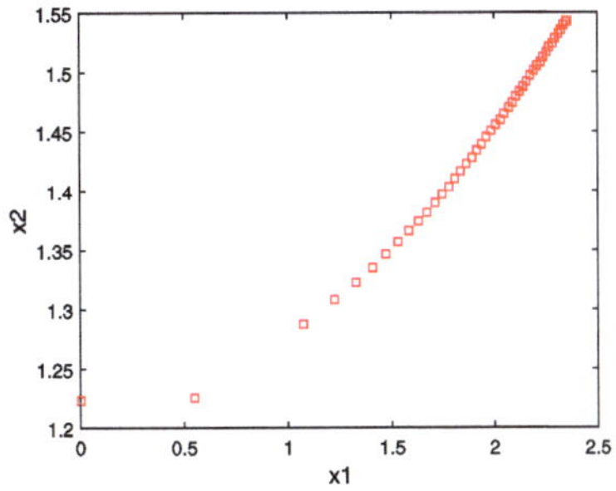

Fig. 14.14 Solutions (decision variables space) of the example

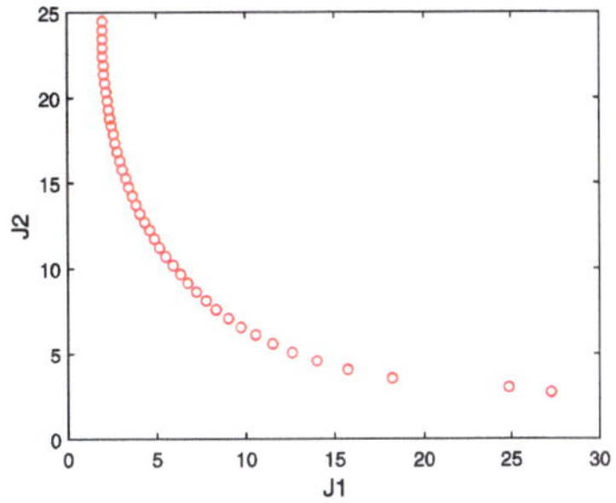

Fig. 14.15 Pareto front of the example

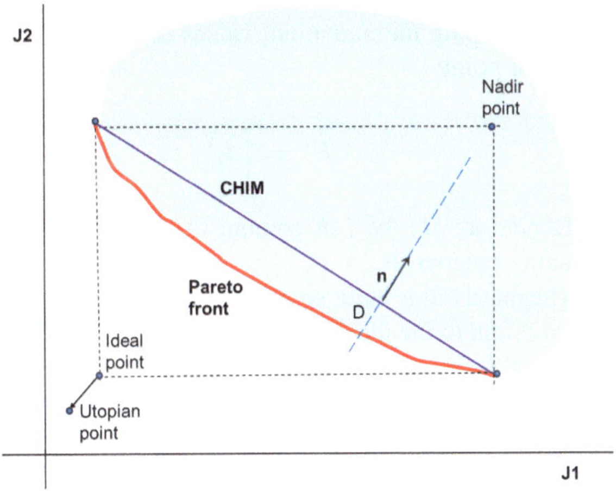

Fig. 14.16 Concept of the NBI method

14.5.2 Normal Boundary Intersection

The normal boundary intersection (NBI) method generates Pareto optimal solutions by means of intersections of "*normal*" lines and the boundary of the viable region. Figure 14.16 illustrates the idea. The normal lines (along vector **n**) are perpendicular to the line CHIM.

The NBI method was introduced by I Das and J.E. Denis in 1998, [37] (2835 cites). Since then, it has gained good popularity.

Next, we shall explain how CHIM is obtained.

Our first task would be to construct the so-called *payoff matrix* Φ.

Let us define a notation as follows. The solution vector that minimizes the *i-th* objective function $J_i(\vec{x})$ is represented as $\vec{x}_i^*$. Then, the minimum value of $J_i(\vec{x})$ at this point is $J_i^*(\vec{x}_i^*)$.

If we compute the value of other objective functions J_k at $\vec{x}_i^*$, we obtain $J_k(\vec{x}_i^*)$, which most probably is not the optimal value of J_k.

Let us build the following matrix M:

$$M = \begin{pmatrix} J_1^*(\vec{x}_1^*) & \ldots & J_k(\vec{x}_1^*) & \ldots J_m(\vec{x}_1^*) \\ \vdots & \ddots & & \\ J_1(\vec{x}_k^*) & \ldots & J_k^*(\vec{x}_k^*) & \ldots J_k(\vec{x}_k^*) \\ \vdots & & \ddots & \\ J_1(\vec{x}_m^*) & \ldots & J_k(\vec{x}_m^*) & \ldots J_m^*(\vec{x}_m^*) \end{pmatrix}$$

With this notation, the ideal point would be:

$$\vec{J}^* = \left[J_1^*(\vec{x}_1^*), \ldots, J_k^*(\vec{x}_k^*), \ldots, J_m^*(\vec{x}_m^*) \right]^T$$

By grouping the maximum values of each objective function, one also obtains the nadir point:

$$\vec{J}^N = \left[J_1^N, \ldots, J_k^N, \ldots, J_m^N \right]^T$$

Denote as M_i the *i-th* column of matrix M. Likewise, denote as Φ_i the *i-th* column of matrix Φ.

The matrix Φ is built, so its columns are $\Phi_i = M_i - \vec{J}^*$ (it corresponds to a shift of the origin to the ideal point).

Recall from the simplex method for linear programming, that vertex points were obtained by linear combinations of the columns of the basis. In a similar form, linear combinations of the columns of Φ would give points of the "*convex hull of individual minima (CHIM)*". This is the line marked as CHIM in Fig. 14.14. The points of that line are obtained with $\Phi\, \vec{w}$, by varying the weight $\vec{w}$ (a column vector).

Notes The original publications that introduced the method, consider that the ideal point and the utopian point are the same (no distance between them). And they call the CHIM line as the utopia line.

Another remark is that some authors introduce further normalizations in the payoff matrix, using lower and upper values of J_k.

In order to illustrate the method, a MATLAB program has been developed (Program 14.5) that obtains and depicts the CHIM line for a bidimensional case (the same example used in precedent sections). The main idea is to explore a range of values of a weight w (between 0 and 1) and obtain CHIM points using $\Phi\, \vec{w}$.

Program 14.5 Example of CHIM line generation

```
% example of CHIM generation for NBI method
% objective functions
J1=@(x)x(1)^4+x(2)^4+x(1)*x(2)-(x(1)^2)*(x(2)^2)-...
9*x(1)^2+ 25;
J2=@(x)x(1)^4+x(2)^4+x(1)*x(2)-(x(1)^2)*(x(2)^2)-...
3*x(2)^2+ 5;
% individual minimization
x0=[1;1];
lb=[0;0];
ub=[3;3];
options=optimoptions('fmincon','Display','off');
[xmJ1,J1optval] = fmincon(J1, x0,[],[],[],[],...
lb,ub,[],options);
[xmJ2,J2optval] = fmincon(J2, x0,[],[],[],[],...
lb,ub,[],options);
% ideal point
Jid = [J1optval; J2optval];
% payoff mmatrix P --------------------
% non-diagonal entries
J2ex= feval(J2, xmJ1);
J1ex= feval(J1, xmJ2);
P = [J1optval-Jid(1) J2ex-Jid(1) ;
J1ex-Jid(2) J2optval-Jid(2)];
% generation of CHIM points
N= 50;
Jim = zeros(N,2);
for i=1:N
  w = [(N-i)/N; i/N];
  Jim(i,:) = P*w;
end;
% find normal direction
% tangent approx.
m=(Jim(N,2)-Jim(1,2))/(Jim(N,1)-Jim(1,1));
nd = -1/m; % perpendicular direction
nX = 5; nY= nd*nX; % perpendicular segment

% arbitrary origin on the CHIM line:
orX = Jim(30,1); orY = Jim(30,2);
%display
figure(1)
plot(Jim(:,1), Jim(:,2),'b-','LineWidth',2); hold on;
% toward origin:
plot([orX orX-nX],[orY orY-nY], 'k-');
% arrow on the normal line:
quiver(orX,orY,-nX,-nY,'LineWidth',2);
xlabel('J1s'); ylabel('J2s');
```

Figure 14.17 was obtained with Program 14.5 and shows the CHIM line on the objective function space (with the origin shifted to the ideal point, as indicated by naming J1s, J2s the axes). It also shows a line segment, which is normal to a certain

Fig. 14.17 Generation of the CHIM line (and its perpendicular)

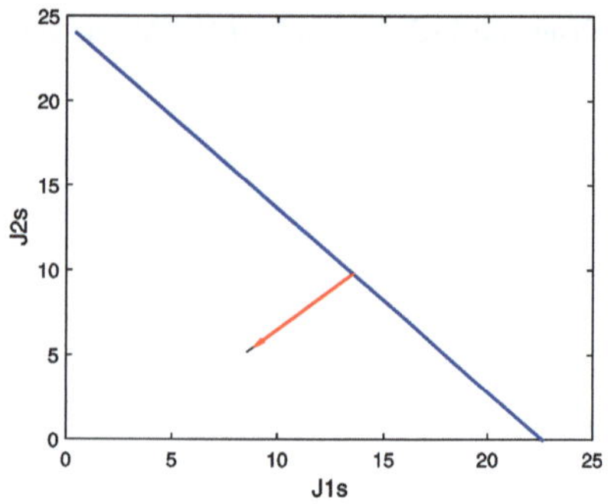

point of the CHIM (as you change the weight $\vec{w}$, you move this point along the CHIM).

Now, the idea is to select a point of the CHIM line, obtain the normal direction, and advance toward the origin until reaching the border of the viable region F.

In more detail: denote as $\hat{n}$ the unit vector normal to CHIM pointing toward the origin (the opposite to $\boldsymbol{n}$ in Fig. 14.14). Then $\Phi\,\vec{w} + D\,\hat{n}$ represents the set of points on the normal line.

The point of intersection between the normal line and the boundary of F (which would be a point of the Pareto front), is found by solving the following optimization problem:

Maximize D
Subject to $\Phi\,\vec{w} + D\hat{n} = F(\vec{x})$

The constraint means that the solution must belong to the viable region F.

Then, the method consists in solving this maximization sub-problem for various weights $\vec{w}$, to obtain a set of Pareto front points.

Program 14.6 is a continuation of the program above, including now the sub-problem optimization at several points of the CHIM. It repeatedly uses *fmincon( )* and a function with the objective function of the optimization sub-problem.

Program 14.6 Pareto front generation with NBI

```
% example of Pareto front generation with NBI
% objective functions
J1=@(x)x(1)^4+x(2)^4+x(1)*x(2)-(x(1)^2)*(x(2)^2)-...
9*x(1)^2+ 25;
J2=@(x)x(1)^4+x(2)^4+x(1)*x(2)-(x(1)^2)*(x(2)^2)-...
3*x(2)^2+ 5;
% individual minimization
x0=[1;1];
lb=[0;0];
ub=[3;3];
options=optimoptions('fmincon','Display','off');
[xmJ1,J1optval] = fmincon(J1, x0,[],[],[],[],...
lb,ub,[],options);
```

```
[xmJ2,J2optval] = fmincon(J2, x0,[],[],[],[],...
lb,ub,[],options);
% ideal point
Jid = [J1optval; J2optval];
% payoff matrix P --------------------
% non-diagonal entries
J2ex= feval(J2, xmJ1);
J1ex= feval(J1, xmJ2);
P = [J1optval-Jid(1) J2ex-Jid(1);
J1ex-Jid(2) J2optval-Jid(2)];
% generation of CHIM points
N= 50;
Jim = zeros(N,2);
Jpareto = zeros(N,2);
xpareto = zeros(N,2);
for i=1:N
  w = [(N-i)/N; i/N];
  Jim(i,:) = P*w;
end;
% find normal direction
% tangent approx.
m=(Jim(N,2)-Jim(1,2))/(Jim(N,1)-Jim(1,1));
nd = -1/m; % perpendicular direction
nX = 5; nY= nd*nX; % perpendicular segment
% arbitrary origin on the CHIM line:
orX = Jim(30,1); orY = Jim(30,2);

% Pareto front generation -------------------------
x0=[1.9;1.7];
lb=[0;0];
ub=[5;5];
for i=6:44
  w = [(N-i)/N; i/N];
  %normal origin on CHIM (it moves with i)
  orig = P*w;
  %line search about D
  %dnew is small until D surpass a limit,
  %this is the frontier
  dnew=0.0; xnew=[0;0]; D=0.0;
  while dnew<0.04
    dold=dnew; xold=xnew;
xnew,dnew] = fmincon(@(x)obj_NBI(x,Jid,orig,D,nd),...
  x0,[],[],[],[],lb,ub,[],options);
    Dold=D;
    D=D+0.1; %decimals
  end
  if Dold == 0,
    D=Dold;
  else
    D=Dold-0.1;
  end
  dnew=0.0; xnew=[0;0];
  while dnew<0.05
    dold=dnew; xold=xnew;
```

```
[xnew,dnew]=fmincon(@(x)obj_NBI(x,Jid,orig,D,nd),...
 x0,[],[],[],[],lb,ub,[],options);
      Dold=D;
      D=D+0.01; %centesimals
    end
    xpareto(i,:)=xold;
    Jp1=J1(xold); Jp2=J2(xold);
    Jpareto(i,:)=[Jp1; Jp2];
end
%display ----------------------------------------
figure(1)
plot(Jim(:,1), Jim(:,2),'b-','LineWidth',2); hold on;
xlabel('J1s'); ylabel('J2s');
plot(Jpareto(:,1),Jpareto(:,2),'k*');
grid;
The program employs the following function:
function f = obj_NBI(x,Jid,orig,D,nd)
% Objective function (get intersection)
lx=orig(1)-D;ly=orig(2)-(nd*D); %motion along normal
distx=(x(1)^4+x(2)^4+x(1)*x(2)-(x(1)^2)*(x(2)^2)-...
9*x(1)^2+ 25 - lx -Jid(1));
disty=(x(1)^4+x(2)^4+x(1)*x(2)-(x(1)^2)*(x(2)^2)-...
3*x(2)^2+ 5 - ly -Jid(2));
f=max(distx,disty);
```

Figure 14.18 shows the result of the program. Notice that Pareto points are evenly distributed: this is one of the advantages of the method.

It should be said that Program 14.6 is somewhat ad hoc, having in view the optimization example being considered and restricting the exploration to a certain range of points on the CHIM. A more general NBI implementation in MATLAB is available from [36].

Another remark to be added is that the original NBI publication [37] introduces the possibility of using a quasi-normal direction, pointing to the origin, instead of the normal direction.

Coming again to the article [37], it conveys a lot of insightful comments on multi-objective optimization and the Pareto fronts. For example, it includes a graphical illustration—similar to Fig. 14.19—that shows a possible version of the NBI method, with the arrow pointing opposite to the origin. This figure shows how

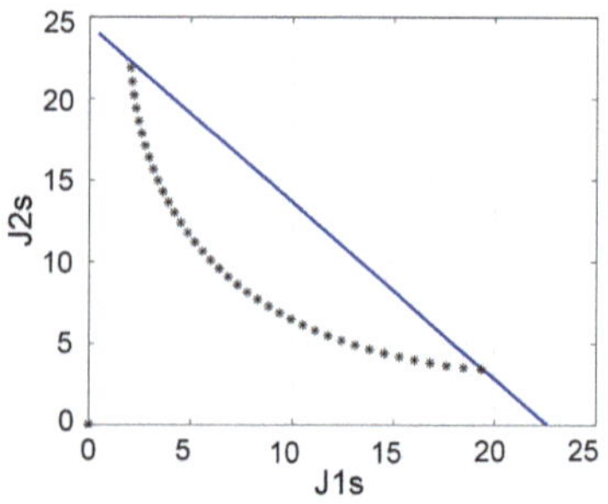

Fig. 14.18 Pareto front obtained with the NBI program

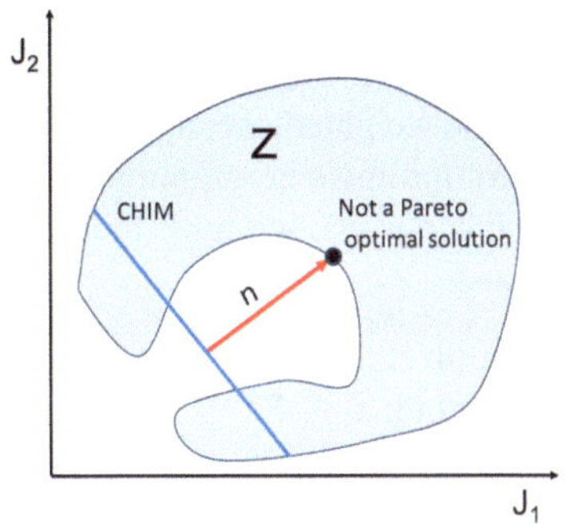

Fig. 14.19 Reaching points in concave boundaries

the method can detect frontier points, which are not Pareto optimal solutions, in concave boundaries of Z.

Another interesting comment of [37] is that in three-dimensional problems, there could exist Pareto optimal points not detectable by NBI. This is one of the issues that has attracted further attention of investigators.

A number of publications have proposed efficient implementations (see for instance [89]) and improvements (like in [141]) of the NBI method.

Let us mention that in [89], its authors consider several chemical industry processes as examples for NBI application.

14.5.3 Weighted Metrics

There are several multi-objective optimization methods based on reference points. For instance, one could choose the ideal point as reference and then try to find *the closest* feasible solution. So, there is a distance—in the objective function space—that can be evaluated with different metrics.

Let us be more specific about metrics. The *Lp-metric* is defined as follows:

$$\|\vec{y}\|_p = \left(\sum |y_k|^p\right)^{1/p}$$

where $1 \leq p \leq \infty$. For p = 1 we have the Manhattan metric. For p = ∞, it is the Chebyshev metric.

Coming to our context, let us consider the following problem:

Minimize: $\left(\sum \left|J_i(\vec{x}) - J_i^*\right|^p\right)^{1/p}$
Subject to: $\vec{x} \in F$

For more general application, we can add weights as follows:

Minimize: $\left(\sum w_i \left|J_i(\vec{x}) - J_i^*\right|^p\right)^{1/p}$ (1)
Subject to: $\vec{x} \in F$

Some authors call this approach *compromise programming*, as M. Zeleny in [168] (1230 cites) (see also [126] and references therein).

The weighted sum approach, already introduced in section 50.4, is a special case of compromise programming with p = 1.

For p = ∞, we have the Chebyshev multi-objective problem, in the following terms:

Minimize: $\max_i \left(w_i \left|J_i(\vec{x}) - J_i^*\right|\right)$
Subject to: $\vec{x} \in F$

Notice that the largest weighted deviation is the only one being considered.

For $1 \leq p < \infty$, the solution obtained by weighted compromise programming (1) is Pareto optimal if this solution is unique, or all weights are nonzero positive. It should be said that for $1 \leq p < \infty$ the method does not necessarily find every Pareto optimal solution.

In turn, the Chebyshev multi-objective approach is able to find every Pareto optimal solution. But, if a solution is not unique, some of the solutions are weakly Pareto optimal.

The article [114] offers a numerical example of application of this method.

In order to identify the Pareto optimal solutions, some authors (like [151]) suggest to aggregate a term to the original problem as follows:

Minimize: $\max_i \left(w_i \left|J_i(\vec{x}) - J_i^*\right|\right) + \rho \sum \left|f_i(\vec{x}) - J_i^*\right|$
Subject to: $\vec{x} \in F$

where ρ is a sufficiently small positive scalar.

This is called *augmented Chebyshev*. Note that with this method, it may be possible not to find some of the Pareto optimal solutions.

14.6 Approaches Based on Goals or Preferences

This section contemplates a series of methods that require the initial intervention of a decision-maker, who specifies goals or preferences.

14.6.1 Lexicographic Method

Frequently, there are most important objectives that should be achieved, on top of other objectives. For example, safety before maximum profit. With the lexicographic method, the objectives are ranked in order of importance—from best to worst—by the decision-maker.

Another name for this method is *preemptive optimization*.

Typically, a sequential algorithm is applied, proceeding as a sequence of single-objective optimization problems. The algorithm starts with the optimization of the most important objective. Then, the optimal value found is added as a constraint for the optimization of the second objective and so on. Therefore, the iterative procedure after the first single-objective optimization could be:

Repeat for $j = 2, \ldots k$:

Minimize: $J_j(\vec{x})$
Subject to: $\vec{x} \in F$

$$J_i(\vec{x}) = J_i^* \quad i = 1, \ldots, j-1$$

(other constraint forms are possible, like for instance inequalities)

The result of the algorithm is a Pareto optimal solution, [44] (book with 5752 cites).

There is a lexicographic simplex algorithm for linear objectives, [76].

And there is a GitHub repository with MATLAB code for lexicographic multi-objective optimization, [54].

14.6.2 Goal-attainment Method

The decision-maker defines targets or goals to attain for each objective. This is given as a goal vector J^{ref}. In addition, the decision-maker specifies a vector of weights $\vec{w}$.

Departing from these requirements, the method proposes the following optimization problem:

Minimize: α
Subject to: $J_i^{ref} + \alpha\, w_i \geq J_i\,(\vec{x})\; i = 1, \ldots, k$

$$\vec{x} \in F$$

where α is a scalar variable.

Figure 14.20 illustrates the idea of this method. By minimizing α a point of intersection with the boundary of Z is obtained, which could be a Pareto optimal point.

Notice that the method is able to touch boundary points in concave segments of the Z border.

The goal-attainment method was introduced by F. Gembicki and Y. Haimes in December 1975, [62]. The method has been used for the optimization of multiresponse systems in [161] and the aircraft control design in [140]. In the year 2019, an enhanced goal-attainment method was proposed by [122], with application in power system operation.

One of the comments of Das and Dennis, in [37], is that the goal attainment method has similarities with the normal boundary intersection (NBI) method. A

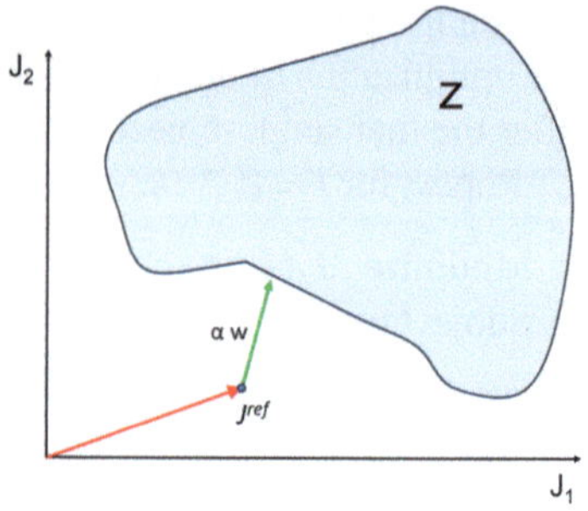

Fig. 14.20 Concept of the goal-attainment method

difference between them is that goal attainment determines one point, while NBI obtains a set of points (this difference, of course, can disappear if you explore changes in the weights and the goal vector when adhering to the goal attainment concept).

The MATLAB Optimization Toolbox includes the function *fgoalattain( )* for using the goal attainment method. See the documentation at

https://mae.ufl.edu/nkim/eas6939/multiobjMATLAB.pdf

A short program (Program 14.7) has been devised for a demonstration on using the function *fgoalattain( )*. The same example considered in precedent programs has been selected for the demonstration.

Program 14.7 Example of goal-attainment method

```
% example of goal attainment method
% objective functions
fun=@(x)[x(1)^4+x(2)^4+x(1)*x(2)-...
  (x(1)^2)*(x(2)^2)- 9*x(1)^2+ 25;...
   x(1)^4+x(2)^4+x(1)*x(2)-(x(1)^2)*(x(2)^2)-...
   3*x(2)^2+5];
Jref = [5,5];
weight = [1,1];
x0 = [1,1];
x = fgoalattain(fun,x0,Jref,weight);
% write results
disp('solution in decision space:')
x
disp('corresponding Pareto point:')
fun(x)
```

14.6.3 *Goal Programming*

This method was introduced by A. Charnes and W.W. Cooper in 1957, [30] (3987 cites). In brief: the decision-maker defines targets or goals to achieve for each objective; this is incorporated in the objective function, which tries to minimize the deviations from targets.

A simple form of the method could be:

Minimize: $\sum |J_i(\vec{x}) - T_i|$
Subject to: $\vec{x} \in F$
where T_i denote the targets.

A more general formulation of the problem, called *generalized goal programming*, is a weighted sum of the *p-th* power of deviations $|J_i(\vec{x}) - T_i|$.

See the book [79], and the presentation [12], for more details. A state of the art of goal programming was described in 1998 by [153].

It is possible to adapt the problem formulation to a linear one being tractable by the simplex algorithm (Linear Programming). With this purpose, new *deviational variables* δ_i^+ and δ_i^- are defined. Both are non-negative scalars. Variable δ_i^+ informs about overachievement, and variable δ_i^- informs about underachievement.

By using these variables, the optimization problem is transformed to:

Minimize: $\sum (\delta_i^+ + \delta_i^-)$
Subject to:

$$f_i(\vec{x}) - \delta_i^+ + \delta_i^- = T_i \quad i = 1, \ldots, k$$

$$\delta_i^+, \delta_i^- \geq 0 \quad i = 1, \ldots, k$$

$$\vec{x} \in F$$

Goal programming can be extended for more flexibility by adding weights. And, what is more, it can take into account a lexicographic ordering of objectives by including priority factors p_i.

Then, a general expression could be the following:

Minimize: $\sum p_i (w_i^+ \delta_i^+ + w_i^- \delta_i^-)$
Subject to:

$$J_i(\vec{x}) - \delta_i^+ + \delta_i^- = T_i \quad i = 1, \ldots, k$$

$$\delta_i^+, \delta_i^- \geq 0 \quad i = 1, \ldots, k$$

$$\vec{x} \in F$$

14.7 Interactive Methods

The search for a satisfactory multi-objective solution could be conveniently aided by the experience (and instinct) of the decision-maker, provided that some interaction was allowed during the computational process.

Let us give more details of some interactive methods.

14.7.1 *Geoffrion-Dyer-Freinberg (GDF) Method*

This method was introduced in 1972 by A. Geoffrion, J. Dyer and A. Feinberg, [63] (1295 cites). It uses a utility function, *only implicitly known*. A direction-finding method is chosen (the authors select the Frank-Wolfe approach), and one starts with the following gradient-based problem:

Maximize: $\nabla_x U(\vec{J}(\vec{x}^h)) \cdot \vec{y}$
Subject to: $\vec{y} \in F$

where $U : \Re^k \rightarrow \Re$ is the utility function, $\vec{x}^h$ is the current point and $\vec{y}$ is the new variable of the problem.

Using the chain rule:

$$\nabla_x U(\vec{J}(\vec{x}^h)) = \sum_{i=1}^{k} \left(\frac{\partial U}{\partial J_i}\right) \nabla_x J_i(\vec{x}^h)$$

Dividing now by $\frac{\partial U}{\partial J_i}$, a reformulation of the problem is obtained:

Maximize: $\left(\sum_{i=1}^{k} -m_i^h \nabla_x J_i(\vec{x}^h)\right) \cdot \vec{y}$
Subject to: $\vec{y} \in F$

where,

$$m_i^h = \left(\frac{\partial U}{\partial J_i}\right) / \left(\frac{\partial U}{\partial J_1}\right) \quad \forall i \,,\, i \neq 1$$

are the *marginal rates of substitution* at $\vec{x}^h$, between objectives J_1 and J_i.

As said in [77], the marginal rate of substitution is the amount of loss on objective J_i that the decision-maker is willing to tolerate in exchange of one unit of gain in objective J_1 (while other objectives remain unchanged).

The interactive GDF procedure involves the following steps, [77]:

1. Provide and initial point $\vec{x}^1 \in F$, set $h = 1$
2. The decision-maker specifies the marginal rates of substitution between $J_1$and the other objectives at the current point $\vec{x}^h$
3. Find the optimal solution $\vec{y}^h$ of the reformulated problem. Set the new search direction $\vec{d}^h = \vec{y}^h - \vec{x}^h$. If $\vec{d}^h = 0$, go to step 5.
4. The decision maker decides the best step-size t^h, to compute the new solution with $\vec{x}^{h+1} = \vec{x}^h + t^h \vec{d}^h$
5. If $\vec{x}^{h+1} = \vec{x}^h$ go to step 5, else set $h = h + 1$ and go to step 1.
6. Return the final solution $\vec{x}^h$.

See [63] for more practical indications in order to follow this approach.

14.7.2 Interactive Weighted Chebyshev Procedure

The interactive weighted Chebyshev procedure was proposed by R.E. Steuer and E-U Choo in 1983, [151]:

Minimize: $\max_i \left\{ (w_i \left|J_i(\vec{x}) - J_i^*\right|), \sum \left|J_i(\vec{x}) - J_i^*\right| \right\}$
Subject to: $\vec{x} \in F$

The procedure follows the next steps, [77]:

1. Determine the ideal point J^*, and set $J^+ = J^* + \vec{\varepsilon}$. Let $\mathbf{W^1}$ be the initial set of weighting vectors. Set $h = 1$
2. Generate a large number (50n) of weighting vectors from $\mathbf{W^h}$
3. Solve the optimization problem (several solutions). Filter the 2P resulting nondominated points to obtain P solutions
4. Show the P solutions to the decision-maker, who selects a preferred one. Let J^h be the selected point
5. If $h = t$ then stop with J^has solution (t is a specified number of iterations).

Else, let $\vec{w}^h$ the weighting vector which generated J^h in step 2; its components are:

$$w_i^h = \frac{1}{J_i^+ - J_i^h} \left[\sum_{j=1}^{k} \frac{1}{J_j^+ - J_j^h} \right] \text{ with } i = 1, \ldots, k$$

Now, determine the reduced set of weighting vectors:

$$\mathbf{W^{h+1}} = \{ \vec{w} \in \Re^k : w_i \in [l_i, u_i], \sum w_i = 1 \}$$

where,

$$[l_i, u_i] = \begin{cases} [0, r^h] & if \ w_i^h \leq r^h/2 \\ [1 - r^h, 1] & if \ w_i^h \geq 1 - r^h/2 \\ [w_i^h - r^h/2, \ w_i^h + r^h/2] & \text{otherwise} \end{cases}$$

and r^h is a specified *convergence factor.*

Set $h = h + 1$ and go to step 1.

14.7.3 Reference Point Methods

In this method, the decision-maker gives a reference point, which represents the aspiration levels for each objective, [160] (1266 cites). Then, an achievement scalarizing function is optimized in relation with the reference point. If the decision-maker is satisfied the interactive process ends; otherwise, the decision-maker provides another reference point.

The achievement scalarizing function is defined as follows: it is a parameterized function $s_{Jref}(J) \ : \ \Re^k \ \rightarrow \ \Re$, where J^{ref} is the reference point given by the decision-maker.

Based on the scalarizing function, the multi-objective problem is transformed to:

Minimize: $s_{Jref}(J)$
Subject to: $J \ \in \ Z$

In most cases, achievement scalarizing functions are based on Chebyshev metrics. An example of this is the augmented weighted Chebyshev scalarizing function:

$$s_\infty(J, J^{ref}) = \max_i \left(\lambda_i \left(J_i - J_i^{ref}\right)\right) \ + \rho \sum_{i=1}^{k} \lambda_i (J_i - J_i^{ref})$$

where $\rho > 0$ sufficiently small, and $\forall \lambda_i \geq 0$, with at least one $\lambda_i > 0$.

Usually, the decision-maker moves the reference point at each iteration, to explore alternatives permitted by the objective space.

In this method, the decision-maker can provide feasible or infeasible reference points. If the reference point is infeasible (outside Z), then the minimum of the Chebyshev scalarizing function gives a point of the Z frontier nearest (based on Chebyshev metrics) to the reference point. If the reference point is feasible (inside Z), then the solution of the minimization problem improves the aspirations marked by the reference point.

Several improvements and modifications have been proposed by the investigators. For instance, the article [40] (1052 cites), introduces an evolutionary algorithm based on reference points, which gives to the decision-maker a set of solution candidates near the reference point.

14.8 Some Extensions of the Problem

Like in the case of single-objective optimization, the research is considering nonlinear scenarios (a large field) and integer or mixed-integer problems.

The purpose of this section is to introduce these topics with opportune references and to include some methods not yet mentioned in this chapter.

14.8.1 Nonlinear Multi-objective Optimization

A most cited book on nonlinear multi-objective optimization (NLMOO) is [104] (8100 cites), by Kaisa Miettinen, Professor of Industrial Optimization at the University of Jyväskylä, Finland.

Figure 14.21 presents an open-air picture of Prof. Kaisa Miettinen, one of the main protagonists of this chapter.

In her article on some methods for NLMOO, [105], K. Miettinen gives the following definitions:

When all the objective and the constraint functions are linear, the problem is called *linear* or an MOLP problem. If at least one of the functions is nonlinear, the problem is a *nonlinear* multiobjective optimization problem. Correspondingly, the problem is *nondifferentiable* is some of the functions is nondifferentiable and *convex* if all the objective functions and the feasible region are convex.

In the review [47] (year 2021) on 20 years of continuous multiobjective optimization in the twenty-first century, some interesting details of recent developments were given. In first place that if one collects multiple scalar-valued objective functions in a vector, then the multi-objective problem can be considered as the optimization of a *vector-valued map*. This is a recent point of view, sustained by several books (like [46], or [7]).

Among other initiatives around the NLMOO problem, it is worth to mention the series of international seminars at Dagstuhl castle in Saarland, Germany, where

Fig. 14.21 Photograph of Kaisa Miettinen, Finland

the Leibnitz Center for Informatics deploys activities from 1990. Seven of these seminars, in recent years, have dealt with multi-objective optimization, multi-criteria decision-making and evolutionary algorithms. Several reports emanated from the seminars, like [REF GRC], which are interesting in order to gauge the recent research trends.

New topics are arising, including two being described in [47]: the multi-objective bilevel optimization, and the robustness in uncertain multi-objective optimization. As said in [47], in bilevel optimization two optimization problems appear, one on upper level, and the other in lower level. The optimization variable of the upper level is a parameter for the lower-level problem. Also, the optimal solutions of the lower-level problem influence the objective function of the upper-level problem. This represents a complicated scenario, which is motivating much research.

People involved in automatic control is familiar with robustness issues. A large airplane should behave well with a few people on board and also well when it is filled with, say, weighty sumo players. The avionic control system should withstand uncertainties. This is assumed in a more general format as robustness in multi-objective optimization in the presence of uncertain aspects.

A valuable section of [47] is devoted to publicly available solvers for multi-objective problems. Some of them, like **Bensolve** (see [91] for code, and [93] for theory), or **inner** (see [34] for code), are based on the Benson's method, which will be introduced below. Others being mentioned are, FLO for facility location problems (FLO is in MATLAB language, [66]), ASMO (also in MATLAB, [49]) or the MOSQP algorithm (there is an implementation in MATLAB, [156]) for nonlinear optimization problems, etc.

14.8.2 *The Benson's Scalarization Method*

Figure 14.21 gives a sketch of the Benson's scalarization method, which is easy to understand.

To obtain a point of the Pareto front, a feasible solution $\vec{x}^k$ is selected at random, its objective vector $\vec{z}^k$ is computed. Then the following scalar single-objective problem is solved:

$$\text{Maximize: } \sum_{m=1}^{M} \max\{0,\ (z_m^k - J_m(\vec{x}))\}$$

$$\text{Subject to: } J_m(\vec{x}) \le z_m^k,\ m = 1, 2, \ldots, M$$

$$g_j(\vec{x}) \le 0, \quad j = 1, 2, \ldots, J$$

$$x_i^L \le x_i \le x_i^U, \quad i = 1, 2, \ldots, n$$

Problem that, in a 2D case, is equivalent to find an "L1-optimal" rectangle touching a point of the Pareto front. This rectangle could be like the one drawn in Fig. 14.22 (note that we had not attempted to precisely draw the optimal one).

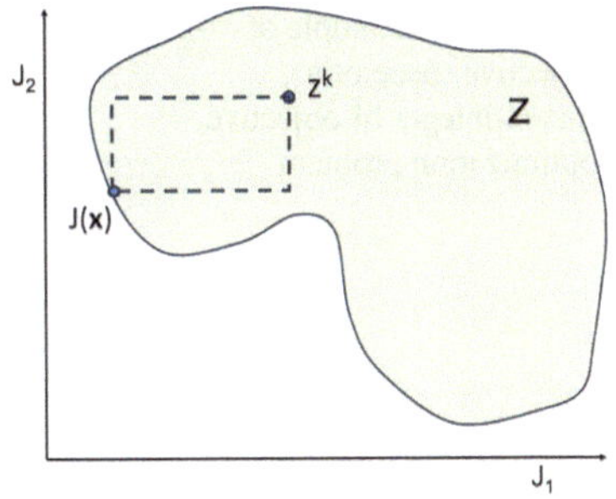

Fig. 14.22 Concept of the Benson's scalarization method

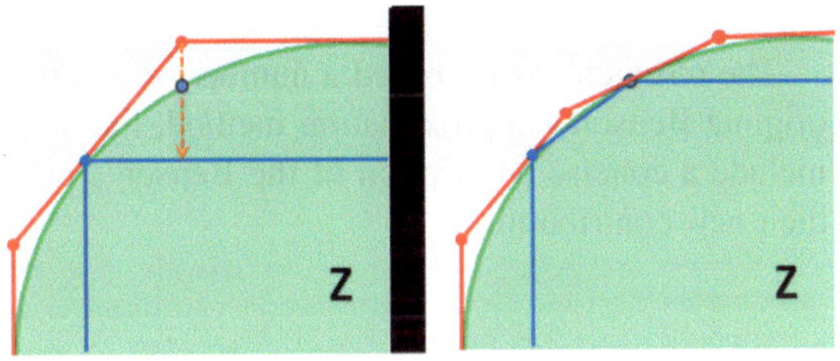

Fig. 14.23 Generating outer and inner closer approximations

The method can be applied repeatedly to obtain a set of Pareto front points.

The advantages of the method are that it is parameter-less and that it can be used for non-convex problems. It is a popular method in the multicriteria decision-making domain. A drawback of the method is that the objective function is nondifferentiable, causing difficulties for gradient-based approaches. The research has proposed changes in the problem formulation to circumvent the issue (see for instance [1]).

This method was introduced in 1978, [18], by Harold P. Benson, when he belonged to the General Motors Research Lab (in 1979 he came as Professor to the University of Florida).

There are other contributions of H.P. Benson. One of these is the **outer approximation algorithm** for generating all efficient extreme points (see [19] (year 1998) for more details). Like the simplex algorithm, this outer algorithm considers a series of vertices. By the way, the article [115] (year 2021), offers an interesting comparison of the outer algorithm and an extended version of multi-objective simplex algorithm.

Some confusion could arise when searching for literature on Benson's method, since several publications refer to the method introduced in 1978 and others to the algorithm published in 1998.

Figure 14.23, based on [42] illustrations, reflects the basic concept about outer approximation applied to our case. All happens in the objective function space, where one has the region **Z**. The plot on the left depicts two polygonal approximations to **Z**, one is an outer approximation (gray lines), and the other is an inner approximation (dark lines). The transition from the left plot to the plot on right shows a basic procedure for getting a closer approximation (increasing the number of vertices). This is the mechanism proposed by Benson for determining nondominated points: a sequence of closer and closer approximations.

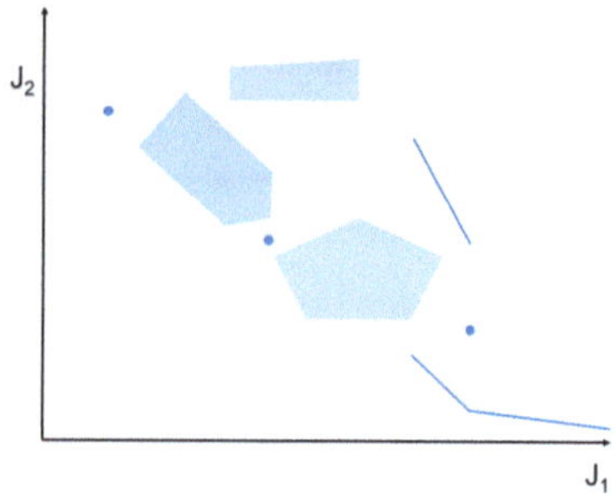

Fig. 14.24 Example of objective space of a mixed-integer bi-objective optimization problem

The research has proposed a number of modifications and improvements of the original Benson's approximation method, like in [42, 92] and [45]. Most of them include a concise description of the Benson's original method and then introduce their new contributions.

14.8.3 Integer and Mixed-Integer Problems

The objective function space of a bi-objective mixed-integer linear optimization problem could look like the case represented in Fig. 14.24, with some patches, some lines, and even isolated points. Normally, the Pareto front of a mixed-integer multi-objective optimization problem would not be a continuous curve.

In the case of multi-objective integer optimization, the objective function space would show just isolated points.

Part of the conclusions of [68], after an extensive review and analysis work on research publications—concerning integer and mixed-integer scenarios—are that:

- For the integer field, most publications propose branch-and-bound, or branch-and-cut algorithms. The second preferred approach is ε-constraint.
- For the mixed-integer field, the winner method is branch-and-bound, closely followed my image space methods.

(by the way, instead of the term "objective function space," some authors use the term "image space").

Let us introduce some bibliographic background, in two parts: first integer optimization and second mixed-integer optimization.

14.8.3.1 Multiobjective Integer Optimization

According to [68], numerous methods for attacking the pure integer problem have been proposed. In fact, this review detects about 70 different algorithms and algorithm variants, able to obtain the whole nondominated set.

We would abbreviate "multiobjective integer linear optimization problem" as MOILP (dropping "optimization").

In general, the initial research contributions on MOILP chose binary problems (variables taking 0–1 values). For example, the doctoral thesis of A.K. Simopoulos proposed in 1977 a tree-search algorithm for binary MOILP, [142]. A branch and bound algorithm for binary MOILP was described in [84], year 1983; the algorithm keeps a search-tree, and nodes are explored one by one to find feasible solutions, and lower and upper bounds are used. Another relevant contribution in 1983 was [41], with an implicit enumeration of all efficient solutions.

Starting from 2005, M. Ehrgott and collaborators (in special X. Gandibleux, and A. Przybylski from 2010) produced many contributions on general MOILP and/or combinatorial optimization. A review of their proposals on multi-objective branch and bound is [121], year 2017. One year before, the three mentioned authors published a book chapter on exact methods for multi-objective combinatorial optimization, [44], covering branch and bound, and scalarization methods. In [60], year 2019, a branch-and-cut algorithm was suggested.

Because its interesting description of algorithms and the graphical illustrations, it is worth to mention the article [118] (year 2019) on branch and bound for bi-objective integer programming.

We would recommend the recent article [59], year 2022, for a contemporary view of multi-objective branch and bound (the article has some materials available from GitHub, [101]).

Matthias Ehrgott, professor at Lancaster University, continues to publish contributions on multi-objective optimization.

Concerning the application of the ε-constraint approach to multi-objective integer optimization, one of the first contributions is [20], year 2009. The target of this article is bi-objective combinatorial optimization problems with integer objective values; one of the objectives is transformed into a constraint. The entire nondominated set is generated by iterative application of the ε-constraint method. Also, in 2009 (with a previous publication in 2008), G. Mavrotas introduced "AUGMECON," an augmented ε-constraint method, [99] (2206 cites). This method was further improved by Mavrotas himself and other investigators.

In 2018, a review of the ε-constraint method and its variants for multi-objective integer optimization was published [135].

14.8.3.2 Multiobjective Mixed-integer Optimization

We shall briefly denote "multiobjective mixed-integer linear optimization problem" as MOMILP.

In order to illustrate, to a certain extent, the peculiar character of mixed-integer problems, one of the examples included by M. de Santis et al. in [38] is being considered. It is a simple bi-objective optimization problem, established as follows:

$$\text{Minimize:} \begin{pmatrix} x_1 + x_2 \\ x_1^2 + x_2^2 \end{pmatrix}$$

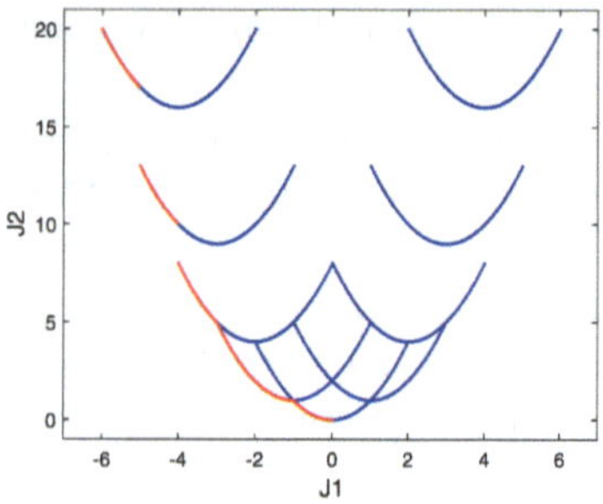

Fig. 14.25 Image of MOMILP example on the objective space, Pareto front in lighter gray

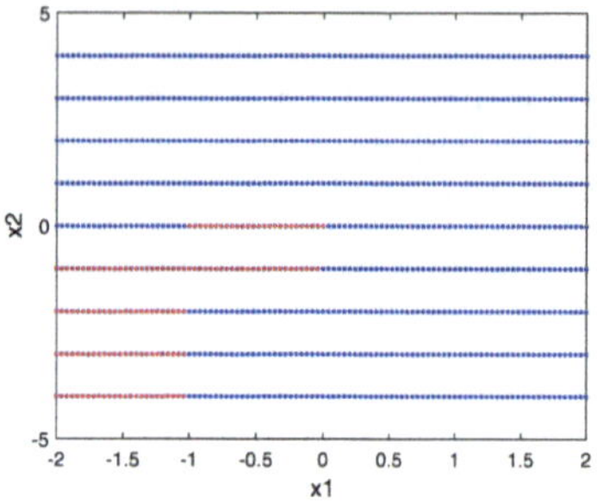

Fig. 14.26 Decision variables of MOMILP example, efficient solutions in lighter gray

Subject to: $x_1 \in [-2, 2]$
$x_2 \in [-4, 4] \cap Z$
where x_1 is continuous variable and x_2 is an integer variable.

Based on the second program in this chapter, another program in MATLAB has been devised to study this case in Pareto style. The program generates two plots, which are presented in the next two figures.

Figure 14.25 depicts the objective function space, in which we see 9 (bowl) curves. The Pareto front is highlighted in lighter gray color.

Figure 14.26 shows the decision variable space. In lighter gray color, the solutions corresponding to the Pareto front (efficient points).

The listing of the program that generated both figures is given below.

Program 14.8 Example of Pareto analysis of mixed-integer problem

```
% example of Pareto analysis on Mixed Integer Problem
% x1 continuous, between -2 and 2 (101 values)
% x2 integer, between -4 and 4
% generation of decision variable space points
L=101; %length of x1 set
N=9*L; %total number of variables
x = zeros(N,2);
i=1;
for k =-4:1:4
  for j =1:L
```

```
    x(i,1)=-2 + ((4*(j-1))/100);
    x(i,2)=k;
    i=i+1;
  end
end
%image of those points due to
% the following functions:
% memory reserve:
J1=zeros(N,1); J2=zeros(N,1); J = zeros(N,2);
for i =1:N
  J1(i)=x(i,1)+x(i,2);
  J2(i)=x(i,1)^2 + x(i,2)^2;
  J(i,1)=J1(i); J(i,2)=J2(i); % to form a matrix
end
% nondominated points detection
% in objective function space
ndmt= J; % nondominated points (initialization)
ndmt_ix = 1:N;
other_ix = [];
% (no other row can be better in all elements)
i = 1; j = 1;
while i <= size(ndmt,1)
  r = size(ndmt,1);
  bb = ones(r,1)*ndmt(i,:)-ndmt;
  bb(i,:) = '';
  if any(all(bb'>=0))
    ndmt(i,:)=[];
    other_ix(j) = ndmt_ix(i);
    ndmt_ix(i) = '';
    j=j+1;
    i=i-1;
  end
  i = i+1;
end
if isempty(ndmt_ix)
  disp('No Pareto points found');
end
figure(1)
plot(J1(:),J2(:),'b.'); hold on
plot(J1(ndmt_ix), J2(ndmt_ix), 'r.');
xlabel('J1'); ylabel('J2');
axis([-7 7 -1 21]);
figure(2)
plot(x(:,1),x(:,2),'b.'); hold on
plot(x(ndmt_ix,1),x(ndmt_ix,2),'r.');
xlabel('x1'); ylabel('x2');
axis([-2 2 -5 5]);
```

Compared to MOILP, the mixed-integer scenario comports a higher degree of difficulty. We had to wait till 1998 to have first algorithms for solving MOMILP problems.

Not many algorithms have been developed from 1998 up to now; the review [68] detects about 15. Due to the short-elapsed time, the number of citations (indicating the acceptance of a proposal) is moderate.

The article from G. Mavrotas and D. Diakoulaki, year 1998, on a branch and bound algorithm for zero-one multiple objective linear programming, [100] (196 cites), opened the way. Later on, the same authors, in 2005, corrected some issues, and other authors have improved the algorithm until recently.

The bi-objective MOMILP case has been treated with branch and bound, by several contributions from P. Belloti, N. Adelgren, B. Soylu, MM. Wiececk, A. Gupte, etc. See for example {$REF ADL] (year 2022), which includes interesting illustrations.

There is an online source of information, called Optimization Online. From this source, it is interesting to access the article [102] (year 2022) with a branch and bound algorithm for bi-objective mixed-integer *quadratic* programs.

The image space usually has a smaller dimension (compared to the decision variable space). This is an advantage for algorithm development, which focus on the structure of image space.

Algorithms based on image space are recent. N. Boland et al. produced from 2014 a series of publications, starting with the introduction of the "triangle splitting method" for bi-objective MOMILP (see [21]). Other proposals from Boland *et. al.* are: the "balanced box method," the "I-shape search method" for tri-objective problems, and the "quadrant shrinking method" (also for tri-objective problems).

Other algorithms are the "one direction search method," proposed by A. Fattahi and M. Türkay, [53] (year 2018) for biojective MOMILP, and the "Gondef algorithm" introduced by S.A.B. Rasmi and M. Türkay, [123] (year 2019) for MOMILP.

Let us mention another algorithm being published in 2019, which is [119] for MOMILP. The authors are W. Pettersson and M. Ozlen; both are involved in new proposals based on parallelization of the algorithms.

A recent and opportune contribution, accessible from Optimization Online, is [48] (year 2022). It introduces a deterministic solver for convex or non-convex MOMILP. This is one of a series of publications from Marianna De Santis and collaborators at the Rome University La Sapienza. A MATLAB implementation of the solver is available from GitHub, [159].

14.9 Multicriteria Decisions

This section will visit a large methodological space. It is a region mainly belonging to multi-criteria decision making (MCDM), or multiple-criteria decision analysis (MCDA). It falls under the interest of business analysis & decision-making, and other socioeconomic and political activities in need of rational decisions.

Sometimes, decisions should be fast, as in Fig. 14.27. Other circumstances may be more elastic and give you time.

An indication of the size of our topic here is that the voice "Multiple-criteria decision analysis" in Wikipedia gives links to over 40 different methods.

In the article [165] (year 2022), on the use of multi-criteria decision-making methods in business analytics (a literature review), a distinction was made between

Fig. 14.27 Decision making

multi-attribute decision-making (MADM) and multi-objective decision-making (MODM). The article detects more than 50 methods under MADM and in its words: *in MODM methods, the number of alternatives is endless.*

There is something special about decision-making. While in optimization problems you have equations and all information on the table, in daily decision-making (business, processes, car driving, etc.), you are surfing on what you may have by now (the reality): probably no equations and no much data about future. Some trial and error (oops!) would be expected, and tools helping for *what-if* exploration on computer are welcome.

The survey article [82] helps to focus on just *interactive approaches* for multi-objective decision-making. The article briefly describes ten methods for interactive continuous multi-objective optimization. Here is a list of those methods:

- STEM (1971)
- GDF (1972)
- Interactive weighted Chebyshev procedure (1983)
- Visual interactive approach, by Korhonen and Laakso (1986)
- Zionts-Wallenius method (1976)
- The sequential proxy optimization technique (SPOT) (1982)
- GUESS (1997)
- STOM (1995)
- NIMBUS (1995)
- Interactive multi-objective evolutionary algorithms (from year 2000, approximately)

This list could be completed with the methods mentioned in [86], in special, the methods allowing for a *progressive articulation of preferences*.

GDF is the Geoffrion-Dyer-Feinberg method, already introduced in this chapter.

The IBM PC was introduced in 1981. This marks a new epoch for calculations and algorithms. Year 1981 could be taken as gross-reference in this section, in order to tell about first methods, and subsequent methodological generations.

Since decision-makers have a key role and should be mentioned frequently from now on, we shall refer to them as DM for short.

14.9.1 First Methods

STEM is an interactive method originally developed for multiobjective linear programming. The method was introduced in 1971 by R. Benayoun et al., [16] (1407 cites). The first three authors of the article (including Benayoun) belonged to the group SEMA-METRA International, Paris (it was the first grey-matter French society).

In STEM, the DM classifies the objective functions at the current iteration as satisfactory or not. A minimum weighted Chebyshev distance to the ideal point is used. The method tries in each iteration to reduce the feasible region, improving some objectives by sacrificing others and converging to the most preferred solution.

The article [127], year 2012, provides a mathematical description of STEM and introduces in detail an improved STEM method.

The **sequential proxy optimization technique (SPOT)** was introduced by Masatoshi Sakawa en 1980, [136] (it was published in a journal in 1982, [137] (95 cites)).

SPOT assumes that the DM uses an implicit value function, which should be maximized over the feasible region. At each iteration, the DM is asked to give preferences (trade-offs) and then being used to find a search direction.

In the mid-1960s, what would become a saga of programs/methods with the common name **ELECTRE** started to develop. The publications about ELECTRE followed the typical research habits: first an internal report of a conference presentation and when a certain maturity was reached, scientific journal articles. In the case of ELECTRE, a first report is [17] (year 1966) by R. Benayoun, B. Roy and N. Sussman, part of the SEMA staff. Later on, in 1968, B. Roy (considered as the father of the method) published an article in French, [128] (2448 cites), which gave publicity to ELECTRE. In 1991, B. Roy published a long article in English, [129] (2998 cites), explaining the foundations of ELECTRE.

The main characteristic of ELECTRE is that it uses an *outranking* approach. Given two alternative decisions and a set of criteria, it is said that one alternative outranks the other if it is at least as good as the later [97] for the majority of the criteria and not significantly worse when other criteria are considered (some criteria vote in favor, other yield a veto). Two types of tests were applied: concordance test and discordance test (opposition between alternatives) to evaluate the situation. The verdict could be that one decision is preferred to the other, or that both decisions are incomparable, or one is indifferent to the other.

A much more detailed description, with mathematical expressions, can be found in [152] (year 2023).

Different ELECTRE versions exist: ELECTRE I, II, III, IV, IS, TRI, TRIB, and TRI-C. ELECTRE I is used for selection problems and ELECTRE II and III for ranking purposes.

An implementation of ELECTRE III in MATLAB is included in the Decision Theory Toolbox (DTT), available from MathWorks File Exchange site.

Another family is PROMETHEE, created by Jean-Pierre Brans et al. [23] (970 cites, year 1982). Brans was able to teach Operations Research both in Flemish and in French in the two universities of Brussels (one for Flemish, the other for French).

The method is also based on outranking and accomplishes a comparison pair per pair of possible decisions along each criterion. PROMETHEE I was developed for partial ranking, PROMETHEE II for complete ranking. They were followed by PROMETHEE III, PROMETHEE IV, PROMETHEE V, PROMETHEE VI (representing the human brain), and then PROMETHEE TRI and PROMETHEE CLUSTER. A recent chapter describing the methodology of this family is [24] (1984 cites, year 2016); two relevant articles on the same topic are [24] (3715 cites, year 1986), and [15] (2049 cites, year 2010).

In the world of decision-making, ELECTRE and PROMETHEE are considered archetypes of the so-called French School.

14.9.2 Second Wave Methods (The 1990s)

The methods to be now introduced are methods intended for progressive information, through an iterative process involving the DM preferences and providing local solutions.

The **GUESS method** is a simple interactive method. It was introduced in 1997 by J.T. Buchanan, [25] (133 cites). The method was presented by Buchanan as a *naïve method*. It is related to the reference point method.

The DM specifies a reference point (a guess) below the nadir point, and then the minimum weighted deviation from the nadir point is maximized. Figure 14.28 depicts the idea.

The DM can obtain a set of candidate solutions by moving the reference point.

The **satisficing trade-off method (STOM)** was introduced by Hirotaka Nakayama in 1995, [113] (131 cites). Professor Nakayama was a faculty member of the Konan University, in Kobe, Japan.

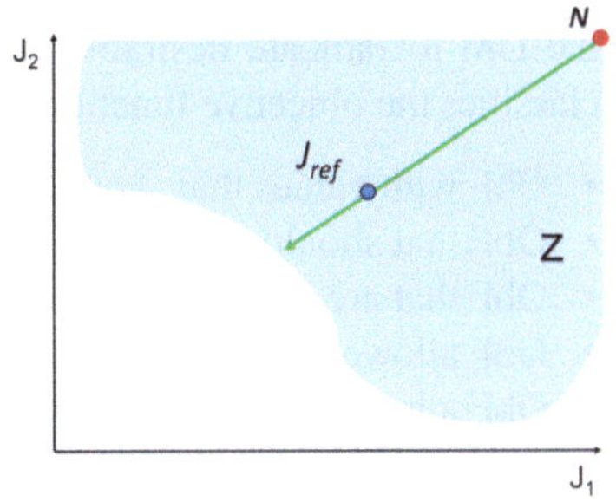

Fig. 14.28 Basic concept of the GUESS method

In brief, STOM works as follows: after a (weakly) Pareto optimal solution has been found with a scalarizing function, that solution is presented to the DM. Then, the DM is asked to classify the objective functions into three classes. The classes are:

1. Unacceptable objective functions, with values to be improved
2. Acceptable objective functions, with values may be relaxed
3. Acceptable objective functions, with values not to be changed

The objective and the constraint functions are assumed to be twice continuously differentiable.

Under some assumptions, trade-off information can be obtained from the KKT multipliers linked to the scalarizing function. This information serves for determining upper bounds for the functions to be relaxed, so the DM only has to specify aspiration levels for functions to be improved: this is called *automatic trade-off.* In the next iteration, the scalarization function is modified, minimized, and the DM is asked again to classify the objective functions at the new solution [97].

The **light beam search method** was described by A. Jaszkiewicz and R Slowski, in [78], year 1999. It combines the reference point approach with multicriteria decision mechanisms. At each iteration, a finite sample of nondominated points is generated. One of those points, called the middle point, was obtained in previous iteration, the other points are neighbors of that middle point. Now, as in ELECTRE, outranking relations are considered. The neighbors that outrank the middle point are called *characteristic neighbors*.

The DM defines three preference thresholds for each objective: the indifferent threshold, the preference threshold and the veto threshold. These thresholds are used to establish the outranking relations.

At each minimization iteration, the DM is asked to choose one of the characteristic neighbors as the next middle point, or update the preferential information, or define a new aspiration (reference) point.

The nondifferentiable interactive multi-objective bundle-based optimization system (NIMBUS) was introduced by K. Miettinen and M.M. Makela in 1995, [103] (209 cites).

As emphasized in its name, the method is able to handle nondifferentiable problems.

NIMBUS is an interactive classification-based method, where it is easy for the DM to indicate desirable improvements and tolerable impairments. The DM classifies the objective functions (ObF) into five classes:

- ObF with values that should be decreased
- ObF that should be decreased to a certain aspiration level
- ObF that are satisfactory at current iteration
- ObF allowed to increase to a certain upper bound
- ObF allowed to change freely

After classification a mini-max problem is formed and solved.

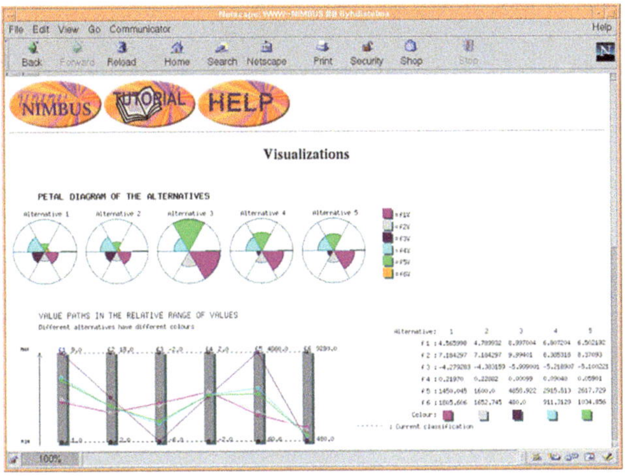

Fig. 14.29 Example of online use of WWW-NIMBUS

With NIMBUS, the DM can explore the (weakly) Pareto optimal set, using non-irreversible classification (the DM can go back).

14.9.3 Methods with Special Interaction Means

Our digital world is being dominated by Internet and graphical visualization. This is marking a trend for new decision-making tools. We shall mention some of these tools, knowing that soon, surely, some more would pop up.

WWW-NIMBUS (access through https://wwwnimbus.it.jyu.fi/) was the first interactive optimization system on the Internet since 1995. It is based on centralized computing and distributed interface. Figure 14.29 shows a visualization of results, once a multi-objective problem was remotely solved. This is a figure that we generated online (choosing one of already on-site available problem examples).

First publications about **NAUTILUS** came in the 2010, [106] (100 cites), with two authors from Finland, K. Miettinen and P. Eskelinen, and two from Spain, F. Ruiz and M. Luque (University of Malaga). NAUTILUS proposes an interactive technique starting a series of iterations from the Nadir point. In each iteration, a solution is obtained on **Z** that dominates the previous one, gaining in all objective functions simultaneously. During iterations the DM is able to direct the search toward the most preferred solution; the DM can reach any Pareto optimal solution. At any iteration, the DM can go back several steps and then take another direction.

Figure 14.30 reproduces a main illustration in [106], where two steps of the exploration have been taken. At each iteration, the DM can modify the direction of the next step. A feature of the method is that the region of interest shrinks as iterations progress, considering that some parts of **Z** will not be reached, unless backward steps were taken.

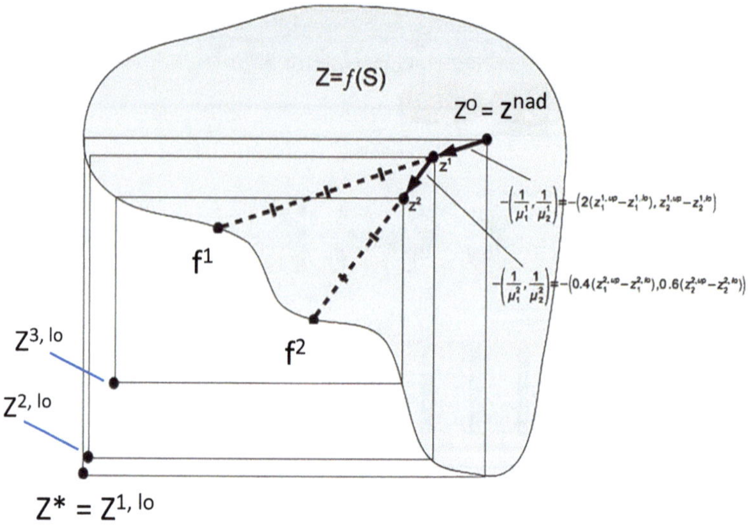

Fig. 14.30 Basic idea of the NAUTILUS method

A number of NAUTILUS variants have been developed, like NAUTILUS 2, [107] (year 2015), and E-NAUTILUS (enhanced-NAUTILUS), [131] (year 2015). In [108] (year 2016), a modular framework is developed, allowing for several variants (the article mention for variants).

In 2019, a research team working with Miettinen in Finland have introduced **DESDEO**, [116]. It is a modular and open-source framework for interactive multi-objective optimization. It can work with scalarization methods or evolutionary methods, or both. The article includes four use cases; one of them the study of river pollution. The modularity of the framework allows for supporting the graphical interface of E-NAUTILUS, for example. See [109] (year 2021) for more details.

Now, we introduce a new aspect of interactive operation. A new breed of tools has appeared for "**Pareto Navigation**". The article [5] compares some initial tools in this line. One of the applications mentioned in this article is multi-objective radiation dose planning in a real radiotherapy scenario. This kind of application has been considered by several medical publications, for example, the article [112] (229 cites) from which Fig. 14.31 reproduces a screenshot.

The screenshot refers to the treatment of a tumor (marked with a white star), achieving high and homogeneous doses in the tumor volumes and low doses in the organs at risk. The spider plot on the left hand side of the figure, depicts the compromise found for six objectives.

The first Pareto navigation tool appeared in 1988. It was created by P. Korhonen and J. Wallenius, Helsinki School of Economics and Business Administration. Prof. Wallenius is a well-known scientist, having published 256 research contributions from 1976 to 2022. The name of the new navigation tool was **Pareto Race**, [87] (349 cites).

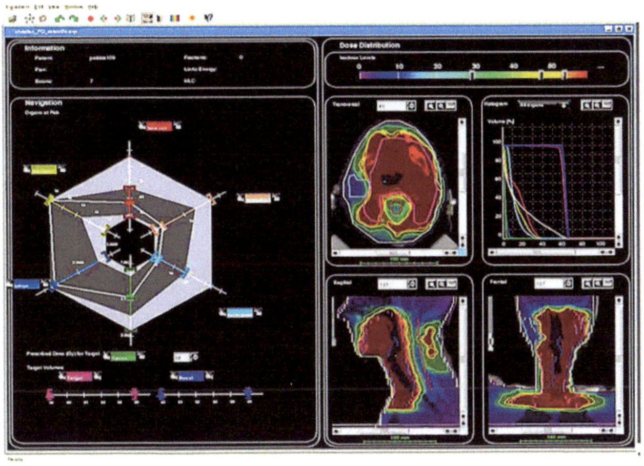

Fig. 14.31 Example of multiobjective Pareto navigation in radiotherapy planning

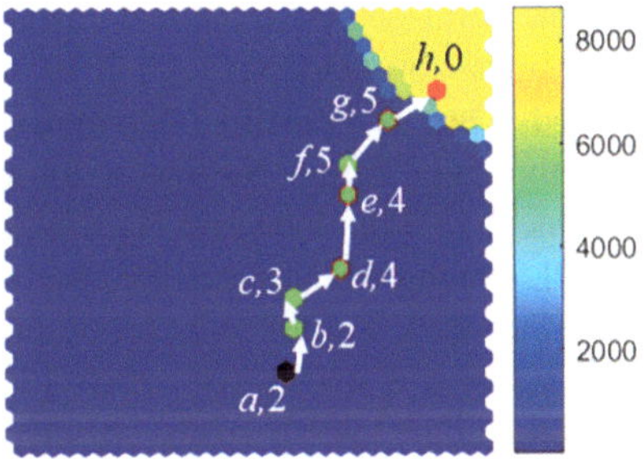

Fig. 14.32 Example of multiobjective Pareto navigation with Pareto race

The Pareto Race algorithm begins with finding in **Z** a nondominated Pareto-optimal start point $\mathbf{z}_0$, based on the DM aspiration level. Next, the DM provides another point **zr**, which defines a reference direction $\mathbf{d} = \mathbf{zr} - \mathbf{z}_0$. A step of length λ is taken in this direction, and the new point is evaluated; with this information, the DM decides a new direction to continue the exploration, etc.

The research has combined the Pareto Race concept with other ingredients. For instance, [164] (year 2023) includes a self-organizing map (SOM) as an aid for the DM during the Pareto race (SOM is an important method of Machine Learning; see https://www.superdatascience.com/blogs/the-ultimate-guide-to-self-organizing-maps-soms). The same authors of [164] published an article with similar topic, [163] (year 2022), from which we would like to extract an illustrative exploration with the Pareto race, as represented in Fig. 14.32.

A second tool was introduced in 2010 by the name **Pareto Navigator**, [51] (82 cites). The authors of the article were P. Eskelinen, K. Miettinen, K. Klamroth and J. Hakanen. This publication was preceded in 2007 by a report of the same author team, [52].

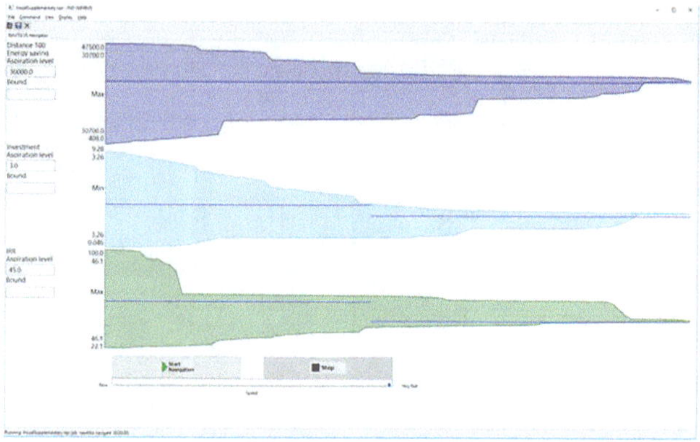

Fig. 14.33 Example of multiobjective Pareto navigation with Nautilus navigator

The recent article [110] (year 2021) describes an application of Pareto navigator for the multi-objective optimal design of a water distribution system.

Almost simultaneously, in 2019, three more Pareto navigation tools were introduced. One is **Nautilus Navigator**, [130], extending the NAUTILUS family. Another is a Pareto Navigator for **nonconvex** problems, developed by M. Hartikainen, K. Miettinen and K. Klamroth, [69]. And the third is **Pareto Explorer**, [139], which contemplates the case of many-objectives optimization problems; so, in a first step, a solution is proposed, and in a second step, a local search around this solution is made. The article [35] (year 2020) specializes the Pareto explorer for finding the knee in many-objectives problems.

By the way, another contribution of Hartikainen et al., [70], is **PAINT**: a popular tool for interpolation in discontinuous Pareto fronts of nonlinear problems.

Figure 14.33 corresponds to Nautilus Navigator. This is one of the iterations toward a satisfactory solution. In words of [130], the progress along the horizontal axis describes the evolution of the reachable ranges for the objective functions; the DM can see how these ranges shrink as the Pareto front is approached and can stop the movement at any time to provide new preferences.

A video containing the Nautilus Navigator solution process can be accessed from https://desdeo.it.jyu.fi/nautilus-navigator.

The publications on Pareto navigation already cited usually contain some application examples that gives an idea of the potential and scope variety of the new techniques. Here is a short list of those examples:

- In [5], radiotherapy treatment design, railway capacity evaluation, vehicle routing, flexible manufacturing system
- In [87], managing disposal of sewage sludge in the New York Bight
- In [164], car side-impact worthiness, multiple clutch brake design, general aviation aircraft (a problem with 10 objectives, 27 decision variables and 18 constraints)
- In [110], water distribution system

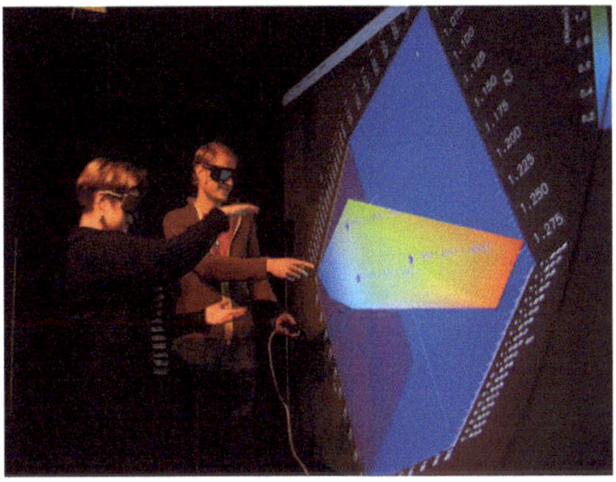

Fig. 14.34 Pareto navigation in 3D with virtual reality

- In [130], auxiliary services in a power plant
- In [69], wastewater treatment plant
- In [139], industrial laundering, plastic injection molding

Much on a contemporary trend, there are visualization proposals pertaining Pareto navigation in 3D, using virtual reality devices. For example, [95] or the impressive doctoral thesis [132]. Figure 14.34 reproduces an illustration of these publications, which shows an interactive exploration on the Pareto objective region, in 3D using virtual reality glasses.

14.9.4 Methods Related to Decision Analysis

In general, when one takes a decision is after analysis of the situation. This analysis could conclude with a ranking of possible alternatives.

Let us add a basic consideration, about the qualitative nature frequently found in decision analysis. In contrast with typical optimization settings, our decisions are not usually governed by numbers. Suppose you are choosing a painting for your office, and you can select among different items with same size and price. Perhaps, you would like figurative art, or abstract; with cold colors; countryside representations or portraits; transpiring serenity, or perhaps chaos; naïve or complex; etc. There are no numbers but characteristics. Indeed, you can still assign numbers (weights) to your preferences and convert the problem to an optimization (that would be subjective).

There are several methods more related to decision analysis that deserve to be introduced, to complete our view of multicriteria handling. For the interested reader, the book entitled *Multiple Criteria Decision Analysis: State of the Art Surveys*, [56] (4294 cites, year 2005), 1011 pages, would give much more information on this topic.

Now, we shall introduce a repertory of well-known methods.

The **Technique for Order of Preference by Similarity to Ideal Solution (TOPSIS)** was introduced in 1981 by C.L. Hwang (Kansas State University) and K. Yoon (Fairleigh Dickinson University), [73] (5718 cites).

It is an easy technique. First the utopia point and the nadir point are determined. Then, objectives are normalized and multiplied by the weight assigned to each objective. Finally, the Euclidean distance of each solution (A_i) in the Pareto front to the utopia point (S_{iu}) and to the nadir point (S_{in}) is calculated and then a score $P_i = S_{in} / (S_{iu} + S_{in})$ is computed. The higher the score, the better the solution. The best compromised solution is the closest to the utopia point and farthest from the nadir point.

Based on the score, it is possible to rank all solutions on the Pareto front, from best to worst.

The name of the next method makes reference to the Utility Theory, which connects with a philosophical way of thinking (the utilitarianism) with old roots. It is related in particular with the ideas of Jeremy Bentham (1748-1832), John Stuart Mill, and others (see [43]). There are several branches and specializations of this theory. The article [125] presents a historical view of utility theory from Bentham to Kahneman more amenable to mathematical treatment.

The **Multi-Attribute Utility Theory (MAUT)** is based on axioms established by Von Neumann and Morgenstern, [157], year 1944. In particular, the Von Neumann-Morgenstern utility theorem is the basis of the **expected** utility theory. The article [57] (731 cites, year 1968) constitutes a kind of compendium of the utility theory (mathematical) aspects that are of interest for decision-making (see references therein for related contributions in the 1960s). This is a publication before small computers came in.

Nowadays, MAUT is a well-established method for solving real-world MCDM problems. The key feature of MAUT is the use of utility functions for modelling attributes. The preference attributes are evaluated on a scale of 0 (undesirable) to 1 (desirable): therefore, different attributes can be compared with a common scaling. The goal of MAUT is to determine the best alternative that maximizes the overall utility or value for the decision-maker.

A book to be recommended for diving into MAUT is [14] (5716 cites), by M.H. Bazerman and D.A. Moore. The book describes utility theory as a formally rational decision process for arriving at a decision with the greatest expected utility, following six steps. In a short note, [13], Max H. Bazerman, University of Harvard, gives an example; it is the case of a senior student that, after applying to several schools for a doctoral program (and being accepted), has to decide where to go. The six (MAUT) steps recommended by Bazerman are literally as follows:

1. Define the problem (selecting the right program)
2. Identify the criteria necessary to judge the multiple options (location, prestige, faculty, etc.)
3. Weight the criteria (rank them in terms of importance to you)
4. Generate alternatives (the schools that admitted you)

5. Rate each alternative on each criterion (rate each school on each criteria that you identified)
6. Compute the optimal decision

The reader could easily imagine how to implement most steps with a series of tables. Anyway, it would be interesting to see [154], where a case of diplomats' selection is treated with numerical details. Also, the web page [94] presents a clear tutorial on MAUT practice.

The **Analytical Hierarchy Process (AHP)** was introduced by Thomas L. Saaty in 1977, [133] (12,956 cites). It is a popular method, with many applications, [124]. The method is oriented to group decision scenarios and uses pairwise comparisons and judgments from experts, in order to derive priority evaluations. AHP consists of three main parts:

1. Decomposing and breaking down the problem into criteria and subcriteria, in a hierarchical way
2. Determining the priorities of criteria and subcriteria
3. Synthesis of priorities, to determine the highest priority

Pairwise comparisons are expressed as matrices with numbers between 1 and 9, estimating the level of importance.

Here is a list of other methods, with convenient references:

- ANP, [134] (2461 cites, year 2013)
- WSM (also known as SAW) method, [143] (year 2014)
- SMART, [144] (123 cites, year 2017)
- VIKOR, [117] (2113 cites, year 2007)
- MACBETH, [10] (318 cites, year 1999)
- RIM, [28] (150 cites, year 2016)
- COPRAS [81] (411 cites, year 2006), [3] (year 2019)
- PROMCALC & GAIA, [22] (year 1994)
- DATA ENVELOPMENT ANALYSIS, [11] (812 cites, year 1989)

(see the survey [6] for structured detailed information on many of the methods we cited).

14.10 Multiobjective Evolutionary Optimization Methods

It has been confirmed by an enormous number of applications that evolutionary algorithms can be quite successfully employed for multi-objective optimization. It has become the preferred approach for most problems.

Indeed, if one chooses a scalarization approach, for example, the heuristic/biological algorithms described in other chapters of our book can be directly used to find a solution for the multi-objective problem. But our horizons can be more ample, having in view the Pareto objective space offering many good compromise points, and so, it is pertinent to devise more ambitious methods.

MATLAB provides the function *gamultiobj( )* for multi-objective optimization based on genetic algorithms. Some other routines in MATLAB are also available for multi-objective evolutionary computation.

This section shall start with the introduction of well-known, archetypical methods in this area. Then, it will continue with a panoramic view of tool generations and the state-of-the-art.

The first implementation of a Multi-Objective Evolutionary Algorithm (MOEA) was made by J.D. Schaffer 1984 (doctoral thesis). This kind of algorithms can generate several points of the Pareto front in a single run and take benefit from any *a priori* knowledge on reasonable compromises.

As a first important example of MOEA, we shall introduce MOGA (Multi-Objective Genetic Algorithm) and then a selection of other algorithms. See [77] for more details on this topic.

MOGA was proposed in 1993 by Carlos M. Fonseca and Peter J. Fleming, [58] (6008 cites). It was one of the first to use Pareto dominance for the ranking of individuals: nondominated individuals receive rank 1. A special fitness scheme is devised to elude solutions being too close one another.

NSGA (Nondominated Sorting Genetic Algorithm) was proposed in 1994 by N. Srinivas and K. Deb, [145] (9914 cites). Before selection, the population is ranked based on nondomination. A stochastic technique is adopted to maintain population diversity.

An improved version, **NSGA-II**, was proposed by K. Deb, [39] (**48,595** cites, year 2000). It employs Evolutionary Operators to create new offspring pools. A crowding distance is computed for each member, which is considered in the selection process to spread the solutions on the Pareto front. It is a popular method.

SPEA (Strength Pareto Evolutionary Algorithm) was introduced by E. Zitzler and L. Thiele, [170] (10,211 cites, year 1999). Mechanisms from other methods were integrated in SPEA: like the use of Pareto dominance ranking. SPEA uses an external archive containing nondominated solutions previously found. The fitness of each individual is computed using the external archive and considers both closeness to the Pareto front and regular distribution of solutions. The external set is partially renewed at each generation, and its size is kept under a threshold.

There is version **SPEA2**, [169] (10,038 cites, year 2001) with a fine-grain fitness calculation for each individual that gauges the number of individuals that dominates it and the number of individuals being dominated by it. The improved method also uses a nearest neighbor density estimation for guiding the search and a better archive truncation method.

PAES (Pareto Archived Evolution Strategy) was due to J.D. Knowles and D. W. Corne, [85] (3115 cites, year 2000). It uses an historical archive with some nondominated solutions found before. The archive is employed for comparison with each mutated individual. To keep diversity, a new crowding procedure is used that divides objective space in a recursive manner, so each solution is placed on a grid location. A map of the grid is maintained.

PESA (Pareto Envelope-based Selection Algorithm) was introduced by D.W. Corne et al., [33] (1418 cites, year 2000] and consists of a small internal population and a larger external population. Selection diversity is maintained with a hypergrid division of phenotype space. It also uses an external archive.

A second version, **PESA-II**, was presented in [32] (1488 cites, year 2001). Selection is region-based and the subject of selection is, instead an individual, an hyperbox. In this way, computational costs are reduced.

Almost all the mentioned methods have been implemented in MATLAB, being available from the MathWorks File Exchange site, and/or the GitHub repositories. In addition, the page

https://neo.lcc.uma.es/emoo/EMOOsoftware.html

supported by the University of Malaga, Spain, offer links to implementations in several languages.

When Google is asked about Multiobjective Evolutionary Optimization, it gives 8,130,000 results. This is an indication that the topic attracts vivid attention. There are also many publications with surveys on the topic. In our case, because is brief and direct, we selected the survey article [72] (year 2019). This survey includes references to 60 contributions.

According to our selected source, the development of MOEAs can be divided into four phases, namely:

- First phase, J.D. Schaffer and other precursors
- Second phase (1990's): MOGA, NSGA, NPGA, etc.
- Third phase (1999–2002): SPEA, PAES, PESA, PESA-II, NPGA, NPGA2, Micro-GA, SPEA2, NSGA II, etc.
- Fourth phase: MOEA/D, NSGA-III, MOPSO, RMEDA, IBEA, SMS-EMOA, etc.

The algorithms of the third phase adopted elitism and better diversity preservation strategies.

During the fourth phase, more routes were explored. Some of the algorithms were indicator-based, or decomposition-based, aiming to many-objective optimization problems. Others included new operators and techniques, considering certain types of dominance and new aspects of the problem. And machine learning is being introduced, together with trends associated to decision-making (DM preferences, interactive operation, etc.).

The survey detects four categories in current MOEAs:

- Domination-based MOEAs: the algorithms focus on better use of Pareto dominance and satisfactory achievement of population diversity. It is difficult for this approach to deal with many-objectives problems, since with the number of objectives the proportion of nondominated solutions increases exponentially, slowing down the evolutionary process.
- Indicator-based MOEAs: performance indicators of solution quality are evaluated, being considered during selection and guiding the search. The first well-known indicator-based MOEA is IBEA. Another is SMS-EMOA. The

selection pressure is higher than Pareto domination, and so the solutions prefer some subregions of **Z**.

- Decomposition-based MOEAs: this is a popular general approach. The basic idea is to decompose the original problem into a number of scalar optimization subproblems. Solutions of each subproblem are optimized by evolutionary operations among its neighboring subproblems. An archetype of this approach is MOEA/D. There are some limitations and drawbacks; for instance, that the population size increases nonlinearly as the number of objectives grows up. The research is actively proposing improvements.
- Hybrid MOEAs: this is a universal trend, the combination of different methods for better performances. For example, a frequent proposal is to combine global search and local search (the search would be divided into several phases). Another pole of interest is to combine various techniques to generate offspring populations.

The final part of [72] presents a list of illustrative applications (with references) on manufacturing, financial data analysis, satellite image processing, resources management, wireless networks, bioinformatics, robotics and control systems, medicine, etc. It also gives a glimpse to future directions. Some of these directions are: many-objectives optimization, interactive MOEAs, deeper insights, algorithm reliability, data mining on solutions, quantum evolutionary algorithms, and creation of relevant benchmark problems.

14.11 Multicriteria Decisions and Soft Computing

Soft computing techniques fit like a glove with multicriteria decisions. Actually, there are thousands of practical confirmations of this fact.

In order to show main traits of this reality, this section will consider each of the fundamental methodologies that constitute soft computing.

14.11.1 Fuzzy Multicriteria Decisions

The point of view of fuzzy logic and methodology suits very naturally to human decision-making. It has been recognized from time ago, and the number of related contributions has notably grown.

There are two main versions of the fuzzy set concepts. One is the classical fuzzy set theory introduced in 1965 by L. Zadeh in a famous article (122,990 cites), an another is the *intuitionistic fuzzy set* theory introduced in 1986 by K.T. Atanassov and S. Stoeva, [9] (21,009 cites).

(Note: other types of fuzzy sets exist, like hesitant fuzzy set, fuzzy multisets, vague sets, etc.)

Adding up data from [80] (year 2015) and [2] (year 2017), which review literature and applications, we obtain the following brief summary about fuzzy multicriteria tools:

Number of publications:

- Based on classical fuzzy sets:
  - Fuzzy TOPSIS, 5630
  - Fuzzy ELECTRE, 1436
  - Fuzzy PROMETHEE, 1164
  - Fuzzy AHP, 10,792
  - Fuzzy ANP 2200
  - Fuzzy VIKOR 1476
- Based on intuitionistic fuzzy sets
  - Fuzzy TOPSIS, 1405
  - Fuzzy ELECTRE, 227
  - Fuzzy PROMETHEE, 154
  - Fuzzy AHP, 908
  - Fuzzy ANP 198
  - Fuzzy VIKOR 494

For each method, a ranking of articles has been made, based on number of citations. We take the five first articles in each ranking and sum their citations. This is the result:

Number of citations (five most cited articles):

- Fuzzy TOPSIS, 877
- Fuzzy ELECTRE, 340
- Fuzzy PROMETHEE, 411
- Fuzzy AHP, 2020
- Fuzzy ANP 585
- Fuzzy VIKOR 477

All these data should be taken as indicative of the situation in 2015–2017. Nowadays, numbers are larger. For instance, the most cited paper on Fuzzy AHP is [29] published by D-Y Chang in 1996: it has now (year 2024) 6470 cites. Likewise, the most cited paper on Fuzzy TOPSIS is [158], published in 2006; it has now 1203 cites.

Some of the findings of [80] are that Fuzzy ELECTRE is preferred for the solution of environmental science problems that Earth and planetary sciences and agricultural and biological sciences only prefer Fuzzy AHP and Fuzzy TOPSIS. Fuzzy VIKOR is mostly used for economics, econometry, and finance areas.

For a detailed bibliographic review for the period 1994–2014, the article [96] would be recommended.

Observe that fuzzy systems are a peculiar version of rule-based systems. Indeed, more conventional rule-based systems are in use, from time ago, for decision-making.

14.11.2 Other Artificial Intelligence Tools and Decision-making

Good decisions usually require good information (even better: good information analysis).

An important role, well suited for artificial intelligence, is data-mining (we could invent here another term: data-distillation or true-information-condensation). Principal component analysis (PCA) or independent component analysis (ICA), which have a number of versions, are important tools in this context.

Data classification is traditionally an important way to acquire perspective. On of the specialties of neural networks is precisely classification.

Querying the crystal ball currently uses models, and this is another specialty of neurons.

From the scientific journal called *Information Fusion*, very interesting contributions can be accessed. We take for example the article [88] (year 2023), which reviews revolutionary research about multi-criterion decision-making. The article includes 177 bibliographic references. The article sees a place for AI in MCDM, as follows: problem-definition, criteria extraction, and parameter determination. It also mentions, as a specific case, the fusion of deep learning with fuzzy systems.

The article [88] is plenty of information, with large tables and illustrations. We will just extract some illustrative contributions:

- PCA, LDA (linear discriminant analysis), neural nets are being used for criteria extraction. Contributions in this sense are [90] with PCA in a large group decision problem; [64] with wavelet neural networks to select criteria weights in a cloud selection service; [111] combines PCA and Fuzzy-TOPSIS in supplier evaluation and selection for the pharmaceutical industry.
- Deep learning and MCDM are combined for Big Data handling, [166] (survey of the topic). An interesting practical contribution is [71] with a deep reinforcement learning based MCDM for optimizing textile chemical process.
- Machine learning methods can help for the determination of parameters. The contribution [8] intends to predict the spatial gully erosion susceptibility, based on COPRAS; the determination of parameters uses Random Forests, Boosted Regression Tree, and Logistic regression. In turn, [162] proposes a MCDM system for the selection of offshore wind turbines, employing PCA and D numbers (which transforms fuzzy and linguistic assessment into reliable quantitative form) to determine weights and affecting factors.
- Machine learning can support the overall process of MCDM, in the form of decision support systems (DSS). In relation with construction industry, [55]

designed a DSS based on deep learning neural networks for the optimization of one-way slabs reducing environmental impact. Other contributions mentioned in [88] propose DSS systems for supplier selection, for scheduling battery use, or an aid for suitable car purchasing.

One of the application fields treated in [88] is sustainable development. The survey detects that neural networks, support-vector machines (SVM) and random forests, are the preferred uses of machine learning in MCDM for this context.

14.12 Some Resources

Since the problems and methods that have been the subject of this chapter are of high interest, in many academic and practical ambits, there exist a good number of resources that help as background material and offer software for personal exploration and for professional uses.

This section would briefly cite some outstanding resource examples.

14.12.1 Books

14.12.1.1 Most Cited

- K. Miettinen, Nonlinear Multiobjective Optimization, 1998 (8124 cites)
- G.H. Tzeng, J.J. Huang, Multiple Attribute Decision Making: Methods and Applications, 2011 (20,785 citcs)
- C.A.C. Coello, Evolutionary Algorithms for Solving Multi-Objective Problems, 2007 (10,197 cites)
- C. Kahraman, Fuzzy Multi-Criteria Decision Making, 2008 (893 cites)

14.12.1.2 Modern

- K. Deb, K. Sindhya, J. Hakanen, Multi-objective Optimization, 2016
- I. Kaliszewski, et al., Multiple Criteria Decision Making by Multiobjective Optimization: A Toolbox, 2018
- K. Deb, Multi-objective Evolutionary Algorithms, 2015
- A.J. Kulkami, Multiple Criteria Decision Making, 2022
- J.J. Thakkar, Multi-criteria Decision Making, 2021

14.12.2 *MATLAB Programs*

14.12.2.1 From MathWorks Site

- See the page on *What is Multiobjective Optimization?*
- See the page on *Performing a Multiobjective Optimization Using the Genetic Algorithm*
- Page about *gamultiobj( )*
- Page about *paretoplot( )*

14.12.2.2 From MathWorks File Exchange Site (See Also GitHub)

(a) The site offers 2963 MATLAB programs relative to “multi-objective”, most of them evolutionary.

  – For example (many downloads):

    - *Multi_objective_optimization_matlab*
    - *GODLIKE: A robust single & multi-objective optimizer*
    - *Multi-objective particle swarm optimization*
    - *PlatEMO*

  (b) The site offers 2449 MATLAB programs relative to “multi-criteria decision,” most of them evolutionary.

  – For example (many downloads):

    - *NGSA-II (46,900 downloads)*
    - *TOPSIS (2000 downloads)*
    - *AHP*
    - *ARMADA data mining tool*
    - *Multi-class SVM*
    - *Decision Theory Toolbox*
    - *MCDM Tools*

14.12.3 *Societies and Web-sites with Links*

- The International Society on MCDM, at https://www.mcdmsociety.org/ (this Society maintains a page on software, with links)
- Links to software at https://neo.lcc.uma.es/emoo/EMOOsoftware.html (Universidad de Malaga, Spain)
- Repository at https://jmetal.sourceforge.net/resources.html

- Repository link at https://www.research.relund.dk/multi-objective-optimization-repository-morepo/

14.12.4 Professional Software, and Books Dealing with Software

(a) Books dealing with software:

- H.R. Weistroffer and Y. Li, Multiple Criteria Analysis Software
- H.R. Weistroffer, C.H. Smith, S.C. Narula, Multiple Criteria Decision Support Software
- A. Ishizaka, P. Nemery, Multi-criteria Decision Analysis: Methods and Software

(b) Information on professional software:

- Web-site of 1000 minds at https://www.1000minds.com/decision-making/what-is-mcdm-mcda
- Page of Medium on five types of decision making software, at https://medium.com/@decioo.com/5-types-of-decision-making-software-cb9169f0384f
- Wikipedia, Decision-making software
- Comparison at https://www.softwareadvice.com/decision-support/

14.13 Exercises

1- In order to buy a car, there are two opportune objectives to consider: fuel efficiency and safety. In a simplistic way, these two aspects could be expressed as:

$f_1(x) = 50 - 0.1\,x$, where x is the car's weight (less weight would mean more fuel efficiency)

$f_2(x) = 0.5\,x + 10$, safety (higher weight would mean larger size and better safety)

Then, we have two objectives to be minimized:

$$\text{Minimize } -f_1(x) \text{ and minimize } -f_2(x)$$
$$\text{suppose } 1000\,\text{kg} \leq x \leq 3000\,\text{kg}$$

Please, visualize the Pareto front and then apply Genetic Algorithms to find a suitable solution, using a weighted sum approach. Add comments on the process and the result.

2- This exercise is related to geometry. It is desired to design a right circular cone, so both the lateral surface area and the cone volume, were minimized. In mathematical terms, we have:

$$S = \sqrt{r^2 + h^2} \text{ (lateral area)}$$
$$V = \tfrac{\pi}{3} r^2 h \text{ (volume)}$$

Constraints: $r \leq 10\,\text{cm},\ h \leq 20\,\text{cm},\ V > 200\,\text{cm}^3$

Please, visualize the Pareto front and then apply the Sine-Cosine algorithm to find a suitable solution, using a ε-constraint method. Add comments on the process and the result.

3- Consider the following bi-objective optimization problem, introduced by J.D. Schaffer:

$$\min\ f_1(x) = x^2$$

$$\min\ f_2(x) = (x-2)^2$$

$$-10 < x < 10$$

Please apply a weighted sum approach, with a Particle Swarm optimization, to find a good compromise. Visualize the Pareto front and add comments.

Also, apply the ε-constraint method for the same problem. Compare.

4- Here is the Tanaka problem:

$$\min\ f_1(x_1, x_2) = x_1$$

$$\min\ f_2(x_1, x_2) = x_2$$

Subject to:

$$g_1(x_1, x_2) = x_1^2 + x_2^2 - 1 - 0.1 \cos\left(16 \arctan \frac{x_1}{x_2}\right) \geq 0$$

$$g_2(x_1, x_2) = (x_1 - 0.5)^2 + (x_2 - 0.5)^2 \leq 0.5$$

$$0 \leq x_1 \leq \pi\ ;\ 0 \leq x_2 \leq \pi$$

Please visualize the Pareto front and then choose an adequate scalarization method for solving the problem. Use Artificial Bee Colony for the optimization. Add comments on the solution process and the results.

5- Let us prepare a lunch diet where you can specify how much you like this or another food. Here is an opportune data table:

Foods	Price	Calories	Protein	Fat	Carbohydrates	Vit_C	Calcium	I like it
Potatoes	0.18	171.5	3.7	0.2	39.9	15.6	22.7	7
Chicken	2.52	277.4	42.2	10.8			21.9	8
Pasta	2.34	358.2	8.2	12.3	58.3	27.9	80.2	7
Scrambled eggs	0.33	99.6	6.7	7.3	1.3	0.1	42.6	5
Beef Farnkfurter	0.81	141.8	5.4	12.8	0.8	10.8	9	1
Apple pie	0.48	67.2	0.5	3.1	9.6	0.9	3.1	9
Bread	0.18	65	2.3	1	11.8		26.2	5
Minimum daily intake		**800**	**100**	**20**	**130**	**400**	**700**	
Maximum daily intake		**1300**	**150**	**50**	**200**	**5000**	**1500**	

Numbers in the last column are between 0 and 10, to express how much I like this food. Of course, you could modify these numbers according to your preferences or needs. Consult [14.1] for other foods and data you may wish to include.

There are two optimization objectives: minimum cost and maximum acceptance (or, in other words: minimum dislike) and a set of constraints.

Please use the weighted sum approach and the Harmony Search method. Visualize the Pareto front. Explain the solution you prefer.

6- Recall from Chap. 5 the knapsack problem. Here is a table of data for an instance of the problem:

Item	1	2	3	4	5	6	7	8	9	10	11	12
Value	79	32	47	18	26	85	33	40	45	59	12	65
Size	85	26	48	21	22	95	43	45	55	52	33	18

The capacity of the knapsack is 120.

State the problem as a 0–1 bi-objective minimization. Visualize the Pareto front. Find a solution with weighted sum and binary Genetic Algorithm. Describe the process evolution and add comments.

7- Consider the following bi-objective optimization problem, proposed by Fonseca and Fleming:

$$\min\ f_1(\vec{x}) = 1 - \exp\left[-\sum_{i=1}^{N}(x_i - \frac{1}{\sqrt{N}})^2\right]$$

$$\min\ f_2(\vec{x}) = 1 - \exp\left[-\sum_{i=1}^{N}(x_i + \frac{1}{\sqrt{N}})^2\right]$$

$$-1 \le x_i \le 1$$

Suppose N $= 6$.

Please, apply goal attainment to find the solution. Visualize the Pareto front and describe the method procedure in this context.

Note: the reader can find more benchmark multi-objective problems, expressed in mathematical terms, in [14.2] (with interesting insights) and [14.3].

14.13.1 Exercise References

14.1 https://www.kdnuggets.com/2019/05/linear-programming-discrete-optimization-python-pulp.html
14.2 https://ro.ecu.edu.au/cgi/viewcontent.cgi?article=3021&context=ecuworks
14.3 https://en.wikipedia.org/wiki/Test_functions_for_optimization

References

1. Abouhawwash, M., Jameel, M., & Deb, K. (2020). A smooth proximity measure for optimality in multi-objective optimization using Benson's method. *Computers & Operations Research*, *117*, 104900. Available from: https://www.egr.msu.edu/~kdeb/papers/c2019003.pdf
2. Afful-Dadzie, E., Oplatkova, Z. K., & Prieto, L. A. B. (2017). Comparative state-of-the-art survey of classical fuzzy set and intuitionistic fuzzy sets in multi-criteria decision making. *International Journal of Fuzzy Systems*, *19*, 726–738. Available from: https://link.springer.com/article/10.1007/s40815-016-0204-y
3. Alinezhad, A., & Khalili, J. (2019). COPRAS method. In *New methods and applications in multiple attribute decision making (MADM)* (pp. 87–91). Available from: https://link.springer.com/chapter/10.1007/978-3-030-15009-9_12
4. Allison, J. (2024). Demonstration of two multi-objective optimization strategies. Available from: https://es.mathworks.com/matlabcentral/fileexchange/64787-demonstration-of-two-multi-objective-optimization-strategies
5. Allmendinger, R., Ehrgott, M., Gandibleux, X., Geiger, J. M., Klamroth, K., & Luque, M. (2017). Navigation in multiobjective optimization methods. *Journal of Multi-Criteria Decision Analysis*, *24*(1–2), 57–70. Available from: https://onlinelibrary.wiley.com/doi/pdf/10.1002/mcda.1599
6. Alvarez, P. A., Ishizaka, A., & Martínez, L. (2021). Multiple-criteria decision-making sorting methods: A survey. *Expert Systems with Applications*, *183*, 115368. Available from: https://sinbad2.ujaen.es/sites/default/files/publications/ESWA2021-PavelAlessioLuis_Sort_survey.pdf

7. Ansari, Q. H., Köbis, E., & Yao, J.-C. (2018). *Vector variational inequalities and vector optimization*. Springer International Publishing AG. Available from: http://library.sadjad.ac.ir/opac/temp/18473.pdf
8. Arabameri, A., Yamani, M., Pradhan, B., Melesse, A., Shirani, K., & Bui, D. T. (2019). Novel ensembles of COPRAS multi-criteria decision-making with logistic regression, boosted regression tree, and random forest for spatial prediction of gully erosion susceptibility. *Science of the Total Environment*, *688*, 903–916. Available from: https://www.sciencedirect.com/science/article/pii/S0048969719327767
9. Atanassov, K. T., & Stoeva, S. (1986). Intuitionistic fuzzy sets. *Fuzzy Sets and Systems*, *20*(1), 87–96. Available from: https://biomed.bas.bg/bioautomation/2016/vol_20.s1/files/20.s1_02.pdf
10. Bana e Costa, C. A., & Vansnick, J.-C. (1999). The MACBETH approach: Basic ideas, software, and an application. In *Advances in decision analysis* (pp. 131–157). Springer. Available from: https://link.springer.com/chapter/10.1007/978-94-017-0647-6_9
11. Banker, R. D., Charnes, A., Cooper, W. W., Swarts, J., & Thomas, D. (1989). An introduction to data envelopment analysis with some of its models and their uses. *Research in Governmental and Nonprofit Accounting*, *5*(1), 125–163. Available from: https://iiif.library.cmu.edu/file/Cooper_box0010c_fld00033_bdl0001_doc0001/Cooper_box0010c_fld00033_bdl0001_doc0001.pdf
12. Barreiro, S. (2020). Goal programming. Available from: https://fenix.isa.ulisboa.pt/downloadFile/563022967873049/GoalProgramming_8-13May.pdf
13. Bazerman, M. H. (2017). Judgment and decision-making. *Noba Textbook Series: Psychology*. Available from: https://www2.psych.ubc.ca/~schaller/308Readings/NOBA-Bazerman.pdf
14. Bazerman, M. H., & Moore, D. A. (2012). *Judgment in managerial decision making*. John Wiley & Sons.
15. Behzadian, M., Kazemzadeh, R. B., Albadvi, A., & Aghdasi, M. (2010). Promethee: A comprehensive literature review on methodologies and applications. *European Journal of Operational Research*, *200*(1), 198–215. Available from: https://www.sciencedirect.com/science/article/pii/S0377221709000071
16. Benayoun, R., De Montgolfier, J., Tergny, J., & Laritchev, O. (1971). Linear programming with multiple objective functions: Step method (STEM). *Mathematical Programming*, *1*(1), 366–375. Available from: https://link.springer.com/article/10.1007/bf01584098
17. Benayoun, R., Roy, B., & Sussman, N. (1966). Manual de reference du programme ELECTRE. *Note de Synthese et Formation*, *25*(79).
18. Benson, H. P. (1978). Existence of efficient solutions for vector maximization problems. *Journal of Optimization Theory and Applications*, *26*, 569–580. Available from: https://link.springer.com/article/10.1007/BF00933152
19. Benson, H. P. (1998). An outer approximation algorithm for generating all efficient extreme points in the outcome set of a multiple objective linear programming problem. *Journal of Global Optimization*, *13*, 1–24. Available from: https://link.springer.com/content/pdf/10.1023/A:1008215702611.pdf
20. Bérubé, J.-F., Gendreau, M., & Potvin, J.-Y. (2009). An exact ε-constraint method for bi-objective combinatorial optimization problems: Application to the traveling salesman problem with profits. *European Journal of Operational Research*, *194*(1), 39–50. Available from: https://www-labs.iro.umontreal.ca/~potvin/berube_2009.pdf
21. Boland, N., Charkhgard, H., & Savelsbergh, M. (2015). A criterion space search algorithm for biobjective mixed integer programming: The triangle splitting method. *INFORMS Journal on Computing*, *27*(4), 597–618. Available from: https://pubsonline.informs.org/doi/abs/10.1287/ijoc.2015.0646
22. Brans, J.-P., & Mareschal, B. (1994). The PROMCALC & GAIA decision support system for multicriteria decision aid. *Decision Support Systems*, *12*(4–5), 297–310. Available from: https://www.sciencedirect.com/science/article/abs/pii/0167923694900485
23. Brans, J.-P., Nadeau, R., & Landry, M. (1982). L'ingénierie de la décision. *Elaboration d'instruments d'aide à la décision. La méthode PROMETHEE. In l'Aide à la Décision: Nature, Instruments et Perspectives d'Avenir* (pp. 183–213).

24. Brans, J.-P., Vincke, P., & Mareschal, B. (1986). How to select and how to rank projects: The PROMETHEE method. *European Journal of Operational Research*, *24*(2), 228–238. Available from: https://edisciplinas.usp.br/pluginfile.php/5261740/mod_resource/content/1/paper%20Brans.pdf
25. Buchanan, J. T. (1997). A naive approach for solving mcdm problems: The GUESS method. *Journal of the Operational Research Society*, *48*, 202–206. Available from: https://www.jstor.org/stable/pdf/3010360.pdf
26. Burachik, R. S., Kaya, C. Y., & Rizvi, M. M. (2022). Algorithms for generating Pareto fronts of multi-objective integer and mixed-integer programming problems. *Engineering Optimization*, *54*(8), 1413–1425. Available from: https://arxiv.org/pdf/1903.07041.pdf
27. Burachik, R. S., Kaya, C. Y., & Rizvi, M. M. (2017). A new scalarization technique and new algorithms to generate Pareto fronts. *SIAM Journal on Optimization*, *27*(2), 1010–1034. Available from: https://epubs.siam.org/doi/abs/10.1137/16M1083967
28. Cables, E., Lamata, M. T., & Verdegay, J. L. (2016). RIM-reference ideal method in multicriteria decision making. *Information Sciences*, *337*, 1–10. Available from: https://www.sciencedirect.com/science/article/pii/S0020025515009007
29. Chang, D.-Y. (1996). Applications of the extent analysis method on fuzzy ahp. *European Journal of Operational Research*, *95*(3), 649–655. Available from: https://www.expertchoice.ir/wp-content/uploads/2017/08/FAHP-Chang-1996.pdf
30. Charnes, A., & Cooper, W. W. (1957). Management models and industrial applications of linear programming. *Management Science*, *4*(1), 38–91. Available from: https://pubsonline.informs.org/doi/abs/10.1287/mnsc.4.1.38
31. Chircop, K., & Zammit-Mangion, D. (2013). On-constraint based methods for the generation of Pareto frontiers. *Journal of Mechanical Engineering and Automation*, *3*(5), 279–289. Available from: https://www.davidpublisher.com/Public/uploads/Contribute/551a4df67d753.pdf
32. Corne, D. W., Jerram, N. R., Knowles, J. D., & Oates, M. J. (2001). PESA-II: Region-based selection in evolutionary multiobjective optimization. In *Proceedings of the 3rd annual conference on genetic and evolutionary computation* (pp. 283–290). Available from: https://dl.acm.org/doi/pdf/10.5555/2955239.2955289
33. Corne, D. W., Knowles, J. D., & Oates, M. J. (2000). The Pareto envelope-based selection algorithm for multiobjective optimization. In *International conference on parallel problem solving from nature* (pp. 839–848). Springer. Available from: https://citeseerx.ist.psu.edu/document?repid=rep1&type=pdf&doi=951117b91bb0221dc2f513f82660be4b1739551e
34. Csirmaz, L. (2019). inner – A multiobjective linear programs solver. Available from: https://github.com/lcsirmaz/inner
35. Cuate, O., & Schütze, O. (2020). Pareto explorer for finding the knee for many objective optimization problems. *Mathematics*, *8*(10), 1651. Available from: https://www.mdpi.com/2227-7390/8/10/1651
36. Das, I., & Dennis, J. E., Jr. (1996). Normal boundary intersection. Available from: https://softlib.rice.edu/NBI.html
37. Das, I., & Dennis, J. E. (1998). Normal-boundary intersection: A new method for generating the Pareto surface in nonlinear multicriteria optimization problems. *SIAM Journal on Optimization*, *8*(3), 631–657. Available from: https://repository.rice.edu/server/api/core/bitstreams/5e4f2ab0-c3db-4077-90ca-c0914852f65a/content
38. De Santis, M., Eichfelder, G., Niebling, J., & Rocktäschel, S. (2020). Solving multiobjective mixed integer convex optimization problems. *SIAM Journal on Optimization*, *30*(4), 3122–3145. Available from: https://iris.uniroma1.it/retrieve/e3835328-56b5-15e8-e053-a505fe0a3de9/DeSantis_postprint_Solving-multiobjective_2020.pdf
39. Deb, K., Pratap, A., Agarwal, S., & Meyarivan, T. A. M. T. (2002). A fast and elitist multiobjective genetic algorithm: NSGA-II. *IEEE Transactions on Evolutionary Computation*, *6*(2), 182–197. Available from: https://www.cp.eng.chula.ac.th/~prabhas/teaching/ec/ec2010/nsga2-ieee-trans-ec-2002.pdf

40. Deb, K., & Sundar, J. (2006). Reference point based multi-objective optimization using evolutionary algorithms. In *Proceedings of the 8th annual conference on genetic and evolutionary computation* (pp. 635–642). Available from: https://dl.acm.org/doi/pdf/10.1145/1143997.1144112
41. Deckro, R. F., & Winkofsky, E. P. (1983). Solving zero-one multiple objective programs through implicit enumeration. *European Journal of Operational Research*, *12*(4), 362–374. Available from: https://www.sciencedirect.com/science/article/abs/pii/0377221783901571
42. Dörfler, D., Löhne, A., Schneider, C., & Weißing, B. (2022). A Benson-type algorithm for bounded convex vector optimization problems with vertex selection. *Optimization Methods and Software*, *37*(3), 1006–1026. Available from: https://arxiv.org/pdf/2006.15600.pdf
43. Driver, J. (2014). The history of utilitarianism. Available from: https://plato.stanford.edu/entries/utilitarianism-history/
44. Ehrgott, M., Gandibleux, X., & Przybylski, A. (2016). Exact methods for multi-objective combinatorial optimisation. In *Multiple criteria decision analysis: State of the art surveys* (pp. 817–850). Available from: https://link.springer.com/chapter/10.1007/978-1-4939-3094-4_19
45. Ehrgott, M., Shao, L., & Schöbel, A. (2011). An approximation algorithm for convex multi-objective programming problems. *Journal of Global Optimization*, *50*, 397–416. Available from: https://link.springer.com/content/pdf/10.1007/s10898-010-9588-7.pdf
46. Eichfelder, G. (2011). *Vector optimization*. Springer. Available from: http://ndl.ethernet.edu.et/bitstream/123456789/77635/1/93.pdf
47. Eichfelder, G. (2021). Twenty years of continuous multiobjective optimization in the twenty-first century. *EURO Journal on Computational Optimization*, *9*, 100014. Available from: https://optimization-online.org/wp-content/uploads/2020/12/8161.pdf
48. Eichfelder, G., Stein, O., & Warnow, L. (2022). A deterministic solver for multiobjective mixed-integer convex and nonconvex optimization. Available from: https://optimization-online.org/?p=18696
49. Eichfelder, G. (2019). ASMO. Available from: https://github.com/GEichfelder/ASMO
50. Emmerich, M. T., & Deutz, A. H. (2018). A tutorial on multiobjective optimization: Fundamentals and evolutionary methods. *Natural Computing*, *17*, 585–609. Available from: https://link.springer.com/content/pdf/10.1007/s11047-018-9685-y.pdf
51. Eskelinen, P., Miettinen, K., Klamroth, K., & Hakanen, J. (2010). Pareto navigator for interactive nonlinear multiobjective optimization. *OR Spectrum*, *32*, 211–227. Available from: https://link.springer.com/content/pdf/10.1007/s00291-008-0151-6.pdf
52. Eskelinen, P., Miettinen, K., Klamroth, K., Hakanen, J., et al. (2007). Interactive learning-oriented decision support tool for nonlinear multiobjective optimization: Pareto navigator. Available from: https://aaltodoc.aalto.fi/server/api/core/bitstreams/1b6db724-8f2b-430e-ba37-cc42903351d4/content
53. Fattahi, A., & Turkay, M. (2018). A one direction search method to find the exact nondominated frontier of biobjective mixed-binary linear programming problems. *European Journal of Operational Research*, *266*(2), 415–425. Available from: https://www.sciencedirect.com/science/article/abs/pii/S0377221717308470?via%3Dihub
54. Ferguson, Z. (2015). Aplmoo. Available from: https://github.com/zfergus/aplmoo/blob/master/notes/matlab_code/lexmin.m
55. Ferreiro-Cabello, J., Fraile-Garcia, E., de Pison Ascacibar, E. M., & Martinez-de Pison, F. J. (2018). Metamodel-based design optimization of structural one-way slabs based on deep learning neural networks to reduce environmental impact. *Engineering Structures*, *155*, 91–101. Available from: https://www.sciencedirect.com/science/article/pii/S0141029617301438
56. Figueira, J., Greco, S., & Ehrgott, M. (Eds.). (2005). *Multiple criteria decision analysis: State of the art surveys*. Springer. Available from: https://link.springer.com/book/10.1007/b100605
57. Fishburn, P. C. (1968). Utility theory. *Management Science*, *14*(5), 335–378. Available from: https://www2.cs.siu.edu/~hexmoor/classes/CS491-F10/Utility.pdf
58. Fonseca, C. M., & Fleming, P. J. (1993). Genetic algorithms for multiobjective optimization: Formulation discussion and generalization. In *ICGA* (Vol. 93, pp. 416–423).

Citeseer. Available from: https://citeseerx.ist.psu.edu/document?repid=rep1&type=pdf&doi=7cc2bc3b1c6a673d3751e26c3c6b0e3feebc1cdb
59. Forget, N., Gadegaard, S. L., Klamroth, K., Nielsen, L. R., & Przybylski, A. (2022). Branch-and-bound and objective branching with three or more objectives. *Computers & Operations Research*, *148*, 106012. Available from: https://www.sciencedirect.com/science/article/pii/S030505482200243X
60. Gadegaard, S. L., Nielsen, L. R., & Ehrgott, M. (2019). Bi-objective branch-and-cut algorithms based on LP relaxation and bound sets. *INFORMS Journal on Computing*, *31*(4), 790–804. Available from: https://pure.au.dk/ws/portalfiles/portal/140604037/IJOCfinal_1.pdf
61. Gass, S., & Saaty, T. (1955). The computational algorithm for the parametric objective function. *Naval Research Logistics Quarterly*, *2*(1–2), 39–45. Available from: https://onlinelibrary.wiley.com/doi/abs/10.1002/nav.3800020106
62. Gembicki, F., & Haimes, Y. (1975). Approach to performance and sensitivity multiobjective optimization: The goal attainment method. *IEEE Transactions on Automatic Control*, *20*(6), 769–771. Available from: https://ieeexplore.ieee.org/document/1101105
63. Geoffrion, A. M., Dyer, J. S., & Feinberg, A. (1972). An interactive approach for multi-criterion optimization, with an application to the operation of an academic department. *Management Science*, *19*(4-part-1), 357–368. Available from: https://pubsonline.informs.org/doi/abs/10.1287/mnsc.19.4.357
64. Gireesha, O., Somu, N., Raman, M. R. G., Reddy, M. S., Kirthivasan, K., & Sriram, V. S. (2020). WNN-EDAS: A wavelet neural network based multi-criteria decision-making approach for cloud service selection. In *Computational intelligence in pattern recognition: Proceedings of CIPR 2019* (pp. 853–865). Springer. Available from: https://link.springer.com/chapter/10.1007/978-981-13-9042-5_73
65. Greenstreet, P. (2023). Issues with weighted sum approach for non-convex sets. Available from: https://www.lancaster.ac.uk/stor-i-student-sites/peter-greenstreet/2020/04/26/issues-with-weighted-sum-approach-for-non-convex-sets/
66. Günther, C. (2021). FLO-Facility location optimizer. Available from: https://es.mathworks.com/matlabcentral/fileexchange/68526-flo-facility-location-optimizer
67. Haimes, Y. (1971). On a bicriterion formulation of the problems of integrated system identification and system optimization. *IEEE Transactions on Systems, Man, and Cybernetics*, *3*, 296–297.
68. Halffmann, P., Schäfer, L. E., Dächert, K., Klamroth, K., & Ruzika, S. (2022). Exact algorithms for multiobjective linear optimization problems with integer variables: A state of the art survey. *Journal of Multi-Criteria Decision Analysis*, *29*(5–6), 341–363. Available from: https://onlinelibrary.wiley.com/doi/pdf/10.1002/mcda.1780
69. Hartikainen, M., Miettinen, K., & Klamroth, K. (2019). Interactive nonconvex Pareto navigator for multiobjective optimization. *European Journal of Operational Research*, *275*(1), 238–251. Available from: https://www.sciencedirect.com/science/article/pii/S0377221718309792
70. Hartikainen, M., Miettinen, K., & Wiecek, M. M. (2012). Paint: Pareto front interpolation for nonlinear multiobjective optimization. *Computational Optimization and Applications*, *52*, 845–867. Available from: https://link.springer.com/article/10.1007/s10589-011-9441-z
71. He, Z., Tran, K.-P., Thomassey, S., Zeng, X., Xu, J., & Yi, C. (2021). A deep reinforcement learning based multi-criteria decision support system for optimizing textile chemical process. *Computers in Industry*, *125*, 103373. Available from: https://www.sciencedirect.com/science/article/pii/S0166361520306072
72. Huang, W., Zhang, Y., & Li, L. (2019). Survey on multi-objective evolutionary algorithms. In *Journal of Physics: Conference series*, *1288*(1), 012057. IOP Publishing. Available from: https://iopscience.iop.org/article/10.1088/1742-6596/1288/1/012057/pdf
73. Hwang, C.-L., & Yoon, K. (1981). *Multiple attribute decision making methods and applications a state-of-the-art survey*. Springer. Available from: https://link.springer.com/book/10.1007/978-3-642-48318-9
74. Ibrahim, A., Rahnamayan, S., Martin, M. V., & Deb, K. (2016). 3D-RadVis: Visualization of Pareto front in many-objective optimization. In *2016 IEEE congress*

on evolutionary computation (CEC) (pp. 736–745). IEEE. Available from: https://ieeexplore.ieee.org/stamp/stamp.jsp?arnumber=7743865&casa_token=eiVLRh7eSPoAAAAA:kvkopMhHZ4RtQCmVgeWeZDqfEspMe9pAdWxqLsrs4Chj69R2uar2FgfjdVwGZog6SG2KdJRbfA
75. Ibrahim, A., Rahnamayan, S., Martin, M. V., & Deb, K. (2018). 3d-radvis antenna: Visualization and performance measure for many-objective optimization. *Swarm and Evolutionary Computation*, *39*, 157–176. Available from: https://www.sciencedirect.com/science/article/pii/S2210650216304758
76. Isermann, H. (1982). Linear lexicographic optimization. *Operations-Research-Spektrum*, *4*(4), 223–228. Available from: https://link.springer.com/article/10.1007/BF01782758
77. Jaimes, A. L., Martınez, S. Z., Coello Coello, C. A., et al. (2009). An introduction to multiobjective optimization techniques. *Optimization in Polymer Processing*, *1*, 29.
78. Jaszkiewicz, A., & Słowiński, R. (1999). The 'light beam search' approach-an overview of methodology applications. *European Journal of Operational Research*, *113*(2), 300–314. Available from: https://www.sciencedirect.com/science/article/pii/S0377221798002185
79. Jones, D., & Tamiz, M. (2010). *Practical goal programming* (Vol. 141). Springer. Available from: https://link.springer.com/book/10.1007/978-1-4419-5771-9
80. Kahraman, C., Onar, S. C., & Oztaysi, B. (2015). Fuzzy multicriteria decision-making: A literature review. *International Journal of Computational Intelligence Systems*, *8*(4), 637–666. Available from: https://link.springer.com/content/pdf/10.1080/18756891.2015.1046325.pdf
81. Kaklauskas, A., Zavadskas, E. K., Raslanas, S., Ginevicius, R., Komka, A., & Malinauskas, P. (2006). Selection of low-e windows in retrofit of public buildings by applying multiple criteria method COPRAS: A lithuanian case. *Energy and Buildings*, *38*(5), 454–462. Available from: http://www.raslanas.lt/uploads/4.pdf
82. Kasimoğlu, F., Amaçlı, Ç., & Verme, K. (2016). A survey on interactive approaches for multi-objective decision making problems. *Defence Science Journal, 15*(1), 231–255. Available from: https://kho.msu.edu.tr/akademik/enstitu/Alp_SAVBEN_dergi/151/9.pdf
83. Keller, A. A. (2017). *Multi-objective optimization in theory and practice I: Classical methods.* Bentham Science Publishers.
84. Kiziltan, G., & Yucaoğlu, E. (1983). An algorithm for multiobjective zero-one linear programming. *Management Science*, *29*(12), 1444–1453. Available from: https://www.jstor.org/stable/pdf/2631029.pdf
85. Knowles, J. D., & Corne, D. W. (2000). Approximating the nondominated front using the Pareto archived evolution strategy. *Evolutionary Computation*, *8*(2), 149–172. Available from: https://www.macs.hw.ac.uk/~dwcorne/newkc.dvi.pdf
86. Koksalan, M. (2018). Multiple criteria decision making: Interactive approaches. Available from: https://mcda-school18.tuc.gr/Slides/Koksalan.pdf
87. Korhonen, P., & Wallenius, J. (1988). A Pareto race. *Naval Research Logistics (NRL)*, *35*(6), 615–623. Available from: https://ideas.repec.org/a/wly/navres/v35y1988i6p615-623.html
88. Liao, H., He, Y., Wu, X., Wu, Z., & Bausys, R. (2023). Reimagining multi-criterion decision making by data-driven methods based on machine learning: A literature review. *Information Fusion*, 101970. Available from: https://www.sciencedirect.com/science/article/pii/S1566253523002865
89. Lim, Y. I., Floquet, P., & Joulia, X. (2001). Efficient implementation of the normal boundary intersection (NBI) method on multiobjective optimization problems. *Industrial & Engineering Chemistry Research*, *40*(2), 648–655. Available from: https://web.archive.org/web/20050906061336id_/http://hknu.ac.kr:80/~limyi/source_documents/I&EC_Lim_NBI_2001.pdf
90. Liu, B., Shen, Y., Zhang, W., Chen, X., & Wang, X. (2015). An interval-valued intuitionistic fuzzy principal component analysis model-based method for complex multi-attribute large-group decision-making. *European Journal of Operational Research*, *245*(1), 209–225. Available from: https://www.sciencedirect.com/science/article/pii/S0377221715001265
91. Löhne, A. (2015). Bensolve. Available from: https://github.com/loehnea/bensolve

92. Löhne, A., Rudloff, B., & Ulus, F. (2014). Primal and dual approximation algorithms for convex vector optimization problems. *Journal of Global Optimization*, *60*(4), 713–736. Available from: https://arxiv.org/pdf/2006.15600.pdf
93. Löhne, A., & Weißing, B. (2017). The vector linear program solver Bensolve-notes on theoretical background. *European Journal of Operational Research*, *260*(3), 807–813. Available from: https://arxiv.org/pdf/1510.04823
94. Luna-Reyes, L. F., Martin, E., & Ivonchyk, M. (2022). Data analytics for public policy and management. Available from: https://pressbooks.pub/decisions/chapter/7-2-mau-model-worked-out-examples/
95. Madetoja, E., Ruotsalainen, H., Monkkonen, V.-M., Hamalainen, J., & Deb, K. (2008). Visualizing multi-dimensional Pareto-optimal fronts with a 3d virtual reality system. In *2008 International multiconference on computer science and information technology* (pp. 907–913). IEEE. Available from: https://web.archive.org/web/20170809145232id_/https://annals-csis.org/proceedings/2008/pliks/90.pdf
96. Mardani, A., Jusoh, A., & Zavadskas, E. K. (2015). Fuzzy multiple criteria decision-making techniques and applications-two decades review from 1994 to 2014. *Expert Systems with Applications*, *42*(8), 4126–4148. Available from: https://e-tarjome.com/storage/btn_uploaded/2020-05-28/1590651998_8646-etarjome%20English.pdf
97. Maurel. (2023). Compute dominate points. Available from: https://es.mathworks.com/matlabcentral/fileexchange/52340-compute-dominate-points
98. Mausser, H. (2006). Normalization and other topics in multi-objective optimization. In *Fields-MITACS industrial problems workshop* (p. 89). Citeseer. Available from: http://miis.maths.ox.ac.uk/miis/233/1/fmipw1-6.pdf
99. Mavrotas, G. (2009). Effective implementation of the ε-constraint method in multi-objective mathematical programming problems. *Applied Mathematics and Computation*, *213*(2), 455–465. Available from: https://www.sciencedirect.com/science/article/pii/S0096300309002574
100. Mavrotas, G., & Diakoulaki, D. (1998). A branch and bound algorithm for mixed zero-one multiple objective linear programming. *European Journal of Operational Research*, *107*(3), 530–541. Available from: https://www.sciencedirect.com/science/article/abs/pii/S0377221797000775
101. MCDMSociety. (2021). MOrepo-Forget21. Available from: https://github.com/MCDMSociety/MOrepo-Forget21
102. Merenchige, P. L. W. J., & Wiecek, M. (2022). A branch and bound algorithm for biobjective mixed integer quadratic programs. *Optimization Online*. Available from: https://optimization-online.org/2022/12/a-branch-and-bound-algorithm-for-biobjective-mixed-integer-quadratic-programs/
103. Miettinen, K., & Mäkelä, M. M. (1995). Interactive bundle-based method for nondifferentiable multiobjective optimization: NIMBUS. *Optimization*, *34*(3), 231–246. Available from: https://www.tandfonline.com/doi/abs/10.1080/02331939508844109
104. Miettinen, K. (1999). *Nonlinear multiobjective optimization* (Vol. 12). Springer Science & Business Media. Available from: https://link.springer.com/book/10.1007/978-1-4615-5563-6
105. Miettinen, K. (2001). Some methods for nonlinear multi-objective optimization. In *International conference on evolutionary multi-criterion optimization* (pp. 1–20). Springer. Available from: https://link.springer.com/chapter/10.1007/3-540-44719-9_1
106. Miettinen, K., Eskelinen, P., Ruiz, F., & Luque, M. (2010). NAUTILUS method: An interactive technique in multiobjective optimization based on the nadir point. *European Journal of Operational Research*, *206*(2), 426–434. Available from: https://www.sciencedirect.com/science/article/pii/S0377221710001785
107. Miettinen, K., Podkopaev, D., Ruiz, F., & Luque, M. (2015). A new preference handling technique for interactive multiobjective optimization without trading-off. *Journal of Global Optimization*, *63*, 633–652. Available from: https://link.springer.com/article/10.1007/s10898-015-0301-8

108. Miettinen, K., & Ruiz, F. (2016). NAUTILUS framework: Towards trade-off-free interaction in multiobjective optimization. *Journal of Business Economics*, *86*, 5–21. Available from: https://link.springer.com/article/10.1007/s11573-015-0786-0
109. Misitano, G., Saini, B. S., Afsar, B., Shavazipour, B., & Miettinen, K. (2021). DESDEO: The modular and open source framework for interactive multiobjective optimization. *IEEE Access*, *9*, 148277–148295. Available from: https://ieeexplore.ieee.org/stamp/stamp.jsp?arnumber=9591595
110. Moazeni, F., & Khazaei, J. (2021). Interactive nonlinear multiobjective optimal design of water distribution systems using Pareto navigator technique. *Sustainable Cities and Society*, *73*, 103110. Available from: https://www.sciencedirect.com/science/article/pii/S2210670721003929
111. Modibbo, U. M., Hassan, M., Ahmed, A., & Ali, I. (2022). Multi-criteria decision analysis for pharmaceutical supplier selection problem using fuzzy TOPSIS. *Management Decision*, *60*(3), 806–836. Available from: https://www.emerald.com/insight/content/doi/10.1108/MD-10-2020-1335/full/html
112. Monz, M., Küfer, K.-H., Bortfeld, T. R., & Thieke, C. (2008). Pareto navigation-algorithmic foundation of interactive multi-criteria imrt planning. *Physics in Medicine & Biology*, *53*(4), 985. Available from: https://iopscience.iop.org/article/10.1088/0031-9155/53/4/011/pdf
113. Nakayama, H. (1995). Aspiration level approach to interactive multi-objective programming and its applications. In *Advances in multicriteria analysis* (pp. 147–174). Available from: https://pure.iiasa.ac.at/id/eprint/4100/1/WP-94-112.pdf
114. Nour, A. R., & Kamali, A. M. (2009). A weighted metric method to optimize multi-response robust problems. *International Journal of Industrial Engineering*. Available from: https://journals.iau.ir/article_511054_8ffc181084440128b4bd3524ad405353.pdf
115. Nyiam, P. B., & Salhi, A. (2021). A comparison of Benson's outer approximation algorithm with an extended version of multiobjective simplex algorithm. *Advances in Operations Research*, *2021*, 1–11. Available from: https://web.archive.org/web/20210713030129id_/https://downloads.hindawi.com/journals/aor/2021/1857030.pdf
116. Ojalehto, V., & Miettinen, K. (2019). DESDEO: An open framework for interactive multi-objective optimization. In *Multiple criteria decision making and aiding: Cases on models and methods with computer implementations* (pp. 67–94). Available from: https://jyx.jyu.fi/bitstream/handle/123456789/60987/1/ojalehtomiettinendesdeo.pdf
117. Opricovic, S., & Tzeng, G.-H. (2007). Extended vikor method in comparison with outranking methods. *European Journal of Operational Research*, *178*(2), 514–529. Available from: https://www.sciencedirect.com/science/article/pii/S0377221706000555
118. Parragh, S. N., & Tricoire, F. (2019). Branch-and-bound for bi-objective integer programming. *INFORMS Journal on Computing*, *31*(4), 805–822. Available from: https://arxiv.org/pdf/1809.06823.pdf
119. Pettersson, W., & Ozlen, M. (2019). Multi-objective mixed integer programming: An objective space algorithm. In *AIP conference proceedings* (Vol. 2070). AIP Publishing. Available from: https://www.researchgate.net/profile/Melih-Ozlen/publication/335713817_Multi-Objective_Mixed_Integer_Programming_An_Objective_Space_Algorithm/links/5d8aeac0a6fdcc255496df26/Multi-Objective-Mixed-Integer-Programming-An-Objective-Space-Algorithm.pdf
120. Pirouz, B., & Khorram, E. (2016). A computational approach based on the ε-constraint method in multi-objective optimization problems. *Advances and Applications in Statistics*, *49*(6), 453–483. Available from: https://www.researchgate.net/profile/Behzad-Pirouz-2/publication/311550947_A_Computational_Approach_Based_on_the_epsilon-Constraint_Method_in_Multi-Objective_Optimization_Problems/links/5b89883492851c1e123f8d42/A-Computational-Approach-Based-on-the-epsilon-Constraint-Method-in-Multi-Objective-Optimization-Problems.pdf
121. Przybylski, A., & Gandibleux, X. (2017). Multi-objective branch and bound. *European Journal of Operational Research*, *260*(3), 856–872. Available from: https://www.sciencedirect.com/science/article/pii/S037722171730067X

122. Rahmani, S., & Amjady, N. (2019). Enhanced goal attainment method for solving multi-objective security-constrained optimal power flow considering dynamic thermal rating of lines. *Applied Soft Computing*, *77*, 41–49.
123. Rasmi, S. A. B., & Türkay, M. (2019). GoNDEF: An exact method to generate all non-dominated points of multi-objective mixed-integer linear programs. *Optimization and Engineering*, *20*, 89–117. Available from: https://link.springer.com/article/10.1007/s11081-018-9399-0
124. Rawat, S. S., Pant, S., Kumar, A., Ram, M., Sharma, H. K., & Kumar, A. (2022). A state-of-the-art survey on analytical hierarchy process applications in sustainable development. *International Journal of Mathematical, Engineering and Management Sciences*, *7*, 883–917. Available from: https://www.proquest.com/docview/2792895950?pq-origsite=gscholar&fromopenview=true&sourcetype=Scholarly%20Journals
125. Read, D. (2004). Utility theory from Jeremy Bentham to Daniel Kahneman. Available from: https://eprints.lse.ac.uk/22750/1/04064.pdf
126. Ringuest, J. L., & Ringuest, J. L. (1992). Compromise programming. In *Multiobjective optimization: Behavioral and computational considerations* (pp. 51–59). Available from: https://link.springer.com/chapter/10.1007/978-1-4615-3612-3_4
127. Roostaee, R., Izadikhah, M., Lotfi, F. H., et al. (2012). An interactive procedure to solve multi-objective decision-making problem: An improvment to STEM method. *Journal of Applied Mathematics*, *2012*. Available from: https://www.hindawi.com/journals/jam/2012/324712/
128. Roy, B. (1968). Classement et choix en présence de points de vue multiples. *Revue Française D'informatique et de Recherche Opérationnelle*, *2*(8), 57–75. Available from: http://www.numdam.org/item/RO_1968__2_1_57_0.pdf
129. Roy, B. (1991). The outranking approach and the foundations of ELECTRE methods. *Theory and Decision*, *31*, 49–73. Available from: https://link.springer.com/article/10.1007/bf00134132
130. Ruiz, A. B., Ruiz, F., Miettinen, K., Delgado-Antequera, L., & Ojalehto, V. (2019). NAUTILUS navigator: Free search interactive multiobjective optimization without trading-off. *Journal of Global Optimization*, *74*(2), 213–231. Available from: https://link.springer.com/article/10.1007/s10898-019-00765-2
131. Ruiz, A. B., Sindhya, K., Miettinen, K., Ruiz, F., & Luque, M. (2015). E-NAUTILUS: A decision support system for complex multiobjective optimization problems based on the NAUTILUS method. *European Journal of Operational Research*, *246*(1), 218–231. Available from: https://ideas.repec.org/a/eee/ejores/v246y2015i1p218-231.html
132. Ruotsalainen, H. (2010). *Interactive multiobjective optimization in model-based decision making with applications* [PhD thesis, Kuopion yliopisto]. Available from: https://erepo.uef.fi/bitstream/handle/123456789/9592/urn_isbn_978-951-27-1462-9.pdf?sequence=1
133. Saaty, T. L. (1977). A scaling method for priorities in hierarchical structures. *Journal of Mathematical Psychology*, *15*(3), 234–281. Available from: https://superdecisions.com/sd_resources/Paper_ScalingMethod.pdf
134. Saaty, T. L., & Vargas, L. G. (2013). *The analytic network process*. Springer. Available from: https://link.springer.com/chapter/10.1007/978-1-4614-7279-7_1
135. Sáez-Aguado, J., & Trandafir, P. C. (2018). Variants of the ε-constraint method for biobjective integer programming problems: Application to p-median-cover problems. *Mathematical Methods of Operations Research*, *87*(2), 251–283. Available from: https://link.springer.com/article/10.1007/s00186-017-0618-9
136. Sakawa, M. (1980). Interactive multiobjective decision making by the sequential proxy optimization technique: SPOT. Available from: https://core.ac.uk/download/pdf/33892874.pdf
137. Sakawa, M. (1982). Interactive multiobjective decision making by the sequential proxy optimization technique: SPOT. *European Journal of Operational Research*, *9*(4), 386–396. Available from: https://pure.iiasa.ac.at/id/eprint/1403/1/WP-80-066.pdf
138. Salian, A. C. V. (2022). A gentle introduction to multi-objective optimisation. Available from: https://codemonk.in/blog/a-gentle-introduction-to-multi-objective-optimization/

139. Schütze, O., Cuate, O., Martín, A., Peitz, S., & Dellnitz, M. (2019). Pareto explorer: A global/local exploration tool for many-objective optimization problems. *Engineering Optimization* Available from: https://www.tandfonline.com/doi/abs/10.1080/0305215X.2019.1617286
140. Shenfield, A., & Fleming, P. J. (2011). Multi-objective evolutionary design of robust controllers on the grid. *IFAC Proceedings Volumes*, *44*(1), 14711–14716. Available from: https://shura.shu.ac.uk/8314/1/2011_ifac_wc.pdf
141. Siddiqui, S., Azarm, S., & Gabriel, S. A. (2012). On improving normal boundary intersection method for generation of Pareto frontier. *Structural and Multidisciplinary Optimization*, *46*, 839–852. Available from: https://systems.jhu.edu/wp-content/uploads/2018/10/2012NBISMO.pdf
142. Simopoulos, A. K. (1977). *Multicriteria integer zero-one programming: A tree-search type algorithm* [PhD thesis, Naval Postgraduate School Monterey, CA]. Available from: https://core.ac.uk/download/pdf/36711449.pdf
143. Singh, A., & Malik, S. K. (2014). Major MCDM techniques & their application—a review. *IOSR Journal of Engineering*, *4*(5), 15–25.
144. Siregar, D., Arisandi, D., Usman, A., Irwan, D., & Rahim, R. (2017). Research of simple multi-attribute rating technique for decision support. *Journal of Physics: Conference Series*, *930*(1), 012015. IOP Publishing. Available from: https://iopscience.iop.org/article/10.1088/1742-6596/930/1/012015/pdf
145. Srinivas, N., & Deb, K. (1994). Muiltiobjective optimization using nondominated sorting in genetic algorithms. *Evolutionary Computation*, *2*(3), 221–248. Available from: http://repository.ias.ac.in/9405/1/308.pdf
146. Stadler, W. (1979). A survey of multicriteria optimization or the vector maximum problem, part I: 1776–1960. *Journal of Optimization Theory and Applications*, *29*, 1–52. Available from: https://link.springer.com/article/10.1007/BF00932634
147. Stadler, W. (1988). *Multicriteria Optimization in Engineering and in the Sciences* (Vol. 37). Springer Science & Business Media. Available from: https://link.springer.com/book/10.1007/978-1-4899-3734-6
148. MathWorks Staff. (2023). Pareto front for multiobjective optimization, problem-based. Available from: https://es.mathworks.com/help/gads/multiobjective-ooptimization-prob-based.html
149. MathWorks Staff. (2023). Plot 3-D Pareto front. Available from: https://es.mathworks.com/help/gads/plot-3-d-pareto-set.html
150. Steuer, R. E. (1986). *Multiple criteria optimization; theory, computation, and application.* Wiley.
151. Steuer, R. E., & Choo, E.-U. (1983). An interactive weighted Tchebycheff procedure for multiple objective programming. *Mathematical Programming*, *26*, 326–344. Available from: https://link.springer.com/article/10.1007/BF02591870
152. Taherdoost, H., & Madanchian, M. (2023). A comprehensive overview of the electre method in multi criteria decision-making. *Journal of Management Science & Engineering Research*, *6*(2). Available from: https://journals.bilpubgroup.com/index.php/jmser/article/download/5637/4885
153. Tamiz, M., Jones, D., & Romero, C. (1998). Goal programming for decision making: An overview of the current state-of-the-art. *European Journal of Operational Research*, *111*(3), 569–581. Available from: https://researchportal.port.ac.uk/en/publications/goal-programming-for-decision-making-an-overview-of-the-current-s
154. Taufik, I., Alam, C. N., Mustofa, Z., Rusdiana, A., & Uriawan, W. (2021). Implementation of multi-attribute utility theory (MAUT) method for selecting diplomats. In *IOP conference series: Materials science and engineering* (Vol. 1098, p. 032055). IOP Publishing. Available from: https://iopscience.iop.org/article/10.1088/1757-899X/1098/3/032055/pdf
155. Ulivieri, V. (2014). Multicriteria optimization: Scalarization tecniques. Available from: https://core.ac.uk/download/pdf/79615211.pdf

156. Vaz, I. (2024). Multi-objective sequential quadratic programming. Available from: https://www.mathworks.com/matlabcentral/fileexchange/58821-multi-objective-sequential-quandratic-programming?s_tid=ta_fx_results
157. Von Neumann, J., & Morgenstern, O. (1947). *Theory of games and economic behavior, 2nd rev*. Princeton University Press.
158. Wang, Y.-M., & Elhag, T. M. S. (2006). Fuzzy TOPSIS method based on alpha level sets with an application to bridge risk assessment. *Expert Systems with Applications*, *31*(2), 309–319. Available from: https://www.sciencedirect.com/science/article/pii/S0957417405002174
159. Warnow, L. (2024). MOMIBB. Available from: https://github.com/LeoWarnow/MOMIBB
160. Wierzbicki, A. P. (1980). The use of reference objectives in multiobjective optimization. In *Multiple criteria decision making theory and application: Proceedings of the third conference Hagen/Königswinter, West Germany, August 20–24, 1979* (pp. 468–486). Springer. Available from: https://pure.iiasa.ac.at/id/eprint/1117/1/WP-79-066.pdf
161. Xu, K., Lin, D. K. J., Tang, L.-C., & Xie, M. (2004). Multiresponse systems optimization using a goal attainment approach. *IIE Transactions*, *36*(5), 433–445. Available from: https://www.stat.purdue.edu/~dkjlin/documents/publications/2004/2004_IIE_QRE.pdf
162. Xu, L., Wang, J., Ou, Y., Fu, Y., & Bian, X. (2022). A novel decision-making system for selecting offshore wind turbines with PCA and D numbers. *Energy*, *258*, 124818. Available from: https://www.sciencedirect.com/science/article/pii/S0544222017212
163. Yadav, D., Ramu, P., & Deb, K. (2022). Visualization-aided multi-criterion decision-making using reference direction based Pareto race. In *2022 IEEE symposium series on computational intelligence (SSCI)* (pp. 125–132). IEEE. Available from: https://www.egr.msu.edu/~kdeb/papers/c2022007.pdf
164. Yadav, D., Ramu, P., & Deb, K. (2023). Interpretable self-organizing map assisted interactive multi-criteria decision-making following Pareto-race. *Applied Soft Computing*, *149*, 111032. Available from: https://www.sciencedirect.com/science/article/pii/S1568494623010505
165. Yalcin, A. S., Kilic, H. S., & Delen, D. (2022). The use of multi-criteria decision-making methods in business analytics: A comprehensive literature review. *Technological Forecasting and Social Change*, *174*, 121193. Available from: https://www.sciencedirect.com/science/article/abs/pii/S0040162521006260
166. Yang, M., Nazir, S., Xu, Q., Ali, S., et al. (2020). Deep learning algorithms and multicriteria decision-making used in big data: A systematic literature review. *Complexity*, *2020*. Available from: https://www.hindawi.com/journals/complexity/2020/2836064/
167. Zadeh, L. (1963). Optimality and non-scalar-valued performance criteria. *IEEE Transactions on Automatic Control*, *8*(1), 59–60.
168. Zeleny, M. (1973). Compromise programming. In *Multiple criteria decision making*. University of South Carolina Press, Columbia.
169. Zitzler, E., Laumanns, M., & Thiele, L. (2001). Spea2: Improving the strength Pareto evolutionary algorithm. *TIK Report*, *103*. Available from: https://www.research-collection.ethz.ch/bitstream/handle/20.500.11850/145755/eth-24689-01.pdf
170. Zitzler, E., & Thiele, L. (1999). Multiobjective evolutionary algorithms: A comparative case study and the strength Pareto approach. *IEEE Transactions on Evolutionary Computation*, *3*(4), 257–271. Available from: https://cs.ijs.si/papa/courses/06_ZitzlerMultiObjective.pdf

Chapter 15
Duality, Games, and Proximal Methods

15.1 Introduction

In his lecture notes on convex analysis and optimization [7], D. P. Bertsekas (M.I.T.) recognizes three main historical epochs:

- Prehistory, 1900–1949. With Caratheodory, Minkowski, Steiniz, and Farkas. The main topics being properties of convex sets and functions
- Fenchel-Rockafellar era, 1949–1980s. With the following main topics:
  - Duality theory
  - Minimax/game theory
  - Subdifferentials, optimality conditions, sensitivity
- Modern era. Paradigm shift. From the mid-1980s to present times. Main topics:
  - Non-smooth analysis
  - Algorithms
  - A change in the assumptions underlying the field

This scheme corresponds to a framework dominated by convexity/non-convexity. In this context, the view of Bertsekas is quite pertinent and indicates a series of topics that we should at least introduce.

In consequence, this chapter will consider duality, minimax/game theory, and non-smooth analysis.

J. M. Giron-Sierra, *Introduction to Optimization with Matlab® Examples*,
https://doi.org/10.1007/978-3-032-05048-9_15

Tic-tac-toe (Japanese)

With no doubt, the most acclaimed topic is game theory. It probably boasts a record of popularity in terms of Google responses. If you ask about "game theory," you receive 916,000,000 answers. Compare this with "theory of relativity," which receives 97,800,000 answers.

In general, the chapter has a mathematical flavor. As it is, it also unveils fundamental ideas that opened the way for important developments and practical applications (for instance, in data and image processing, artificial intelligence, and business).

15.2 Duality

It is usual, when using interior-point methods, to deal with dual problems. This is one of the reasons that have motivated much interest on duality issues. In this context, it is illuminating to consider the Fenchel's duality theorem, which we introduce now by means of two simple figures.

The situation consists of a convex and a concave curve. The problem is to find the minimum vertical distance between the two curves. Figure 15.1 depicts the case.

A second problem is to find two separated parallel tangents, so their distance is maximal. Figure 15.2 depicts this other case.

The *Fenchel's duality theorem* establishes that the points having the minimal vertical separation are also the tangency points for the maximally separated parallel tangents.

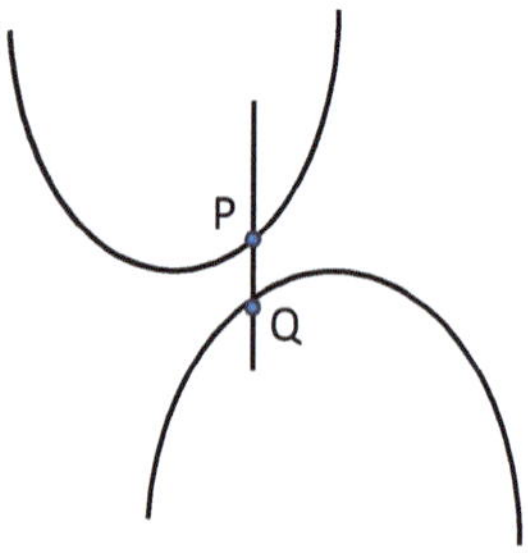

Fig. 15.1 Minimal vertical distance between two curves

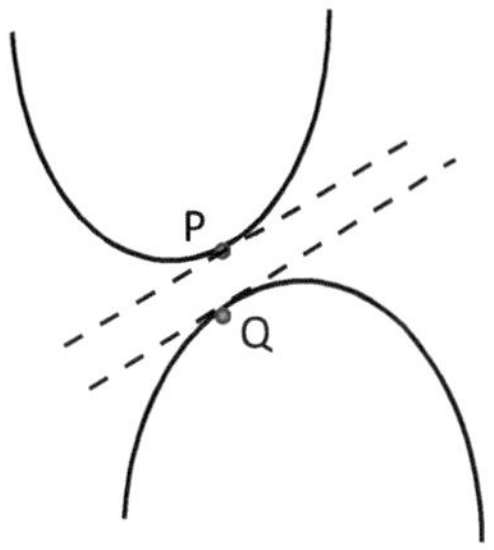

Fig. 15.2 Maximally separated parallel tangents

Let us devote some efforts to formalize this result. Of course, what makes it interesting for us is because of primal-dual solutions and duality gap issues.

Some Applications of Fenchel's Duality

Evidently, Fenchel's duality is directly related to the Support Vector Machine concepts.

Motivated by *web navigators'* needs, the recent *rankSVM* algorithm is a learning retrieval function that employs SVM pairwise ranking methods. There are several variations of this type of algorithm, for example, *FenchelRank*, which is based on Fenchel duality; see [45, 48], and references therein.

15.2.1 The Legendre-Fenchel Transform

A first important ingredient is the Legendre-Fenchel (LF) transform, which is a generalization of the Legendre transform. The Legendre transform applies for convex, differentiable functions, while the LF transform can be applied to a continuous but not necessarily differentiable function [89].

The concept of the Legendre transform can be intuitively introduced by considering the exercise represented in Fig. 15.3. A certain slope a is chosen. Now, one moves a line with slope a until touching the curve $f(x)$. This tangent intersects the vertical axis at b.

The value of the Legendre transform $f^*(a)$ will be $-b$ (notice the change of sign).

Given a convex, differentiable function $f(x)$, its Legendre transform is a continuous set of values $-b$ obtained for different values of a. The definition of the Legendre transform is

$$f^*(a) = \max_x (a\,x - f(x)).$$

The LF transform of a continuous function $f(x)$ is defined as follows:

$$f^*(a) = \sup_x (a\,x - f(x))$$

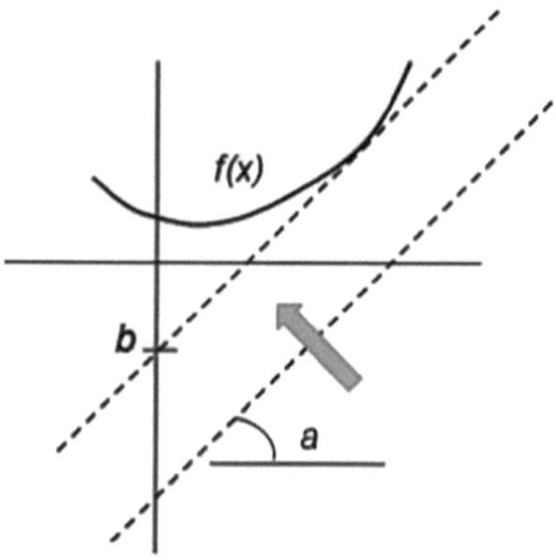

Fig. 15.3 Obtaining a value of $f^*(a)$ *(the Legendre transform)*

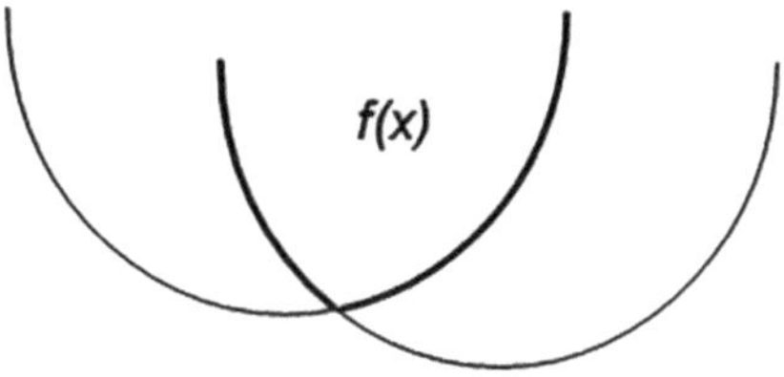

Fig. 15.4 Example of non-smooth function

The LF transform of $f^*(a)$ is

$$f^{**}(x) = \sup_a (ax - f^*(a)).$$

The reader may suspect that mathematical difficulties may arise when functions are not convex and/or non-differentiable. Of course, it is not our mission here to try a complete study of this question, but it is convenient to include some hints.

For instance, Fig. 15.4 shows how the intersection of two differentiable functions may give a non-differentiable (non-smooth) function.

Given a function $f(x)$, it is said that it admits a *supporting line* at x if there exists a parameter p such that

$$f(y) \geq f(x) + p(y - x), \quad \forall y \in \Re.$$

A supporting line is strictly supporting if

$$f(y) > f(x) + p(y - x), \quad \forall y \neq x.$$

For differentiable functions, a supporting line is also a tangent line.

Figure 15.5 depicts a curve with a number of interesting points. Two supporting lines have been plotted (dash-point-dash lines).

Let us make the following remarks:

- The point (a) admits a strictly supporting line.
- The point (b) does not admit any supporting line.

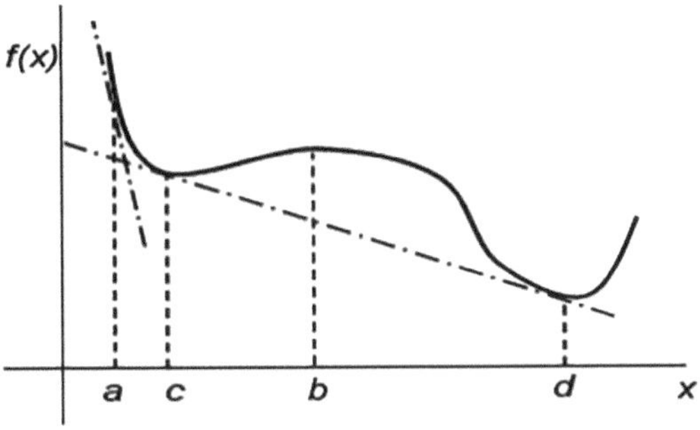

Fig. 15.5 An example of curve with some points of interest

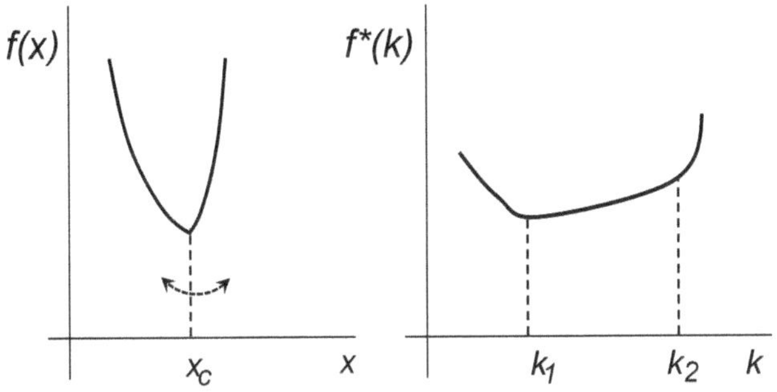

Fig. 15.6 LF transform of a non-differentiable point

- The point (c) admits a non-strictly supporting line, as it touches another point (d) of $f(x)$.

For the general cases, it can be shown that if $f(x)$ admits a supporting line at x with slope k, then $f^*(k)$ admits a supporting line at k with slope x. Moreover, if $f(x)$ admits a strictly supporting line at x with slope k, then $f^*(k)$ admits a tangent line at k with slope $f^{*'}(k) = x$ (therefore, $f^*(k)$is differentiable).

- It also can be shown that $f(x) = f^{**}(x)$ iff $f(x)$ admits a supporting line at x.
- In addition, if $f^*(k)$ is differentiable at k, then $f(x) = f^{**}(x)$ at $x = f^{*'}(k)$.
- Indeed, if $f^*(k)$ is everywhere differentiable, then $f(x) = f^{**}(x)$ for all x.

An important property of the LF transform is that $f^*(k)$ and $f^{**}(x)$ are convex functions (they are U-shaped).

Other properties of interest are the following:

- A convex function can always be written as the LF transform of another function.
- $f^{**}(x)$ is the largest convex function that satisfies $f^{**}(x) \leq f(x)$.

And finally, concerning non-differentiable points, suppose that $f(x)$ has one of these points (like the one represented in Fig. 15.4) and denote it as x_C. This point admits infinitely many supporting lines, with slopes in the range $k_1 \ldots k_2$. The corresponding $f^*(k)$ would be a line of constant slope x_C in the interval $k_1 \ldots k_2$. Figure 15.6 depicts an example.

See [89] for more details on the LF transform.

15.3 The Fenchel's Duality Theorem

Let us express the Fenchel's duality theorem using the formalism already introduced. Suppose that $f(\vec{x})$ is a proper convex function, $g(\vec{x})$ is a proper concave function, and regularity conditions are satisfied. Then,

$$\min_{\vec{x}}(f(\vec{x}) - g(\vec{x})) = \max_{\vec{a}}(g_*(\vec{a}) - f^*(\vec{a})),$$

where

$$f^*(\vec{a}) = \sup_{\vec{x}}(\vec{a}^T \vec{x} - f(\vec{x}))$$

$$g_*(\vec{a}) = \inf_{\vec{x}}(\vec{a}^T \vec{x} - g(\vec{x})).$$

The function $f^*(\vec{x})$ is called the *convex conjugate* of $f(\vec{x})$, and the function $g_*(\vec{x})$ is called the *concave conjugate* of $g(\vec{x})$.

15.3.1 Other Relationships

From the very definition of LF transform, it is possible to write the following:

$$f(\vec{x}) + f^*(\vec{a}) \geq \vec{a}^T \vec{x}$$

This is called the *Fenchel's inequality*.

Under certain regularity conditions, if $f(\vec{x}) \geq g(\vec{x})$, $\forall x \in \Re^n$, then there exists a *separation plane* between $f(\vec{x})$ and $g(\vec{x})$ given by

$$f(\vec{x}) \geq \vec{a}^T \vec{x} + b \geq g(\vec{x})$$

(this result is established by the hyperplane separating theorem [10]).

If $f(x)$ is convex and differentiable, one can consider the Legendre transform

$$f^*(a) = \max_{x}(a\,x - f(x))$$

and compute the maximization by taking the derivative

$$\frac{d}{dx}(a\,x - f(x)) = a - \frac{df(x)}{dx} = 0.$$

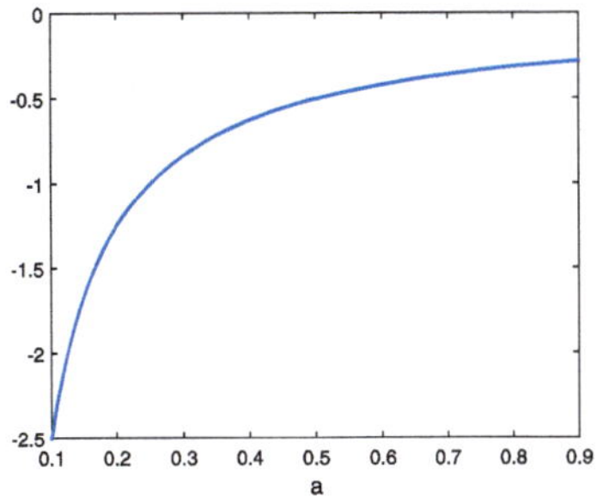

Fig. 15.7 LF transform of sqrt( )

The usual procedure is to write

$$a = \frac{df(x)}{dx} \tag{15.1}$$

and to obtain from this equation x as a function of a. Then,

$$f^*(a) = a\,x(a) - f(x(a)).$$

A program in MATLAB, Program 15.1, has been developed for computing and visualizing the LF transform of a function $f(x)$, based on the procedure just sketched. Note that Eq. (15.1) implies that $f(x)$ is differentiable.

Figure 15.7 has been generated by our simple program and corresponds to the LFT of the function $sqrt(x)$. Any other single variable function can be tested.

The Program 15.1 is written in symbolic MATLAB. Our experience in testing the LF transform of several functions is that there are troubles when double alternatives arise inside the procedure (for instance, check trigonometric functions). Most probably, the program could be improved based on recent MATLAB versions.

Program 15.1 Visualization of the LFT of a function

```
% Visualization of the LFT of a function
% (symbolic MATLAB)
syms x a
%insert below the function to be transformed
f = sqrt(x); %(edit)
Df = diff(f,x);
g = finverse(Df);
h=subs(g,a); %this is x(a)
m=subs(f,h); %this is f(x(a))
L= (a*h)-m;
seq=char(L); %convert symbolic to string
%display--------
fplot(seq, [0.1 0.9]);
xlabel('a');
```

Another definition of Legendre transform says that $f(x)$ and $f^*(a)$ are Legendre transforms of each other if $D(f(x)) = (D(f^*(a))^{-1}$ (where D means derivative). Actually, from the definition, it can be deduced that

$$a = \frac{df(x)}{dx}; \quad x = \frac{df^*(a)}{da}.$$

These expressions make you recall Hamiltonian mechanics and other main topics of Physics.

Notice that based on (15.1), in the n-dimensional case, one can write

$$f^*(\nabla f(\vec{x})) = \nabla f(\vec{x})^T \vec{x} - f(\vec{x}).$$

An interpretation of the LF transform in the context of Economics Theory is that in

$$f^*(a) = \sup_{\vec{x}}(a\,x - f(x))$$

$f^*(a)$ can represent the gain, a is the price, x is the level of production, and $f(x)$ is the cost of this production. It may happen that the price depended on x, so our formula should be modified. This is in accordance with the initiative of Moreau, who extended the Fenchel's conjugation using *coupling functions* (we will introduce Moreau's contributions later on):

$$f^*(a) = \sup_{x}\,(c\,(a, x) - f(x))$$

(*Moreau's conjugate*)

15.3.2 Saddle Value

The term *saddle* is used in several types of mathematical problems. The typical 3D case would be represented as in Fig. 15.8 (the shape is similar to a seat for riding a horse).

Specifically, the function represented in Fig. 15.8 is

$$z = x^2 - y^2.$$

Figure 15.8 has been generated with Program 15.2. It is almost the same program we included in the first chapter, in Sect. 1.3.2.

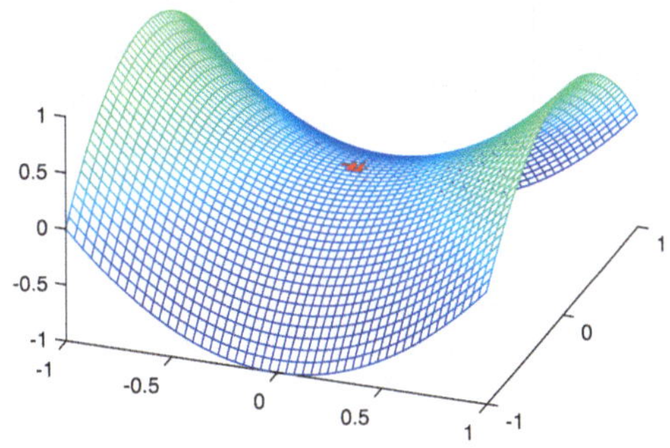

Fig. 15.8 A 3D function surface and a saddle point

Program 15.2 A saddle point in 3D

```
%A 3D saddle point
[x1, x2] = meshgrid(-1:0.04:1);
z= (x1.^2)-(x2.^2);
mesh(x1,x2,z); hold on;
plot(0,0,'r*','MarkerSize',10,'LineWidth',4);
colormap('winter');
view(20,40);
```

A saddle point, in mathematics, is a point on the surface of the graph of a function where derivatives in orthogonal directions are all zero but which is not a local extremum of the function. When gradient-based algorithms are used in search of extremum points, they can get stuck in saddle points. In order to discern whether a point is or not a saddle point, a key tool is the Hessian, which has both positive and negative eigenvalues at saddle points. See [28] for more details.

Let us return to the topic of this section: duality.

An important reference concerning duality is the book of Rockafellar [74] (1623 cites). It introduces relevant mathematical concepts and terminology. The central problem started with a certain function $K(x, y)$ and the following pair of functions:

$$f(x) = \sup_y K(x, y)$$

$$g(y) = \inf_x K(x, y)$$

The variables x and y belong to certain sets X and Y. Two optimization problems were considered:

minimize $f(x)$ over all $x \in X$
maximize $g(x)$ over all $y \in Y$

Clearly,

$$f(x) \geq K(x, y) \geq g(y)$$

and then

$$\inf_x f(x) \geq \sup_y g(y).$$

If equality holds, the common value is called the *saddle value* of K, which exists if there is a *saddle point* of K. A point (x_s, y_s) is a saddle point of K if

$$K(x, y_s) \geq K(x_s, y_s) \geq K(x_s, y), \quad \forall x \in X, \; \forall y \in Y$$

the saddle point solves the two optimization problems.

According to [Ro2], many different functions $K(x, y)$ can be proposed, and also given an optimization problem, many duals could be constructed: the question is how to obtain desired properties.

The principal theorems about the existence of the saddle value of K require that $K(x, y)$ be convex in x and concave in y or almost so [Ro2].

A lot of scientific contributions have been made (many of them related to image processing) on primal-dual optimization and the saddle value of an appropriate $K(x, y)$. As it is done in [74], a Lagrangian can be used for this $K(x, y)$ function.

15.3.3 Examples of Conjugate Functions

A rapid overview of conjugate function examples may help for an evaluation of the duality approach potential.

Let us start with the indicator function and norms.

Given a set C, its indicator function $i_C(\vec{x})$ is 0 if $\vec{x} \in C$ and ∞ otherwise. The conjugate of $i_C(\vec{x})$ is $\sup_{x \in C}(\vec{a}^T \vec{x})$ which is the support function of C and is sometimes denoted as $\|a\|_*$.

Figure 15.9 shows the norm $\|x\|$ and its conjugate, which is 0 for $\|a\| \leq 1$, and ∞ otherwise. Note that $\|x\|$ is not differentiable at the origin.

In the case of

$$f(x) = \frac{1}{p} \|x\|^p \;, \; 1 < p < \infty,$$

the conjugate is

$$f^*(a) = \frac{1}{q} \|a\|^q$$

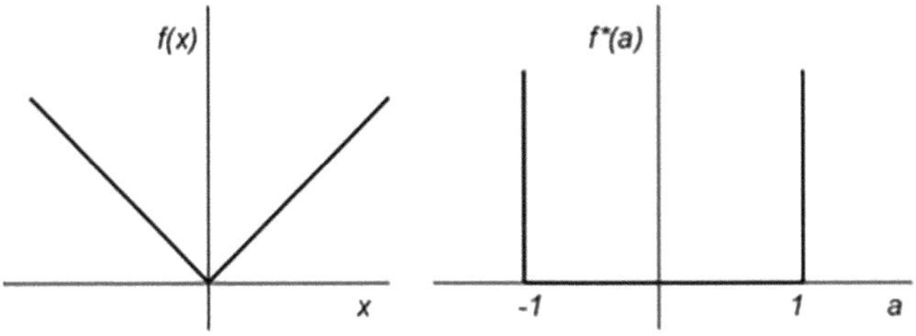

Fig. 15.9 The norm $\|x\|$ and its conjugate

Fig. 15.10 The conjugate of a parabola

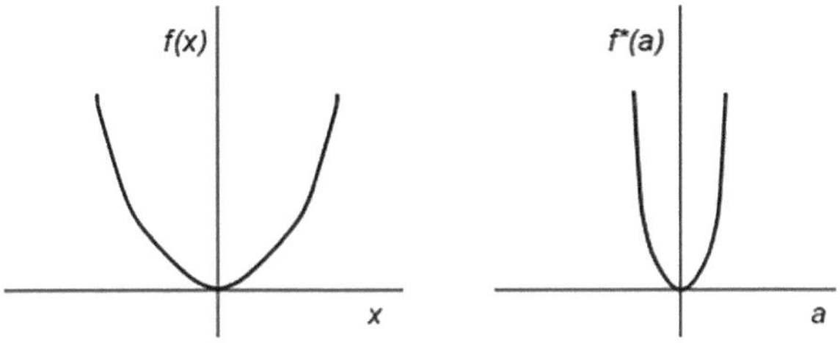

with

$$\frac{1}{p}+\frac{1}{q}=1.$$

Now, let us devote some space for curves.

Suppose that $f(x)$ is a parabola: $f(x)=x^2$. Its conjugate is $f^*(a)=\frac{1}{4}a^2$. Figure 15.10. visualizes this case.

If $f(x)$ is a general quadratic curve

$$f(\vec{x})=\frac{1}{2}\vec{x}^T A\vec{x},$$

the conjugate is

$$f^*(\vec{a})=\frac{1}{2}\vec{a}^T A^{-1}\vec{a}.$$

A more general case could be

$$f(\vec{x})=\frac{1}{2}\vec{x}^T A\vec{x}+\vec{c}^T\vec{x}+b.$$

The conjugate is

$$f^*(\vec{a})=\frac{1}{2}(\vec{a}-\vec{c})^T A^{-1}(\vec{a}-\vec{c})+\vec{c}^T A^{-1}(\vec{a}-\vec{c})+b.$$

The conjugate of the exponential function $f(x) = e^x$ is

$$f^*(a) = a\,(\ln a - 1).$$

And the conjugate of $f(x) = -\log x$ is

$$f^*(a) = -1 - \log(-a)$$

(only valid if $a<0$).

Coming to optimization matters, let us consider

$$\min_x f(x) + g(x)$$

which is equivalent to

$$\min_{x,y} f(x) + g(y) \quad \text{subject to } x = y.$$

Then, the dual problem would be

$$\max_u \; - f^*(a) - g^*(-a).$$

Examples of Application of the LFT

The report [34] describes a series of applications of the Legendre-Fenchel transform (LFT) to computer vision problems, like image denoising, image deconvolution, optic flow, and super-resolution.

The article [3] (year 2015) introduces a method for classification of hyperspectral camera images, with these images being improved by LFT.

A variety of applications are cited in [49], including robot navigation, medical imaging, electrical networks, thermodynamics, etc.

15.3.4 The Fenchel-Rockafellar Duality

The importance of Fenchel's contributions was realized by Rockafellar, with fruitful results concerning duality understanding. Figure 15.11 presents photographs of Prof. R. Tyrrell Rockafellar, University of Washington, and Werner Fenchel, who, born in Berlin, had to flee from Nazi's prosecution, finding refuge in Sweden and becoming a professor at the University of Copenhagen.

Rockafellar extended the Fenchel's duality theorem considering as primal problem the following,

$$\min_{\vec{x}} f(\vec{x}) + g(A\vec{x}),$$

Fig. 15.11 From left to right, W. Fenchel and R. T. Rockafellar

and writing the corresponding dual problem as

$$\min_{\vec{y}} f^*(-A^\dagger \vec{y}) + g^*(\vec{y})$$

where $A^\dagger$ is the adjoint. Denote

$$\mu = \inf_{\vec{x}} f(\vec{x}) + g(A\vec{x})$$

$$\mu^* = \inf_{\vec{y}} f^*(-A^\dagger \vec{y}) + g^*(\vec{y}).$$

Then, $\mu \geq -\mu^*$.

Moreover, if a (primal) *constraint qualification* is satisfied, such as $0 \in int\,(dom\, g - A \cdot dom f)$, then, $\mu = -\mu^*$. The unique solution $\vec{x}_M$ of the primal problem can be derived from a (nonnecessarily unique) solution $\vec{y}_M$ of the dual problem as

$$\vec{x}_M = \nabla f^*(-A^\dagger \vec{y}_M).$$

In [51], this duality was further extended to the context of saddle functions and dual minimax problems.

Examples of Application of the Fenchel-Rockafellar Duality

Reinforcement learning (RL) is one of the key methods in machine learning. Most of the RL algorithms adopt the dynamic programming paradigm; this is, for instance, the case of Q-learning. An alternative paradigm is based on linear programming, which is now receiving renewed interest, since it can more easily afford large-scale problems (by using duality). A recent contribution [58] develops a method for RL based on LP via Fenchel-Rockafellar duality. It is an interesting proposal, being now considered by several investigators.

Children use to imitate. It is a way of learning. Robots could imitate, why not? The document [92] (year 2023) introduces diverse offline imitation procedures, via Fenchel duality. It is a matter of unsupervised skill discovery, interacting with expert demonstrations (not expressly directed to teaching).

The Fenchel-Rockafellar duality is a main ingredient of the contribution [81] (year 2020), which showed that the problem of support vector regression (related to SVM) can be explicitly solved with the introduction of a new type of kernel (of tensorial type).

15.4 Minimax Results: Games

As the reader surely had noticed, some of the previous expressions involve a chaining of min (or inf) and max (or sup). Actually, this is connected with minimax results, which in turn are representative of game theory [93] (book, 47.700 cites, commemorative edition).

This important topic gives us the occasion to render tribute to John von Neumann, who contributed so much to the present era of digital processing based on computers. Figure 15.12 presents a photograph of him.

15.4.1 Minimax Results

Figure 15.13 visualizes one of the situations that may happen in linear programming, considering both the primal and the dual problem.

This situation may also correspond to a company trying to get maximal benefits from a product (upper curve) and a client trying to pay not much (lower curve). Initially, as shown by the figure, there is a gap: no deal. However, perhaps, after

Fig. 15.12 John von Neumann

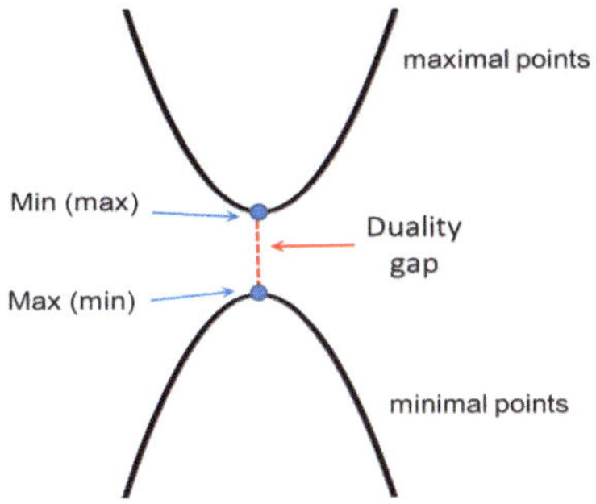

Fig. 15.13 A basic duality scenario

some bargaining, the gap could become zero, and the product would come to the client's hands. In any case, notice that we have minimax and maximin points.

Of course, the bargaining just mentioned could be seen as a game (a real-life game).

Soon, we will use a generic two-variable function $g(x, y)$ to express important results; this function may well correspond to something like Fig. 15.8 with a saddle point or any other scenario.

Although the famous book of von Neumann and Morgenstern was published in 1944, marking the official start of game theory, the minimax results to be briefly consigned below were published by von Neumann in 1928 (an article in French and another in German). See [43] for a detailed history.

Let us select some theorems.

The first (weak duality) says that under certain conditions (continuity, compact domain)

$$\min_x \max_y g(x, y) \geq \max_y \min_x g(x, y)$$

A second theorem (strong duality) establishes that given a continuous $g(x, y)$, convex on $x \in X$ and concave on $y \in Y$, with X and Y convex and compact, then,

$$\min_x \max_y g(x, y) = \max_y \min_x g(x, y).$$

The conditions can be relaxed a bit, so a third theorem (strong duality) due to Sion in 1958, establishes that given a $g(x, y)$, lower semicontinuous quasi-convex on $x \in X$ and upper semicontinuous quasi-concave on $y \in Y$, with X and Y convex and one of them compact, then,

$$\min_x \max_y g(x, y) = \max_y \min_x g(x, y).$$

There are several theorems and mathematical results in connection with the minimax context. For instance, [21] includes up to 14 min-max theorems, showing that they form an equivalent chain.

15.4.2 Introduction to Game Theory (Concepts and Examples)

Games can be expressed in normal or extensive forms.

A standard two-person game is the *prisoner's dilemma*, which can be represented in the following *normal form*:

dilemma		Prisoner B (column)	
		Confess	Don't
Prisoner A (row)	Confess	-10, -10	0, -20
	Don't	-20, 0	-1, -1

In each cell, the payoff of each prisoner is given in terms of year's reduction.

Equivalently, the game can be expressed as a *bimatrix*, as follows:

$$\begin{pmatrix} (-10, -10) & (0, -20) \\ (-20, 0) & (-1, -1) \end{pmatrix}$$

Figure 15.14 depicts the prisoner's dilemma game in an extensive form. The dashed line means imperfect information.

The prisoner's dilemma is an example of noncooperative game (the two players were kept in separate rooms). There exist cooperative games, like joint ventures of companies.

An important concept is *strategic dominance*. Players try to select the best strategy. Given two strategies, A or B, that a player can choose, there are the following situations:

- B is equivalent to A: Both give the same outcome, no matter what the other players do (will abbreviate it as NoMPd).

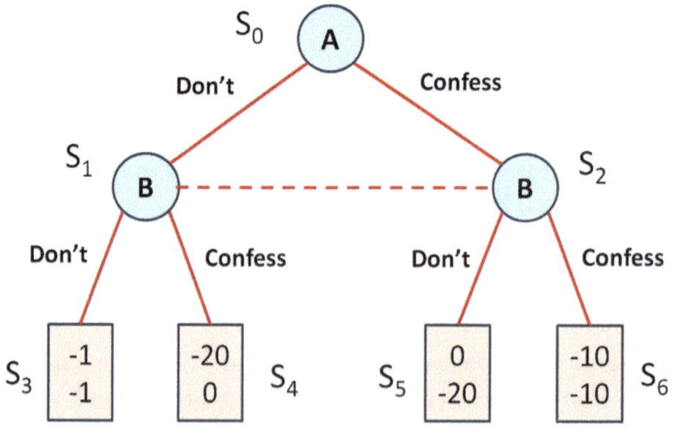

Fig. 15.14 Prisoner's dilemma in an extensive form

- B strictly dominates A: B always gives better outcome than A, NoMPd.
- B weakly dominates A: B always gives better or equal outcome as A gives, with at least one case in which B is better, NoMPd.
- B and A are intransitive, A is better in some cases, and B is better in other cases, depending on what the opponent plays.

Changing roles, B could be strictly dominated by A or B weakly dominated by A.

Nash equilibrium in a noncooperative game is an outcome in which no player's expected outcome can be improved by changing one's own strategy.

Consider the following popular example:

	A	B
A	1,1	0,0
B	0,0	0,0

There are two strategies, A and B. You are the player on the left of the table; the opponent is on top:

- If you play A, you can get 1 if the opponent plays A (getting 1) or 0 if the opponent plays B (getting 0).
- If you play B, both you and the opponent get 0, NoMPd.

There is one case in which it is better for me to play A, instead of B (the corner with 1, 1). Hence, A weakly dominates B.

If both players choose B, neither player would improve by unilaterally switching his strategy. The bottom-right corner is a Nash equilibrium. Likewise, the top-left corner is another Nash equilibrium.

The fact is that any dominant strategy equilibrium is a Nash equilibrium but not every Nash equilibria are dominant strategy equilibria.

For illustrative purposes, a MATLAB program (Program 15.3) has been devised, adapting the code included in the academic report [59].

Given the normal form description of a two-player game, the program finds the Nash equilibria of the game and gives the coordinates of these equilibria in the form.

The game example considered in the program is the game of chicken, which is described as follows:

	Swerve	Straight
Swerve	0,0	-1,1
Straight	1,-1	-10,10

The game has two Nash equilibria that had been highlighted in yellow color. The coordinates of these equilibria are (1, 2) and (2, 1).

When we say "best response," it refers to a strategy (strategy 1, or strategy 2).

As you can easily observe, the program is a classifier with many *if* instructions:

- If both players' first best responses were one, then one Nash equilibrium is (1, 1).
- If both players' second best responses were two, then one Nash equilibrium is (2, 2).
- If both players' first best responses were two and player 1 second best response was one, a Nash equilibrium is (1, 2).
- If both players' second best responses were one and player 1 first best response was two, a Nash equilibrium is (2, 1).

Variables h_{11}, h_{12}, h_{21}, and h_{22} have values 1 or 2. The output of the program is text: numbers indicating the coordinates of Nash equilibria.

Program 15.3 Finding Nash equilibrium nodes

```
% Find Nash equilibrium nodes
% two players
% Payoff matrices (two matrices)
% Example (game of chicken)
P1=[0 -1;
1 -10];
P2=[0 1;
-1 -10];
% initialization --------------------------------
C1=P1(:,1); D1=P1(:,2); %player 1
C2=P2(1,:); D2=P2(2,:); %player 2
% best response functions:
[br11 h11]=max(C1); [br12 h12]=max(D1); %player 1
[br21 h21]=max(C2); [br22 h22]=max(D2); %player 2
% Finding Nash Equilibria ------------------------
if (h11==h12) % P1 has a strictly dominant strategy
  if (h21==h22) % P2 has a strictly dominant strategy
    nash=[h11 h21];
  % P2 does not have a strictly dominant strategy:
  elseif (h21~=h22)
    if (h11==1), nash=[h11 h21];
    elseif (h11==2), nash=[h11 h22];
    end;
  end;
% P1 does not have a strictly dominant strategy:
elseif (h11~=h12)
  if (h21==h22) % P2 has a strictly dominant strategy
    if (h21==1), nash=[h11 h21];
    elseif (h21==2), nash=[h12 h21];
    end;
  % P2 does not have a strictly dominant strategy:
  elseif (h21~=h22)
    if (h11==h21)
      if (h11==1), nash1=[h11 h21];
      elseif (h11==2),
```

```
            if (h12==1), nash1=[h12 h21];
            elseif (h12==2), nash1=[0 0];
            end;
          end;
        elseif (h11~=h21), nash1=[0 0];
        end;
        if (h12==h22)
          if (h12==2), nash2=[h12 h22];
          elseif (h12==1)
            if (h11==2), nash2=[h11 h22];
            elseif (h11==1), nash2=[0 0];
            end;
          end;
        elseif (h12~=h22), nash2=[0 0];
        end;
        if (nash1==nash2), nash=nash1;
        elseif (nash1~=nash2), nash=[nash1;nash2];
        end;
      end;
    end;
    % Display results ---------------
    nash
```

15.4.2.1 Game Simplification via Dominance

Rational players would never play strictly dominated strategies. Therefore, dominated strategies can be eliminated. For example, in the following game, player 1 (rows) does not have any dominated strategy, but in the case of player 2 (columns), the strategy C is dominated by strategy R (compare cell by cell: 1 <2; 4 <6; 6 <8):

	L	C	R
U	4,3	5,1	6,2
M	2,1	8,4	3,6
D	5,9	9,6	2,8

Then, we can eliminate column C, obtaining the following simplified game:

	L	R
U	4,3	6,2
M	2,1	3,6
D	5,9	2,8

Now, it happens that strategy M of player 1 is dominated by strategy U (2 <4; 3 <6). Then, we can eliminate row M, getting the next simplified game:

	L	R
U	4,3	6,2
D	5,9	2,8

Strategy R of player 2 is now dominated by strategy L (2 <3; 8 <9). Let us eliminate column R, obtaining the most simplified game:

	L
U	4,3
D	5,9

(this is also called *a reduced game*).

See [85] for more details on game simplification.

15.4.2.2 Zero-Sum Games

The main initial focus of von Neumann was put on zero-sum games.

The sum of payoffs in each cell of a *zero-sum game* is zero. That means in a two-person game, that if one player wins 100, then the other player loses 100. This type of games can be represented by just one payoff matrix (the other having term by term the opposite sign).

Take, for example, the case of a negotiation between a football player and his employer:

contract		Employer (column)	
		A	B
Player (row)	1	35000	40000
	2	30000	20000

Usually, zero-sum tables put on the left the agent that wants to maximize the outcome (choosing between strategies 1 or 2) and on top the agent that wants to minimize it (selecting strategy A or B). If the player chooses 1, the employer will choose A. If the player chooses 2, the employer will choose B. The best for the athlete would be 1-A. In mathematical terms,

$$\max_i \min_j P = 35{,}000,$$

where P is a matrix with the four profit entries of the table.

From the point of view of the employer, if he chooses A, the player will choose 1. Instead, if he chooses B, the player will choose also 1. The best for the employer would be A-1. In mathematical terms,

$$\min_j \max_i P = 35{,}000.$$

And so, a minimax result has been found. It is said that 35,000 (equally found by maximin() or minmax()) is *the value of the game*.

The entry 1-A, with the value 35,000, is a *saddle point* of the game. Because of the saddle point, neither player can take advantage of the rival's strategy; it is said that one has a *stable solution* (or *equilibrium solution*).

15.4.2.3 Game Simplification via Dominance Again

As other games, zero-sum games can be simplified based on dominance. An example of game simplification has already been given, using tables with two entries in each cell. For zero-sum games, tables are simpler, with only one entry per cell.

For the simpler tables, the rule for row elimination—corresponding to player 1—is the same as before: if all elements of a row X are *less than or equal to* the corresponding elements of another row Y, you can eliminate row X.

However, for columns—corresponding to player 2—the rule for elimination is different: if all elements of column A are *greater than or equal to* the corresponding elements of another column B, you can eliminate column A.

Let us work on the academic example given in [88]. Observe first that row 4 is dominated by row 3.

	A	B	C	D	E
1	4	6	5	10	6
2	7	8	5	9	10
3	8	9	11	10	9
4	6	4	10	6	4

We can eliminate row 4, obtaining the following game:

	A	B	C	D	E
1	4	6	5	10	6
2	7	8	5	9	10
3	8	9	11	10	9

Row 1 is weakly dominated by row 3. We can eliminate this row, and so we obtain

	A	B	C	D	E
2	7	8	5	9	10
3	8	9	11	10	9

Column D is dominated by column A. We eliminate column D, so obtaining

	A	B	C	E
2	7	8	5	10
3	8	9	11	9

Also, column E is dominated by column A. Column E is eliminated.

	A	B	C
2	7	8	5
3	8	9	11

Next steps would be to eliminate column B (dominated by column A) and then to eliminate row 2 (dominated by row 3):

	A	C
2	7	5
3	8	11

	A	C
3	8	11

Finally, column A dominates column C. C is eliminated. We obtain an equilibrium. The value of the game is 8:

	A
3	8

15.4.2.4 Equivalence Between Games and Linear Programming

Recall the standard formulation of the linear programming problem

- min $\vec{c}^T\vec{x}$ subject to $A\vec{x} \geq \vec{b},\ \vec{x} \geq 0$

  and the dual

- max $\vec{b}^T\vec{y}$ subject to $A^T\vec{y} \leq \vec{c},\ \vec{y} \geq 0$.

The question is whether the LP problem is equivalent to a game.

In a very enjoyable article [20], Dantzig himself describes a historic encounter of von Neumann and Dantzig in 1947 at Princeton. Those were the times when it was conjectured that close relationships existed between game theory, duality, and linear programming. Now, it is well known (see, for instance, [1]) that any zero-sum game can be reduced to a linear programming problem (and it can be used to prove the minimax theorem based on a strong duality), and vice-versa: a linear programming problem can be reduced to a zero-sum game.

In fact, the study of games that represent complexity difficulties, not easy for game theory analysis, can be tackled with linear programming.

Dantzig suggested to reduce the pair primal-dual of LP problems to a symmetric zero-sum game by using the following payoff matrix,

$$P = \begin{pmatrix} 0 & A & -\vec{b} \\ -A^T & 0 & \vec{c} \\ \vec{b}^T & -\vec{c}^T & 0 \end{pmatrix}$$

(when P is skew-symmetric, i.e., $P^T = -P$, the game is called a symmetric game).

This reduction may present some flaws, as discussed in [1]. See the recent article [13] on this issue and how to solve it.

15.4.2.5 Mixed Strategies

Here is a common, simple game, *matching pennies*:

	Heads	Tails
Heads	1,-1	-1,1
Tails	-1,1	1,-1

Key observation: There is no (pure strategy) Nash equilibrium in this game.

If we play this game, we should be "unpredictable," that is, we would randomize our behavior by *mixing strategies* in a random (but optimal) manner.

Let us analyze for best response. Player 1 would play heads with probability q, while player 2 would play heads with probability p. Then,

	Heads (prob p)	Tails (prob 1-p)	*Expected payoff*
Heads (prob q)	1,-1	-1,1	*2p-1*
Tails (prob 1-q)	-1,1	1,-1	*1-2p*
Expected payoff	*1-2q*	*-1+2q*	

If $2p - 1 > 0$, player 1 will choose playing heads. If $2p - 1 < 0$, player 1 will choose playing tails. Only when player 2 uses $p = 1/2$, it is indifferent for player 1 to choose one or another strategy, and he will be willing to randomize (*equality of expected payoffs*).

Since the matching pennies game is completely symmetric, both (rational) players shall choose heads with 50% probability and tails with 50% probability. Here, we have a *mixed-strategy Nash equilibrium*.

Figure 15.15 visualizes the situation.

In general, players should avoid a predictable strategy, which may suppose an advantage for the opponent. Hence, each player should choose at random among their alternatives according to some probability distributions $\vec{x} = (x_1, x_2, \ldots, x_n)$ of player 1 and $\vec{y} = (y_1, y_2, \ldots, y_m)$ of player 2.

Take again the game of chicken (we changed the values on the right-bottom corner for visualization purposes in the next figure):

	Swerve	Straight
Swerve	0,0	-1,1
Straight	1,-1	-4,4

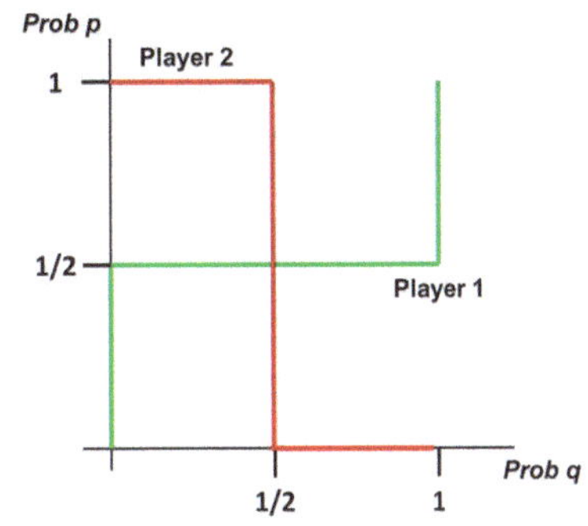

Fig. 15.15 Reaction functions in matching pennies game

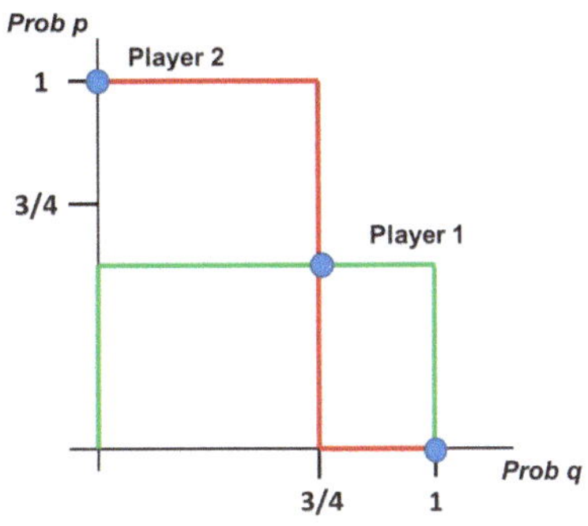

Fig. 15.16 Reaction functions in chicken game

The reaction functions are as depicted in Fig. 15.16.

Notice in Fig. 15.16 that the three Nash equilibrium points of this game have been marked with a circle.

Let N be the number of (pure strategy) Nash equilibria and M the number of mixed-strategy Nash equilibria. It can be shown that if $Q = N + M$ is finite, then Q is odd, for example, N+M could be 1, 3, etc.

Historical Note

The classic book of von Neumann and Morgenstern, published in 1944, extended the work on zero-sum games and also started cooperative game theory. First contributions from John Forbes Nash started in 1949, proving that an equilibrium exists in every game. He extended the study to non-zero-sum games and also began with bargaining. In 1957, R. Luce and H. Raiffa published the book *Games and Decisions: Introduction and Critical Survey*, giving impulse to the popularization of game theory. During 1967–1968, John Harsanyi created methods to study games of incomplete information (this was important for extending the applications field).

John Nash received the Nobel Prize in 1994. A total of 18 scientists with research on game theory have received the Noble Prize from 1972 to 2020.

As an *excursus* around a historical event connected with game theory, we wish to summarize what happened in the Battle of Bismarck Sea, February–March 1943, based on [35]. A Japanese convoy was sailing toward New Guinea and had to surround the New Britain island. The Japanese commander had two choices:

he could sail North of the island or sail South of the island. General Kenney, commander of air forces, received orders from MacArthur to intercept the convoy.

With the northern route, the convoy would be exposed to bombing at most 2 days. With the southern route, the risk of bombing increases to 3 days. Japanese had to choose North for minimum risk.

Kenney, the maximizing player, concentrated the surveillance on the North: If Japanese were there, they would be detected (allowing for 2 days of bombing). If Japanese were South, the scarce surveillance would perhaps miss 1 day, leaving still 2 for bombing.

Under the perspective of game theory, the situation can be reflected with the following table:

		Japanese strategies		Minimum of row
		North	South	
Kenney strategies	North	2 days	2 days	2 days (maximin)
	South	1 day	3 days	1 day
Maximum of column		2 days (minimax)	3 days	

Both commanders played well, choosing the saddle point (but there were chances and factors escaping the abstract view of theory). More details and one more battle can be reached in [35].

15.4.2.6 Resources

Game theory and MATLAB

There is a MATLAB Game Theory Toolbox, named *MatTuGames*. It is available on the Internet from the GitHub repository:

https://github.com/himeinhardt/MatTuGames.

The article [52] offers abundant information on *MatTuGames*, including examples.

The MathWorks File Exchange site give access to MATLAB programs related to game theory at the following links:

https://es.mathworks.com/matlabcentral/fileexchange/51572-game-theory
https://es.mathworks.com/matlabcentral/fileexchange/51151-game-theory

Likewise, there is a GitHub repository on game theory at
https://github.com/giakou4/game_theory.

Introductory Guides and Tutorials
There are numerous tutorial publications on game theory, most of them easy to access from the Internet.

We could cite, for instance,

- The Lecture Notes [39] (37 pages) from National Taiwan University, which concisely covers utility theory, two-person games (including Nash equilibrium and Pareto optimality), extensive games, mixed strategies, and many examples.
- There is also [37] (76 pages) from the University of California, Berkeley, which starts with combinatorial games, then two-person zero-sum games and general sum games, then social choice (voting), and auctions.
- In addition, we have [83] (157 pages) from the University of Wisconsin, with normal form games, extensive form games, Bayesian games, and repeated games.

Bibliography on Game Theory and Applications
Game theory represents a large and multiform field, many books, and much more scientific publications.

Books that deserve a special mention are [24] (year 1991, 16.126 cites), [63] (year 1994, 10.998 cites), [57] (year 2013, 9.054 cites), [71] (year 1989, 4.495 cites), and [4] (year 1998, 8.137 cites) on dynamic noncooperative game theory and [96] (year 1997, 6.439 cites) on evolutionary game theory, etc.

Together with Alex Newell and Cliff Shaw, Herbert A. Simon developed a computer program for chess playing. This was the first chess program winning a human. Years after, in 1978, H.A. Simon received the Noble Prize in economy—economy and games: a profound relationship.

Books on game theory and economy are many. This happens from the very start, the seminal book of Von Neumann and Morgenstern [93]. Other well-known books are [23] (year 1986, 1.815 cites), [29] (year 1991, 3.554 cites), [70] (year 1993, 7.716 cites), etc.

Many books connect game theory and social sciences, political action, defense, agriculture, etc. In particular, the book [6] presents a notable variety of engineering applications.

Of special interest are the articles on game theory and economic matters from L. Samuelson [82] (year 2016) or [22] (year 1991, 2.437 cites) on evolutionary games. In other order of things, machine learning is also showing much interest on game theory, like in adversarial/antagonist neural networks for generative AI [53] (year 2023) or in learning techniques, like, for example, the contribution [47] (year 2017) on a game theoretic approach to multiagent reinforcement learning.

15.4.3 The Farkas' Lemma

It is quite frequent to see references to the Farkas's lemma in the scientific and academic publications, in particular when dealing with linear programming and related topics.

The Farkas' lemma can be expressed in several ways [40, 78]. In geometrical terms, it says that a vector is either in a given convex cone or there is a hyperplane separating the vector from the cone. The lemma can be regarded as a particular case of the separating hyperplane theorem.

Another formulation is the following: let A be a $m \times n$ matrix and $\vec{b}$ a column vector of size n. Only one of the next alternatives holds:

- There exists $\vec{x} \in \Re^n$ such that $A\,\vec{x} = \vec{b}$ and $\vec{x} \geq 0$.
- There exists $\vec{y} \in \Re^m$ such that $A^T\,\vec{y} \geq 0$ and $\vec{b}^T\,\vec{y} < 0$.

Note that this second alternative is equivalent to the following:

- There exists $\vec{y} \in \Re^m$ such that $A^T\,\vec{y} \leq 0$ and $\vec{b}^T\,\vec{y} > 0$.

An important application of the lemma is for the analysis of linear programming problems, concerning in particular feasibility aspects.

For example, in the case of the LP problem

$$\min\ \vec{c}^T\vec{x} \text{ subject to } A\vec{x} \geq \vec{b},\ \vec{x} \geq 0$$

and the dual

$$\max\ \vec{b}^T\vec{y} \text{ subject to } A^T\vec{y} \leq \vec{c},\ \vec{y} \geq 0,$$

the Farkas' lemma implies that exactly one of the following cases occurs:

- Both the primal and the dual have optimal solutions $\vec{x}*$ and $\vec{y}*$ with equal values $\vec{c}^T\,\vec{x}* = \vec{b}^T\,\vec{y}*$.
- The dual is unfeasible, and the primal is unbounded ($\vec{c}^T\vec{x} \rightarrow \infty$).
- The primal is unfeasible, and the dual is unbounded ($\vec{b}^T\vec{y} \rightarrow \infty$).
- Both primal and dual are unfeasible.

The lemma can be used as a **certificate** of infeasibility [2]. In particular, the primal problem

$$\min\ \vec{c}^T\vec{x} \text{ subject to } A\vec{x} \geq \vec{b},\ \vec{x} \geq 0$$

is unfeasible if and only if

$$\text{there exists } \vec{y} \in \Re^m \text{ such that } A^T\,\vec{y} \leq 0 \text{ and } \vec{b}^T\vec{y} > 0.$$

Some New Applications Related to the Farkas's Lemma

Some recent research is considering how to incorporate prior knowledge for the benefit of machine learning methods. In general, the results were encouraging. One of the recent contributions in this sense is [8] (year 2021), with the following title: *Farkas' theorem of the alternative for prior knowledge in deep networks*. This contribution includes interesting illustrations and bibliographic references.

Time before, in 2011, the article [100] proposed a knowledge-incorporated multiple criteria classifier (for data mining, etc.). It quotes the use of Farkas' theorem in their development (a linear programming approach).

In a similar trend, there are proposals about knowledge-based SVMs, which use the Farkas' theorem, for example, [25] and [46].

In 2019, the document [69] introduced the so-called "*Farkas layers*." It means an improvement for the training of neural networks, guaranteeing that at least one neuron is active at a given layer of the network. There is a GitHub repository about this method (FarkasLayers).

15.4.4 *Types of Functions: Bounds*

Given an objective function to be minimized, it is important to establish upper and lower bounds.

15.4.4.1 Upper Bounds

Let us start with an upper bound, which is based on Lipschitz functions. A function $f(\vec{x})$ is said to satisfy a Lipschitz condition if a constant $L>0$ exists with

$$\|f(\vec{x}) - f(\vec{y})\|_2 \leq L \|\vec{x} - \vec{y}\|_2 , \; \forall \vec{x}, \vec{y} \in \Re^n .$$

This expression says that the slope of the line connecting the points $f(\vec{x})$ and $f(\vec{y})$ is no greater than L.

Functions satisfying this condition are called *Lipschitz functions*; also, colloquially, one could say that a function $f(\vec{x})$ is Lipschitz.

The Lipschitz condition is stronger than continuity. Indeed, Lipschitz functions are continuous functions.

The function $|\vec{x}|$ is Lipschitz, but not differentiable. The function $\sqrt{x}$ is a continuous function but it is not a Lipschitz function because it becomes infinitely steep as $x \rightarrow 0$. Likewise, the function x^2 is a continuous function but is not Lipschitz, since it becomes arbitrarily steep as $x \rightarrow \infty$. See [64] for more examples.

In the case of differentiable functions, gradients are defined and so they can be object of study. The gradient of a function is Lipschitz if a constant $L>0$ exists with

$$\|\nabla f(\vec{x}) - \nabla f(\vec{y})\|_2 \leq L \|\vec{x} - \vec{y}\|_2 , \; \forall \vec{x}, \vec{y} \in \Re^n .$$

On the basis of the Cauchy-Schwarz inequality, it can be shown (lemma) that if the gradient is Lipschitz, then

$$|f(\vec{y}) - f(\vec{x}) - \nabla f(\vec{x})^T (\vec{y} - \vec{x})| \leq \frac{L}{2} \|\vec{y} - \vec{x}\|_2^2 .$$

Put in other way,

$$f(\vec{y}) \leq f(\vec{x}) + \nabla f(\vec{x})^T (\vec{y} - \vec{x})| + \frac{L}{2} \|\vec{y} - \vec{x}\|_2^2 .$$

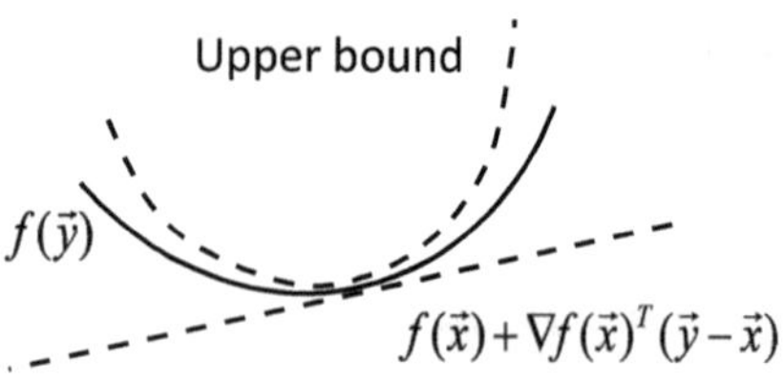

Fig. 15.17 Upper bound corresponding to the gradient being Lipschitz

A consequence of this lemma is that

$$\frac{L}{2}\,\|\vec{x}\|_2^2 \,-\, f(\vec{x})$$

is convex.

For twice differentiable $f(\vec{x})$, if the gradient is Lipschitz, $\nabla^2 f(\vec{x}) \preceq L\,I$. Figure 15.17 depicts the upper bound in case the gradient of $f(\vec{x})$ is Lipschitz.

15.4.4.2 Lower Bounds

Now, for lower bounds, one considers some more definitions related to convexity:

- A function $f(\vec{x})$ is *convex* iff

$$f(\alpha\,\vec{x} \,+\, (1-\alpha)\,\vec{y}) \,\leq\, \alpha\, f(\vec{x}) \,+\, (1-\alpha)\, f(\vec{y})\,, \;\; \forall \vec{x}, \vec{y}\,;\; 0 \leq \alpha \leq 1.$$

- The function $f(\vec{x})$ is *strictly convex* if the equality holds only when $\vec{x} \,=\, \vec{y}$:

$$f(\alpha\,\vec{x} \,+\, (1-\alpha)\,\vec{y}) \,<\, \alpha\, f(\vec{x}) \,+\, (1-\alpha)\, f(\vec{y})\,, \;\; \forall \vec{x}, \vec{y}\,;\; \vec{x} \neq \vec{y}\,;\; 0 \leq \alpha \leq 1$$

- The function $f(\vec{x})$ is *strongly convex* (with modulus m) if

$$f(\alpha\,\vec{x} \,+\, (1-\alpha)\,\vec{y}) \,\leq\, \alpha\, f(\vec{x}) \,+\, (1-\alpha)\, f(\vec{y}) \,-\, \frac{m}{2}(\alpha\,(1-\alpha))\,\,\|x - y\|_2^2\,,$$

$$\forall \vec{x}, \vec{y},\; 0 \leq \alpha \leq 1.$$

Again, if $f(\vec{x})$ is differentiable, it is possible to consider gradients, obtaining some more inequalities corresponding to convexity:

- The function $f(\vec{x})$ is *convex* iff

$$f(\vec{y}) \,\geq\, f(\vec{x}) \,+\, \nabla f(\vec{x})^T\,(\vec{y} - \vec{x})\,, \;\; \forall \vec{x}, \vec{y}.$$

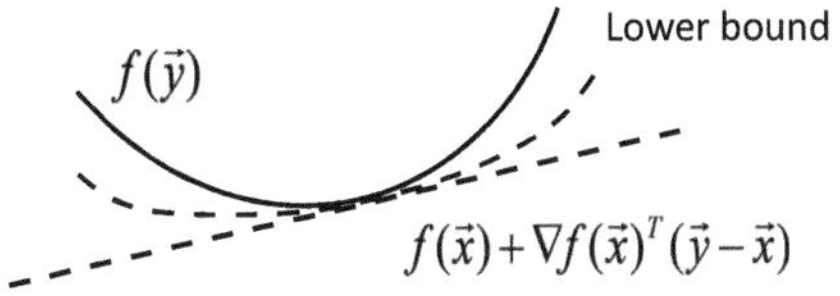

Fig. 15.18 Lower bound corresponding to differentiable strongly convex $f(\vec{x})$

- The function $f(\vec{x})$ is *strictly convex* if the equality holds only when $\vec{x} = \vec{y}$:

$$f(\vec{y}) > f(\vec{x}) + \nabla f(\vec{x})^T(\vec{y} - \vec{x}), \quad \forall \vec{x}, \vec{y};\ \vec{x} \neq \vec{y}$$

- The function $f(\vec{x})$ is *strongly convex* (with modulus m) if

$$f(\vec{y}) \geq f(\vec{x}) + \nabla f(\vec{x})^T(\vec{y} - \vec{x}) + \frac{m}{2}\|\vec{y} - \vec{x}\|_2^2, \quad \forall \vec{x}, \vec{y}.$$

For twice differentiable $f(\vec{x})$, the function is convex iff $\nabla^2 f(\vec{x}) \geq 0$ and strongly convex iff $\nabla^2 f(\vec{x}) \geq mI$. The condition $\nabla^2 f(\vec{x}) > 0$ is sufficient for $f(\vec{x})$ to be strictly convex, but it is not a necessary condition (e.g., $f(\vec{x}) = x^4$).

Figure 15.18 depicts the lower bound in case of $f(\vec{x})$ being differentiable and strongly convex.

15.4.4.3 A Sandwich

Notice that in the case of a differentiable and strongly convex $f(\vec{x})$ with the gradient being Lipschitz, one has the following "sandwich,"

$$C(\vec{x}, \vec{y}) + \frac{m}{2}\|\vec{y} - \vec{x}\|_2^2 \leq f(\vec{y}) \leq C(\vec{x}, \vec{y}) + \frac{L}{2}\|\vec{y} - \vec{x}\|_2^2$$

where

$$C(\vec{x}, \vec{y}) = f(\vec{x}) + \nabla f(\vec{x})^T(\vec{y} - \vec{x}).$$

15.4.4.4 Monotone Gradients

In order to complete our recital of inequalities, let us consider the *monotonicity of gradient*, which happens when

$$(\nabla f(\vec{y}) - \nabla f(\vec{x}))^T(\vec{y} - \vec{x}) \geq 0\ ;\ \forall \vec{x}, \vec{y}.$$

A differentiable $f(\vec{x})$is convex iff it enjoys monotonicity of gradient.
The gradient of a differentiable strictly convex function is strictly monotone:

$$(\nabla f(\vec{y}) - \nabla f(\vec{x}))^T(\vec{y} - \vec{x}) > 0 \; ; \; \forall \vec{x}, \vec{y}$$

And the gradient of a differentiable strongly convex function is strongly monotone or coercive:

$$(\nabla f(\vec{y}) - \nabla f(\vec{x}))^T(\vec{y} - \vec{x}) \geq m \; \|\vec{y} - \vec{x}\|_2^2 \; ; \; \forall \vec{x}, \vec{y}$$

A function$f(\vec{x})$ is said to be *coercive* if the limit of $f(\vec{x})$ when $\|\vec{x}\| \to \infty$ is ∞. A coercive function must increase without limit on any path on $\vec{x}$ going to infinity. For example, in 3D, a coercive function looks like a bowl.

A function$f(\vec{x})$ is said to be *co-coercive* with modulus m if

$$(\vec{y} - \vec{x})^T(f(\vec{y}) - f(\vec{x})) \geq m \; \|f(\vec{y}) - f(\vec{x})\|_2^2 \; ; \; \forall \vec{x}, \vec{y}.$$

In the case of differentiable $f(\vec{x})$ with gradient being Lipschitz,

$$(\nabla f(\vec{y}) \; - \; \nabla f(\vec{x}))^T(\vec{y} - \vec{x}) \geq \frac{1}{L} \; \|\nabla f(\vec{y}) \; - \; \nabla f(\vec{x})\|_2^2 \quad \forall \vec{x}, \vec{y}$$

and, so, the gradient $\nabla f(\vec{x})$ is co-coercive with modulus $1/L$.
Strong convexity and Lipschitz gradient are related by Fenchel duality. According to the lemma 3.10 (pg.100 of the unpublished book) in [86],

1. If $f(.)$ is strongly convex with modulus m, the conjugate $f^*(.)$ has Lipschitz gradient with $L = 1/m$.
2. If$f(.)$ is convex and has Lipschitz gradient, the conjugate $f^*(.)$ is strongly convex with modulus $m = 1/L$.

For the reader wishing to know more, the survey [79] (Stanford University) would be recommended. A very short lecture on coercive functions, with examples, is given by [33].

About Image Processing Applications
Lipschitz and coercive functions frequently appear in certain image processing applications, for example, convex and Lipschitz functions in [17] (a contribution about image segmentation), coercive functions in the chapter [62] on image processing based on energy, and more convex analysis and coercive functions in the article [9] (2.317 cites) about image restoration.

More details of optimization in image processing shall be covered in the next chapter.

15.5 Non-smooth Analysis and Optimization

It has been found that it is not always possible to have differentiable objective functions in optimization contexts. The article [31] includes a long section describing several sources of non-differentiability in optimization problems; likewise, the book [84] (2421 cites) includes, in Chap. 5, an extensive reference of this kind of problems. Standard analytical methods for optimization, which are based on derivatives, are difficult to apply in such cases.

This section is devoted to the treatment of non-differentiable optimization problems. Important concepts to be introduced are those of *subgradient* and *subdifferential*, which prolong to a certain extent the classical concepts of gradient and differential. The section focuses on minimization problems.

15.5.1 Non-smooth Analysis

Some initial examples would be helpful to illustrate why we are concerned with non-differentiability.

A typical case, which is found in image processing, involves an objective function like $z = |x|$, which is differentiable for all $x \neq 0$, but it is non-differentiable at $x = 0$. Figure 15.19 shows a plot of this function. It happens that the minimum of the function takes place at $x = 0$.

A similar situation is depicted in Fig. 15.20, where two curved components define a convex function that is non-differentiable at x_0. It also happens that this is the point where the function reaches its minimum.

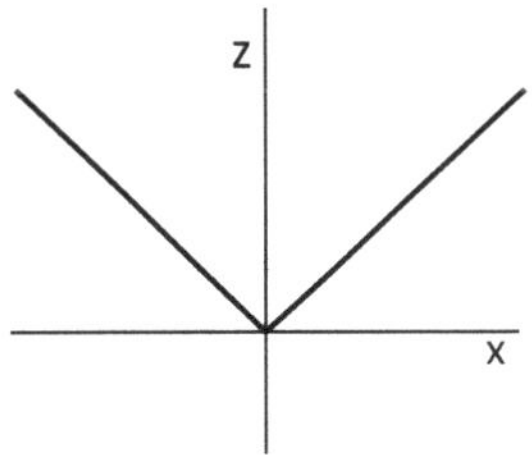

Fig. 15.19 The function $z = |x|$

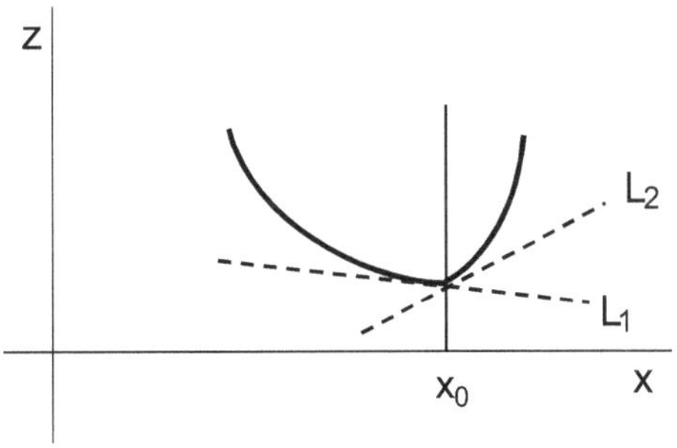

Fig. 15.20 Two "subtangent" lines

Something has to be devised in order to keep, as much as possible, the traditional methodology, which determines the minimum by equaling derivatives to zero. With this in mind, a repertory of new concepts has been proposed. Let us now introduce some of these concepts.

Considering in particular Fig. 15.20, there would be infinite lines that go through the point $(x_0,\ f(x_0))$ and which are either touching or below the graph of the function. The slope of any of these lines is a *subderivative*. The two lines, L_1 and L_2, on this figure are two examples of these lines.

In more formal terms, a subderivative at x_0 is a real number c such that:

$$f(x) - f(x_0) \geq c\,(x - x_0)$$

The set $[a, b]$ of all subderivatives is called the '*subdifferential*' of the function $f()$ at x_0. The subdifferential is denoted as $\partial f(x_0)$.

These concepts can be generalized to functions of several variables. A vector $\vec{g}$ is called a *subgradient* of the function $f()$ at $\vec{x}_0$ if

$$f(\vec{x}) - f(\vec{x}_0) \geq \vec{g}^T \cdot (\vec{x} - \vec{x}_0). \tag{15.2}$$

The set of all subgradients is called the *subdifferential* of the function $f()$ at $\vec{x}_0$. The subdifferential is denoted as $\partial f(\vec{x}_0)$.

Note that Eq. (15.2) can be written as

$$f(\vec{x}) \geq f(\vec{x}_0) + \vec{g}^T \cdot (\vec{x} - \vec{x}_0). \tag{15.3}$$

A point $\vec{x}_0$ is a *global minimum* of a convex function if zero is contained in the subdifferential at $\vec{x}_0$.

Returning to the function $z = |x|$, Fig. 15.21 shows its subdifferential $\partial f(x)$ as a function of x. The subdifferential at the origin is the interval $[-1, 1]$.

Notice that according to Fig. 15.21, $sign(x) \in \partial f(x)$.

Perhaps the reader could find (for instance, in the literature about Filippov's differential inclusion) that some authors, like [Lo, Dan], do use a *set-valued*

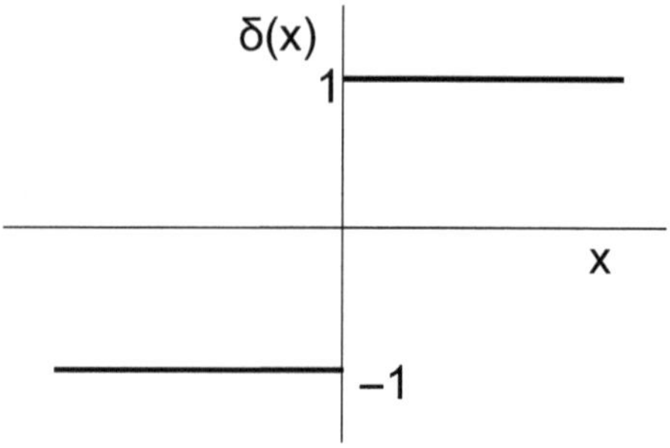

Fig. 15.21 Subdifferential of the $z = |x|$ function

signum function, which has values $[-1, 1]$ for $x = 0$ (so it is coincident with the subdifferential depicted in Fig. 15.21).

See [12] (1308 cites) for more details on subgradients and related concepts. A recommended lecture presentation on this topic would be [27].

Historical Note

The book of Rockafellar [76], in 1970 with the title *Convex Analysis*, provided momentum for new theory developments (the reedition of the book in 2015 has now 32.805 cites). It was realized that many convex optimization problems do not have a derivative at the optimum, and this motivated the introduction of subdifferentials. The term *non-smooth analysis* was introduced by Clarke [19] (14,697 cites), 1983, who was the second student of Rockafellar and who extended the study to locally Lipschitz functions. New important contributions in order to further extensions have been done by Mordukhovich [54, 55] in 2006.

Optimization Methods for Non-smooth Problems

A basic methodology for non-smooth optimization is based on subgradients. Another type of methods is related to cutting planes, resulting in faster convergence.

A third type of methods, which is attracting a lot of attention, is based on the proximal mapping.

Of course, it is always possible to use direct searching methods [44] or heuristic methods. However, we prefer to put now the focus on the more specific methods already mentioned.

The next three subsections are devoted to optimization via subgradients, or cutting planes, or proximal mapping.

The last section of the article [97] discuss with emphasis on stochastic subgradients, various alternatives that the research has explored, frequently in search of faster methods.

Two important application examples are the support vector machine (there are non-smooth versions) and the LASSO problem (*least absolute shrinkage and selection operator*) in connection with compressed sensing and sparse representation (matters to be treated in the next chapter).

Recall from a precedent chapter that the SVM problem is to find support vectors such as

Minimize $\vec{w}^T\,\vec{w}$
Subject to

$$y_i(\vec{w}^T\vec{x}_i + b) \geq 1 \tag{15.4}$$

where $y_i = 1 \; if \; \vec{x}_i \in A$ or $y_i = -1 \; if \; \vec{x}_i \in B$ and $\vec{x}_i$ are data points.

In turn, the formulation of the LASSO problem is the following:

Minimize $\left\| A\vec{x} - \vec{b} \right\|_2^2$

Subject to $\|\vec{x}\|_1 < c$

Based on the Lagrange approach, the LASSO problem can be also stated as follows:

Minimize $\left\|A\vec{x} - \vec{b}\right\|_2^2 + \lambda\ \|\vec{x}\|_1$

The second term $\lambda\ \|\vec{x}\|_1$ is a penalty term for regularization purposes.

This last expression is usually found in the context of regressions, and it is called *LASSO regression*.

There is another type of regression, called *ridge regression*, in which the term $\lambda\ \|\vec{x}\|_1$ is replaced by $\lambda\ \|\vec{x}\|_2^2$.

The survey articles [41] and [94] give abundant information on many types of loss functions and regressions with different purposes.

15.5.2 Subgradient Methods

The subgradient method uses the following simple iteration,

$$\vec{x}_{k+1} = \vec{x}_k - \alpha_k\, \vec{g}_k,$$

where $\vec{g}_k$ is any subgradient of $f()$ at $\vec{x}_k$. The positive parameter α_k is the step size. When $f()$is differentiable at $\vec{x}_k$, there is only one possible subgradient, which is the gradient. When it is not differentiable, the chosen subgradient would *not* necessarily lead to a descent of the function value; for this reason, one usually retains a memory of the best point found along the iterations. The number of iterations could be high (perhaps millions).

The convergence of the method has been subject of much study. Some relevant results are summarized in [12], starting from the assumption that there is a bound G such that $\|\vec{g}_k\|_2 \le G$, which is, for instance, the case when $f()$ satisfies the Lipschitz condition:

$$|f(u) - f(v)| \le G\ \|u - v\|_2$$

Considering in particular the difference $D = f_k^{best} - f^*$, then, under the previous assumption, it can be established that

- For constant step size $\alpha_k = h$, then $D \to G^2 h/2$ as $k \to \infty$.
- For constant step length $\alpha_k = h/\ \|\vec{g}_k\|_2$, then $D \to G h/2$ as $k \to \infty$.

There are many other step size alternatives that have been proposed, for example the following nonsummable, i.e., $\sum_k \alpha_k = \infty$, step sizes:

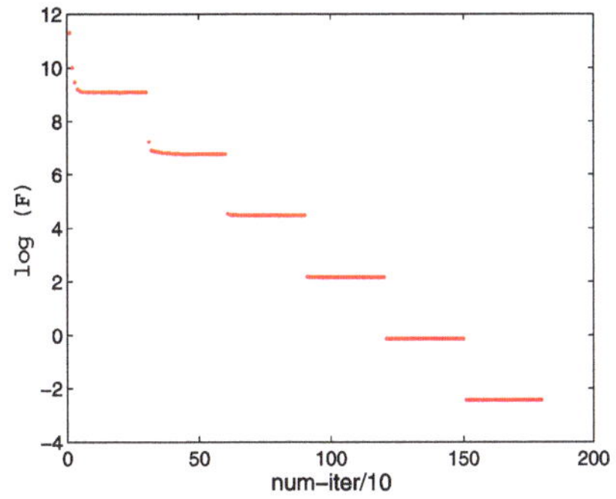

Fig. 15.22 Evolution of the LASSO minimization

- Square summable:

$$\sum_k \alpha_k^2 < \infty$$

(typical example: $\alpha_k = a/(b+k)$ with positive a and b)

- Diminishing step size:

$$\lim_{k\to\infty} \alpha_k = 0$$

(typical example: $\alpha_k = a/\sqrt{k}$ with positive a)

- And also, diminishing step length:

$$\alpha_k = \gamma_k \,/\, \|\vec{g}_k\|_2\,; \quad \lim_{k\to\infty} \gamma_k = 0; \quad \sum_k \gamma_k = \infty$$

In these three alternatives, the procedure converges to the optimum, that is, $D \to 0$ as $k \to \infty$.

In order to illustrate the use of subgradient iteration, based on the code available from [99], a program in MATLAB (Program 15.4) has been devised.

The program considers a LASSO problem and applies changes of parameter λ every 300 iterations (this number can be modified).

Figure 15.22 visualizes how the objective function decreases along the process (notice that vertical axis is logarithmic).

The program uses two functions, one for computing the value of the objective function and another for computing the subgradient of the LASSO problem as follows:

$$\partial F = A^T A\,\vec{x} - A^T\,\vec{b} + \lambda \cdot sign(\vec{x})$$

The reader is invited to explore the effect of changing parameters in the program.

Program 15.4 Using subgradient iterations for a LASSO problem

```
% example of subgradient iterations
% for a LASSO problem
% generate data
n = 1024; m = 512;
A = randn(m,n);
u = sprandn(n,1,0.1);
b = A*u;
%parameters
s = 0.00028;
maxiter = 300;
count_num = 6;
count_alpha = 10;
%init
Frec=zeros(1,maxiter*count_num);
lambda = 1e-3;
x0 = rand(n,1);
ATA=A'*A;
ATb=A'*b;
lambdai = lambda*count_alpha^(count_num-1);
x = x0; Ni=1;
%main loop ------------------------------
for i=1:count_num
  for j=1:maxiter
    x=x-s*g(x,ATA,ATb,lambdai);
    Frec(Ni) = F(x,A,b,lambdai);
    Ni=Ni+1;
  end
  if i<count_num
    lambdai = lambdai/count_alpha;
  end
end
%Results ---------------------------------
figure(1)
plot(log(Frec(1:10:end)), 'r.');
xlabel('num-iter/10'); ylabel('log (F)');
out.val = F(x,A,b,lambdai);
out.s = s;
out.lambda = lambdai;
out
```

The program uses the following two functions:

Function 15.1 Objective function for the LASSO problem

```
function [Fxx]=F(xx,A,b,lambdai)
Fxx = 0.5*norm(A*xx-b)^2+lambdai*norm(xx,1);
end
```

Function 15.2 The subgradient in the LASSO problem

```
function [gxx]=g(xx,ATA,ATb,lambdai)
gxx = ATA*xx-ATb+lambdai*sign(xx);
end
```

The Polyak's Step Size

When the optimal value f° is known, it is convenient to use the Polyak's step size, which is given by

$$\alpha_k = \gamma_k \frac{f(\vec{x}_k) - f^\circ}{\|\vec{g}_k\|_2^2}$$

where $0 < \gamma_k < 2$.

Indeed, there are some optimization problems for which f° is known, like, for example, the *feasibility problem* of finding a point $\vec{x}_c$ satisfying a particular system of inequalities.

Some modifications of the Polyak's step size have been proposed to circumvent the use of f°.

15.5.2.1 Projected Subgradient Method

The *projected subgradient method* is a simple extension of the basic method that is suitable for the following convex constrained problem:

Minimize $f(x)$
Subject to $x \in C$
(C is a convex set)

The projected subgradient method is given by

$$\vec{x}_{k+1} = P(\vec{x}_k - \alpha_k \, \vec{g}_k)$$

where P is a projection on C.

The step sizes can be chosen as in the basic method, with similar convergence properties.

Here is a short list of projections that could be useful [67, 91, 95]:

- For half-space $C = \{\vec{x} : \ \vec{a}^T\vec{x} \leq \vec{b}\}$,

  $P_C(\vec{x}) = \vec{x} + \frac{\vec{b}-\vec{a}^T\vec{x}}{\|\vec{a}\|_2^2} \ if \ \vec{a}^T\vec{x} > \vec{b} \quad and$
  $P_C(\vec{x}) = \vec{x} \ if \ \vec{a}^T\vec{x} \leq \vec{b}.$
- For non-negative orthant, the projection would be $P(\vec{x}_i) = \max(\vec{x}_i, 0)$.

- For a rectangle $C = \{\vec{x} \in \Re^n : \vec{l} \leq \vec{x} \leq \vec{u}\}$,

$$P_C(x_i) = \begin{cases} l_i & x_i \leq l_i \\ x_i & l_i \leq x_i \leq u_i \\ u_i & x_i \geq u_i \\ . \end{cases}$$

- For a linear system such that $C = \{\vec{x} : A\vec{x} = \vec{b}\}$, the projection would be

$$P_C(\vec{x}_k) = \vec{x}_k - A^T (A\,A^T)^{-1} (A\vec{x}_k - \vec{b}).$$

In this last case, the subgradient update would be

$$\vec{x}_{k+1} = \vec{x}_k - \alpha_k \,(I - A^T (A\,A^T)^{-1} A)\,\vec{g}_k$$

(note that $A\vec{x}_k = \vec{b}$).

Comments from literature are coincident in that the projected subgradient method usually has faster convergence.

One of the typical applications of the method is *l1* minimization, for example, consider the following problem [12]:

- Minimize $z = \|\vec{x}\|_1$
- Subject to $A\vec{x} = \vec{b}$

A subgradient of z is given by $g = sgn(x)$ (the conventional signum function). Therefore, the optimization problem can be solved with the following projected subgradient iterations:

$$\vec{x}_{k+1} = \vec{x}_k - \alpha_k \,(I - A^T (A\,A^T)^{-1} A)\, sgn(\vec{x}_k)$$

Another popular application is for problems like the following,

- Minimize $\frac{1}{2} \left\| A\,\vec{x} - \vec{b} \right\|_2^2$

where all components of $\vec{x}$ should be nonnegative (so $\vec{x}$ belongs to the positive orthant).

Based on [36], a coding in MATLAB (Program 15.5) has been devised to illustrate the application of projected subgradient iterations to this last problem.

Figure 15.23 shows the evolution of the objective function along the iterative process. And Fig. 15.24 shows the components of the solution $\vec{x}$ reached at the end of the iterations.

Both figures have been generated by Program 15.5.

The program uses a diminishing step size.

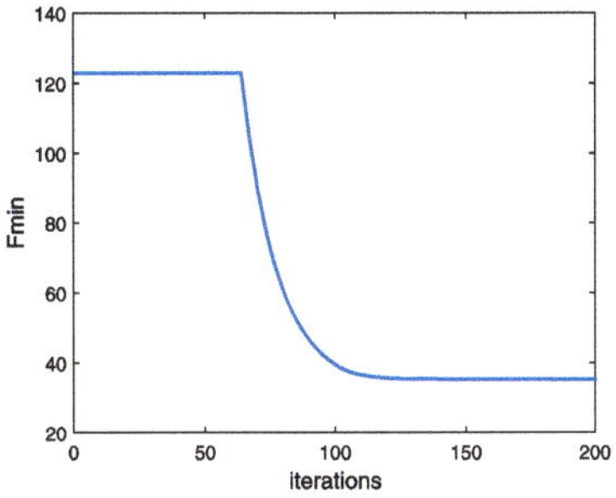

Fig. 15.23 Evolution of the subgradient projection minimization

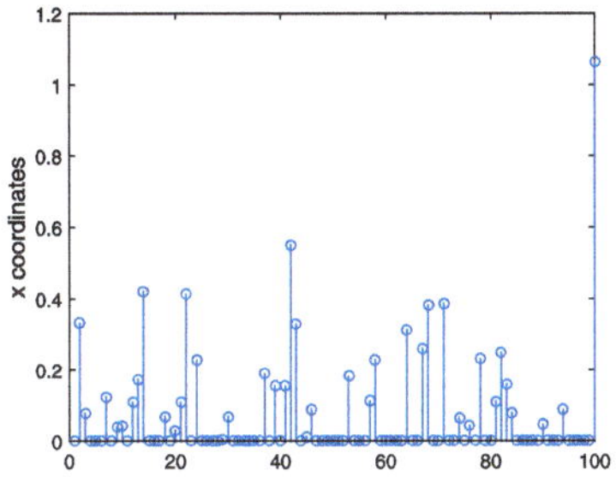

Fig. 15.24 Components of the solution vector

Program 15.5 Example of subgradient projection

```
% example of Subgradient Projection
% generate data
n = 100; m = 100;
A = randn(m,n);
u = sprandn(n,1,0.1);
b = A*u;
% init
xcur=zeros(n,1);
maxiter=200;
%------------------------------------------------
fvals=0.5*norm(b)^2; fmin=fvals; xmin=xcur;
% save objective function values:
fminvals=zeros(maxiter,1);
% main loop ------------------
for i=1:maxiter
  alpha=0.025/sqrt(i);
  res = (A*xcur-b);
  subg = A'* res; % (sub)-gradient of objective
  % Projection onto positive orthant:
  xcur = max(0,xcur - alpha*subg);
  %value of objective function
  fcur = 0.5*norm(res)^2;
  if(fcur<fmin)
    fmin=fcur; xmin=xcur;
  end
  fminvals(i) = fmin;
end
```

```
% results ------------------
figure(1)
plot(fminvals);
xlabel('iterations'); ylabel('Fmin');
figure(2)
stem(xmin);
ylabel('x coordinates');
```

Historical Note

The subgradient method was developed in 1962 by N.Z. Schor. Other authors, like Ermoliev and Polyak, contributed to consolidate the method. See [30] for more technical and historical details, including main references. The method belongs to a large class of methods known as *first-order methods*, which are based on subgradients and the like; second-order methods usually employ Hessians.

Subgradient methods are very easy to implement and can be applied to many different problems. The main drawback is that they can be very slow, sometimes with a zigzag behavior.

15.5.2.2 Some Variants of the Subgradient Method

The subgradient method can be extended [97] to solve problems with inequalities:

Minimize $f_0(x)$
Subject to $f_i(\vec{x}) \leq 0, \; i = 1, 2, \ldots, m$

where $f_i(.)$ are convex.
In $\vec{x}_{k+1} = \vec{x}_k - \alpha_k \, \vec{g}_k$, the subgradient is chosen as follows,

$$\vec{g}_k = \begin{cases} \partial f_0(\vec{x}_k) \; if \; f_i(\vec{x}_k) \leq 0, \; \forall i \\ \partial f_j(\vec{x}_k) \; for \; some \, j \; such \; that \; f_j(\vec{x}_k) > 0 \end{cases}$$

(if the current point is feasible, one uses $\partial f_0(.)$; if not, one uses a subgradient of any violated constraint).

As said before, subgradients are not guaranteed to be directions of descent. Some modifications have been proposed to detect good directions. One is the *conjugated subgradient method*, which includes a minimization task (to find the best direction) in each iteration. Another technique is called *ε-subgradient method*, which uses approximated subgradients (see [73]).

15.5.2.3 Stochastic Subgradient Methods

There are reasons, like, for instance, the large size of the optimization problem at hand or its involved randomness, which can recommend the use of *stochastic*

Fig. 15.25 The hinged loss function

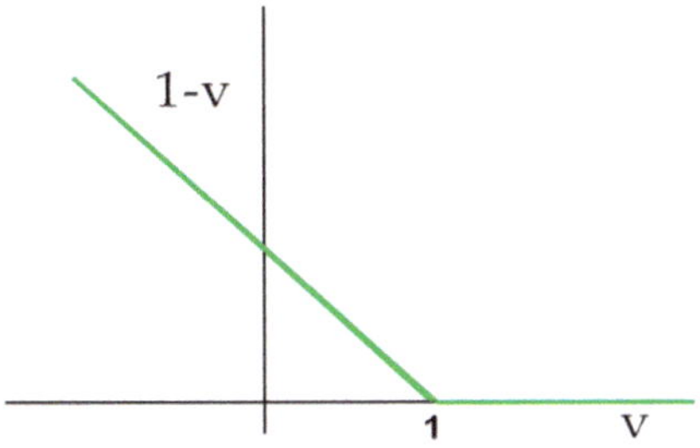

subgradient methods [11, 97]. This method is almost the same as the subgradient method but using noisy subgradients and a reduced set of step size alternatives.

It is said that $\tilde{g}$ is a *noisy subgradient* of $f(.)$ at $\vec{x}$ (which is random) if $E(\tilde{g} \mid \vec{x}) \in \partial f(\vec{x})$ (where $E(.)$ is the expected value). One could consider the noisy subgradient as the sum of $\vec{g}$ and a zero-mean random variable.

The *stochastic subgradient iteration* is

$$\vec{x}_{k+1} = \vec{x}_k - \alpha_k \, \tilde{g}_k$$

where $\tilde{g}_k$ is the noisy subgradient at $\vec{x}_k$.

Let us consider again the support vector machine (SVM). In case of not linearly separable data, we cannot maintain the original statement of the problem (as given in (15.4)). Instead, the SVM problem can be restated as follows [15],

$$\text{Minimize } \frac{\lambda}{2} \, \|\vec{w}\|_2^2 + \frac{1}{m} \sum_{i=1}^{m} l(y_i, < \vec{w}, \vec{x}_i >,$$

where one inserts a loss function $l(y_i, < \vec{w}, \vec{x}_i >)$.

The parameter λ is used to trade between misclassification penalizing and margin maximization.

A commonly used loss function is

$$l(y, < \vec{w}, \vec{x} >) = \max(0, \; 1 - y < \vec{w}, \vec{x} >)$$

which is called *hinged loss function*. It is not a differentiable function, motivating the use of subgradients. Figure 15.25 depicts this function.

The subgradient of the hinged loss function is [16]

$$\vec{g}_i = \begin{cases} 0, & if \; y_i(\vec{w}\vec{x}_i + b) > 1 \\ y_i \vec{x}_i, & otherwise \end{cases}$$

In case of too many data points, a subset S_k of these points can be obtained by sampling, and a smaller problem can be formulated as

$$\text{Minimize } \frac{\lambda}{2} \, \|\vec{w}\|_2^2 + \frac{1}{|S_k|} \sum_{i \in S_k}^{l(y_i, <\vec{w}, \vec{x}_i>)} .$$

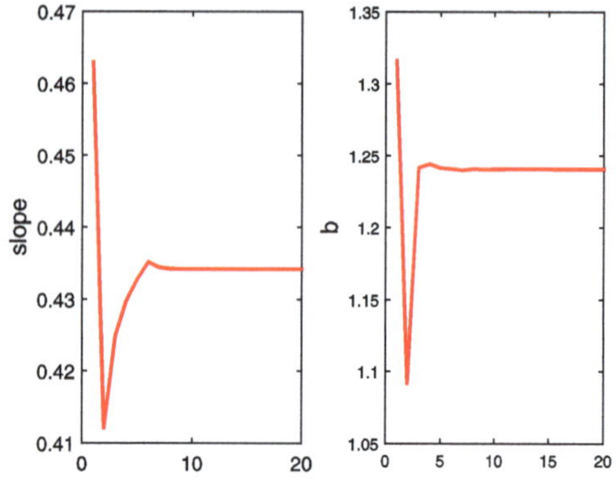

Fig. 15.26 Evolution of slope and b parameters, along stochastic gradient iterations

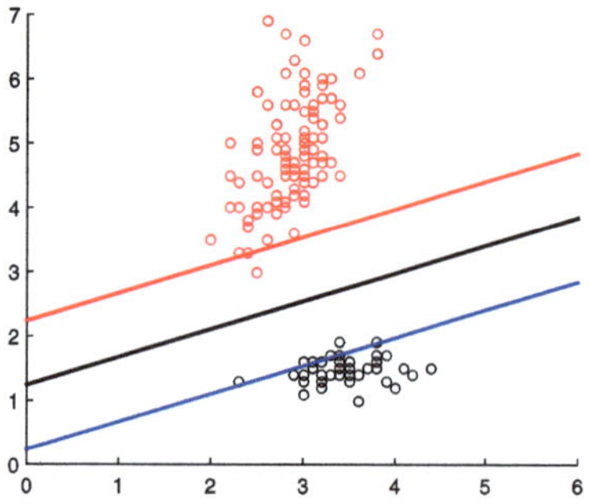

Fig. 15.27 Components of the solution vector

In order to solve this problem, the following stochastic subgradient iteration can be applied:

$$\vec{w}_{k+1} = \vec{w}_k - \alpha_k \, \tilde{g}_k$$

A simple implementation in MATLAB code has been developed, Program 15.6, to illustrate the application of the stochastic subgradient method for the SVM problem. The program generates two figures.

The solution of the problem is given in terms of an optimal separating line, with a certain slope and b parameters. Figure 15.26 depicts the evolution of those parameters as the iterative process searches for their optimal values. Not much iterations were necessary.

As in other programs included in previous chapters, we employed the iris flower data as the case to be considered.

Figure 15.27 shows the result of the stochastic gradient program. It is not perfect, but it could be satisfactory enough, and its computation is fast (and no quadratic optimization solver is needed).

Here is the listing of the program, with the application of the stochastic subgradient.

Program 15.6 Applying stochastic subgradient method for SVM-based classification

```
% stochastic subgradient example
% SVM for classification
% IRIS data
% x=sepal width, y=petal length
%read IRIS data
D=dlmread('iris.data'); %columns
%there are 50 points for class A
% and 100 points for class B
X=[D(:,2), D(:,3)]; %two columns (x,y)
Y=ones(150,1); Y(1:50)=-Y(1:50); %labels:1 or -1
lambda=0.27;
alpha=2.7;
maxiter=20;
wx=1; wy=1; b=1;
Vslop=zeros(maxiter,1);
Vb=zeros(maxiter,1);
% loop ----------------------------------------
for nn=1:maxiter,
  gx=0; gy=0; gb=0; %subgradient
  for i=1:150,
    x=X(i,1); y=X(i,2);
    % subgradient application
    if ((Y(i)*(wx*x+wy*y+b)) <=1)
      gx=gx+(Y(i)*x); gy=gy+(Y(i)*y);
      gb=gb+Y(i);
    end
  end
  gx=gx/150; gy=gy/150; gb=gb/150;
  krd=0.2;
  gx=gx+(lambda)*wx + krd*(rand(1)-0.5);
  gy=gy+(lambda)*wy + krd*(rand(1)-0.5);
  gb=gb+(lambda)*b + krd*(rand(1)-0.5);
  alpha=alpha/sqrt(nn);
  wx=wx-(alpha*gx);
  wy=wy-(alpha*gy);
  b=b-(alpha*gb);
  slop=wy/wx;
  Vslop(nn,1)=slop;
  Vb(nn,1)=b;
end
%display ------------------------------------
figure(1)
subplot(1,2,1);
plot(Vslop(:,1),'r');
ylabel('slope');
subplot(1,2,2);
plot(Vb(:,1),'r');
ylabel('b');
figure(2)
```

```
%the data
scatter(X(1:50,1),X(1:50,2),32,'k'); hold on;
scatter(X(51:150,1),X(51:150,2),32,'r');
% the lines
slop=wy/wx;
vx=[0:0.1:6];
vy=(slop*vx + b);
% optimal separating line:
plot(vx, vy, 'k', 'LineWidth',2);
vx=[0:0.1:6];
vy=(slop*vx + b +1);
% class 1 border line:
plot(vx, vy, 'r', 'LineWidth',2);
vy=(slop*vx + b -1);
% class 2 border line:
plot(vx, vy, 'b', 'LineWidth',2);
```

15.5.2.4 Random Coordinate Descent Methods

A recently popularized idea [60] (1537 cites), [72] (849 cites), [90] (902 cites) is the *random coordinate descent method.* At each step k of the iteration, a coordinate index j is chosen at random (with uniform PDF), and then the update is done as follows,

$$(x_j)_{k+1} \;=\; (x_j)_k \;-\; \alpha_k\,(\vec{g}_j)_k$$

$$(x_i)_{k+1} \;=\; (x_i)_k \;\; for\;\; i \neq j,$$

where (x_i) denotes the *i-component* (or coordinate) of $\vec{x}$.

A program in MATLAB has been developed in order to visualize the searching concept (Program 15.7). Figure 15.28 depicts the result for a 2D problem that was considered in Chap. 2. The noticeable zigzag corresponds to advancing through one coordinate and then through another coordinate.

Notice that for this case the program uses directly gradients (the subgradients would be the same).

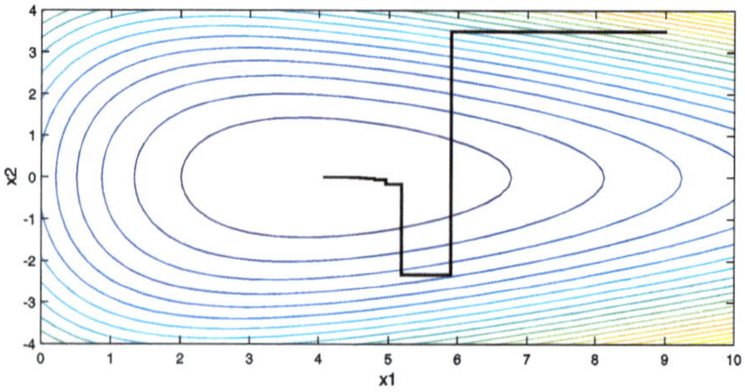

Fig. 15.28 Coordinate descent trajectory

Program 15.7 Example of coordinate descent method

```
% Example of Coordinate Descent
% (based on gradients)
%
[vx1,vx2]=meshgrid(0:0.2:10, -4:0.2:4);
z=(vx1.^2)+(vx2.^2);
y=(exp(-(vx1-3)/2))+(exp((vx1+4*vx2)/10))+...
(exp((vx1-4*vx2)/10));
%
% function display-------------
figure(1)
contour(vx1,vx2,y,20); hold on;
xlabel('x1'); ylabel('x2');
%
% searching ------------------
N=30;
% for trajectory saving:
x1rec=zeros(1+2*N,1); x2rec=zeros(1+2*N,1);
% initial point
x1=9; x2=3.5;
x1rec(1,1)=x1; x2rec(1,1)=x2;
alphai=3;
k=2;
for j=1:N,
  % coordinate x1
  % gradient
 g1=(-0.5*exp(-(x1-3)/2))+(0.1*exp((x1+4*x2)/10))+...
  (0.1*exp((x1-4*x2)/10));
  % update
  alpha=alphai/sqrt(k-1);
  x1=x1-(alpha*g1); x1rec(k,1)=x1; x2rec(k,1)=x2;
  k=k+1;
  % coordinate x2
  % gradient
  g2=(0.4*exp((x1+4*x2)/10))-(0.4*exp((x1-4*x2)/10));
  % update
  alpha=alphai/sqrt(k-1);
  x2=x2-(alpha*g2); x1rec(k,1)=x1; x2rec(k,1)=x2;
  k=k+1;
end
% display -------------------------
plot(x1rec(1:k-1,1), x2rec(1:k-1,1), 'k');
```

15.5.3 Bundle Methods

In order to introduce the bundle methods, it is important to make two observations:

The first is that for convex functions a first-order Taylor approximation is a lower bound. The same can be said about subgradients, based on Eq. (15.3).

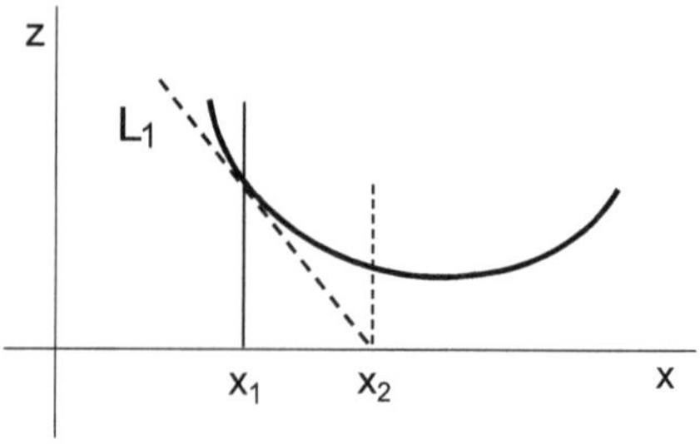

Fig. 15.29 First step of the modeling process

The second is that when running the subgradient methods described above, almost no record is kept on the exploration history. It would be more convenient to take into account the knowledge acquired about the objective function as the iterative optimization proceeds.

By using a combination of first-order linear approximations, it is possible to build a piecewise approximation of the objective function, which can be considered as model $f^M(.)$ of the objective function $f(.)$. This was the basis of the cutting-plane method proposed in [18] and [42] (2302 cites), which considers a series of subgradients $\vec{g}_1,\ \vec{g}_2,\ \ldots \vec{g}_n$ evaluated at points $\vec{x}_1,\ \vec{x}_2,\ \ldots \vec{x}_n$ and uses it for building the model

$$f_n^M(\vec{x}) = \max_{1 \le k \le n} \{f(\vec{x}_k) + (\vec{g}_k^T(\vec{x} - \vec{x}_k)\}.$$

This model is a lower bound of the objective function $f_n^M(\vec{x}) \le f(\vec{x})$.

The information on the current model is kept in a *bundle*:

$$B_n = \{(\vec{x}_k,\ f(\vec{x}_k),\ \vec{g}_k),\ k = 1,\ 2,\ \ldots,n\}$$

At iteration $n + 1$, a new point is added. This point is chosen as

$$\vec{x}_{n+1} = \arg\min_{\vec{x}} f_n^M(\vec{x}).$$

Let us illustrate how the piecewise model is built. Figure 15.29 shows a convex objective function $f(.) \ge 0$. The problem is to find its minimum. A first point $\vec{x}_1$ has been chosen, and a first model of the objective function has been found: the line L_1. The next point, $\vec{x}_2$, to be added to the exploration is simply found at the intersection of L_1 with the horizontal axis.

Taking $f(\vec{x}_2)$ as new point for a linear approximation, which is plotted as the line L_2, it is possible to determine the next point to be added, $\vec{x}_3$. Now, the model consists of two linear pieces (a "V"), taken from the two lines. Figure 15.30 depicts this step.

Figure 15.31 shows the result of a third step. Notice that now the next point to be added, $\vec{x}_4$, is found at the intersection of lines L_2 and L_3. The piecewise model, made of segments of L_1, L_2, and L_3, is getting better.

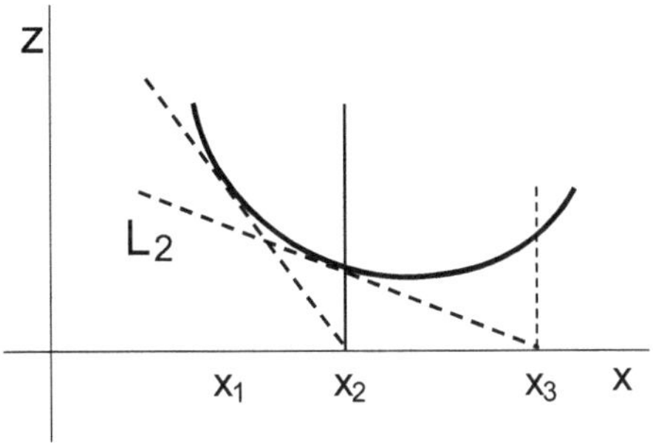

Fig. 15.30 Second step of the modeling process

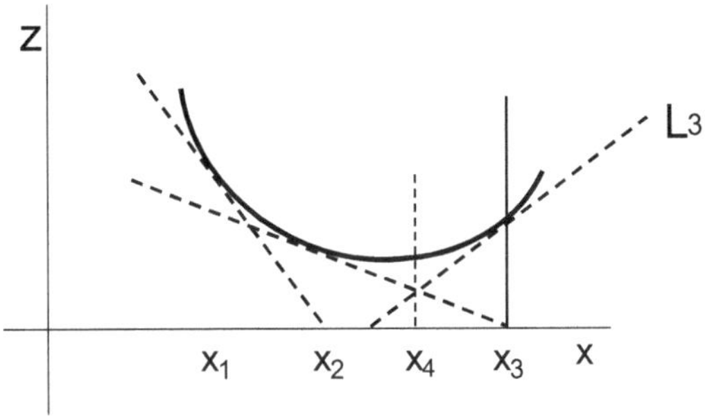

Fig. 15.31 Third step of the modeling process

Along the modeling iterations, the cutting-plane method monitors the following quantity, which monotonically decreases

$$\varepsilon = \min_{0<k<n} \{f(\vec{x}_k) - f_n^M(\vec{x}_n)\}.$$

Once this quantity becomes smaller than a certain specified threshold, the iterative process terminates.

A series of observed drawbacks of the cutting-plane basic scheme has been reported by the scientific literature: instability, etc.

See [80] for details.

15.5.4 Proximal Methods

Suppose that the function to be minimized has several points where it is not differentiable, and so the plot of the function has kinks (informally speaking) where smoothness is lost. An interesting idea to recover good mathematical conditions is to use *"envelopes,"* that is, to use a smooth approximation of the function, so instead of kinks you have arcs. Figure 15.32 tries to illustrate the idea.

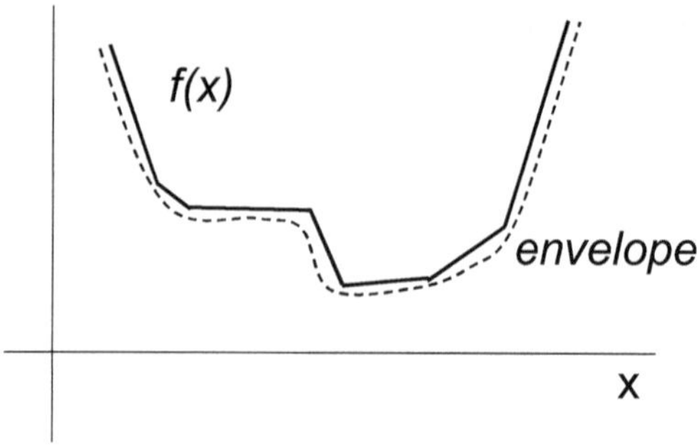

Fig. 15.32 An example of envelope

An important example of envelope is the *Moreau's envelope*, defined by

$$M_{\lambda f}(\vec{x}) = \inf_{\vec{u}} \left(f(\vec{u}) + \frac{1}{2\lambda} \|\vec{u} - \vec{x}\|_2^2 \right)$$

(Moreau employed $\lambda = 1$).

Directly related to this envelope, the *proximal mapping* is defined by

$$\text{Prox}_{\lambda f}(\vec{x}) = \arg\min_{\vec{u}} \left(f(\vec{u}) + \frac{1}{2\lambda} \|\vec{u} - \vec{x}\|_2^2 \right) \tag{15.5}$$

where $f(x)$ is a proper lower semi-continuous function (*plsc*).

In general, the Moreau's envelope approximates $f(\vec{x})$ from below, becoming a better approximation as λ is decreased. The envelope is a finite, continuous function. The lower bounds of $f(\vec{x})$ and $M_{\lambda f}(\vec{x})$ are equal:

$$\inf f(\vec{x}) = \inf M_{\lambda f}(\vec{x})$$

Therefore, the original problem of minimizing a possibly non-differentiable $f(\vec{x})$ can be replaced with the minimization of the smooth function $M_{\lambda f}(\vec{x})$.

If $f(\vec{x})$ is convex and plsc, the gradient of the envelope would be

$$\nabla M_{\lambda f}(\vec{x}) = \frac{1}{\lambda}(I - \text{Prox}_{\lambda f}(\vec{x})).$$

Also, if $f(\vec{x})$ is convex and plsc, the function inside the parenthesis in (15.5) is strongly convex because of the added Euclidean norm term.

Moreau introduced the envelope and the proximal mapping in 1962 [56] (in French, 807 cites) as a way of regularizing and approximating a convex function. Actually, it was called *prox-regularization*, and it can be regarded as analogue to the Tikhonov regularization [38]. The book [77] (several reeditions, 11633 cites) places the envelope and the proximal mapping in a rigorous mathematical context.

The Moreau's envelope is an example of infimal convolution. The *infimal convolution* of closed proper convex functions $f(\vec{x})$ and $g(\vec{x})$ is defined as

$$(f \otimes g)(\vec{v}) = \inf_{\vec{x}}(f(\vec{x}) + g(\vec{v} - \vec{x})).$$

There are a number of interpretations [68] of the approach initiated by Moreau, for instance, it can be considered as a penalty method for sequential unconstrained minimization [14]. In what follows, some attention is paid to this aspect.

15.5.4.1 Penalty Functions

The problem is to minimize a function $f(\vec{x})$. The *sequential unconstrained minimization* (SUM) algorithms use a set of auxiliary functions $p_k(\vec{x})$ so that at the kth step one minimizes

$$P_k(\vec{x}) = f(\vec{x}) + g_k(\vec{x})$$

to get a minimizer $\vec{x}^k$.

The functions $p_k(\vec{x})$ are chosen so that the infinite sequence $\{\vec{x}^k\}$ converges to a solution of the $f(\vec{x})$ minimization problem.

In the context of constrained minimization, where a feasible set exists, one could distinguish two approaches. Barrier-function methods (for instance, interior-point methods) require the minimizers $\{\vec{x}^k\}$ to belong to the feasible set. This is not the case with penalty-function methods, where constraint violations are discouraged but not prohibited (hence, sometimes they are called *exterior-point* methods).

Some examples of penalty functions, like the absolute value, the Courant-Bertrami, the quadratic-loss, etc., can be found in [14]. Also, in [14], connections of penalty-function methods with the minimization of cross-entropy, the regularized least squares, and the Lagrangian approach in optimization are described.

As highlighted in [14], the proximal approach can be used for a sequential minimization algorithm (this will be introduced later on).

15.5.4.2 Some Examples of Envelopes and Proximal Maps

A first and important example is $f(x) = |x|$. Its Moreau's envelope is

$$M_{\lambda f}(x) = \begin{cases} \frac{1}{2\lambda} x^2 & x \in [-\lambda, \lambda] \\ |x| - \frac{\lambda}{2} & otherwise, \end{cases}$$

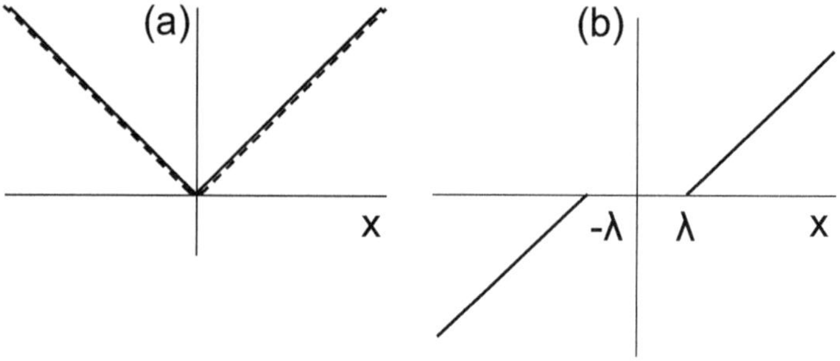

Fig. 15.33 (**a**) Moreau's envelope (dashes) and (**b**) proximal mapping of $|x|$

and the proximal mapping is

$$\text{Prox}_{\lambda f}(x) = \begin{cases} 0 & x \in [-\lambda, \lambda] \\ x - \lambda & x > \lambda \\ x + \lambda & x < \lambda. \end{cases} \tag{15.6}$$

Figure 15.33 shows the Moreau's envelope and the proximal mapping of $|x|$.

In the case of the $l1$ norm of a vector, $\|\vec{x}\|_1 = \sum_i |x_i|$, the ith coordinate of $\text{Prox}_{\lambda f}(\vec{x})$ is given by (15.6) replacing x with x_i.

The proximal mapping of the $l1$ norm is also called the *soft thresholding operator*.

For $\lambda = 1$, the Moreau's envelope of $|x|$ would be a particular case of the *Huber penalty function*:

$$L_\lambda(x) = \begin{cases} \frac{1}{2}x^2 & x \in [-\lambda, \lambda] \\ \lambda\left(|x| - \frac{\lambda}{2}\right) & otherwise \end{cases}$$

Another example is the quadratic case: $f(\vec{x}) = \frac{1}{2}\vec{x}^T A\,\vec{x} + \vec{b}^T\vec{x} + c$. Its proximal mapping is

$$\text{Prox}_{\lambda f}(\vec{x}) = (I + \lambda A)^{-1}(\vec{x} - \lambda\vec{b}).$$

There are three special cases of this mapping:

- If $f(\vec{x}) = \vec{b}^T\vec{x} + c$, then $\text{Prox}_{\lambda f}(\vec{x}) = \vec{x} - \lambda\vec{b}$.
- If $f(\vec{x}) = c$, then $\text{Prox}_{\lambda f}(\vec{x}) = \vec{x}$ (the proximal operator is the identity).
- If $f(\vec{x}) = \frac{1}{2}\|\cdot\|_2^2$, then $\text{Prox}_{\lambda f}(\vec{x}) = (\frac{1}{1+\lambda})\,\vec{x}$.

Other example of interest, in connection with logarithmic barriers, is the case of $f(x) = -\log x$. Its proximal mapping is

$$\text{Prox}_{\lambda f}(\vec{x}) = \frac{x + \sqrt{x^2 + 4\lambda}}{2}.$$

The next case is of mathematical interest, for minimization problems where solutions are constrained to belong to a certain set.

If $f(\vec{u}) = i_C(\vec{u})$, where $i_C(\vec{u})$ is the indicator function of the set C

$$i_C(\vec{u}) = \begin{cases} 0 & x \in C \\ \infty & x \notin C, \end{cases}$$

then $\text{Prox}_{\lambda f}(\vec{x}) = \arg\min_{\vec{u}} \|\vec{u} - \vec{x}\|_2^2 = P_C(\vec{x})$, which is the projection over set C.

And, finally, the case of the Euclidean norm $f(x) = \|\vec{x}\|_2$. Its proximal mapping is

$$\text{Prox}_{\lambda f}(\vec{x}) = \begin{cases} \left(\frac{1-\lambda}{\|\vec{x}\|_2}\right)\vec{x} & \|\vec{x}\|_2 \geq \lambda \\ 0 & otherwise \end{cases}.$$

See [66] for more details.

15.5.4.3 Characterization and Properties of the Proximal Mapping

It can be easily shown [66] that if the point $\vec{x}^*$minimizes $f(\vec{x})$, then $\text{Prox}_{\lambda f}(\vec{x}^*) = \vec{x}^*$.

Also, the point $\vec{x}^*$ minimizes

$$\left(f(\vec{x}) + \frac{1}{2\lambda}\|\vec{x} - \vec{v}\|_2^2\right)$$

if and only if

$$0 \in \partial f(\vec{x}^*) + \frac{1}{\lambda}(\vec{x}^* - \vec{v}). \tag{15.7}$$

Taking $\vec{x}^* = \vec{v}$, it follows that $0 \in \partial f(\vec{x}^*)$, so $\vec{x}^*$minimizes $f(\vec{x})$.
Therefore, from the above results, the point $\vec{x}^*$ minimizes $f(\vec{x})$ iff

$$\vec{x}^* = \text{Prox}_{\lambda f}(\vec{x}^*).$$

Then, the point $\vec{x}^*$ is a fixed point of the operator $\text{Prox}_{\lambda f}$.

The expression (15.7) is the *subgradient characterization* of the proximal mapping.

The proximal mapping of a separable function such that $f(x, y) = \alpha(x) + \beta(y)$ would be

$$\text{Prox}_{\lambda\, f}(v,\, w) \;=\; (\text{Prox}_{\lambda\, \alpha}(v), \text{Prox}_{\lambda\, \beta}(w)).$$

Thus, the proximal mapping of $f(x, y)$ is done with the proximal mapping of each of the separable parts. In the general case of a fully separable $f(\vec{x})$, the proximal mapping would consist of proximal mappings of scalar functions.

15.5.4.4 The Proximal Point Algorithm

As announced before, when dealing with penalty functions, there is a sequential minimization algorithm, based on the proximal mapping. The algorithm is simply expressed as follows:

$$\vec{x}^{k+1} \;=\text{Prox}_{\lambda\, f}(\vec{x}^{k})$$

Convergence of the sequence $\{\vec{x}^k\}$ to the set of minimizers of $f(\vec{x})$ is guaranteed for values of λ that satisfy $\lambda^k > 0$ and $\sum_{i=1}^{\infty} \lambda^k = \infty$.

The convergence of the algorithm has been studied in detail in [75] (4643 cites) and [32] (846 cites). It may be quite slow. In [87], a modification of the algorithm was proposed in order to obtain strong convergence.

For illustrative purposes, a MATLAB program (Program 15.8) has been devised for using the proximal algorithm in the LASSO problem (already introduced in Sect. 15.5.1). It is based on scripts from Stanford University [65].

The program generates two figures: Fig. 15.34 depicts the evolution of the objective function along iterations.

Figure 15.35 shows the components of the final solution $\vec{x}^*$.

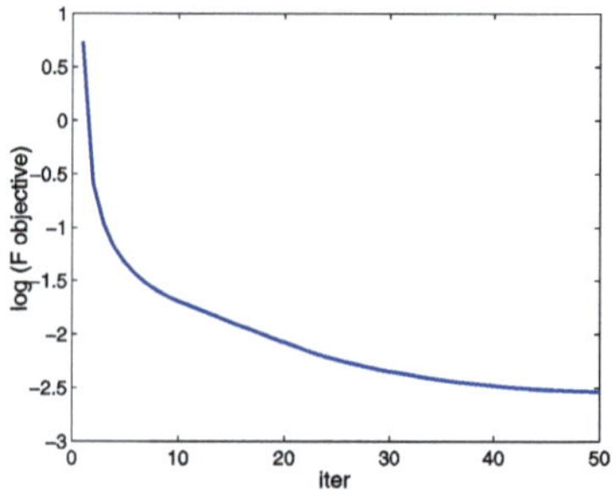

Fig. 15.34 Convergence of proximal iterations

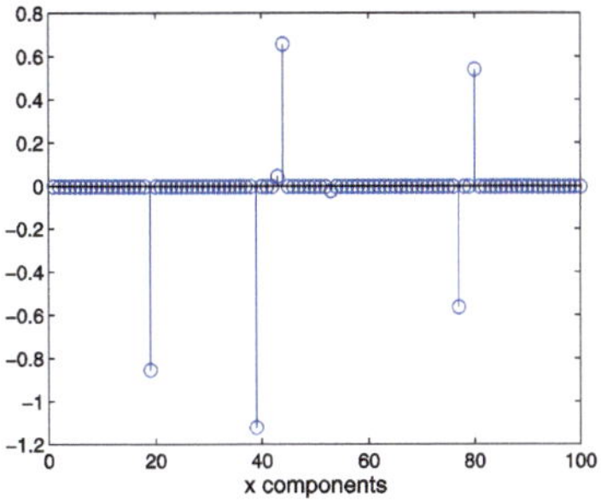

Fig. 15.35 Components of the solution vector

The program uses two functions.

Program 15.8 Example of proximal method

```
% example of proximal method
m = 50; % number of examples
n = 100; % number of features
% generate dATA
x0 = sprandn(n,1,0.05);
A = randn(m,n);
% normalize columns:
A = A*spdiags(1./sqrt(sum(A.^2))',0,n,n);
v = sqrt(0.001)*randn(m,1);
b = A*x0 + v;
gamma_max = norm(A'*b,'inf');
gamma = 0.1*gamma_max;
ATA = A'*A;
ATb = A'*b;
maxiter = 50;
% trayectory of objective function:
Fty=zeros(maxiter,1);
%--------------------
lambda = 1;
beta = 0.5;
x = zeros(n,1);
for k = 1:maxiter
  while 1
    grad_x = ATA*x - ATb;
    %proximal:
    z = prox_l1(x - lambda*grad_x, lambda*gamma);
    qaux=(z-x).^2;
    if pf(A,z,b) <= pf(A,x,b) + grad_x'*(z - x) +...
     (1/(2*lambda))*sum(qaux)
      break;
    end
    lambda = beta*lambda;
  end
  Fty(k,1)=pf(A,x,b);
  x = z;
```

```
end
% display result --------------
figure(1)
plot(log(Fty(:,1)));
xlabel('iter'); ylabel('log (F objective)');
figure(2)
stem(x)
xlabel('x components');
(proximal operator)
function x = prox_l1(v, lambda)
% proximal operator of the l1 norm.
x = max(0, v - lambda) - max(0, -v - lambda);
(objective function)
function [f]=pf(A,x,b)
aux=(A*x-b);
f = 0.5*sum(aux.^2);
```

Part of the literature uses the term proximal fixed-point algorithm, emphasizing that the solution would be a fixed point of the proximal operator. The article [66] remarks that if $\mathrm{Prox}_{\lambda\, f}$ was a contraction (Lipschitz with $\lambda < 1$), the repeated application of the proximal mapping would find a fixed point. However, $\mathrm{Prox}_{\lambda\, f}$ needs not to be a contraction. It turns out that it has a different property which is sufficient for fixed-point iteration. This property is called *firm nonexpansiveness*:

$$\| \mathrm{Prox}_{\lambda\, f}(\vec{x}) - \mathrm{Prox}_{\lambda\, f}(\vec{y}) \|_2^2 \;\leq\; (\vec{x} - \vec{y})^T (\mathrm{Prox}_{\lambda\, f}(\vec{x}) - Prox_{\lambda\, f}(\vec{y}))$$

Now, a short digression seems opportune. Firmly nonexpansive operators are a particular case of nonexpansive operators, which are Lipschitz with $\lambda = 1$. Iterations of a nonexpansive operator need not converge, like, for instance, in the case of rotation operators. Anyhow, given a nonexpansive operator F, the sequence

$$\vec{x}^{k+1} \;=\; (1-\alpha)\,\vec{x}^k + \alpha\, F(\vec{x}^k),$$

with $0 < \alpha < 1$, will converge to one of the fixed points of F.

Operators like T are called α- averaged operators. The iteration $\vec{x}^{k+1} \;=\; T(\vec{x}^k)$ converges to a fixed point of T.

Firmly nonexpansive operators are averaged. In particular, the proximal operator is an averaged operator, and so the proximal point algorithm converges to a fixed point.

By the way, given a set of closed proper convex functions, $f_1,\, f_2, \ldots, f_n$, one has

$$\frac{1}{m} \sum_{i=1}^{n} \mathrm{Prox}_{\lambda\, f_i} \;=\; \mathrm{Prox}_{\lambda\, g}.$$

The function g is called the *proximal average* of the set of functions, not to be confused with the concept of averaged operator.

Let us enter into another aspect. The already cited article [75] connects *monotone operators* and the proximal mapping. In the case of functions, the concept of monotonicity is clear: a monotonically increasing function $f(x)$ would never decrease, or vice-versa. In the case of operators, it is said that a set-valued operator M (mapping x to a set $M(x)$) is *monotone* if

$$(\vec{u} - \vec{v})^T(\vec{x} - \vec{y}) \geq 0\ , \quad \forall \vec{x},\ \vec{y},\ \vec{u} \in M(\vec{x}),\ \ \vec{v} \in M(\vec{y}).$$

The subdifferential of a closed convex function is a monotone operator.

The *resolvent* of a set-valued mapping M is the mapping:

$$R_\lambda = (I + \lambda M)^{-1}$$

If M is monotone, then R_λ is firmly nonexpansive, so R_λ is single-valued and Lipschitz.

The resolvent of ∂f is the proximal mapping:

$$(I + \lambda\, \partial f)^{-1}(\vec{x}) = \text{Prox}_{\lambda f}(\vec{x})$$

15.5.4.5 The Linear Proximal SVM

The linear proximal SVM (LPSVM) was introduced by G. Fung and O.L. Mangasarian in the Knowledge Discovery and Data Mining Conference, San Francisco, year 2001 [26]. It had large resonance, with 1405 cites (year 2024).

Instead of using the SVM standard approach (that would require a solver), a proximal version is chosen, which classifies points depending on proximity to one of two parallel planes.

The SVM optimization problem can be formulated as follows [50],

$$\begin{aligned}&\text{Minimize } \tfrac{1}{2}(\vec{w}^T\, \vec{w} + b^2) + \tfrac{1}{2}\, C\, \left\|\vec{\xi}\right\|^2 \\ &\text{Subject to } D\,(A\,\vec{w} - \vec{e}\, b) + \vec{\xi} \geq \vec{e}\end{aligned}$$

where D is a diagonal matrix with the data labels (+1 or $-$ 1) in the diagonal, C is a positive constant, and $\vec{\xi}$ is introduced for constraint relaxing in view of non-separable data [98].

With a simple reformulation, the problem can be treated as a proximal SVM:

$$\begin{aligned}&\text{Minimize } \tfrac{1}{2}(\vec{w}^T\, \vec{w} + b^2) + \tfrac{1}{2}\, C\, \left\|\vec{\xi}\right\|^2 \\ &\text{Subject to } D\,(A\,\vec{w} - \vec{e}\, b) + \vec{\xi} = \vec{e}\end{aligned}$$

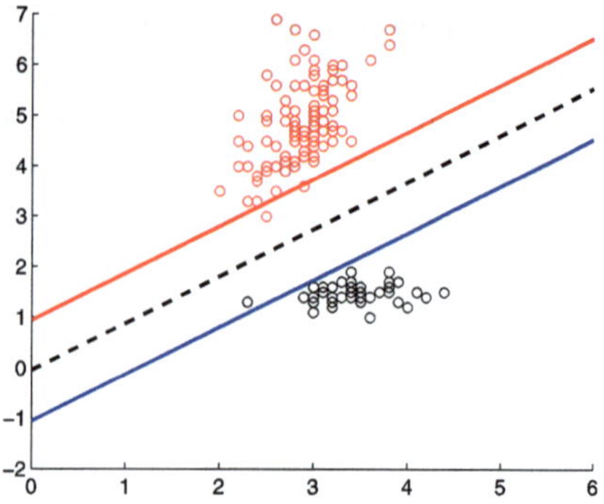

Fig. 15.36 Result of the LPSVM application to IRIS data

The corresponding Lagrangian would be

$$L = \frac{1}{2}(\vec{w}^T\,\vec{w} + b^2) + \frac{1}{2}C\,\left\|\vec{\xi}\right\|^2 - \lambda\,(D\,(A\,\vec{w} - \vec{e}\,b) + \vec{\xi} - \vec{e}.$$

Based on KKT conditions, it was found that

$$\lambda = \left(\frac{I}{C}D\,(A\,A^T + \vec{e}\,\vec{e}^T)\,D\right)^{-1}\,\vec{e} = (W)^{-1}\,\vec{e}$$

where $W = \frac{I}{C} + H\,H^T$ and $H = D\,(A - \vec{e})$.

The original contribution [26] offers a short MATLAB program (six lines of code) that implements in a simple manner the proximal idea. On the basis of this code, a program has been devised (Program 15.9) for testing the LPSVM. Figure 15.36 shows the result (again not perfect but quite satisfactory).

The authors of the algorithm report that it is more than 1000 times faster than the standard SVM procedure.

Program 15.9 Example of linear proximal SVM

```
% example of Linear Proximal SVM
% IRIS data
% x=sepal width, y=petal length
%read IRIS data
DAT=dlmread('iris.data'); %columns
%there are 50 points for class A
% and 100 points for class B
A=[DAT(:,2), DAT(:,3)]; %two columns (x,y)
Y=ones(150,1); Y(1:50)=-Y(1:50); %labels:1 or -1
D=diag(Y);
nu=0.05;
% code from paper
[m,n] = size(A);
e = ones(m,1);
```

```
H = D*[A -e];
r=sum(H)'; %r=H'*e;
r=(speye(n+1)/nu+H'*H)\r; %solve
u=nu*(1-H*r);
s=D*u;
w=(s'*A)'; %w=A'*D*u
gamma=sum(s); %gamma=-e'*D*u
%display -----------------------------------
figure(1)
%the data
scatter(A(1:50,1),A(1:50,2),32,'k');
hold on;
scatter(A(51:150,1),A(51:150,2),32,'r');
% the lines
slop=-w(1)/w(2);
b=gamma;
vx=[0:0.1:6];
vy=(slop*vx + b +1);
% class 1 border line:
plot(vx, vy, 'r', 'LineWidth',2);
vx=[0:0.1:6];
vy=(slop*vx + b);
%optimal separating line:
plot(vx, vy, 'k--', 'LineWidth',2);
vy=(slop*vx +b -1);
% class 2 border line:
plot(vx, vy, 'b', 'LineWidth',2);
```

15.5.4.6 Forward-Backward Algorithm

Now, consider the following case,

$$M(\vec{x}) = A(\vec{x}) + B(\vec{x}),$$

with A and B monotone, $A(\vec{x})$ single-valued, and B with easy-to-compute resolvent. An important algorithm is the following:

$$\vec{x}^{k+1} = (I + \lambda_k B)^{-1} (I - \lambda_k A) (\vec{x}^k)$$

It is called the *forward-backward* algorithm, since it combines a "forward operator" $(I - \lambda_k A)$ and a "backward operator" $(I + \lambda_k B)^{-1}$.
Particular cases of this algorithm are as follows:

- The *proximal gradient method*,

$$\vec{x}^{k+1} = \text{Prox}_{\lambda f}(\vec{x}^k - \lambda_k \nabla g(\vec{x}^k))$$

with $A(\vec{x}) = \nabla g(\vec{x})$ and $B(\vec{x}) = \partial f(\vec{x})$

- The *projection method for variational inequality*:

$$\vec{x}^{k+1} = P_C(\vec{x}^k - \lambda_k F(\vec{x}^k))$$

with $A(\vec{x}) = F(\vec{x})$ and $B(\vec{x}) = N_C$ (the normal cone operator, which is 0 for $\vec{x} \notin C$ and $\{\vec{w} \mid \vec{w}^T(\vec{z} - \vec{x}) \leq 0,\ \forall \vec{z} \in C\}$ for $\vec{x} \in C$).

Let us study next, in more details, the proximal gradient method.

15.5.4.7 Composite Functions and the Proximal Gradient Method

Consider the minimization of

$$f(\vec{x}) = g(\vec{x}) + h(\vec{x})$$

where $g(\vec{x})$ and $h(\vec{x})$ are closed, proper and convex, and $g(\vec{x})$ is differentiable. The function $h(\vec{x})$ can be extended-valued, so $h(\vec{x}) : \Re^n \rightarrow \Re \cup \{+\infty\}$.

The proximal gradient method for the minimization problem is

$$\vec{x}^{k+1} = \text{Prox}_{\lambda\, h}(\vec{x}^k - \lambda_k \nabla g(\vec{x}^k)).$$

When $\nabla g(\vec{x})$ is Lipschitz with constant L, the method converges with rate $O(1/k)$ if $\lambda_k \in (0, 1/L)$. If L was not known, the step size λ_k can be found by line search [66].

The iterative method can be interpreted as a fixed-point iteration. A point $\vec{x}^*$ is a solution of the minimization problem iff

$$0 \in \nabla g(\vec{x}^*) + \partial h(\vec{x}^*).$$

This condition is satisfied if and only

$$\vec{x}^* = (I + \lambda \partial h)^{-1} (I - \lambda \nabla g)(\vec{x}^*)$$

or equivalently

$$\vec{x}^* = \text{Prox}_{\lambda\, h}(\vec{x}^* - \lambda \nabla g(\vec{x}^*)).$$

The point $\vec{x}^*$ would be a fixed point of the forward-backward operator $(I + \lambda\, \partial\, h)^{-1} (I - \lambda \nabla g)$.

Another interpretation is that the method is an example of *majorization-minimization algorithm* [66]. This class of algorithms embodies many important instances, like the gradient descent, the Newton's method, the EM algorithm, etc.

Given a function $g(\vec{x})$, its minimization could be done with the following iterative algorithm,

$$\vec{x}^{k+1} = \arg\min_{\vec{x}} \hat{g}(\vec{x}, \vec{x}^k),$$

where $\hat{g}(\vec{x})$ is a convex upper bound to $g(\vec{x})$ satisfying that $\hat{g}(\vec{x}, \vec{x}^k) \geq g(\vec{x})$ and $\hat{g}(\vec{x}, \vec{x}) = g(\vec{x})$.

This generic algorithm iteratively exerts a majorization with the upper bound and then minimizes the majorization.

An example of upper bound could be the following:

$$\hat{g}(\vec{x}, \vec{y}) = g(\vec{y}) + \nabla g(\vec{y})^T (\vec{x} - \vec{y}) + \frac{1}{2\lambda} \|\vec{x} - \vec{y}\|_2^2$$

Consider now the function

$$q(\vec{x}, \vec{y}) = \hat{g}(\vec{x}, \vec{y}) + h(\vec{x}).$$

This function is a surrogate for $g(\vec{x}) + h(\vec{x})$, with fixed $\vec{y}$. It can be shown that the majorization-minimization iteration

$$\vec{x}^{k+1} = \arg\min_{\vec{x}} q(\vec{x}, \vec{x}^k)$$

is equivalent to the proximal gradient iteration.

The minimization of $q(\vec{x}, \vec{x}^k)$ can be expressed as the minimization of

$$\frac{1}{2} \left\| \vec{x} - (\vec{x}^k - \lambda \nabla g(\vec{x}^k)) \right\|_2^2 + \lambda\, h(\vec{x}).$$

Therefore, the solution tries to balance the gradient step with the weighted contribution of $h(\vec{x})$.

Nesterov [61] (5508 cites) introduced and accelerated proximal gradient method, with the following scheme,

$$\vec{y}^{k+1} = \vec{x}^k + w_k (\vec{x}^k - \vec{x}^{k-1})$$

$$\vec{x}^{k+1} = \text{Prox}_{\lambda h}(\vec{y}^{k+1} - \lambda_k \nabla g(\vec{y}^{k+1}))$$

where $w_k \in [0, 1)$ is an extrapolation parameter. A choice of this parameter could be

$$w_k = \frac{k}{k+3}.$$

When $\nabla g(\vec{x})$ is Lipschitz with constant L, the accelerated method converges with rate $O(1/k^2)$ if $\lambda_k \in (0, 1/L)$. Notice the difference with the standard method, which converges with rate $O(1/k)$. Also like before, if L is not known, the step size λ_k can be found by line search

An extensive treatise on monotone operators and proximal methods can be found in [5].

15.6 Exercises

1- [15.1] Next table corresponds to the bar hopping game:

	2	4	5
2	10,10	14,12	14,15
4	12,14	20,20	28.15
5	15,14	15,28	25,25

Please, simplify the game. Describe the process. Then, change one or two of the numbers in the table, for changing the final winner cell: explain your idea.

2- [15.2] Two companies, GG and UU, competitors in the soda market, are deciding how much to invest in advertising, supposing that profits depend directly on this advertising. Next table describes the situation:

	Spend 5M$	Spend 10M$	Spend15M$
Spend 5M$	45,45	28,57	20,60
Spend 10M$	57,28	40,4	30.45
Spend 15M$	60,20	45,30	35,35

Please simplify the game, and find what rational decisions should happen.

3- [15.3] [15.4] It is typical of machine learning to deal with large datasets. The basic processes should be as fast as possible, since they will be repeated millions of times. This is why two fast versions of SVM were included in Chap. 15.

In order to feel the effect of processing speed, a convenient dataset would be banknote authentication [15.3], with 1372 instances and 4 features. The two first features show that the data can somehow be divided into two classes.

Please, use the two SVM versions of Chap. 15 for this dataset. In addition, use a kernel-SVM like the one in Chap. 7 for the same dataset. Compare processing times and the quality of results.

Indeed, other bigger datasets could be used for this comparison, for example, the HIGGS dataset (see page [15.4]), with 11,000,000 instances and 28 features.

4- [15.5] Subgradient methods are frequently used in network optimization. An interesting example is given in [15.5] for the network represented in the next figure:

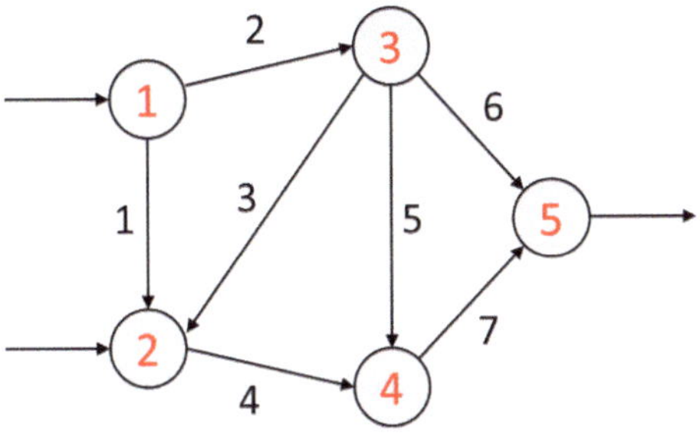

Please, obtain the optimal flow of the network using a subgradient method. Follow the guide provided by [15.5], where the problem is treated in the dual form. Visualize the evolution of iterations

The mentioned reference introduces an intuitive analogy with electrical networks. Some other publications align with the subgradient approach for wireless communication scenarios and situations with queues (like in the case of message handling by digital devices, Wi-Fi networks, etc.).

5- [15.6] Like SVM, the logistic regression is used for classification. The target of the classification is to obtain a function $f(\vec{x}) = \vec{w}^T\vec{x} + b$, to divide the data into two classes (in the 2D case, $f(\vec{x})$ is a straight line). The parameters of the function are obtained, using the labelled data (x_i, y_i), as follows,

$$\min_{\vec{w},b} \sum_{i=1}^{N} \log[1 + \exp(-y_i f(\vec{x}_i)]$$

where $y_i \in \{-1, +1\}$

(a term $\lambda \; \|\vec{w}\|^2$could be added for regularization).

Please, use logistic regression minimized with the subgradient projection method for classification of the UCI ionosphere dataset (already used in a previous exercise) in two classes. Describe how you apply the method. Visualize the classification result. Add comments.

6- Consider the Goldstein-Price function:

$$J = [1 + (x + y + 1)^2(19 - 14x + 3x^2 - 14y + 6xy + 3y^2)] \cdot$$
$$\cdot [30 + (2x - 3y)^2(18 - 32x + 12x^2 + 48y - 36xy + 27y^2)]$$

Please, apply coordinate descent to find the minimum. Visualize four cases, starting from four different points. Add comments.

(Coordinate descent can be also applied for SVM or logistic regression.)

7- [15.7] There are several alternatives for solving LASSO problems, like, for instance, the proximal point algorithm we used in Program 15.8. The target of this exercise is to explore the use of CVX.

Please, apply CVX and MATLAB for the problem considered in Program 15.8. Compare convergence behaviors.

15.6.1 Exercise References

[15.1] http://slantchev.ucsd.edu/courses/gt/04-strategic-form.pdf
[15.2] https://medium.com/game-theory-101/chapter-2-dominance-solvable-games-83990f998a08
[15.3] https://archive.ics.uci.edu/dataset/267/banknote+authentication
[15.4] https://www.csie.ntu.edu.tw/~cjlin/libsvmtools/datasets/
[15.5] https://web.stanford.edu/class/ee392o/subgrad_method.pdf
[15.6] https://www.robots.ox.ac.uk/~az/lectures/ml/2011/lect4.pdf
[15.7] https://brgaines.github.io/presentations/cvx/cvxDemo/cvxDemo.html#15

References

1. Adler, I. (2013). The equivalence of linear programs and zero-sum games. *International Journal of Game Theory, 42*, 165–177. Available from: https://link.springer.com/content/pdf/10.1007/s00182-012-0328-8.pdf.
2. Andersen, E. D., & ApS, M. (2013). How to use Farkas's lemma to say something important about infeasible linear problems. Technical report, MOSEK. Available from: https://docs.mosek.com/whitepapers/infeas.pdf.
3. Aswathy, C., Sowmya, V., & Soman, K. (2015). Admm based hyperspectral image classification improved by denoising using Legendre Fenchel transformation. *Indian Journal of Science and Technology, 8*(24), 1. Available from: https://sciresol.s3.us-east-2.amazonaws.com/IJST/Articles/2015/Issue-24/Article6.pdf.
4. Başar, T., & Olsder, G. J. (1998). *Dynamic noncooperative game theory*. SIAM.
5. Bauschke, H., & Combettes, P. (2011). *Convex analysis and monotone operator theory in Hilbert spaces*, vol. 10. CMS Books in Mathematics. Available from: https://link.springer.com/book/10.1007/978-3-319-48311-5.
6. Bauso, D. (2016). *Game theory with engineering applications*. SIAM.
7. Bertsekas, D. P. (2012). Convex analysis and optimization. Available from: https://ocw.mit.edu/courses/6-253-convex-analysis-and-optimization-spring-2012/resources/mit6_253s12_lec01/.
8. Bhat, S., Kline, J., & Fung, G. (2021). Farkas's theorem of the alternative for prior knowledge in deep networks. In *OPT2021: 13th Annual Workshop on Optimization for Machine Learning*. Available from: https://opt-ml.org/oldopt/papers/2021/paper16.pdf.
9. Bioucas-Dias, J. M., & Figueiredo, M. A. (2007). A new twist: Two-step iterative shrinkage/thresholding algorithms for image restoration. *IEEE Transactions on Image Processing, 16*(12), 2992–3004. Available from: http://www.lx.it.pt/~mtf/IEEE_TIP_2007_Dias_Figueiredo.pdf.
10. Border, K. C. (2010). *Separating hyperplane theorems*. Notes, California Institute of Technology. Available from: https://healy.econ.ohio-state.edu/kcb/Notes/SeparatingHyperplane.pdf.

11. Boyd, S., & Mutapcic, A. (2008a). Stochastic subgradient methods. In *Lecture Notes for EE364b, Stanford University*, 9. Available from: https://see.stanford.edu/materials/lsocoee364b/04-stoch_subgrad_notes.pdf.
12. Boyd, S., & Mutapcic, A. (2008b). Subgradient methods. In *Lecture Notes for EE364b, Stanford University*. Available from: https://see.stanford.edu/materials/lsocoee364b/02-subgrad_method_notes.pdf.
13. Brooks, B., & Reny, P. J. (2023). A canonical game—75 years in the making—showing the equivalence of matrix games and linear programming. *Economic Theory Bulletin, 11*(2), 171–180. Available from: https://link.springer.com/content/pdf/10.1007/s40505-023-00252-8.pdf.
14. Byrne, C. L. (2013). Continuous optimization. Available from: https://faculty.uml.edu/cbyrne/optfirst.pdf.
15. Caramanis, C., & Sanghavi, S. (2012). Lecture 20, large scale optimization. Available from: https://users.ece.utexas.edu/~cmcaram/EE381V_2012F/Lecture_20_Scribe_Notes.final.pdf.
16. Carpuat, M. (2017). (sub)gradient descent, lecture 12. Available from: http://www.cs.umd.edu/class/spring2017/cmsc422/slides0101/lecture12.pdf.
17. Chan, R., Lanza, A., Morigi, S., & Sgallari, F. (2018). Convex non-convex image segmentation. *Numerische Mathematik, 138*, 635–680. Available from: https://link.springer.com/article/10.1007/s00211-017-0916-4.
18. Cheney, E. W., & Goldstein, A. A. (1959). Newton's method for convex programming and tchebycheff approximation. *Numerische Mathematik, 1*(1), 253–268.
19. Clarke, F. H. (1990). *Optimization and nonsmooth analysis*. SIAM. Available from: https://epubs.siam.org/doi/pdf/10.1137/1.9781611971309.fm.
20. Dantzig, G. B. (2002). Linear programming. *Operations Research, INFORMS, 50*(1), 42–47. Available from: https://courses.cs.duke.edu/spring07/cps296.2/papers/LinearProgramming_article.pdf.
21. Frenk, J. B. G., Kassay, G., & Kolumbán, J. (2004). On equivalent results in minimax theory. *European Journal of Operational Research, 157*(1), 46–58. Available from: https://citeseerx.ist.psu.edu/document?repid=rep1&type=pdf&doi=6b765b3c4f18b96809d0c91786bd0cef98e172bb.
22. Friedman, D. (1991). Evolutionary games in economics. *Econometrica: Journal of the Econometric Society*, 637–666. Available from: http://www.dklevine.com/archive/refs4392.pdf.
23. Friedman, J. W. (1986). *Game theory with applications to economics*. Oxford University Press.
24. Fudenberg, D., & Tirole, J. (1991). *Game theory*. MIT Press.
25. Fung, G., Mangasarian, O., & Shavlik, J. (2002). Knowledge-based support vector machine classifiers. *Advances in Neural Information Processing Systems*, 15. Available from: https://proceedings.neurips.cc/paper_files/paper/2002/file/934b535800b1cba8f96a5d72f72f1611-Paper.pdf.
26. Fung, G., & Mangasarian, O. L. (2001). Proximal support vector machine classifiers. In *Proceedings of the Seventh ACM SIGKDD International Conference on Knowledge Discovery and Data Mining*, pp. 77–86. Available from: https://dl.acm.org/doi/pdf/10.1145/502512.502527.
27. Gärtner, B., & He, N. (2022). Optimization for data science. ETH Zürich, FS 261-5110-00l. Available from: https://odi.inf.ethz.ch/teaching/ODS/Handout/handout05.pdf.
28. Ge, R. (2016). Escaping from saddle points. Available from: https://www.offconvex.org/2016/03/22/saddlepoints/.
29. Gibbons, R. S. (1992). *Game theory for applied economists*. Princeton University Press.
30. Goffin, J.-L. (2012). *Subgradient optimization in nonsmooth optimization (including the Soviet revolution)*. Groupe d'études et de recherche en analyse des décisions. Available from: https://www.math.uni-bielefeld.de/documenta/vol-ismp/43_goffin-jean-louis.pdf.

31. Goffin, J.-L., & Vial, J.-P. (2002). Convex nondifferentiable optimization: A survey focused on the analytic center cutting plane method. *Optimization Methods and Software, 17*(5), 805–867. Available from: https://citeseerx.ist.psu.edu/document?repid=rep1&type=pdf&doi=7686928a365297800a9d162173258903a9fa5b42.
32. Güler, O. (1991). On the convergence of the proximal point algorithm for convex minimization. *SIAM Journal on Control and Optimization, 29*(2), 403–419. Available from: https://epubs.siam.org/doi/abs/10.1137/0329022.
33. Guo, X. (2010). Coercive functions and global min. Available from: https://classes.engineering.wustl.edu/2010/spring/ese415/ch1-4.pdf.
34. Handa, A., Newcombe, R. A., Angeli, A., & Davison, A. (2011). Applications of Legendre-Fenchel transformation to computer vision problems. Available from: https://www.doc.ic.ac.uk/~ahanda/lfreport.pdf.
35. Haywood Jr, O. G. (1954). Military decision and game theory. *Journal of the Operations Research Society of America, 2*(4), 365–385. Available from: https://pubsonline.informs.org/doi/abs/10.1287/opre.2.4.365.
36. Hein, J.-P. M. (2010). Convex optimization and modeling, solution of exercise sheet 11. Available from: https://www.ml.uni-saarland.de/Lectures/CVX-SS10/CvxSolution11.pdf.
37. Henriquez-Auba, R. (2019). Stat155—game theory lecture notes. Available from: https://henriquezauba.com/assets/files/Teaching/summaries/STAT155_GameTheory_LectureNotes.pdf.
38. Hiriart-Urruty, J.-B., & Le, H. Y. (2013). From Eckart and Young approximation to Moreau envelopes and vice versa. *RAIRO-Operations Research-Recherche Opérationnelle, 47*(3), 299–310. Available from: http://www.numdam.org/article/RO_2013__47_3_299_0.pdf.
39. Hu, T.-W. (2011). Game theory: Lecture 1. Available from: https://homepage.ntu.edu.tw/~josephw/TWHgametheory-lecture1.pdf.
40. Jaćimović, M. (2011). Farkas' lemma of alternative. *The Teaching of Mathematics, 25*(27), 77–86. Available from: http://elib.mi.sanu.ac.rs/files/journals/tm/27/tm1422.pdf.
41. Jadon, A., Patil, A., & Jadon, S. (2022). A comprehensive survey of regression based loss functions for time series forecasting. *arXiv preprint arXiv:2211.02989*. Available from: https://arxiv.org/pdf/2211.02989.pdf.
42. Kelley, Jr, J. E. (1960). The cutting-plane method for solving convex programs. *Journal of the society for Industrial and Applied Mathematics, 8*(4), 703–712. Available from: https://www.jstor.org/stable/pdf/2099058.pdf.
43. Kjeldsen, T. H. (2001). John von Neumann's conception of the minimax theorem: A journey through different mathematical contexts. *Archive for History of Exact Sciences, 56*(1), 39–68. Available from: https://web.archive.org/web/20170809192303id_/http://web.math.ucsb.edu/~crandall/math201b/vnminimax.pdf.
44. Kolda, T. G., Lewis, R. M., & Torczon, V. (2003). Optimization by direct search: New perspectives on some classical and modern methods. *SIAM Review, 45*(3), 385–482. Available from: https://www.ii.uib.no/~lennart/drgrad/Kolda2003.pdf.
45. Krasotkina, O., & Mottl, V. (2015). A bayesian approach to sparse learning-to-rank for search engine optimization. In *Machine Learning and Data Mining in Pattern Recognition: 11th International Conference, MLDM 2015, Hamburg, Germany, July 20-21, 2015, Proceedings 11*, pp. 382–394. Springer. Available from: https://link.springer.com/chapter/10.1007/978-3-319-21024-7_26.
46. Kunapuli, G., Maclin, R., & Shavlik, J. (2011). Advice refinement in knowledge-based svms. *Advances in Neural Information Processing Systems*, 24. Available from: https://proceedings.neurips.cc/paper_files/paper/2011/file/d79aac075930c83c2f1e369a511148fe-Paper.pdf.
47. Lanctot, M., Zambaldi, V., Gruslys, A., Lazaridou, A., Tuyls, K., Pérolat, J., Silver, D., & Graepel, T. (2017). A unified game-theoretic approach to multiagent reinforcement learning. *Advances in Neural Information Processing Systems*, 30. Available from: https://core.ac.uk/download/pdf/146488735.pdf.

48. Laporte, L., Flamary, R., Canu, S., Déjean, S., & Mothe, J. (2013). Nonconvex regularizations for feature selection in ranking with sparse svm. *IEEE Transactions on Neural Networks and Learning Systems, 25*(6), 1118–1130. Available from: https://arxiv.org/pdf/1507.00500.pdf.
49. Lucet, Y. (2010). What shape is your conjugate? a survey of computational convex analysis and its applications. *SIAM Review, 52*(3), 505–542. Available from: https://citeseerx.ist.psu.edu/document?repid=rep1&type=pdf&doi=94ba39925cc84c0b3f320e7ce00419c29b52411e.
50. Mangasarian, O. L., & Musicant, D. R. (2001). Lagrangian support vector machines. *Journal of Machine Learning Research, 1*(Mar), 161–177. Available from: https://www.jmlr.org/papers/volume1/mangasarian01a/mangasarian01a.pdf.
51. McLinden, L. (1974). An extension of Fenchel's duality theorem to saddle functions and dual minimax problems. *Pacific Journal of Mathematics, 50*(1), 135–158. Available from: https://msp.org/pjm/1974/50-1/pjm-v50-n1-p15-s.pdf.
52. Meinhardt, H. (2013). The matlab game theory toolbox mattugames version 0.4: an introduction, basics, and examples. Available from: https://www.researchgate.net/profile/Holger-Meinhardt/publication/295513836_The_Matlab_Game_Theory_Toolbox_MatTuGames_Version_04_An_Introduction_Basics_and_Examples/links/56cadb4208ae5488f0da48ce/The-Matlab-Game-Theory-Toolbox-MatTuGames-Version-04-An-Introduction-Basics-and-Examples.pdf.
53. Mohebbi Moghaddam, M., Boroomand, B., Jalali, M., Zareian, A., Daeijavad, A., Manshaei, M. H., & Krunz, M. (2023). Games of GANs: Game-theoretical models for generative adversarial networks. *Artificial Intelligence Review*, 1–37. Available from: https://link.springer.com/article/10.1007/s10462-023-10395-6.
54. Mordukhovich, B. S. (2006a). *Variational analysis and generalized differentiation I: Basic theory*, vol. 330. Springer.
55. Mordukhovich, B. S. (2006b). *Variational analysis and generalized differentiation II: Applications*, vol. 331. Springer.
56. Moreau, J. J. (1962). Fonctions convexes duales et points proximaux dans un espace hilbertien. *Comptes Rendus Hebdomadaires des Séances de l'Académie des Sciences, 255*, 2897–2899.
57. Myerson, R. B. (2013). *Game theory*. Harvard University Press.
58. Nachum, O., & Dai, B. (2020). Reinforcement learning via fenchel-rockafellar duality. *arXiv preprint arXiv:2001.01866*. Available from: https://arxiv.org/pdf/2001.01866.pdf.
59. Najjar, N. (2009). Game theory portfolio, a computational exploration of the Nash equilibrium. Available from: https://www.uvic.ca/socialsciences/economics/assets/docs/student/GameTheoryNajjar.pdf.
60. Nesterov, Y. (2012). Efficiency of coordinate descent methods on huge-scale optimization problems. *SIAM Journal on Optimization, 22*(2), 341–362. Available from: https://dial.uclouvain.be/pr/boreal/object/boreal%3A29221/datastream/PDF_01/view.
61. Nesterov, Y. E. (1983). A method of solving a convex programming problem with convergence rate $\mathcal{O}\frac{1}{(k^2)}$. *Doklady Akademii Nauk, 269*(3), 543–547. Available from: https://www.researchgate.net/publication/257291640_A_method_of_solving_a_convex_programming_problem_with_convergence_rate_O1k2.
62. Nikolova, M. (2011). Energy minimization methods. Available from: https://hal.science/file/index/docid/841479/filename/MinEnergy-chapter.pdf.
63. Osborne, M. J., & Rubinstein, A. (1994). *A course in game theory*. MIT Press.
64. O'Searcoid, M. (2006). *Metric spaces*. Springer Science & Business Media. Available from: https://link.springer.com/book/10.1007/978-1-84628-627-8.
65. Parikh, N., & Boyd, S. (2013). Matlab scripts for proximal methods. Available from: https://web.stanford.edu/~boyd/papers/prox_algs/index.html.
66. Parikh, N., Boyd, S., et al. (2014). Proximal algorithms. *Foundations and Trends® in Optimization, 1*(3), 127–239. Available from: https://web.stanford.edu/~boyd/papers/pdf/prox_algs.pdf.

67. Pilanci, M. (2023). EE364b, subgradient methods for constrained problems. Available from: https://web.stanford.edu/class/ee364b/lectures/constr_subgrad_slides.pdf.
68. Planiden, C. D. (2018). *Theory and algorithmic applications of the proximal mapping and Moreau envelope*. PhD thesis, University of British Columbia. Available from: https://open.library.ubc.ca/media/stream/pdf/24/1.0363957/4.
69. Pooladian, A.-A., Finlay, C., & Oberman, A. M. (2019). Farkas layers: don't shift the data, fix the geometry. *arXiv preprint arXiv:1910.02840*. Available from: https://arxiv.org/pdf/1910.02840.pdf.
70. Rabin, M. (1993). Incorporating fairness into game theory and economics. *The American Economic Review*, 1281–1302.
71. Rasmusen, E. (1989). *Games and information*, vol. 13. Basil Blackwell Oxford.
72. Richtárik, P., & Takáč, M. (2014). Iteration complexity of randomized block-coordinate descent methods for minimizing a composite function. *Mathematical Programming, 144*(1-2), 1–38. Available from: https://link.springer.com/content/pdf/10.1007/s10107-012-0614-z.pdf.
73. Robinson, S. M. (1999). Linear convergence of epsilon-subgradient descent methods for a class of convex functions. *Mathematical Programming, 86*, 41–50. Available from: https://link.springer.com/content/pdf/10.1007/s101070050078.pdf.
74. Rockafellar, R. T. (1974). *Conjugate duality and optimization*. SIAM. Available from: https://epubs.siam.org/doi/pdf/10.1137/1.9781611970524.fm.
75. Rockafellar, R. T. (1976). Monotone operators and the proximal point algorithm. *SIAM Journal on Control and Optimization, 14*(5), 877–898. Available from: https://citeseerx.ist.psu.edu/document?repid=rep1&type=pdf&doi=240c2cb549d0ad3ca8e6d5d17ca61e95831bbe6d.
76. Rockafellar, R. T. (2015). *Convex analysis:(PMS-28)*. Princeton University Press. Available from: https://www.degruyter.com/document/doi/10.1515/9781400873173/html.
77. Rockafellar, R. T., & Wets, R. J.-B. (2009). *Variational analysis*, vol. 317. Springer Science & Business Media. Available from: https://sites.math.washington.edu/~rtr/papers/rtr169-VarAnalysis-RockWets.pdf.
78. Rudnev, M. (2011). Farkas alternative and duality theorem. Available from: https://people.maths.bris.ac.uk/~maxmr/opt/farkas.pdf.
79. Ryu, E. K., & Boyd, S. (2016). Primer on monotone operator methods. *Applied and Computational Mathematics, 15*(1), 3–43. Available from: https://web.stanford.edu/~boyd/papers/pdf/monotone_primer.pdf.
80. Säljö, R. (2004). *Implementation of a bundle algorithm for convex optimization*. Chalmers tekniska högsk. Available from: https://www.math.chalmers.se/~mipat/LATEX/Reine_exjobb.pdf.
81. Salzo, S., & Suykens, J. A. (2020). Generalized support vector regression: duality and tensor-kernel representation. *Analysis and Applications, 18*(01), 149–183. Available from: https://arxiv.org/pdf/1603.05876.pdf.
82. Samuelson, L. (2016). Game theory in economics and beyond. *Journal of Economic Perspectives, 30*(4), 107–130. Available from: https://pubs.aeaweb.org/doi/pdf/10.1257/jep.30.4.107.
83. Sandholm, W. H. (2014). Lecture notes on game theory. Available from: https://citeseerx.ist.psu.edu/document?repid=rep1&type=pdf&doi=9d594f6da38fbf1e888bf025c7073d8f49b42a7a.
84. Shor, N. Z. (2012). *Minimization methods for non-differentiable functions*, vol. 3. Springer Science & Business Media. Available from: https://link.springer.com/book/10.1007/978-3-642-82118-9.
85. Slantchev, B. L. (2008). Game theory: Dominance, Nash equilibrium, symmetry. *Department of Political Science, University of California–San Diego*. Available from: https://pages.ucsd.edu/~bslantchev/courses/gt/04-strategic-form.pdf.
86. Smola, A., & Wishwanathan, S. V. N. (2008). *Introduction to machine learning*. Smola. Available from: https://alex.smola.org/drafts/thebook.pdf.

87. Solodov, M. V., & Svaiter, B. F. (2000). Forcing strong convergence of proximal point iterations in a hilbert space. *Mathematical Programming, 87*, 189–202. Available from: https://link.springer.com/content/pdf/10.1007/s101079900113.pdf.
88. Staff, G&G. (2024). Game theory (normal form game): Set 4 (dominance property-pure strategy. Available from: https://www.geeksforgeeks.org/game-theory-normal-form-game-set-4-dominance-property-pure-strategy/.
89. Touchette, H. (2005). Legendre-Fenchel transforms in a nutshell. Available from: http://www.maths.qmul.ac.uk/~ht/archive/lfth2.pdf.
90. Tseng, P., & Yun, S. (2009). A coordinate gradient descent method for nonsmooth separable minimization. *Mathematical Programming, 117*, 387–423. Available from: https://link.springer.com/content/pdf/10.1007/s10107-007-0170-0.pdf.
91. Vandenberghe, L. (2022). Lecture 3, ECE236C subgradient method. Available from: http://www.seas.ucla.edu/~vandenbe/236C/lectures/sgmethod.pdf.
92. Vlastelica, M., Kolev, P., Cheng, J., & Martius, G. (2023). Diverse offline imitation via Fenchel duality. *arXiv preprint arXiv:2307.11373*. Available from: https://arxiv.org/pdf/2307.11373.pdf.
93. Von Neumann, J., & Morgenstern, O. (2007). *Theory of games and economic behavior (60th Anniversary Commemorative Edition)*. Princeton University Press.
94. Wang, Q., Ma, Y., Zhao, K., & Tian, Y. (2022). A comprehensive survey of loss functions in machine learning. *Annals of Data Science*, 1–26. Available from: https://faculty.ist.psu.edu/vhonavar/Courses/ds310/lossfunc.pdf.
95. Wang, Y.-X., & Liu, Y. (2019). Lecture 5, CS292A convex optimization: Gradient methods and online learning. Available from: https://sites.cs.ucsb.edu/~yuxiangw/classes/CS292A-2019Spring/Lectures/Lecture5_scribe.pdf.
96. Weibull, J. W. (1997). *Evolutionary game theory*. MIT Press.
97. Weng, L., & Chen, Y. (2012). Stochastic subgradient method. *Bren School of Information and Computer Science, University of California, Irvine*. Available from: https://ics.uci.edu/~xhx/courses/ConvexOpt/projects/stochastic_subgradient_methods_report.pdf.
98. Xu, X., Ye, N., & Ye, Q. (2008). Nonlinear proximal support vector machine classifiers aiming at large scale classification problems. In *11th Joint International Conference on Information Sciences*, pp. 627–633. Atlantis Press. Available from: https://www.atlantis-press.com/proceedings/jcis2008/1771.
99. Wang, Y. (2018). LASSO. Available from: https://github.com/YiifeiWang/LASSO.
100. Zhang, D., Tian, Y., & Shi, Y. (2011). A group of knowledge-incorporated multiple criteria linear programming classifiers. *Journal of Computational and Applied Mathematics, 235*(13), 3705–3717. Available from: https://www.sciencedirect.com/science/article/pii/S0377042711000252.

Chapter 16
Applications in Image and Data Processing

16.1 Introduction

Many thinkers and observers consider that we are in the era of digitalization. In particular, image and video digital processing and transmission are now of particular relevance in terms of social and economic impact.

This is an immense field where optimization finds many applications, perhaps not consciously perceived by users, but very real.

In this chapter, we just suggest with examples some selected lines of application. It is not pretended to cover the full area, but most of the main ideas would be introduced.

In view of the large size of digital data—in particular, if related with images—optimization algorithms were required to be fast and capable. This is why a new wave of simple approaches is being developed. In part, this chapter is justified by the intent of introducing some of these algorithms.

The programs included in this chapter make use of the MATLAB Image Processing Toolbox when dealing with pictures.

For extensive treatment of signal and image digital processing theory, our book (three volumes) on this topic could be recommended [35].

16.2 Image Processing on Computers: Basics

Digital cameras and other image capture devices give the results as arrays of values. The elements of these arrays, when visually represented, are *pixels*. MATLAB manages those arrays as matrices .

There is a set of MATLAB functions for image processing. When MATLAB displays an image A (which corresponds to a matrix A), the first element of matrix

J. M. Giron-Sierra, *Introduction to Optimization with Matlab® Examples*,
https://doi.org/10.1007/978-3-032-05048-9_16

Fig. 16.1 Visualizing an image from a file

A—which is a_{11}—goes to the top-left corner of the image. In other words, with images, the origin of coordinates is that top-left corner.

Frequently, the studies on image processing use gray scale, so each pixel corresponds to a value between 0 and 255 (0 is black, 255 is white, and the intermediate values are gray color, more or less dark). Color images usually require three matrices for each image, for instance, when using R-G-B format, one matrix corresponds to red, another for green, and the third for blue.

Program 16.1 shows in few MATLAB lines how to read an image from a file and to visualize it (Fig. 16.1).

Program 16.1 Displaying grayscale pictures

```
% Display grayscale picture
%read the image file into a matrix:
cpic=imread('APnewyork.jpg');
figure(1)
imshow(cpic); %display the photo
```

One of the data types that MATLAB can handle is *uint8* with integer values between 0 and 255. The entries of the arrays used to represent images and obey to this data type.

High-pass, low-pass, and other classes of filters can be applied to images.

Figure 16.2 shows the effects of passing the image through a low-pass filter. Borders and details were lost. Color masses (uniform areas) were emphasized. The

Fig. 16.2 The picture after low-pass filtering

Fig. 16.3 The picture after high-pass filtering

general impression is of a blurred image. The filtering is based on a pixel-by-pixel strategy, which averages values of a number of neighbors for each pixel.

The low-pass filtering was achieved by the following program (program 16.2) in MATLAB. The important function is *filter2( )*.

Program 16.2 Low-pass filtering of a grayscale picture

```
% Display filtered grayscale picture
% average filter

%read the image file into a matrix:
cpic=imread('APnewyork.jpg');
fil=fspecial('average',[10,10]); %filter molecule
fc=filter2(fil,cpic);
bfc=uint8(round(fc)); %convert to unsigned 8-bit
figure(1)
imshow(bfc);
```

Figure 16.3 shows the effects of a high-pass filter (also based on neighbors of each pixel). The high frequencies are on the edges, borders, discontinuities, etc.; those are the details that were emphasized.

The high-pass filtering was achieved by Program 16.3, very similar to the previous program.

Program 16.3 High-pass filtering of a grayscale picture

```
% Laplacian filter

%read the image file into a matrix:
cpic=imread('APnewyork.jpg');
fil=fspecial('laplacian',0.3); %filter molecule
fc=filter2(fil,cpic);

%convert to unsigned 8-bit:
bfc=uint8(round(abs(fc)));
abfc=imadjust(bfc,[0.1 0.7],[]); %image adjust
figure(1)
imshow(abfc);
```

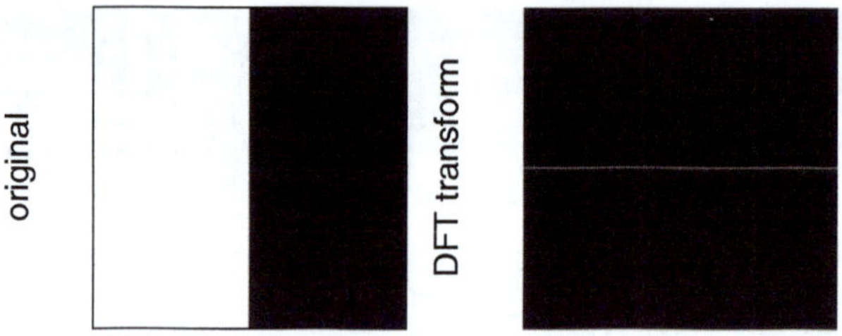

Fig. 16.4 2D DFT of an edge

16.2.1 Bidimensional Fourier Transform

The 2D DFT (discrete Fourier transform) of a matrix is another matrix with the same size. The 2D DFT can be applied to matrices corresponding to images. The Fourier transform takes you to the frequency domain. It is customary to denote the 2D DFT of a matrix A, as $A(\omega)$.

We have seen that image edges correspond to high frequencies. It is opportune to confirm it by using the 2D DFT over images with evident edges.

Figure 16.4 shows on the left the image selected for testing the 2D DFT (the image has a distinctive edge). The result is shown on the right, where the horizontal white line corresponds to the edge. It is well known that the 2D DFT causes a 90° rotation.

Figure 16.4 has been generated by Program 16.4, which takes good benefit of the *fft2( )* function of MATLAB, which is a fast and quite powerful tool.

Program 16.4 Discrete 2D Fourier transform of an edge

```
% DFT of a simple edge
fedg=[ones(256,128) zeros(256,128)]; %simple edge
Ffg=fftshift(fft2(fedg)); %Fourier transform
M=max(max(Ffg)); % the one maximum value
sFfg=(256*Ffg/M); %normalization
figure(1)
subplot(1,2,1)
%plots the binary image with a simple edge:
imshow(fedg);
ylabel('original','FontSize',14);
subplot(1,2,2)
imshow(abs(sFfg)); %plots the Fourier transform
ylabel('DFT transform','FontSize',14);
```

Original images can be recovered by using the inverse 2D DFT . This is demonstrated in Fig. 16.5, which shows an original image, its 2D DFT transform, and the recovery of the image by inverse transform (using the function *ifft2( )*).

Figure 16.5 has been generated with Program 16.5.

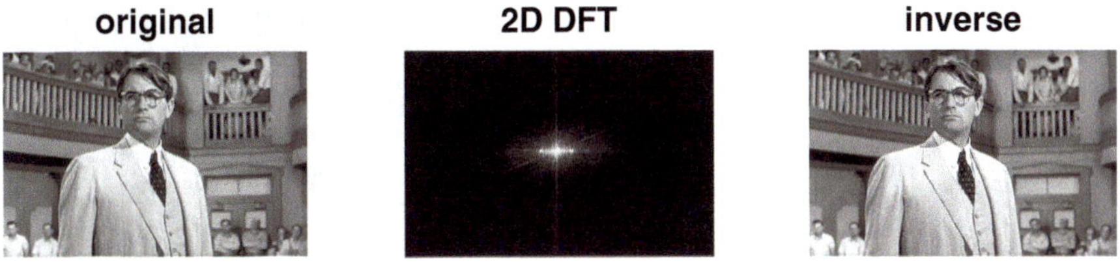

Fig. 16.5 2D DFT of a picture and recovery by inverse 2D DFT

Program 16.5 2D DFT of a picture and recovery with inverse 2D DFT

```
% 2D DFT of a picture, and recovery
% with inverse 2D DFT

%read the image file into a matrix:
cpic=imread('kukoo.jpg');
Fcpic=fftshift(fft2(cpic)); %Fourier transform
M=max(max(Fcpic)); % the one maximum value
sFcpic=(256*Fcpic/M); %normalization
IFc=ifft2(Fcpic); %inverse transform
uIFc=uint8(abs(IFc)); %convert to unsigned 8-bit
figure(1)
subplot(1,3,1)
imshow(cpic); %display the photo
title('original');
subplot(1,3,2)
imshow(abs(sFcpic)); %Fourier transform of the photo
title('2D DFT');
subplot(1,3,3)
imshow(uIFc); %inverse of the Fourier transform
title('inverse');
```

Filtering of images can easily be done in the frequency domain. Given an image A and a filter D, the result of filtering would be

$$Z(\omega) = D(\omega) \cdot A(\omega).$$

The filtered image can be recovered by inverse 2D DFT of $Z(\omega)$.

In the original image domain, the same filtering can be done by convolution of filter D and image A. Convolution in the original domain is equivalent to the product in the frequency domain.

Low frequencies go to the center of the 2D DFT matrix. A simple low-pass filter $D(\omega)$ could be just a white disk surrounded by black pixels, like the one represented in Fig. 16.6 (this image is used as a mask to let pass the low frequencies).

Fig. 16.6 Example of a low-pass filter $D(\omega)$

16.2.2 Bayesian Estimation and Images

When you see an unfocused photograph, you immediately detect that something is wrong (unless the loss of focus was on purpose). Something like "a priori" expectation tells you there was a problem. This evidently sounds to Bayes and priors.

As a matter of fact, the first images sent by the Hubble telescope were unfocused. It was necessary to refocus the photographs by image processing on computer. Later on, the telescope was repaired in a space mission.

An important application of image processing is to deal with degraded images. A simple model of image degradation would be:

$$Z(\omega) = D(\omega) \cdot A(\omega) + V(\omega)$$

where we find, again, a filter $D(\omega)$ and noise $V(\omega)$.

The *optimization objective* is to recover, as best as possible, the original image. In general, this would be a case belonging to the broad family of *inverse problems*.

In Bayesian terms, one could write

$$P(A|Z) = \frac{P(Z|A)\, P(A)}{P(Z)}$$

where $P(A)$ is the *prior* and $P(Z|A)$ corresponds to image degradation. Usually, $P(Z)$ is considered a normalization factor.

The term $P(Z|A)$ is the **likelihood** of A given Z.

The term $P(A|Z)$ is the posterior probability of A given Z. Therefore,

$$posterior \propto prior \times likelihood.$$

In terms of probability density functions (PDFs),

$$f_X(A|Z) \propto f_X(A) \times L(A|Z),$$

where $L(A|Z)$ is the likelihood of A given Z.

The original image A can be estimated as the maximum of $P(A|Z)$: this is the maximum a posteriori (MAP) estimation.

If the prior is flat, then the MAP estimation is the same as the maximum likelihood (ML) estimation.

Note that if the prior is not flat, it may take the MAP away from the ML.

16.2.3 Bayesian Estimation and Unidimensional Signals

To give a more specific idea of Bayesian methodology, it is convenient to consider unidimensional (scalar) signals, like, for example, measurements of temperature, or pressure, taken from sensors.

Our model would be

$$y = x + v.$$

Suppose that x and v are zero-mean Gaussian random variables. Specifically,

$$f_x(x) = \frac{1}{\sigma_x \sqrt{2\pi}} \exp(-x^2/2\sigma_x^2)$$

$$f_v(v) = \frac{1}{\sigma_v \sqrt{2\pi}} \exp(-v^2/2\sigma_v^2).$$

Suppose one observation of y is obtained. In Bayesian terms, one has the following:

- Prior:
  $f_x(x) = \frac{1}{\sigma_x \sqrt{2\pi}} \exp(-x^2/2\sigma_x^2)$
- Likelihood:
  $L(x|y) = \frac{1}{\sigma_v \sqrt{2\pi}} \exp(-y - x)^2/2\sigma_v^2)$
- Posterior:
  $f_x(x|y) \propto \exp[(-x^2/2\sigma_x^2) - (-y - x)^2/2\sigma_v^2)]$

  Writing the posterior as

$$f_x(x|y) \propto \exp(-(x - \mu)^2/2\sigma^2)$$

and by developing and comparing the exponents of the posteriors, it can be deduced that

$$\sigma = \frac{\sigma_x^2 \sigma_v^2}{\sigma_x^2 + \sigma_v^2}$$

$$\mu = \frac{\sigma_x^2}{\sigma_x^2 + \sigma_v^2} \cdot y$$

(the value of μ is the Bayesian estimate of x).

The expression of the variance can be written as follows:

$$\sigma = \frac{1}{\frac{1}{\sigma_x^2} + \frac{1}{\sigma_v^2}}$$

Notice that this is similar to the resistance of two resistors in parallel. Clearly,

$$\sigma < \min(\sigma_x, \sigma_v)$$

(the MAP estimate is more trustful than the priors).

In case the mean of x is nonzero, the corresponding Bayesian estimate is

$$\mu = \frac{\sigma_v^2 \mu_x + \sigma_x^2 y}{\sigma_x^2 + \sigma_v^2} = [\frac{\mu_x}{\sigma_x^2} + \frac{y}{\sigma_v^2}]\ \sigma.$$

This expression can be interpreted as a weighted compromise between the expected prior mean and the observed value y.

To put a daily life example, consider the car driver assistance given by a TomTom. This device uses GPS to take measurements and a mathematical model (a prior) of your motion. When you are in a tunnel, with no GPS, the device can still use the model. Outside the tunnel, the model is also useful, since it helps to counteract GPS fluctuations.

16.3 The Lucy-Richardson Algorithm for Image Deblurring

Several factors can complicate the capture of precise images, with the result of blurred pictures. A number of image processing methods have been created to remedy, to a certain extent, this problem. One important example of those methods is the Lucy-Richardson algorithm (RLA) .

The RLA was independently proposed by Richardson (1972) and Lucy (1974) for astronomy image applications (see [29] and references therein). It was derived using Bayesian arguments.

Later on, the algorithm was derived by other investigators for medical applications (emission tomography). Nowadays, the algorithm and related variants are used for microscopy, cameras, and medical devices.

People involved in optical systems frequently use *point-spread functions* (PSF). In the case of a telescope, the PSF describes the two-dimensional distribution of light in the telescope focal plane, the light coming from astronomical point sources.

The PSF can model many image degradation phenomena. Some of these models are now briefly introduced:

(a) Uniform Out-of-Focus Blur

The image of any point source is a circle, denoted as *circle of confusion* (COC), with a certain radius R. A simple PSF model would be

$$d(x, y) = \begin{cases} \frac{1}{\pi R^2}, \ if \ \sqrt{x^2 + y^2} < R \\ 0, \ elsewhere \end{cases}$$

where x and y are Euclidean coordinates.

(b) Atmospheric Turbulence Blur

Several models have been proposed for atmospheric blur. A frequently used PSF model is Gaussian

$$d(x, y) = C \exp(-\frac{x^2 + y^2}{2\sigma^2})$$

where σ determines the spread of the blur.

(c) Linear Motion Blur

The length of motion L is velocity multiplied by exposure time, the velocity as seen by the camera. The PSF is

$$d(x, y) = \begin{cases} \frac{1}{L}, \ if \ \sqrt{x^2 + y^2} \leq L/2 \\ 0, \ elsewhere \\ . \end{cases}$$

The variables x and y are related, $y = x \cdot \tan \alpha$, where α is the angle of the linear motion.

The RLA Algorithm

RLA considers photon counting in the CCD sensor pixels, so the noise is (mostly) of Poisson-type. Astronomical objects are supposed equi-probable, so the prior is flat (this is called an uninformative prior). Therefore, the MAP estimation is a maximum likelihood (ML) estimation.

Supposing a two-dimensional array of pixels, with indexes $iandj$, the image model would be

$$y(i, j) = d(i, j) * x(i, j) + n(i, j)$$

where * denotes convolution, $d(i, j)$ is the PSF, $n(i, j)$ is noise, $x(i, j)$ is the original image, and $y(i, j)$ is the observed image.

The Poisson likelihood would be

$$P(y|x) = \prod_{i,j} (\frac{1}{y(i, j)!} ((d * x)(i, j))^{y(i,j)} \exp(-(d * x)(i, j)).$$

Taking logarithms,

$$\ln P(y|x) = const. - \sum_{i,j} ((d * x)(i, j)) + \sum_{i,j} y(i, j) \ln((d * x)(i, j)) .$$

By taking the negative logarithm of the likelihood, one obtains the Kullback-Leibler divergence:

$$D_{KL}(x) = -\ln P(x|y).$$

The ML estimate is given by the following optimization:

$$\hat{x}_{ML} = \arg\min (D_{KL}(x)).$$

Let us obtain this argument:

$$\frac{\partial \ln (P(y|x)(i, j))}{\partial x(i, j)} = 0$$

From the Karush-Kuhn-Tucker (KKT) conditions, which are necessary and sufficient for the previous equation, one deduces that

$$\left[\frac{y}{d * x} * d^A\right](i, j) = 1$$

where d^A is the adjoint: a spatially reversed version of d that is *d (-i,-j).*

This equation can be solved with the following iteration:

$$x^{(n+1)}(i, j) = \left[\frac{y}{d * x^{(n)}} * d^A\right](i, j) \cdot x^{(n)}(i, j)$$

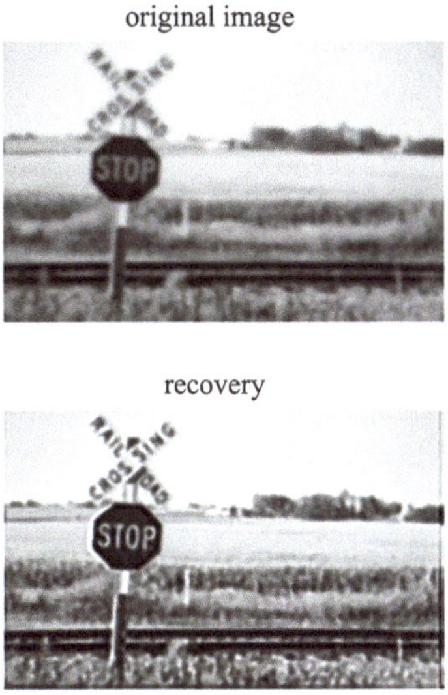

Fig. 16.7 Example of image deblurring with RLA

This is the Lucy-Richardson algorithm. Using text, it can be written as follows:

$$image^{(n+1)} = image^{(n)} \cdot \left[\frac{original\, image}{image^{(n)} * PSF} * PSF^{A} \right] = image^{(n)} \cdot CF$$

Usually, the convolutions are computed using Fourier transforms. The factor *CF* can be seen as a *correction factor*.

Let us consider an application example. Suppose one has captured the blurred image displayed on top of Fig. 16.7.

The RLA algorithm has been implemented by Program 16.6 in MATLAB. It downloads the blurred image, process this information, and returns an improved image shown at the bottom of Fig. 16.7.

The result of RLA is not perfect but is considerably better than the original blurred image.

More details of optical aspects can be found in my book on digital signal processing [35].

Here is the listing of the program with LRA for image files.

Program 16.6 Image deblurring with Lucy-Richardson algorithm

```
% Lucy-Richardson example
%load the image
ip=imread('APtrenBlur1.jpg');
op=im2double(ip); %convert to float
%create the PSF
[oL,oC]=size(op);
```

```
L=oL-1; C=oC-1; %number-1 of rows and columns
sigma=1.2;
[x,y]=meshgrid(-C/2:C/2, -L/2:L/2);
argm=-((x.^2)+(y.^2))/(2*sigma*sigma);
d=exp(argm);
md=sum(d(:));
d=d/md; %the normalized PSF
%the degradation OTF
D=fft2(d);
%preparation for LRA
mp=medfilt2(op); %median filtering
ep=mp; %initial estimated image
%the LRA iterations
for nn=1:5,
  %denominator (convolution using Fourier)
  EP=fft2(ep);
  BEP=D.*EP;
  bep=abs(ifftshift(ifft2(BEP)));
  %num/den
  r=ep./(0.0001+bep); %simple regularization
  %correction vector (convolution using Fourier)
  R=fft2(r);
  CV=D.*R;
  cv=abs(ifftshift(ifft2(CV)));
  %next estimate
  ep=cv.*ep;
end;
%display----------------
figure(1)
subplot(2,1,1)
imshow(op); title('original image')
subplot(2,1,2)
imshow(ep); title('recovery');
```

16.4 Image Decompositions

For detailed study of image contents, it is convenient to decompose images into parts, for example, in medical imaging, it is useful to delineate the perimeter of tumors, vessels, organs, etc. or to add colors to certain regions of interest... Many other examples could be mentioned, related to visualization techniques in engineering and scientific fields.

16.4.1 *Texture + Cartoon*

Yves Meyer, one of the progenitors of wavelet theory, proposed in a seminal publication [43] (year 2001, 1472 cites) the decomposition of images into texture and cartoon parts. This approach has originated an active research on edges, textures, and separation into parts.

The idea can be simply expressed as a decomposition of the image f into cartoon u and texture v, as follows:

$$f = u + v$$

The u term is also called, by some authors, the geometry or the structure term.

The v term could include constant color regions, textures that could be modeled with oscillatory functions, and others. Textiles can usually be modeled with oscillatory functions.

Although the decomposition seems to be like the separation made by low-pass and high-pass filters (as in Figs. 16.2 and 16.3), this is not the case, because both cartoon and texture contain high frequencies. A linear filtering will not obtain our desired decomposition. Meyer proposed instead a variational approach.

In general, variational methods try to obtain u as a function that minimizes a certain criterion, which is expressed as an image model.

The recent history of variational decompositions is summarized in [11]. In 1989, an important proposal was done by Mumford and Shah [45] (7515 cites). It is worthwhile to consider this contribution in some detail.

16.4.1.1 Mumford-Shah and Image Segmentation

It may well happen that in a certain photograph one wants to separate objects of interest, for instance, blood cells, roads, faces, etc. This is the type of applications addressed by image segmentation. It leads typically to the detection of edges, and so it has many things in common with $f = u + v$ decomposition.

Suppose a rectangular picture with multiple objects $O_1, O_2, \ldots, O_n$. There would be a set of regions $\Omega_1, \Omega_2, \ldots, \Omega_n$ in the image corresponding to these objects.

Denote as Γ the set of smooth arcs that make up boundaries for the regions Ω_i. In total, we would consider $\Omega = \Omega_1 \cup \Omega_2 \cup \ldots \cup \Omega_n \cup \Gamma$.

Our main goal is to capture the boundaries, being so that textures do not vary much inside each object. The image f could in this case be approximated by a piecewise-smooth function u.

Information is related to changes. With a certain analogy, one could speak of energies in the sense of information content. Using the Sobolev's H^1 norm, **the**

energy of a region Ω_i would be

$$E(\Omega_i) = \int_{\Omega_i} |\nabla u(\vec{x})|^2 d\vec{x}$$

(the function u must belong to the space H^1 of functions whose derivative is square-integrable).

It is known that functions belonging to H^1 cannot present discontinuities across lines, such as boundaries. Therefore, u alone cannot model the boundaries, and so, it is opportune to consider the energy of the boundaries, as follows:

$$E(\Gamma) = Length\,(\Gamma)$$

(more formally, this would be the Hausdorff measure of Γ).

In many applications, there is noise and blurring to be taken into account. A *fidelity term* can be used for this purpose:

$$E(error) = \int_{\Omega} (f(\vec{x}) - u(\vec{x}))^2\, d\vec{x}$$

In their seminal article [45], Mumford and Shah approximated the image f by a piecewise-smooth function u and a set Γ that solves the following problem:

$$\arg\min_{u,\,\Gamma} \left\{ \mu\, Length(\Gamma) + \int_{\Omega-\Gamma} |\nabla\, u(\vec{x})|^2\, d\vec{x} + \lambda \int_{\Omega} (f(\vec{x}) - u(\vec{x}))^2 d\vec{x} \right\}$$

Clearly, it is an **energy minimization**. It can be shown that removing any of the three terms gives a trivial, not suitable, solution.

The observed result of the minimization is usually a simplified version of the original image, similar to a cartoon. It is not a perfect tool, for instance, shadows, reflections, and gross textures may cause difficulties.

16.4.1.2 The Chan-Vese Model

Mainly because the term with Γ, the minimization problem is not easy to solve. Many methods and approximations have been proposed. There is a related popular segmentation method based on the **Chan-Vese model** (which is described in [31]).

An example of image segmentation using the Chan-Vese method will soon be given. In preparation for the example, let us include a brief summary of this method. An important simplification is that $u(\vec{x})$ can take only two values: c_{in} for $\vec{x}$ inside Γ and c_{out} for $\vec{x}$ outside Γ. These constants can be regarded as average gray values:

$$c_{in} = \frac{\int_{in\,\Gamma} f\, d\vec{x}}{\int_{in\,\Gamma}\, d\vec{x}} \qquad c_{out} = \frac{\int_{out\,\Gamma} f\, d\vec{x}}{\int_{out\,\Gamma}\, d\vec{x}}$$

Instead of a direct manipulation of Γ, it is represented as zero crossings of a *level set* function φ. This function φ is positive inside Γ and negative outside. This idea was introduced in [50] (19720 cites).

The **energy to be minimized** would be written as follows:

$$E = \mu \int_{\Omega} \delta(\varphi(\vec{x}))\, |\nabla \varphi(\vec{x})|\, d\vec{x} \,+\, \lambda_{in} \int_{in\Gamma} (f(\vec{x}) - c_{in})^2 d\vec{x} +$$

$$+\, \lambda_{out} \int_{out\Gamma} (f(\vec{x}) - c_{out})^2 d\vec{x}$$

The first integral in the previous equation corresponds to the length of Γ.

A semi-implicit gradient descent could be applied for the minimization. This is done with an iterative evolution of

$$\frac{\partial \varphi}{\partial t} = \delta\,(\varphi) \left[\frac{\varphi_{xx}\,\varphi_y^2 \;-\; 2\,\varphi_x \varphi_y \varphi_{xy} \;+\; \varphi_{yy} \varphi_x^2}{(\varphi_x^2 \;+\; \varphi_y^2)^{3/2}} \;-\; \lambda_{in}(f - c_{in})^2 + \right.$$

$$\left. + \lambda_{out}(f - c_{out})^2 \right]$$

(where subindexes represent partial derivatives).

The $\delta(\varphi)$ can be approximated with

$$\delta\,(\varphi) \;=\; \frac{\varepsilon}{\pi(\varepsilon^2 + t^2)}.$$

In each iteration, the values of c_{in} and c_{out} were updated.

Now, let us introduce an example. A gray scale picture has been obtained from a painting of Hooper, as visualized in Fig. 16.8. This is the original image to be processed.

A program in MATLAB has been devised (Program 16.7) for demonstration of the Chan-Vese method. Note that the MathWorks File Exchange site (MATLAB programs) offers several versions of this image segmentation method.

Figure 16.9 displays in pseudo-colors the level set obtained by Chan-Vese corresponding to the original image.

By applying the *contour( )* function of MATLAB to the level set, the Chan-Vese segmentation of the image can be emphasized, as shown by Fig. 16.10 in gray scale.

Here is the listing of the MATLAB program. The reader is invited to explore the effects of *lambda* and *epl.*

Fig. 16.8 Original image

Fig. 16.9 Level set

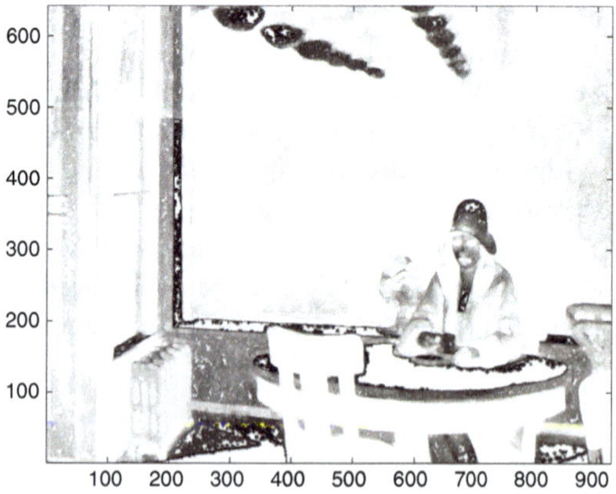

Fig. 16.10 Chan-Vese segmentation

Program 16.7 Image segmentation with Chan-Vese algorithm

```
% Example of Chan-Vese algorithm
I=imread('hooper1.jpg'); %read b&w image
I = double(I);
[m,n] = size(I);
lambda=100; %parameter
%Set initial box=1, with border=-1
Phi=ones(m,n);
Phi(1,1:n)=-1; Phi(m,1:n)=-1;
Phi(1:m,1)=-1; Phi(1:m,n)=-1;
%Prepare loop
disp('working...');
a = 0.01; %to avoid division by zero.
epl = 0.2; %small value
T = 5;
dt = 0.1;
for t = 0:dt:T
ax2=2*Phi;
P1=Phi(:,[2:n,n]); P2=Phi([2:m,m],:);
P3=Phi(:,[1,1:n-1]); P4=Phi([1,1:m-1],:);
%partial derivatives (approx.)
Phi_x = (P1 - P3)/2;
Phi_y = (P2 - P4)/2;
Phi_xx = P1 - ax2 + P3;
Phi_yy = P2 - ax2 + P4;
Q1=Phi([2:m,m],[2:n,n]); Q2=Phi([1,1:m-1],[1,1:n-1]);
Q3=Phi([1,1:m-1],[2:n,n]); Q4=Phi([2:m,m],[1,1:n-1]);
Phi_xy = (Q1 + Q2 - Q3 - Q4)/4;
%TV term
Num = (Phi_xx.*Phi_y.^2 )- (2*Phi_x.*Phi_y.*Phi_xy) +
(Phi_yy.*Phi_x.^2);
Den = (Phi_x.^2 + Phi_y.^2).^(3/2) + a;
%Compute averages
c_in = sum([Phi>0].*I)/(a+sum([Phi>0]));
c_out = sum([Phi<0].*I)/(a+sum([Phi<0]));
%Update
aux=( Num./Den - lambda*(I-c_in).^2 +...
 lambda*(I-c_out).^2);
Phi = Phi + dt*epl./(pi*(epl^2+Phi.^2)).*aux;
end;
% display of results --------------
figure(1)
imagesc(I);
title('Original image');
colormap gray;
figure(2)
imagesc(Phi);
title('Level Set');
colormap cool;
figure(3)
title('Chan-Vese Segmentation');
```

```
aux=zeros(m,n);
for i=1:m,
    for j=1:n;
    aux(i,j)=Phi(m+1-i,j);
    end;
end;
contour(aux,12,'LineWidth',2);
colormap gray;
```

16.4.2 Total Variation and BV Functions

The analysis of images can be enriched by studying the gradients inside an image. A gross quantitative parameter can be obtained by adding the absolute values of those gradients, which results in the so-called **total variation (TV).**

For a formal definition of TV, some authors (like [20] (2071 cites)) suggest to use some test functions $\vec{\phi}$ belonging to the set $C_0^1(\Omega\ ,\ \Re^2)$ of continuously differentiable vector functions of compact support contained in Ω, such as

$$\left\|\vec{\phi}\right\|_\infty \leq 1.$$

These functions are employed for the definition of TV as follows:

$$TV(u) \ = \sup\{ \int_\Omega u\, div\, \vec{\phi}\, d\vec{x}\ ,\ \forall \vec{x} \in \Omega\, \}$$

Given a differentiable function u defined on a bounded open set Ω, its total variation is

$$TV(u) \ = \ \int_\Omega |\nabla\, u(\vec{x})\, |\, d\vec{x}.$$

If the TV of the differentiable function u is $TV(u) \ < \ +\infty$, then $u \ \in \ BV$, where BV is the space of *bounded variation functions*.

The set of bounded variation functions is a Banach space with the norm

$$\|u\|_{BV} \ = \ \|u\|_{L^1} \ + \ TV(u).$$

Bounded variation functions can have sharp edges. Actually, the norm $\|\cdot\|_{BV}$ takes into account the number of edges.

The BV functions play an important role in the cartoon + texture decomposition.

16.4.2.1 The Rudin, Osher, and Fatemi (ROF) Model

In 1992, Rudin, Osher, and Fatemi [54] (19344 cites) proposed the application of TV for *image denoising* in a variational framework with the following expression,

$$\inf_u \left\{ \int_\Omega |\nabla u(\vec{x})| \, d\vec{x} + \lambda \int_\Omega (f(\vec{x}) - u(\vec{x}))^2 d\vec{x} \right\},$$

where $u \in BV$.

Evidently, the ROF model (enclosed in braces) is composed of a TV term and a fidelity term.

The TV term removes noise or small details while preserving edges. The authors of the ROF model give in [54] an algorithm for computing the adequate value of λ if the noise level was known. In other cases, this value has to be chosen, considering that it determines in some sense the smallest image feature to be kept.

From the point of view of optimization, the good news is that the ROF denoising problem is convex, so the solution exists in BV and is unique. A detailed study of image recovery via TV minimization is [20]. The field of TV in imaging, including algorithms, is reviewed by [17].

The ROF variational method can be adapted to different types of noise, as introduced in [32]. This paper is plenty of practical numerical and analytical details and contains a link to one implementation in C code.

16.4.2.2 Meyer's Approach

Meyer suggested to replace the $l2$ norm in the fidelity term with a weaker norm more adequate for modeling textures or oscillatory patterns. Hence, he proposed the following **minimization** problem ,

$$\inf_u \left\{ \int_\Omega |\nabla u(\vec{x})| \, d\vec{x} + \lambda \, \|f(\vec{x}) - u(\vec{x})\|_* \right\},$$

where $u \in BV$.

Continuing with his approach, Meyer defined the space G, which is the Banach space of all generalized functions v that can be written as $v = div(\vec{g})$, where $\vec{g} = (g_1, g_2)$ and $g_1, g_2 \in l_\infty(\Omega)$. The space G is endowed with the G-norm, which is defined as the lower bound of all $l_\infty(\Omega)$ norms of the functions $|\vec{g}|$, with the infimum being computed over all decompositions of v .

The space G is the dual of the closed subspace BV of BV. When applying the G-norm in the minimization problem, the second term is $\lambda \, \|f(\vec{x}) - u(\vec{x})\|_G$. Meyer also defined two more spaces: E and F (see [43]). The spaces are related as follows: $BV \subset G \subset E \subset F$.

G-functions may have large oscillations and nevertheless small norms, which is suitable for the minimization to preserve textures.

Meyer did not propose any numerical procedure for the decomposition. Some algorithms were contributed by other investigators afterward. In [11], a simple conversion of a linear filter pair into a nonlinear filter pair was proposed, obtaining a fast separation of cartoon and texture.

16.4.2.3 TV-l1 Approach

It was suggested in [47] (843 cites) to replace the $l2$ term in the ROF model by a $l1$ term. This article, year 2004, contains interesting references from the 1990s about the fidelity term and how to avoid outliers. According to this approach, the functional to minimize is

$$\inf_u \left\{ \int_\Omega |\nabla u(\vec{x})| \, d\vec{x} + \lambda \int_{|f(\vec{x})-u(\vec{x})|} d\vec{x} \right\}.$$

As shown in [47], the $l1$ norm is well suited to remove salt and pepper noise. Further analysis by [21] (933 cites) and [67] shows that the model enjoys interesting properties of morphological invariance and texture extraction by scale, so geometrical features are better preserved. A first, fast algorithm for solving the optimization problem was presented in [4] (812 cites).

All three models, Meyer, Vese-Osher, and TV-$l1$, were compared by [68] using a uniform computing approach. The three models were solved as second-order cone programs (SCOP). The comparison refers to 1D signals and 2D images. Also, [68] contains a detailed history, with references, of alternatives for solving the three models.

See [39] for a practical treatment of the TV-$l1$decomposition, including various examples and a link to software from a web site.

16.5 Bregman-Related Algorithms

Lev M. Bregman proposed in 1967 a convenient iterative method for finding extrema of convex functions [10] (3525 cites). The method was part of his Ph.D. thesis in Leningrad University (in the 1990s, Bregman migrated to a famous university in Israel).

Later on, in 2005, it was shown by Osher et al. [49] (2236 cites) that this method was very appropriate for image processing (in particular for total variation).

The method is based on the Bregman divergence, and it can be employed in its basic iterative version or as split Bregman iteration [36] (5149 cites).

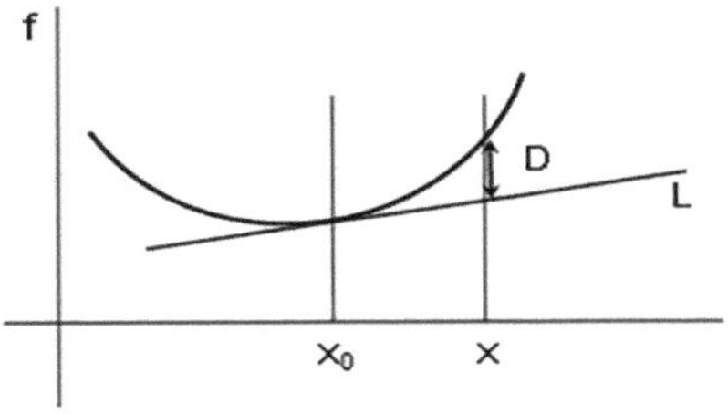

Fig. 16.11 Example of Bregman distance

16.5.1 Bregman Divergence

An expression of the Bregman divergence between two points $\vec{x}$ and $\vec{x}_0$ would be the following:

$$D(\vec{x}, \vec{x}_0) = f(\vec{x}) - f(\vec{x}_0) - \nabla f(\vec{x}_0)^T \cdot (\vec{x} - \vec{x}_0)$$

This concept was introduced by Bregman for differentiable convex functions and was given the name *Bregman distance* by Censor and Lent [19] in 1981.

Figure 16.11 gives a graphical interpretation of this distance D, sitting above the tangent L.

Notice that a first-order Taylor expansion to approximate $f(\vec{x})$ would be $f_a(\vec{x}) = f(\vec{x}_0) + \nabla f(\vec{x}_0)^T(\vec{x} - \vec{x}_0)$, and therefore the distance D is the difference: $D(\vec{x} - \vec{x}_0) = f(\vec{x}) - f_a(\vec{x})$.

Distances are nonnegative functions. A metric distance has symmetry, identity, and triangle inequality properties. Bregman distances are not necessarily symmetric, nor supporting triangle inequality.

Particular cases of Bregman distance are the Euclidean distance, with the choice $f(\vec{x}) = \|\vec{x}\|^2$; the Kullback-Leibler divergence, with $f(\vec{x}) = \sum -x_i \log x_i$; and the Itakura-Saito distance, with $f(\vec{x}) = \sum -\log x_i$. The Mahalanobis distance is also a particular case.

Other equivalent expression of the Bregman distance is the following:

$$D(\vec{x}, \vec{x}_0) = f(\vec{x}) - f(\vec{x}_0) - < \nabla f(\vec{x}_0), \cdot \vec{x} - \vec{x}_0 >$$

Recently, some important authors, like in [36], used the following expression,

$$D(\vec{x}, \vec{x}_0) = f(\vec{x}) - f(\vec{x}_0) - < p, \cdot \vec{x} - \vec{x}_0 >$$

where p is the subgradient at $\vec{x}_0$.

16.5.2 Bregman Iteration

Consider the following **constrained minimization** problem,

$$\min_u J(\vec{u}) \,, \text{ subject to } H(\vec{u}) = 0,$$

where J and H are defined in $\Re^n$ and convex. The associated unconstrained problem is

$$\min_u J(\vec{u}) + \lambda\, H(\vec{u}).$$

The Bregman distance of J between $\vec{u}$ and $\vec{v}$ would be

$$D(\vec{u},\, \vec{v}) = J(\vec{u}) - J(\vec{v}) - < p, \cdot\, \vec{u} - \vec{v} > .$$

The *Bregman iteration* for solving the minimization problem is

$$\vec{u}^{(k+1)} = \arg\min_u D^{(k)}(\vec{u},\, \vec{u}^{(k)}) + \lambda\, H(\vec{u}).$$

Suppose that H is differentiable. In this case, the Bregman iteration would be

$$\begin{aligned} \vec{u}^{(k+1)} &= \arg\min\nolimits_u D^{(k)}(\vec{u},\, \vec{u}^{(k)}) + \lambda\, H(\vec{u}) \\ p^{(k+1)} &= p^{(k)} - \lambda\, \nabla H(\vec{u}^{(k+1)}). \end{aligned}$$

For the particular problem of TV-based image denoising [49], one could take

$$H(\vec{u},\, \vec{f}) = \frac{1}{2} \left\| A\, \vec{u} - \vec{f} \right\|_2^2 ;$$

then, the iteration can be expressed in the following simplified form:

$$\vec{u}^{(k+1)} = \arg\min_u J(\vec{u}) + \lambda\, H(\vec{u},\, \vec{f}^{(k)})$$

$$\vec{f}^{(k+1)} = \vec{f}^{(k)} + (\vec{f} - A\, \vec{u}^{(k+1)})$$

It has been verified that these iterations converge very quickly.

16.5.2.1 Linearized Bregman Iteration

In appropriate cases, it would be possible to use the following approximation:

$$H(\vec{u}) \approx H(\vec{u}^{(k)}) + \nabla H(\vec{u}^{(k)}) \cdot (\vec{u} - \vec{u}^{(k)})$$

For a better approximation a penalty quadratic term could be added, and then:

$$\vec{u}^{(k+1)} = \arg\min_{u} J(\vec{u}) + \left\langle \lambda\, H(\vec{u}^{(k)}) - p^{(k)},\, \vec{u} \right\rangle + \frac{1}{2\delta} \left\| \vec{u} - \vec{u}^{(k)} \right\|_2^2$$

16.5.2.2 Split Bregman Iteration

A crucial aspect of split Bregman method is that it could separate the typical $l1$ and $l2$ portions of important image processing approaches, for instance, suppose that the minimization problem is

$$\min_{u} \left\| \vec{d} \right\|_1 + H(\vec{u}) + \frac{\lambda}{2} \left\| \vec{d} - \vec{\phi}(\vec{u}) \right\|_2^2.$$

The split Bregman iteration for solving this problem is

$$\begin{aligned} \vec{u}^{(k+1)} &= \arg\min\nolimits_u\ H(\vec{u}) + \tfrac{\lambda}{2} \left\| \vec{d}^{(k)} - \vec{\phi}(\vec{u}) - \vec{b}^{(k)} \right\|_2^2 \\ \vec{d}^{(k+1)} &= \arg\min\nolimits_d\ \left\| \vec{d} \right\|_1 + \tfrac{\lambda}{2} \left\| \vec{d} - \vec{\phi}(\vec{u}^{(k+1)}) - \vec{b}^{(k)} \right\|_2^2 \\ \vec{b}^{(k+1)} &= \vec{b}^{(k)} + (\vec{\phi}(\vec{u}^{(k+1)}) - \vec{d}^{(k+1)}). \end{aligned}$$

As you can see, the method involves two optimization steps and a simple update of $\vec{b}$. The first optimization step is differentiable, and so it can be solved by a variety of methods, like Gauss-Seidel or conjugate gradient, or even in the Fourier domain. The second optimization step can be computed by shrinkage, that is,

$$\vec{d}_j^{(k+1)} = shrink(\phi(\vec{u})_j + \vec{b}_j^{(k)},\ 1/\lambda),$$

where the shrinkage would be

$$shrink(\vec{x}, \lambda) = \frac{\vec{x}}{|\vec{x}|} * \max(\,|\vec{x}| - \lambda), 0).$$

16.5.3 *Using Split Bregman for ROF-TV Image Denoising*

The already mentioned article [36] shows how to apply the split Bregman method for total variation (TV) image denoising, based on the ROF model. The method is simple and efficient. It handles two-dimensional variables.

There are two denoising formulations: the anisotropic problem and the isotropic problem. In the *anisotropic* problem, one has to solve

$$\min_u |\nabla_x u| + |\nabla_y u| + \frac{\mu}{2} \parallel u - f \parallel_2^2 .$$

Let us replace $\nabla_x u$ by d_x and $\nabla_y u$ by d_y. In order to strengthen the constraints, two penalty terms were added:

$$\min_u \ |d_x| + |d_y| + \frac{\mu}{2} \parallel u - f \parallel_2^2 + \frac{\lambda}{2} \parallel d_x - \nabla_x u \parallel + \frac{\lambda}{2} \parallel d_y - \nabla_y u \parallel$$

Coming now to the split Bregman algorithm, the first optimization step would be

$$u^{(k+1)} = \arg\min_u \frac{\mu}{2} \parallel u - f \parallel_2^2 + \frac{\lambda}{2} \left\| d_x^{(k)} - \nabla_x u - b_x^{(k)} \right\| +$$
$$+ \frac{\lambda}{2} \left\| d_y - \nabla_y u - b_y^{(k)} \right\| .$$

Because the system is strictly diagonal, it is recommended in [36] to get the solution with the Gauss-Seidel method:

$$u_{i,j}^{(k+1)} = G_{i,j}^{(k)} = \frac{\mu}{\mu+4\lambda} f_{i,j} + \frac{\lambda}{\mu+4\lambda} \{u_{i+1,j}^{(k)} + u_{i-1,j}^{(k)} + u_{i,j+1}^{(k)} + u_{i,j-1}^{(k)} +$$
$$+ d_{x,i-1,j}^{(k)} - d_{x,i,j}^{(k)} + d_{y,i,j-1}^{(k)} - d_{y,i,j}^{(k)} - b_{x,i-1,j}^{(k)} + b_{x,i,j}^{(k)} - b_{y,i,j-1}^{(k)} + b_{y,i,j}^{(k)} \}$$

Then, the complete split Bregman algorithm, to be iterated, is

$$u^{(k+1)} = G^{(k)}$$

$$d_x^{(k+1)} = shrink(\nabla_x u^{(k+1)} + b_x^{(k)},\ 1/\lambda)$$

$$d_y^{(k+1)} = shrink(\nabla_y u^{(k+1)} + b_y^{(k)},\ 1/\lambda)$$

$$b_x^{(k+1)} = b_x^{(k)} + (\nabla_x u^{(k+1)} - d_x^{k+1})$$

$$b_y^{(k+1)} = b_y^{(k)} + (\nabla_y u^{(k+1)} - d_y^{k+1}).$$

In the case of *isotropic* denoising, the minimization problem is

$$\min_u \sum \sqrt{(\nabla_x u)_i^2 + (\nabla_x u)_i^2} + \frac{\mu}{2} \parallel u - f \parallel_2^2 .$$

Like before, the problem is transformed to

$$\min_u \sum_{i,j} \sqrt{d_{x,i,j}^2 + d_{y,i,j}^2} + \frac{\mu}{2} \| u - f \|_2^2 + \frac{\lambda}{2} \| d_x - \nabla_x u \| + \frac{\lambda}{2} \| d_y - \nabla_y u \| .$$

And the split Bregman algorithm, to be iterated, is the following,

$$u^{(k+1)} = G^{(k)}$$

$$d_x^{(k+1)} = \max(s^k - 1/\lambda, 0) \frac{\nabla_x u^{(k)} + b_x^{(k)}}{s^k}$$

$$d_y^{(k+1)} = \max(s^k - 1/\lambda, 0) \frac{\nabla_y u^{(k)} + b_y^{(k)}}{s^k}$$

$$b_x^{(k+1)} = b_x^{(k)} + (\nabla_x u^{(k+1)} - d_x^{k+1})$$

$$b_y^{(k+1)} = b_y^{(k)} + (\nabla_y u^{(k+1)} - d_y^{k+1}),$$

where

$$s^k = \sqrt{\left|\nabla_x u^{(k)} + b_x^{(k)}\right|^2 + \left|\nabla_y u^{(k)} + b_y^{(k)}\right|^2}$$

According to [12], the shrinking steps could be approximated as follows:

$$d_x^{(k+1)} = \frac{s^k \lambda (\nabla_x u^{(k)} + b_x^{(k)})}{s^k \lambda + 1}$$

$$d_y^{(k+1)} = \frac{s^k \lambda (\nabla_y u^{(k)} + b_y^{(k)})}{s^k \lambda + 1}$$

Figure 16.12 shows an example of ROF-TV isotropic denoising. The image on the left is contaminated by Poisson noise (a typical noise in medical and astronomical imaging). The good effect of the denoising algorithm is shown in the image on the right.

Figure 16.12 has been generated with the Program 16.8. Its MATLAB code is based on the implementation offered by [61] (University College London), accessible from the MathWorks File Exchange site, complemented with a similar coding from [44] accessible from GitHub. Both sources do use the Kronecker product for compact coding.

Notice the easy addition of Poisson noise by using the *imnoise( )* function (included in the MATLAB Image Processing Toolbox).

Fig. 16.12 ROF total variation denoising using split Bregman

As a measure of noise elimination, an accumulated pixel-by-pixel difference of images is computed and printed in each iteration of the algorithm, algorithm that only requires a few iterations.

Program 16.8 Denoising with split Bregman algorithm

```
%Split Bregman Isotropic Total Variation(TV)denoising
% initialize parameters
lambda=0.5;
mu=0.4;
P=imread('ActressH.jpg');
[M,N]=size(P);
% (note: it is a square image, M=N)
A=imnoise(P,'poisson');
f=reshape(A,[],1); % one column (noisy image)
n=length(f);
F=double(f);
% Split Bregman TV denoising algorithm
% initialization -------------
d= zeros(2*n,1);
b= d;
% gradient operators -------------------------
D= spdiags([-ones(N,1) ones(N,1)],[0,1],N,N+1);
D(:,1)=[]; D(1,1)=0;
del_x = kron(speye(N),D);
del_y= kron(D, speye(N));
B= [del_x;del_y];
Bt=B';
BtB=Bt*B;
% loop ----------------
for iter=1:5,
```

```
  % image update
  aux1=speye(n)+BtB;
  aux2=lambda*Bt*(b-d);
  [u_update,r]= cgs(aux1,F-aux2,1e-5,100);
  % d and b update
  Bub=(B*u_update)+b;
  s=sqrt(Bub(1:n).^2 + Bub(n+1:end).^2);
  q=mu/lambda;
  d=[max(s-q,0).*Bub(1:n)./s ;
  max(s-q,0).*Bub(n+1:end)./s];
  b=Bub-d;
  % substract original-denoised:
  acuDif=sum(F-u_update)
end
% result display ------------------
figure; imshow(A); title('Noisy Image');
u_den=reshape(u_update,N,N);
figure; imshow(uint8(u_den));
title('Denoised Image');
```

16.6 Douglas-Rachford Splitting

The Douglas-Rachford algorithm is one of the most successful optimization methods, able to tackle non-smooth objective functions. The method has also been applied to nonconvex problems. An interesting series of applications is cited in [3], including image retrieval, graph coloring, protein studies, matrix completion, and even nonograms (a kind of Japanese puzzles).

The optimization problem is

Minimize $f(\vec{x}) + g(\vec{x})$.

Using the formulation given in [Com], the algorithm is simply

$$\vec{x}_k = \mathrm{Pr}\,ox_{\lambda\, g}(\vec{y}_k)$$

$$\vec{y}_{k+1} = \vec{y}_k + \lambda_k(\mathrm{Pr}\,ox_{\lambda\, f}(2\vec{x}_k - \vec{y}_k) - \vec{x}_k)$$

where $\lambda_k \in [\varepsilon\,,\, 2-\varepsilon]$.

See the Thesis [24] for an extensive treatment of splitting methods. The history of the Douglas-Rachford algorithm and other aspects of interest are concisely included in [22] (3027 cites). There is a web page focusing on Douglas-Rachford (https://carmamaths.org/DRmethods/).

The Douglas-Rachford method will be used for examples in the next section.

16.7 Matrix Completion

A matrix could be regarded as a set of signal samples, for instance, it may correspond to the measurements taken at a given time by a network of spatially distributed sensors, in which some sensors could fail.

The consideration of matrices has originated a fruitful new research area, which is very active and expansive. This section introduces some fundamental aspects of this area.

16.7.1 Introduction to Matrix Completion

Emmanuel Candès and Benjamin Recht introduced in a celebrated article [14] (year 2012, 6952 cites) a topic of considerable interest: the recovery of a data matrix from a sampling of its entries.

A number m of entries are chosen uniformly at random from a matrix M, and the question is whether it is possible to entirely recover the matrix M from these m entries.

Candès and Recht found that if the matrix M has low rank or approximately low rank, then accurate and even exact recovery from random sampling is possible by nuclear norm minimization.

By the way, it is now opportune to quote a certain set of norms, denoted as *Schatten-p norms*, for $p = 1, 2$ or ∞. In particular,

- Spectral norm:

$$\text{The largest singular value: } \|X\|_S = \max \sigma_i(X) = \|\sigma(X)\|_\infty$$

- Nuclear norm:

$$\text{The sum of singular values: } \|X\|_* = \sum_{i=1}^{N} \sigma_i(X) = \|\sigma(X)\|_1$$

- Frobenius norm:

$$\|X\|_F = \left(\sum_{i=1}^{N} \sum_{j=1}^{M} X_{ij}^2 \right)^{1/2} = \left(\sum_{i=1}^{t} \sigma_i^2(X) \right)^{1/2} = \|\sigma(X)\|_2$$

Using a simple example of a $n \times n$ matrix M that has all entries equal to zero except for the first row, it is noted in [14] that this matrix cannot be recovered from a subset of its entries. There are also more pathological cases as well. In general, we need the singular vectors of M to be spread across all coordinates. If this happens, the recovery could be done by solving a **rank minimization problem** :

Minimize $rank(X)$, subject to $X_{i\,j} = M_{i\,j}$, $(i, j) \in \Omega$
where X is the recovered matrix.

Most frequently, it is preferred to consider the **minimization of the nuclear norm**:

Minimize $\|X\|_*$, subject to $X_{i\,j} = M_{i\,j}$, $(i, j) \in \Omega$

The contribution from Candes and Recht shows that most matrices can be recovered provided that the number of samples m obeys,

$$m \geq C\, n^{6/5}\, r\ \log n$$

where C is some positive constant and r is the rank.

The nuclear norm minimization problem can be solved in many ways. A special mention should be done of the singular value thresholding algorithm introduced in [13]. This algorithm is a de facto reference for comparison with other algorithms.

Actually, the proximal operator corresponding to the minimization of $\|X\|_*$ is the singular value shrinkage operator that shrinks toward zero the singular values of X by subtracting a certain threshold value [51].

An example of matrix sampling and recovery is given below. The Douglas-Rachford splitting is used [52] based on two proximal operators: one corresponds to the indicator function and the other to the nuclear norm. In the case of the indicator function, the proximal operator is a projection that accumulates repeated samples of the same matrix entry. The algorithm is implemented in the Program 16.9, which generates two figures.

Figure 16.13 shows the evolution of the nuclear norm as the iterations go on. It converges in less than 100 iterations.

In order to show how accurate the recovery, we obtain A-X (where A is the original random matrix and X is the recovered matrix) and plot this difference. Also, for comparison purposes, we generated another random matrix B and obtained A-B.

Figure 16.14 shows on the left A-B. As expected, there is a mix of gray pixels indicating differences. On the right, we visualize A-X, which is white: the recovery is practically exact (it has an accumulated error of 2.7894e-004).

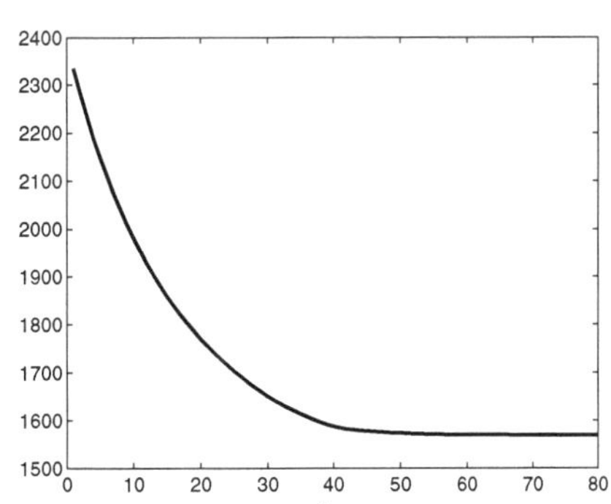

Fig. 16.13 Evolution of nuclear norm

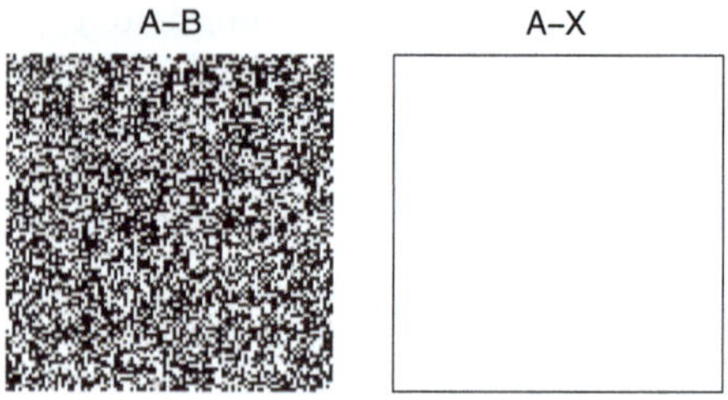

Fig. 16.14 A test of reconstruction quality (see text)

Figures 16.13 and 16.14 have been generated with Program 16.9. In principle, for A-X=0, a black square would be visualized (pixels being zero); however, for better emphasis of results, we employed the function *imcomplement()* to give a white square (the complement of black).

Program 16.9 Matrix completion based on Douglas-Rachford procedure

```
% Example of Matrix Completion
% (using Douglas-Rachford)
clear all;
disp('working...');
% a random matrix of rank r
n=100;
r=16; %rank
A=randn(n,r)*randn(r,n); % Original random matrix
disp('nuclear norm of original matrix:');
[u,d,v]=svd(A);
nuc_N = sum(diag(d));
% a subset of A is measured by
% sampling at random a number p of entries of A
p=round(n*log(n)*r); % see theory
aux=randperm(n*n);
ix=aux(1:p)'; %extract p integer random numbers
% b is a column vector:
b=A(ix); %retain a subset of entries of A
% start the algorithm ----------------------------
X=zeros(n,n); %will become the reconstructed matrix
Y=zeros(n,n);
L=n*n;
niter=80; %number of iterations
lambda=1; gamma=2;
rnx=zeros(niter,1);
for nn=1:niter,
  % X update ------
  O=zeros(L,1);
  % accumulating repeated entries
  for j=1:p;
    ox=ix(j);
```

```
    O(ox)=O(ox)+(b(j)-Y(ox));
  end;
  Q=reshape(O,[n n]);
  % proxF (indicator function)
  X=Y+Q; %X update
  % Y update ------
  %proxG (soft thresholding of singular values):
  P=(2*X)-Y;
  [U,D,V]=svd(P);
  for j=1:n,
    aux=D(j,j);
    if abs(aux)<=gamma,
      D(j,j)=0;
    else
     if aux>gamma, D(j,j)=aux-gamma; end
     if aux<-gamma, D(j,j)=aux+gamma; end;
    end;
  end;
  S=U*D*V'; % result of thresholding
  Y=Y+ (lambda*(S-X));
  % recording
  [u,d,v]=svd(X);
  rnx(nn)=sum(diag(d)); %nuclear norm
end
%display ----------------------------
% evolution of nuclear norm
figure(1)
plot(rnx,'k');
xlabel('niter');
% see estimation accuracy
figure(2)
% X is the reconstructed matrix
% generate another random matrix for comparison
B=randn(n,r)*randn(r,n);
subplot(1,2,1)
imshow(A-B); %see differences between A and B
title('A-B', 'FontSize',14);
subplot(1,2,2)
DI=A-X; J=imcomplement(DI);
imshow(J); %see differences between A and X
title('A-X', 'FontSize',14);
% a measure of error
er=sum(sum(A-X))
```

16.7.2 Decomposition of Matrices

E. Candès and Y. Plan made in [16] (2132 cites) the following observation: in real-world applications, the measured entries would be corrupted by noise (perhaps outliers, [64]).

This remark originated a new type of problem, in which one has to find a decomposition of the observed matrix M into a low-rank matrix L and a sparse matrix S. The problem was identified as a robust principal component analysis (PCA) in [15] (7572 cites), being stated as follows:

Minimize $(\|L\|_* + \lambda\,\|S\|_1)$, subject to $M = L + S$

Again, the new problem led to a broad range of research efforts, which have found many interesting applications.

Note About PCA

The principal component analysis (PCA) is the most important statistical methodology for detecting significant variables, like, for instance, in disease symptoms or in data mining (could be in big data contexts). An effective, concise tutorial on PCA is [57].

Simple Example of $L + S$ Decomposition

Let us build a simple example by adding a low-rank random matrix and a sparse matrix (just a cross-diagonal matrix). Figure 16.15 shows images corresponding to those matrices.

The result of adding the previous two matrices is shown in Fig. 16.16.

Now, the problem is to recover L and S from M. One of the methods that can be used is an adaptation of the Douglas-Rachford algorithm for this case. According to [30], it can be formulated as follows:

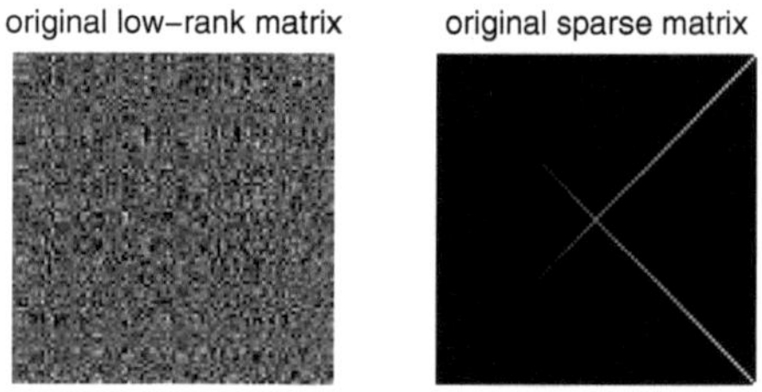

Fig. 16.15 A low-rank matrix and a sparse matrix

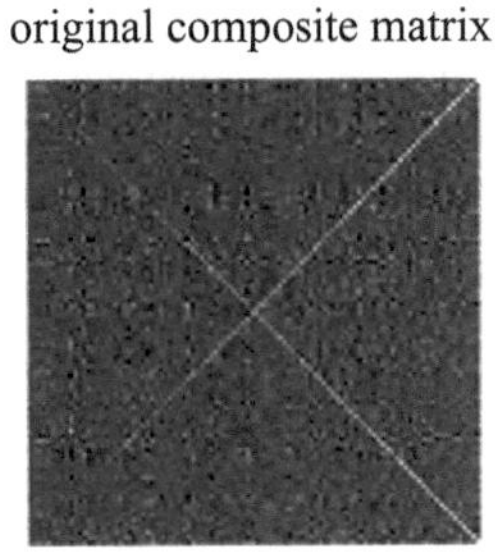

Fig. 16.16 The observed matrix $M = L + S$

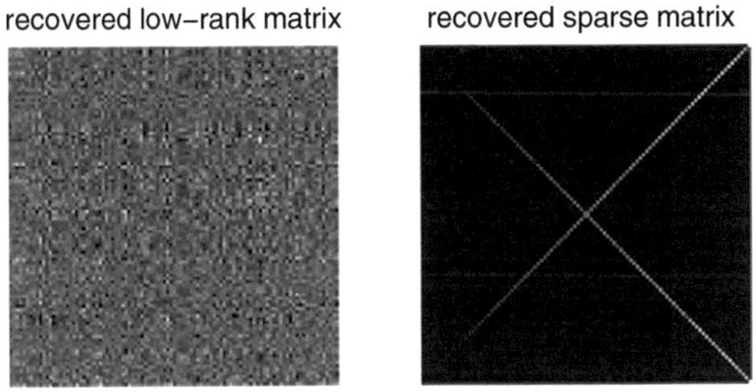

Fig. 16.17 The recovered matrices L and S

Repeat

$$Le = (M + L^{(k)} - S^{(k)})/2$$

$$Se = (M - L^{(k)} + S^{(k)})/2$$

$$L^{(k+1)} = L^{(k)} + t_k(shrink(2Le - L^{(k)}) - Le)$$

$$S^{(k+1)} = S^{(k)} + t_k(soft\_threshold(2Se - S^{(k)}) - Se)$$

until convergence.

The shrinkage corresponds to the proximity operator for the nuclear norm, and the soft-thresholding corresponds to the proximity operator for the $l1$ norm.

An implementation of this algorithm is provided by Program 16.10. The result is satisfactory as shown in Fig. 16.17.

Figure 16.18 visualizes the recovery accuracy concerning the low-rank part. It was just a matter of subtracting the original and the recovered low-rank part. We employed again the MATLAB function *imcomplement( )* in order to visualize the differences as gray or black pixels (white means zero difference). Notice some slight traces of the cross.

Fig. 16.18 Difference between L original and L recovered

Here is the MATLAB program that made the matrix decomposition and generated all Figs. 16.15, 16.16, 16.17, 16.18.

Program 16.10 Decomposition into low-rank and sparse matrices

```
% decomposition into low-rank (L)
% and sparse (S) matrices
% Douglas-Rachford
% a random matrix of rank r
n=100;
r=8; %rank
L0=randn(n,r)*randn(r,n); % a low-rank matrix
% a sparse (cross) matrix
nn=1:100;
aux=diag(0.2*nn,0);
S0=aux+aux(end:-1:1,:);
% composite original matrix
M=L0+S0;
% parameter settings
lambda=0.2;
tk=1;
Th=1; %Threshold
niter=40; % number of iterations
rnx=zeros(niter,1);
L=zeros(n,n); S=zeros(n,n);
% start the algorithm --------------------
for nn=1:niter,
  Le=0.5*(M+L-S); Se=0.5*(M-L+S);
  % shrinking---------
  aux1=(2*Le)-L;
  [U D V]=svd(aux1);
  for j=1:n,
    D(j,j)=max(D(j,j)-Th,0);
  end;
  aux=U*D*V';
  L=L+(tk*(aux-Le));
  % soft_threshold---------
  aux1=(2*Se)-S;
```

```
  aux=sign(aux1).*max(0, abs(aux1)-lambda);
  S=S+(tk*(aux-Se));
  [u,d,v]=svd(L);
  rnx(nn)=sum(diag(d)); %nuclear norm, record
end;
% display -----------------------------
figure(1)
subplot(1,2,1)
imshow(L0,[]);
title('original low-rank matrix', 'FontSize',14);
subplot(1,2,2)
imshow(S0,[]);
title('original sparse matrix', 'FontSize',14)
figure(2)
imshow(M,[]);
title('original composite matrix', 'FontSize',14);
figure(3)
subplot(1,2,1)
imshow(L,[]);
title('recovered low-rank matrix', 'FontSize',14)
subplot(1,2,2)
imshow(S,[]);
title('recovered sparse matrix', 'FontSize',14)
figure(4)
DI=L0-L; J=imcomplement(DI);
imshow(J); %see differences between A and X
title('Low-rank Original-Recovered', 'FontSize',14);
```

A number of methods for matrix decomposition have been proposed. One of the factors that promote the popularity of certain methods is the public availability of code [41] (3565 cites), [42] (632 cites), and [69].

16.7.3 Picture Reconstruction Based on Matrix Completion

A direct example of matrix completion is the case of taking at random some pixels of a picture and see if it is possible to recover the picture from these samples.

An image with evident redundancies has been chosen (Fig. 16.19). It could be expected that this image has approximately low rank. Anyway, a certain value of p was tried.

The sampling and recovery experiment was done with the Program 16.11, which is just an adaptation of the Program 16.9 (under Fig. 16.14).

Figure 16.20 shows the evolution of the nuclear norm along iterations.

And Fig. 16.21 shows the good result of the image recovery.

Fig. 16.19 Original image

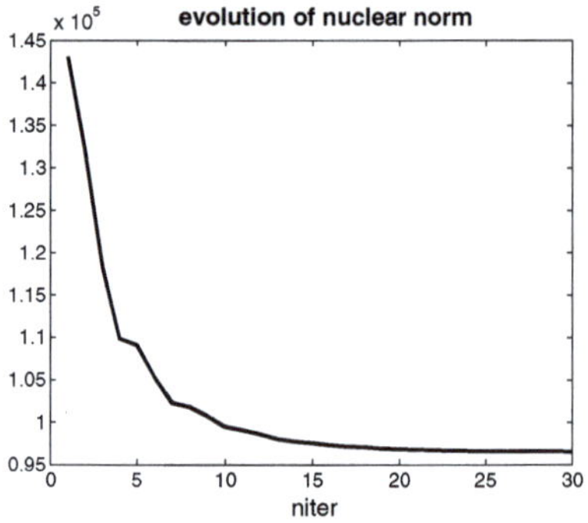

Fig. 16.20 Evolution of nuclear norm

Fig. 16.21 Image reconstruction by matrix completion

Program 16.11 Picture completion

```
% Example of Picture Completion
% (using Douglas-Rachford)
clear all;
disp('working...')
% original picture
figu=imread('patron.jpg'); %read picture
F=double(figu);
n=225;
```

```
A=F(1:n,1:n); %crop;
aux=mean(mean(A));
A=A-aux;
disp('nuclear norm of original matrix:');
[u,d,v]=svd(A);
nuc_N = sum(diag(d))
% a subset of A is measured by
% sampling at random a number p of entries of A
p=12000;
aux=randperm(n*n);
ix=aux(1:p)'; %extract p integer random numbers
% b is a column vector:
b=A(ix); %retain a subset of entries of A
% start the algorithm--------------------------
X=zeros(n,n);
Y=zeros(n,n);
L=n*n;
niter=30;
lambda=1.8; gamma=1000; %(notice the value of gamma)
rnx=zeros(niter,1);
for nn=1:niter,
  % X update
  % R(b-M(Y)) term
  O=zeros(L,1);
  % accumulating repeated entries
  for j=1:p;
    ox=ix(j);
    O(ox)=O(ox)+(b(j)-Y(ox));
  end;
  Q=reshape(O,[n n]);
  % proxF (indicator function)
  X=Y+Q;
  % Y update
  %proxG (soft thresholding of singular values):
  P=(2*X)-Y;
  [U,D,V]=svd(P);
  for j=1:n,
    aux=D(j,j);
    if abs(aux)<=gamma,
      D(j,j)=0;
    else
      if aux>gamma, D(j,j)=aux-gamma; end
      if aux<-gamma, D(j,j)=aux+gamma; end;
    end;
  end;
  S=U*D*V'; % result of thresholding
  Y=Y+ (lambda*(S-X));
  % saving
  [u,d,v]=svd(X);
  rnx(nn)=sum(diag(d)); %nuclear norm
end
```

```
aux=mean(mean(X));
X=X-aux;
%display ------------------------------
rnx(end)
% evolution of nuclear norm
figure(1)
plot(rnx,'k','LineWidth',2);
title('evolution of nuclear norm')
xlabel('niter');
figure(2)
imshow(A);
title('original picture')
figure(3)
imshow(X);
title('reconstructed picture')
```

An opportune article about robust matrix completion is [59] (year 2020), with a good description of the state of the art in its introductory section.

16.7.4 Text Removal

Suppose you have a photograph and someone has written some text on it. It would be good to remove that text. Given this situation, matrix decomposition could be helpful, as far as the picture has low rank, and the text corresponds to a sparse matrix of pixels.

In order to explore this application, a synthetic problem has been fabricated: some text has been added to a wall of bricks. It is a crude simulation of a graffiti. Figure 16.22 shows the original image.

Since we wanted to explore alternating minimization schemes, we selected a fast scheme proposed by [53] where the matrix decomposition problem is treated as

Minimize $(\frac{1}{2} \|L + S - M\|_F + \lambda \|S\|_1)$, subject to $rank(L) = t$.

Fig. 16.22 Simulated graffiti image

Fig. 16.23 Image decomposition into low-rank and sparse matrices

The problem is solved with the following alternating minimization:

$$L^{(k+1)} = \arg\min_{L}\left(\left\| L + S^{(k)} - M \right\|_F\right), \text{ subject to } rank(L) = t$$

$$S^{(k+1)} = \arg\min_{S}\left(\left\| L^{(k+1)} + S - M \right\|_F + \lambda \, \|S\|_1\right)$$

These two subproblems are solved as follows:

- The first subproblem is solved with a partial SVD of $S^{(k)} - M$. Only t components of the SVD are selected (corresponding to the t largest singular values). This is done with the *lansvd( )* routine included in the PROPACK library [40], which is commonly found in the public available codes.
- The second subproblem is an element-wise shrinkage of $M - L^{(k)}$.

Program 16.12 presents a simple implementation of the algorithm. Figure 16.23 shows the result, which is not perfect but gives an idea of the approach.

Program 16.12 Text removal by alternating minimization (ADMM) procedure

```
% decomposition into low-rank (L)
% and sparse (S) matrices
% Alternating Minimization
% load image
figu=imread('grafiti1.jpg'); %read picture
F=double(figu);
n=350;
M=F(1:n,1:n); %crop;
aux=mean(mean(M));
M=M-aux;
% parameter settings
lambda=0.1;
lambF=1.01;
```

```
Th=1; %Threshold
rank0=1; %intial rank guess
irk=1; %for rank increments
nnL=1000; % number of loops
rank=rank0; %current rank
% start the algorithm ------------------------------
%
[UL SL VL] = lansvd(M, rank, 'L'); %partial SVD
L1=UL*SL*VL'; %initial low-rank approximation
aux=M-L1;
S1= sign(aux).*max(0,abs(aux)-lambda); %shrinkage
for nn=2:nnL,
  if irk==1,
    % lambda is modified in each iteration:
    lambda = lambda * lambF;
    rank = rank + irk; % rank is increased " " "
  end;
  [UL SL VL] = lansvd(M-S1, rank, 'L'); %partial SVD
  L1=UL*SL*VL'; %current low-rank approximation
  aux=M-L1;
  S1= sign(aux).*max(0,abs(aux)-lambda); %shrinkage
  % change rank increment when appropriate
  vv=diag(SL);
  rho=vv(end)/sum(vv(1:end-1));
  if rho<Th,
    irk=0;
  else
    irk=1;
  end;
end;
% display ------------------------------
figure(1)
imshow(M,[]);
title('original picture','FontSize',14)
figure(2)
subplot(1,2,1)
imshow(L1,[]);
title('low-rank component','FontSize',14)
subplot(1,2,2)
imshow(S1,[]);
title('sparse component','FontSize',14)
```

Background extraction or subtraction is a subject of considerable interest (for instance, in video surveillance).

The web site of T. Bouwmans
https://sites.google.com/site/thierrybouwmans/computer-vision-research/background-subtraction-research:

offers several surveys on this topic.

16.8 Sparsity Optimization

This section is related with compressed sensing [23] (34168 cites), [38] and mostly 1D signals.

Sparse representations intend to represent signals with as few as possible significant coefficients. This is important for many applications, like, for instance, **compression**. Then, a lot of interest is being paid to finding sparsest representations. This turns out to be a not-so-standard optimization problem.

To start with, consider the following well-known expression:

$$A\vec{x} = \vec{b}$$

The target is to obtain sparse solutions.

First of all, looking at the $m \times n$ matrix A and the corresponding system of linear equations, there are three cases: $m > n$, $m = n$, and $m < n$.

The case of more equations than unknowns is typical of statistics and regression problems.

On the other hand, the undetermined system is the basic scenario considered in *compressive sensing* (formed by sensors sending compressed information).

Usually, in order to get sparse solutions, the question is stated in an optimization framework.

A matrix or a vector is said to be **sparse** if it has only a few (compared to the total size) nonzero entries. Of course, one needs to formalize this initial concept.

In the case of a vector, its $l0$ norm, $\|\vec{x}\|_0$, is simply the number of nonzero entries in the vector.

Continuing with vectors, the $l1$ norm, $\|\vec{x}\|_1$, is the sum of the absolute value of the entries in a vector. The $l2$ norm of a vector is its Euclidean length: $\|\vec{x}\|_2 = \left(\sum_{i=1}^{n} |x_i|^2\right)^{1/2}$.

From the point of view of sparsity, $l0$ is the most appropriate norm, while $l2$ says almost nothing. On the other hand, from the point of view of mathematical analysis, finding the sparsest solution is an optimization problem, and in this context, the $l2$ norm is a well-known terrain while $l0$ is unusual.

As we shall see, our efforts will concentrate on using the $l1$ norm, which is linked to sparsity and is easier for mathematical work.

This section focuses on methods for solving sparsity optimization problems. The basic literature is [9, 26] (4055 cites), and [62] (1391 cites).

16.8.1 The Central Problem and Basic Approaches

Consider the problem of finding a sparse $n \times 1$ vector $\vec{x}$ such that

$$A\vec{x} = \vec{b}$$

where the $m \times 1$ vector $\vec{b}$ and the $m \times n$ matrix A are given.

Recall that in the case of regression problems, with $m > n$, this is an overdetermined system that usually has no solution. But one could be interested on establishing a model, based on the following problem:

Minimize $\left\| A\vec{x} - \vec{b} \right\|_2^2$

In general, this minimization would not enforce sparsity. Some pertinent terms should be added, as we shall see.

In the case of undetermined system, with $m < n$, there would be in principle infinite candidate solutions. A typical proposal is to choose the solution with the minimum $l2$ norm, so the problem is

Minimize $\|\vec{x}\|_2^2$, subject to $A\vec{x} = \vec{b}$.

Using Lagrange multipliers and taking derivatives, it is possible to analytically get the solution:

$$\vec{x} = (A^T A)^{-1} A^T \vec{b}$$

Again, in general, this solution would be not sparse. The problem should be stated with other terms.

But why it would be interesting to obtain sparse solutions?

If you are trying to establish a model in a regression problem, a sparse solution will reveal the pertinent variables as those with nonzero values. This helps to identify and select the correct variables.

When dealing with signals, it is sometimes desired to transform a set of measurements $\vec{b}$ into a sparse vector $\vec{x}$. Later on, the vector $\vec{x}$ could be used to exactly reproduce $\vec{b}$, or, in certain applications, it would be sufficient with an approximate recovery. Then, it would make sense to consider an approximation of the form

$$\left\| A\vec{x} - \vec{b} \right\| < \varepsilon$$

where some of the entries of $\vec{x}$ were set to zero. This is a sparse approximation.

Suppose that it is an undetermined system, with infinite candidate solutions. One wants the sparsest solution:

- In formal terms, the problem of sparsest representation can be written as follows:

  Minimize $\|\vec{x}\|_0$, subject to $A\vec{x} = \vec{b}$

- There is a method, called *basis pursuit (BP)*, which replaces $l0$ with $l1$, so now the problem is

  Minimize $\|\vec{x}\|_1$, subject to $A\,\vec{x} = \vec{b}$.

  Under certain conditions, BP will find a sparse solution or even the sparsest one.

Another method (already mentioned), called the *Lasso*, states the following problem:

Minimize $\left\| A\vec{x} - \vec{b} \right\|_2^2$, subject to $\|\vec{x}\|_1 < c$

This method can be used in the regression context and for undetermined systems to find sparse approximations. Like BP, under certain conditions, it will find the sparsest solution.

As regards sparsity, a pertinent definition is the following:

A vector $\vec{x}$ is *s-sparse* if at most s of its entries are nonzero.

Suppose you have obtained a solution $\vec{x}$ for $A\,\vec{x} = \vec{b}$ with a few nonzero entries. Then, you may ask yourself if it is the sparsest solution. Here, it becomes relevant to consider the following definition:

The *spark* of a matrix A is the smallest number of columns of A that are linearly dependent.

According to this definition, the vectors in the null space of the matrix (i.e., $A\,\vec{x} = 0$) must satisfy $\|\vec{x}\|_0 \geq spark(A)$.

Now, there is a theorem that establishes the following: if $A\,\vec{x} = \vec{b}$ has a solution $\vec{x}$ obeying $\|\vec{x}\|_0 < spark(A)\,/2$, this solution is necessarily the sparsest possible [26].

16.8.2 Basis Pursuit and Lasso Problems

As said before, the basis pursuit method considers the problem:

Minimize $\|\vec{x}\|_1$, subject to $A\,\vec{x} = \vec{b}$

This problem can be solved with linear programming. The reader can find in the page (https://math.stackexchange.com/questions/1639716/how-can-l-1-norm-minimization-with-linear-equality-constraints-basis-pu) how to use the MATLAB *linprog( )* function for this case. Indeed, interior point algorithms can be also employed.

In basic regression scenarios, the term

$$\left\| A\vec{x} - \vec{b} \right\|_2^2$$

is minimized. In practical applications, this approach may have numerical difficulties or unsatisfactory results (like, for instance, overfitting). By adding penalty terms, both regularization and suitable fitting could be obtained.

To this aim, the so-called *ridge regularization* substitutes the minimization of $\left\| A\vec{x} - \vec{b} \right\|_2^2$ with the minimization of

$$\left(\left\| A\vec{x} - \vec{b} \right\|_2^2 + \lambda \, \|\vec{x}\|_2^2 \right).$$

An $l2$ penalty was added.

One of the good effects of ridge regularization is a certain shrinking of the solution components. According to [55], higher values of λ force these components to be more similar to each other.

In 1996, Robert Tibshirani [60] (58,333 cites) introduced the method *least absolute shrinkage and selection operator (LASSO)*. An $l1$ penalty is added.

The Lasso method can be expressed in two equivalent forms:

- Penalized form:

  Minimize $\left(\left\| A\vec{x} - \vec{b} \right\|_2^2 + \lambda \, \|\vec{x}\|_1 \right)$

- Constraint form:

  Minimize $\left\| A\vec{x} - \vec{b} \right\|_2^2$, subject to $\|\vec{x}\|_1 < c$

The $l1$ penalty has the advantage of promoting sparsity. Contrary to the $l2$ penalty, the $l1$ penalty leads to solutions with many zero components, and this helps to determine the relevant variables. This is a main reason of Lasso popularity.

With respect to analytical solutions, recall that the least squares problem in undetermined systems has the following solution:

$$\vec{x} = (A^T A)^{-1} A^T \, \vec{b}$$

The solution for the ridge regularization would be

$$\vec{x} = (A^T A + \lambda \, I)^{-1} A^T \, \vec{b}.$$

In the case of Lasso, finding the solution is more involved, and many methods have been proposed. A popular one is *LARS* (*least order regression*), which was introduced in 2004 [25] (12,423 cites). It is a stepwise iterative algorithm that starts with $c = 0$ (constraint form) and detects the most influential variables as c increases (see my book [35] for more details and MATLAB code).

In the special case of A being orthonormal (so $A^T A = I$), the ordinary least square (OLS) regression has the following solution, which we will denote as $\hat{x}$:

$$\hat{x} = A^T \vec{b}$$

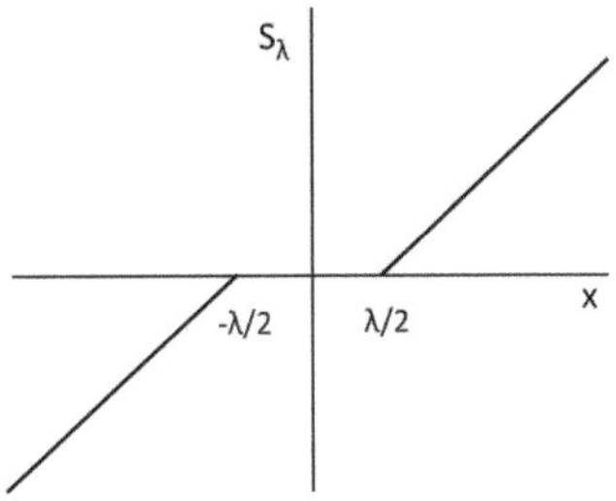

Fig. 16.24 Soft-thresholding operator

- The solution for the ridge regression would be

$$\vec{x} = \frac{1}{1+\lambda}\,\hat{x}.$$

- And the solution for the Lasso would be

$$x_i = S_{\lambda/2}(\hat{x}_i)\ ,\ i = 1,\, 2,\, \ldots, n.$$

In this last expression, the *soft-thresholding operator* $S_{\lambda/2}$ is used, its values being

$$S_{\lambda/2}(x) = \begin{cases} x + (\lambda/2)\,, & if\ x \leq -\lambda/2 \\ 0\,, & if\,|x| < \lambda/2 \\ x - (\lambda/2), & if\ x \geq \lambda/2 \end{cases}$$

The action of this operator is expressed in Fig. 16.24. This operator appears in other optimization contexts.

In a most cited paper, H. Zou and T. Hastie [70] (20,586 cites) introduced the ***elastic net***, which is a two-step method. For the first step, the paper introduces the *naïve elastic net* as the following problem:

$$\text{Minimize}\ \left(\left\|A\vec{x} - \vec{b}\right\|_2^2 + \lambda_1 \left\|\vec{x}\right\|_1 + \lambda_2 \left\|x\right\|_2^2\right)$$

Denote as $\tilde{x}$ the estimate obtained by the first step. The second step is just the following correction:

$$\vec{x} = \frac{\tilde{x}}{\sqrt{1+\lambda_2}}$$

The elastic net can be regarded as compromise between ridge regression and Lasso. Because of the quadratic term, the elastic net penalty is strictly convex.

16.8.3 First-Order Methods

Most sparse approximation problems can be tackled with LP or QP algorithms. However, there are applications in which, due to the size or the characteristics of the problem, computational difficulties appear [5] (13,342 cites). Some of these difficulties are detailed in [28] (4336 cites). Therefore, other solution alternatives have been explored, including first-order methods.

An example of gradient-based method is the ***gradient projection sparse representation (GPSR)*** proposed in [28]. Positive and negative variables are separated, and the problem

$$\text{Minimize } \left(\frac{1}{2}\left\|\vec{b} - A\vec{x}\right\|_2^2 + \lambda\ \|\vec{x}\|_1\right)$$

is rewritten as

$$\text{Minimize } Q(\vec{x}) = \left(\frac{1}{2}\left\|\vec{b} - [A, -A][\vec{x}_+;\ \vec{x}_-]\right\|_2^2 + \lambda\, 1^T\,(\vec{x}_+ + \vec{x}_-)\right)$$

with $\vec{x}_+ \geq 0;\ \vec{x}_- \geq 0.$

Now, denote

$$\vec{z} = [\vec{x}_+; \vec{x}_-],\ \vec{c} = \lambda\, 1 + [-A^T\vec{b};\ A^T\vec{b}], \text{ and:}$$

$$B = \begin{pmatrix} A^T A & -A^T A \\ -A^T A & A^T A \end{pmatrix}.$$

Hence,

$$Q(\vec{z}) = \frac{1}{2}\vec{z}^T B\,\vec{z} + \vec{c}^T\vec{z},$$

and the gradient $\nabla_z\, Q(\vec{z}) = B\,\vec{z} + \vec{c}$.

Then, a steepest descent algorithm can be enounced:

$$\vec{z}_{k+1} = \vec{z}_k - \alpha_k\, \nabla_z\, Q(\vec{z}_k)$$

It was shown in [28] that the computation of the matrix B can be done more economically than its size suggests and that the gradient of Q requires one multiplication each by A and A^T. No inversion of matrices is needed.

16.8.3.1 ISTA Method

There is an important family of methods in connection with the *iterative shrinkage/thresholding (IST)* algorithm. Initially, IST was introduced as an expectation-maximization (EM) algorithm for image deconvolution [48]. According to [28], it can be used for Lasso problems and only requires matrix-vector multiplications involving A and A^T. Let us describe the idea; suppose you have the following composite problem:

Minimize $f(\vec{x}) + \lambda\, g(\vec{x})$

Consider the next iterative approximation of the solution,

$$\vec{x}_{k+1} = \arg\min_x\{f(\vec{x}_k) + (\vec{x} - \vec{x}_k)^T \nabla f(\vec{x}_k) + \tfrac{1}{2}\, \|\vec{x} - \vec{x}_k\|_2^2\, \nabla^2 f(\vec{x}_k) + \lambda g(\vec{x})\}$$
$$\approx \arg\min_x\{(\vec{x} - \vec{x}_k)^T \nabla f(\vec{x}_k) + \tfrac{\alpha_k}{2}\, \|\vec{x} - \vec{x}_k\|_2^2 + \lambda\, g(\vec{x})\}$$

(the Hessian $\nabla^2 f(\vec{x}_k)$ is approximated by a diagonal matrix $\alpha_k I$).

Denote

$$\vec{u}_k = \vec{x}_k - \frac{1}{\alpha_k} \nabla f(\vec{x}_k).$$

Then, the iterative procedure can be written as

$$\vec{x}_{k+1} = \arg\min_x\{\ \frac{1}{2}\ \|\vec{x} - \vec{u}_k\|_2^2 + \frac{\lambda}{\alpha_k}\, g(\vec{x})\}$$

(the $\approx$ was replaced by =).

In the case of $g(\vec{x})$ being $\|\vec{x}\|_1$, which is separable, then there is a component-wise closed-form solution,

$$(\vec{x}_{k+1})_i = \arg\min_x\{\ \frac{1}{2}\ ((\vec{x})_i - (\vec{u}_k)_i)^2 + \frac{\lambda}{\alpha_k}\, |(\vec{x})_i|\} = S_T((\vec{u}_k)_i)$$

where we used $(\vec{x})_i$ to represent the *i-th* component of $\vec{x}$. The last, right-hand term includes a soft-thresholding operator $S_T(y)$:

$$S_T(y) = \begin{cases} y + T\,, & if\ y \le -T \\ 0\,, if\ |y| < T \\ y - T, & if\ y \ge T \end{cases}$$

with $T = \frac{\lambda}{\alpha_k}$.

Therefore, a simple iterative shrinkage/thresholding algorithm, named ISTA, has been devised [5]. There are two parameters to specify, λ and α_k. The research has proposed a number of strategies for this specification.

Another equivalent form of expressing the soft-thresholding operator is

$$S_T(y) = sgn(y) \cdot (|y| - T)_+.$$

In case the composite problem was of Lasso-type,

Minimize $\left(\frac{1}{2}\left\|\vec{b}-A\vec{x}\right\|_2^2+\lambda\,\|\vec{x}\|_1\right)$,

then, $\nabla f(\vec{x}_k)=-A^T(\vec{b}-A\vec{x}_k)$.

And so, the iterative algorithm would be

$$\vec{x}_{k+1} = S_T\left(\vec{x}_k+\frac{1}{\alpha_k}A^T(\vec{b}-A\,\vec{x}_k)\right) \qquad (1)$$

(the operator should be component-wise applied).

The algorithm is clearly related with the ***Landweber algorithm***, introduced in the 1950s to solve ill-posed linear inverse problems and that has the following general expression,

$$\vec{x}_{k+1} = \vec{x}_k+\lambda\,A^T M(\vec{b}-A\,\vec{x}_k)$$

(where M is a positive symmetric matrix and λ is a parameter).

An example of ISTA coding in MATLAB will be given in the next subsection. Another implementation example can be found in [56].

16.8.3.2 FISTA Method

Since the ISTA algorithm involves a gradient descent, it is a good idea to use the Nesterov's accelerated gradient descent. The result is the *fast iterative shrinkage/thresholding algorithm (FISTA)*, which was introduced in 2009 [5].

While ISTA can be expressed as

$$\vec{x}_{k+1} = S_{\lambda/\alpha_k}(\vec{x}_k-\frac{1}{\alpha_k}\nabla f(\vec{x}_k)),$$

FISTA can be expressed as the next three steps:

$$\vec{x}_{k+1} = S_{\lambda/L}(\vec{y}_k-\frac{1}{L}\nabla f(\vec{y}_k))$$

$$t_{k+1} = \frac{1+\sqrt{1+4t_k^2}}{2}$$

$$\vec{y}_{k+1} = \vec{x}_k+\left(\frac{t_k-1}{t_{k+1}}\right)(\vec{x}_{k+1}-\vec{x}_k)$$

There are direct generalizations of the ISTA and FISTA algorithms by replacing the soft-thresholding operator with the proximal operator. This is frequent in the most recent literature. In particular, ISTA is equivalent to the proximal gradient method.

Another idea of Nesterov was to substitute non-smooth objective functions by suitable smooth approximations [46] (3173 cites). One application of this idea for our problem is to consider a smooth approximation of the $l1$ norm minimization as follows:

- First, the minimization is written as minimax:

$$\min_x \|\vec{x}\|_1 = \min_x \max_u \vec{u}_T\vec{x} \ ; \quad -1 \leq u_i \leq 1$$

- Second, the smoothed approximation is introduced:

$$smoothed \min_x \|\vec{x}\|_1 = \min_x \max_u \vec{u}_T\vec{x} \ + \frac{\mu}{2} \|\vec{u}\|_2^2$$

Based on this approach, a popular algorithm called NESTA was introduced in [6] (1445 cites). Other relevant algorithms are TwIST [7] (2315 cites), SALSA [2] (1454 cites), and SpaRSA [63] (2271 cites).

16.8.4 Alternating Directions

Let us now focus on the *alternating direction method of multipliers (ADMM)*. According to [66] (1454 cites), this approach was not widely employed in the field of image and signal processing until recently, from 2009. It is an iterative method.

Consider the problem

Minimize $f(\vec{x}) + g(\vec{x})$

where $f(\vec{x})$ and $g(\vec{x})$ are closed proper convex functions. Both functions can be non-smooth.

The ADMM method can be expressed as follows [51]:

$$\vec{x}_{k+1} = \mathrm{Pr}\,ox_{\lambda\, f}(\vec{z}_k - \vec{u}_k)$$

$$\vec{z}_{k+1} = \mathrm{Pr}\,ox_{\lambda\, g}(\vec{x}_{k+1} + \vec{u}_k)$$

$$\vec{u}_{k+1} = \vec{u}_k + \vec{x}_{k+1} - \vec{z}_{k+1}$$

Typically, $g(\vec{x})$ encodes the constraints and $\vec{z}_k \in dom\ g(\vec{x})$ (the domain of $g(\vec{x})$). On the other hand, $\vec{x}_k \in dom\ f(\vec{x})$. Along the iterations, $\vec{x}_k$ and $\vec{z}_k$ converge to each other. The ADMM method is helpful when the proximal mappings of $f(\vec{x})$ and $g(\vec{x})$ are easy to compute but the proximal mapping of $f(\vec{x}) + g(\vec{x})$ is not so.

Equivalently, the method can be expressed in terms of an ***augmented Lagrangian***, starting from the following problem,

Minimize $f(\vec{x}) + g(\vec{z})$ subject to $\vec{x} - \vec{z} = 0$

(the problem is expressed in *consensus form*).

An augmented Lagrangian can be written

$$L(\vec{x}, \vec{y}, \vec{z}) = f(\vec{x}) + g(\vec{z}) + \vec{y}^T (\vec{x} - \vec{z}) + \frac{\rho}{2} \|\vec{x} - \vec{z}\|_2^2 ,$$

and the ADMM would be

$$\vec{x}_{k+1} = \arg\min_x L(\vec{x}, \vec{z}_k, \vec{y}_k)$$

$$\vec{z}_{k+1} = \arg\min_z L(\vec{x}_{k+1}, \vec{z}, \vec{y}_k)$$

$$\vec{y}_{k+1} = \vec{y}_k + \rho (\vec{x}_{k+1} - \vec{z}_{k+1}).$$

It is shown in [51] that these three equations can be translated to the proximal form introduced above.

One of the applications of ADMM treated in [8] (22162 cites) is the basis pursuit (BP) problem. This problem can be written as

Minimize $f(\vec{x}) + \|\vec{z}\|_1$

Subject to $\vec{x} - \vec{z} = 0$

where $f(\vec{x})$ is the indicator function of the set $C = \{\vec{x} \mid A\vec{x} = \vec{b}\}$. In this setting, the ADMM iterations are the following,

$$\vec{x}_{k+1} = P_C(\vec{z}_k - \vec{u}_k)$$

$$\vec{z}_{k+1} = S_{(1/\rho)}(\vec{x}_{k+1} + \vec{u}_k)$$

$$\vec{u}_{k+1} = \vec{u}_k + \vec{x}_{k+1} - \vec{z}_{k+1}$$

where $S_{(1/\rho)}$ is the soft-thresholding operator and the projection P_C can be computed as follows:

$$\vec{x}_{k+1} = (I - A^T (A\,A^T)^{-1} - A)(\vec{z}_k - \vec{u}_k) + A^T (A\,A^T)^{-1}\,\vec{b}$$

Program 16.13 provides an implementation example of ADMM for the BP problem, closely related to the examples given in an ADMM web page (https://web.stanford.edu/$\sim$boyd/papers/admm/). The program assigns random values to the matrix A and to $\vec{b}$. Then, it finds the sparsest solution to $A\vec{x} = \vec{b}$ (undetermined system). The program has a second part in which ISTA is applied to the same problem.

Each time the program is run, different matrix A and vector $\vec{b}$ are generated, and different sparsest solutions are obtained.

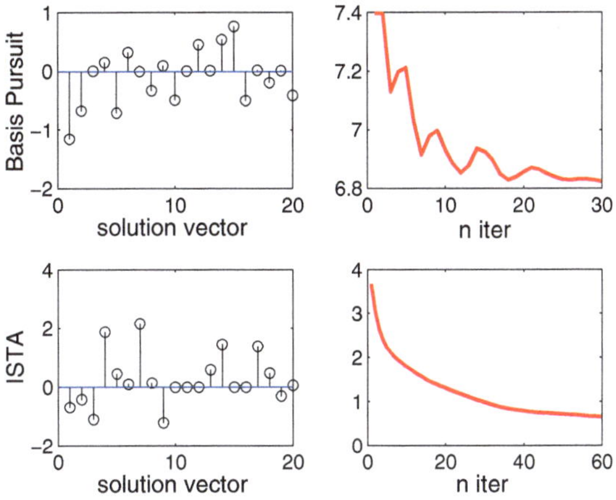

Fig. 16.25 Results of ADMM-BP and ISTA

Figure 16.25 has been generated by Program 16.13 and visualizes the results with ADMM-BP and with ISTA. The two algorithms give different solutions to the problem. Actually, ADMM-BP solves more accurately the equation $A\vec{x} = \vec{b}$.

Notice that the evolution of the objective function in the case of ISTA is given in logarithms.

The program also yields numerical results about $\|\vec{x}\|_1$ ($l1$ norm of the solution) and about $\left\| \vec{b} - A\vec{x} \right\|_2$.

For example, the numerical results in the specific case visualized in Fig. 16.26 were the following:

ADMM-BP, norm1(x) = 3.6388, norm2(b-Ax) = 8.28 e-015
ISTA (Lasso), norm1(x) = 0.9268, norm2(b-Ax) = 1.9729

Note (Program 16.13)

The complete listing of the program 16.13, about ADMM, has been included in the Appendix for Long Programs.

16.9 Image Processing Experiments

This final section comes back to images, having paid attention to ADMM in the precedent pages.

We shall include two experiments, one related with denoising and the other to photo restoration.

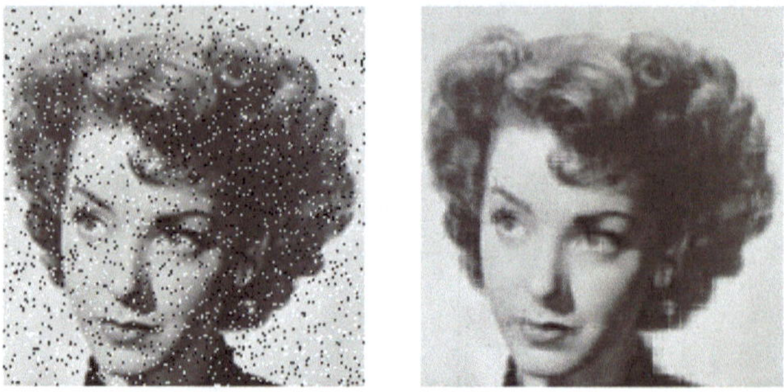

Fig. 16.26 Salt & pepper denoising with ADMM

16.9.1 Salt and Pepper Noise

Salt & pepper noise is a representative example of impulsional noise. It is formed at random by completely black or completely white pixels (or perhaps almost completely B or W). It is an extreme case, difficult to eradicate.

In our experiment, salt & pepper noise was added to a photograph, and then ADMM was applied for denoising by solving the following optimization problem,

Minimize $\|L\|_* + \lambda \ \|S\|_1$

Subject to $L + S = M$,

where $\|L\|_*$represents the nuclear norm of the low-rank matrix L, S is a sparse matrix, and the matrix M is the noisy observation. All this is related with matrix completion.

Figure 16.26 shows on the left the original noisy image we prepared for the experiment and on the right the result of the denoising.

The denoising has been obtained with Program 16.14, which is based on the MATLAB implementation developed by [1], available from GitHub web site.

Program 16.14 Salt and pepper noise elimination with ADMM

```
% ADMM denoising example
% Elimination of salt & pepper noise
% source image and parameters -----------
F = imread('martaBW.jpg');
sigma = 0.1; %density of the noise
M = imnoise(F,'salt & pepper',sigma);
M = double(M);
lamb = 0.1;
rho = 0.01;
niter=70;
L = zeros(size(M)); % Low rank matrix
S = zeros(size(M)); % Sparse matrix
```

```
Mu = zeros(size(M)); % Lagrange multipliers
% main loop -------------------------------
for i = 1:niter
  % L subproblem, singular value thresholding
  X = M + Mu./rho;
  [U, P, V] = svd(X-S,'econ');
  L = U*shrink1(P,1/rho)*V';
  % S subproblem. soft thresholding
  S = shrink1(X-L, lamb/rho);
  % Lagrange multipliers update
  Mu = Mu - rho*(L+S-M);
end
% display --------------------
figure(1)
subplot(1,2,1)
imshow(uint8(M));
title('original noisy image');
subplot(1,2,2)
imshow(uint8(L));
title('restored image');
function Z = shrink1(X,r) %Shrinkage
Z = sign(X).*max(abs(X)- r,0);
end
```

16.9.2 Inpainting

Image restoration is a popular topic, with a wide application perspective covering damaged old photographs, paintings, or even movies. An established term for the topic is *image inpainting* . For the interested reader, the surveys [37] and [58] would be useful. The academic thesis [18] presents a good rationale of key techniques, including MATLAB programs. And [27] gives an interesting survey on variational approaches on this topic.

The MATLAB Signal Processing Toolbox includes the *inpaintExemplar( )* and *inpaintCoherent( )* functions for two inpainting methods. GitHub offers 13 repositories of MATLAB software for image inpainting. Nowadays, deep learning (which also is based on optimization) is obtaining notable results on this field. The MathWorks File Exchange site offers a repository with five different inpainting methods:

https://es.mathworks.com/matlabcentral/fileexchange/55326-matlab-codes-for-the-image-inpainting-problem

Among many old photographs of my family, a notably damaged one was selected for this experiment. This image is shown on the left-hand side of Fig. 16.27.

Several inpainting implementations are available from the Internet. After testing of alternatives, an example from [65] was selected as a basis in order to devise

Fig. 16.27 Image restoring with inpainting algorithm

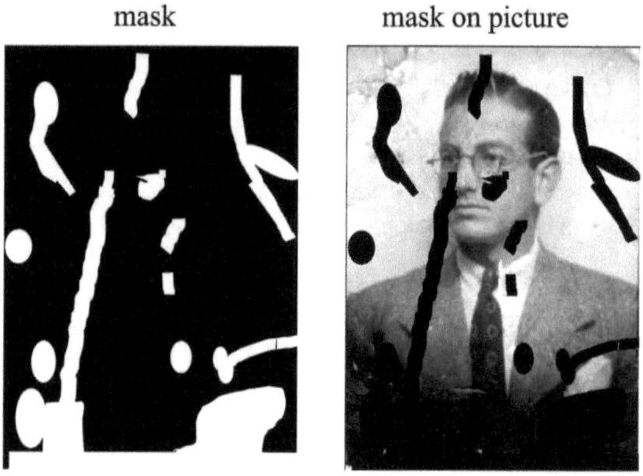

Fig. 16.28 Mask preparation and checking

Program 16.15 in MATLAB. It is an application of the split Bregman iterations to a total variation L1-norm analysis of images, as proposed by Pascal Getreuer in [33].

The result of the program is shown on the right of Fig. 16.27. The most significant defects were eliminated. Minor details were left for manual retouching (perhaps with a magic wand).

As said by Getreuer in [34], inpainting is a matter of image completion, inferring the image content in corrupted regions. In principle, it is an interpolation problem, where reconstruction has to agree with the boundaries of problematic regions.

A main idea of the procedure is to prepare a mask indicating the regions to be restored. In our specific experiment, we prepared the mask shown on the left of Fig. 16.28. In part, this mask corresponds to the application of a gray-level threshold to the image, so pixels become black or white according to the threshold. Some elliptical patches were also added.

The mask is converted to its negative and then applied to the photograph. After a few iterations of the restoring algorithm, it is possible to check the correct positioning of the mask on the image to be restored. The mask appears in black. After hundreds of iterations, the mask becomes more and more faded, taking color from borders.

The program uses five functions. Notice that 4000 iterations were run, taking about 10 minutes.
Note (Program 16.15)

The complete listing of the program 16.15 has been included in the Appendix for Long Programs.

Quote:

"You don't take a photograph, you make it"
Ansel Adams.

16.10 Exercises

1- [16.1] [16.2] [16.3] Contrast enhancement of a photograph seems to be an easy task, only that it is a subjective matter. The research is proposing automatic ways for getting a better contrast. This exercise is simply to use a piece of software from GitHub, in [16.1], and check how well it works on the next photograph (from [16.2]). The software combines evolutionary algorithms. The page [16.3] could be useful for better understanding of the problem and its history. Here is the image to be improved:

Please apply the cited software to the image. Analyze the behavior of the evolutionary algorithms. Perhaps some modifications of the code or parameters could be opportune. You can also try [16.4], with a particle swarm optimization algorithm.
2- [16.5] [16.6] Edge detection in images is quite important for pattern recognition, some robotic functions, etc. Heuristic algorithms can be applied for this task. We selected in [16.7] the next photograph as subject of study.

Please apply ant colony method for edge detection in this photograph. The article [16.6] can serve as a suitable guide. There is an implementation in MATLAB that can be accessed from [16.5]. Probably some parameters should be adjusted. Visualize the results, and add comments on the detection quality.

3- [16.8] [16.9] [16.10] First photographs taken by the Hubble Space Telescope in 1990 turned up to be blurred. A shape flaw in the primary mirror of the telescope was causing spherical aberration. Here is the image of Saturn (see [16.8]) as captured by the Hubble:

Please apply the Lucy-Richardson algorithm for image deblurring. Use the uniform-out-of-focus PSF, and try changes in the parameters of the algorithm. Comment the results.

Perhaps you could obtain from the Internet a more suitable PSF.

4- [16.11] [16.12] Medical image processing makes an extensive use of segmentation to delineate tumors, lesions, vessels, etc. The next image has been obtained by magnetic resonance and corresponds to a brain seen from the top. Some tumors can be noticed.

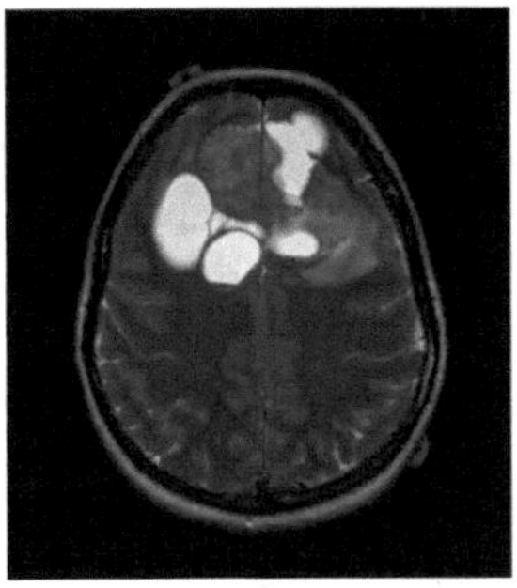

The image (named glioma(110).jpeg) was taken from a Kaggle dataset [16.11].

Please apply the Chan-Vese model, with the program included in Chap. 16, in order to highlight the tumors. Since the central problem is the minimization of an "energy," it is also possible to employ an evolutionary algorithm. Apply a fast bio-inspired algorithm for image segmentation, and compare the result and the processing time with the previous approach.

5- [16.13] Photographs of moving objects, like cars, horses, birds, etc., can be blurred. Experts in photography matters would tell in this respect about camera shutter speed. Sometimes, those experts do explicitly intend to capture motion by getting blurred images. Our experiment in this exercise is to use a motion-blurred picture and try to deblur it. Here is a photograph (see [16.10]) that seems to have motion blur:

Please apply the Lucy-Richardson algorithm for image deblurring, with the PSF for motion blur. Try changes in the PSF (motion direction?) and algorithm parameters. Comment results.

6- [16.14] [16.15] The ADMM method can also be used for SVM training. In the GitHub page [16.14], it is clearly described how this can be done. The page includes MATLAB code and a large dataset for discovering spam messages on the Internet.

More information on spam and related problems can be found in [16.15].

Please follow the guide offered by [16.14] for spam classification. Visualize the results, and add comments.

Most probably you could find on the web other spam datasets for comparison purposes.

7- [16.16] [16.17] Apart from the Lucy-Richardson algorithm, there are other methods that can be employed for image deblurring. In this exercise, we propose to use ISTA and total variation (TV), following the guide offered by [16.16] and [16.17]. We shall produce a blurred photograph, and then we shall try to recover the original.

Motion blur can be caused by camera shaking. The MATLAB function *fspecial( )*, provides a simple way to simulate motion blur whatever the cause is.

This is the original picture, [16.18], where you can read some texts:

Please create a motion-blurred version of this picture. Then, apply ISTA+TV for image deblurring. Comment the result (can you read the texts?) (how about the horse, etc.).

16.10.1 Exercise References

[16.1] https://github.com/pouryahoseini/Nature-Inspired-Image-Enhancement/blob/master/im-enhance/imenhance.m
[16.2] https://www.naturettl.com/black-and-white-safari-photography/
[16.3] https://github.com/Gdahuks/Image_contrast_enhancement
[16.4] https://github.com/dleliuhin/PSO-Image-Enhancementer/blob/master/mainPso.m
[16.5] https://es.mathworks.com/matlabcentral/fileexchange/20997-image-edge-detection-using-ant-colony-optimization
[16.6] https://www.math.md/files/csjm/v23-n3/v23-n3-(pp343-359).pdf
[16.7] https://photocontest.smithsonianmag.com/photocontest/latest/
[16.8] https://www.researchgate.net/figure/Raw-image-of-planet-Saturn-obtained-with-the-WF-PC-camera-of-the-Hubble-Space-Telescope_fig6_46772094
[16.9] https://www.researchgate.net/publication/46772094_Image_restoration_in_astronomy_-_A_Bayesian_perspective

[16.10] https://www.stsci.edu/hst/instrumentation/focus-and-pointing/focus/tiny-tim-hst-psf-modeling
[16.11] https://www.kaggle.com/datasets/fernando2rad/brain-tumor-mri-images-17-classes
[16.12] https://es.mathworks.com/matlabcentral/fileexchange/106050-pso-sa-image-segmentation
[16.13] https://bikepacking.com/plog/riders-lens-will-keating/
[16.14] https://github.com/tianrui-qi/ADMM-for-SVM?tab=readme-ov-file
[16.15] https://link.springer.com/article/10.1007/s10462-022-10195-4
[16.16] https://htmlpreview.github.io/?https://github.com/Yunhui-Gao/ISTA/blob/master/docs/README.html
[16.17] https://htmlpreview.github.io/?https://github.com/Yunhui-Gao/ISTA/blob/master/docs/README.html#example-2-image-deblurring
[16.18] https://www.shorpy.com/node?page=960

References

1. Adam, T. (2019). Image-inverse-problems-algorithms. Available from: https://github.com/YW81/Image-Inverse-Problems-Algorithms/blob/master/RPCA%20Image%20Inpainting/RPCA_ADMM.m.
2. Afonso, M. V., Bioucas-Dias, J. M., & Figueiredo, M. A. (2010). Fast image recovery using variable splitting and constrained optimization. *IEEE Transactions on Image Processing, 19*(9), 2345–2356. Available from: http://www.lx.it.pt/~mtf/Afonso_BioucasDias_Figueiredo_twocolumn_v7.pdf.
3. Artacho, F. J. A., Borwein, J. M., & Tam, M. K. (2013). Recent results on Douglas-Rachford methods. *Serdica Mathematical Journal, 39*(3-4), 313p–330p. Available from: http://sci-gems.math.bas.bg/jspui/bitstream/10525/3442/1/2013-313-330.pdf.
4. Aujol, J.-F., Gilboa, G., Chan, T., & Osher, S. (2006). Structure-texture image decomposition–modeling, algorithms, and parameter selection. *International Journal of Computer Vision, 67*, 111–136. Available from: https://link.springer.com/content/pdf/10.1007/s11263-006-4331-z.pdf.
5. Beck, A., & Teboulle, M. (2009). A fast iterative shrinkage-thresholding algorithm for linear inverse problems. *SIAM Journal on Imaging Sciences, 2*(1), 183–202. Available from: https://www.math.mcgill.ca/yyang/comp/papers/FISTA.pdf.
6. Becker, S., Bobin, J., & Candès, E. J. (2011). Nesta: A fast and accurate first-order method for sparse recovery. *SIAM Journal on Imaging Sciences, 4*(1), 1–39. Available from: https://arxiv.org/pdf/0904.3367.pdf.
7. Bioucas-Dias, J. M., & Figueiredo, M. A. (2007). A new twist: Two-step iterative shrinkage/thresholding algorithms for image restoration. *IEEE Transactions on Image Processing, 16*(12), 2992–3004. Available from: http://www.lx.it.pt/~mtf/IEEE_TIP_2007_Dias_Figueiredo.pdf.
8. Boyd, S., Parikh, N., Chu, E., Peleato, B., Eckstein, J., et al. (2011). Distributed optimization and statistical learning via the alternating direction method of multipliers. *Foundations and Trends® in Machine Learning, 3*(1), 1–122. Available from: https://members.cbio.mines-paristech.fr/~jvert/svn/bibli/local/Boyd2011Distributed.pdf.
9. Breen, P. (2009). Algorithms for sparse approximation. *School of Mathematics, University of Edinburgh, Year*, 4.

10. Bregman, L. M. (1967). The relaxation method of finding the common point of convex sets and its application to the solution of problems in convex programming. *USSR Computational Mathematics and Mathematical Physics, 7*(3), 200–217.
11. Buades, A., Le, T. M., Morel, J.-M., & Vese, L. A. (2010). Fast cartoon+ texture image filters. *IEEE Transactions on Image Processing, 19*(8), 1978–1986. Available from: https://www.researchgate.net/profile/Jean-Michel-Morel-2/publication/42638728_Fast_Cartoon_Texture_Image_Filters/links/5da6f96c4585159bc3d0dab7/Fast-Cartoon-Texture-Image-Filters.pdf.
12. Bush, J. (2011). Bregman algorithms. *Senior Thesis. University of California, Santa Barbara.* Available from: https://ccs.math.ucsb.edu/senior-thesis/Jacqueline-Bush.pdf.
13. Cai, J.-F., Candès, E. J., & Shen, Z. (2010). A singular value thresholding algorithm for matrix completion. *SIAM Journal on Optimization, 20*(4), 1956–1982. Available from: https://arxiv.org/pdf/0810.3286.
14. Candes, E., & Recht, B. (2012). Exact matrix completion via convex optimization. *Communications of the ACM, 55*(6), 111–119. Available from: https://dl.acm.org/doi/pdf/10.1145/2184319.2184343.
15. Candès, E. J., Li, X., Ma, Y., & Wright, J. (2011). Robust principal component analysis? *Journal of the ACM (JACM), 58*(3), 1–37. Available from: https://members.cbio.mines-paristech.fr/~jvert/svn/bibli/local/Candes2011Robust.pdf.
16. Candes, E. J., & Plan, Y. (2010). Matrix completion with noise. *Proceedings of the IEEE, 98*(6), 925–936. Available from: https://web.archive.org/web/20170815094135id_/http://authors.library.caltech.edu/18601/1/Candes2010p10178P_Ieee.pdf.
17. Caselles, V., Chambolle, A., Novaga, M., et al. (2015). Total variation in imaging. *Handbook of Mathematical Methods in Imaging, 1*(2), 3. Available from: https://people.dm.unipi.it/novaga/papers/chambolle/TVHandbook/TVHandbook6.pdf.
18. Cebollero, C. M. (2020). Numerical resolution of inpainting methods. Available from: https://zaguan.unizar.es/record/97973/files/TAZ-TFG-2020-3227.pdf?version=1.
19. Censor, Y., & Lent, A. (1981). An iterative row-action method for interval convex programming. *Journal of Optimization Theory and Applications, 34*(3), 321–353. Available from: https://link.springer.com/article/10.1007/BF00934676.
20. Chambolle, A., & Lions, P.-L. (1997). Image recovery via total variation minimization and related problems. *Numerische Mathematik, 76*, 167–188. Available from: https://link.springer.com/content/pdf/10.1007/s002110050258.pdf.
21. Chan, T. F., & Esedoglu, S. (2005). Aspects of total variation regularized l 1 function approximation. *SIAM Journal on Applied Mathematics, 65*(5), 1817–1837. Available from: https://dept.math.lsa.umich.edu/~esedoglu/Papers_Preprints/tvL1_chan_esedoglu_revised.pdf.
22. Combettes, P. L., & Pesquet, J.-C. (2011). Proximal splitting methods in signal processing. *Fixed-point Algorithms for Inverse Problems in Science and Engineering*, 185–212. Available from: https://arxiv.org/pdf/0912.3522.pdf.
23. Donoho, D. L. (2006). Compressed sensing. *IEEE Transactions on Information Theory, 52*(4), 1289–1306. Available from: https://www.cmor-faculty.rice.edu/~yzhang/caam699/Image%20papers/CompSensing.pdf.
24. Eckstein, J. (1989). *Splitting methods for monotone operators with applications to parallel optimization.* PhD thesis, Massachusetts Institute of Technology.
25. Efron, B., Hastie, T., Johnstone, I., & Tibshirani, R. (2004). Least angle regression. *The Annals of Statistics, 32*(2), 407–499. Available from: https://members.cbio.mines-paristech.fr/~jvert/svn/bibli/local/Efron2004Least.pdf.
26. Elad, M. (2010). *Sparse and redundant representations: from theory to applications in signal and image processing.* Springer Science & Business Media. Available from: https://link.springer.com/book/10.1007/978-1-4419-7011-4.
27. Fidaner, I. B. (2008). *A survey on variational image inpainting, texture synthesis and image completion.* Bogazici University. Available from: https://fidaner.wordpress.com/wp-content/uploads/2013/10/survey_draft3.pdf.
28. Figueiredo, M. A., Nowak, R. D., & Wright, S. J. (2007). Gradient projection for sparse reconstruction: Application to compressed sensing and other inverse problems. *IEEE Journal*

of Selected Topics in Signal Processing, 1(4), 586–597. Available from: https://optimization-online.org/wp-content/uploads/2007/06/1704.pdf.
29. Fish, D., Brinicombe, A., Pike, E., & Walker, J. (1995). Blind deconvolution by means of the Richardson-Lucy algorithm. *JOSA A, 12*(1), 58–65. Available from: https://prancer.physics.louisville.edu/classes/650/deconvolution/fish_blind_lucy_richardson_deconvolution_josa1995.pdf.
30. Gandy, S., & Yamada, I. (2010). Convex optimization techniques for the efficient recovery of a sparsely corrupted low-rank matrix. *Journal of Math-for-Industry, 2*, 147–156. Available from: https://catalog.lib.kyushu-u.ac.jp/opac_download_md/18491/JMI2010B-5.pdf.
31. Getreuer, P. (2012a). Chan-Vese segmentation. *Image Processing On Line, 2*, 214–224. Available from: https://www.ipol.im/pub/art/2012/g-cv/article.pdf.
32. Getreuer, P. (2012b). Rudin-Osher-Fatemi total variation denoising using split Bregman. *Image Processing On Line, 2*, 74–95. Available from: https://www.ipol.im/pub/art/2012/g-tvd/article.pdf.
33. Getreuer, P. (2012c). Total variation inpainting using split Bregman. *Image Processing On Line, 2*, 147–157.
34. Getreuer, P. (2012d). Total variation inpainting using split Bregman. Available from: https://getreuer.info/papers/getreuer2012tvinpainting/index.html.
35. Giron-Sierra, J. M. (2017). *Digital signal processing with matlab examples, Volume 3*. Springer. Available from: https://link.springer.com/book/10.1007/978-981-10-2537-2.
36. Goldstein, T., & Osher, S. (2009). The split Bregman method for L1-regularized problems. *SIAM Journal on Imaging Sciences, 2*(2), 323–343. Available from: https://hannes.nickisch.org/code/glm-ie/pls/goldsteinOsher10splitBregman.pdf.
37. Jam, J., Kendrick, C., Walker, K., Drouard, V., Hsu, J. G.-S., & Yap, M. H. (2021). A comprehensive review of past and present image inpainting methods. *Computer Vision and Image Understanding, 203*, 103147. Available from: https://drive.google.com/file/d/1_8xa65kNwLpeygQbopg6Ky5gy_ZgxfY2/view.
38. Kutyniok, G. (2013). Theory and applications of compressed sensing. *GAMM-Mitteilungen, 36*(1), 79–101. Available from: https://www.ai.math.uni-muenchen.de/publications/survey/survey_cs_revisions.pdf.
39. Le Guen, V. (2014). Cartoon+ texture image decomposition by the TV-L1 model. *Image Processing On Line, 4*, 204–219. Available from: https://hal.science/hal-03041129/document.
40. Lin, Z. (2011). Some software packages for partial SVD computation. *arXiv preprint arXiv:1108.1548*. Available from: https://arxiv.org/pdf/1108.1548.pdf.
41. Lin, Z., Chen, M., & Ma, Y. (2010). The augmented Lagrange multiplier method for exact recovery of corrupted low-rank matrices. *arXiv preprint arXiv:1009.5055*. Available from: https://arxiv.org/pdf/1009.5055.pdf.
42. Lin, Z., Ganesh, A., Wright, J., Wu, L., Chen, M., & Ma, Y. (2009). Fast convex optimization algorithms for exact recovery of a corrupted low-rank matrix. *Coordinated Science Laboratory Report no. UILU-ENG-09-2214, DC-246*. Available from: https://people.eecs.berkeley.edu/~yima/matrix-rank/Files/rpca_algorithms.pdf.
43. Meyer, Y. (2001). *Oscillating patterns in image processing and nonlinear evolution equations: the fifteenth Dean Jacqueline B. Lewis memorial lectures*, volume 22. American Mathematical Soc. Available from: https://dl.acm.org/doi/abs/10.5555/516207.
44. Mukherjee, S. (2019). Image_denoising_split_Bregman. Available from: https://github.com/sovanlal/Image_Denoising_split_Bregman.
45. Mumford, D. B., & Shah, J. (1989). Optimal approximations by piecewise smooth functions and associated variational problems. *Communications on Pure and Applied Mathematics*. Available from: https://dash.harvard.edu/bitstream/handle/1/3637121/Mumford_OptimalApproxPiece.pdf?sequence=1.
46. Nesterov, Y. (2005). Smooth minimization of non-smooth functions. *Mathematical Programming, 103*, 127–152. Available from: https://link.springer.com/content/pdf/10.1007/s10107-004-0552-5.pdf.

47. Nikolova, M. (2004). A variational approach to remove outliers and impulse noise. *Journal of Mathematical Imaging and Vision, 20*(1), 99–120. Available from: https://link.springer.com/content/pdf/10.1023/B:JMIV.0000011326.88682.e5.pdf.
48. Nowak, R. D., & Figueiredo, M. A. (2001). Fast wavelet-based image deconvolution using the em algorithm. In *Conference Record of Thirty-Fifth Asilomar Conference on Signals, Systems and Computers (Cat. No. 01CH37256)*, volume 1, pages 371–375. IEEE. Available from: http://www.lx.it.pt/~mtf/Asilomar_2001.pdf.
49. Osher, S., Burger, M., Goldfarb, D., Xu, J., & Yin, W. (2005). An iterative regularization method for total variation-based image restoration. *Multiscale Modeling & Simulation, 4*(2), 460–489. Available from: https://search.iczhiku.com/paper/Rjc92fh0U2S7cS2z.pdf.
50. Osher, S., & Sethian, J. A. (1988). Fronts propagating with curvature-dependent speed: Algorithms based on hamilton-jacobi formulations. *Journal of Computational Physics, 79*(1), 12–49. Available from: https://ntrs.nasa.gov/api/citations/19880001113/downloads/19880001113.pdf.
51. Parikh, N., Boyd, S., et al. (2014). Proximal algorithms. *Foundations and Trends® in Optimization, 1*(3), 127–239. Available from: https://web.stanford.edu/~boyd/papers/pdf/prox_algs.pdf.
52. Peyre, G. (2014). Matrix completion with nuclear norm minimization. Available from: https://github.com/gpeyre/numerical-tours/tree/master/matlab/solutions/sparsity_3_matrix_completion.
53. Rodriguez, P., & Wohlberg, B. (2013). Fast principal component pursuit via alternating minimization. In *2013 IEEE International Conference on Image Processing*, pp. 69–73. IEEE. Available from: https://brendt.wohlberg.net/publications/pdf/rodriguez-2013-fast.pdf.
54. Rudin, L. I., Osher, S., & Fatemi, E. (1992). Nonlinear total variation based noise removal algorithms. *Physica D: Nonlinear Phenomena, 60*(1-4), 259–268. Available from: https://members.cbio.mines-paristech.fr/~jvert/svn/bibli/local/Rudin1992Nonlinear.pdf.
55. Schmidt, M. (2005). Least squares optimization with L1-norm regularization. *CS542B Project Report, 504*(2005), 195–221. Available from: https://people.duke.edu/~hpgavin/SystemID/References/Schmidt-LASSO-2005.pdf.
56. Selesnick, I. W. (2009). Sparse signal restoration. *Connexions*, 1–13. Available from: https://citeseerx.ist.psu.edu/document?repid=rep1&type=pdf&doi=0ea9bd6b0f62c151d30a5bc3bf8a2bf7a1fd506c.
57. Shlens, J. (2014). A tutorial on principal component analysis. *arXiv preprint arXiv:1404.1100*. Available from: https://www.cs.cmu.edu/~elaw/papers/pca.pdf.
58. Shu, C., & Liu, Y. (2016). Inpainting: Survey and experiments. *International Journal of Signal Processing, Image Processing and Pattern Recognition, 9*(10), 21–36. Available from: https://article.nadiapub.com/IJSIP/vol9_no10/3.pdf.
59. Tang, L., & Guan, W. (2020). Robust matrix completion with complex noise. *Multimedia Tools and Applications, 79*(3), 2703–2717. Available from: https://link.springer.com/article/10.1007/s11042-019-08430-2.
60. Tibshirani, R. (1996). Regression shrinkage and selection via the Lasso. *Journal of the Royal Statistical Society Series B: Statistical Methodology, 58*(1), 267–288. Available from: https://www.math.cuhk.edu.hk/~btjin/math6221/p7.pdf.
61. Tremoulheac, B. (2012). Split Bregman method for total variation denoising. Available from: https://es.mathworks.com/matlabcentral/fileexchange/36278-split-bregman-method-for-total-variation-denoising.
62. Tropp, J. A., & Wright, S. J. (2010). Computational methods for sparse solution of linear inverse problems. *Proceedings of the IEEE, 98*(6), 948–958. Available from: https://core.ac.uk/download/pdf/216136386.pdf.
63. Wright, S. J., Nowak, R. D., & Figueiredo, M. A. (2009). Sparse reconstruction by separable approximation. *IEEE Transactions on Signal Processing, 57*(7), 2479–2493. Available from: https://optimization-online.org/wp-content/uploads/2007/10/1813.pdf.

64. Yan, M., Yang, Y., & Osher, S. (2013). Exact low-rank matrix completion from sparsely corrupted entries via adaptive outlier pursuit. *Journal of Scientific Computing, 56*, 433–449. Available from: https://link.springer.com/article/10.1007/s10915-013-9682-3.
65. Yang, C. (2019). L1-norm-total variation-inpainting-using-split-Bregman-iteration. Available from: https://github.com/ycynyu007/L1-norm-total-variation-inpainting-using-Split_Bregman-Iteration/tree/master/test2.
66. Yang, J., & Zhang, Y. (2011). Alternating direction algorithms for ℓ_1 -problems in compressive sensing. *SIAM Journal on Scientific Computing, 33*(1), 250–278. Available from: https://arxiv.org/pdf/0912.1185.pdf.
67. Yin, W., Goldfarb, D., & Osher, S. (2005). Total variation based image cartoon-texture decomposition. *Columbia University CORC Report TR-2005-01, UCLA CAM Report*, pp. 05–27. Available from: https://apps.dtic.mil/sti/pdfs/ADA478457.pdf.
68. Yin, W., Goldfarb, D., & Osher, S. (2007). A comparison of three total variation based texture extraction models. *Journal of Visual Communication and Image Representation, 18*(3), 240–252. Available from: https://repository.rice.edu/server/api/core/bitstreams/0c2882d4-9415-4389-9c1a-40bb2826c661/content.
69. Yuan, X., & Yang, J. (2009). Sparse and low-rank matrix decomposition via alternating direction methods. Preprint, 12(2). Available from: https://optimization-online.org/wp-content/uploads/2009/11/2447.pdf.
70. Zou, H., & Hastie, T. (2005). Regularization and variable selection via the elastic net. *Journal of the Royal Statistical Society Series B: Statistical Methodology, 67*(2), 301–320. Available from: https://www.stat.purdue.edu/~tlzhang/mathstat/ElasticNet.pdf.

Appendix A
Long Programs

A.1 Introduction

Some of the programs developed for this book are long. In favor of shortening the chapters, those programs have been moved to the present Appendix.

Small versions of the figures generated by the programs have been added, to help identifying each one.

A.2 Chapter 2: Searching Methods

A.2.1 Methods Based on Hessian Approximation (2.4.3)

Method of Davidon-Fletcher-Powell

Program 2.23-DFP

In Sect. 2.4.3, the Broyden's method was introduced, including Program 2.23 and Fig. 2.44 (for a given example). Now, the DFP method was applied for the same example, obtaining the result shown in Fig. A.1, which is clearly similar to the Broyden's result.

J. M. Giron-Sierra, *Introduction to Optimization with Matlab® Examples*,
https://doi.org/10.1007/978-3-032-05048-9

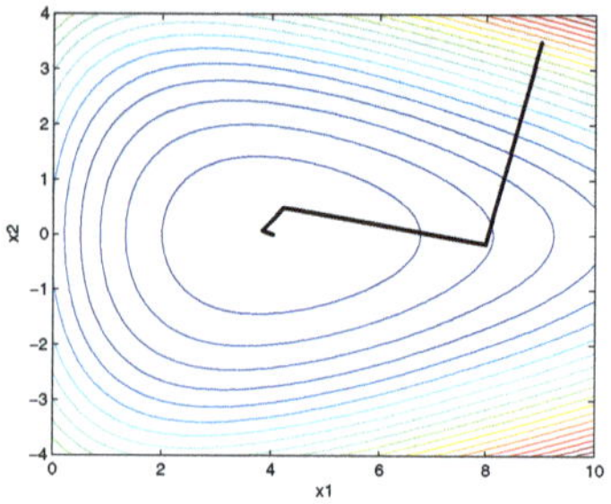

Fig. A.1 Example of DFP method

A.1 (P2.23-DFP) Example of DFP method

```
% Example of DFP method
%
% function y=(exp(-(x1-3)/2))+(exp((x1+4*x2)/10))...
% +(exp((x1-4*x2)/10));
%
[vx1,vx2]=meshgrid(0:0.1:10, -4:0.1:4);
y=(exp(-(vx1-3)/2))+(exp((vx1+4*vx2)/10))...
+(exp((vx1-4*vx2)/10));
%
% function display-------------
figure (1)
contour(vx1,vx2,y,20); hold on;
xlabel('x1'); ylabel('x2');
%
% initial point
x1i=9; x2i=3.5;
%
a=(exp(-(x1i-3)/2));
b=(exp((x1i+4*x2i)/10));
c=(exp((x1i-4*x2i)/10));
yi=a+b+c;
%
% gradient
aux1=(-0.5*a)+(0.1*b)+(0.1*c);
aux2=(0.4*b)-(0.4*c);
g=[aux1;aux2];
%
% matrix A
A=eye(2); %matrix I
%
% first step -------------------------
delta=-A*g;
% sign of delta along x1:
s=1;
if delta(1)<0, s=-1; end;
%
% first search segment
m=delta(2)/delta(1); %slope
```

```
flag=1; yref=yi;
x1=x1i;
while flag==1,
  x2=m*(x1-x1i)+x2i;
  a=(exp(-(x1-3)/2));
  b=(exp((x1+4*x2)/10));
  c=(exp((x1-4*x2)/10));
  y=a+b+c;
  if y>yref,
    flag=0;
  else
    yref=y;
    plot(x1,x2,'k.');
    x1o=x1; x2o=x2;
    x1=x1+(s*0.01);
  end;
end;
%
x1=x1o; x2=x2o;
%
% next steps ------------------------
for M=1:4,
  a=(exp(-(x1-3)/2));
  b=(exp((x1+4*x2)/10));
  c=(exp((x1-4*x2)/10));
  aux1=(-0.5*a)+(0.1*b)+(0.1*c);
  aux2=(0.4*b)-(0.4*c);
  go=g; %old value of the gradient
  g=[aux1;aux2]; %new gradient
  dg=g-go;
  dx=[x1-x1i; x2-x2i];
  Ao=A; %old value of A
  %QA (CODE FRAGMENT BEG)
    %new value of A:
    A=Ao+((dx*dx')/(dx'*dg))- ...
    ((Ao*dg*dg'*Ao)/(dg'*Ao*dg));
  %QB (CODE FRAGMENT END)
  x1i=x1; x2i=x2; %new start point
  %
  % line search
  delta=-A*g;
  % sign of delta along x1:
  s=1;
  if delta(1)<0, s=-1; end;
  %
  m=delta(2)/delta(1); %slope
  flag=1;
  x1=x1i;
  while flag==1,
    x2=m*(x1-x1i)+x2i;
    a=(exp(-(x1-3)/2));
```

```
      b=(exp((x1+4*x2)/10));
      c=(exp((x1-4*x2)/10));
      y=a+b+c;
      if y>yref,
        flag=0;
      else
        yref=y;
        plot(x1,x2,'k.');
        x1o=x1; x2o=x2;
        x1=x1+(s*0.01);
      end;
    end;
    x1=x1o; x2=x2o;
end;
```

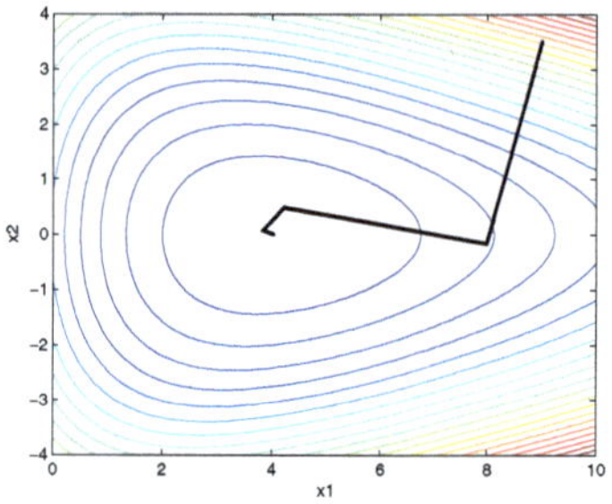

Fig. A.2 Example of BFGS method

Method of Broyden-Fletcher-Goldfarb-Shanno

Program 2.23-BFGS

Here we continue with the same case considered in Program 2.23 and Fig. 2.44, corresponding to the Broyden's method. Now, the BFGS method was applied for the example, obtaining the result shown in Fig. A.2, which is again similar to Fig. 2.44.

A.2 (P2.23-BFGS) Example of BFGS method

```
% Example of BFGS method
%
% function y=(exp(-(x1-3)/2))+(exp((x1+4*x2)/10))...
% +(exp((x1-4*x2)/10));
%
[vx1,vx2]=meshgrid(0:0.1:10, -4:0.1:4);
y=(exp(-(vx1-3)/2))+(exp((vx1+4*vx2)/10))...
+(exp((vx1-4*vx2)/10));
%
% function display-------------
figure (1)
contour(vx1,vx2,y,20); hold on;
xlabel('x1'); ylabel('x2');
%
% initial point
```

```
x1i=9; x2i=3.5;
%
a=(exp(-(x1i-3)/2));
b=(exp((x1i+4*x2i)/10));
c=(exp((x1i-4*x2i)/10));
yi=a+b+c;
%
% gradient
aux1=(-0.5*a)+(0.1*b)+(0.1*c);
aux2=(0.4*b)-(0.4*c);
g=[aux1;aux2];
%
% matrix A
A=eye(2); %matrix I
%
% first step -------------------------
delta=-A*g;
% sign of delta along x1:
s=1;
if delta(1)<0, s=-1; end;
%
% first search segment
m=delta(2)/delta(1); %slope
flag=1; yref=yi;
x1=x1i;
while flag==1,
  x2=m*(x1-x1i)+x2i;
  a=(exp(-(x1-3)/2));
  b=(exp((x1+4*x2)/10));
  c=(exp((x1-4*x2)/10));
  y=a+b+c;
  if y>yref,
    flag=0;
  else
    yref=y;
    plot(x1,x2,'k.');
    x1o=x1; x2o=x2;
    x1=x1+(s*0.01);
  end;
end;
%
x1=x1o; x2=x2o;
%
% next steps -------------------------
for M=1:4,
  a=(exp(-(x1-3)/2));
  b=(exp((x1+4*x2)/10));
  c=(exp((x1-4*x2)/10));
  aux1=(-0.5*a)+(0.1*b)+(0.1*c);
  aux2=(0.4*b)-(0.4*c);
  go=g; %old value of the gradient
  g=[aux1;aux2]; %new gradient
```

```
  dg=g-go;
  dx=[x1-x1i; x2-x2i];
  Ao=A; %old value of A
  %QA (CODE FRAGMENT BEG)
    %new value of A
    A=Ao+(((dx*dx')/(dx'*dg))*...
    (1+(dg'*Ao*dg)/(dx'*dg)))- ...
    ((Ao*dg*dx')/(dx'*dg))-((dx*dg'*Ao)/(dx'*dg));
  %QB (CODE FRAGMENT END)
  x1i=x1; x2i=x2; %new start point
  %
  % line search
  delta=-A*g;
  % sign of delta along x1:
  s=1;
  if delta(1)<0, s=-1; end;
  %
  m=delta(2)/delta(1); %slope
  flag=1;
  x1=x1i;
  while flag==1,
    x2=m*(x1-x1i)+x2i;
    a=(exp(-(x1-3)/2));
    b=(exp((x1+4*x2)/10));
    c=(exp((x1-4*x2)/10));
    y=a+b+c;
    if y>yref,
      flag=0;
    else
      yref=y;
      plot(x1,x2,'k.');
      x1o=x1; x2o=x2;
      x1=x1+(s*0.01);
    end;
  end;
  x1=x1o; x2=x2o;
end;
```

A.2.2 Trust-Region Methods (2.5)

Based on Cauchy Points

Program 2.24

The trust-region method using Cauchy points has been used for a non-quadratic example (Fig. A.3).

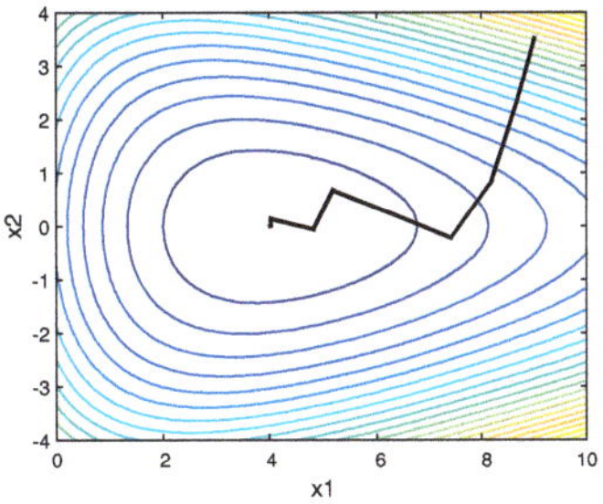

Fig. A.3 Example of trust-region method

A.3 (P2.24) Trust-region method using Cauchy points

```
% Example of trust-region method
% via Cauchy point
%
% function y=(exp(-(x1-3)/2))+...
(exp((x1+4*x2)/10))+(exp((x1-4*x2)/10));
%
[vx1,vx2]=meshgrid(0:0.1:10, -4:0.1:4);
y=(exp(-(vx1-3)/2))+(exp((vx1+4*vx2)/10))+...
(exp((vx1-4*vx2)/10));
%
% function display-------------
figure(1)
contour(vx1,vx2,y,20); hold on;
xlabel('x1'); ylabel('x2');
%
% initial point
x1i=9; x2i=3.5;
%
% initial trust-region  radius
rr=0.4;
% trust rules parameters
LaL=0.25;
LaH=0.75;
% end flag
EF=0;
% step accept flag
SF=0;
%
% steps
while EF==0,
a=(exp(-(x1i-3)/2));
b=(exp((x1i+4*x2i)/10));
c=(exp((x1i-4*x2i)/10));
yi=a+b+c;
%
% gradient
aux1=(-0.5*a)+(0.1*b)+(0.1*c);
aux2=(0.4*b)-(0.4*c);
g=[aux1;aux2];
%
```

```
% Hessian
qH11=(0.25*a)+(0.01*b)+(0.01*c);
qH12=(0.04*b)-(0.04*c);
qH21=(0.04*b)-(0.04*c);
qH22=(0.16*b)+(0.16*c);
H=[qH11 qH12; qH21 qH22];
%
% find Cauchy point
aux1=g'*H*g;
if aux1>0,
   aux2=((norm(g,2))^3)/(rr*aux1);
   if aux2>1,
      tau=1;
   else,
      tau=aux2;
   end;
else,
   tau=1;
end;
%step to Cauchy:
delta=-((tau*rr)/norm(g,2))*g;
x1a=x1i+delta(1);
x2a=x2i+delta(2);
%
% see reduction ------------
%  function
a=(exp(-(x1a-3)/2));
b=(exp((x1a+4*x2a)/10));
c=(exp((x1a-4*x2a)/10));
yia=a+b+c;
Ra=yi-yia;
%  model
aux=(g'*delta)+0.5*(delta'*H*delta);
Rp=-aux;
%
rho=Ra/Rp;
% trust rules
if rho<LaL,
   rr=rr/2; %radius decrease
else,
   SF=1; %step accept ON
   if rho>LaH,
      aux=norm(delta,2);
      if (2*aux>rr),
         rr=2*aux; %radius increase
      end;
   end;
end;
%
% decisions
if SF==1,  %accept step
   SF=0;
```

```
   x1f=x1i+delta(1);
   x2f=x2i+delta(2);
   plot([x1i x1f], [x2i x2f],'k');
   %
   x1i=x1f; x2i=x2f;
end;
%
% end when reduction was insignificant
if Ra<0.0001,
   EF=1;%end flag ON
end;
% end of program cycle
end
```

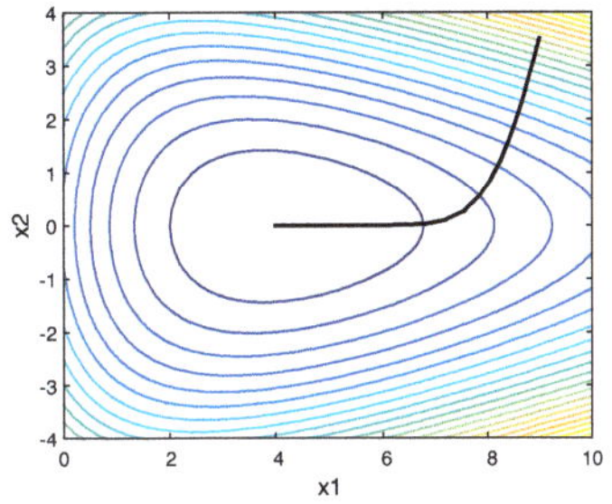

Fig. A.4 Example of trust-region dog-leg method

Dog-Leg Method

Program 2.25

The dog-leg method has been used for the same non-quadratic example as before (Fig. A.4).

A.4 (P2.25) Trust-region dog-leg method

```
% Example of trust-region method
% via Dog-leg
%
% function y=(exp(-(x1-3)/2))+...
(exp((x1+4*x2)/10))+(exp((x1-4*x2)/10));
%
[vx1,vx2]=meshgrid(0:0.1:10, -4:0.1:4);
y=(exp(-(vx1-3)/2))+(exp((vx1+4*vx2)/10))+...
(exp((vx1-4*vx2)/10));
%
% function display-------------
figure(1)
contour(vx1,vx2,y,20); hold on;
xlabel('x1'); ylabel('x2');
%
% initial point
x1i=9; x2i=3.5;
```

```
%
% initial trust-region  radius
rr=0.4;
% steps
for nn=1:20,
a=(exp(-(x1i-3)/2));
b=(exp((x1i+4*x2i)/10));
c=(exp((x1i-4*x2i)/10));
yi=a+b+c;
%
% gradient
aux1=(-0.5*a)+(0.1*b)+(0.1*c);
aux2=(0.4*b)-(0.4*c);
g=[aux1;aux2];
%
% Hessian
qH11=(0.25*a)+(0.01*b)+(0.01*c);
qH12=(0.04*b)-(0.04*c);
qH21=(0.04*b)-(0.04*c);
qH22=(0.16*b)+(0.16*c);
H=[qH11 qH12; qH21 qH22];
%
% find full step
fus=-H\g;
aux=(g'*fus)+0.5*(fus'*H*fus);
Rn=-aux;  % corresponding reduction
Ln=norm(fus,2); %full step length
%
if Rn>0, % the full step can yield reduction
   % if the full step is inside trust region,
   % use it
   if Ln<=rr,
      delta=fus;
   else,
      % find Cauchy point
      aux1=g'*H*g;
      if aux1>0,
         aux2=((norm(g,2))^3)/(rr*aux1);
         if aux2>1,
           tau=1;
         else,
           tau=aux2;
         end;
      else,
        tau=1;
      end;
      %step to Cauchy
      delta=-((tau*rr)/norm(g,2))*g;
      aux=norm(delta,2);
      %Cauchy is not inside trust region
      if aux>=rr,
```

```
            % go to intersection
            delta=-rr*(g/norm(g,2));
        else, % dog-leg
            % segment from Cauchy to Newton
            sgm=fus-delta;
            M=sgm'*sgm;
            % find intersection with
            % trust border
            b=2*sgm'*delta;
            c=(delta'*delta)-rr^2;
            q=(-b+sqrt(b^2-4*M*c))/2/M;
            delta=delta+(q*sgm);
        end;
    end;
end;
x1f=x1i+delta(1);
x2f=x2i+delta(2);
plot([x1i x1f], [x2i x2f],'k');
%
x1i=x1f; x2i=x2f;
% end of program cycle
end
```

A.3 Chapter 3: Linear Programming

A.3.1 The Tableau Format (3.3.2)

Example of MATLAB Coding of the Simplex Method

Program 3.6

This is a program in MATLAB that generates the series of tableaux corresponding to the simplex method. A basic optimization problem was considered. The image below shows the feasible region of this problem (Fig. A.5).

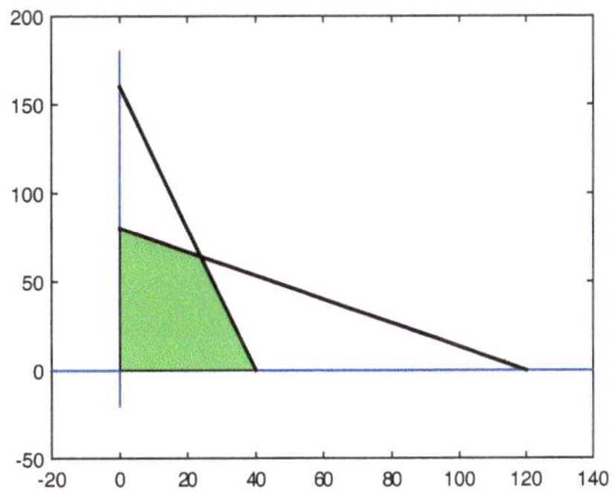

Fig. A.5 Feasible region of the problem considered

A.5 (P3.6) A MATLAB coding of the simplex method

```
% simplex with tableaux
% example 1:
% maximize z=8x1+6x2
% subject to:
% 4x1+x2<=160
% 2x1+3x2<=240
%
% Problem specification:
% (2 slack variables are added)
c=[8; 6; 0; 0];
A=[4 1 1 0;
   2 3 0 1];
b=[160; 240];
%
% algorithm preparation:
ib=[3,4]; %indexes of initial basis
% initial zj row
z=zeros(1,4);
p1=c(ib(1)); p2=c(ib(2));
for nn=1:4,
   z(nn)=p1*A(1,nn)+p2*A(2,nn);
end;
q=p1*b(1)+p2*b(2);
% cj-zj row
u=c-z';
ic=0; ir=0; %pivot pointers
pv=0; %pivot value
% basis names (cell)
bs={'x1', 'x2', 's1', 's2'};
%
% tableau contents
C=cell(4,1);
R=cell(2,4);
Z=cell(4,1);
U=cell(4,1);
B=cell(2,1);
% spacing pads
P1=['  |  '];
P2=['  |'];  %for - sign
%
for nn=1:4,
   C{nn}=num2str(c(nn));
   aux=num2str(z(nn),'%2.2f');
   Z{nn}=[aux,P1];
   sg=sign(u(nn)); aux2=P1;
   aux1=num2str(u(nn),'%2.2f');
   if sg<0, aux2=P2; end;
   U{nn}=[aux2,aux1];
end;
% central part
r=A; %initial values
for i=1:2,
   for j=1:4,
      sg=sign(r(i,j)); aux2=P1;
      aux1=num2str(r(i,j),'%2.2f');
      if sg<0, aux2=P2; end;
      R{i,j}=[aux2,aux1];
   end;
end;
% b column
for nn=1:2,
   B{nn}=num2str(b(nn),'%3.1f');
end;
Q=num2str(q,'%2.2f');
%
% spacing and rulers
O=['      |  '];
LL=['-']; LM=['_'];
for n=1:60,
   LL=[LL,'-'];
   LM=[LM,'_'];
end;
V1=bs{ib(1)}; V2=bs{ib(2)}; %current basis
W1=C{ib(1)}; W2=C{ib(2)}; %current c coeffs
%
% initial tableau
```

```
str1=['| cj-> |           | ',C{1},O,C{2},O,C{3},O,...
C{4},O,'    ',O];
str2=['|      | Basis   | x1    | x2    | s1    | s2...
    | b      |'];
str3=['| ',W1,'       | ',V1,'      ',R{1,1},R{1,2},...
R{1,3},R{1,4},P1,B{1},P1];
str4=['| ',W2,'       | ',V2,'      ',R{2,1},R{2,2},...
R{2,3},R{2,4},P1,B{2},P1];
str5=['| ','      | zj      | ',Z{1},Z{2},Z{3},Z{4},...
Q,P1];
str6=['| ','      | cj-zj ',U{1},U{2},U{3},U{4},P1];
%
% display
disp(LM);
disp('*** initial tableau ***');
disp(LL); disp(str1);
disp(LL); disp(str2);
disp(LL); disp(str3);
disp(LL); disp(str4);
disp(LM); disp(str5);
disp(LL); disp(str6);
disp(LM);
%
% ---------------------------------------------------
% next tableaux
%
nT=1; %tableau number
%iteration of tableaux:
K=1; %flag
while K==1,
%
% pivot finding
[mx,ic]=max(u); %pivot column
  %prevent division by zero
   vv=0.00001;
aux=b./(vv+r(:,ic));
% get smallest non-negative result
[Y,I]=sort(aux);
fl=0; nn=1;
while fl==0,
   if Y(nn)<0,
      nn=nn+1;
   else
      ir=I(nn); %pivot row
      fl=1;
   end;
end;
%
pv=r(ir,ic); %pivot value
ib(ir)=ic;   %basis change
V1=bs{ib(1)}; V2=bs{ib(2)};
%
% central part
% new pivot row
r(ir,:)=r(ir,:)/pv;
b(ir)=b(ir)/pv;
% new remaining row(s)
for nn=1:2,
   if nn~=ir,
      op=r(nn,ic); %entry in pivot column
      vaux=op*r(ir,:);
      r(nn,:)=r(nn,:)-vaux; %new row
      b(nn)=b(nn)-(op*b(ir));
   end;
end;
%
% zj row
p1=c(ib(1)); p2=c(ib(2));
for nn=1:4,
   z(nn)=p1*r(1,nn)+p2*r(2,nn);
end;
q=p1*b(1)+p2*b(2);
% cj-zj row
u=c-z';
%
% tableau contents
%
for nn=1:4,
```

```
    aux=num2str(z(nn),'%2.2f');
    Z{nn}=[aux,P1];
    sg=sign(u(nn)); aux2=P1;
    aux1=num2str(u(nn),'%2.2f');
    if sg<0, aux2=P2; end;
    U{nn}=[aux2,aux1];
end;
% central part
for i=1:2,
    for j=1:4,
       sg=sign(r(i,j)); aux2=P1;
       aux1=num2str(r(i,j),'%2.2f');
       if sg<0, aux2=P2; end;
       R{i,j}=[aux2,aux1];
    end;
end;
% b column
for nn=1:2,
    B{nn}=num2str(b(nn),'%3.1f');
end;
Q=num2str(q,'%2.2f');
%
V1=bs{ib(1)}; V2=bs{ib(2)}; %current basis
W1=C{ib(1)}; W2=C{ib(2)}; %current c coeffs
%
% tableau
str1=['| cj-> |          | ',C{1},0,C{2},0,C{3},...
0,C{4},0,'   ',0];
str2=['|       | Basis   | x1    | x2    | s1    | s2...
   | b     |'];
str3=['| ',W1,'      | ',V1,'      ',R{1,1},R{1,2},...
R{1,3}, R{1,4},P1,B{1},P1];
str4=['| ',W2,'      | ',V2,'      ',R{2,1},R{2,2},...
R{2,3},R{2,4},P1,B{2},P1];
str5=['| ','        | zj      | ',Z{1},Z{2},Z{3},Z{4},...
Q,P1];
str6=['| ','        | cj-zj ',U{1},U{2},U{3},U{4},P1];
%
% display
disp(' ');
disp(LM);
msg=['*** tableau number ',num2str(nT),' ***'];
disp(msg);
disp(LL); disp(str1);
disp(LL); disp(str2);
disp(LL); disp(str3);
disp(LL); disp(str4);
disp(LM); disp(str5);
disp(LL); disp(str6);
disp(LM);
%
nT=nT+1;
% the last?
[aux,I]=max(u);
if aux<=0,
    K=0; %exit
end;
% prevent cycling
if nT>30,
    K=0; %exit
end;
end;
```

A.3.2 A Simple Diet (3.6.1.1)

Series of Simplex Tableaux for Optimal Simple Diet

Program 3.13

Given a set of ingredients, we want to find a simple optimal diet. The simplex method is applied, going step by step to a final solution, using a series of tableaux. The program computes and visualizes each of the tableaux (Fig. A.6).

Fig. A.6 Final simplex tableau for the simple diet

*** tableau number 3 ***

cj->								
		8	1.5	0	0	1000	1000	
	Basis	x1	x2	s1	s2	xa1	xa2	b
0	s1	5.58	0.00	1.00	0.01	1.00	0.00	6.0
1.5	x2	1.09	1.00	0.00	-0.01	0.00	1.00	3.1
	zj	1.64	1.50	0.00	-0.02	1000.00	1000.00	4.64
	cj-zj	6.36	0.00	0.00	0.02	0.00	0.00	

A.6 (P3.13) Simple diet series of tableaux

```
% simplex with tableaux
% diet 0:
% minimize cost: z=8x1+1.5x2
% x1, cheese; x2, potato
% subject to:
% 7x1+1.3x2<=10
% 106x1+97x2>=300
%
% Problem specification:
% (2 slack and 2 surplus variables are added)
M=1000; %large value (edit)
c=[8; 1.5; 0; 0; M; M];
A=[7 1.3 1 0 1 0;
   106 97 0 -1 0 1];
b=[10; 300];
%
% algorithm preparation:
ib=[5,6]; %indexes of initial basis
% initial zj row
z=zeros(1,6);
p1=c(ib(1)); p2=c(ib(2));
for nn=1:6,
   z(nn)=p1*A(1,nn)+p2*A(2,nn);
end;
q=p1*b(1)+p2*b(2);
% cj-zj row
u=c-z';
ic=0; ir=0; %pivot pointers
pv=0; %pivot value
% basis names (cell)
bs={'x1', 'x2', 's1', 's2', 'xa1', 'xa2'};
%
% tableau contents
C=cell(6,1);
R=cell(2,6);
Z=cell(6,1);
U=cell(6,1);
B=cell(2,1);
% spacing pads
P1=[' | '];
P2=[' |']; %for - sign
%
for nn=1:6,
   C{nn}=num2str(c(nn));
   aux=num2str(z(nn),'%2.2f');
   Z{nn}=[aux,P1];
   sg=sign(u(nn)); aux2=P1;
   aux1=num2str(u(nn),'%2.2f');
   if sg<0, aux2=P2; end;
   U{nn}=[aux2,aux1];
end;
% central part
r=A; %initial values
for i=1:2,
   for j=1:6,
      sg=sign(r(i,j)); aux2=P1;
```

```
            aux1=num2str(r(i,j),'%2.2f');
            if sg<0, aux2=P2; end;
            R{i,j}=[aux2,aux1];
      end;
end;
% b column
for nn=1:2,
      B{nn}=num2str(b(nn),'%3.1f');
end;
Q=num2str(q,'%2.2f');
%
% spacing and rulers
O=['     |  '];
LL=['-']; LM=['_'];
for n=1:80,
      LL=[LL,'-'];
      LM=[LM,'_'];
end;
V1=bs{ib(1)}; V2=bs{ib(2)}; %current basis
W1=C{ib(1)}; W2=C{ib(2)}; %current c coeffs
%
% initial tableau
str1=['| cj-> |         | ',C{1},O,C{2},O,C{3},O,...
C{4},O,C{5},O,C{6},O,'   ',O];
str2=['|       | Basis   | x1    | x2    | s1    | s2...
   | xa1  | xa2  | b      |'];
str3=['|  ',W1,'     |  ',V1,'     ',R{1,1},R{1,2},...
R{1,3},R{1,4},R{1,5},R{1,6},P1,B{1},P1];
str4=['|  ',W2,'     |  ',V2,'     ',R{2,1},R{2,2},...
R{2,3},R{2,4},R{2,5},R{2,6},P1,B{2},P1];
str5=['|  ','        | zj       |  ',Z{1},Z{2},Z{3},Z{4},...
Z{5},Z{6},Q,P1];
str6=['|  ','        | cj-zj  ',U{1},U{2},U{3},U{4},...
U{5},U{6},P1];
%
% display
disp(LM);
disp('*** initial tableau ***');
disp(LL); disp(str1);
disp(LL); disp(str2);
disp(LL); disp(str3);
disp(LL); disp(str4);
disp(LM); disp(str5);
disp(LL); disp(str6);
disp(LM);
%
% -------------------------------------------
% next tableaux
%
nT=1; %tableau number
%iteration of tableaux:
K=1; %flag
while K==1,
%
% pivot finding
[mx,ic]=min(u); %pivot column
  %prevent division by zero
    vv=0.00001;
aux=b./(vv+r(:,ic));
% get smallest non-negative result
[Y,I]=sort(aux);
fl=0; nn=1;
while fl==0,
      if Y(nn)<0,
            nn=nn+1;
      else
            ir=I(nn); %pivot row
            fl=1;
      end;
end;
%
pv=r(ir,ic); %pivot value
ib(ir)=ic;    %basis change
V1=bs{ib(1)}; V2=bs{ib(2)};
%
% central part
% new pivot row
r(ir,:)=r(ir,:)/pv;
b(ir)=b(ir)/pv;
```

```
% new remaining row(s)
for nn=1:2,
   if nn~=ir,
      op=r(nn,ic); %entry in pivot column
      vaux=op*r(ir,:);
      r(nn,:)=r(nn,:)-vaux; %new row
      b(nn)=b(nn)-(op*b(ir));
   end;
end;
%
% zj row
p1=c(ib(1)); p2=c(ib(2));
for nn=1:4,
   z(nn)=p1*r(1,nn)+p2*r(2,nn);
end;
q=p1*b(1)+p2*b(2);
% cj-zj row
u=c-z';
%
% tableau contents
%
for nn=1:4,
   aux=num2str(z(nn),'%2.2f');
   Z{nn}=[aux,P1];
   sg=sign(u(nn)); aux2=P1;
   aux1=num2str(u(nn),'%2.2f');
   if sg<0, aux2=P2; end;
   U{nn}=[aux2,aux1];
end;
% central part
for i=1:2,
   for j=1:4,
      sg=sign(r(i,j)); aux2=P1;
      aux1=num2str(r(i,j),'%2.2f');
      if sg<0, aux2=P2; end;
      R{i,j}=[aux2,aux1];
   end;
end;
% b column
for nn=1:2,
   B{nn}=num2str(b(nn),'%3.1f');
end;
Q=num2str(q,'%2.2f');
%
V1=bs{ib(1)}; V2=bs{ib(2)}; %current basis
W1=C{ib(1)}; W2=C{ib(2)}; %current c coeffs
%
% tableau
str1=['| cj-> |          | ',C{1},O,C{2},O,C{3},O,...
C{4},O,C{5},O,C{6},O,'  ',O];
str2=['|      | Basis   | x1    | x2    | s1    | s2...
   | xa1  | xa2  | b     |'];
str3=['| ',W1,'     | ',V1,'     ',R{1,1},R{1,2},...
R{1,3},R{1,4},R{1,5},R{1,6},P1,B{1},P1];
str4=['| ',W2,'     | ',V2,'     ',R{2,1},R{2,2},...
R{2,3},R{2,4},R{2,5},R{2,6},P1,B{2},P1];
str5=['| ','      | zj      | ',Z{1},Z{2},Z{3},...
Z{4},Z{5},Z{6},Q,P1];
str6=['| ','      | cj-zj ',U{1},U{2},U{3},...
U{4},U{5},U{6},P1];
%
% display
disp(' ');
disp(LM);
msg=['*** tableau number ',num2str(nT),' ***'];
disp(msg);
disp(LL); disp(str1);
disp(LL); disp(str2);
disp(LL); disp(str3);
disp(LL); disp(str4);
disp(LM); disp(str5);
disp(LL); disp(str6);
disp(LM);
%
nT=nT+1;
% the last?
[aux,I]=min(u);
if aux>=0,
```

```
    K=0;  %exit
end;
% prevent cycling
if nT>30,
    K=0;  %exit
end;
end;
```

Fig. A.7 A more complete diet

A.3.3 A More Complete Diet (3.6.1.2)

Series of Simplex Tableaux for a More Complete Optimal Diet

Program 3.14

The number of ingredients was increased, wishing to find a more complete, optimal diet (Fig. A.7).

The program does not display the series of tableaux (they have large size); it just gives the numerical final result.

A.7 (P3.14) A more complete diet; primal method

```
%% simplex with invisible tableaux
% diet 1:
%
% Problem specification:
N=4; %number of variables
L=4; %number of constraints
W=N+(2*L); %total number of variables+added variables
Oc=[2; 8; 1.5; 11];
OA=[4 7 1.3 8;
    1 9 0.1 7;
    15 0.4 22.6 0;
    90 106 97 130];
Ob=[10; 8; 10; 300];
c=zeros(W,1); A=zeros(L,W); p=zeros(L,1);
%
% (L slack and L surplus variables are added)
M=1000; %large value (edit)
c(1:N)=Oc; c(N+L+1:end)=M*ones(L,1);
A(:,1:N)=OA;
A(:,N+1:N+L)=[1 0 0 0;
              0 -1 0 0;
              0 0 -1 0;
              0 0 0 -1]; %(edit according
                         % with constraints)
A(:,N+L+1:end)=eye(N);
```

```
b=Ob;
%
% algorithm preparation:
z=zeros(1,W);
q=0;
r=A; % the matrix to be manipulated
ic=0; ir=0; %pivot pointers
pv=0; %pivot value
ib=zeros(1,L);
ib=[N+L+1:W]; %indexes of initial basis
% initial zj row
for nn=1:L,
   p(nn)=c(ib(nn));
end;
for nn=1:W,
   aux=0;
   for m=1:L,
      aux=aux+p(m)*r(m,nn);
   end;
   z(nn)=aux;
end;
q=0;
for m=1:L,
   q=q+p(m)*b(m);
end;
% cj-zj row
u=c-z';
% ---------------------------------------
% next (invisible) tableaux
%
nT=1; %tableau number
%iteration of invisible tableaux:
K=1; %flag
while K==1,
%
% pivot finding
[mx,ic]=min(u); %pivot column
% avoid division difficulties
aux=zeros(1,L);
vv=0.001; %small value
for j=1:L,
   if b(j)==0,
      b(j)=0.01; %small perturbation
   end;
   if r(j,ic)==0,
      aux(j)=b(j)/vv; %large value
   else
      aux(j)=b(j)/r(j,ic);
   end;
end;
% get smallest non-negative result
[Y,I]=sort(aux);
fl=0; nn=1;
while fl==0,
   if Y(nn)<0,
      nn=nn+1;
   else
      ir=I(nn); %pivot row
      fl=1;
   end;
end;
%
pv=r(ir,ic); %pivot value
ib(ir)=ic;   %basis change
% avoid pivot=0
if pv==0,
   pv=vv;
end;
%
% central part
% new pivot row
r(ir,:)=r(ir,:)/pv;
b(ir)=b(ir)/pv;
% new remaining row(s)
for nn=1:L,
   if nn~=ir,
      op=r(nn,ic); %entry in pivot column
      vaux=op*r(ir,:);
      raux=r(nn,:);
```

```
         r(nn,:)=raux-vaux; %new row
         b(nn)=b(nn)-(op*b(ir));
      end;
end;
%
% zj row
for nn=1:L,
   p(nn)=c(ib(nn));
end;
for nn=1:W,
   aux=0;
   for m=1:L,
      aux=aux+p(m)*r(m,nn);
   end;
   z(nn)=aux;
end;
q=0;
for m=1:L,
   q=q+p(m)*b(m);
end;
% cj-zj row
u=c-z';
%
nT=nT+1
% the last?
[aux,I]=min(u);
if aux>=0,
   K=0;   %exit
end;
% prevent cycling
if nT>20,
   K=0;   %exit
end;
ib
end;
% result
ib
b
q
```

A.3.4 A More Complete Diet, Dual Method (3.6.1.2)

Dual Method with a Series of Tableaux, a More Complete Diet

Program 3.15

In order to confirm the results for the more complete diet, the dual problem is solved (Fig. A.8).

The program does not show the series of tableaux; it just gives the numerical final result.

Fig. A.8 Dual method for a more complete diet

A.8 (P3.15) A more complete diet; dual method

```
% simplex with invisible tableaux
% diet 1:
% dual problem
%
% Problem specification:
N=4; %number of variables
L=4; %number of constraints
W=N+(2*L); %total number of variables+added variables
Oc=[2; 8; 1.5; 11];
OA=[4 7 1.3 8;
    1 9 0.1 7;
    15 0.4 22.6 0;
    90 106 97 130];
Ob=[-10; 8; 10; 300]; %notice the minus sign
c=zeros(W,1); A=zeros(L,W); p=zeros(L,1);
%
% (L slack and L surplus variables are added)
M=-1000; %large negative value (edit)
c(1:N)=Ob; c(N+L+1:end)=M*ones(L,1); %dual
A(:,1:N)=OA'; %dual
A(:,N+1:N+L)=[-1 0 0 0;
              0 1 0 0;
              0 0 1 0;
              0 0 0 1]; %(edit according
                        % with constraints)
A(:,N+L+1:end)=eye(L);
b=Oc; %dual
%
% algorithm preparation:
z=zeros(1,W);
q=0;
r=A; % the matrix to be manipulated
ic=0; ir=0; %pivot pointers
pv=0; %pivot value
ib=zeros(1,L);
ib=[N+L+1:W]; %indexes of initial basis
% initial zj row
for nn=1:L,
   p(nn)=c(ib(nn));
end;
for nn=1:W,
   aux=0;
   for m=1:L,
      aux=aux+p(m)*r(m,nn);
   end;
   z(nn)=aux;
end;
q=0;
for m=1:L,
   q=q+p(m)*b(m);
end;
% cj-zj row
u=c-z';

% ----------------------------------------
% next (invisible) tableaux
%
nT=1; %tableau number
%iteration of invisible tableaux:
K=1; %flag
while K==1,
%
% pivot finding
[mx,ic]=max(u); %pivot column
% avoid division difficulties
aux=zeros(1,L);
vv=0.001; %small value
for j=1:L,
   if b(j)==0,
      b(j)=0.01; %small perturbation
   end;
   if r(j,ic)==0,
      aux(j)=b(j)/vv; %large value
   else
      aux(j)=b(j)/r(j,ic);
   end;
```

```
end;
% get smallest non-negative result
[Y,I]=sort(aux);
fl=0; nn=1;
while fl==0,
   if Y(nn)<0,
      nn=nn+1;
   else
      ir=I(nn); %pivot row
      fl=1;
   end;
end;
%
pv=r(ir,ic); %pivot value
ib(ir)=ic;   %basis change
% avoid pivot=0
if pv==0,
   pv=vv;
end;
%
% central part
% new pivot row
r(ir,:)=r(ir,:)/pv;
b(ir)=b(ir)/pv;
% new remaining row(s)
for nn=1:L,
   if nn~=ir,
      op=r(nn,ic); %entry in pivot column
      vaux=op*r(ir,:);
      raux=r(nn,:);
      r(nn,:)=raux-vaux; %new row
      b(nn)=b(nn)-(op*b(ir));
   end;
end;
%
% zj row
for nn=1:L,
   p(nn)=c(ib(nn));
end;
for nn=1:W,
   aux=0;
   for m=1:L,
      aux=aux+p(m)*r(m,nn);
   end;
   z(nn)=aux;
end;
q=0;
for m=1:L,
   q=q+p(m)*b(m);
end;
% cj-zj row
u=c-z';
%
nT=nT+1
% the last?
[aux,I]=max(u);
if aux<=0,
   K=0;  %exit
end;
% prevent cycling
if nT>20,
   K=0;  %exit
end;
ib
end;
% result
ib
b
q
```

Fig. A.9 Illustration of the blending problem

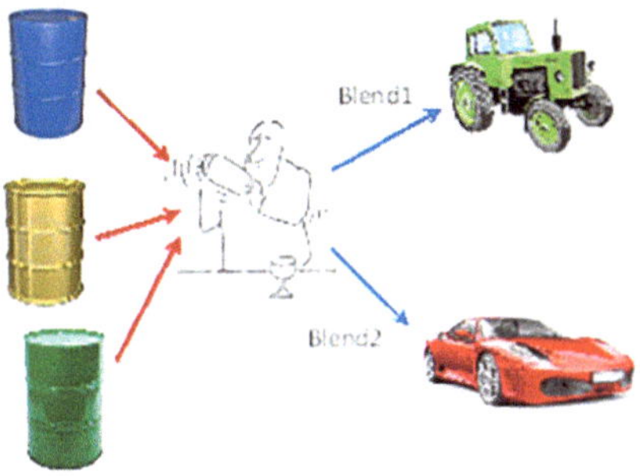

A.3.5 *Blending (3.6.2)*

Example of Blending Optimization

Program 3.16

Blending or mixing is a frequent process in industries like oil&gas, food, paint, etc. The optimization target is to maximize the profit (Fig. A.9).

A.9 (P3.16) Blending optimization

```
% simplex with invisible tableaux
% blend0
%
% Problem specification:
N=6; %number of variables
L=11; %number of constraints
W=N+(2*L); %total number of variables+added variables
Oc=[0.8; 0.7; 0.62; 0.9; 0.8; 0.72];
OA=[-0.7 0.3 0.3 0 0 0;
   -0.5 0.5 -0.5 0 0 0;
   0.3 0.3 -0.7 0 0 0;
   0 0 0 0.6 -0.4 -0.4;
   0 0 0 0.35 -0.65 0.35;
   0 0 0 -0.4 -0.4 0.6;
   1 0 0 1 0 0;
   0 1 0 0 1 0;
   0 0 1 0 0 1;
   1 1 1 0 0 0;
   0 0 0 1 1 1];
Ob=[0; 0; 0; 0; 0; 0; 6000; 10000; 12000;...
 10000; 10000];
c=zeros(W,1); A=zeros(L,W); p=zeros(L,1);
%
% (L slack and L surplus variables are added)
M=-1000; %large negative value (edit)
c(1:N)=Oc; c(N+L+1:end)=M*ones(L,1);
A(:,1:N)=OA;
%(edit according with constraints):
A(:,N+1:N+L)=eye(L);
A(L-1,N+1:N+L)=[0 0 0 0 0 0 0 0 0 -1 0];
A(L,N+1:N+L)=[0 0 0 0 0 0 0 0 0 0 -1];
A(:,N+L+1:end)=eye(L);
b=Ob;
%
% algorithm preparation:
z=zeros(1,W);
q=0;
r=A; % the matrix to be manipulated
ic=0; ir=0; %pivot pointers
pv=0; %pivot value
```

```
ib=zeros(1,L);
ib=[N+L+1:W]; %indexes of initial basis
% initial zj row
for nn=1:L,
   p(nn)=c(ib(nn));
end;
for nn=1:W,
   aux=0;
   for m=1:L,
      aux=aux+p(m)*r(m,nn);
   end;
   z(nn)=aux;
end;
q=0;
for m=1:L,
   q=q+p(m)*b(m);
end;
% cj-zj row
u=c-z';
% -----------------------------------------
% next (invisible) tableaux
%
nT=1; %tableau number
%iteration of invisible tableaux:
K=1; %flag
while K==1,
%
% pivot finding
[mx,ic]=max(u); %pivot column
% avoid division difficulties
aux=zeros(1,L);
vv=0.001; %small value
for j=1:L,
   if b(j)==0,
      b(j)=0.01; %small perturbation
   end;
   if r(j,ic)==0,
      aux(j)=b(j)/vv; %large value
   else
      aux(j)=b(j)/r(j,ic);
   end;
end;
% get smallest non-negative result
[Y,I]=sort(aux);
fl=0; nn=1;
while fl==0,
   if Y(nn)<0,
      nn=nn+1;
   else
      ir=I(nn); %pivot row
      fl=1;
   end;
end;
%
pv=r(ir,ic); %pivot value
ib(ir)=ic;   %basis change
% avoid pivot=0
if pv==0,
   pv=vv;
end;
%
% central part
% new pivot row
r(ir,:)=r(ir,:)/pv;
b(ir)=b(ir)/pv;
% new remaining row(s)
for nn=1:L,
   if nn~=ir,
      op=r(nn,ic); %entry in pivot column
      vaux=op*r(ir,:);
      raux=r(nn,:);
      r(nn,:)=raux-vaux; %new row
      b(nn)=b(nn)-(op*b(ir));
   end;
end;
%
% zj row
for nn=1:L,
   p(nn)=c(ib(nn));
end;
```

```
for nn=1:W,
    aux=0;
    for m=1:L,
        aux=aux+p(m)*r(m,nn);
    end;
    z(nn)=aux;
end;
q=0;
for m=1:L,
    q=q+p(m)*b(m);
end;
% cj-zj row
u=c-z';
%
nT=nT+1
% the last?
[aux,I]=max(u);
if aux<=0,
    K=0;   %exit
end;
% prevent cycling
if nT>20,
    K=0;   %exit
end;
ib
end;
% result
ib
b
q
```

A.4 Chapter 7: Quadratic Programming

A.4.1 Gradient Projection Methodology (7.5.5)

Example of Projected Gradient

Program 7.5

Gradient projection methods propose a piecewise search for the optimal solution, using projections of the objective function gradients onto the feasible set boundary (Fig. A.10).

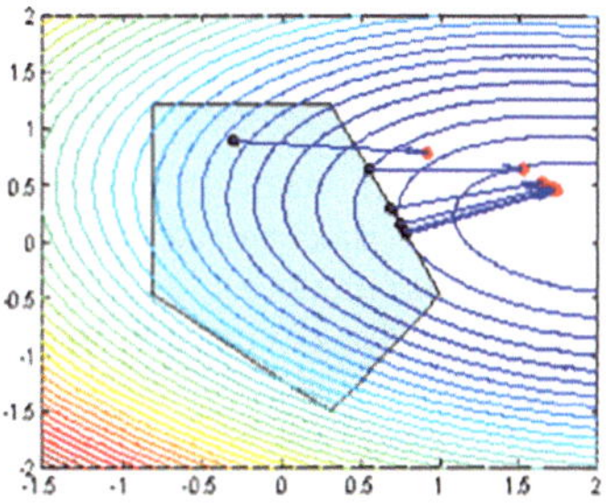

Fig. A.10 Search done with the gradient projection program

A.10 (P7.5) Example of projected gradient

```
%Example of Projected Gradient
Q=[0.30 0.08
0.08 0.7];
b=[-1.2 ; -0.5];
x0=[-0.3; 0.9]; %Initial point
%contour data
[vx1,vx2] = meshgrid(-1.5:0.05:2, -2:0.05:2);
y=Q(1,1)*(vx1.^2)+(Q(2,1)+...
Q(1,2))*vx1.*vx2+Q(2,2)*(vx2.^2)+b(1)*vx1+b(2)*vx2;
%-----------------------------------------
%Define a polygonal feasible set (Ax=c)
% number of polygon sides
nply = 5;
A = zeros(nply,2);
c = zeros(nply,1);
% find vertices (inside a circle)
theta_ply = (0:(2*pi)/nply:2*pi);
xply = cos(theta_ply);
yply = 1.5*sin(theta_ply-pi/10);
for i = 1:nply
  % compute edge directions p = x2-x1
  p = [xply(i+1)-xply(i)
  yply(i+1)-yply(i)];
  % get the normal direction
  % rotate 90 degrees counter-clockwise
  a = [p(2),-p(1)];
  a = a/norm(a);
  A(i,:) = a;
  % offset of edges
  c(i) = a*[xply(i); yply(i)];
end
%Algorithm -----------------------------------
% anitialization
maxiter=50;
miner=0.005;
mina=1e-6;
cc=0.5; %Armijo, parameter
alphak =1; %Armijo, step size
rx=zeros(maxiter,2);
rd=zeros(maxiter,2); %search directions
rg=zeros(maxiter,2); %gradients
% start ------------------------
xk=PG_proj(x0,A,c); %first projection onto F
for k=1:maxiter
  Jk=0.5*xk'*Q*xk + b'*xk;
  dJk=Q*xk + b; %gradient
  dk=PG_proj(xk-dJk,A,c)-xk; %search direction
  if (dJk'*dk > 0)
    disp('terminate: non descent direction');
```

```
    break;
  end
  %Armijo linesearch (backtracking)
  xk1 = xk + alphak * dk;
  Jk1 = 0.5*xk1'*Q*xk1 + b'*xk1;
  while (Jk1>Jk+cc*alphak*dJk'*dk) && (alphak > mina)
    %alphak backstep
    alphak = 0.5 * alphak;
    %function evaluation at new point
    xk1 = xk + alphak * dk;
    Jk1 = 0.5*xk1'*Q*xk1 + b'*xk1;
  end
  %save iteration data
  rx(k,:)= xk;
  rd(k,:)= dk;
  rg(k,:)= dJk;
  %iteration update
  xk = xk1;
  %stopping
  if (norm(dk,2)<miner)
    disp('too small step direction');
  break
  end
  if (alphak $<$ mina)
    disp('too small step size');
    break
  end
end
%solution:
x = xk
k
figure(1)
plot(xply,yply,'k','LineWidth',2); hold on
patch(xply,yply,[0.8 0.95 0.95]);
contour(vx1,vx2,y,35);
plot(rx(1:k,1),rx(1:k,2),'ko',...
'MarkerFaceColor','k')
plot(rx(1:k,1)-rg(1:k,1),rx(1:k,2)-rg(1:k,2),'ro',...
'MarkerFaceColor','r')
quiver(rx(1:k,1),rx(1:k,2),-rg(1:k,1),-rg(1:k,2),0)
%Function
function x = PG_proj(y,A,c)
n = length(y);
H = eye(n);
x0 = y;
options = optimset('Display','off');
x = quadprog(H,-y,A,c,[],[],[],[],x0,options);
return
```

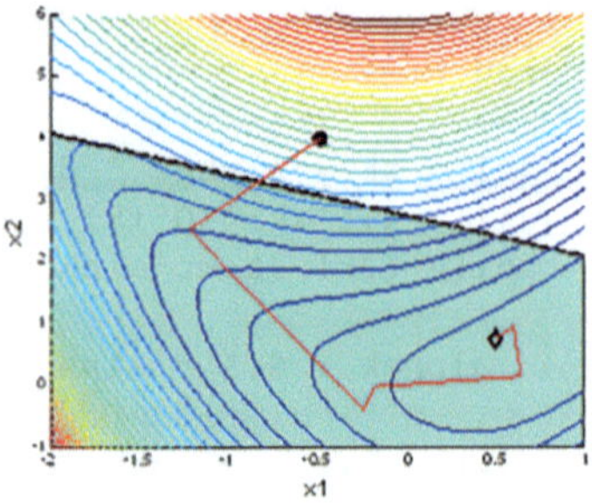

Fig. A.11 Evolution of SQP iterations for the example considered

A.4.2 Sequential Quadratic Programming (SQP) (7.6)

Example of SQP Algorithm
Program 7.8

Sequential programming methods belong to the most powerful algorithms for solving differentiable nonlinear problems. SQP is specialized on quadratic programming, being quite popular (Fig. A.11).

A.11 (P7.8) Example of SQP iterations

```
%Example of SQP algorithm
clear all; close all; clc;
%initialization
vars = 2; % number of variables
cons = 1; % number of constraints
maxIter=100; % max iterations
xrecord = zeros(vars,maxIter+1);
x = [-0.5;4]; % initial guess point
xrecord(:,1)= x;
lambda =0; % LagrangeMultipliers vector
H = eye(vars,vars); % Initial Hessian matrix
%Evaluation at initial point
fEval= sqpf(x);
gEval = sqpg(x);
[gViol,lViol] = Viols(gEval,lambda);
gradfEval = gradf(x);
gradgEval = gradg(x);
P = Penalty(fEval,gViol,lViol);
%Algorithm iterations----------------------
for iter=1:maxIter
  % Solve KKT (optimum of
  %  the quadratic approximation)
  sol = SolveKKT(gradfEval,gradgEval,gEval,H);
  xSol = sol(1:vars);
  lSol = sol(vars+1:vars+cons);
  % If Lagrange multiplier negative, set it to zero
  for j = 1:length(lSol)
```

```
    if lSol(j)<0
      sol= H$-gradfEval)';
      xSol = sol(1:vars);
      lSol(j)=0;
    end
  end
  % Evaluate at new candidate point
  xNew = x + xSol;
  lNew = lSol;
  fEvalNew = sqpf(xNew);
  gEvalNew = sqpg(xNew);
  gradfEvalNew = gradf(xNew);
  gradgEvalNew = gradg(xNew);
  [gViolNew,lViolNew] = Viols(gEvalNew,lNew);
  PNew = Penalty(fEvalNew,gViolNew,lViolNew);
  % Lower step by half if penalty function increased
  while PNew-P$>$1e-4
    xSol = 0.5*xSol;
    xNew = x + xSol;
    fEvalNew = sqpf(xNew);
    gEvalNew = sqpg(xNew);
    gradfEvalNew = gradf(xNew);
    gradgEvalNew = gradg(xNew);
    [gViolNew,lViolNew] = Viols(gEvalNew,lNew);
    PNew = Penalty(fEvalNew,gViolNew,lViolNew);
  end
  % Stopping criterion
  if norm(xNew(1:vars)-x(1:vars))<=1e-2
    break
  end
  % Hessian update
  % lnew not l!!!
  gradLEval=gradLagr(gradfEval,gradgEval,lNew,vars);
  gradLEvalNew=gradLagr(gradfEvalNew,gradgEvalNew,...
  lNew,vars);
  Q = gradLEvalNew-gradLEval;
  dx = xNew-x;
  HNew = UpdateH(H,dx,Q);
  % general update for next iteration
  H = HNew;
  fEval = fEvalNew;
  gEval = gEvalNew;
  gradfEval = gradfEvalNew;
  gradgEval = gradgEvalNew;
  P = PNew;
  x = xNew;
  xrecord(:,iter+1) = x;
end
% display of results ------------------------------
[vx1,vx2] = meshgrid(-2:0.1:1, -1:0.1:6);
y = (vx1.^4) - (2*vx2.*(vx1.^2))+ (vx2.^2)+...
(vx1.^2)-(2*vx1)+5;
figure(1)
```

```
patch([-2,-2,1,1],[-1,4.0833,2.0833,-1],...
[0.6 0.9 0.85]);
hold on;
contour(vx1,vx2,y,30,'LineWidth',1.5); hold on;
%constraint
plot([-2,1],[4.0833,2.0833],'k--','LineWidth',2.5);
%initial point
plot(xrecord(1,1),xrecord(2,1),'ko','LineWidth',4);
%final solution
plot(x(1),x(2),'kd','MarkerSize',8,'LineWidth',3);
%SQP iterations
plot(xrecord(1,1:iter),xrecord(2,1:iter),'r',...
'LineWidth',2);
xlabel('x1'); ylabel('x2');
disp(['x(1)= ',num2str(x(1))]);
disp(['x(2)= ',num2str(x(2))]);
disp(['iterations = ',num2str(iter)]);
---------------------
function y = sqpf(x)
y = x(1)^4 - 2*x(2)*x(1)^2 + x(2)^2 + x(1)^2 - ...
2*x(1)+5;
end
---
function y = sqpg(x)
y(1) = -(x(1)+0.25)^2+0.75*x(2);
end
---
function y = gradf(x)
y(1) = 2*x(1)-4*x(1)*x(2)+4*x(1)^3-2;
y(2) = -2*x(1)^2 + 2*x(2);
end
---
function y = gradg(x)
y(1,1) = -2*x(1)-1/2;
y(1,2) = 3/4;
end
---
function y = gradLagr(gradfEval,gradgEval,l,n)
y = gradfEval';
sum = zeros(n,1);
for i = 1:length(l)
  sum = sum -l(i)*gradgEval(i:n)';
end
y = y + sum;
end
---
function y = SolveKKT(gradfEval,gradgEval,gEval,...
Hessian)
A = [Hessian -gradgEval';gradgEval 0];
b = [-gradfEval -gEval]';
y = A\b;
end
---
function y = UpdateH(H,dx,gamma)
term1=(gamma*gamma') / (gamma'*dx);
term2 = ((H*dx)*(dx'*H)) / (dx'*(H*dx));
y = H + term1-term2;
```

```
end
---
function [gViol,lViol] = Viols(gEval,l)
gViol=[];
lViol=[];
for i = 1:length(gEval)
  if gEval(i)<0
    gViol(i)=gEval(i);
    lViol(i)=l(i);
  end
end
end
---
function y = Penalty(fEval,gViol,lViol)
sum = 0;
y = fEval;
for i = 1:length(gViol)
  sum = sum + lViol(i)*abs(gViol(i));
end
y = y + sum;
end
```

A.4.3 *The Support Vector Machine (7.8)*

Kernel-SVM Classification Example (Breast Cancer)

Program 7.13

One of the most important data classification techniques is the support vector machine (SVM). By means of kernels, the technique can be adapted for difficult to separate data clouds. The program execution time would require some patience. The case considered is the classification of breast cancer into benign or malign (Fig. A.12).

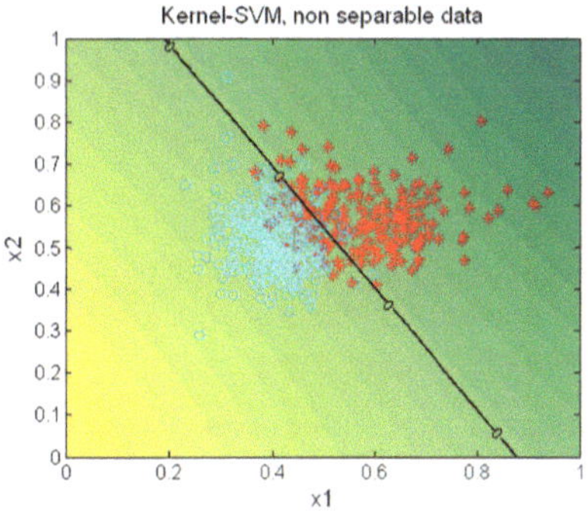

Fig. A.12 Using Kernel-SVM for classification of intermingled data

A.12 (P7.13) Kernel-SVM for breast cancer classification

```
% kernel-SVM classification example
% Wisconsin Diagnostic Breast Cancer
%read cancer data
D=dlmread('wisc_dbc.data'); %columns
X=[D(:,3), D(:,7)]; %two columns (x,y)
N=size(D,1);
%I edited data, replacing M by -1, and B by 1
Y=D(:,2);
%normalize
X(:,1) = X(:,1)/30;
X(:,2) = X(:,2)/0.18;
%----------------------------------------------------
%prepare quadratic optimization
C=8.1; threshold = 0.05; %for soft margin
%labels
Dy = diag(Y);
% Gaussian kernel
ps=X*X'; %scalar product
normx=sum(X.^2,2);
[nps,mps]=size(ps);
aux=-2*ps+repmat(normx,1,mps)+repmat(normx',nps,1);
K=exp(-aux/2); %gaussian kernel
H =(Dy * K * Dy);
H = (H+H')/2; %make it symmetric
f = -ones(N,1);
Aeq=Y';
ceq = 0.0;
%bounds
cL= zeros(N,1); cU= repmat(C,N,1);
%optimizer
options = optimset('LargeScale','off','Display',...
'off','MaxIter',600);
alpha = quadprog(H,f,[],[],Aeq,ceq,cL,cU,[],options);
%index of alphas satisfying conditions
serial_n = (1:size(X,1))';
ser_sv = serial_n((alpha > threshold) & ...
(alpha < (C -0.05)));
nVectors = size(ser_sv,1);
%support vectors
xsup=X(ser_sv,:);
ysup=Y(ser_sv);
w=alpha(ser_sv).*ysup;
baux=0;
maux=0;

for ii = 1:nVectors
  baux = baux + Y(ser_sv(ii));
  maux= sum(alpha(ser_sv(ii))*Y(ser_sv(ii))*...
   K(ser_sv,ser_sv(ii)));
  baux = baux - maux;
end
b = baux/size(ser_sv,1); %decision function coeff
```

```
%---------------------------------------------------------
% Prepare for visualization a grid of test points
[xg1,xg2]=meshgrid([0:0.01:1],[0:0.01:1]);
[nl,nc]=size(xg1); q=nl*nc;
xt1=reshape(xg1,1,q); xt2=reshape(xg2,1,q);
xt=[xt1;xt2]'; %set of test points
xtest=xt;
%using gaussian kernel
ps=xt*xsup'; %scalar product
normxt=sum(xt.^2,2);
normxsup=sum(xsup.^2,2);
[nps,mps]=size(ps);
yt=zeros(nps,1);
aux=-2*ps+repmat(normxt,1,mps)+...
repmat(normxsup',nps,1);
Kt=exp(-aux/2); %gaussian kernel
yt(:)=yt(:)+Kt*w(1:mps);
yt=yt+b;
yt2d=reshape(yt,nl,nc);
%---------------------------------------------------------
%Display
figure(1)
%background levels
colormap('summer')
contourf(xg1,xg2,yt2d,20);shading flat; hold on
%data
for ii=1:N,
  if Y(ii)<0
    plot(X(ii,1), X(ii,2), 'r*', 'MarkerSize', 8);
  else
    plot(X(ii,1), X(ii,2), 'co', 'MarkerSize', 6);
  end
end
%separating line and margins
[L,A]=contour(xg1,xg2,yt2d,[-1 0 1],'k');
clabel(L,A); %plot of lines with labels
xlabel('x1');ylabel('x2');
title('Kernel-SVM, non separable data');
axis([0 1 0 1]);
```

A.5 Chapter 11: Evolutionary Optimization: Introduction

A.5.1 Ant Colonies (11.8)

TSP Solving by Ant Colony

Program 11.5

Based on indirect communication between ants, through pheromone, which originates a self-reinforcement process, the ant colonies are able to find the shortest path toward food (Fig. A.13).

Fig. A.13 The best tour found by ants for a TSP example

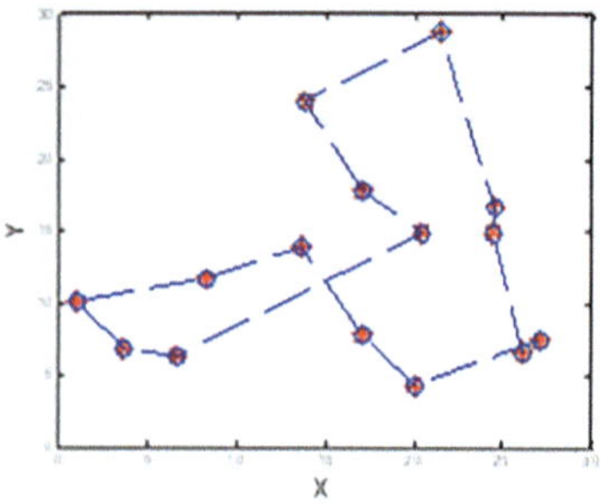

A.13 (P11.5) Example of TSP optimization by ant colony

```
%simple Ant Colony optimization example
% solving a TSP problem
%this main program uses 3 auxiliary functions
%TSP initialization -------------------------------
Nc=15; %number of cities
x=zeros(Nc,1);
y=zeros(Nc,1);
%generate coordinates of cities, at random
for jj=1:Nc
  x(jj)=rand*30;
  y(jj)=rand*30;
end
%plot the TSP map
figure(1)
plot(x,y,'r*', 'MarkerSize', 12); hold on
xlabel('x'); ylabel('y');
%generate matrix of distance between cities
D=zeros(Nc,Nc);
for i=1:Nc
  for j=1:Nc
    D(i,j)=sqrt((x(i)-x(j))^2 + (y(i)-y(j))^2);
  end
end
%start ACO ----------------------------------------
%initialize
S=zeros(Nc,Nc); %sight matrix
for i=1:Nc
  for j=1:Nc
    if (D(i,j)==0)
      S(i,j)=0;
    else
      S(i,j)=1/D(i,j);
    end
  end
end
Na=30; %number of ants
niter=20; %number of iterations
```

```
alpha=1; %effect of ants' sight
beta=4; %trace's effect
t=0.0001*ones(Nc); %primary traces
elim=0.97; %common distance elimination
ev=0.15; %evaporation coeff
Ip=zeros(Na,Nc); %for ant positions on cities
%main iteration----------
for jj=1:niter
  %initial ant positions (cities)
  for i=1:Na
    Ip(i,1)=fix(1+rand*(Nc-1)); %ant position
  end
  %generate tours for each ant
  % (using function antTour)
  [tour] = antTour(Ip,Na,Nc,S,t,alpha,beta);
  %concatenation of tours
  tour=horzcat(tour,tour(:,1));
  %compute total distances (using function calcDist)
  [dist,f] = calcDist(Na,Nc,D,tour,elim);
  %trace updating (using function traceUpdt)
  [t]=traceUpdt(Na,Nc,t,tour,f,ev);
  mean_dist(jj)=mean(dist);
  %extract the best route
  [min_dist(jj),best_index]=min(dist);
  best_tour(jj,:)=tour(best_index,:);
  itercount(jj)=jj;
end
%show results
%plot best route
[K,L]=min(min_dist);
for i=1:(Nc+1)
  aux=best_tour(L,i);
  X(i)=x(aux);
  Y(i)=y(aux);
end
plot(X,Y,'--bo');
xlabel('X'); ylabel('Y');
disp('total distance')
num2str(K)
%history of the search
figure(2)
plot(itercount, mean_dist, 'm');
xlabel('iteration'); ylabel('average distance');
% Function
%find ant tours
function [new_places] = antTour(Ip,Na,Nc,S,t,...
alpha,beta)
for i=1:Na
  mS=S;
  for j=1:(Nc-1)
    c=Ip(i,j);
    mS(:,c)=0;
```

```
    aux=(t(c,:).^beta).*(mS(c,:).^alpha);
    sm=(sum(aux));
    p=(1/sm).*aux;
    r=rand;
    sm=0;
    for k=1:Nc
      sm=sm+p(k);
      if r<=sm
        Ip(i,j+1)=k;
       break
      end
    end
  end
end
new_places=Ip;
% Function
%distance calculation
function [dist,f]= calcDist(Na,Nc,D,atour,elim)
for i=1:Na
  aux=0;
  for j=1:Nc
    aux=aux+D(atour(i,j),atour(i,j+1));
  end
  f(i)=aux;
end
dist=f;
f=f-(elim*min(f));
% Function
%update the trace
function [t]= traceUpdt(Na,Nc,t,tour,f,ev)
for i=1:Na
  for j=1:Nc
    dt=1/f(i);
    aux1=tour(i,j); aux2=tour(i,j+1);
    t(aux1,aux2)=dt+((1-ev)*t(aux1,aux2));
  end
end
```

A.6 Chapter 12: Evolutionary Optimization: Bio-Inspired Methods

A.6.1 Krill Herd Algorithm (12.5)

Simple Coding of Krill Herd Algorithm

Program 12.2

The Krill Herd algorithm is metaheuristic swarm intelligence optimization method, with crossover and mutation mechanisms. The algorithm is based on the motion, foraging activity and physical diffusion of krill (Fig. A.14).

Fig. A.14 Krills (food for many sea inhabitants)

A.14 (P12.2) Simple coding of Krill Herd algorithm

```
%Simple coding of Krill Herd algorithm
format long
NR = 10; % number of runs
NK = 25; % number of Krills
MI = 50; % number of iterations
C_flag = 1; % crossover flag [Yes=1]
% Bounds
UB = 10*ones(1,10);
LB = -10*ones(1,10);
NP = length(LB); % number of parameter(s)
Dt = mean(abs(UB-LB))/2; % scale factor
F = zeros(NP,NK);D = zeros(1,NK);N = zeros(NP,NK);
Vf = 0.02; Dmax = 0.005; Nmax = 0.01; Sr = 0;
% main loop
for nr = 1:NR
  %Initial Krill positions
  for z1 = 1:NP
    X(z1,:) = LB(z1) + (UB(z1) - LB(z1)).*rand(1,NK);
  end
  for z2 = 1:NK
    K(z2)=cost(X(:,z2));
  end
  Kib=K;
  Xib=X;
  [Kgb(1,nr), A] = min(K);
  Xgb(:,1,nr) = X(:,A);
  for j = 1:MI
    % Virtual food
    for ll = 1:NP;
```

```
    Sf(ll) = (sum(X(ll,:)./K));
  end
  Xf(:,j) = Sf./(sum(1./K)); %food location
  % see bounds:
  Xf(:,j) =findlimits(Xf(:,j)',LB,UB,Xgb(:,j,nr)');
  Kf(j) = cost(Xf(:,j));
  if 2<=j
    if Kf(j-1)<Kf(j)
      Xf(:,j) = Xf(:,j-1);
      Kf(j) = Kf(j-1);
    end
  end
  Kw_Kgb = max(K)-Kgb(j,nr);
  w = (0.1+0.8*(1-j/MI));
  for i = 1:NK
    % Calculation of distances
    Rf = Xf(:,j)-X(:,i);
    Rgb = Xgb(:,j,nr)-X(:,i);
    for ii = 1:NK
      RR(:,ii) = X(:,ii)-X(:,i);
    end
    R = sqrt(sum(RR.*RR));
    %Movement ----------
    % Best Krill effect on movements
    if Kgb(j,nr) < K(i)
      alpha_b = -2*(1+rand*(j/MI))*(Kgb(j,nr)-...
       K(i)) /Kw_Kgb/sqrt(sum(Rgb.*Rgb)) * Rgb;
    else
      alpha_b=0;
    end
    % Neighbors Krill effects
    nn=0;
    ds = mean(R)/5;
    alpha_n = 0;
    for n=1:NK
      if and(R<ds,n~=i)
        nn=nn+1;
        if and(nn<=4,K(i)~=K(n))
          alpha_n=alpha_n-(K(n)-K(i))/...
           Kw_Kgb/R(n)*RR(:,n);
        end
      end
    end
    %
    % Induced movement
    N(:,i) = w*N(:,i)+Nmax*(alpha_b+alpha_n);
    %Foraging ----------
    % Food attraction
    if Kf(j) < K(i)
      Beta_f=-2*(1-j/MI)*(Kf(j) - K(i)) /Kw_Kgb/...
       sqrt(sum(Rf.*Rf)) * Rf;
    else
      Beta_f=0;
    end
```

```
      % Best position attraction
      Rib = Xib(:,i)-X(:,i);
      if Kib(i) < K(i)
        Beta_b=-(Kib(i) - K(i)) /Kw_Kgb/...
         sqrt(sum(Rib.*Rib)) *Rib;
      else
        Beta_b=0;
      end
      % Foraging motion
      F(:,i) = w*F(:,i)+Vf*(Beta_b+Beta_f);
      %Diffusion ----------
      D =
       Dmax*(1-j/MI)*floor(rand+(K(i)-...
       Kgb(j,nr))/Kw_Kgb)*(2*rand(NP,1)-ones(NP,1));
      % Motion process ----------
      DX = Dt*(N(:,i)+F(:,i));
      % Crossover -----------
      if C_flag ==1
        C_rate = 0.8 + 0.2*(K(i)-Kgb(j,nr))/Kw_Kgb;
        Cr = rand(NP,1) < C_rate ;
        % Random selection of Krill number
        % for Crossover
        NK4Cr = round(NK*rand+.5);
        % Crossover scheme
        X(:,i)=X(:,NK4Cr).*(1-Cr)+X(:,i).*Cr;
      end
      % Update the position ---------
      X(:,i)=X(:,i)+DX;
      % see bounds:
      X(:,i)=findlimits(X(:,i)',LB,UB,Xgb(:,j,nr)');
      K(i)=cost(X(:,i));
      if K(i)<Kib(i)
        Kib(i)=K(i);
        Xib(:,i)=X(:,i);
      end
    end
    % Update the current best
    [Kgb(j+1,nr), A] = min(K);
    if Kgb(j+1,nr)<Kgb(j,nr)
      Xgb(:,j+1,nr) = X(:,A);
    else
      Kgb(j+1,nr) = Kgb(j,nr);
      Xgb(:,j+1,nr) = Xgb(:,j,nr);
    end
  end
end
%-------------------------
% Post-processing
[Best, Ron_No] = min(Kgb(end,:))
Xgb(:,end,Ron_No)
% Convergence plot of the best values
figure(1)
plot(1:MI+1,Kgb(:,Ron_No),'m');
xlabel('iter')
```

```
ylabel('best value')
%Function
%objective function
function f=cost(X)
% Ackley function.
n = 10;
a = 20; b = 0.2; c = 2*pi;
s1 = 0; s2 = 0;
for i=1:n;
  s1 = s1+X(i)^2;
  s2 = s2+cos(c*X(i));
end
f = -a*exp(-b*sqrt(1/n*s1))-exp(1/n*s2)+a+exp(1);
%Function
%see limits
function [ns]=findlimits(ns,Lb,Ub,best)
% Evolutionary Boundary Constraint Handling Scheme
n=size(ns,1);
for i=1:n
  ns_tmp=ns(i,:);
  I=ns_tmp<Lb;
  J=ns_tmp>Ub;
  A=rand;
  ns_tmp(I)=A*Lb(I)+(1-A)*best(I);
  B=rand;
  ns_tmp(J)=B*Ub(J)+(1-B)*best(J);
  ns(i,:)=ns_tmp;
end
```

A.6.2 Cuckoo Search (12.10)

Vectorized Implementation of Cuckoo Search Algorithm

Program 12.7

The cuckoo search algorithm is inspired by the brooding parasitism of some cuckoo species. They lay their eggs in the nests of other host birds; each of these eggs represents a candidate solution (Fig. A.15).

Fig. A.15 Cuckoos and parasitic breeding on other's nest

A.15 (P12.7) Implementation of cuckoo search

```
%Cuckoo search
%vectorized implementation
%parameters
Nn = 25; %number of nests
rd = 0.25; %discovery rate
tol = 0.001; %tolerance
%bounds
Nd = 2; %problem dimension
bl = -5*ones(1,Nd);
bu = 5*ones(1,Nd);
%objective function visualization
[vx1,vx2]=meshgrid(-5:0.1:5, -5:0.1:5);
y=(vx1.^2) + (vx2.^2)+ (25*((sin(vx1)).^2)) +
(25*((sin(vx2)).^2));
figure(1)
colormap('winter');
mesh(vx1,vx2,y); hold on;
contour(vx1,vx2,y);
%initial solutions at random
for jj=1:Nn
  nest(jj,:) = bl + (bu-bl).*rand(size(bl));
end
%get current best
fitness = (10^10)*ones(Nn,1);
[OFV,bestnest,nest,fitness] =...
 bestNest(nest,nest,fitness);
%start the algorithm ------------------------------
niter=0;
while(OFV>tol)
  %get new solutions (keeping current best)
  new_nest = getCucks(nest,bestnest,bl,bu);
  [Fnew,best,nest,fitness] =...
   bestNest(nest,new_nest,fitness);
  %
  niter=niter+Nn; %update counter
  %discovery
  new_nest=emptyNests(nest,bl,bu,rd);
  %evaluate recent new solutions
  [Fnew,best,nest,fitness] =...
   bestNest(nest,new_nest,fitness);
  %
  niter=niter+Nn; %update counter again
  %find best objective at present
  if(Fnew<OFV)
    OFV=Fnew;
    bestnest=best;
  end
end
%results ---------------
OFV
bestnest
```

```
figure(2)
contour(vx1,vx2,y,20); hold on;
for i=1:Nn
  plot(nest(i,1),nest(i,2),'ro');
end
%----------------------------
%Function
%get cuckoos
function nest = getCucks(nest,best,bl,bu);
N=size(nest,1);
beta=3/2;
aux1=gamma(1+beta); aux2=gamma((1+beta)/2);
aux3=2^((beta-1)/2);
sigma=(aux1*sin(pi*beta/2) /...
 (aux2*beta*aux3))^(1/beta);
%Levy flights
for i=1:N
  s=nest(i,:);
  aux=size(s);
  u=sigma*randn(aux);
  v=randn(aux);
  step=u./abs(v).^(1/beta);
  stepsize=0.01*step.*(s-best);
  s=s+stepsize.*randn(aux);
  nest(i,:)=useBounds(s,bl,bu);
end
%Function
%find the best nest
function [OFV,bnest,nest,fitness] =...
 bestNest(nest,newnest,fitness)
%evaluate new solutions
for i=1:size(nest,1)
  aux=newnest(i,:);
  %egg-crate function
  x1=aux(1); x2=aux(2);
  fnew=(x1^2)+(x2^2)+(25*((sin(x1))^2))+...
   (25*((sin(x2))^2));
  if fnew<=fitness(i)
    fitness(i)=fnew;
    nest(i,:)=newnest(i,:);
  end
end
[OFV,ix]=min(fitness);
bnest=nest(ix,:);
%Function
%replace some nets by new ones
function new_nest=emptyNests(nest,bl,bu,rd);
%some bad nets are discovered wit probability rd
N=size(nest,1);
K=rand(size(nest))>rd;
%new solution
stepsize=rand*(nest(randperm(N),:)-...
nest(randperm(N),:));
```

```
new_nest=nest+stepsize.*K;
for i=1:size(new_nest,1)
  aux=new_nest(i,:);
  new_nest(i,:)=useBounds(aux,bl,bu);
end
%Function
%simple use of bounds
function s=useBounds(s,bl,bu);
%apply lower bound
qaux=s;
ix=qaux<bl;
qaux(ix)=bl(ix);
%apply upper bound
jx=qaux>bu;
aux(jx)=bu(jx);
s=qaux;
```

A.7 Chapter 13: Evolutionary Optimization: Other Metaphors

A.7.1 Imperialist Competitive Algorithm (13.2.2)

Example of Imperialist Competitive Algorithm (ICA)

Program 13.3

A few countries in the world are imperialist, and the others can be considered as colonies. Stronger empires tend to grow (Fig. A.16).

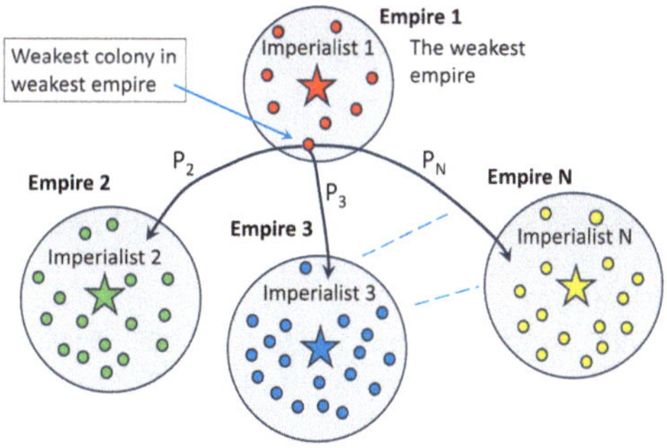

Fig. A.16 Stronger empires tend to grow

A.16 (P13.3) An implementation of the imperialist competitive algorithm

```
%Example of Imperialist Competitive Algorithm
clear all;
%Number of Population & Dimensions
pop_size = 100;
dim = 2;
%Iteration Condition
max_iteration = 500;

%memory of best results along iterations
Best_fit=zeros(max_iteration,1);
%Variable limits
from = 0;
to = 10;
%Nfe Condition
nfe = 0;
max_nfe = 40000;
%Initializing Random Population
Countries=from+((to-from)*rand(pop_size,dim));
%Number of Empires
n_empire = ceil(0.1 * pop_size);
Empires = zeros(n_empire,dim+1);
%Number of Colonies
n_col = pop_size - n_empire;
Colonies = zeros(n_col,dim+2);
%Minimum Colonies for each Empire
min_col = 3;
%Revolution_Rate
rev_rate = 0.2;
%Calculating Fitness & Sorting
[sorted_Countries, nfe, F_result] =
ICA_Sort(Countries,pop_size,nfe,dim,max_nfe);
%Initializing Empires
Empires(:,1:dim) = sorted_Countries(1:n_empire,:);
Empires(:,dim+1) = F_result(1,1:n_empire)';
%Initializing Colonies
Colonies(:,1:dim)=sorted_Countries(n_empire+1:end,:);
Colonies(:,dim+1) = F_result(1,n_empire+1:end)';
%Plot of initial empires ---------------------
figure(1)
[vx1,vx2]= meshgrid(0:0.1:10, 0:0.1:10);
y=(vx1.*sin(4*vx1))+1.1*(vx2.*sin(2*vx2));
colormap('cool');
contour(vx1,vx2,y,10); hold on;
plot(Empires(:,1), Empires(:,2),'k*',...
'MarkerSize',14);
xlabel('x'); ylabel('y');
%Normalized Fitness of each Empire
emptyE = zeros(1,n_empire);
cn = emptyE;
Cn = emptyE;
```

```
cn(1,:) = Empires(:,dim+1)';
Cn(1,:) = max(cn) - cn(1,:);
Best_fit(1)=min(Empires(:,dim+1)); %initial best fit
%Relative Power of each Empire
pn = emptyE;
pn(1,:) = abs(Cn(1,:) / sum(Cn));
%Number of Primitive Colonies for each Empire
NCn = emptyE;
NCn(1,:) = min_col;
NCn(1,:) = NCn(1,:) + (round( pn(1,:) * (n_col -
(n_empire*min_col))) );
%Shuffling the Colonies
randx = randperm(n_col);
Colonies = Colonies(randx,:);
%Assign Colonies to the Empires
j = 1;
k = NCn(1,1);
for i = 1:n_empire
  Colonies(j:k,dim+2) = i;
  j = j + NCn(1,i);
  if i+1 > n_empire
    break;
  end
  k = k + NCn(1,i+1);
end
%Control the Assignation Term
if size(Colonies,1) > n_col
Colonies = Colonies(1:n_col,:);
end
for i = 1:n_col
  if Colonies(i,dim+2) == 0
    rand_empire = 1+round((n_empire-1)*rand);
    Colonies(i,dim+2) = rand_empire;
  end
end
%Starting the Main Loop ----------------------------
for iteration = 1:max_iteration
  %Compute Euclidean Distance Between the Empires
  %and their Colonies
  distance = zeros(n_col,dim);
  for i = 1:n_empire
    for j = 1:n_col
      if Colonies(j,dim+2) == i
        distance(j,:)=sqrt(sum((Empires(i,1:dim)-...
         Colonies(j,1:dim)).^2));
      end
    end
  end
  %Move the Colonies
  beta = 2;
  teta= (-pi/4) + ((pi/2)*rand(n_col, dim));
  x = zeros(n_col,dim);
  aux=beta*distance(:,:); [Mx,Nx]=size(aux);
```

```
x(:,:)=aux.*rand(Mx,Nx);
X = teta .* x;
Colonies(:,1:dim) = X + Colonies(:,1:dim);
%Control the Domain of the Colonies
for i = 1:n_col
  for j = 1:dim
    r_c = rand;
    if (Colonies(i,j) > to)
      Colonies(i,j) = to - r_c;
    elseif (Colonies(i,j) < from)
      Colonies(i,j) = from + r_c;
    end
  end
end
%Fitness of New Colonies
Colonies(:,dim+1) = ImpObj(Colonies(:,1:dim));
nfe = nfe + n_col;
%Termination Condition
if(nfe >= max_nfe)
  break;
end
%Exchange the Positions of Empire and Colony
%(if Fitness of Colony is better than Empire)
for i = 1:n_empire
  for j = 1:n_col
    if Colonies(j,dim+2) == i
      if Colonies(j,dim+1) < Empires(i,dim+1)
          temp = Empires(i,1:dim+1);
          Empires(i,1:dim+1) = Colonies(j,1:dim+1);
          Colonies(j,1:dim+1) = temp;
      end
    end
  end
  %Calculate the Average Fitness of the Colonies
  % of each Empire
  ind = find(Colonies(:,dim+2)==i);
  if(isempty(ind) == 0)
    Empires(i,dim+2) =...
     mean(Colonies(ind(1):ind(end),dim+1));
  end
end
%The Whole Power of each Empire
zeta = 0.05;
TCn = Empires(:,dim+1) + (zeta * Empires(:,dim+2));
%Normalized Whole Power of each Empire
NTCn = max(TCn) - TCn;
%Probability of Takeover for each Empire
%insert 1 to avoid division by zero:
Ppn = abs(NTCn / (1+sum(NTCn)));
%Colonial Competition
if (size(Empires,1) > 1)
  R = rand(n_empire,1);
  D = Ppn - R;
  %Minimization
```

```
    weak_emp_ind = find(NTCn==min(NTCn));
weak_colonies=sortrows(Colonies(Colonies(:,dim+2)...
     == weak_emp_ind(1),:),(dim+1));
    colonies_takeover_size=...
     ceil(size(weak_colonies,1)*0.5);
    takeover_ind = find(D==max(D));
    for i = 1:colonies_takeover_size
      aux1=(Colonies(:,dim+2)==weak_emp_ind(1));
aux2=(Colonies(:,dim+1) == weak_colonies(i,dim+1));
      fx=find(aux1 & aux2);
      weakest_colonies_ind=fx(1);
Colonies(weakest_colonies_ind,dim+2) = takeover_ind;
    end
  end
  %Fall of the Weakest Empire
  if(size(weak_colonies,1)-colonies_takeover_size)==0
    if n_empire - 1 ~= 0
      n_col = n_col + 1;
      n_empire = n_empire - 1;
      Colonies(n_col,1:dim+1) =...
       Empires(weak_emp_ind(1),1:dim+1);
      rand_empire = 1+round((n_empire-1)*rand);
      Colonies(n_col,dim+2) = rand_empire;
      Empires(weak_emp_ind(1),:)=[]; %remove row
      %Reindex
      for index = n_empire+1:-1:weak_emp_ind(1)
        if index - 1 ~= 0
 Colonies(Colonies(:,dim+2)==index,dim+2) = index-1;
        end
      end
    end
  end
  %Revolution Process
  y=from+((to-from)*rand(floor(rev_rate*n_col),dim));
  for i = 1:size(y,1)
    rand_rev = 1+round((n_col-1)*rand);
Colonies(rand_rev,1:dim)=Colonies(rand_rev,1:dim)+...
     y(i,:);
  end
  Best_fit(iteration+1)=min(Empires(:,dim+1));
end
x=Empires(1,1)
y=Empires(1,2)
optim. value = Empires(1,3)
figure(2)
y=(vx1.*sin(4*vx1))+1.1*(vx2.*sin(2*vx2));
colormap('winter');
contour(vx1,vx2,y,10); hold on;
plot(Colonies(:,1), Colonies(:,2),'mx',...
'MarkerSize',12,'LineWidth',2);
xlabel('x'); ylabel('y');
figure(3)
nn=1:iteration;
plot(nn,Best_fit(nn),'m');
```

```
xlabel('iter'); ylabel('best result');
% Function
function [Out, nfe, fitness_result] =
ICA_Sort(X,n_size,nfe,dim,max_nfe)
s = size(X,1);
F_result = zeros(1, s);
for i = 1:s
  F_result(1,i) = ImpObj(X(i,:));
  nfe = nfe + 1;
  if(nfe >= max_nfe)
    break;
  end
end
[F_result, sorted_index] = sort(F_result,2);
sorted = zeros(n_size,dim);
for j=1:n_size
  sorted(j,:) = X(sorted_index(1,j),:);
end
Out = sorted;
fitness_result = F_result;
% Function (Schwefel 2.26)
function [Out] = ImpObj(X)
x=X(1); y=X(2);
Out=x*sin(4*x) + 1.1*y*sin(2*y);
```

A.7.2 Biogeography-Based Optimization (13.5.1)

Solving N-Queen Problem with Biogeography-Based Optimization

Program 13.6 (Fig. A.17)

Fig. A.17 The N-queen problem

A.17 (P13.6) Solving N-queen problem with biogeography-based algorithm

```
%Example of Biogeography-based optimization
% Solving n-queen problem
clear all;
% number of queens (edit)
nQueen=12;
%-----------------------------------------------
nVar=nQueen; % Decision Variables
VarSize=[1 nVar]; % Decision Variables Matrix Size
VarMin=0; % Lower Bound of Variables
VarMax=1; % Upper Bound of Variables
%% BBO Parameters
MaxIt = 50; % Iterations
nHb = 100; % Number of Habitats (Population Size)
KeepRate = 0.2; % Keep Rate
% Number of Kept Habitats:
nKeep = round(KeepRate*nHb);
nNew = nHb-nKeep; % Number of New Habitats
% Migration Rates
mu = linspace(1, 0, nHb); % Emmigration Rates
lambda = 1-mu; % Immigration Rates
alpha = 0.9;
pMutation = 0.4;
sigma = 0.02*(VarMax-VarMin);
%-----------------------------------------------
% Empty Habitat
habitat.Position = [];
habitat.Cost = [];
habitat.Sol = [];
% Create Habitats Array
Hpop = repmat(habitat, nHb, 1);
% Habitats
for i = 1:nHb
  aux=(VarMax-VarMin)*rand(1,nVar) + VarMin;
  Hpop(i).Position = aux;
  [Hpop(i).Cost Hpop(i).Sol]=CostF(Hpop(i).Position);
end
% Sort Population
[aux, SortOrder] = sort([Hpop.Cost]);
Hpop = Hpop(SortOrder);
% Best Solution
Final = Hpop(1);
% For recording Best Costs
CostValue = zeros(MaxIt, 1);
%% BBO main loop ------------------------------
for it = 1:MaxIt
  newHpop = Hpop;
  for i = 1:nHb
    for k = 1:nVar
      % Migration
```

```
      if rand <= lambda(i)
        % Emigration Probabilities
        EP = mu;
        EP(i) = 0;
        EP = EP/sum(EP);
        % Select Source Habitat
        j = RWS(EP);
        % Migration
      newHpop(i).Position(k)=Hpop(i).Position(k) ...
   +alpha*(Hpop(j).Position(k)-Hpop(i).Position(k));
      end
      % Mutation
      if rand <= pMutation
   newHpop(i).Position(k)=newHpop(i).Position(k)+...
        sigma*randn;
      end
    end
    % Apply Lower and Upper Bound Limits
    newHpop(i).Position = max(newHpop(i).Position,...
     VarMin);
    newHpop(i).Position = min(newHpop(i).Position,...
     VarMax);
    % Evaluation
    [newHpop(i).Cost newHpop(i).Sol] =...
     CostF(newHpop(i).Position);
  end
  % Sort New Population
  [aux, SortOrder] = sort([newHpop.Cost]);
  newHpop = newHpop(SortOrder);
  % Select Next Iteration Population
  Hpop = [Hpop(1:nKeep)
  newHpop(1:nNew)];
  % Sort Population
  [aux, SortOrder] = sort([Hpop.Cost]);
  Hpop = Hpop(SortOrder);
  % Update Best Solution
  Final = Hpop(1);
  % Store Best Cost
  CostValue(it) = Final.Cost;
  % Plot Best Solution ---------------------------
  figure(1);
  ShowRes(Final.Sol);
  if CostValue(it)==0
   break;
  end
end
%% Best result evolution -----------------------
figure(2);
plot(CostValue, 'r-');
xlabel('iter'); ylabel('Cost Value');
%function
function i=RWS(P)
r=rand;
C=cumsum(P);
```

```
ix=find(r<=C);
i=ix(1);
%function
function ShowRes(sol)
X=sol.X-0.5; Y=sol.Y-0.5; Hit=sol.Hit; z=sol.z;
n=prod(size(X));
for i=1:n-1
  for j=i+1:n
    if Hit(i,j)==1
   plot([X(i) X(j)],[Y(i) Y(j)],'b:','LineWidth',2);
      hold on;
    end
  end
end
plot(X,Y,'k^','MarkerSize',12,'MarkerFaceColor','r');
strTitle=[num2str(n) '-Queens: '];
if z==0
  title([strTitle 'No Hits']);
elseif z==1
  title([strTitle 'Just one Hit']);
else
  title([strTitle num2str(z) ' Hit(s)']);
end
ax = gca;
ax.FontSize = 12; ax.FontWeight='bold';
grid on;
axis square;
set(gca,'XTick',0:n); set(gca,'YTick',0:n);
xlim([0 n]); xlim([0 n]);
hold off;
%function
function [z, sol]=CostF(s)
n=prod(size(s));
[aux, X]=sort(s);
Y=1:n;
Hit=zeros(n,n);
z=0;
for i=1:n-1
  for j=i+1:n
    if abs(X(i)-X(j))==abs(Y(i)-Y(j))
      Hit(i,j)=1; Hit(j,i)=1;
      z=z+1;
    end
  end
end
sol.X=X; sol.Y=Y;
sol.Hit=Hit;
sol.z=z;
```

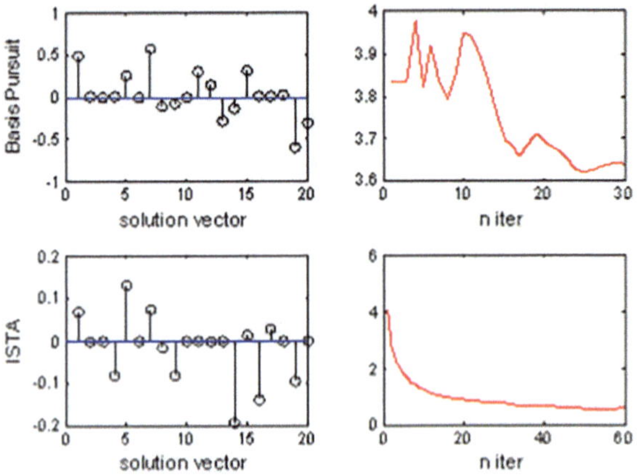

Fig. A.18 Results of ADMM-BP and ISTA

A.8 Chapter 16: Applications in Image and Data Processing

A.8.1 Alternating Directions (16.8.4)

Finding Sparsest Solution with ADMM

Program 16.13

The alternating direction method of multipliers (ADMM) is a recent iterative method being used in data and image processing (Fig. A.18).

A.18 (P16.13) Sparsest solution reached with ADMM

```
% Finding sparsest solution to Ax=b
% with ADMM for BP, and with Lasso
n=20; m=14;
A=randn(m,n);
b=randn(m,1);
%
% -------------------------------------------------
% ADMM for BP algorithm, to find sparsest solution
niter=30;
x=zeros(n,1); z=zeros(n,1); u=zeros(n,1);
rob=zeros(niter,1);
alpha=1.0; mu=1.0;
uA=inv(A*A');
aux1=eye(n)-(A'*uA*A);
aux2=A'*uA*b;
for nn=1:niter,
  % x update
  x=(aux1*(z-u))+aux2;
  % z update
  zo=z;
  xe=(alpha*x)+((1-alpha)*zo);
```

```
  aux=xe+u; imu=1/mu;
  z=max(0,aux-imu)-max(0,-aux-imu); %shrinkage
  % u update
  u=u+(xe-z);
  % recording
  rob(nn)=norm(x,1);
end
BPx = x; %solution with ADMM-BP
BProb = rob; %evolution of l1-norm with ADMM-BP
%results
disp('ADMM for BP: -------');
disp('norm1(x): --');
norm(x,1)
disp('norm2(b-Ax): --');
V=b-(A*x);
norm(V,2)
%
% --------------------------------------------------
% Lasso-ISTA algorithm for finding sparsest solution
niter=60;
J=zeros(1,niter); %objective function
x=5*randn(n,1); %initialize x
lambda=0.1; %parameter
alpha=40; %parameter
T=lambda/2; %threshold
for k=1:niter
  q=A*x;
J(k)=sum(abs(b(:)-q(:)).^2) + lambda*sum(abs(x(:)));
  p=x+(A'*(b-q))/alpha;
  %soft thresholding (component-wise)
  for j=1:n
    if abs(p(j)) <= T,
      x(j)=0;
    else
      if p(j)>T,
        x(j)=p(j)-T;
      else
        x(j)=p(j)+T;
      end;
    end;
  end;
end
Lx = x; %solution with Lasso
Lrob = J; %evolution of J with Lasso
%results
disp('ISTA: --------');
disp('norm1(x): --');
norm(x,1)
disp('norm2(b-Ax): --');
V=b-(A*x);
norm(V,2)
%
```

```
% display --------------------------
figure(1)
subplot(2,2,1)
stem(BPx,'k'); hold on;
plot([0 20],[0 0],'b');
xlabel('solution vector');
ylabel('Basis Pursuit');
subplot(2,2,2)
plot(BProb,'r','LineWidth',2);
xlabel('n iter');
subplot(2,2,3)
stem(Lx,'k'); hold on;
plot([0 20],[0 0],'b');
xlabel('solution vector');
ylabel('ISTA');
subplot(2,2,4)
plot(log10(Lrob),'r','LineWidth',2);
xlabel('n iter');
```

A.8.2 *Inpainting (16.9.2)*

Split Bregman Iterations for Inpainting

Program 16.15

A split Bregman iteration procedure for total variation L1 norm analysis of images is applied for a portrait inpainting example (Fig. A.19).

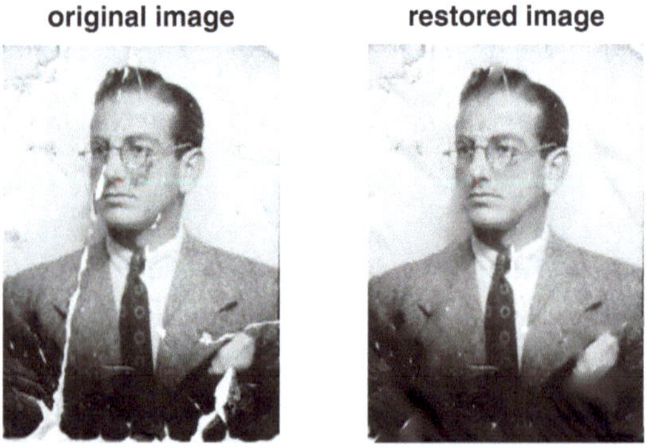

Fig. A.19 Image restoring with inpainting algorithm

A.19 (P16.15) Image inpainting with split Bregman iterations

```
% Split Bregman inpainting example
% Photo restoring
% source image and parameters ----------------
global uk
u = double(imread('TTo.jpg')); %original photo
[height,width]=size(u);
mmk = double(imread('MK1.jpg')); %edited mask
Mask = mmk(1:height, 1:width); %crop if needed
lambda = 255;
gamma = 0.5;
niter = 4000;
%convert mask to negative
Lambda = zeros(size(u));
for i=1:height,
  for j=1:width,
    if Mask(i,j) > 220,
      Lambda(i,j) = 0;
    else
      Lambda(i,j) = lambda;
    end
  end
end
b_x = zeros(height,width);
b_y = zeros(height,width);
uk = zeros(height,width);
e1 = ones(width,1); e2 = ones(height,1);
D1 =spdiags([-e1,e1],0:1,width,width);
D1(width,1) = 1;
D1 = D1';
D2 = spdiags([-e2,e2],0:1,height,height);
D2(height,1) = 1;
% main loop ------------------------------
for nn = 1:niter
  % d subproblem
  [x_grad,y_grad] = u_grad(uk,D1,D2);
  [d_x,d_y] = shrink2(x_grad,y_grad,b_x,b_y,gamma);
  % u subproblem, Gauss-Seidel
  Gk =
  gau_seid1(d_x,d_y,b_x,b_y,Lambda,gamma,u,uk,...
  height,width);
  dif = (uk-Gk).^2;
  uk = Gk;
  % updates
  b_x = b_x + x_grad - d_x;
  b_y = b_y + y_grad - d_y;
end
%display ------------------------
figure(1)
subplot(1,2,1)
imshow(uint8(u))
```

```
title('original image')
subplot(1,2,2)
imshow(uint8(uk))
title('restored image')

function [x_grad,y_grad] = u_grad(u,D1,D2)
x_grad = u*D1;
y_grad = D2*u;

function [d_x,d_y] = shrink2(x_grad,y_grad,...
b_x,b_y,gamma)
x1 = (x_grad + b_x)./abs(x_grad + b_x);
m1 = abs(x_grad + b_x)-1/gamma;
m1 = max(m1,0);
d_x = x1.*m1;
d_x(isnan(d_x)) = 0;
x2 = (y_grad + b_y)./abs(y_grad + b_y);
m2 = abs(y_grad + b_y)-1/gamma;
m2 = max(m2,0);
d_y = x2.*m2;
d_y(isnan(d_y)) = 0;

function Gk =
gau_seid1(d_x,d_y,b_x,b_y,lambda,gamma,u,uk,...
height,width)
D_prod = lambda .* u;
lambda_4g = lambda + 4*gamma;
coef1 = D_prod./lambda_4g;
coef2 = gamma./lambda_4g;
lapla = [0,1,0;1,0,1;0,1,0];
uk_sum = conv2(padarray(uk,[1,1],'replicate',...
'both'),lapla,'same');
uk_sum = uk_sum(2:height+1,2:width+1);
db_sum = diff_row(d_x) + diff_col(d_y)-...
diff_row(b_x)-diff_col(b_y);
Gk = coef1 + coef2.*(uk_sum + db_sum);

function x_diff = diff_row(x_data)
[row,aux] = size(x_data);
x_data_rowless = [x_data(1,:);x_data(1:row-1,:)];
x_diff = x_data_rowless - x_data ;

function y_diff = diff_col(y_data)
[aux,col] = size(y_data);
y_data_rowless = [y_data(:,1),y_data(:,1:col-1)];
y_diff = y_data_rowless - y_data ;
```

Index

J. M. Giron-Sierra, *Introduction to Optimization with Matlab® Examples*,
https://doi.org/10.1007/978-3-032-05048-9

MIX
Papier aus verantwortungsvollen Quellen
Paper from responsible sources
FSC® C105338

If you have any concerns about our products,
you can contact us on
ProductSafety@springernature.com

In case Publisher is established outside the EU,
the EU authorized representative is:
Springer Nature Customer Service Center GmbH
Europaplatz 3, 69115 Heidelberg, Germany

Printed by Libri Plureos GmbH
in Hamburg, Germany